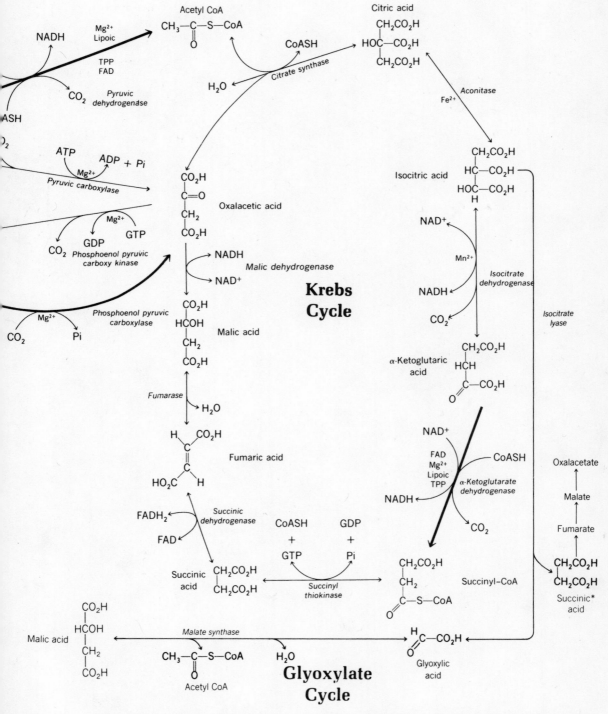

Krebs Cycle

Glyoxylate Cycle

* Succinic acid produced by isocitrate lyase is converted back to oxalacetate in order to keep the glyoxylate cycle in operation.

OUTLINES
OF
BIOCHEMISTRY

OUTLINES OF BIOCHEMISTRY 5 / E

ERIC E. CONN
PAUL K. STUMPF
GEORGE BRUENING
ROY H. DOI

University of California at Davis

JOHN WILEY & SONS

New York
Chichester
Brisbane
Toronto
Singapore

Production Supervised by Linda R. Indig
Cover & Interior designed by Carolyn Joseph
Cover Credit; Space-filling model of a portion
    of the surface structure of Mengo virus,
    showing the association of four different
    protein molecules. Based upon the research
    of Ming Luo, Gerrit Vriend, Greg Kamer,
    Tom Smith, Michael G. Rossmann and others
    at Purdue University. Image produced in part
    by graphics program ANIMOL as modified by
    M. G. Webb and R. P. Millane.
Illustrations by John Balbalis with the assistance
    of the Wiley Illustration Department.
Manuscript edited by Patricia Brecht with the
    supervision of Glenn Petry.

Library of Congress Cataloging in Publication Data:

Conn, Eric E.
    Outlines of biochemistry.

    Includes index.
    1.   Biological chemistry.   I.   Title.
QP514.2.C65   1987   574.19′2   86-24688
ISBN 0-471-05288-4

Printed in the United States of America

10 9 8 7 6 5 4 3 2

In the quarter century since the appearance of the first edition of Conn and Stumpf's *Outlines of Biochemistry,* the science of biochemistry, and its technological implications, have changed in ways its pioneers could not have imagined. The fifth edition required extensive revision to properly reflect current biochemistry, especially in the rapidly expanding areas of gene expression and gene manipulation, as well as in the related topics of nucleic acid and protein structure and function. To achieve the desired breadth of coverage in this edition, the original authors invited George Bruening and Roy Doi to participate as coauthors. We have also endeavored to bring the student a text that provides the basic tenets of modern biochemistry in a volume of reasonable size and balance.

Based upon our experience in teaching general biochemistry courses at the University of California, Davis, the general plan of earlier editions of *Outlines* has been retained with three sections emphasizing the structure of biologically important molecules, the metabolism of small molecules, and the metabolism of macromolecules and the regulation of gene expression.

An extensive revision of the first chapter gives an overview of biochemistry, presenting some central themes, such as a common genetic code and common metabolic intermediates for all forms of life. The aim is to avoid losing these important "take-home lessons" of biochemistry in what often seems to the student as a thicket of structures, metabolic pathways, and restriction endonuclease maps. We also present the general experimental philosophies that are characteristic of biochemistry, such as reductionist and nonreductionist approaches and their respective merits and the use of complex *in vitro* systems that integrate, for example, transcription and translation.

This fifth edition of *Outlines* retains the metabolic orientation of the former editions and their straightforward presentation of biochemical principles. But this edition also describes some important recent advances in biochemistry, especially the ability to manipulate and transfer genes and to study the biochemistry of particular molecules introduced into intact cells. These experimental approaches are conveyed to the student because they have given biochemists much greater insight into the operation of the cell and have reduced the need to extrapolate results from cell extracts, as the biochemist attempts to understand the intact organism. New instrumentation is also emphasized in this edition, not only for the analysis of biochemical systems but also for the precise chemical synthesis of such critical molecules as peptides and DNA fragments, which are then ready for tests of biological activity *in vitro* and *in vivo*. The combination of traditional and modern biochemical techniques has brought us a new and far more complete understanding of the regulation of metabolism, especially through the regulation of enzyme synthesis. The regulation of enzyme synthesis and the processing and transport of proteins to their sites of function in the cell receive particular emphasis in the fifth edition.

An entirely new last chapter describes extracellular nucleic acids, both in the form of experimental recombinant DNA molecules and naturally occurring plasmids and viruses. The methodological appendix found in the earlier editions has been abandoned in favor of separate subsections—set off from the main body of the chapter—that describe methods such as amino acid analysis, gas–liquid chromatography, and fundamental topics such as the tautomeric forms of the nucleic acid bases. In this way biochemistry as an experimental science, and the roots of biochemistry in chemistry, are emphasized without breaking the flow of the text. Throughout the text, the implications of biochemistry for medicine, agriculture, and industry are emphasized.

In preparing this edition, as well as the earlier editions, we are indebted to our students and colleagues who have used the text in their courses. In particular, we thank the students who have been enrolled in Biochemistry 101AB at Davis for providing many helpful suggestions over the years. And the comments of Dr. Larry Sprechman, who has taught undergraduate biochemistry courses for many years, have been especially valuable.

E. E. Conn
P. K. Stumpf
G. Bruening
R. H. Doi

Davis, California

CONTENTS

PART 1 STRUCTURES AND FUNCTIONS OF BIOLOGICAL MOLECULES

PART 1

STRUCTURES AND FUNCTIONS OF BIOLOGICAL MOLECULES

WHAT BIOCHEMISTS STUDY

The goal of biochemistry is to understand the chemical basis of biological phenomena. This introductory chapter begins our presentation of the current state of that understanding by covering three main subjects.

We first give a perspective on the field of biochemistry, including some information on the methodologies used. The realm of biochemistry is the biosphere, which consists of organisms of all types and the environments in which they live. Our intent is to describe the vantage point from which biochemists view the biosphere as they attempt to explain the operation of organisms and their interactions with their environments in chemical terms.

Second, we summarize the contents of this book and its organization into sections that present structural, functional, and informational aspects of biologically important molecules.

Finally, we focus on one important compound. Of all the compounds in the biosphere, water is the most abundant. Because of its abundance and the fact that water is essential to cells and their environments and accounts for 70 to 90% of the mass of most cells, we devote the end of this chapter to the properties of this amazing substance, the solvent for biochemical reactions.

1.1 CELL AND ORGANISM AS BIOCHEMICAL ENTITIES

Biology in its broadest sense may be regarded simply as the study of the biosphere. What are the most characteristic properties of the organisms that populate the biosphere? A particular organism may be characterized by its form and essential processes such as reproduction, development, and energy transduction. That organisms take particular forms and mediate specific processes reflects and requires order. Reducing disorder, or entropy, inevitably requires energy. Organisms excel in utilizing energy to create order, and complex organisms do so on a continuous basis. What we know about how organisms use energy to create orderly structures and carry out orderly processes comes primarily from years of effort on the part of thousands of biochemists. One of the most important contributions of biochemistry to biology is the understanding that has been attained about how the chemical processes can create order.

The process that most definitely sets organisms apart from other objects is reproduction. Both cellular and noncellular organisms give rise to more organisms of similar design as part of a life cycle that may be simple or elaborate. For cellular organisms, cell division is, of course, an essential event in reproduction. Energy is used to manufacture new molecules that are incorporated into new structures in a precise developmental process that leads to the formation of two new cells. In contrast to most cells, the viruses and other even less complex noncellular organisms all are parasitic or symbiotic; they depend on specific host cells to provide an environment in which replication can occur by processes that do not overtly resemble cell division. For all organisms then, cellular and noncellular, the site of action is the cell, and the cell is the entity that is essential to reproduction.

The biochemist views the cell as an exceedingly complex and refined machine. These machines have capabilities that are far beyond those of current man-made machines. A given cell may have a very large repertoire of responses to environmental changes. The cells of a multicellular organism must be able to interact to give the even greater repertoire of responses that characterize higher organisms. "Adaptability" is a characteristic of the cell-machine. The complex organism detects and processes stimuli from the environment and from within the organism with sophistication. Movement, secretion, cell division, and so on are accomplished in a precisely controlled fashion to allow the entire organism to grow and reproduce itself.

The biochemist must be aware of the abilities of organisms to change and adapt, not only in the time frame of an individual organism but also on an evolutionary time scale. It is not enough to consider any noncellular, single cell, or multicellular organism in isolation or only as it exists today. Organisms can be classified as belonging to species or similar taxonomic or functional groupings. Organisms exchange, or at least transmit, symbolically encoded, controlling information. That is, they have genetic systems. Capabilities of the progeny reflect the capabilities of the parent(s), and the progeny follow definite plans of development. Those plans, though definite, are not rigid. They are modified by the environment and time. Organisms cannot be understood without a consideration of what their antecedents must have been like. Organisms have evolved and adapted to changing conditions on a geological time scale and continue to do so. Thus, biochemists seek chemical explanations of how organisms adapt to their environment both in the short term and over the eons.

The cell carries out the most elaborate of chemical transformations while maintaining its internal environment within definitely controlled limits. Cells have boundaries and, within those boundaries, compartments. There, thousands of types of small molecules and macromolecules, in an aqueous (water) environment, undergo chemical transformations and exchange materials and heat with the exterior of the cell. The elaborate macroscopic behavior of living systems anticipates a great variety of microscopic catalytic and regulatory mechanisms. Only a small fraction of these are now understood in molecular detail.

Fortunately, the results currently available show a startling unity of biology from a biochemist's point of view. Organisms, from the simplest noncellular entity to the most complex vertebrates and higher plants, are seen to have in common many chemical structures and reactions. Thus, the biochemist studying a particular organism may be able to anticipate its biochemical reactions from what is already known from other, even distantly related, organisms. This grand guideline, that organisms show a unity at the chemical level that is unsuspected when one views their biological diversity, strongly influences the thinking of biochemists.

1.2 THE BIOCHEMIST'S POINT OF VIEW

Biochemistry has its roots in medicine, nutrition, agriculture, fermentation, and natural products chemistry. Today, it is principally concerned with the chemistry of molecules found in and associated with living systems, especially the chemistry of the interactions of these molecules. Most biochemists operate in a particular philosophical framework. The usually unspoken assumption is that the activities of cells, for all of their complexity, must be susceptible to explanation as chemical and physical phenomena. Developing this understanding has required and will continue to require the careful application of chemical and physical laws and methods *in combination* with the careful biological manipulation of the systems under study.

Biochemistry is in large measure, but not wholly, a reductionist science. Biochemists routinely disrupt cells and fractionate what has been released into the extract for the purpose of obtaining one or another component of the cell in isolation. The biochemist's intent in preparing extracts and fractions is to reduce the number of variables that remain uncontrolled and to obtain a more clear-cut and reliable result. The cell has capabilities beyond those of its subsystems, but what we know about the cell has come largely from studying isolated parts of the cell, even individual molecules.

A common example of the reductionist approach is an *in vitro* enzyme assay. Enzymes are the principal *in vivo* catalysts of the cell. An enzyme will effectively facilitate, in the test tube, complex chemical transformations that would occur at only an imperceptible rate in the absence of the enzyme. The investigator selects conditions of temperature, concentration, and so on that can be achieved in the test tube and allow the enzyme to transform biologically authentic substrates into biologically authentic products. However, the conditions that are convenient for the investigator may resemble only slightly the conditions of the intact cell. Interpretation of the results in the context of the intact cell requires care and considerable attention to the detailed physiology of the system. Often, the results must be correlated with other biochemical, physiological, and/or genetic data before the puzzle can be solved.

Several modern biochemical approaches are not reductionist but take advantage of technological advances to study intact systems. We complete this subsection with a discussion of four such approaches. Each allows the investigator to deal directly with the intact organism and yet learn with great precision the course of a particular reaction, the consequences of introducing a specific change into a particular structure, and so forth.

Radioisotopes are tools that allow biochemists to observe metabolic transformations in the intact organism. The radioactive forms of specific molecules, often with only the atoms at a specific position in the molecules labeled, allow the radioactive molecules to be located against a background of almost unlimited numbers of other molecules. This sensitivity and specificity are achieved because atomic events, such as radioactive decay, generate energetic, readily detected subatomic particles. Atomic events typically have 10^5 or more times the energy of chemical events. Thus, acetic acid may be prepared with one or both of its carbon atoms radioactive, that is, with unstable, radioactive ^{14}C atoms replacing the far more abundant, nonradioactive ^{12}C atoms. Some of the radioactive acetate, when submitted to an intact organism, will be incorporated into compounds, such as lipids, that are being synthesized by the organism at that time (see, e.g., Chapter 13). The chemical difference between acetate with ^{14}C atoms and acetate with ^{12}C atoms is so small that the two forms of acetate are interchangeable in biochemical reactions. Subsequent analysis of extracts prepared from the organism or its secretions will reveal which compounds, those that are radioactive, have been formed from the radioactive acetate. Although this analysis, like the reductionist approaches described above, requires disruption of cells, the incorporation itself occurred in the intact organism. Thus, if the experiment is carried out properly, the results should reflect the metabolism of the intact organism.

A second approach is the use of nuclear magnetic resonance spectroscopy, abbreviated NMR. NMR allows specific chemical reactions to be monitored in undisrupted cells. For example, the nuclear magnetic resonance signals from inorganic phosphate ions in a cell allow the hydrogen ion concentration (see Section 1.6) in the cell to be estimated.

Recombinant DNA technology (see Chapter 22) allows genes to be moved, *in vitro,* from one location to another in a chromosome. Genes even may be synthesized and/or rearranged *in vitro.* These can then be used to transform organisms and produce new genotypes. The biochemical consequences to the intact organism, of changing only a single gene, then can be assessed.

The fourth example of a nonreductionist, minimally intrusive approach is the transfer of macromolecules or even subcellular organelles from one organism to another. Microinjection and other technologies for accomplishing this now are well developed, at least for transfer to single cells. Often, the macromolecules or organelles will function in the foreign cell, allowing the effects of different, and nearly undisturbed, cellular backgrounds on the functioning of a particular enzyme or protein, for example, to be assessed.

1.3   ARRANGEMENT OF THIS BOOK

This fifth edition has three sections.

Part 1, Structures and Functions of Biological Molecules, is concerned with one of the oldest endeavors of biochemistry: chemical characterization of the macromolecules and the low molecular weight molecules of living systems. Although the cell is the premiere structure of living systems, we reserve an in-depth discussion of it for Chapter 8, the first chapter of Part 2. Chapter 8 is intended to build on preceding descriptions of the components of cells so that the cell, its membranes and compartments, can be presented in biochemical terms.

Part 2 is Energy Metabolism and Biosynthesis of Small Molecules. The highly specific and controlled chemical reactions by which small molecules and some macromolecules are synthesized, interconverted, and metabolized for energy production are described.

Part 3 is Genes, Gene Expression, and the Metabolism of Informational Macromolecules. The genetic control of the cell, as it is mediated by the synthesis of biological macromolecules, is the primary topic.

As an introduction to Part 1, we point out that most of the organic molecules of living systems fall into one of four general classes: carbohydrates, proteins, nucleic acids, and lipids. Only six elements account for almost the entire mass of these organic molecules: carbon, hydrogen, oxygen, nitrogen, phosphorus, and sulfur. The carbohydrates, compounds principally of carbon, hydrogen, and oxygen, serve mainly as sources of energy and structural materials. However, they also control the polarity of proteins and lipids to which they may be attached and often serve as recognition sites. As such, the carbohydrate portions of biological macromolecules may control their subcellular location and other important characteristics. We consider carbohydrates as the first of the four classes because they illustrate simply and directly certain important chemical characteristics of biological molecules, such as molecular asymmetry.

Proteins are polymers of amino acids and are the biological macromolecules that show the greatest diversity of function, including serving as structures and structural components, catalysts, and hormones.

Nucleic acids, the informational macromolecules, are essential contributors to that most characteristic properties of living systems: reproduction and genetic control. Low molecular weight forms of nucleic acids participate in energy metabolism and biosynthesis.

Lipids serve as energy stores and the principal building blocks of cell membranes. Certain lipids serve as hormones and other types of biological signals; others are key metabolic intermediates.

The organic and most of the inorganic constituents of the cell, as crucial as they are to the structure and functioning of the cell, constitute only a small fraction of the mass of a typical cell. The most abundant compound of the cell is, of course, water.

1.4 SOME IMPORTANT PROPERTIES OF WATER

Undoubtedly, no factor has had more influence on the development of life on earth than the peculiar properties of water and its abundance on our planet. Consider the group of compounds listed in Table 1.1. These compounds may be compared with H_2O because they all have protons bound to oxygen or some other electronegative atom. As can be seen, H_2O has the highest boiling point, the highest specific heat of vaporization, and, by far, the highest melting point of all these compounds. Pauling has expressed the anomalous behavior of H_2O in another way by comparing it with the hydrides of other elements in Group VI of the periodic table: H_2S, H_2Se, and H_2Te. When this is done, we would predict that H_2O should have boiling point of $-100°C$, instead of its $+100°C$ boiling point!

The water molecule is highly polarized. That is, its electrons are not evenly distributed over it. The electronegative oxygen atom tends to draw electrons away from the hydrogen atoms, leaving a net positive charge surrounding the

TABLE 1.1   **Some Physical Properties of Water and Other Compounds**

SUBSTANCE	MELTING POINT (°C)	BOILING POINT (°C)	HEAT OF VAPORIZATION (cal/g)	HEAT CAPACITY (cal/g)	HEAT OF FUSION (cal/g)
Water	0	100	540	1.00	80
Ethanol	−114	78	204	0.58	25
Methanol	−98	65	263	0.60	22
NH_3	−78	−33	327	1.12	84
H_2S	−83	−60	132	—	17
HF	−92	19	360	—	55

proton. Because of this polarization, water molecules behave like dipoles and can be oriented toward both positive and negative ions. This property, in turn, accounts for the unusual ability of water to act as a solvent for polar compounds. Positive or negative ions in a crystal lattice can be approached by dipolar water molecules and brought into solution. Once in solution, ions of both positive and negative charge will be surrounded by protective layers of water molecules, and further interaction between those ions of opposite charge subsequently will be decreased.

The high boiling and melting points of H_2O and its high heat of vaporization are the result of an interaction between adjacent water molecules known as hydrogen bonding. Briefly put, the term **hydrogen bond** refers to the interaction of a hydrogen atom that is covalently bonded to one electronegative atom with the electrons of a second electronegative atom to which it is not directly covalently bonded. There is a tendency for the hydrogen atom to associate with the second electronegative atom by sharing the nonbonded electron pair of that atom, and a weak bond of approximately 4.5 kcal/mole can exist. In biological material, the two atoms most commonly involved in hydrogen bonding are nitrogen and oxygen. In ice, the water molecules are nearly completely hydrogen-bonded. Even in liquid water, small transient chains of water molecules will occur due to this interaction.

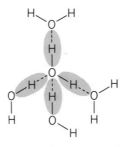

The energy necessary to disrupt even the most stable hydrogen bonds (4–10 kcal/mole) is much less than that required to break most covalent bonds. In an aqueous solution, hydrogen bonds are broken and formed readily. The cumulative effect of hydrogen bonding in water is a major factor in explaining many of the unusual properties of H_2O. Thus, the extra energy required to boil water and melt ice may be attributed largely to extensive hydrogen bonding.

Other unusual properties of water make it an ideal medium for living organisms. The specific heat capacity of H_2O—the number of calories required to raise the temperature of 1 g of water from 15 to 16°C—is 1.0 and is unusually high among several of the hydrogen bonding solvents considered in Table 1.1. Only a few solvents, such as liquid ammonia, have greater heat capacities than water. The greater the specific heat of a substance, the less the change in temperature that results when a given amount of heat is absorbed by that substance. Thus, H_2O is well-designed for keeping the temperature of a living organism relatively constant.

The heat of vaporization of water, as already mentioned, is unusually high. Expressed as the specific heat of vaporization (calories absorbed per gram vaporized), the value for water is 540 at its boiling point and even greater at lower temperatures. This high value is very useful in helping the living organism keep its temperature constant, since a large amount of heat can be dissipated by vaporization of H_2O.

The high heat of fusion of water (Table 1.1) is also of significance in stabilizing the biological environment. Although cellular water rarely freezes in higher living forms, the heat released by H_2O on freezing is a major factor in slowing the temperature drop of a body of water during the winter. Thus, a gram of H_2O must give up 80 times as much heat in freezing at 0°C as it does in being lowered from 1 to 0°C just before freezing.

One final example of a property of H_2O that is of biological significance must be cited: H_2O expands on solidifying, and ice is less dense than water. Only a few other substances expand on freezing. The importance of this property for biology has long been recognized. If ice were heavier than liquid water, it would sink to the bottom of its container on freezing. This would mean that oceans, lakes, and streams would freeze from the bottom to the top and, once frozen, would be extremely difficult to melt. Such a situation would obviously be incompatible with those bodies of water serving as the habitat of many living forms, as they do. As it is, however, the warmer, liquid water falls to the bottom of any lake and the ice floats on top where heat from the external environment can reach and melt it.

Additional properties of water such as high surface tension and a high dielectric constant have significance in biology. However, we refer the student to the classical publication by Henderson, *The Fitness of the Environment,* in which this subject is discussed in more detail. Instead, in the next section, we consider the ability of water to dissociate into hydrogen (H^+) and hydroxyl (OH^-) ions and the mechanisms by which the concentrations of hydrogen and hydroxyl ions in aqueous solutions are controlled, both experimentally and in the living organism. To do this, we review first the *law of mass action* and the *ion product of water.*

1.5 THE LAW OF MASS ACTION

For the reaction

$$A + B \rightleftharpoons C + D \tag{1.1}$$

in which two reactants A and B interact to form two products C and D, we may write the expression

$$K_{eq} = \frac{C_C \cdot C_D}{C_A \cdot C_B} \tag{1.2}$$

This is an expression of the **law of mass action,** applied to Reaction 1.1, which states that *at equilibrium, the product of the concentrations of the product divided by the product of the concentrations of the reactants is a constant known as the equilibrium constant, K_{eq}.* This constant is fixed for any given temperature. If the concentration of any single component of the reaction is varied, it follows that the concentration of at least one other component must also change in order to meet the conditions of the equilibrium as defined by K_{eq}.

To be precise, we should distinguish between the concentration of the reactants and products in this reaction and the **activity** or **effective concentration** of these reactants and products. It has long been recognized that the concentration of a substance does not precisely reflect its reactivity in a chemical reaction. Moreover, these discrepancies in behavior are appreciable when the concentration of reactant is large. At high concentrations, the individual reactant molecules may exert a mutual attraction on each other or exhibit alterations in their interactions with the solvent in which the reaction occurs. On the other hand, in dilute solutions, chemical reactivity may be proportional to reactant concentration, indicating that such interactions are negligible. In order to correct for the difference between concentration and effective concentration, the activity coefficient γ was introduced. Thus,

$$a_A = C_A \times \gamma \tag{1.3}$$

where a_A refers to the activity and C_A to the concentration of the substance. The activity coefficient is not a fixed quantity but varies in value depending on the solution composition. At very dilute concentrations, the activity coefficient of the solute approaches unity, because there is little if any solute-solute interaction. At infinite dilution, the activity and the concentration are the same. For the purpose of this book, we usually will not distinguish between activities and concentrations; rather, we use the latter term. Often, this does not cause serious error, since the reactants in many biochemical reactions are quite low in concentration, including the H^+ ion concentration, the subject of the next subsection.

1.6 DISSOCIATION OF WATER AND ITS ION PRODUCT, K_w

Water is far from inert in its service as a solvent for biochemical reactions. In addition to its abilities as a solvent, water is chemically reactive. It is a weak electrolyte that dissociates only slightly to form H^+ and OH^- ions.

$$H_2O \rightleftharpoons H^+ + OH^- \tag{1.4}$$

The equilibrium constant for this dissociation reaction has been accurately measured, and at 25°C, it has the value 1.8×10^{-16} mole/liter. That is,

$$K_{eq} = \frac{C_{H^+}C_{OH^-}}{C_{H_2O}} = 1.8 \times 10^{-16}$$

The concentration of H_2O (C_{H_2O}) in pure water may be calculated to be 1000 g/liter divided by 18 g/mole, giving 55.5 moles/liter. Since the concentration of H_2O in dilute aqueous solutions is essentially unchanged from that in pure H_2O, this figure may be taken as a constant. It is, in fact, usually incorporated into the expression for the dissociation of water, to give

$$C_{H^+}C_{OH^-} = 1.8 \times 10^{-16} \times 55.5 = 1.01 \times 10^{-14}$$
$$= K_w = 1.01 \times 10^{-14} \tag{1.5}$$

at 25°C.

This new constant K_w, termed the **ion product of water,** expresses the relation between the concentration of H^+ and OH^- ions in dilute aqueous solutions; for example, this relation may be used to calculate the concentration of H^+ in pure water. To do this, let x equal the concentration of H^+. Since in pure water, one OH^- is produced for every H^+ formed on dissociation of a molecule of H_2O, x must also equal the concentration of OH^-. Substituting in Equation 1.5, we have

$$(x)(x) = 1.01 \times 10^{-14} = x^2$$

Therefore,
$$x = 1 \times 10^{-7}$$
$$= C_{H^+} = C_{OH^-} = 1.0 \times 10^{-7} \text{ mole/1}$$

1.7 pH

In 1909, Sorensen introduced the term **pH** as a convenient manner of expressing the concentration of H^+ ion by means of a base 10 logarithmic function; pH may be defined as

$$pH = \log \frac{1}{a_{H^+}} = -\log a_{H^+} \tag{1.6}$$

where a_{H^+} is defined as the activity of H^+. If the activity coefficient is assumed to be 1, then

$$pH = \log \frac{1}{[H^+]} = -\log [H^+] \tag{1.7}$$

In this equation, to indicate that we are dealing with concentrations, we use brackets. Thus, the concentration of H^+, (C_{H^+}), is represented as $[H^+]$.

The hydrogen ion is one of the few substances for which estimates of both the concentration and the activity routinely are obtained in the biochemical laboratory. The pH meter is an electrochemical instrument that is connected to two electrodes. These electrodes are in contact with a solution, one directly and one through a special glass membrane that is far more permeable to protons than it is to most other cations. The potential difference between the two electrodes, which is measured by the pH meter, is related logarithmically to the hydrogen ion activity and hence to the pH as defined by Equation 1.6. (Note that in some modern pH meter systems, the two electrodes are combined into one "concentric" electrode.)

The distinction between activity and concentration is indicated by the following example. The pH of $0.1M$ HCl when measured with a pH meter is 1.09. This value can be substituted in Equation 1.6, as the pH meter measures activities.

$$1.09 = \log \frac{1}{a_{H^+}}$$
$$a_{H^+} = 10^{-1.09}$$
$$= 8.1 \times 10^{-2} \text{ moles/liter}$$

The concentration of hydrogen ion is expected to be 0.1 moles/liter in $0.1M$ HCl since HCl is a strong, fully dissociated acid. The activity coefficient may be calculated as

$$\gamma = a_{H^+}/[H^+]$$
$$= 0.081/0.1$$
$$= 0.81$$

Thus, even at a concentration of $0.1M$, the activity coefficient for hydrogen ions in an HCl solution is only slightly less than 1. It approaches 1 more closely at the lesser hydrogen ion concentrations that are characteristic of most biochemical reactions.

It is important to stress that the pH is a logarithmic function; thus, when the pH of a solution is decreased one unit from 5 to 4, the H^+ concentration has increased tenfold from $10^{-5}M$ to $10^{-4}M$. When the pH has increased three-tenths of a unit from 6 to 6.3, the H^+ concentration has decreased, by a factor of about 2, from $10^{-6}M$ to $5 \times 10^{-7}M$.

If we now apply the term of pH to the ion product expression for pure water, we obtain another useful expression

$$[H^+] \times [OH^-] = 1.0 \times 10^{-14}$$

We take the logarithms of this equation

$$\log [H^+] + \log [OH^-] = \log (1.0 \times 10^{-14})$$
$$= -14$$

and multiply by -1

$$-\log [H^+] - \log [OH^-] = 14$$

If we now define $-\log [OH^-]$ as pOH, a definition similar to that of pH, we have an expression relating the pH and pOH in any aqueous solution

$$pH + pOH = 14 \tag{1.8}$$

1.8   BRÖNSTED ACIDS

A most useful definition of acids and bases in biochemistry is that proposed by Brönsted. He defined *an acid as any substance that can donate a proton,* and *a base as a substance that can accept a proton.* Although other definitions of acids, notably one proposed by Lewis, are even more general, the Brönsted concept should be thoroughly understood by students of biochemistry.

The following substances shown in orange are examples of Brönsted acids:

$$HCl \longrightarrow H^+ + Cl^-$$
$$CH_3COOH \longrightarrow H^+ + CH_3COO^-$$
$$NH_4^+ \longrightarrow NH_3 + H^+$$

and the generalized expression would be

$$HA \longrightarrow H^+ + A^-$$

The corresponding bases are now shown reacting with a proton:

$$Cl^- + H^+ \longrightarrow HCl$$
$$CH_3COO^- + H^+ \longrightarrow CH_3COOH$$
$$NH_3 + H^+ \longrightarrow NH_4^+$$

Note that the first of these three reactions does not occur to an appreciable extent in aqueous solution because HCl is a strong acid.

The corresponding base for the generalized weak acid HA is

$$A^- + H^+ \longrightarrow HA$$

It is customary to refer to the acid–base pair as follows: HA (or HA^+ in the case of protonated amines) is the **Brönsted acid** because it can furnish a proton; the anion A^- (or neutral A in the case of amines) is called the **conjugate base** because it can accept the proton to form the acid HA.

1.9 IONIZATION OF WEAK ACIDS

Hydrochloric acid (HCl), a familiar mineral acid, is completely dissociated in H_2O.

$$HCl \longrightarrow H^+ + Cl^- \tag{1.9}$$

Note that the dissociation is represented as an irreversible reaction. The same is true of strong bases, such as NaOH, and the salts of strong bases and strong acids, such as NaCl. A weak acid, in contrast to a strong acid, is only partially ionized in aqueous solution. This is represented by a reversible ionization of the generalized weak acid, HA.

$$\begin{array}{cccc} HA & + & H_2O & \rightleftharpoons & H_3O^+ & + & A^- \\ \text{(Conjugate} & & \text{(Conjugate} & & \text{(Conjugate} & & \text{(Conjugate} \\ \text{acid, 1)} & & \text{base, 2)} & & \text{acid, 2)} & & \text{base, 1)} \end{array}$$

or the simpler   (1.10)

$$HA \rightleftharpoons H^+ + A^-$$

The proton donated by HA is accepted by H_2O to form the hydronium ion, H_3O^+. Although it is strictly correct to recognize that the proton is hydrated in solution, other ions are as well, and the formulas for ions generally are written in the simple, unhydrated form. Hence, we can write the equilibrium constant as

$$K_{eq} = \frac{[H^+][A^-]}{[HA]} = K_a \tag{1.11}$$

where K_a is the dissociation constant for the weak acid.

1.10 IONIZATION OF WEAK BASES

One way to consider the ionization of a weak base is by direct analogy with the ionization of a strong base, writing the equation as reversible rather than the irreversible $NaOH \rightarrow Na^+ + OH^-$. That is, both a strong base and a weak base may be defined as a substance that furnishes OH^- ions on dissociation.

$$BOH \rightleftharpoons B^+ + OH^-$$

$$K_{eq} = K_b = \frac{[B^+][OH^-]}{[BOH]} \tag{1.12}$$

For ammonium hydroxide (NH_4OH), for example, the K_b is given in chemical handbooks as 1.8×10^{-5}. Therefore, the extent of dissociation of NH_4OH is identical with that of acetic acid (CH_3COOH; $K_a = 1.8 \times 10^{-5}$). The important difference, of course, is that NH_4OH dissociates to form hydroxyl ions (OH^-), whereas CH_3COOH dissociates to form protons (H^+), and that the pH of $0.1M$ solutions of these two substances is by no means similar, the former being above pH 7 and the latter, acetic acid, below pH 7.

One of the most common types of weak base encountered in biochemistry is the group called organic amines (e.g., the amino groups of amino acids). Such compounds, when represented with the general formula $R-NH_2$, do not contain hydroxyl groups that can dissociate as in Reaction 1.12. Note that even ammonium hydroxide itself can be written in this form, as HNH_2, or more simply, NH_3. Such compounds can ionize in H_2O to produce hydroxyl ions, according to an equation that has a form similar to that of Equation 1.10.

$$RNH_2 \quad + \quad H_2O \quad \rightleftharpoons \quad OH^- \quad + \quad RNH_3^+$$

(Conjugate	(Conjugate	(Conjugate	(Conjugate
base, 1)	acid, 2)	base, 2)	acid, 1)

or, alternatively, $\tag{1.13}$

$$RNH_2 \quad + \quad H_3O^+ \quad \rightleftharpoons \quad H_2O \quad + \quad RNH_3^+$$

(Conjugate	(Conjugate	(Conjugate	(Conjugate
base, 1)	acid, 2)	base, 2)	acid, 1)

which simplifies to

$$RNH_2 + H^+ \rightleftharpoons RNH_3^+$$

In the first representation of this reaction, H_2O serves as an acid to contribute a proton to the base RNH_2. The corresponding ionization constant is

$$K_{ion} = \frac{[RNH_3^+][OH^-]}{[RNH_2][H_2O]} \tag{1.14}$$

that by analogy with Equation 1.12 can be rewritten as

$$K_b = [H_2O]K_{ion} = \frac{[RNH_3^+][OH^-]}{[RNH_2]} \tag{1.15}$$

The simplified version of Equation 1.13 can be written in reverse, as we have done for NH_4^+ in Section 1.8. This is the dissociation of a Brönsted weak acid.

$$RNH_3^+ \rightleftharpoons RNH_2 + H^+$$

The corresponding dissociation equilibrium constant is defined by

$$K_a = \frac{[RNH_2][H^+]}{[RNH_3^+]} \tag{1.16}$$

Thus, it is possible, and convenient, to treat both weak acids and weak bases as Brönsted acids by simply writing the dissociation of the protonated form of the base. Both tables of pK_b and of pK_a values are published for weak bases. Note that these two quantities are simply related. Given one, the other can be calculated. By multiplying Equations 1.15 and 1.16 together, we obtain

$$(K_a)(K_b) = [H^+][OH^-] = K_w$$

or   (1.17)

$$pK_a + pK_b = -\log K_w = 14$$

1.11 THE HENDERSON–HASSELBALCH EQUATION

Henderson and Hasselbalch have rearranged the **mass law** equation, as it applies to the ionization of weak acids, into a useful expression known as the Henderson–Hasselbalch equation. Consider the ionization of a generalized weak acid HA.

$$HA \rightleftharpoons H^+ + A^-$$
$$K_{ion} = K_a = \frac{[H^+][A^-]}{[HA]}$$

Rearranging terms, we have

$$[H^+] = K_a \frac{[HA]}{[A^-]}$$

Taking logarithms, we find

$$\log [H^+] = \log K_a + \log \frac{[HA]}{[A^-]}$$

and multiplying by -1,

$$-\log [H^+] = -\log K_a - \log \frac{[HA]}{[A^-]}$$

If $-\log K_a$ is defined as pK_a and $\log [A^-]/[HA]$ is substituted for $-\log [HA]/[A^-]$, we obtain

$$pH = pK_a + \log \frac{[A^-]}{[HA]}$$ (1.18)

Note that when $[A^-] = [HA]$, $pH = pK_a$ because the logarithm of 1 is zero.

This form of the Henderson–Hasselbalch equation can be written in a more general expression, in which we replace $[A^-]$ with the term "conjugate base" and $[HA]$ with "conjugate acid."

$$pH = pK_a + \log \frac{[\text{Conjugate base}]}{[\text{Conjugate acid}]}$$ (1.19)

This expression may then be applied not only to weak acids such as acetic acid, but also to the ionization of ammonium ions and those substituted amino groups found in amino acids. In this case, NH_4^+ ions or the protonated amino groups RNH_3^+ are the conjugate acids that dissociate to form protons and the conjugate bases NH_3 and RNH_2, respectively.

$$NH_4^+ \rightleftharpoons NH_3 + H^+$$
$$RNH_3^+ \rightleftharpoons RNH_2 + H^+$$

Applying Equation 1.19 to the protonated amine, we have

$$pH = pK_a + \log \frac{[RNH_2]}{[RNH_3^+]} \tag{1.20}$$

Biochemistry handbooks often list the K_a (or pK_a) for the conjugate acids of substances we normally consider as bases (e.g., NH_4OH, amino acids, organic amines). If they do not, the K_b (or pK_b; see Equation 1.17) for the ionization of the weak base will certainly be listed, and the K_a or (pK_a) must first be calculated before employing the generalized Henderson-Hasselbalch equation. Although more care must be taken to identify correctly the conjugate acid–base pairs in that expression, its use leads directly to the pH of mixtures of weak bases and their salts.

1.12   TITRATION CURVES AND BUFFERING ACTION

The titration curve obtained when 100 ml of $0.1N\,CH_3COOH$ is titrated with $0.1N\,NaOH$ is shown in Figure 1.1. This curve can be obtained experimentally in the laboratory by measuring the pH of $0.1N\,CH_3COOH$ before and after the addition of different aliquots of $0.1N\,NaOH$. The curve may also be calculated by the Henderson–Hasselbalch equation for all the points except the first, for which no NaOH has been added, and the last, for which a stoichiometric amount (100 ml) of $0.1N\,NaOH$ has been added. Clearly, the Henderson-Hasselbalch equation cannot be used to determine the pH at the limits of the titration, when the ratio of salt to acid is either zero or infinite.

The pH of the acetic acid solution, before any NaOH has been added, may be calculated from Equation 1.11. Equal concentrations of acetate ion and protons are generated by the dissociation of pure acetic acid in water. If we set the concentration of acetate ion and protons equal to x and the concentration of undissociated acetic acid equal to $(C - x)$, Equation 1.11 becomes

$$K_a = x^2/(C - x)$$

or $\hspace{10cm}$ (1.21)

$$x^2 = K_a(C - x)$$

which is a quadratic equation. This can be solved for x by the standard methods for quadratic equations. Alternatively, we can simplify the equation for a limited extent of dissociation by recognizing that $(C - x)$ is approximately equal to C. Then,

$$x = \sqrt{K_a C} \tag{1.22}$$

Since the K_a for acetic acid is 1.8×10^{-5} and the concentration of acetic acid in our example is $0.1M$, $x = 1.3 \times 10^{-3}$. Since x is small compared to C ($0.1M$), our assumption in making the above approximation is justified, and the same

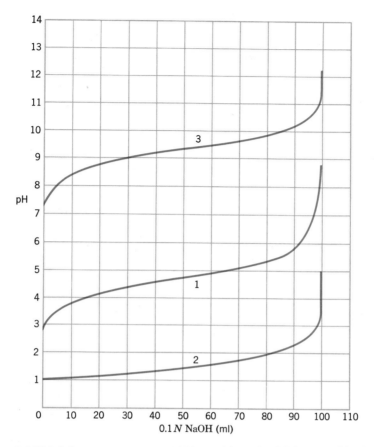

FIGURE 1.1 Titration curve of 100 ml of 0.1N CH$_3$COOH (1), 100 ml 0.1N HCl (2), and 100 ml of 0.1N NH$_4$Cl (3) with 0.1N NaOH.

answer would, in fact, be obtained by solution of the quadratic equation. For $x = [H^+] = 0.0013M$, pH $= 2.88$, in agreement with curve 1 of Figure 1.1.

At the end of the titration of acetic acid (curve 1, Figure 1.1), we will have a 200 ml solution of 0.05M sodium acetate. The reaction that describes the equilibrium at this point in the titration, the titration end point, is

$$Ac^- + H_2O \rightleftharpoons HAc + OH^-$$

We write an equilibrium constant as

$$K = \frac{[HAc][OH^-]}{[Ac^-][H_2O]}$$

From this, we derive

$$K_{Hydr} = K[H_2O] = \frac{[HAc][OH^-]}{[Ac^-]}$$

because the concentration of water remains approximately constant. The subscript "Hydr" recognizes that the acetate anion is undergoing a hydrolysis reaction. By analogy with the equation developed in the previous paragraph, we set $[OH^-] = [HAc] = x$. Then,

$$x = \sqrt{CK_{Hydr}}$$

using the approximation that the concentration of acetate anion, C, will not be changed significantly from $0.05M$ by the hydrolysis reaction. The student will be able to show that $K_{Hydr} = K_w/K_a$. The result is $[OH^-] = 5.3 \times 10^{-6}M$, which corresponds to a pH of 8.7, again in agreement with curve 1 of Figure 1.1.

In considering the shape of the titration curve of acetic acid, we see that the change in pH per unit of alkali added is greatest at the beginning and end of the titration, whereas the smallest change in pH for each unit of alkali is obtained when the titration is half-complete. In other words, an equimolar mixture of sodium acetate and acetic acid shows less change in pH initially, when acid or alkali is added, than a solution consisting mainly of either acetic acid or sodium acetate. We refer to the ability of a solution to resist a change in pH as its **buffering action,** and it can be shown that a buffer exhibits its *maximum* action when the titration is half-complete. This corresponds to $[HA] = [A^-]$ in the Henderson-Hasselbalch equation, that is, when the pH is equal to the pK_a (Equation 1.18). In Figure 1.1, the point of maximum buffering action for acetic acid and acetate anions is at the pH of 4.74, the value of the pK_a of acetic acid.

Another way of representing the condition that exists when the pH of a mixture of acetic acid and sodium acetate is at the pK_a is to state that the acid at this pH is half-ionized. That is, half the "total acetate" species is present as undissociated CH_3COOH, while the other half is in the form of acetate ion, CH_3COO^-. Since at its pK_a any weak acid will be half-ionized, this is one of the most useful ways of distinguishing between individual weak acids. The pK_a is also a characteristic property of each acid, because the ionization constant is a function of the inherent properties of the weak acid.

The titration curve of $0.1N$ HCl is also represented in Figure 1.1. The Henderson-Hasselbalch equation is of no use in calculating the curve for HCl, since it applies only for weak electrolytes. However, the pH at any point on the HCl curve can be calculated by determining the milliequivalents of HCl remaining and correcting for the volume. Thus, when 30 ml of $0.1N$ NaOH has been added, 7.0 meq of HCl will remain in a volume of 130 ml. The concentration of H^+ will therefore be 7.0/130 or $0.054M$. If the activity coefficient is neglected, the pH may be calculated from Equation 1.7 as 1.27.

Curve 3 in Figure 1.1 is the titration curve obtained when 100 ml of $0.1N$ NH_4Cl is titrated with $0.1N$ NaOH. In this titration, the protons contributed by NH_4^+ are neutralized by the OH^- ions provided by the NaOH.

$$NH_4^+ + OH^- \longrightarrow NH_3 + H_2O$$

Again, the Henderson-Hasselbalch equation is of no value when calculating the pH of the solution of NH_4Cl before any NaOH has been added. Instead, we can use an equation of the form of Equation 1.21 or 1.22, with NH_4^+ as the acid and NH_3 as the conjugate base. However, the Henderson-Hasselbalch equation can be employed to determine any of the other points on the curve, so long as the NH_4Cl has not been completely titrated.

Up to this point, we have considered only monobasic acids, such as acetic acid. Polybasic or **polyprotic** acids, which commonly are encountered in biochemistry, are acids capable of ionizing to yield more than one proton per molecule of acid. Each step of the dissociation has associated with it a K_{ion} or K_a. In the case of phosphoric acid (H_3PO_4), three protons may be furnished on complete ionization of a mole of this acid.

$$H_3PO_4 \rightleftharpoons H^+ + H_2PO_4^- \qquad K_{a1} = 7.5 \times 10^{-3} \qquad pK_{a1} = 2.12$$
$$H_2PO_4^- \rightleftharpoons H^+ + HPO_4^{2-} \qquad K_{a2} = 6.2 \times 10^{-8} \qquad pK_{a2} = 7.21$$
$$HPO_4^{2-} \rightleftharpoons H^+ + PO_4^{3-} \qquad K_{a3} = 4.8 \times 10^{-13} \qquad pK_{a3} = 12.32$$

This means that at the pH of 2.12, the first ionization of H_3PO_4 is half-complete; the pH must be 12.32, however, before the third and final ionization of H_3PO_4 is 50% complete. At the pH of 7.0, which is frequently encountered in the cell, the second proton of phosphoric acid ($pK_{a2} = 7.21$) will be about half-dissociated. At this pH, the monoanion $H_2PO_4^-$ and the dianion HPO_4^- of phosphoric acid will be present in approximately equal concentrations. A phosphoric acid ester, such as α-glycerol phosphate, has only two ionizations. These correspond, approximately, to the first two pK_a's of phosphoric acid, so that at pH 7, the most abundant species, indicated below, are present in approximately equal concentrations. The phosphate ester thus

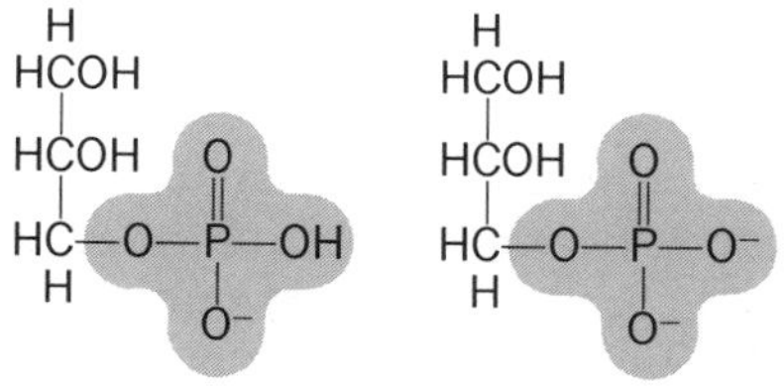

```
    H                      H
   HCOH                   HCOH
    |                      |
   HCOH   O               HCOH   O
    |     ‖                |     ‖
   HC—O—P—OH              HC—O—P—O⁻
    |     |                |     |
    H     O⁻               H     O⁻
```

appears to be "missing" the "last" pK_a. This is due in large measure to electrostatics. It is more difficult to remove a positively charged proton from a phosphate monoanion or phosphate ester monoanion than from unionized phosphoric acid or phosphoric acid ester. Thus, for both compounds, the first ionization corresponds to that of a similarly strong acid and the second to that of a similarly weak acid, since the same ionic charges are involved for phosphoric acid and its monoester.

Many of the common organic acids encountered in intermediary metabolism are polyprotic; for example, succinic acid ionizes according to the following scheme:

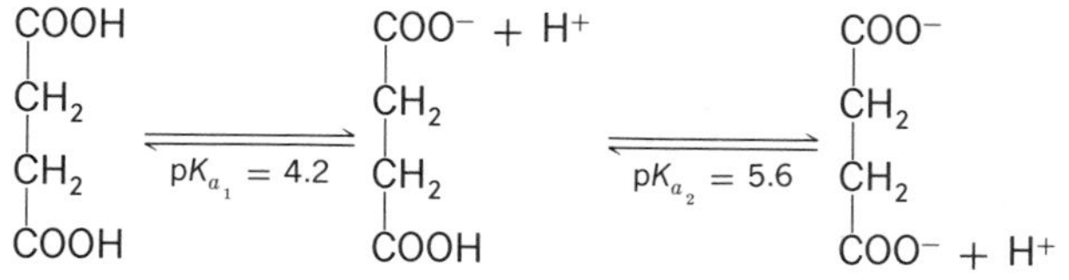

```
  COOH              COO⁻ + H⁺            COO⁻
   |                 |                    |
  CH₂               CH₂                  CH₂
   |      pKₐ₁=4.2    |      pKₐ₂=5.6      |
  CH₂    ⇌          CH₂      ⇌           CH₂
   |                 |                    |
  COOH              COOH                 COO⁻ + H⁺
```

At pH 7.0 in the cell, succinic acid will exist predominantly as the dianion $^-OOC—CH_2—CH_2—COO^-$, commonly referred to as succinate. Most of the organic monoacids that serve as metabolites in the cell, for example, palmitic acid, lactic acid, and pyruvic acid, are acids of strength equal to or greater than that of acetic acid. Thus, like succinate, they will be present as their anions. This has led to the common use of the names of the anions (palmitate, lactate, and pyruvate) when these compounds are discussed in biochemistry. In writing chemical reactions, however, it will be the practice in this text to use the formulas for the undissociated acid. Table 1.2 lists the pK_a's for several of the organic acids commonly encountered in intermediary metabolism.

TABLE 1.2   The pK_a of Some Organic Acids

	pK_{a1}	pK_{a2}	pK_{a3}
Acetic acid (CH_3COOH)	4.74		
Acetoacetic acid (CH_3COCH_2COOH)	3.58		
Citric acid ($HOOCCH_2C(OH)(COOH)CH_2COOH$)	3.09	4.75	5.41
Formic acid ($HCOOH$)	3.62		
Fumaric acid ($HOOCCH{=}CHCOOH$)	3.03	4.54	
DL-Glyceric acid ($CH_2OHCHOHCOOH$)	3.55		
DL-Lactic acid ($CH_3CHOHCOOH$)	3.86		
DL-Malic acid ($HOOCCH_2CHOHCOOH$)	3.40	5.26	
Pyruvic acid ($CH_3COCOOH$)	2.50		
Succinic acid ($HOOCCH_2CH_2COOH$)	4.18	5.56	

1.13   DETERMINING pK_a VALUES

The pK_a of any dissociable groups is a characteristic property of a molecule and one that is relatively easy to determine. The pK_a may be determined in the laboratory by preparing a titration curve experimentally, using a pH meter, a buret, and titrant. As known amounts of alkali or acid are added to a solution of the unknown, the pH is determined, and the titration curve can be plotted. From this curve, the inflection point (pK_a) may be determined. This is possible for molecules with single titratable groups, such as acetic acid, and for molecules such as phosphoric acid that undergo multiple ionizations. However, when the pK_a's are closely spaced, for example, for citric acid (Table 1.2), the pK_a values do not correspond exactly to inflection points, and a more complex mathematical analysis is required to determine them.

At this point, we urge the student to review chemical stoichiometry. The meanings of gram molecular weight (mole) and gram equivalent weight (equivalent) and the significance of molarity, molality, and normality must be thoroughly understood. Biochemistry is a quantitative science, and the student must recognize immediately such terms as millimole and micromole. In connection with titrations, for example, we must be aware that the H^+ concentrations of $0.1 N\,H_2SO_4$ and $0.1 N\,CH_3COOH$ are by no means similar, although 1 liter of each of these solutions contains the same amount of total titratable acid.

1.14   BUFFERS

Control of pH is an essential property of biological systems. The pH of human blood plasma is maintained within 0.2 pH unit of 7.2 to 7.3; values outside this range are not compatible with life. The enzymes (Chapter 4) that are responsible for the catalysis of reactions both within and outside of cells frequently exhibit their maximum catalytic action at some definite pH, and some have significant catalytic activity only within a narrow pH range. Metabolic reactions constantly produce Brönsted acids and bases, often not in balanced amounts. Any imbalance that tends to shift the pH outside the cell's pH range for optimum activity must be compensated. The control of pH in natural biochemical systems is accomplished both by buffers, the subject of this section, and by energy-requiring, active mechanisms.

Biological systems contain dissolved proteins, organic substrates, and inorganic salts, many of which can act as buffers. In the laboratory, the biochemist wishes to examine reactions *in vitro* under conditions where the change in pH is minimal. He or she most often obtains these conditions by using efficient buffers, preferably inert ones, in the reactions under investigation. The buffers may include weak acids such as phosphoric and acetic, or maleic acids or weak bases such as ammonia, pyridine, and tris-(hydroxymethyl)amino methane.

With a thorough understanding of the ionization of weak electrolytes, it is possible to discuss buffered solutions. A *buffered solution is one that resists a change in pH on the addition of acid or alkali.* Most commonly, the buffer solution consists of a mixture of a weak Brönsted acid and its conjugate base; for example, mixtures of acetic acid and sodium acetate or of ammonium hydroxide and ammonium chloride are buffer solutions.

Let us consider the mechanism by which a buffer solution exerts control over large pH changes. When alkali (e.g., NaOH) is added to a mixture of acetic acid (CH_3COOH) and potassium acetate (CH_3COOK), the following reaction occurs:

$$OH^- + CH_3COOH \longrightarrow CH_3COO^- + H_2O$$

This reaction states that OH^- ion reacted with protons furnished by the dissociation of the weak acid and formed H_2O.

$$CH_3COOH \rightleftharpoons CH_3COO^- + H^+$$
$$\qquad\qquad\qquad\qquad\quad |\!\!-\!\!OH^-$$
$$\qquad\qquad\qquad\qquad\quad \downarrow$$
$$\qquad\qquad\qquad\qquad\quad H_2O$$

On the addition of alkali, there is a further dissociation of the available CH_3COOH to furnish additional protons and thus to keep the H^+ concentration or pH unchanged.

When acid is added to an acetate buffer, the following reaction occurs:

$$H^+ + CH_3COO^- \longrightarrow CH_3COOH$$

The protons added (in the form of HCl, e.g.) combine exceedingly rapidly with the CH_3COO^- anion present in the buffer mixture (as potassium acetate) to form the undissociated weak acid CH_3COOH. Consequently the resulting pH change is much less than would occur if the conjugate base were absent.

In discussing the quantitative aspects of buffer action, we should point out that two factors determine the effectiveness or **capacity** of a buffer solution. Clearly, the molar concentration of the buffer components is one of them. The buffer capacity is directly proportional to the concentration of the buffer components. Here, we encounter the convention used in referring to the concentration of buffers. The concentration of a buffer may be defined as the sum of the concentration of the weak acid and its conjugate base. Thus, a $0.1M$ acetate buffer could contain 0.05 mole of acetic acid and a 0.05 mole of sodium acetate in 1 liter of H_2O. It could also contain 0.065 mole of acetic acid and 0.035 mole of sodium acetate in 1 liter of H_2O.

The second factor influencing the effectiveness of a buffer solution is the

ratio of the concentration of the conjugate base to the concentration of the weak acid. Quantitatively, it should seem evident that the most effective buffer would be one with **equal concentrations** of basic and acidic components, since such a mixture could furnish **equal quantities** of basic or acid components to react, respectively, with acid or alkali. An inspection of the titration curve for acetic acid (Figure 1.1) similarly shows that the minimum change in pH resulting from the addition of a unit of alkali (or acid) occurs at the pK_a for acetic acid. That this is so can be proved mathematically. At this pH, we have already seen that the ratio of CH_3COO^- to CH_3COOH is 1. On the other hand, at values of pH far removed from the pK_a (and therefore, at ratios of conjugate base to acid greatly differing from unity), the change in pH for each added unit of acid or alkali is much larger, that is, the buffer capacity is much less.

Having stated the two factors that influence the buffer capacity, we may consider the decisions involved in selecting a buffer to be effective at the desired pH value, for example, pH = 5. Clearly, it would be most desirable to select a weak acid having a pK_a of 5.0. If this cannot be done, the weak acid whose pK_a is closest to 5.0 is the first choice. In addition, it is evident that we would want to use as high a concentration as is compatible with other features of the system. Too high a concentration of salt frequently inhibits the activity of enzymes or

TABLE 1.3   **Buffers**

COMPOUND	pK_{a1}	pK_{a2}	pK_{a3}	pK_{a4}
N-(2-acetamido-)iminodiacetic acid (ADA)	6.6			
Acetic acid	4.7			
Ammonium chloride	9.3			
Carbonic acid	6.1	10.3		
Citric acid	3.1	4.7	5.4	
Diethanolamine	8.9			
Ethanolamine	9.5			
Fumaric acid	3.0	4.5		
Glycine	2.3	9.6		
Glycylglycine	3.1	8.1		
Histidine	1.8	6.0	9.2	
N-2-Hydroxyethylpiperazine-N'-2-ethanesulfonic acid (HEPES)	7.6			
Maleic acid	2.0	6.3		
2-(N-morpholino)-ethanesulfonic acid (MES)	6.2			
Phosphoric acid	2.1	7.2	12.3	
Pyrophosphoric acid	0.9	2.0	6.7	9.4
Triethanolamine	7.8			
Tris-(hydroxymethyl)aminomethane (Tris)	8.0			
N-Tris(hydroxymethyl)methyl-2-amino-ethanesulfonic acid (TES)	7.5			
Sodium diethylbarbiturate	8.0			
Ethylenediaminetetraacetic acid	2.0	2.7	6.2	10.3

other physiological systems, however. The solubility of the buffer components may also limit the concentration that can be employed.

Table 1.3 lists the pK_a for some buffers commonly employed in biochemistry.

1.15 PHYSIOLOGICAL BUFFERS

The ability to ionize is an essential property of many biological compounds. Organic acids, amino acids, proteins, nucleotides, and other phosphate esters are examples of biochemicals that are ionized to varying degrees in biological systems. Since the pH of most biological fluids is near 7, the extent of dissociation of some of these compounds may be complete there. For example, the first ionization of H_3PO_4 will be complete; the second ionization ($pK_{a2} = 7.2$) will be approximately half-complete. Other examples will be encountered later in the book. One may ask which of these are physiologically significant as buffers in the intact organism. The answer is dependent on several factors including those listed in the preceding section; that is, the molar concentration of the buffer components and the ratio of the concentration of the conjugate base to that of the weak acid. The former of these factors would appear to rule out many of the compounds encountered in intermediary metabolism in which the concentrations of such metabolites are seldom large. This would include the phosphate esters of glycolysis, the organic acids of the Krebs cycle, and the free amino acids. In plants, however, certain of the organic acids — malic, citric, and isocitric — can accumulate in the vacuoles and, in that case, play a major role in determining the pH of that part of the cell. Yeasts can also accumulate relatively large concentrations of phosphate esters during glycolysis.

In animals, a complex system of pH control, with both buffering and active pH controlling elements, is found in the circulating blood. The components of this system include CO_2 and HCO_3, NaH_2PO_4 and Na_2HPO_4, the oxygenated and nonoxygenated forms of hemoglobin, and the plasma proteins. Two of these components deserve further comment. Since the pK_{a1} for H_2CO_3 is 6.1, the ratio of conjugate base to weak acid is approximately 20:1 in the normal pH range of 7.35 to 7.45 of blood. Consequently, one would expect that the H_2CO_3–HCO_3^- pair is not very effective as a buffer. Nevertheless, the carbonic acid–bicarbonate pair does have an important role in pH control, because blood is subject to CO_2 exchange with tissues and the atmosphere. The weak acid H_2CO_3 is in rapid equilibrium with dissolved CO_2 in the plasma (Equation 1.23). This equilibrium is catalyzed by the enzyme **carbonic anhydrase** that is found in red blood cells.

$$H_2CO_3 \rightleftharpoons CO_{2,\text{diss}} + H_2O \qquad (1.23)$$

The dissolved CO_2 is, in turn, in equilibrium with CO_2 in the atmosphere and, depending on the partial pressure of CO_2 in the gas phase, will either escape into the air phase (as in the lungs where CO_2 is expired) or will enter the blood (as in the peripheral tissues where CO_2 is produced by respiring cells). Thus, the H_2CO_3–HCO_3^- buffer system is less important than the rapid $CO_2 \rightleftharpoons H_2CO_3$ equilibrium that makes CO_2 into what is essentially a very volatile acid, the total amount of which is controlled in the blood to be within certain limits in order to maintain the ratio of 20:1 for conjugate base (HCO_3^-) to weak acid (H_2CO_3).

The two forms of hemoglobin found in the blood (oxygenated hemoglobin, $HHbO_2$, and unoxygenated hemoglobin, HHb) constitute the other major pH control system of blood. The buffering capacity of the hemoglobins, the most abundant protein in blood, is due in large measure to the imidazole group of histidine residues (see Section 3.14.3). The hemoglobin in 1 liter of blood can buffer 27.5 meq of H^+ (i.e., theoretically 27.5 meq of acid would be required to change the pH by one unit). In contrast, the plasma proteins can buffer only 4.24 meq of H^+. Active pH control by hemoglobins occurs because the two forms of hemoglobin differ in their pK_a's; $HHbO_2$ is the stronger acid and dissociates with a pK_{a1} of 6.2.

$$HHbO_2 \rightleftharpoons H^+ + HbO_2^-, \qquad pK_{a1} = 6.2$$

Therefore, in the lungs where the partial pressure of O_2 is high, $HHbO_2$ will predominate over the unoxygenated form and the blood tends to become more acidic. In the peripheral tissues where the partial pressure O_2 is relatively lower, HHb with the higher pK_{a1} of 7.7 will predominate, and the pH will tend to increase, since the equilibrium shown below shifts to the left.

$$HHb \rightleftharpoons H^+ + Hb^-, \qquad pK_{a1} = 7.7$$

The two effects compensate for the low concentration of CO_2 in the lungs relative to that in the peripheral tissues, and the two effects working together provide for a minimum change in pH. The pH control of the blood, then, is one example of the biochemical machinery (see Section 1.1) that is absolutely essential to the proper functioning of biological systems.

REFERENCES

1. *L. J. Henderson,* The Fitness of the Environment. *Boston: Beacon Press, 1958.*
2. *I. H. Segel,* Biochemical Calculations, *2nd ed. New York: Wiley, 1976.*

REVIEW PROBLEMS

1. What constitutes the earth's biosphere?

2. What are some of the potential advantages and disadvantages of studying a particular biochemical reaction in isolation, with purified cell components in a test tube, rather than with and in intact cells?

3. What properties of water make it particularly suited to the support of life on our planet?

4. Approximately what would be the pH of a solution that was prepared by dissolving 0.02 moles of formic acid and 0.012 moles of NaOH in water to give a final volume of 100 ml?

5. Calculate the compositions of two buffers. Each buffer is to be $0.10M$ in ethanolamine ($H_2N—CH_2—CH_2—OH$) and to contain acetic acid. One buffer is to be pH 5.0, the other 9.2. Which of these two buffers will have the greater buffering capacity?

CHAPTER 2

CARBOHYDRATES

In the next few chapters of this text, we shall describe the more important building blocks of the biosphere — the simple sugars, fatty acids, amino acids, and mononucleotides — that are assembled into the biopolymers of the cell — the polysaccharides, lipids, proteins, and nucleic acids. In the present chapter, we shall examine the simple sugars, the storage carbohydrates, and the structural polysaccharides. The subject of stereochemistry will be reviewed and the various ways of representing carbohydrates with different structural formulas will be presented. Physical and chemical properties of simple sugars will be described, and the structures of some of the more complex storage and structural polysaccharides will be given. Complex carbohydrates play not only a structural role in the cell but may serve as a reservoir of chemical energy to be enlarged and decreased as fits the needs of the organism. Two examples of structural carbohydrates are cellulose, the major structural component of plant cell walls, and the peptidoglycans of bacterial cell walls. The storage carbohydrates include starch and glycogen, polysaccharides that may be produced and consumed in line with the energy needs of the cell.

Carbohydrates may be defined as polyhydroxy aldehydes or ketones, or as substances that yield one of these compounds on hydrolysis. Many carbohydrates have the empirical formula $(CH_2O)_n$ where n is 3 or larger. This formula obviously contributed to the original belief that this group of compounds could be represented as **hydrates of carbon.** It became clear that this definition was not suitable when other compounds were encountered that had the general properties of carbohydrates but contained nitrogen, phosphorus, or sulfur in addition to carbon, hydrogen, and oxygen. Moreover, the important simple sugar deoxyribose, found in every cell as a component of deoxyribonucleic acid, has the molecular formula $C_5H_{10}O_4$ rather than $C_5H_{10}O_5$.

2.1 CLASSIFICATION OF CARBOHYDRATES

Carbohydrates can be classified into three groups based on the number of sugar units they contain: monosaccharides, oligosaccharides, and polysaccharides. Monosaccharides are simple sugars that cannot be hydrolyzed into smaller units under reasonably mild conditions. The simplest monosaccharides fitting our definition and empirical formula are the **aldose** glyceraldehyde and its isomer, the **ketose** dihydroxy acetone. Both of these sugars are trioses because they contain three

$$CH_2OH—CHOH—C\overset{H}{\underset{O}{\diagdown}} \qquad CH_2OH—\overset{}{\underset{\overset{\|}{O}}{C}}—CH_2OH$$

Glyceraldehyde   Dihydroxy acetone

carbon atoms. In addition to the functional groups (aldehyde and ketone) that are used in describing these two sugars, note that they contain alcoholic hydroxyls and, in the case of glyceraldehyde, an asymmetric carbon atom.

Oligosaccharides are hydrolyzable polymers of monosaccharides that contain from two to six molecules of simple sugars. The disaccharides, which have two monosaccharide units, are the most abundant; trisaccharides also occur free in nature. Oligosaccharides with more than three subunits are usually found bound as side chains in glycoproteins.

Polysaccharides are polymers, frequently insoluble, consisting of hundreds or thousands of monosaccharide units; they may be either linear or branched in structure. If the polymer is made up from a single monosaccharide, the polysaccharide is called a *homo*polysaccharide. If two or more different monosaccharides are found in the polymer, it is called a *hetero*polysaccharide. Some of the monosaccharides that are bound together by glycosidic bonds to form polysaccharides are glucose, xylose, and arabinose.

2.2 STEREOISOMERISM

The study of carbohydrates requires an understanding of isomerism, especially stereoisomerism. The subject of isomerism may be divided into **structural isomerism** and **stereoisomerism.** Structural isomers have the same molecular formula but differ from each other by having different structures; that is, they differ in the order in which their atoms are bonded together. Stereoisomers have the same molecular formula and the same structure, but they differ in **configuration,** that is, in the arrangement of their atoms in space. Structural isomers, in turn, can be of three types. One type is the **chain isomers,** in which the isomers have different arrangements of the carbon atoms. As an example, *n*-butane is a chain isomer of

$$\underset{n\text{-Butane}}{H—\overset{\overset{H}{|}}{\underset{\underset{H}{|}}{C}}—\overset{\overset{H}{|}}{\underset{\underset{H}{|}}{C}}—\overset{\overset{H}{|}}{\underset{\underset{H}{|}}{C}}—\overset{\overset{H}{|}}{\underset{\underset{H}{|}}{C}}—H} \qquad \underset{\text{Isobutane}}{H—\overset{\overset{H}{|}}{\underset{\underset{H}{|}}{C}}—\overset{\overset{H}{|}}{\underset{\underset{CH_3}{|}}{C}}—\overset{\overset{H}{|}}{\underset{\underset{H}{|}}{C}}—H}$$

isobutane. Other examples of structural isomers are the **positional isomers,** *n*-propyl chloride and isopropyl chloride, in which the two compounds involved have the same carbon chain but differ in the position of a substituent group. The third type of

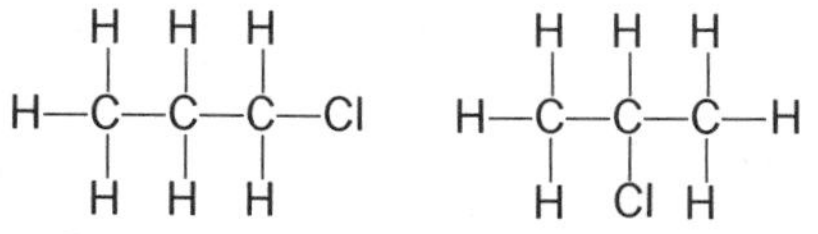

$$
\begin{array}{ccc}
& H & H & H \\
& | & | & | \\
H—& C—& C—& C—Cl \\
& | & | & | \\
& H & H & H
\end{array}
\qquad
\begin{array}{ccc}
& H & H & H \\
& | & | & | \\
H—& C—& C—& C—H \\
& | & | & | \\
& H & Cl & H
\end{array}
$$

n-Propyl chloride   Isopropyl chloride

structural isomers is the **functional group isomers,** in which the compounds have different functional groups. Examples are *n*-propanol and methylethyl ether.

$H_3C—CH_2—CH_2OH$ $H_3C—CH_2—O—CH_3$

n-Propanol   Methylethyl ether

The subject of stereoisomerism can be divided into the smaller areas of **optical** isomerism and **geometrical** (or *cis-trans*) isomerism. The latter type of isomerism is illustrated by the *cis-trans* pair, fumaric and maleic acids.

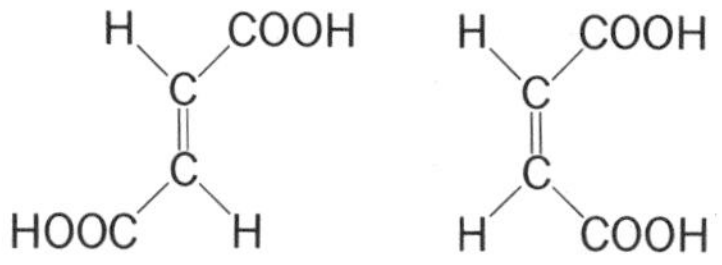

Fumaric acid   Maleic acid
(*trans*)   (*cis*)

2.2.1 OPTICAL ISOMERISM

This is the type of isomerism commonly found in carbohydrates; it is usually encountered when a molecule contains one or more **chiral** (Greek cheir = hand) or asymmetric carbon atoms. The subject of stereoisomerism was extensively developed after van't Hoff and LeBel introduced the concept of the **tetrahedral carbon atom.** This atomic structure has four covalent bonds or bond axes that extend out from the central carbon nucleus to the corners of a tetrahedron (Structure 2.1).

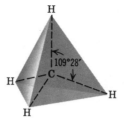

STRUCTURE 2.1

When four different groups are attached to those bonds, the carbon atom in the center of the molecule is said to be a chiral center (or a chiral **carbon atom**). This is indicated in Structure 2.2 in which the compound C(ABDE), containing a single chiral carbon atom, is represented as having the four groups A, B, D, and E attached. These groups may be arranged in space in two different ways so that two different compounds are formed. These compounds are obviously different; that is, they cannot be superimposed on each other. Instead, one compound is related to the other as a right hand is related to a left hand. Such

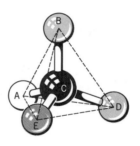

STRUCTURE 2.2

chiral molecules are said to possess "handedness" and are therefore mirror images of each other; if one molecule is held before a mirror, the image in the mirror corresponds to the other molecule (Structure 2.3).

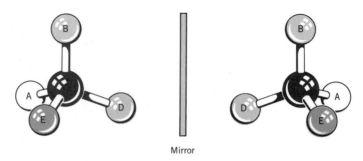

Mirror

STRUCTURE 2.3

These mirror image isomers constitute an **enantiomeric pair;** one member of the pair is said to be the **enantiomer** of the other.

In molecules that contain more than one chiral carbon atom, more than two stereoisomers can exist. Any two that are mirror image isomers, and therefore constitute an enantiomeric pair, will be related to the remaining stereoisomers as **diastereomers.** They are stereoisomeric with, but are not mirror image isomers of, the remaining compounds.

2.2.2   OPTICAL ACTIVITY

Almost all the properties of the two members of an enantiomeric pair are identical: they have the same boiling point, the same melting point, the same solubility in various solvents. They also exhibit optical activity; in this property, they differ in one important manner. One member of the enantiomeric pair will rotate a plane of polarized light in a clockwise direction and is therefore said to be *dextro*rotatory. Its mirror image isomer or enantiomer will rotate the plane of polarized light to the same extent, but in the opposite or counterclockwise direction. This isomer is said to be *levo*rotatory. It must be noted, however, that not all compounds possessing a chiral center are chiral or exhibit optical activity. On the other hand, a molecule may possess chirality, exhibit optical activity, and not contain a chiral center.

2.2.3   PROJECTION AND PERSPECTIVE FORMULAS

In the study of carbohydrates, many examples of optical isomerism are encountered, and it is necessary to have a means for representing the different possible isomers. One way of representing them is to use the **projection formula** introduced in the nineteenth century by the illustrious German organic chemist, Emil Fischer. The projection formula represents the four groups attached to the carbon atom as being projected onto a plane. This projection can be represented for the asymmetric molecule depicted previously as shown in Structure 2.4. In the Fischer

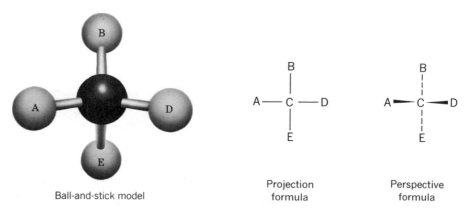

Ball-and-stick model   Projection formula   Perspective formula

STRUCTURE 2.4

projection formula, the horizontal bonds are understood to be in front of the plane of the paper, while the vertical bonds are behind. This relationship is seen more clearly in the **perspective formula**. Here, dashed lines indicate bonds extending behind the plane of the page, while solid wedges identify bonds projecting forward out of the plane of the page. The projection and perspective formulas can be used to distinguish between the compound shown in Structure 2.4 and its mirror image isomer below. These two pairs of formulas together with a **simplified version** of the ball-and-stick model constitute three different ways of writing formulas to represent the enantiomeric pairs.

Simplified ball-and-stick model   Projection formula   Perspective formula

In one's imagination, the perspective formula can be rotated in all planes without fear of confusing the two enantiomers. Caution is required in the use of the Fischer formulas; although they may be rotated a full 180° in the plane of the paper, rotation of only 90° results in the enantiomer because of the convention that the horizontal bonds are represented as being in front of the plane of the paper. The Fischer formula cannot, in the mind's eye, be removed from the plane of the paper.

2.2.4   D-GLYCERALDEHYDE AS A REFERENCE COMPOUND

With the existence of a large number of optical isomers in carbohydrates, it is also necessary to have a reference compound. The simplest monosaccharide that possesses an asymmetric carbon atom has been chosen as the reference standard; this compound is the triose **glycerose** or **glyceraldehyde**. Since this compound has a chiral center, it can exist as two stereoisomers. These may be represented by their Fischer projection formulas as well as the simplified ball-and-stick models and perspective formulas.

Fischer formulas

$$\begin{array}{c} \text{CHO} \\ | \\ \text{H—C—OH} \\ | \\ \text{CH}_2\text{OH} \end{array} \qquad \begin{array}{c} \text{CHO} \\ | \\ \text{HO—C—H} \\ | \\ \text{CH}_2\text{OH} \end{array}$$

Ball-and-stick models

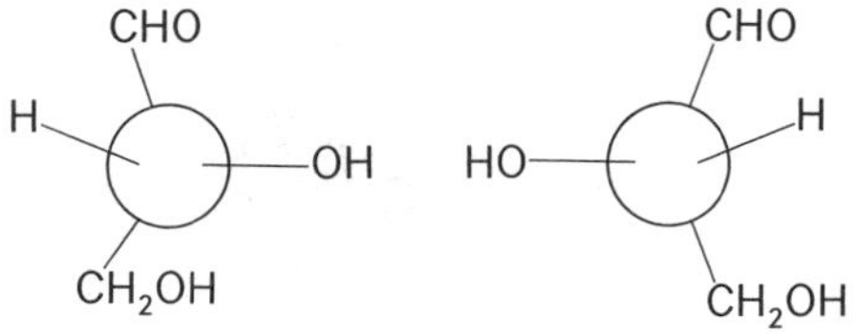

Perspective formulas

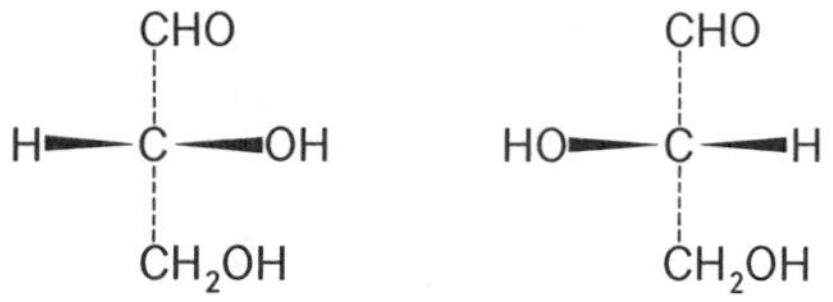

$$\begin{array}{c} \text{CHO} \\ | \\ \text{H} \blacksquare\text{—C}\text{—}\blacktriangleleft\text{OH} \\ | \\ \text{CH}_2\text{OH} \end{array} \qquad \begin{array}{c} \text{CHO} \\ | \\ \text{HO}\blacksquare\text{—C}\text{—}\blacktriangleleft\text{H} \\ | \\ \text{CH}_2\text{OH} \end{array}$$

These three pairs of structures are related to each other as mirror image isomers. Although they will have the same melting point, boiling point, and solubility in H_2O, they will differ in the direction in which they rotate plane-polarized light. The isomer that rotates light in the clockwise direction is identified with the symbol (+) to indicate that it is the dextrorotatory enantiomer. At the turn of this century, that isomer was also assigned the Fischer formula in which the hydroxyl group appears on the right when the aldehyde group is at the top. Moreover, it was agreed that this form should be designated as D(+)-glyceraldehyde. For clarification, both the projection formula and the frequently seen ball-and-stick representations are given.

D(+)-Glyceraldehyde

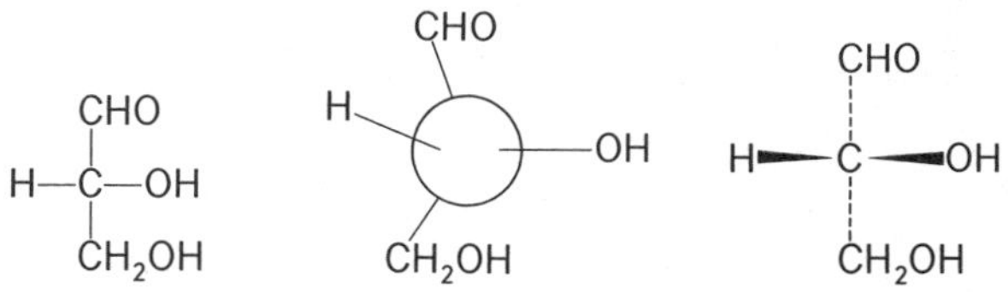

$$\begin{array}{c} \text{CHO} \\ | \\ \text{H—C—OH} \\ | \\ \text{CH}_2\text{OH} \end{array}$$

Fischer projection   Ball-and-stick model   Perspective formula

This assignment of the above structures to the glyceraldehyde isomer that rotated polarized light in the clockwise direction had only a 50:50 chance of being correct. Fifty years later, research involving x-ray diffraction studies on tartaric acid showed that Fischer's choice was the correct one.

D(+)-Glyceraldehyde serves an important role as a reference compound not only for carbohydrates, but also for hydroxy and amino acids encountered in biochemistry. The D and L notation has been particularly useful in relating groups of carbohydrates (the naturally occurring D-sugars) and amino acids (the naturally occurring L-amino acids). However, the D and L system cannot be used for all compounds containing chiral centers, since, in theory, this would call for converting the compound of interest into D- or L-glyceraldehyde or a compound known to be related by stereochemistry to these reference standards. Therefore, a new system called the Cahn–Ingold–Prelog "sequence rule" has been devised to describe, in an absolute manner, the configuration of individual chiral centers. In brief, the method is based on orienting the molecule so that the smallest ranked group attached to the chiral center is facing away and then observing the direction, clockwise (R for rectus) or counterclockwise (s for sinister), that one's eye takes as it moves in an assigned **order of rank** from the highest atomic number of the substituent atoms to the lowest. According to the sequence rule, D(+)-glyceraldehyde would be designated as (R)-glyceraldehyde. Although the sequence rule is employed in many areas of organic chemistry, it is not convenient to use with carbohydrates or amino acids for which the "local systems" based on D(+)-glyceraldehyde and L_s-serine are retained (see Sections 2.2.5 and 4.2).

2.2.5 CYANOHYDRIN SYNTHESIS

As an illustration of the use of D-glyceraldehyde as a reference compound, consider the formation of tetrose sugars from a triose by the **Kiliani–Fischer** synthesis. This synthesis is a process by which the chain length of an aldose is increased by one carbon atom, and two new aldoses are formed (Figure 2.1). In this synthesis, the chirality of the asymmetric carbon in D-glyceraldehyde is not disturbed, but the aldehyde carbon is converted into a new chiral center. Therefore, two four-carbon compounds with opposite configuration at that center are formed. These new compounds are D-erythrose and D-threose; the D-prefix specifically denotes their relationship to D-glyceraldehyde. If the cyanohydrin synthesis were applied to L-glyceraldehyde, L-erythrose and L-threose would be formed.

When the Kiliani–Fischer synthesis is applied to the two D-aldotetroses shown in Figure 2-1, four aldopentoses are formed. Elongation of these four pentoses results in a total of eight aldohexoses including the familiar D-glucose, D-mannose, and D-galactose.

At this point, it is informative to consider the stereochemical relationships between the eight aldohexoses and the reference compound D-glyceraldehyde. First, the eight hexoses have the same structural formula, CH_2OH—CHOH—CHOH—CHOH—CHOH—CHO and therefore are stereoisomers rather than structural isomers. Second, with regard to their stereoisomerism, they are optical rather than geometric isomers. Third, these D-aldohexoses are structurally related to D-glyceraldehyde because in theory they could all have been produced by application of the cyanohydrin synthesis to that triose. This relationship is also shown in the projection formulas for the eight hexoses in which the hydroxyl (—OH) group on the **reference carbon atom,** the highest-numbered chiral carbon, is on the right as in D-glyceraldehyde (the carbons in aldoses are numbered starting with the aldehyde carbon). Fourth, these aldohexoses are related to each other as diastereoisomers; they cannot be enantiomeric since their mirror-image isomers, for example, L-glucose, the mirror image of D-glucose, is not shown. Fifth, four pairs of **epimers,** diastereomers differing in the configuration at a **single** carbon atom, exist among the aldohexoses (e.g., D-allose and D-altrose). Finally, the fact that these sugars are all "D-sugars" derived from D-glyceraldehyde bears no relationship to whether the sugar is dextrorotatory or levorotatory.

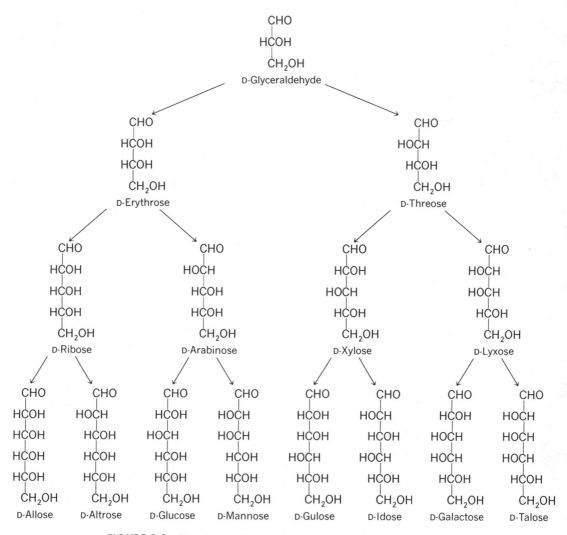

FIGURE 2.1   Structures of the D-aldoses.

As the number of asymmetric carbon atoms increases in a carbohydrate molecule, the number of optical isomers also increases. Van't Hoff established that 2^n represents the number of possible optical isomers, where n is the number of asymmetric carbon atoms. Thus, in the trioses where n is 1, there are two optical isomers; in the tetroses where n is 2, four optical isomers, known as D-erythrose, L-erythrose, D-threose, and L-threose, occur. In the aldohexoses, where there are four asymmetric carbon atoms, there are 16 optical isomers, eight of which are the D-sugars shown in Figure 2.1. In the ketohexoses, when n is 3, there are eight possible isomers.

Although Figure 2.1 shows eight aldohexoses, only three of these are frequently encountered in nature. Their projection formulas are repeated here together with the structure of D-fructose, an important ketohexose. The following statements can be made about

```
      CHO              CHO              CHO             CH₂OH
    HCOH             HOCH             HCOH              C=O
    HOCH             HOCH             HOCH             HOCH
    HCOH             HCOH             HOCH             HCOH
    HCOH             HCOH             HCOH             HCOH
    CH₂OH            CH₂OH            CH₂OH            CH₂OH
 D(+)-Glucose    D(+)-Mannose    D(+)-Galactose    D(−)-Fructose
```

their isomerism: All four sugars are D-sugars because they have the same configuration as D-glyceraldehyde on the highest number chiral atom; the use of the
term D has no bearing on whether these sugars are dextro- or levorotatory.

The three aldohexoses are stereoisomers—more specifically, optical
isomers. Because no one of the three is an enantiomer of either of the other two,
they are related as *diastereomers.* As such, they have different melting points,
different boiling points, different solubilities, different specific rotations, and,
in general, different chemical properties.

D(+)-Glucose may be said to be an *epimer* of D(+)-mannose because these
compounds differ from each other by their configuration on a single asymmetric carbon atom. Similarly, D(+)-glucose is an epimer of D(+)-galactose. On the
other hand, there is no epimeric relationship between D(+)-mannose and D(+)-
galactose.

D-Fructose is a structural isomer of the other three hexoses. Although it has
the same molecular formula ($C_6H_{12}O_6$), it has a different functional group; it is
a ketose rather than an aldose.

2.3 THE STRUCTURE OF GLUCOSE

Emil Fischer received the Nobel Prize in Chemistry for his studies on the
structure of glucose, more specifically for establishing the configuration of the
four asymmetric carbon atoms in that aldohexose relative to D(+)-glyceralde-
hyde. From Fischer's work, chemists were able to write the projection and
ball-and-stick formulas for D- and L-glucose (Structure 2.5).

If the ball-and-stick model just represented is actually constructed and the
−CHO and −CH₂OH groups are held so that they extend away from the holder
(behind the plane of the paper), the remainder of the carbon atoms will tend to
form a ring extending toward the holder, and the H and OH groups will project
out toward the holder.

Although the aldohexoses have been considered polyhydroxy aldehydes or
ketones to this point, there is abundant evidence to indicate that other forms (of
glucose, e.g.) exist and, indeed, predominate both in the solid phase and solution. For instance, aldohexoses undergo the Kiliani–Fischer synthesis with
difficulty, although cyanohydrin formation with simple aldehydes is usually
rapid. Glucose and other aldoses fail to give the Schiff test for aldehydes. Solid
glucose is quite inert to oxygen, but aldehydes are notoriously autoxidizable.
Finally, it is possible to prepare two crystalline forms of D-glucose. When
D-glucose is dissolved in water and allowed to crystallize out by evaporation of
the water, a form designated as α-D-glucose is obtained. If glucose is crystallized

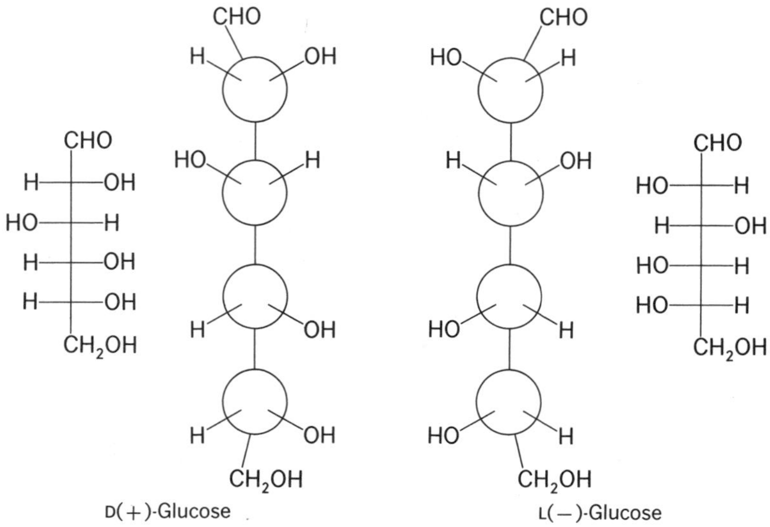

STRUCTURE 2.5

from acetic acid or pyridine, another form, β-D-glucose, is obtained. These two forms of D-glucose show the phenomenon of **mutarotation.** A freshly prepared solution of α-D-glucose has a specific rotation $[\alpha]_D^{20}$ of $+113°$; when the solution is left standing, it changes to $+52.5°$. A fresh solution of β-D-glucose, on the other hand, has an $[\alpha]_D^{20}$ of $+19°$; on standing, it also changes to the same value, $+52.5°$.

The existence of the two forms of glucose, as well as the other anomalous properties described, was explained by the English chemist, W. N. Haworth, who showed that aldohexoses and other sugars react internally to form cyclic hemiacetals. Hemiacetal formation is a characteristic reaction between aldehydes and alcohols.

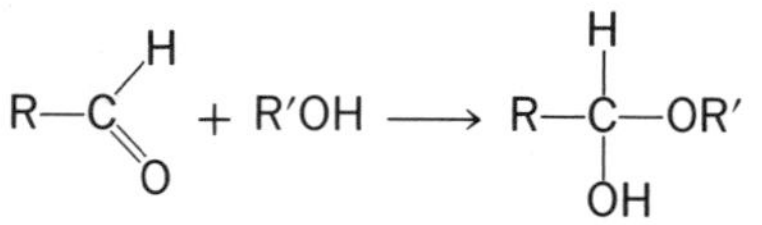

The formation of the cyclic hemiacetal structure for D-glucose is shown in Figure 2.2.

Rotation of the bond between carbon atoms 4 and 5 moves the C-5 hydroxyl group into a postiion where it can react with the aldehyde group on C-1. This places the —CH₂OH group on C-5 above the ring. As the hemiacetal ring is formed, note that C-1 becomes a chiral center. Therefore, two diastereomeric molecules are possible. These isomers are the α and β forms of glucose; they are diastereomers, however, rather than enantiomers, for the α form differs from the β form only in the configuration at a single carbon. Since the cyclic forms of the aldohexoses have five asymmetric carbon atoms, there are 32 optical isomers of the cyclic aldohexoses consisting of 16 pairs of enantiomers.

As noted above, the angles of the tetrahedral carbon atom bend the glucose molecule into a ring. When the C-5 hydroxyl group reacts, as shown in Figure 2-2, a six-membered ring is formed. If the C-4 hydroxyl were to react, a five-membered ring would result; a seven-membered ring is too strained to permit the C-6 hydroxyl of an aldohexose to form a hemiacetal. The six-membered

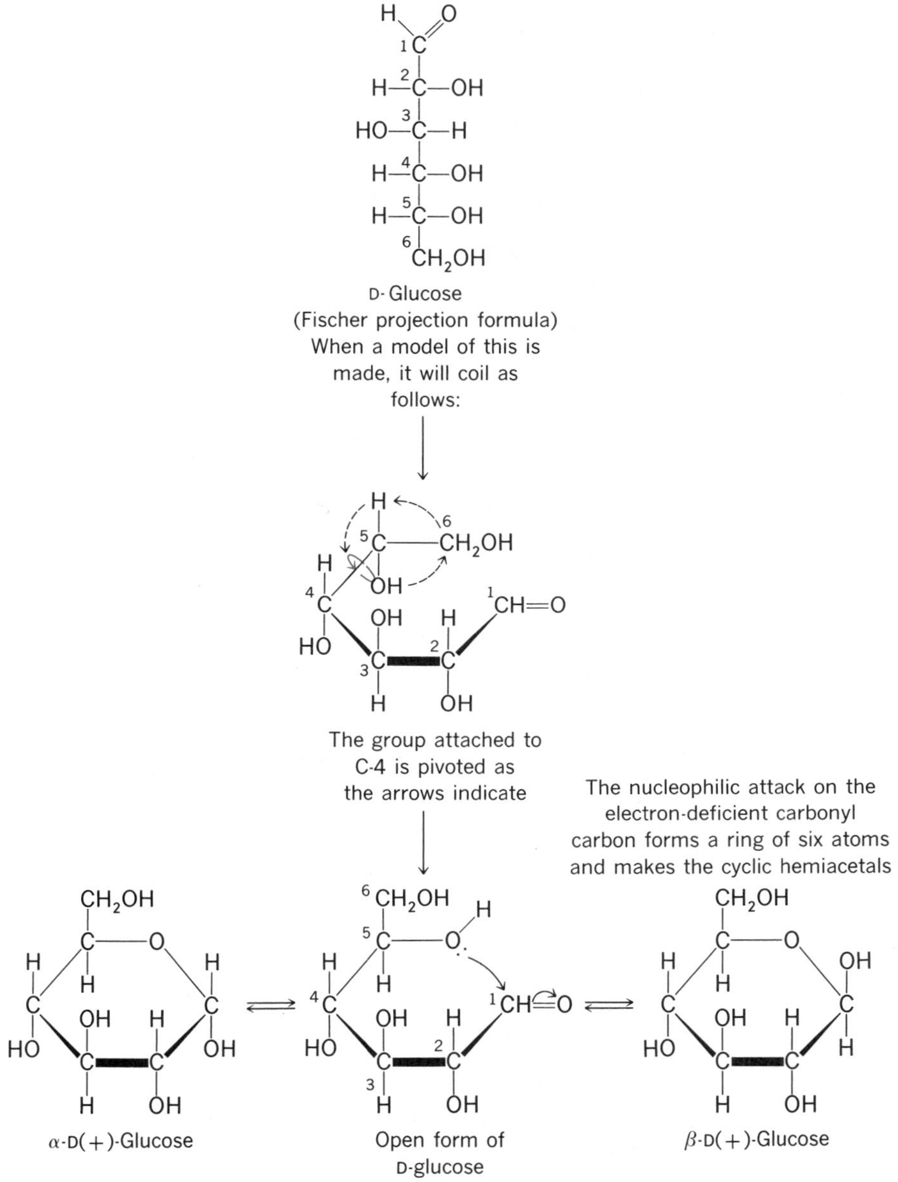

FIGURE 2.2 Scheme depicting the formation of the hemiacetal forms of D-glucose. Note that an equilibrium exists between the α and β forms and the open-chain form.

ring sugars may be considered derivatives of pyran, whereas the five-membered rings are considered relatives to furan.

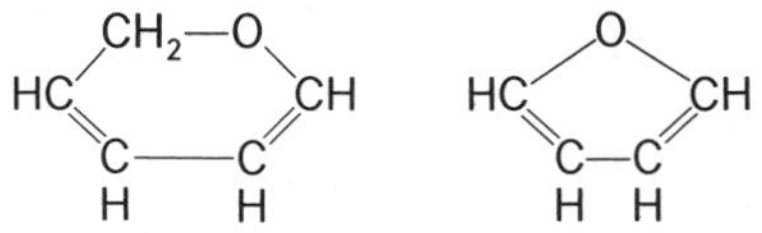

α-Pyran Furan

Hence, it is customary to refer to the **pyranose** or **furanose** form of the monosaccharide. Furanose forms of the hexoses are less stable than the pyranose forms in solution; combined forms of furanose sugars (as in the fructose unit of sucrose) are found in nature, however.

Haworth proposed that these cyclic hemiacetal forms of glucose and other sugars be represented as a hexagonal ring in a plane perpendicular to the plane of the paper. The side of the hexagon that is nearer to the reader would then be indicated by a thickened line. When this is done, the substituents on the carbon atom then will extend above or below the plane of the six-membered ring. Carbon atom 6, a substituent on C-5, will therefore be above the plane of the ring. The Haworth formulas for α-D(+)-glucose and β-D(+)-glucose may then be compared with the Fischer projection formulas for these diastereomers (Structure 2.6).

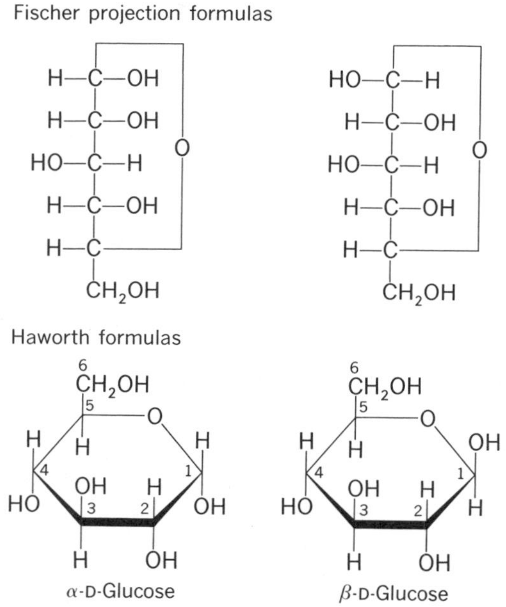

Fischer projection formulas

Haworth formulas

α-D-Glucose β-D-Glucose

STRUCTURE 2.6

With respect to assigning structures to the α- and β-anomers, Fischer originally suggested that, in the D-series, the more dextrorotatory compound be called the α-anomer whereas, in the L series, the α-anomer would be the more levorotatory substance. Later, Freudenberg proposed that the α- and β- anomers be classified with respect to their configuration rather than sign or magnitude of rotation. The relationship of the anomeric hydroxyl to the reference carbon atom is easy to see when Fischer projection formulas for the ring structures are used. In these projections, the α-anomer is the isomer in which the anomeric hydroxyl is on the same side (*cis*) of the carbon chain as the hydroxyl group on the reference carbon atom. If the reference hydroxyl group happens, as it does in α-D-glucopyranose, to be involved in ring formation, then the anomeric hydroxyl of the α-isomer is on the same side as the ring structure formed by the oxygen bridge. In the β-anomer, the hemiacetal hydroxyl group is *trans* to the hydroxyl on the reference carbon atom.

The assignment of configuration to the anomeric carbon atom is less readily

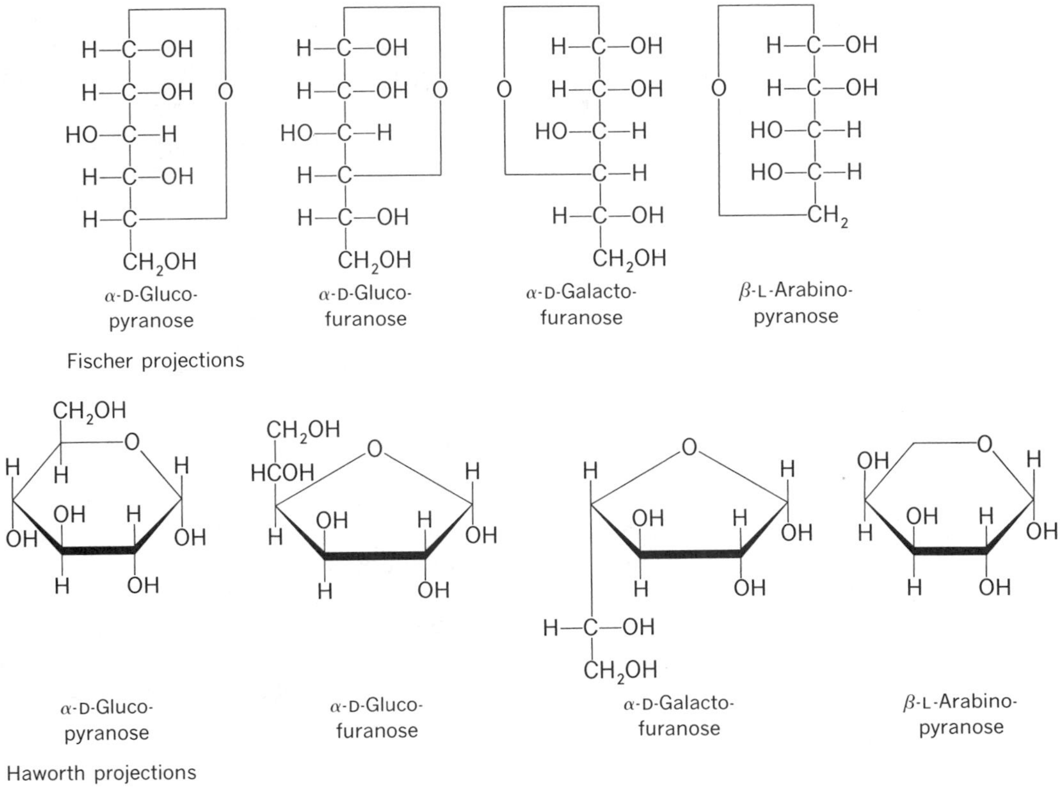

α-D-Gluco-
pyranose

α-D-Gluco-
furanose

α-D-Galacto-
furanose

β-L-Arabino-
pyranose

Fischer projections

α-D-Gluco-
pyranose

α-D-Gluco-
furanose

α-D-Galacto-
furanose

β-L-Arabino-
pyranose

Haworth projections

FIGURE 2.3 Fischer and Haworth formulas of some common monosaccharides.

seen with the Haworth formulas. In the case of D-hexoses and D-pentoses, the α-anomer has the anomeric hydroxyl written below the plane of the ring. The β-anomer then has the anomeric hydroxyl above the plane of the ring. Examples are given in Figure 2.3.

There is still one final aspect of the structure of glucose to be mentioned; this is its **conformation.** Because the C—O—C bond angle of the hemiacetal ring (111°) is similar to that of the C—C—C ring angles (109°) in cyclohexane, the pyranose ring of glucose, rather than forming a true plane, is puckered in much the same way as cyclohexane. Like cyclohexane, glucopyranose can exist in two conformations: the **chair** and **boat** forms. The chair conformation of glucose minimizes torsional strain and further, the conformational structure in which a maximum number of bulky groups ($-OH$ and $-CH_2OH$) are **equatorial** rather than **axial** to an axis passing through the ring is preferred. The following diagram

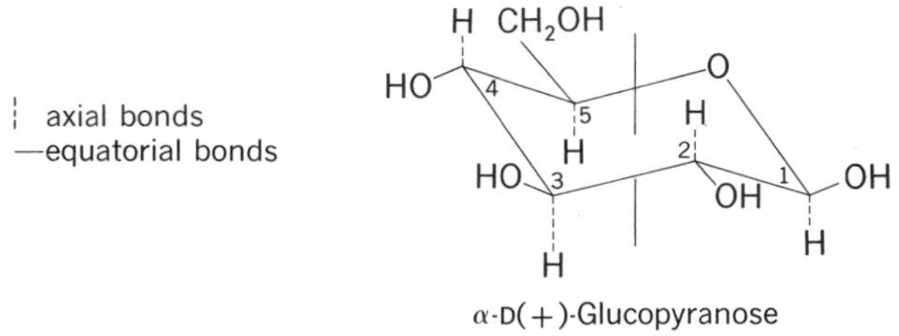

axial bonds
—equatorial bonds

α-D(+)-Glucopyranose

shows that β-D(+)-glucopyranose can achieve a conformation in which all bulky groups are equatorial (or perpendicular) to an axis passing through the plane of the ring. This conformation is thermodynamically more stable than that in which the hydroxyls and the $-CH_2OH$ are axial (parallel to the axis shown). α-D-Glucopyranose can have a conformation in which all bulky groups *except* the anomeric hydroxyl are equatorial, and the preferred structure for this form may be represented as

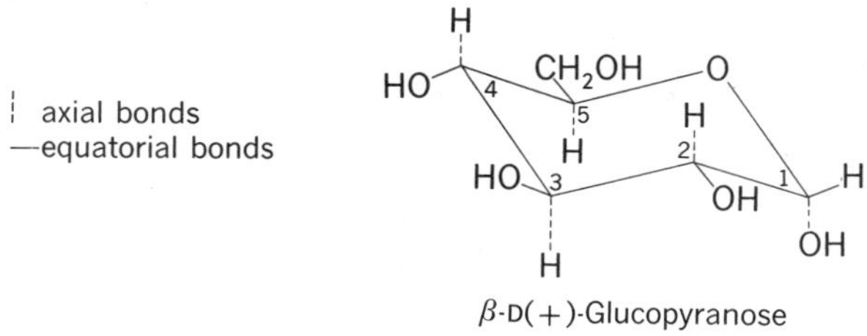

:   axial bonds
—equatorial bonds

β-D(+)-Glucopyranose

Therefore, one of the two anomers, namely the β-anomer with *all* bulky groups equatorial, should predominate in a solution over the α-isomer with one axial group, the anomeric hydroxyl. Thus, in aqueous solution, β-D(+)-glucopyranose is present to the extent of about 63% after mutarotation, whereas α-D(+)-glucopyranose comprises about 36%. The linear polyhydroxy aldehyde form accounts for less than 1% of the total carbon present as glucose (see Figure 2.2).

2.4   STRUCTURES OF OTHER MONOSACCHARIDES

Pyranose forms for the other aldohexoses mentioned on page 33 may be written by the proper arrangement of the hydroxyl groups on C-2, C-3, and C-4. Similarly, the Haworth formulas for α-D-fructopyranose and β-D-fructopyranose may be written as shown in Structure 2.7. Note, however, that the five-member furanose structure is the one encountered for fructose when the hemiketal (from the ketone group of the ketohexoses) group is substituted as in sucrose (see Section 2.6.1) and fructosans.

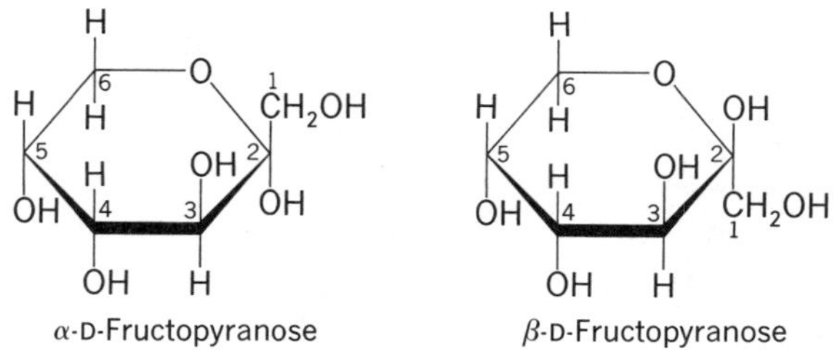

α-D-Fructopyranose β-D-Fructopyranose

STRUCTURE 2.7

The ubiquitous pentose, D-ribose, a component of ribonucleic acid, exists as a furanose; 2-deoxy-D-ribose, a component of 2-deoxyribonucleic acid, is also a furanose sugar. Both α- and β-isomers can exist in solution, but the β-isomer is the one that is found in the nucleic acids.

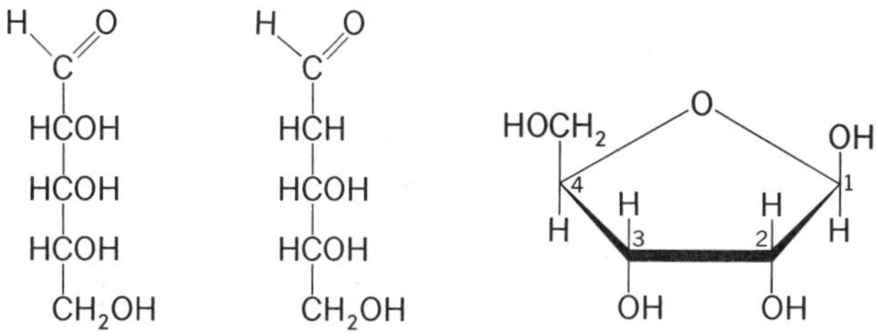

D-Ribose   2-Deoxy-D-ribose   β-D-Ribofuranose

STRUCTURE 2.8

Four other monosaccharides that play important roles in the metabolism of carbohydrates during photosynthesis are the aldotetrose, D-erythrose; the keto-pentoses, D-xylulose and D-ribulose; and the ketoheptose, D-sedoheptulose.

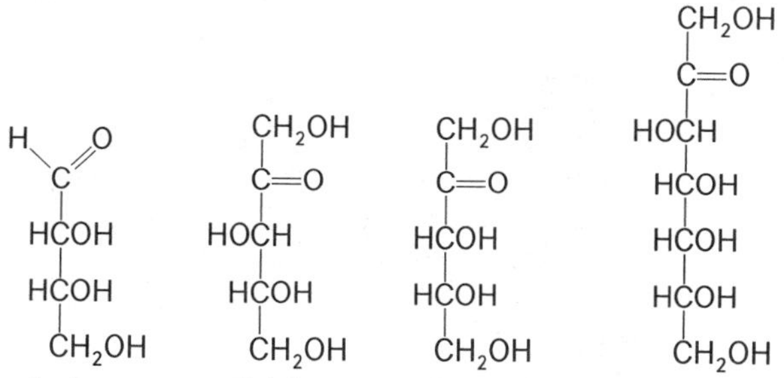

D-Erythrose   D-Xylulose   D-Ribulose   D-Sedoheptulose

Although five-membered hemiacetal (erythrose) or hemiketal (xylulose, ribulose) structures of these monosaccharides may be written, the metabolically active forms are the phosphate esters in which the primary alcohol ($-CH_2OH$) group has been esterified with H_3PO_4, thereby preventing its participation in a ring structure.

Two other deoxy sugars are found in nature as components of cell walls. Those are L-rhamnose (6-deoxy-L-mannose) and L-fucose (6-deoxy-L-galactose).

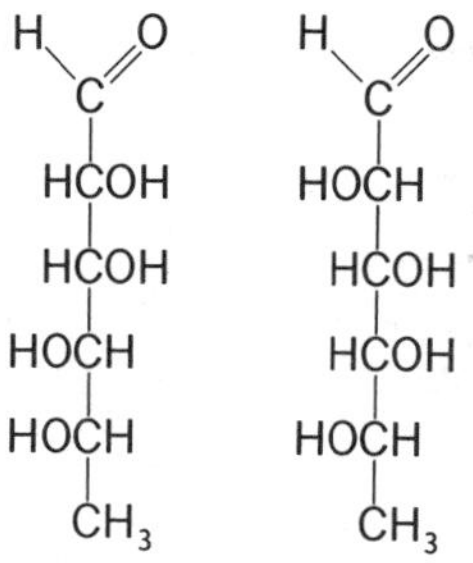

L-Rhamnose   L-Fucose

Two amino sugars, D-glucosamine and D-galactosamine, exist in which the hydroxyl group at C-2 is replaced by an amino group. The former is a major component of chitin, a structural polysaccharide found in insects and crustaceans. D-Galactosamine is a major component of the polysaccharide of cartilage. Their hemiacetal forms are shown here. The derived forms of the amino sugars will be described later.

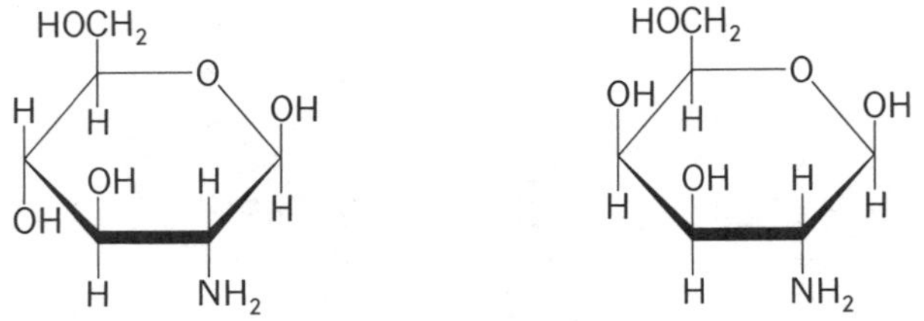

2-Deoxy-2-amino-β-D-glucopyranose
(β-D-Glucosamine)

2-Deoxy-2-amino-β-D-galactopyranose
(β-D-Galactosamine)

2.5 PROPERTIES OF MONOSACCHARIDES

2.5.1 MUTAROTATION

We have already referred to the phenomenon of mutarotation exhibited by the anomeric forms of D-glucopyranose. Mutarotation is a property exhibited by the hemiacetal and ketal forms of sugars that are free to form the open-chain sugar. As pointed out in Figure 2.2, the open-chain polyhydroxy aldehyde or ketone is an intermediate in the interconversion of the α and β forms during mutarotation.

2.5.2 ENOLIZATION

When glucose is exposed to dilute alkali for several hours, the resulting mixture contains both fructose and mannose. If either of these sugars is treated with dilute alkali, the equilibrium mixture will contain the other sugar as well as glucose. This reaction, known as the Lobry de Bruyn–von Ekenstein transformation, is due to the enolization of these sugars in the presence of alkali. **Enediol** intermediates that are common to all three sugars are responsible for the establishment of the equilibrium. At higher concentrations of alkali, the monosaccharides are generally unstable and undergo oxidation, degradation, and polymerization.

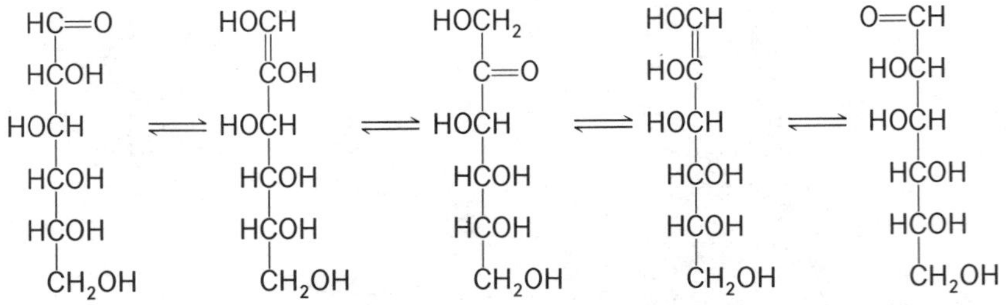

| D-Glucose | *trans*-Enediol | D-Fructose | *cis*-Enediol | D-Mannose |

Isomerization in dilute alkali

The alkaline enolization shown has its enzymatic counterparts in that there are **isomerases** and **epimerases** that catalyze the interconversion of phosphorylated forms of these three hexoses (Sections 10.3 and 10.6).

2.5.3 OXIDATION–REDUCTION

Carbohydrates may be classified as either reducing or nonreducing sugars. The reducing sugars, which are more common, are able to function as reducing agents because free, or potentially free, aldehyde groups, as in the cyclic hemiacetal forms, are present in the molecule. This aldehyde group is readily oxidized to the carboxylic acid at neutral pH by mild oxidizing agents and enzymes. This property is utilized in detecting and quantitating monosaccharides, especially glucose, in biological fluids such as blood or urine. The monocarboxylic acid that is formed is known as an aldonic acid (e.g., gluconic acid from glucose). The structures of several of these are shown.

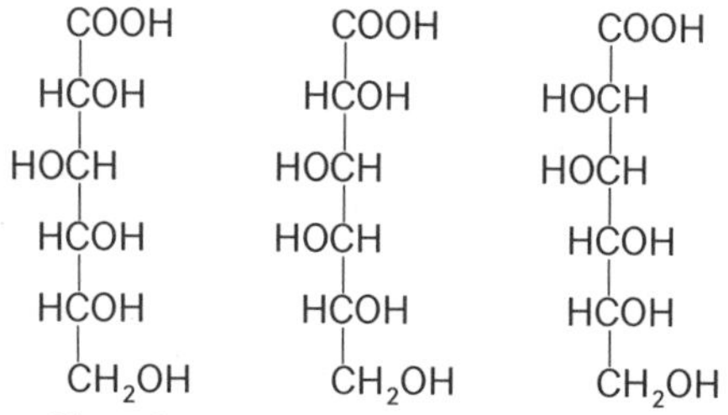

```
    COOH          COOH          COOH
    HCOH          HCOH          HOCH
    HOCH          HOCH          HOCH
    HCOH          HOCH          HCOH
    HCOH          HCOH          HCOH
    CH₂OH         CH₂OH         CH₂OH
  D-Gluconic    D-Galactonic   D-Mannonic
     acid           acid          acid
```

In the presence of a strong oxidizing agent like HNO_3, both the aldehyde and the primary alcoholic function will be oxidized to yield the corresponding dicarboxylic or aldaric acid (e.g., galactaric acid).

One of the more important oxidation products of monosaccharides is the monocarboxylic acid obtained by the oxidation of only the primary alcoholic group, usually by specific enzymes, to yield the corresponding uronic acid (e.g., galacturonic acid). Such acids are components of important heteropolysaccharides found in nature.

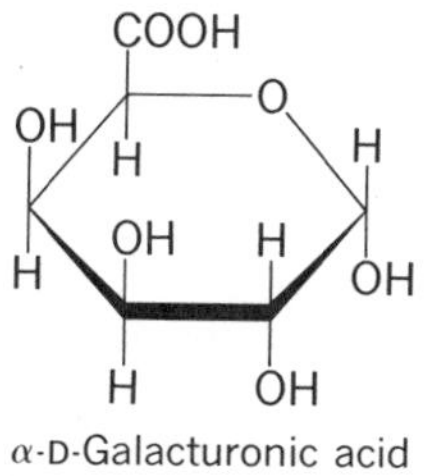

α-D-Galacturonic acid

The aldehyde and ketone groups of monosaccharides may be reduced nonenzymatically (with hydrogen or $NaBH_4$) or with enzymes to yield the corresponding sugar alcohols. Thus, D-glucose when reduced yields D-sorbitol, and D-mannose produces D-mannitol. Sorbitol is found in the berries of many higher plants, especially in the *Rosaceae;* it is a crystalline solid at room temperature but has a low melting point. D-Mannitol is found in algae and fungi. Both compounds are soluble in H_2O and have a sweet taste.

2.5.4 GLYCOSIDE FORMATION

One of the most important properties of monsaccharides is their ability to form glycosides. Consider as an example the formation of the methyl glycoside of glucose. When D-glucose in solution is treated with methanol and HCl, two compounds are formed. Determination of their structure has shown that these two compounds are the diastereomeric methyl α- and β-D-glucosides. These glucosides, and glycosides in general, are acid labile but are relatively stable at neutral and alkaline pH. Since the formation of the methyl glycoside converts the hemiacetal into an acetal group, the glycoside is not a reducing sugar and does not show the phenomenon of mutarotation.

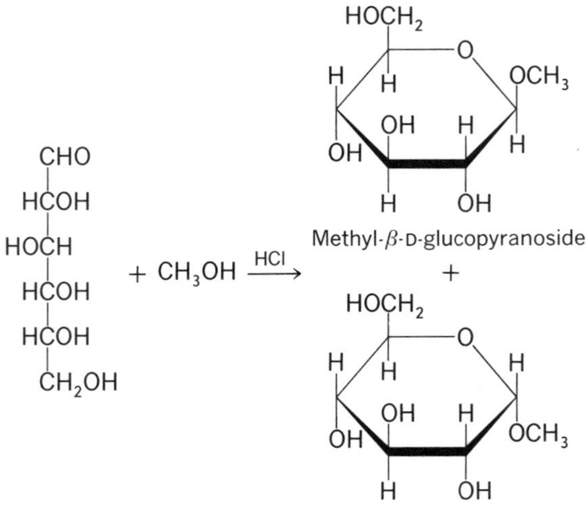

Methyl-β-D-glucopyranoside

Methyl-α-D-glucopyranoside

When an alcoholic hydroxyl group on a second sugar molecule reacts with the hemiacetal (or hemiketal) hydroxyl, the resulting glycoside is a disaccharide. The bond between the two sugars is known as a **glycosidic bond**. Polysaccharides are formed by linking together a large number of monosaccharide units with glycosidic bonds.

Although the anomeric hydroxy group of sugars may be methylated with ease, as in the formation of methyl glycosides just described, methylation of the remaining hydroxyl functions requires much stronger methylating agents. Nevertheless, the remaining four hydroxyl groups of methyl-α-D-glucopyranoside can be reacted with methyl iodide or dimethyl sulfate to yield the pentamethyl derivative. Such compounds, in turn, are useful in determining the ring structure of the parent sugar as in the following example:

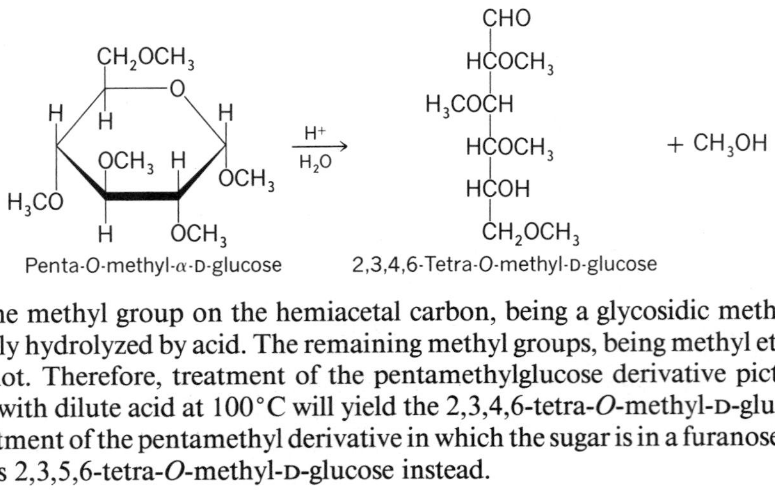

Penta-O-methyl-α-D-glucose   2,3,4,6-Tetra-O-methyl-D-glucose

The methyl group on the hemiacetal carbon, being a glycosidic methyl, is readily hydrolyzed by acid. The remaining methyl groups, being methyl ethers, are not. Therefore, treatment of the pentamethylglucose derivative pictured here with dilute acid at 100°C will yield the 2,3,4,6-tetra-O-methyl-D-glucose. Treatment of the pentamethyl derivative in which the sugar is in a furanose ring yields 2,3,5,6-tetra-O-methyl-D-glucose instead.

2.5.5 ESTER FORMATION

The alcoholic groups of carbohydrates may be esterified in both nonenzymatic and enzymatic reactions. Thus, when α-D-glucopyranose is treated with acetic anhydride, all the hydroxyl functions are acetylated to yield the penta-*O*-acetyl glucose pictured here. These acetyl groups, being esters, can be hydrolyzed either in acid or alkali.

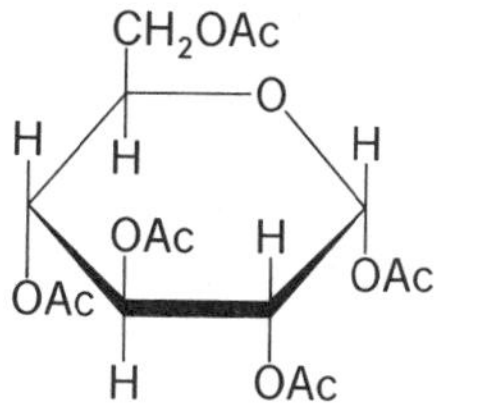

Penta-*O*-acetyl-α-D-glucose
($Ac = CH_3$—C—)
      O

An important type of carbohydrate ester encountered in intermediary metabolism is the phosphate ester. Such compounds are formed by the reaction of the carbohydrate with biological phosphorylating reagents such as adenosine triphosphate (ATP) in the presence of an appropriate enzyme. An example is fructose-1,6-bisphosphate.

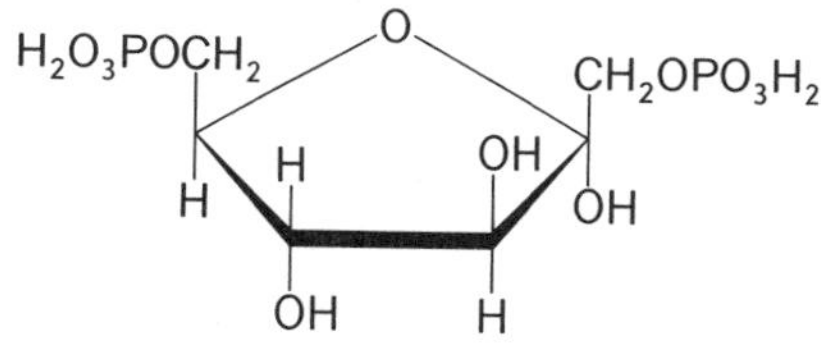

α-D-Fructose-1,6-bisphosphoric acid

The correct name of the nonionized form of this compound is α-D-fructo-furanose-1,6-bisphosphoric acid. Such phosphate esters are relatively strong acids with values of approximately 2.1 and 7.2 for pK_{a1} and pK_{a2}. Thus, at neutral pH, the sugar phosphates are anions and are normally referred to by the name of the anion, that is, fructose-1,6-bisphosphate.

2.5.6 ALDOL CONDENSATION

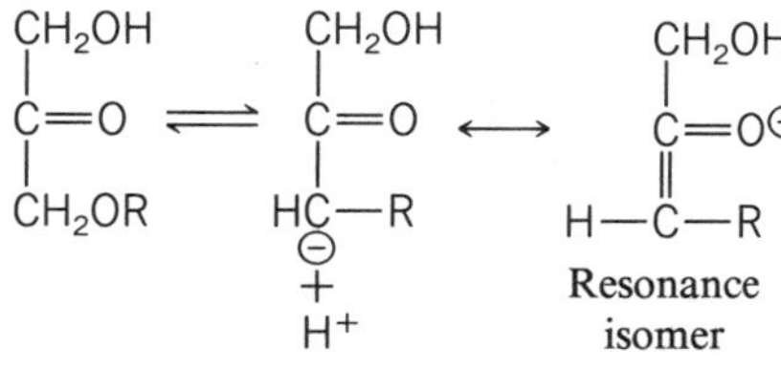

Another important reaction that is typical of carbohydrates and occurs frequently in biochemistry is the aldol condensation (or its reverse, the aldol cleavage). This reaction depends on the acidity of the hydrogens (α-hydrogens) on the carbon atom adjacent to a carbonyl group and the ability of the ionized ion to be stabilized by resonance. The enolate ion is then capable of acting as a nucleophile and can attack the aldehyde group of a second sugar molecule.

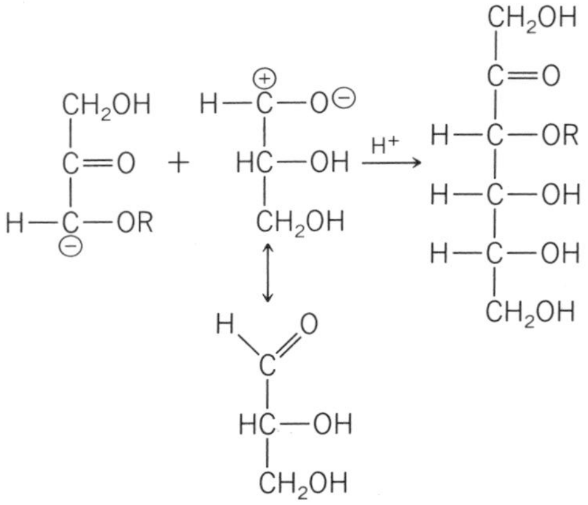

As shown above, two trioses can condense in an aldol condensation to yield a hexose. In the process, a new chiral center is produced. In theory, two hexoses with either an R or S configuration on carbon 4 would be produced. In metabolism, an enzyme assists in the formation of the enolate ion, and only one of the two possible diasteromers would be formed.

2.6 DISACCHARIDES

The oligosaccharides (see page 26 for definition) most frequently encountered in nature are disaccharides that on hydrolysis yield two moles of monosaccharides. Among the disaccharides encountered is the sugar maltose; this sugar is obtained as an intermediate in the hydrolysis of starch by enzymes known as amylases. In maltose, one molecule of glucose is linked through the hydroxyl group on the C-1 carbon atom in a glycosidic bond to the hydroxyl group on the C-4 of a second molecule of glucose.

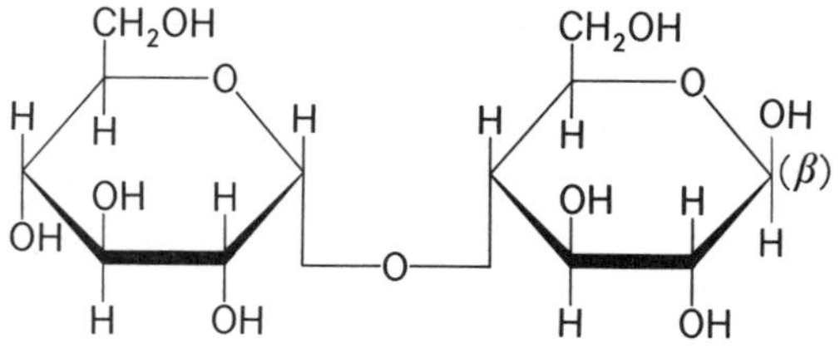

Maltose

The glycosidic linkage between the two glucose residues is designated as $\alpha(1 \rightarrow 4)$ to specify that the anomeric carbon involved in the glycosidic bond has the α-configuration and that it is linked to the 4-position of the second glucose molecule. This second glucose moiety possesses a free anomeric hydroxyl that can exist in either the α- or β-configuration (the β-isomer is shown); this free anomeric hydroxyl thus confers the property of mutarotation on maltose, and the disaccharide is a reducing sugar. That maltose has the structure shown was determined by analyzing the two products obtained on acid hydrolysis of its *octa*methyl derivative. The fully methylated maltose yields 2,3,4,6-tetra-*O*-methyl-D-glucose and 2,3,6-tri-*O*-methyl-D-glucose on hydrol-

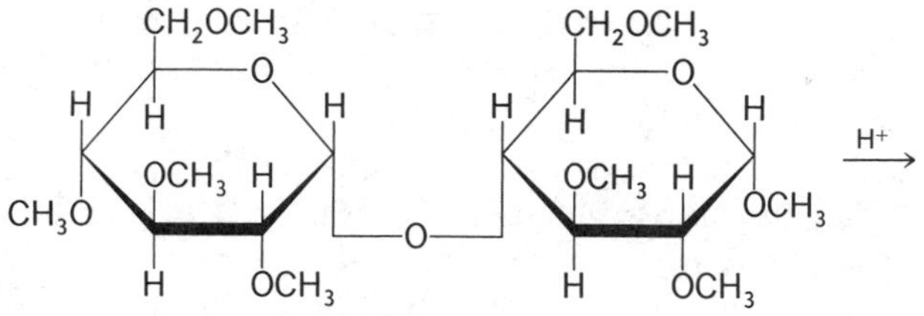

Octamethyl-D-maltose

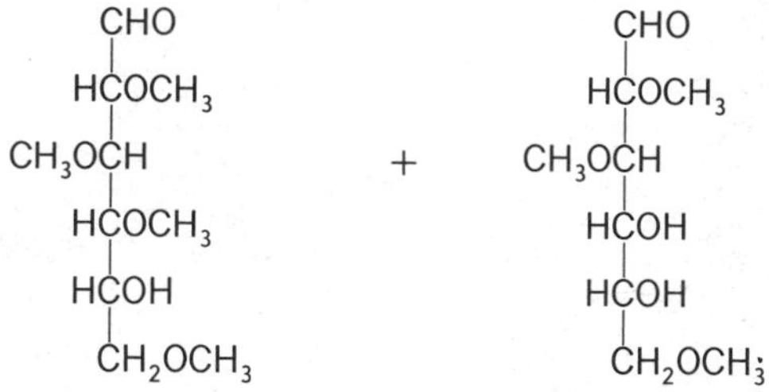

$\xrightarrow{\text{H}^+}$

2,3,4,6-Tetra-O-methyl-D-glucose

+

2,3,6-Tri-O-methyl-D-glucose

ysis. While the anomeric carbon of maltose is methylated on treatment of the disaccharide with dimethyl sulfate, this *O*-methyl glycosidic bond as well as the glycosidic bond between the two glucose units of the disaccharide is acid labile charide is acid labile and both are cleaved on hydrolysis with acid.

The disaccharide cellobiose is identical with maltose except that the former compound has a $\beta(1 \rightarrow 4)$ glycosidic linkage. Cellobiose is a disaccharide formed during the acid hydrolysis of

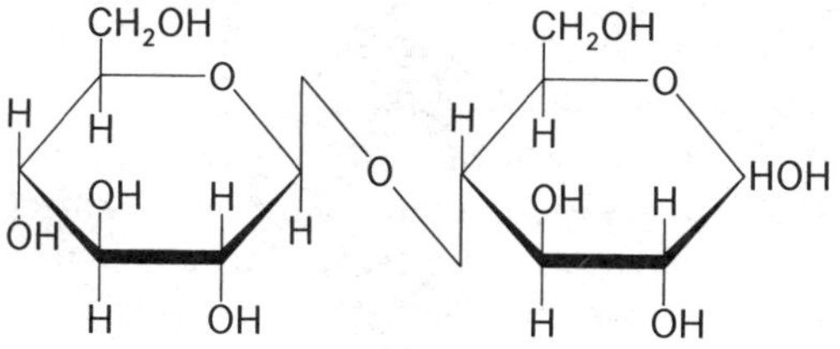

Cellobiose

cellulose. It is a reducing sugar and undergoes mutarotation. Treatment of cellobiose with dimethyl sulfate would also yield an octamethylated sugar, and acid hydrolysis would yield the same products that were obtained from octa-methyl maltose.

Isomaltose, another disaccharide obtained during the hydrolysis of certain polysaccharides, is similar to maltose

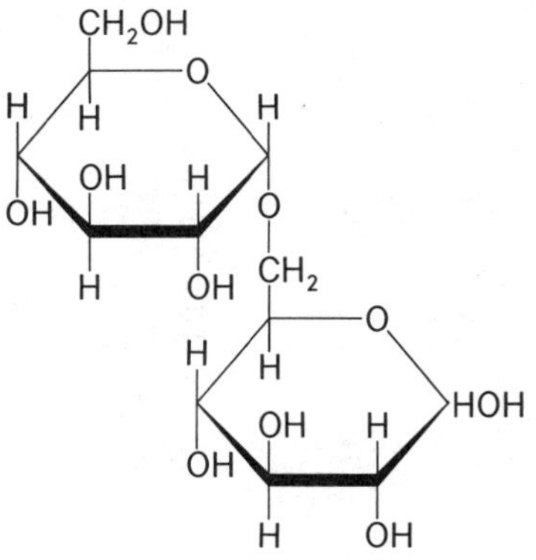

Isomaltose

except that it has an $\alpha(1 \rightarrow 6)$ linkage. Exhaustive methylation and acid hydrolysis of octamethyl isomaltose would yield 2,3,4,6-tetra-*O*-methyl-D-glucose and 2,3,4-tri-*O*-methyl-D-glucose.

Lactose is a disaccharide found in milk; on hydrolysis, it yields one mole each of D-galactose and D-glucose. It possesses a $\beta(1 \rightarrow 4)$ linkage, is a reducing sugar, and can undergo mutarotation. α-Lactose has the following formula in which the configuration at the reducing end of the disaccharide is shown as α.

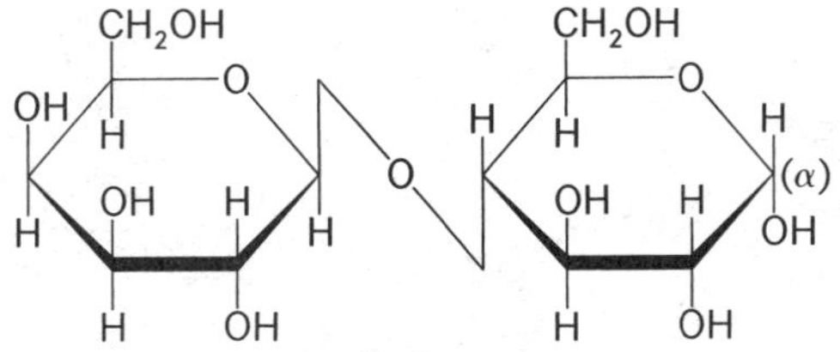

Lactose

Being the major carbohydrate in milk, lactose is extremely important in the nutrition of young mammals. Most of the world's human population relies on this sugar as a major form of energy during the first years of life. Lactose itself cannot be absorbed into the blood, but must first be hydrolyzed to its constituent monosaccharides by intestinal lactase. This enzyme is abundant in nursing infants but tends to disappear with age. Only Northern Europeans and a few other African peoples retain the enzyme in adulthood. Most other human groups have little intestinal lactase as adults and some, especially Mediterranean peoples and Orientals, may exhibit an intolerance to the sugar. In these people, high dietary intake of lactose results in intestinal disturbance in the form of diarrhea and pain.

Sucrose, the sugar of commerce, is produced by higher plants; sugar beets and sugar cane are major commercial sources. On hydrolysis, sucrose yields one molecule each of glucose and fructose, but, in contrast to all the other mono- and disaccharides described previously, sucrose is not a reducing sugar. This means that the reducing groups in both of the monosaccharide components must be involved in the glycosidic linkage between the two sugars. That is, the C-1 and C-2 carbon atoms, respectively, of the glucose and fructose moieties must be covalently linked in a glycoside bond. Permethylation studies of sucrose has shown that it must have the following structure with an α-configuration on the glucose subunit and a β-configuration on the fructose moiety.

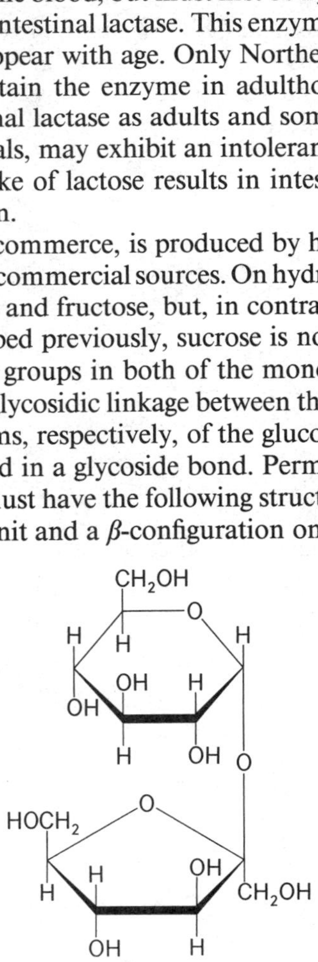

Sucrose

Sucrose is a major product of plant photosynthesis (Section 11.2.2 and Chapter 15). Sucrose is also the form in which carbohydrate, produced in the leaves by photosynthesis, is transported into storage organs such as developing seeds, tubers, or roots. It has been suggested that sucrose has an advantage, both as a storage product and a transport form of carbohydrate, over glucose and the other common sugars since both of its anomeric carbon atoms are protected from oxidative attack.

Sucrose, and maltose to a lesser extent, are important carbohydrate components of the human diet. However, they cannot be directly absorbed into the body and, like lactose, must first be hydrolyzed by specific enzymes, sucrase and maltase, found in the intestinal mucosa. Sucrose is used extensively as a sweetening agent in the food industry. It is readily available and is sweeter than the other common sugars maltose, lactose, and glucose. Only fructose is sweeter, and today enzymically produced mixtures of glucose and fructose, obtained from corn and other plant starches, are replacing sucrose as a commercial sweetener. Such mixtures are nutritionally equivalent to sucrose on a weight basis and significantly sweeter.

2.7 POLYSACCHARIDES

The polysaccharides found in nature either serve a structural function or play an important role as a stored form of energy. All polysaccharides can be hydrolyzed with acid or enzymes to yield monosaccharides and/or monosaccharide derivatives. Those polysaccharides that on hydrolysis yield only a single type of monosaccharide molecule are termed **homopolysaccharides. Heteropolysaccharides** on hydrolysis yield a mixture of constituent monosaccharides and derived products.

2.7.1 STORAGE POLYSACCHARIDES

Starch is a storage homopolysaccharide produced by plants (see Section 10.6.5). Whereas all green plants produce starch as an end product of photosynthesis, the cereal crops (wheat, rice, maize, and sorghum) are noted for the high starch content of their seeds. Indeed, these cereals together with a few other crops like potato and cassava, where the starch is stored in an underground tuber, provide the majority of the calories for nearly all of mankind. Starch consists of two components, amylose and amylopectin, that are present in varying amounts. The amylose component consists of D-glucose units linked in a linear fashion by $\alpha(1 \rightarrow 4)$ linkages; it has a nonreducing end and a reducing end (Structure 2.9). Its molecular weight can vary from a few thousand to 150,000. Amylose gives a characteristic blue color with iodine due to the ability of the halide to

Amylose

STRUCTURE 2.9

occupy a position in the interior of a helical coil of glucose units that is formed when amylose is suspended in water (Structure 2.10).

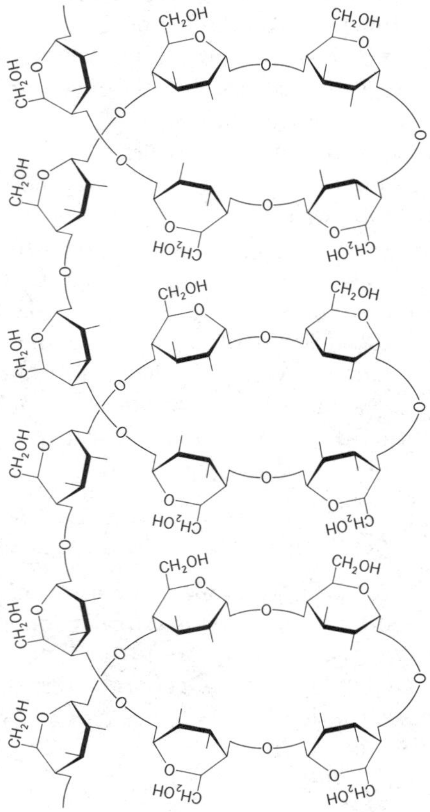

STRUCTURE 2.10

Amylopectin is a branched polysaccharide; in this molecule, shorter chains (about 30 units) of glucose units linked $\alpha(1 \rightarrow 4)$ are also joined to each other by $\alpha(1 \rightarrow 6)$ linkings (from which isomaltose can be obtained). (See Structure 2.11.) The molecular weight of potato amylopectin varies greatly and may be 500,000 or larger. Amylopectin produces a purple to red color with iodine.

Much has been learned about the structure of starch not only from studies

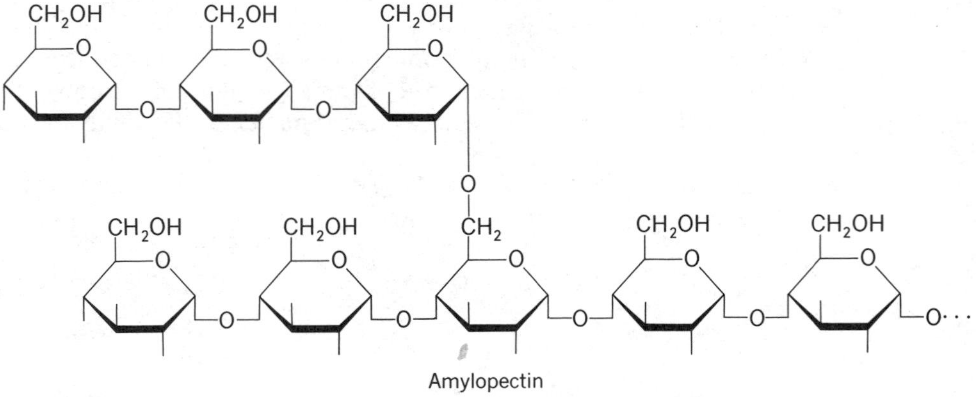

Amylopectin

STRUCTURE 2.11

with exhaustive methylating and oxidizing agents, but also by the action of enzymes on the polysaccharide. One enzyme, α-amylase, found in the digestive tract of animals (in saliva and the pancreatic juice), hydrolyzes the linear amylose chain by attacking $\alpha(1 \rightarrow 4)$ linkages at random throughout the chain to produce a mixture of maltose and glucose. β-Amylase, an enzyme found in plants, attacks the nonreducing end of amylose to yield successive units of maltose. (The prefixes α and β used with the amylases do not refer to glycosidic linkage, but simply designate these two enzymes.)

Amylopectin can also be attacked by α- and β-amylase, but the $\alpha(1 \rightarrow 4)$ glycosidic bonds near the branching point in amylopectin and the $\alpha(1 \rightarrow 6)$ bond itself are not hydrolyzed by these enzymes. A separate "debranching" enzyme, an $\alpha(1 \rightarrow 6)$ glucosidase, can hydrolyze the bond at the branch point. Therefore, the combined action of α-amylase and the $\alpha(1 \rightarrow 6)$ glucosidase will hydrolyze amylopectin ultimately to a mixture of glucose and maltose.

The storage polysaccharide of animal tissues is glycogen; it is similar in structure to amylopectin in that it is a branched homopolysaccharide composed of glucose units. It is more highly branched than amylopectin, however, having branch points about every 8 to 10 glucose units. Like amylopectin, glycogen is hydrolyzed by α- and β-amylases to form glucose, maltose, and a limit dextrin.

A final example of a nutrient polysaccharide will suffice. This is inulin, a storage carbohydrate found in the bulbs of many plants (e.g., dahlias and Jerusalem artichokes). Inulin consists chiefly of fructofuranose units joined together by $\beta(2 \rightarrow 1)$ glycosidic linkages.

2.7.2 STRUCTURAL POLYSACCHARIDES

The most abundant carbon compound in the world is the structural polysaccharide cellulose. Cellulose is found in the cell walls of plants where it contributes in a major way to the physical structure of the organism (see Section 8.1.2). Lacking a skeleton of bone onto which organs and specialized tissues may be organized, the higher plant relies on its cell walls to bear its own weight whether it is a sunflower or a sequoia. The wood of trees is an insoluble, organized structure composed of cellulose and another polymer called **lignin** derived from the amino acid phenylalanine.

Cellulose is a linear homopolymer of D-glucose units linked by $\beta(1 \rightarrow 4)$ glycosidic bonds (Structure 2.12). The seemingly small difference in structure

Cellulose

STRUCTURE 2.12

from amylose, however, confers very different and important properties on cellulose. Instead of forming a coiled helix, cellulose forms a structure of parallel chains that are cross-linked by hydrogen bonding.

In contrast to starch, the $\beta(1 \rightarrow 4)$ linkages of cellulose are highly resistant to acid hydrolysis; strong mineral acid is required to produce D-glucose; partial hydrolysis yields the reducing disaccharide, cellobiose. The $\beta(1 \rightarrow 4)$ linkages of cellulose are not hydrolyzed by the amylases found in the digestive tracts of humans or most other higher animals. Consequently, man and most animals cannot utilize the energy present in this glucose polymer. Ruminants are an important exception, however, since the bacteria that reside within the rumen secrete cellulase, a β-glucosidase, that catalyzes the hydrolysis of cellulose. These bacteria and others resident in the rumen then metabolize the glucose produced in a remarkable fermentation that is beneficial to the host animal. Snails and wood rotting fungi also secrete cellulase that can degrade cellulose. Termites can also degrade cellulose because their digestive tract contains a parasite that secretes cellulase.

Other examples of structural polysaccharides in plants are known. Plants contain pectins and hemicelluloses. The latter are not cellulose derivatives, but rather are polysaccharides enriched in D-xylose (xylans), D-mannose (mannans), and galactose (galactans) and linked by $\beta(1 \rightarrow 4)$ or $\beta(1 \rightarrow 3)$ glycosidic bonds. Pectins contain arabinose, galactose, and galacturonic acid. Pectic acid is a homopolymer of the methyl ester of D-galacturonic acid.

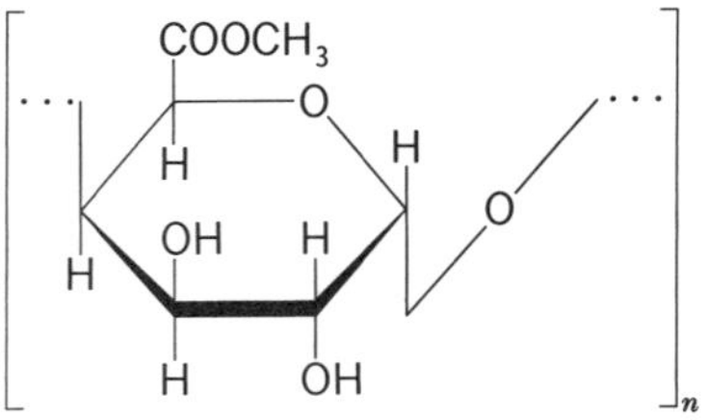

Pectic acid

Chitin, a homopolymer of *N*-acetyl-D-glucosamine, linked $\beta(1 \rightarrow 4)$ is the structural polysaccharide that constitutes the shell of crustaceans and the exoskeleton of insects.

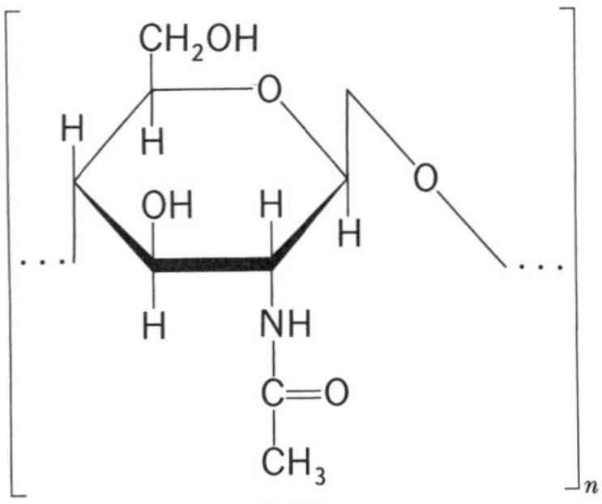

Chitin

2.8 POLYSACCHARIDES IN CELL WALLS

Animal cells do not possess a well-defined cell wall but have a **cell coat,** visible in the electron microscope, that plays an important role in the interaction with adjacent cells. These cell coats contain glycoproteins, glycolipids, and mucopolysaccharides. The chemical nature of the first two will be discussed later (Sections 2.9 and 7.7). The mucopolysaccharides are gelatinous substances of high molecular weights (up to 5×10^6) that both lubricate and serve as a sticky cement. One common mucopolysaccharide is hyaluronic acid, a heteropolysaccharide composed of alternating units of D-glucuronic acid and *N*-acetyl-D-glucosamine. The two different monosaccharides are linked by a $\beta(1 \rightarrow 3)$ bond to form a disaccharide that is linked $\beta(1 \rightarrow 4)$ to the next repeating unit. Hyaluronic acid, found in the vitreous humor of the eye and the umbilical cord, is water soluble but forms viscous solutions.

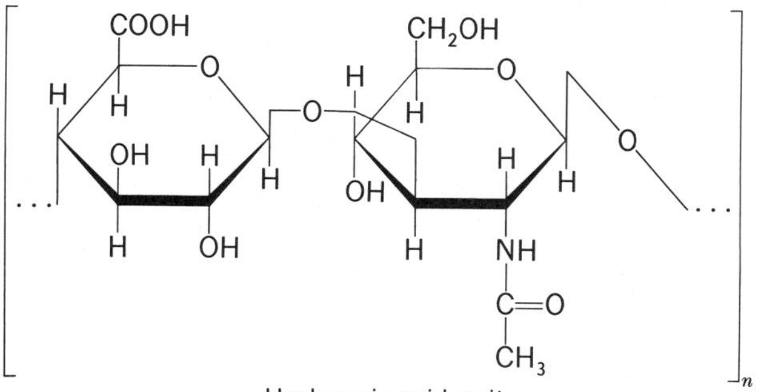

Hyaluronic acid unit

Chondroitin, similar in structure to hyaluronic acid except that the amino sugar is *N*-acetyl-D-galactosamine, is also a component of cell coats. Sulfate

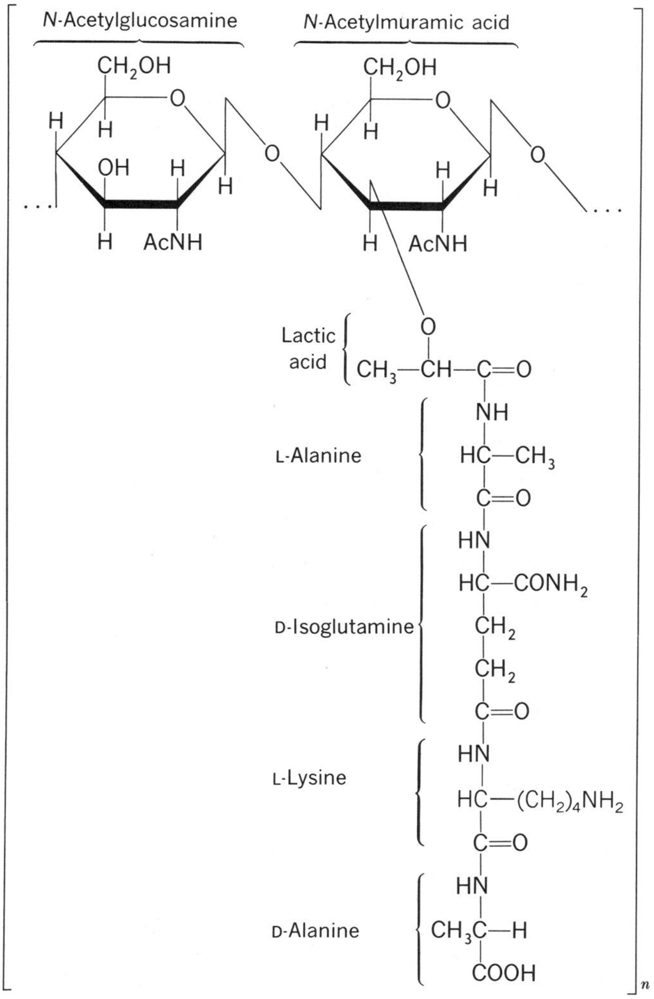

Repeating unit of peptidoglycan
of *Staphylococcus aureus*

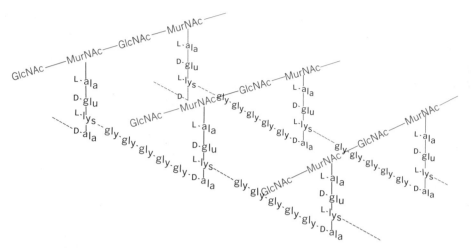

FIGURE 2.4   The linear chains of the peptidoglycan are cross-linked by glycine penta-peptides.

esters (at the C-4 or C-6 positions of the amino sugar) of chondrotin are major structural components of cartilage, tendons, and bones.

Bacterial cell walls, which determine many of the physiological characteristics of the organism they enclose, contain a heteropolysaccharide linked to a short chain of amino acids. Since the individual chains of amino acids are not as long as in proteins, such polymers have been termed peptidoglycans rather than glycoproteins. The heteropolysaccharide is an alternating chain of N-acetyl-D-glucosamine (GlcNAc) and N-acetylmuramic acid (MurNAc) joined by a $\beta(1 \rightarrow 4)$glycosidic bond. N-Acetylmuramic acid consists of a N-acetyl glucosamine unit that has its C-3 hydroxyl group joined to the α-hydroxyl group of lactic acid by an ether linkage. In the peptidoglycan, the carboxyl group of each lactic acid unit is linked, in turn, to a tetrapeptide that usually contains both D-alanine and L-alanine. Other amino acids found in the tetrapeptide may include D-glutamine, D-*iso*-glutamine, L-lysine, or diaminopimelic acid.

The linear polysaccharide chain of the peptidoglycan has a tetrapeptide branch at every second hexoseamine unit that is cross-linked to adjacent, parallel, polysaccharide chains. In the cross-linking (Figure 2.4), the carboxyl group of the terminal D-alanine moiety is attached to a pentaglycine residue that, in turn, is attached to the ε-amino group of lysine in the next adjacent glycan unit.

The antibiotic activity of penicillin is due to its ability to inhibit the last step in the biosynthesis of bacterial peptidoglycans. With the synthesis of this essential component of the cell wall inhibited, the bacteria are unable to grow or replicate.

2.9   GLYCOPROTEINS

Most of the oligo- and polysaccharides in the animal and plant cell are linked covalently to protein or lipid molecules known as glycoproteins or glycolipids. In many glycoproteins, the amide group of asparagine is linked through an N-glycosyl bond to a core trisaccharide consisting of one molecule of mannose (Man) and two molecules of N-acetyl glucosamine (GlcNAc).

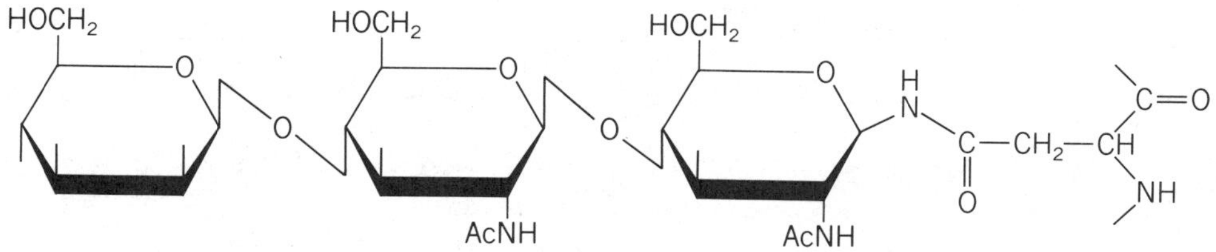

Manβ1,4GlcNAcβ1,4GlcNacβ-Asn

The mannose moiety of the core polysaccharide constitutes a branch point where two more mannose molecules are linked in $\alpha(1 \rightarrow 3)$ and $\alpha(1 \rightarrow 6)$ linkages. Either or both of the two mannose units may then serve as additional branch points for further enlargement of the polysaccharide component. In a highly branched example, galactose, sialic acid, and *N*-acetyl glucose are additional components of the branched polysaccharide.

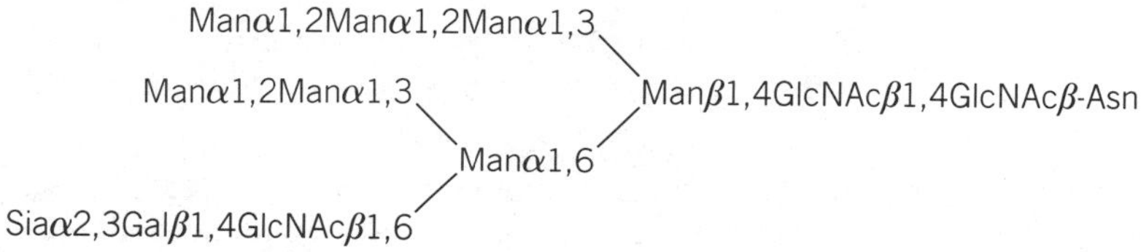

A sialic acid is a ketose containing nine carbon atoms (ketononose) that may be acylated with acetic or glycolic acid. *N*-acetyl-D-neuraminic acid is a specific example of a sialic acid.

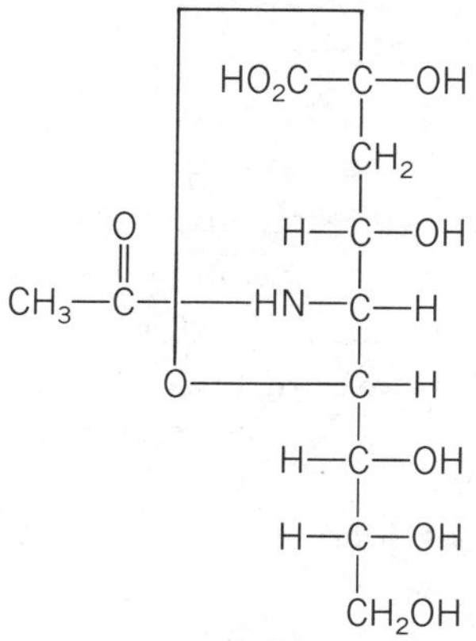

N-Acetyl-D-neuraminic acid

The core polysaccharide can also be linked to the protein component through an *O*-glycosyl bond to the hydroxyl group of serine instead of the asparagine amide group. Hydroxylysine can substitute for serine, and other sugars such as xylose and galactose can substitute for the GlcNAc. The latter are found in such glycoproteins as collagen and proteoglycan. It is apparent that glycoproteins, which are widely distributed in all living matter, can show considerable diversity.

Much remains to be learned of the structure of cell walls before we can completely understand such important phenomena as the immune response and cellular growth and differentiation.

REFERENCES

1. *R. T. Morrison and R. N. Boyd,* Organic Chemistry, *4th ed. Boston: Allyn and Bacon, 1983.*

2. *W. Pigman and D. Horton, eds.,* The Carbohydrates, *2nd ed. New York: Academic Press, 1970, 1972, 1980 (4 volumes).*

3. *R. Barker,* Organic Chemistry of Biological Compounds. *Englewood Cliffs, N.J.: Prentice-Hall, 1971.*

4. *R. L. Whistler and M. L. Wolfrom, eds.,* Methods in Carbohydrate Chemistry. *New York: Academic Press, 1962–1980 (8 volumes).*

5. *M. L. Wolfrom and R. S. Tipson, eds.,* Advances in Carbohydrate Chemistry and Biochemistry. *New York: Academic Press, 1945–1984 (42 volumes).*

REVIEW PROBLEMS

1. If heptoses (7-carbon sugars) are synthesized by Kiliani synthesis from a given 4-carbon sugar, how many isomers would be obtained?

2. An equilibrium mixture of α- and β-D-galactose has an $[\alpha]_D^{25°}$ of $+80.2°$. The specific rotation of pure α-D-galactose is $+150.7°$. The specific rotation of pure β-D-galactose is $+52.8°$. Calculate the proportions of α- and β-D-galactose in the equilibrium mixture.

3. Draw the structure of any β-D-aldoheptose in the pyranose ring form, using the Fischer projection or the Haworth ring structure, and answer the following questions:

   (a) How many asymmetric carbon atoms does the above sugar have?

   (b) How many stereoisomers of the above sugars are theoretically possible?

   (c) Draw the structure of the *anomer* of the above β-D-aldoheptose.

   (d) Draw the structure of the *enantiomer* of the above β-D-aldoheptose.

   (e) Draw the structure of an *epimer* (other than the anomer) of the above sugar.

   (f) Draw the structure of a *diastereoisomer* of the above β-aldoheptose.

   (g) Draw the structure of a *structural isomer* of the above β-aldoheptose.

   (h) Draw the structures of the two different sugars that you would obtain if you used the aldoheptose drawn initially as the starting material for a Kiliani synthesis (involving HCN addition, etc.).

   (i) Why does the Kiliani synthesis yield two different sugars starting from a single precursor?

   (j) Draw the structures of two different sugars that would yield the same osazone as the β-D-aldoheptose drawn initially.

   (k) Draw the structure of the same β-D-aldoheptose drawn initially in the furanose ring form.

4. An unknown disaccharide was purified from bacteria. Equal amounts of D-glucose and D-galactose were obtained after acid hydrolysis of the disaccharide and the two sugars were found to be linked by an α-glycosidic linkage. Exhaustive methylation of the disaccharide produced equal amounts of 2,3,4,6-tetramethylgalactose and 2,4,6-trimethylglucose. Using the Haworth formula, draw the structure of the disaccharide suggested by the above information and show clearly the linkage between the sugars.

CHAPTER 3

AMINO ACIDS AND PROTEINS

As was described in Chapter 1, cells utilize energy to overcome the general tendency toward disorder. No class of molecules in the cell is more important than the proteins in mediating the reactions and forming the structures that generate order in organisms. Proteins are the workhorses of the cell.

The Swedish chemist Berzelius suggested the name "protein," derived from the Greek noun *protios,* meaning "holding the first place." His suggestion was inspired by observations and deductions made by the Dutch chemist Gerardus Mulder. Mulder had been able to extract very similar nitrogen-rich substances from both animal and plant sources and had guessed that these substances, whatever their other properties, must be important simply because their occurrence is so widespread and they are so abundant in tissues. Half and more of the dry weight of many cells is, in fact, protein.

3.1 SOME FUNCTIONS OF PROTEINS

Most of the functions of proteins fall into the categories of binding, catalysis, conduction or transport, contraction, nutrition, and/or structure. Often, a protein will have more than one of these functions, and the categories are in any case not mutually exclusive. Consider some examples of binding. A protein that binds to a specific portion of a nucleic acid molecule may be able to control the expression of the genetic information encoded in nearby regions of that molecule. A cell surface protein receptor that binds insulin molecules may be able, in effect, to let the cell "sense" the concentration of insulin in its environment, providing a link between the hormone and the hormone's action. Serum antifreeze proteins that are crucial to the survival of certain cold water fishes appear to act by binding to the surface of ice crystals. The binding of oxygen and protons by the protein hemoglobin was discussed in Chapter 1.

The tremendous variety of chemical reactions within the cell must all be carried out within the narrow ranges of temperature, pH, and so on, under which the cell is active. The cell does not have the option of using extremes of temperature or pH to facilitate a reaction. Instead, catalysts of specific reactions are used. Virtually all of the important catalysts of the cell are proteins: the enzymes. Enzymes, of course, bind the molecules on which they act, but they also transform them, breaking and making covalent and noncovalent bonds. Without enzymes, clearly there would be no metabolism. Chapter 4 of this book is devoted to enzymes, the catalytic proteins.

Some proteins facilitate the movement of compounds through cell membranes. The movement may be passive, a conduction of the substance through the membrane from a concentrated to a less concentrated solution. However, often the transport is active. In an energy-requiring process, a substance moves against a gradient of its concentration. Conduction and transport involve binding and contribute to such processes as nutrient uptake and nerve conduction. We discuss some of these processes in greater detail in Chapter 8.

Contraction is, of course, a property of muscles. Muscles are principally protein, and it is the sliding motion of muscle proteins, one relative to the other, that is responsible for contraction of muscles. Some proteins contribute to motion and coordination directly through their contraction.

The storage proteins of seeds supply much of the necessary nitrogen and energy to support growth of the plant until it can be supported by photosynthesis. Prolamins, the most abundant class of cereal storage proteins, make up a significant fraction of the protein consumed by man and domestic animals. Collagen, a fibrous protein of skin and bone, is just one example of the many structural proteins. Thus, proteins influence almost every facet of the cell's activity.

3.2   PROTEINS AS POLYMERS OF AMINO ACIDS

The abundance of proteins and their relatively high content of nitrogen, roughly 15 to 18% by weight, brought them to the attention of pioneering biochemists. Nitrogen is an essential constituent of proteins because proteins are polymers of **amino acids.** Protein amino acids have a central carbon atom, designated the α-carbon, to which is bonded both a carboxyl group and an amino group, the α-amino group. The general formula of a naturally occurring amino acid may be represented with a modified ball-and-stick formula or the Fischer projection formula (Structure 3.1).

Ball-and-stick model   Fischer projection formula

STRUCTURE 3.1

Because the amino group is on the carbon atom adjacent to the carboxyl group, the amino acids having this general formula are known as α-amino acids. It is also apparent that if R in this structure is not equal to H, the α-carbon atom is asymmetric. That is, like the sugars described in Section 2.2, the α-amino acids are **chiral** or "optically active" compounds. It is well-known that all the naturally occurring amino acids found in proteins have the same configuration.

With respect to the reference compound for carbohydrates, D-glyceralde-hyde, the amino acids that occur in proteins have the opposite or L-configura-tion. This relationship is shown in Structure 3.2 where, in the ball-and-stick model and the Fischer projection, the amino group of L-serine is on the left when the carboxyl group is written at the top of the formula. An early accom-plishment in biochemistry was the conversion of L-serine into L-glyceraldehyde by a series of chemical reactions that did not modify the configuration of the α-carbon atom. In this way, the absolute configuration of L-serine was estab-lished. Other amino acids are compared, for absolute configuration about the α-carbon atom, to L-serine as the reference compound. (When reference is made to the absolute configuration of L-serine rather than to the actual optical rotation of the amino acid, the notation L_s often is used.)

Ball-and-stick model

Fischer projection formula

L-Serine   D-Glyceraldehyde

STRUCTURE 3.2

Careful comparison of the general formula in Structure 3.1 with those in Structure 3.2 will disclose that the amino acid represented in Structure 3.1 also has the L-configuration; if $R = -CH_2OH$, the general formula becomes L-ser-ine. Note that the amino group is below the α-carbon atom in the structure of an L-amino acid when the carboxyl group is written to the right in the projection formula.

As with the carbohydrates, it is important to stress that the use of L and D conventions refers only to the relative configuration of these compounds and does not provide any information regarding the direction in which these opti-cally active compounds rotate polarized light.

Note that the amino acids are represented in different ionic forms in Struc-tures 3.1 and 3.2. As is described in Box 3.A, the zwitterionic form of the amino acid, as represented in Structure 3.2, most closely represents the state of ioniza-tion of the amino acids in solutions of neutral pH.

BOX 3.A   AMINO ACIDS AS ZWITTERIONS

Consider the amino acid alanine, in which the R group is a methyl group. Since alanine contains both a carboxyl and an amino group, it should react with acids and alkalies. Such compounds are referred to as **amphoteric substances.** If a solid sample of alanine is dissolved in water, the pH of this solution will be approximately neutral. If electrodes are placed in solution and a difference in potential is placed across the electrodes, the amino acid will not migrate in the electric field. These results are in keeping with the representation of the amino acid as a neutral, uncharged molecule. However, the same is true if alanine were represented as the **zwitterion.** This formula, first proposed by Bjerrum in 1923, depicts the carboxyl group as being dissociated while the amino group is protonated. The name zwitterion is derived from the German "zwitter," meaning "hybrid." Thus, the zwitterion is a hybrid of positive and negative ionic groups, as shown.

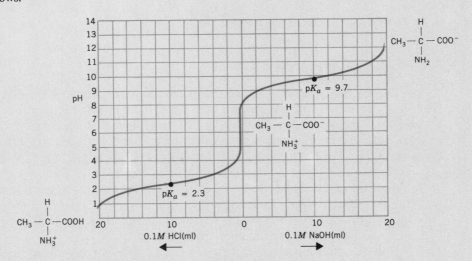

In theory, both structures could give rise to the same titration curves (see Sections 1.9 and 1.10 for a discussion of titration curves of carboxylic acids and alkyl amines). Certainly, the same forms will be seen at the pH extremes: the carboxylate anion and unprotonated amine at high pH, both amino and carboxyl groups protonated at low pH. If 20 ml of $0.1M$ alanine (2 mmole) in solution is titrated with $0.1M$ NaOH, a curve with a pK_a at 9.7 is obtained. When 10 ml of $0.1M$ NaOH (1 mequiv, corresponding to half the number of moles of alanine) has been added, the pH should correspond to the pK_a. This signifies that, at pH 9.7, some group capable of furnishing protons to react with the added alkali is half-neutralized. Similarly, if $0.1M$ HCl is added to the solution of alanine, the other half of the titration curve is obtained and the pH corresponding to a pK_a of 2.3 is reached when 10 ml of $0.1M$ HCl has been added. The titration curve for the zwitterion form of alanine is represented as follows:

The next formulas represent the ionization reactions of the titration.

$$CH_3-CH-COOH \xleftarrow{H^+} CH_3-CH-COO^- \xrightarrow{OH^-} CH_3-CH-COO^- + H_2O$$
$$\quad\;\; ^+NH_3 \qquad\qquad\quad\;\; ^+NH_3 \qquad\qquad\qquad NH_2$$

Note that in this scheme, the carboxylate group can be considered to be titrated by acid, the protonated amino group by base. The pK_a's observed in the titration curve are consistent with this representation because, as noted in Chapter 1, the carboxyl groups of organic acids dissociate in the pH range of 3 to 5, while protonated alkyl amines are weak acids with pK_a's in the range of 9 to 11. Of course, protons are capable of ready migration, and both the zwitterion and the corresponding unionized alanine structures must be considered to contribute to the structure of alanine in solution at neutral pH, with the zwitterion making by far the greater contribution.

Several other properties of amino acids are consistent with the zwitterion structure at neutral pH. Additional evidence for the protonated α-amino group at neutral pH was obtained by titrating the amino acid in formaldehyde. Formaldehyde reacts with the uncharged amino group to produce a mixture of the mono- and dihydroxymethyl derivatives.

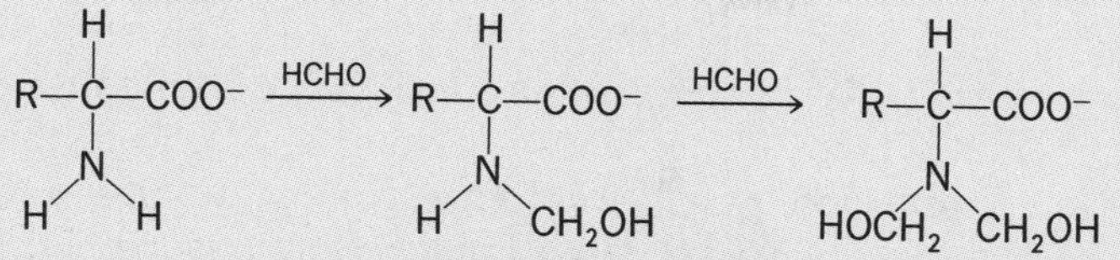

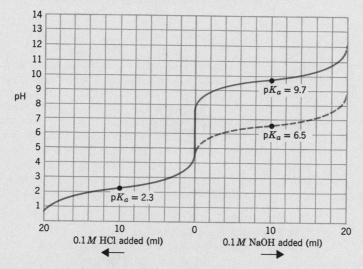

The hydroxymethyl groups of these secondary and tertiary amines are electron-withdrawing. Thus, they are weaker bases (= stronger acids) than the corresponding underivatized amines. This is manifested in the titration curve by a lowering of the pK_a for the amino group in the presence of formaldehyde, as indicated by the dashed titration curve shown here.

Note that only the high pH side of the titration curve is affected. Formaldehyde is not known to react with carboxyl groups under the conditions of the titration. Hence, the amino group rather than the carboxyl group is titrated at high pH, as required by the zwitterionic structure at neutral pH.

Amino acids, with certain exceptions, are generally soluble in water and are quite insoluble in nonpolar organic solvents such as ether, chloroform, and acetone. This observation is not in keeping with the known properties of protonated carboxylic acids and unprotonated organic amines. Aliphatic and aromatic carboxylic acids, particularly those having several carbon atoms, have limited solubility in water but are readily soluble in organic solvents. Similarly, the higher amines are usually soluble in organic solvents but not water. Crystalline amino acids have high melting points that often result in decomposition, whereas the melting points of the solid carboxylic acids and amines are usually low and well defined. Thus, the crystalline amino acids melt as if they are ionic rather than neutral. The high water solubility of most amino acids and the melting behavior of the crystalline solids are consistent with the zwitterionic form predominating both in solution and the solid.

Other evidence of the zwitterionic nature of all amino acids is found in their spectroscopic properties, effects on the dielectric constant of aqueous solutions, and titrations in organic solvents.

That the fundamental structural unit of proteins is the α-amino acid is easily demonstrated by hydrolyzing purified proteins by either chemical or enzymatic procedures. For example, a protein may be hydrolyzed completely or nearly completely to its constituent amino acids in a period of 18 to 24 hours by the action of 6N HCl at 110°C in a sealed tube. Under these conditions, 17 of the 20 common protein amino acids are released in good yield. Isolating 17 distinct amino acids from a hydrolyzate by crystallization and other classical techniques of organic chemistry was a considerable accomplishment of pioneering biochemists. Box 3.B describes a current approach to the microanalysis of amino acids by ion exchange chromatography. An amino acid analysis, which gives the relative amounts of the amino acids that survive the hydrolysis procedure, is an important step in the preliminary characterization of a newly purified protein.

BOX 3.B   AMINO ACID ANALYSIS

A routine and often essential part of the characterization of a protein is the determination of the relative molar amounts of the amino acid residues in that protein—an amino acid analysis. Current technology for amino acid analysis is destructive of the sample because it requires that the protein be hydrolyzed to its constituent amino acids. A requirement for a useful hydrolysis method is that it is vigorous enough to leave only an insignificant portion of the peptide bonds of the protein intact but gentle enough to preserve the released amino acids. No hydrolysis procedure fully lives up to these requirements. The carboxamido groups of asparagine and glutamine generally are more sensitive to hydrolysis (to the corresponding carboxylic acid) than peptide bonds, so subsequent analysis gives only the sum of aspartic acid and asparagine residues and the sum of glutamic acid and glutamine residues in the protein, not the amounts of these amino acid residues individually.

In a typical hydrolysis condition, aliquots of the protein sample in 6N HCl are heated at 110° for 6 to 96 hours in sealed tubes. Several time periods of hydrolysis are used to compensate for the strong resistance to hydrolysis exhibited by peptide bonds of aliphatic amino acid residues such as leucine and isoleucine and the sensitivity to destruction of serine and threonine. The values obtained from subsequent analyses of the amino acids are extrapolated to long times of hydrolysis for the aliphatic amino acids and to short times of hydrolysis for serine and threonine. Tryptophan is destroyed in 6N HCl at 110° unless a reducing agent such as thioglycolic acid is added during hydrolysis.

The HCl is removed by evaporation under vacuum, leaving the amino acids as their hydrochloric acid salts. These are dissolved and applied to a chromatography column for separation. A chromatography column is a tube, usually much longer than its diameter, that is equipped with a porous disk at one or both ends as indicated in the diagram. The column is packed with very small beads (the "chromatography resin") composed of a material that will show a differential affinity for the substances that are to be separated, amino acids in this case. The diagram shows a bed of resin packed in a chromatography column. Arrows indicate the direction of flow of eluting solution. Contrary to what is illustrated, actual chromatography resins are very fine, so that the bead diameters are very much smaller than the column diameter.

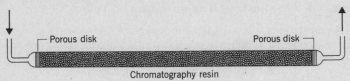
Chromatography resin

Amino acids most often are separated on a chromatography resin that is sulfonated polystyrene. The chromatography is carried out in buffers of low pH, from 3 to 6.5. This maintains the amino acids as cations because both the amino and the carboxyl groups are fully or partially protonated. Since sulfonic acids are strong acids, the sulfonyl groups, —SO$_3^-$, retain their negative charge even below pH 3. A representation of a portion of the structure of a cross-linked, insoluble, sulfonated polystyrene is:

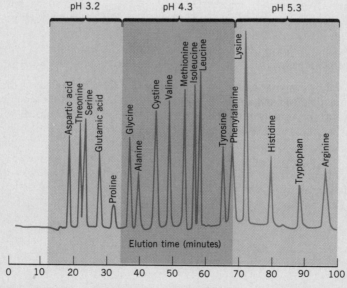

The amino acid cations bind to the resin electrostatically. The process of eluting the amino acids (or other ions) from an ionic resin by competition with buffered salt solutions passed through the column is referred to as "ion exchange chromatography." However, because both the polystyrene matrix of the resin and the R groups of many of the amino acids are nonpolar, the process of separation of amino acids is influenced not only by ion exchange chromatography, but also by absorption to the resin matrix.

An "amino acid analyzer" is an integrated apparatus that combines the chromatography column, pumps, valves, timers, photometric and/or fluorometric detectors, electronics, and chart paper recorder to automatically record and even perform preliminary calculations of the separation. Here, we present an example of a "chromatogram" produced by an amino acid analyzer, for the chromatography of a protein hydrolysate on a sulfonated polystyrene resin.

The abscissa is time after the sample was introduced into the stream of liquid that is being pumped through the column. Since the pump operates at a constant flow rate, the abscissa also is proportional to the volume that has passed through the column after the sample was applied. The ordinate is proportional to the concentration of the amino acid being eluted, the proportionality being similar but not identical for most of the amino acids. The proportionality between signal (ordinate) and amount of each amino acid, and the time required to elute each amino acid are established by the application of standards, known amounts of amino acids applied singularly or in small groups to the column in separate runs.

The chromatography often is carried out with citrate buffers as the elution solution. Typically, two or three buffers are used, the first with a pH as low as 3 and the last with a pH as high as 6.5. At the lowest pH, all of the protein amino acids will be bound to the column. The basic amino acids, lysine, histidine, and arginine, will be bound so tightly that they will not elute in a reasonable period of time if the eluting buffer is not changed. Note that the order of elution of the amino acids can be rationalized, at least in part, by consideration of the amino acid charge and polarity. Aspartic and glutamic acids are acidic, polar amino acids and hence should have a low affinity for the acidic, nonpolar resin. The neutral amino acids, beginning with threonine and serine, elute generally in order of decreasing polarity. The basic amino acids are eluted late, only after the pH of the eluting buffer has been increased.

To conserve valuable protein samples, considerable effort has been expended in developing sensitive methods for detecting amino acids. The ninhydrin reagent is the basis of a quantitative colorimetric analysis for amino acids. The dehydrated form of ninhydrin is able to form a Schiff base with the α-amino group of amino acids and, in hot, weakly acidic solution, to promote the decarboxylation of the amino acid portion of the Schiff base. Dissociation of an aldehyde leaves a primary amine that can react with the reduced form of ninhydrin, hydrindantin, to give the intensely blue-colored "Ruhemann's purple."

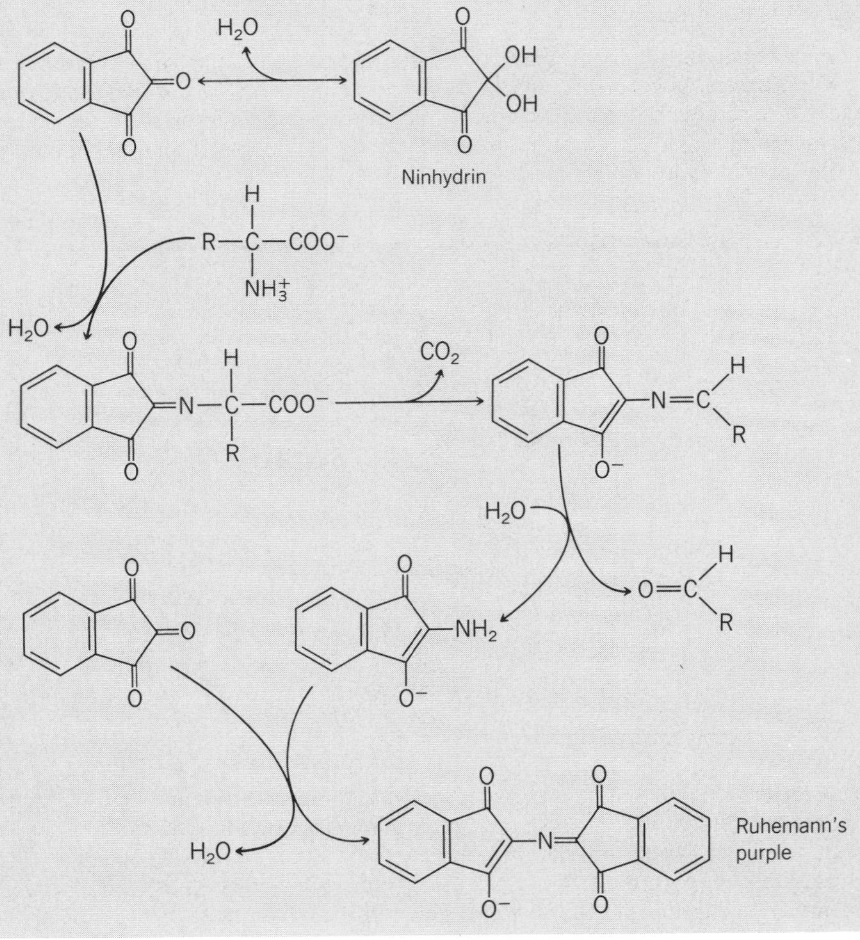

In an amino acid analyzer, the stream that emerges from the chromatography column is mixed with another stream of liquid, a solution of ninhydrin and hydrindantin, catalysts of the ninhydrin reaction and buffer. In a process known as postcolumn derivatization, the combined, flowing streams are passed through a long coil of fine tubing in a heated water bath where the ninhydrin reaction occurs. The stream that emerges from the coil is passed through a "flow cell" with a long optical path length. The flow cell is in a special photometer that is designed to quantitatively detect the blue derivative with high sensitivity. The intense blue product is generally characteristic of those amino acids having α-amino groups. However, proline and the rarer amino acid hydroxyproline, which are secondary amines, yield yellow products in the ninhydrin reaction. These imino acids require a separate detector set at a wavelength of 440 nm, whereas the reaction products of amino acids with ninhydrin are detected at 570 nm. Since primary amines, as well as α-amino acids, react with the ninhydrin reagent, lysine gives a more intense color than is given by other protein amino acids.

Two other reagents that are widely used for sensitive detection of amino acids are o-phthalaldehyde and fluorescamine. Each of these reagents produces a fluorescent derivative when it reacts with aliphatic primary amines. Detection, of course, requires a fluorometer.

o-Phthalaldehyde   Fluorescamine

With current instrumentation, the amino acid composition of less than a nanomole (10^{-9} mole) of protein can be determined.

Early investigators realized, from certain physical properties of proteins, that they must be high molecular weight substances. High molecular weight substances already were well known in the chemistry laboratory. The tars that accumulate in some reactions and the colloidal substances formed by an electric arc between two metal electrodes under water are examples. Such high molecular weight substances are said to be "polydisperse" because they are, in fact, mixtures of compounds of related chemical structure but highly variable size. Tars and colloidal substances, as well as hemoglobin and certain other oxygen-transporting proteins, all were examined by applying strong centrifugal fields to their solutions or suspensions. The relatively large size of the molecules in tars and proteins or the aggregates in colloidal substances caused them to move much more rapidly than low molecular solutes, such as salts, in the centrifugal field. However, purified proteins, such as hemoglobin, behaved very differently from tars and colloids. Only the proteins migrated in the centrifugal field in such a way as to give a sharp, discrete boundary. This was observed because proteins are high molecular weight **compounds,** meaning that they have a definite atomic composition. Each constituent of a tar or colloid will have its characteristic rate of sedimentation in the centrifugal field. However, since the constituents are of different sizes, no sharp, discrete boundary could be observed for these substances.

The atomic composition of hemoglobin, stripped of its oxygen-bearing heme groups, is

$$C_{2796}H_{4592}O_{832}N_{812}S_8$$

This atomic composition is typical of protein molecules, which usually are 30 to 33% C, roughly 50% H, 9 to 11% O and 7 to 9% N, with small amounts of S. The student should contrast these atomic percentages with those for carbohydrates, Chapter 2.

3.3   PRIMARY STRUCTURE OF PROTEINS

The covalent structure of a protein is essentially linear. Amino acids are connected together to form a chain, the connection being a peptide bond. The peptide bond is simply an amide bond between the carbonyl carbon of one amino acid and the amino nitrogen of another.

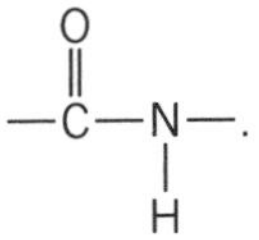

$$-\overset{\overset{\displaystyle O}{\|}}{C}-\underset{\underset{\displaystyle H}{|}}{N}-\,.$$

Conceptually, the formation of the peptide bond may be considered to result from the removal of two protons and one oxygen atom (i.e., one molecule of water) from a pair of amino acids. In the reaction shown below, the oxygen atom may be considered to have been removed from the carboxylate group of amino acid 1 and two protons from the alkyl ammonium group of amino acid 2.

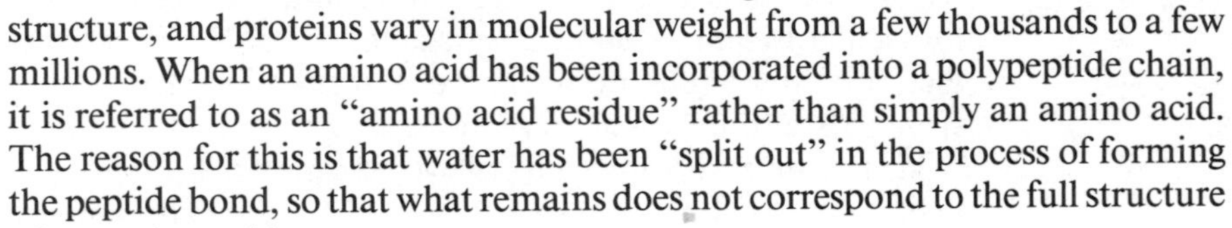

$$^+H_3N-\overset{\overset{\displaystyle R_1}{|}}{CH}-COO^- + {}^+H_3N-\overset{\overset{\displaystyle R_2}{|}}{CH}-COO^- \longrightarrow$$

$$H_2O + {}^+H_3N-\overset{\overset{\displaystyle R_1}{|}}{CH}-\overset{\overset{\displaystyle O}{\|}}{C}-NH-\overset{\overset{\displaystyle R_2}{|}}{CH}-COO^-$$

The **polypeptide** character of a protein is shown in Figure 3.1. Obviously, there is no theoretical limit to the molecular weight attainable with such a chain structure, and proteins vary in molecular weight from a few thousands to a few millions. When an amino acid has been incorporated into a polypeptide chain, it is referred to as an "amino acid residue" rather than simply an amino acid. The reason for this is that water has been "split out" in the process of forming the peptide bond, so that what remains does not correspond to the full structure of an amino acid.

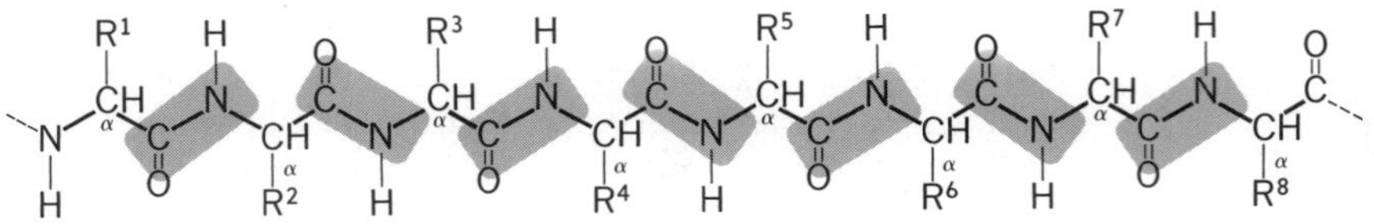

FIGURE 3.1   A generalized structure of a polypeptide chain showing the linkage of adjacent amino acid residues through peptide bonds.

Twenty different amino acid residues account for the vast majority of protein structures. With the exception of the **imino** acid proline (whose structure follows), these all are α-amino acids, which means that they differ only in the R group. The average weight of a protein amino acid residue is approximately 110, which means that proteins have from a few tens to about ten thousand amino acid residues.

The primary structure of a protein is simply the order of amino acid residues in the polypeptide chain. In the remainder of this section, we present the structures of the 20 common protein amino acids and of a few less common protein amino acids. These could be classified according to the chemical nature (aliphatic, aromatic, hetercyclic) of their R groups into appropriate subclasses. More meaningful, however, is a classification based on the polarity of the R group or residue because it emphasizes the possible functional roles that the different amino acids can play in proteins and their possible contributions to the folding of the polypeptide chain. In this classification, the 20 amino acids commonly found in proteins may be described as

1. Nonpolar or hydrophobic.
2. Polar but uncharged.
3. Polar because of a negative charge at the physiological pH of 7.
4. Polar because of a positive charge at physiological pH.

Hydrophobic means "water hating." It is a term used to describe aliphatic and aromatic hydrocarbon compounds or portions of molecules, or other chemical groups, that share the property of having only very limited solubility in water.

1. *Amino acids with nonpolar or hydrophobic R groups.* This group contains amino acids with both aliphatic (alanine, valine, leucine, isoleucine, methionine) and aromatic (phenylalanine and tryptophan) residues that are understandably hydrophobic in character. One of the amino acids, proline, is unusual in that its nitrogen atom present is as a *secondary* rather than a primary amine. Thus, it is in fact an *imino* acid rather than an *amino* acid.

L$_s$-Alanine

L$_s$-Valine

L$_s$-Leucine

L$_s$-Isoleucine

L$_s$-Proline

L$_s$-Phenylalanine

L$_s$-Tryptophan

L$_s$-Methionine

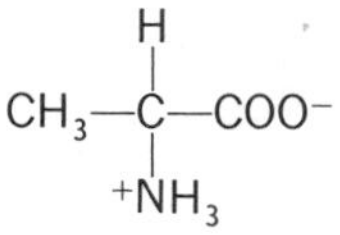

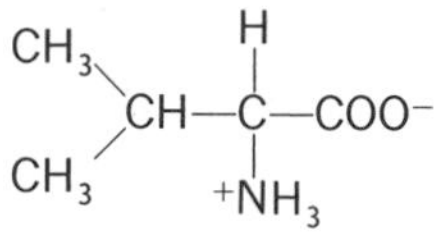

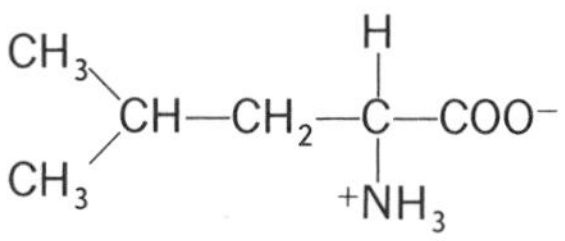

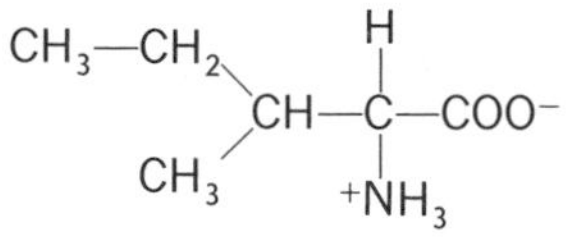

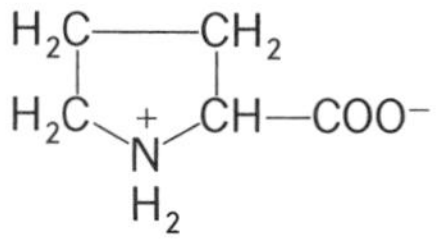

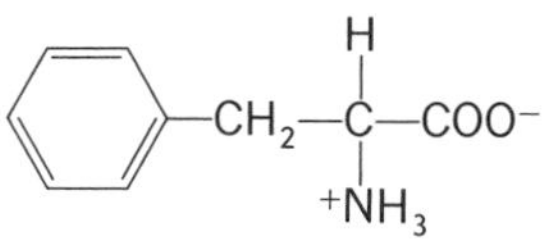

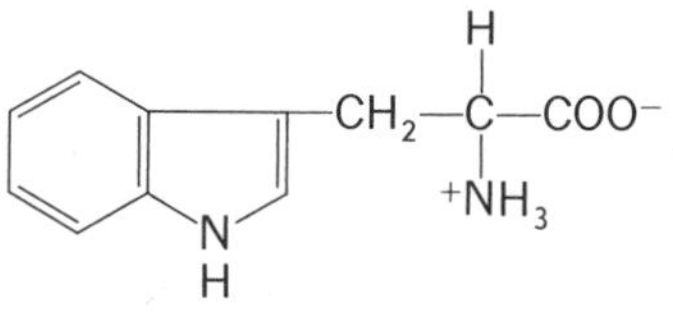

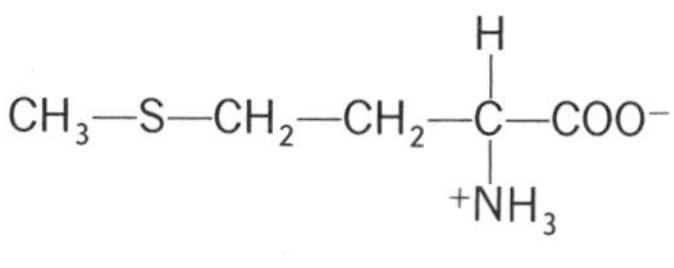

2. *Amino acids with polar, but uncharged R groups.* Most of these amino acids contain polar R residues that can participate in hydrogen bond formation. A hydrogen bond is a highly ionic bond that results when two heteroatoms, usually oxygen or nitrogen, but occasionally sulfur, share one proton (see Sections 1.4 and 3.9). Several of the amino acids in this group possess a hydroxyl group (serine, threonine, and tyrosine) or sulfhydryl group (cysteine), while two (asparagine and glutamine) have amide groups. Glycine, which lacks an R group, is included in this grouping because of its definite polar nature, a property it possesses because its carbonyl and amino nitrogen groups constitute such a large proportion of the mass of the glycine molecule. Both aliphatic and aromatic (tyrosine) compounds are included in this group.

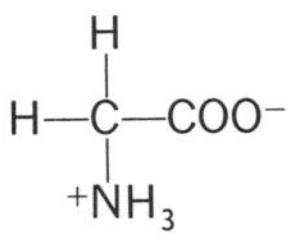

Glycine 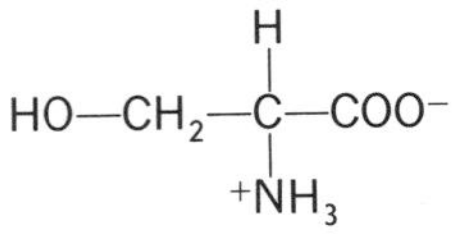   Serine 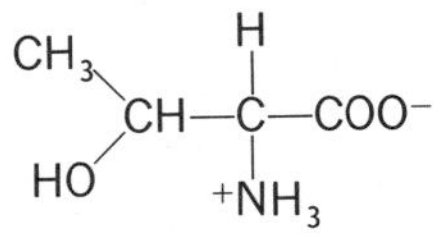 L_s-Threonine

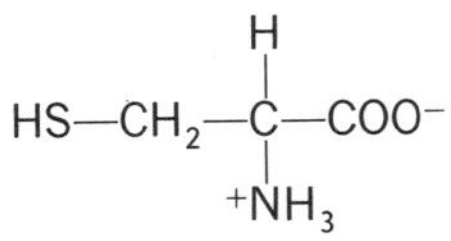

L_s-Cysteine 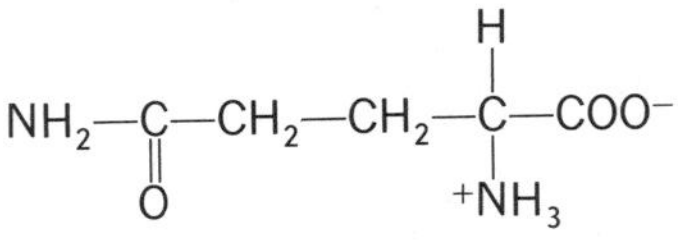 L_s-Glutamine

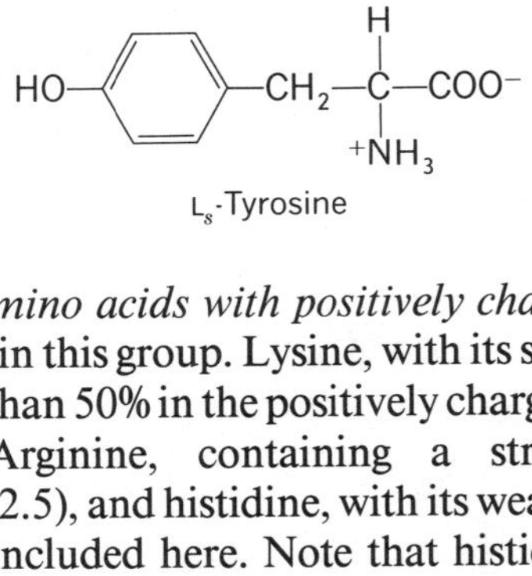

L_s-Tyrosine 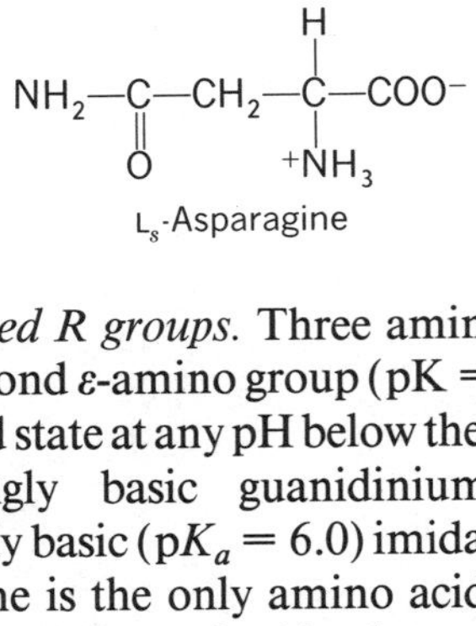 L_s-Asparagine

3. *Amino acids with positively charged R groups.* Three amino acids are included in this group. Lysine, with its second ε-amino group (pK = 10.5), will be more than 50% in the positively charged state at any pH below the pK_a of that group. Arginine, containing a strongly basic guanidinium function (pK_a = 12.5), and histidine, with its weakly basic (pK_a = 6.0) imidazole group, are also included here. Note that histidine is the only amino acid that has a proton that dissociates in the neutral pH range. It is this characteristic that allows certain histidine residues to play an important role in the catalytic activities of some enzymes.

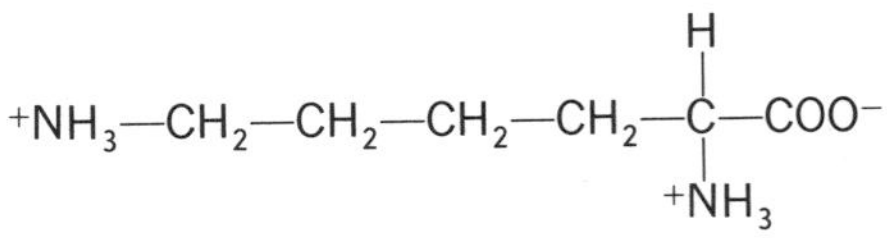

L_s-Lysine

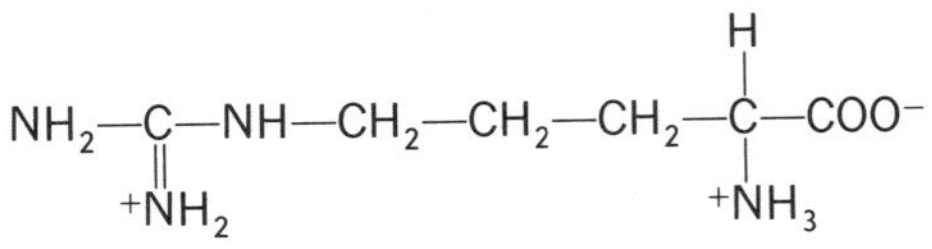

L_s-Arginine 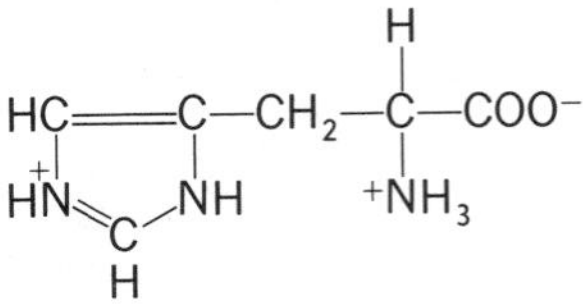 L_s-Histidine

4. *Amino acids with negatively charged R groups.* This group includes the two dicarboxylic amino acids aspartic acid and glutamic acid. At neutral pH, their second carboxyl groups with pK_{a2}'s of 3.9 and 4.3, respectively, dissociate, giving a net charge of -1 to these compounds.

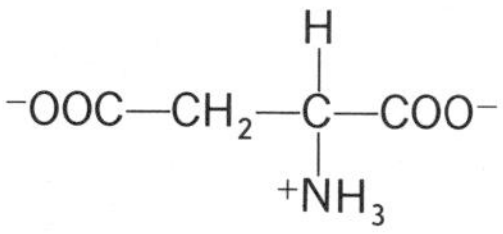

$$^-OOC-CH_2-\overset{\overset{\displaystyle H}{|}}{\underset{\underset{\displaystyle {}^+NH_3}{|}}{C}}-COO^-$$

L_s-Aspartic Acid

$$^-OOC-CH_2-CH_2-\overset{\overset{\displaystyle H}{|}}{\underset{\underset{\displaystyle {}^+NH_3}{|}}{C}}-COO^-$$

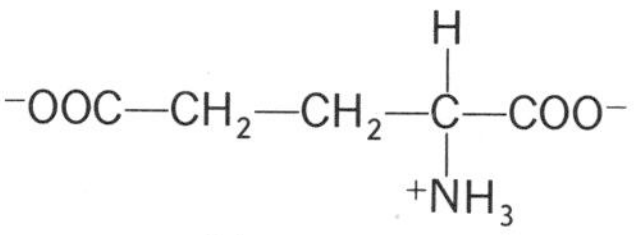

L_s-Glutamic Acid

These 20 amino acids constitute the bulk of the amino acids in the proteins of bacteria, plants, and animals, illustrating the unity of living systems on our planet. The 20 side chains apparently provide a sufficient diversity of chemical reactivities and conformations of the polypeptide chain so that only these amino acids are required in most proteins. The 20 common protein amino acids have a standard set of three-letter and one-letter abbreviations for use in writing amino acid sequences in polypeptide chains (Table 3.1).

TABLE 3.1 Notations for 20 Standard Protein Amino Acids

AMINO ACID	THREE-LETTER SYMBOL	ONE-LETTER SYMBOL
Alanine	Ala	A
Valine	Val	V
Leucine	Leu	L
Isoleucine	Ile	I
Proline	Pro	P
Phenylala-nine	Phe	F
Tryptophan	Trp	W
Methionine	Met	M
Glycine	Gly	G
Serine	Ser	S
Threonine	Thr	T
Cysteine	Cys	C
Glutamine	Gln	Q
Asparagine	Asn	N
Tyrosine	Tyr	Y
Lysine	Lys	K
Arginine	Arg	R
Histidine	His	H
Aspartate	Asp	D
Glutamate	Glu	E

3.4   NATURALLY OCCURRING MODIFICATIONS OF AMINO ACIDS IN PROTEINS

Less widely distributed but nevertheless important are a number of other amino acids that occur in only a few proteins, but occasionally as a large proportion of the residues in those few proteins. These additional amino acids are formed by modifications of the 20 common protein amino acids, usually after the polypeptide chain has been synthesized (see Part III).

The sulfhydryl group of cysteine undergoes reactions typical of the $-SH$ group both in the free amino acid and proteins. The most common of these is the reversible oxidative reaction with another molecule of cysteine to form the disulfide derivative **cystine.** Cystine readily can be formed by exposing a solution of cysteine to oxygen. Cysteine is regenerated by supplying an appropriate reducing agent to the cystine solution. Disulfide linkages between two cysteine residues in a polypeptide chain are a frequent occurrence in protein structures, especially in proteins that are secreted into the extracellular environment where requirements of stability may be particularly stringent. The ribonucleic acid degrading enzyme bovine pancreatic ribonuclease, for example, contains four disulfide bonds between four pairs of cysteine residues. The structure to follow shows a disulfide bond connecting two portions of a polypeptide chain (or even two polypeptide chains):

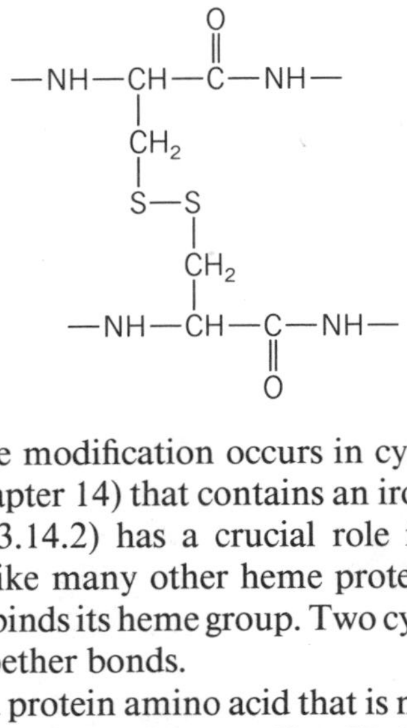

A second type of cysteine modification occurs in cytochrome c, an electron-transporting protein (Chapter 14) that contains an iron chelating heme group. Cytochrome c (Section 3.14.2) has a crucial role in biological reduction–oxidation reactions. Unlike many other heme proteins such as hemoglobin, cytochrome c covalently binds its heme group. Two cysteine residues are linked to the heme through thioether bonds.

A second example of a protein amino acid that is modified in some proteins is proline, which is converted to **hydroxyproline.** Hydroxyproline has a limited distribution in nature, but constitutes more than 12% of the structure of collagen, an important structural protein of animals. Hydroxyproline also is an

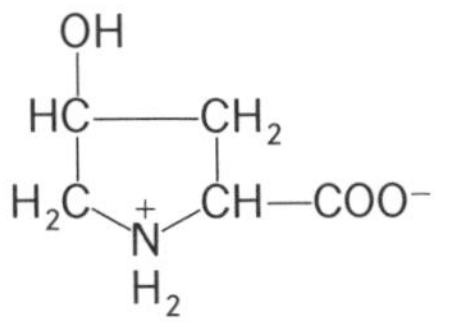

L$_s$-Hydroxyproline
(*erythro*-4-Hydroxy-L$_s$-proline)

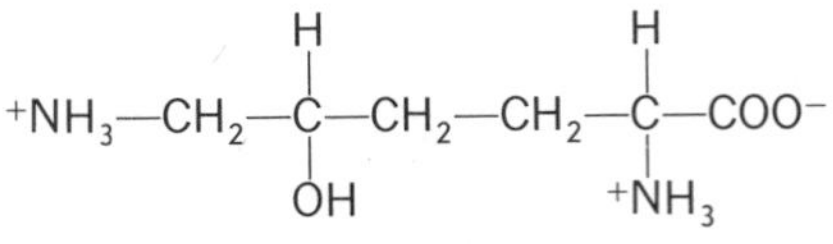

L$_s$-Hydroxylysine
(*erythro*-5-Hydroxy-L$_s$-lysine)

important constituent of the plant cell wall protein extensin and certain other plant proteins that appear to be involved in plant responses to pathogenic agents. The second amino acid shown here, **hydroxylysine,** also is a component of collagen.

Specific serine, threonine or tyrosine residues, and even arginine or histidine residues, of certain proteins can be **phosphorylated** in enzyme catalyzed reactions. The examples of a phosphoserine residue and a phosphohistidine residue, as part of polypeptide chains, are shown in the following:

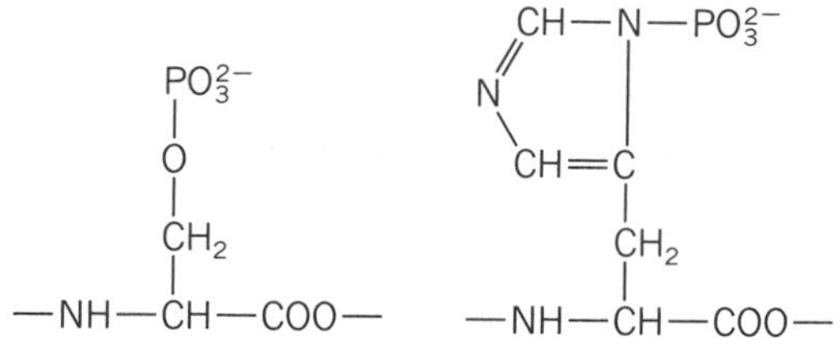

$$
\begin{array}{cc}
\text{PO}_3^{2-} & \text{CH} - \text{N} - \text{PO}_3^{2-} \\
| & \diagup \quad\quad\quad | \\
\text{O} & \text{N} \\
| & \diagdown \quad\quad | \\
\text{CH}_2 & \text{CH} = \text{C} \\
| & | \\
-\text{NH} - \text{CH} - \text{COO}- & \text{CH}_2 \\
& | \\
& -\text{NH} - \text{CH} - \text{COO}-
\end{array}
$$

P—N bonds generally are unstable, and a phophohistidine residue is not an exception; it is an unstable intermediate in certain enzymically catalyzed phosphorylation reactions. P—O bonds generally are much more stable. When a specific serine residue of the glycogen degrading enzyme glycogen phosphorylase is phosphorylated, by introduction of the ionized phosphoryl group, $-\text{OPO}_3^{2-}$, the enzyme is converted to a form that is much more active under physiological conditions. Many other examples are known of the regulation of the enzymic or other activity of proteins by phosphorylation, including proteins that bind to deoxyribonucleic acid and the glycolytic enzyme phosphoglucomutase. The milk protein casein contains many phosophorylated serine residues.

Acetylation is another chemical modification that may result in the alteration of protein function. An acetylated lysine residue is presented here.

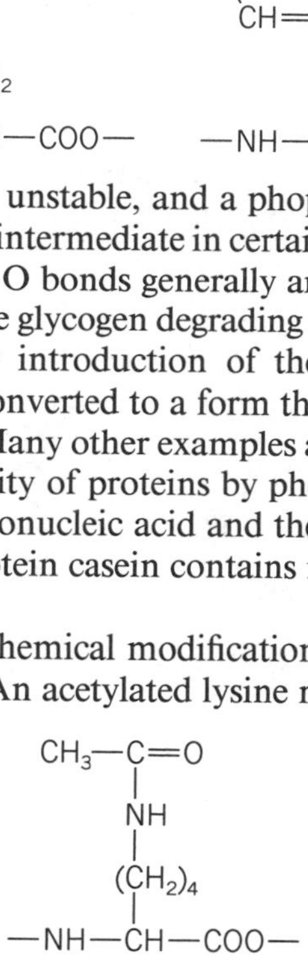

$$
\begin{array}{c}
\text{CH}_3 - \text{C} = \text{O} \\
| \\
\text{NH} \\
| \\
(\text{CH}_2)_4 \\
| \\
-\text{NH} - \text{CH} - \text{COO}-
\end{array}
$$

Some proteins are acetylated at the amino terminus.

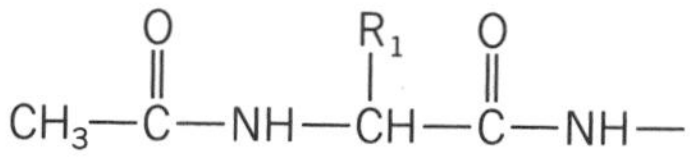

$$
\begin{array}{ccc}
\text{O} & \text{R}_1 & \text{O} \\
\| & | & \| \\
\text{CH}_3 - \text{C} - \text{NH} - \text{CH} - \text{C} - \text{NH} -
\end{array}
$$

Similarly, the polypeptide chain of some bacterial proteins begins with a formyl-methionine residue, which is a modification that occurs at the time of polypeptide synthesis rather than after (Chapter 20). Methylation of amino groups also is an important modification of certain proteins.

One of the most widely occurring and important of protein amino acid modifications is of a significantly more complex character. It is **glycosylation,** the covalent attachment of monosaccharides and oligosaccharides (Sections 2.5 and 2.6) to proteins. The resulting conjugate, in which the protein : oligosaccharide mass ratio generally is greater than one, is referred to as a **glycoprotein.** The carbohydrate portion of the glycoprotein can impart to the protein special properties, such as the localized polar regions of cell surface proteins. The carbohydrate moieties of the antifreeze proteins of certain Antarctic fish are essential to their function. Glycoproteins serve to impart viscosity and lubricating properties to body fluids and joints. Arabinose, fucose, galactose, glucose, mannose, and xylose; the acetylated amino sugars *N*-acetylgalactoseamine, *N*-acetylglucose-amine, and *N*-acetylneuraminic acid; and several uronic acids have been demonstrated in glycoproteins. Because of their complexity, determining the exact structures of glycoproteins is among the most difficult technical problems in modern biochemistry.

We do not go into the details of glycoprotein structure in this text, but give only some examples of the known types of linkages between protein and carbohydrate. The examples in Figure 3.2 are of (*a*) human red cell ABO blood group substances, (*b*) immunoglobulins (Section 3.14.4), and (*c*) the plant cell wall protein extensin. Note that in most instances, other monosaccharide residues will be bonded to the monosaccharide shown to create oligosaccharide side chains and that only certain serine, threonine, asparagine, and other such residues of a given glycoprotein actually will bear such an oligosaccharide side chain.

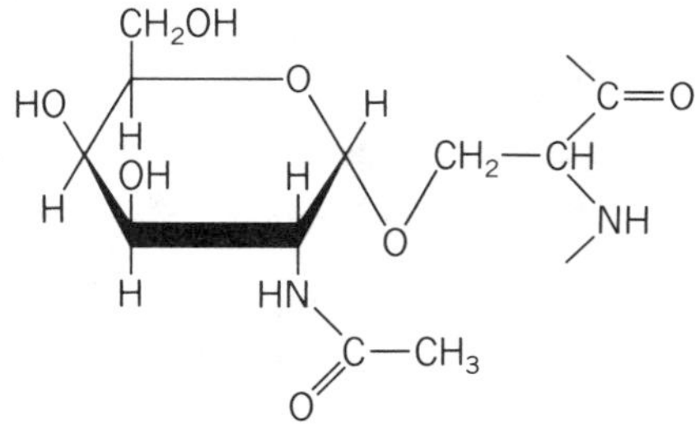

(*a*) α-D-*N*-Acetylgalactopyranosyl-seryl residue

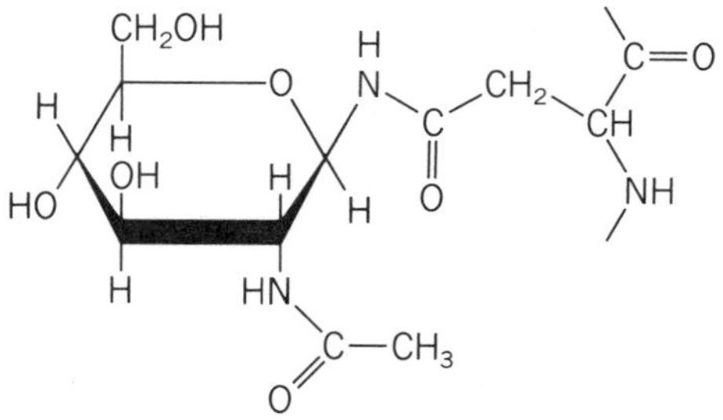

(*b*) β-D-*N*-Acetylglucopyranosyl-asparaginyl residue

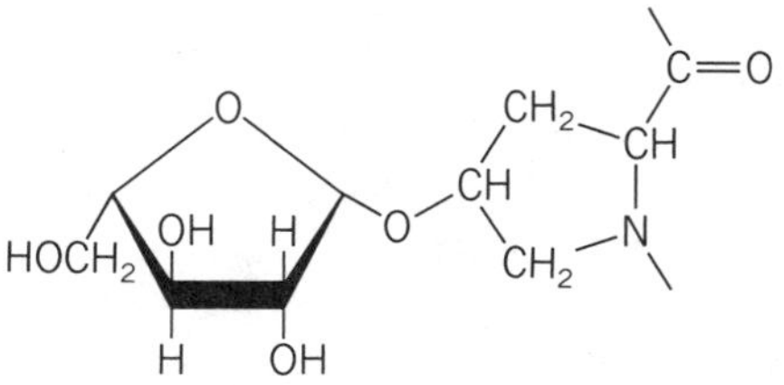

(*c*) β-L-Arabinofuranosyl-hydroxyprolyl residue

FIGURE 3.2 Examples of carbohydrate-amino acid covalent linkages that have been found in glycoproteins. (*a*) An O-glycosidic bond between *N*-acetylgalactosamine and a serine residue, (*b*) an N-glycosidic bond between *N*-acetylglucosamine and an asparagine residue, (*c*) an O-glycosidic bond between arabinose and a hydroxyproline residue.

3.5 NONPROTEIN AMINO ACIDS

Amino acids having the D configuration also exist in peptide linkage in nature, but not as components of large protein molecules. Their occurrence appears limited to smaller, cyclic peptides or as components of **peptidoglycans** (covalent complexes of polypeptides and polysaccharides in which the latter form the bulk of the conjugate) of bacterial cell walls. The antibiotic gramicidin-S (Structure 3.3) is an example of a peptide that contains two residues of a D-amino acid, D-phenylalanine (also see Section 20.4).

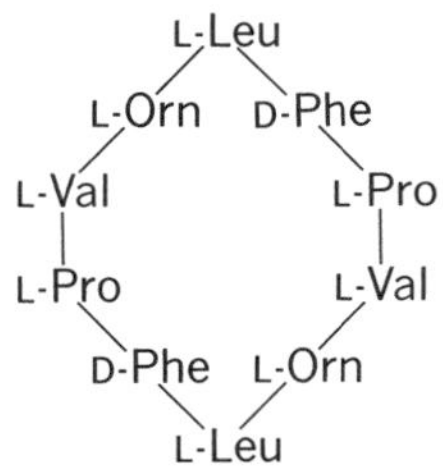

L-Leu
L-Orn   D-Phe
L-Val   L-Pro
L-Pro   L-Val
D-Phe   L-Orn
L-Leu

Gramicidin-S

STRUCTURE 3.3

Gramicidin-S also contains the nonprotein amino acid L-ornithine, which is an important metabolic intermediate in the synthesis of amino acids and other compounds.

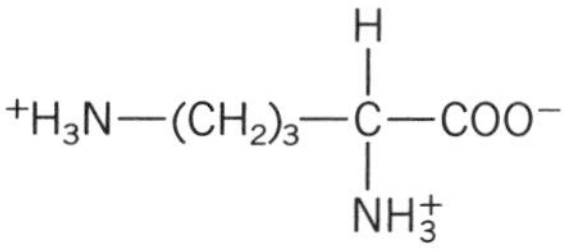

$$^+H_3N—(CH_2)_3—\overset{\overset{\displaystyle H}{|}}{\underset{\underset{\displaystyle NH_3^+}{|}}{C}}—COO^-$$

L-Ornithine

D-Valine occurs in actinomycin-D, a potent inhibitor of RNA synthesis, and D-alanine and D-glutamic acid are found in the peptidoglycan of the cell wall of gram-positive bacteria (Section 2.8).

An isomer of alanine, β-alanine, occurs free in nature and as a component of the vitamin pantothenic acid, coenzyme A, and acyl carrier protein (Chapter 5). The quaternary amine creatine, a derivative of glycine, plays a fundamental role in the energy storage process in vertebrates, where it is phosphorylated and converted to creatine phosphate (Chapter 9).

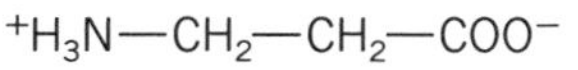

$$^+H_3N—CH_2—CH_2—COO^-$$
β-Alanine

$$H_2N—C{=}NH_2^+$$
$$CH_3—\overset{|}{N}—CH_2—COO^-$$
Creatine

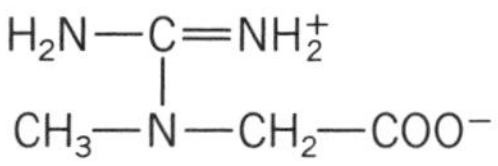

In addition to these nonprotein amino acids, for which metabolic roles have been described, several hundred other nonprotein amino acids have been detected as natural products. Higher plants are a particularly rich source of these amino acids. In contrast to the amino acids previously described, however, these compounds do not occur widely, but may be limited to a single species or only a few species within a genus. These nonprotein amino acids are usually related to the protein amino acids as homologs or substituted derivatives. Thus,

L-azetidine-2-carboxylic acid, a homolog of proline, may account for 50% of the nitrogen present in the rhizome of Solomon's seal (*Polygonatum multiflorum*). Orcylalanine (2,4-dihydroxy-6-methyl phenyl-L-alanine), found in the seed of the corncockle *Agrostemma githago,* may be considered as a substituted phenylalanine or tyrosine.

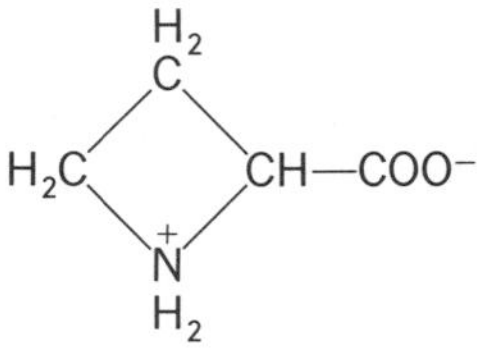

Azetidine-2-carboxylic
acid

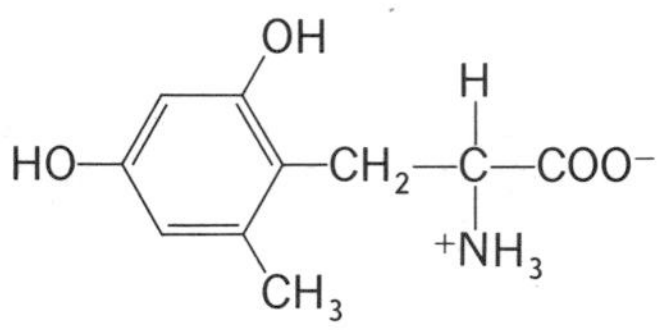

Orcyl-L$_s$-alanine

A particularly interesting group of nonprotein amino acids is found among the opines, which are formed in the tumors induced on plants by the crown gall bacterium, *Agrobacterium tumefaciens* (see Box 22.A). Some opines are conjugates of protein amino acids and α-keto acids. Octopine, for example, is a conjugate of pyruvic acid and arginine. It has two chiral centers: one derived from arginine in the L-configuration and the other from pyruvic acid in the D-configuration.

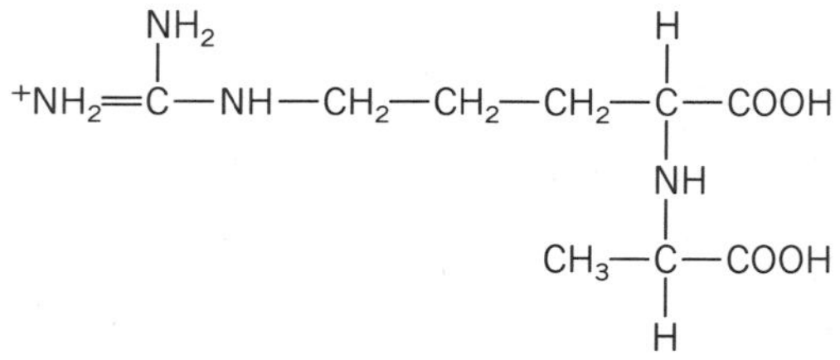

Octopine

These and the many other nonprotein amino acids that occur in nature are presently being studied in order to learn more about the conditions under which they arise and their function in the organism in which they occur. Other aspects of nonprotein amino acids are discussed in Chapter 17.

3.6   CHEMISTRY OF AMINO ACIDS AND THE POLYPEPTIDE CHAIN

In the formation of a polypeptide chain, each peptide bond can be considered to consume one α-amino group and one α-carboxyl group. Since a polypeptide chain that has no terminal blocking group has one fewer peptide bond than it has amino acid residues, one α-amino group and one α-carboxyl group remain at the chain termini. These define the **amino terminus** or the **amino end** and the **carboxyl terminus** or **carboxyl end** of the polypeptide chain, respectively. By convention, the polypeptide chain is written with the amino terminus on the left and the carboxyl terminus on the right. In order to simplify the representation, three-letter abbreviations of the amino acid residues are used. For example,

$$H_2N—Gly—Asp—Tyr—Ser—COOH$$

or simply

$$Gly—Asp—Tyr—Ser$$

for which the corresponding full structure is

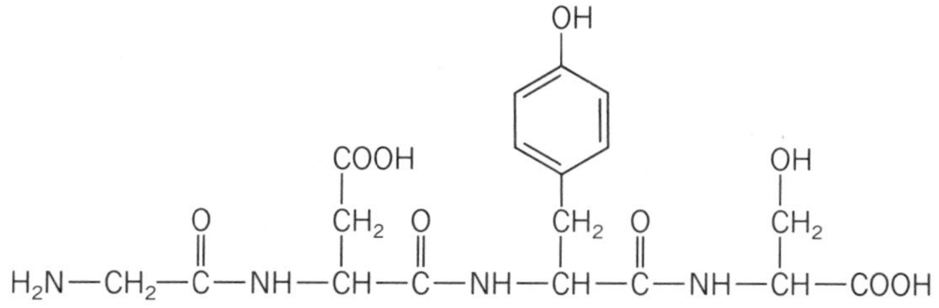

$$H_2N-CH_2-\overset{\overset{\displaystyle O}{\|}}{C}-NH-\underset{\underset{\displaystyle CH_2}{|}}{\overset{\overset{\displaystyle COOH}{|}}{CH}}-\overset{\overset{\displaystyle O}{\|}}{C}-NH-\underset{\underset{\displaystyle CH_2}{|}}{CH}-\overset{\overset{\displaystyle O}{\|}}{C}-NH-\underset{\underset{\displaystyle CH_2}{|}}{CH}-COOH$$

The general term for a short polypeptide chain of the type indicated above is an **oligopeptide.** We consider some biologically active oligopeptides in Section 3.7, and the cyclic oligopeptide gramicidin S has already been introduced in Section 3.5. Such a cyclic oligopeptide has no carboxyl or amino terminus, but the amino-to-carboxyl polarity nevertheless is inherent in the structure of gramicidin S. It corresponds to the clockwise direction in Structure 3.3. The student may find it instructive to draw out a more complete structural formula for gramicidin S.

3.6.1   IONIZATION OF AMINO ACID SIDE CHAINS

Box 3.A describes the proton dissociation and association reactions of the free α-carboxyl and α-amino groups of amino acids. The α-carboxyl and α-amino groups of free amino acids generally have pK_a values in the range of 1.7 to 2.4 and of 9 to 11, respectively. When these groups participate in the formation of a peptide bond, they, of course, no longer are ionized in the neutral pH range, and only the terminal amino and carboxyl residues of the polypeptide chain will be available for titration in the pH 1 to pH 12 range. However, some amino acids (Section 3.3) have additional ionizable groups.

Those amino acids that have more than one carboxyl or amino group will have corresponding pK_a values, and the side chain ionizable residues may contribute to the ionic charge of a peptide or polypeptide that contains that residue. Thus, the pK_a for the α-carboxyl group of aspartic acid is 2.1, while the pK_a for the β-carboxyl is 3.9. The pK_a for the amino group is 9.8. The titration curve for aspartic acid is shown in Figure 3.3. Depending on the pH, different aspartic acid species can predominate. Three examples are shown below. The left most species will predominate at pH 1, while the middle species will predominate at pH 3. However, the situation is slightly more complicated than it, at first, may appear. At pH 3, the right-hand species will be a minor but perceptible proportion of the aspartic acid.

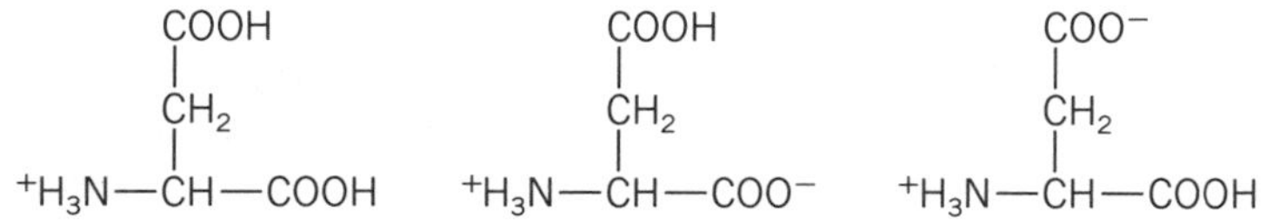

$$\underset{^+H_3N-CH-COOH}{\underset{\underset{\displaystyle CH_2}{|}}{\overset{\overset{\displaystyle COOH}{|}}{}}}\qquad\qquad\underset{^+H_3N-CH-COO^-}{\underset{\underset{\displaystyle CH_2}{|}}{\overset{\overset{\displaystyle COOH}{|}}{}}}\qquad\qquad\underset{^+H_3N-CH-COOH}{\underset{\underset{\displaystyle CH_2}{|}}{\overset{\overset{\displaystyle COO^-}{|}}{}}}$$

The student is urged to write structures for the aspartic acid species expected to predominate at pH 7 and at pH 11. Glutamic acid behaves similarly, with a pK_a value of approximately 4.3 for the γ-carboxyl group. Both aspartic acid and glutamic acid thus have an ionizable carboxyl group that is available when these amino acids are in a polypeptide chain.

The ε-amino group of lysine has a pK_a value of approximately 10.8. The

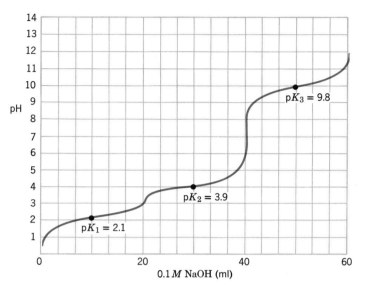

FIGURE 3.3   Titration curve obtained when 20 ml of 0.1M aspartic acid hydrochloride are titrated with 0.1M NaOH.

nitrogen atoms of the heterocyclic ring of histidine behave as a Brönsted base. The pK_a for the histidine ring protonation in free histidine is 6.0. An organic guanidino group, such as that of arginine, is very basic and therefore is completely protonated under physiological conditions. The arginine guanidinium pK_a is 12.5.

$$
\begin{array}{ccc}
H_2N & & H_2N \\
| & & | \\
C=NH_2^+ & \longleftrightarrow & C=NH + H^+ \\
| & & | \\
HN & & HN \\
| & & | \\
R & & R
\end{array}
$$

Other ionizable groups of amino acid side chains are the phenolic hydroxyl (pK_a = 10.4) of tyrosine and the sulfhydryl group of cysteine (pK_a of 8.2).

$$
\begin{array}{ccc}
COO^- & & COO^- \\
| & & | \\
H_3\overset{+}{N}-C-H & \rightleftharpoons & H_3\overset{+}{N}-C-H + H^+ \\
| & & | \\
CH_2 & & CH_2 \\
| & & | \\
SH & & S^-
\end{array}
$$

The pK_a values quoted here are for the ionizable groups of the amino acid side chains of the *free* amino acids. These, of course, are altered when the amino acid is incorporated into a polypeptide chain, as is discussed in Section 3.8.

3.6.2   OTHER REACTIONS OF AMINO ACID SIDE CHAINS

Amino acids act not only as electrolytes, but also in reactions that are characteristic of the organic functional groups of the amino acid side chains and the α-amino and α-carboxyl groups. Biochemists often take advantage of the reactivity of protein organic functional groups in order to learn about protein structure and function. A historically very important protein-derivatizing reagent is 1-fluoro-2,4-dinitrobenzene. The reagent is abbreviated FDNB. It often is referred to as Sanger's reagent, after F. Sanger who used it in his determination of the amino acid sequence insulin, the first protein to be so analyzed. The reaction of FDNB with a simple, free amino acid is

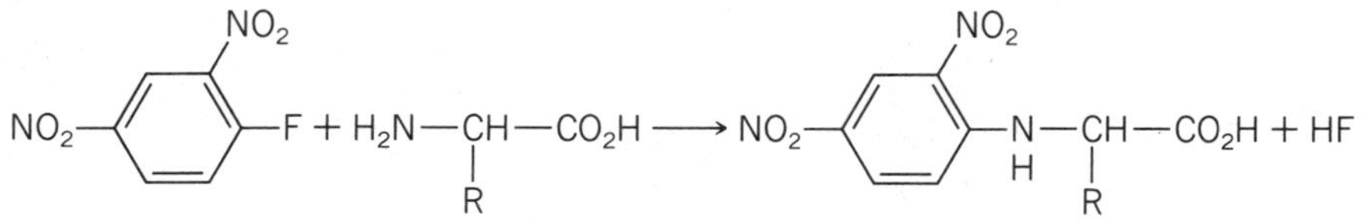

In this reaction, the colored dinitrobenzene nucleus is attached to the nitrogen atom of the amino acid to yield a yellow derivative, the 2,4-dinitrophenyl derivative or DNP-amino acid. The compound FDNB will react with the free amino group on the amino end of a polypeptide as well as the amino groups of free amino acids. The C—N bond that is formed generally is far more stable than a peptide bond. Thus, by reacting a native protein or intact polypeptide with FDNB, hydrolyzing the protein in acid, and isolating the colored DNP-amino acids, one can identify the amino terminal amino acid in a polypeptide chain. The ε-amino group of lysine and certain other side chain functional groups also will react with FDNB. After hydrolysis, however, only the derivative of the original amino terminal amino acid will have its α-amino group blocked, and such α-DNP amino acids can be separated from the other DNP derivatives by simple extraction procedures. Any of several chromatographic methods will identify the α-DNP amino acids.

The amino groups of both free amino acids and peptide chains will react with another reagent for identifying the amino terminal amino acid: **dansyl chloride** (5-dimethylamino-naphthalene-1-sulfonyl chloride). The derivative, a **dansyl amino acid,** is highly fluorescent and therefore detectable in minute amounts. The great sensitivity of detection makes dansyl chloride preferable to FDNB when only very small amounts of protein or peptide are available for analysis.

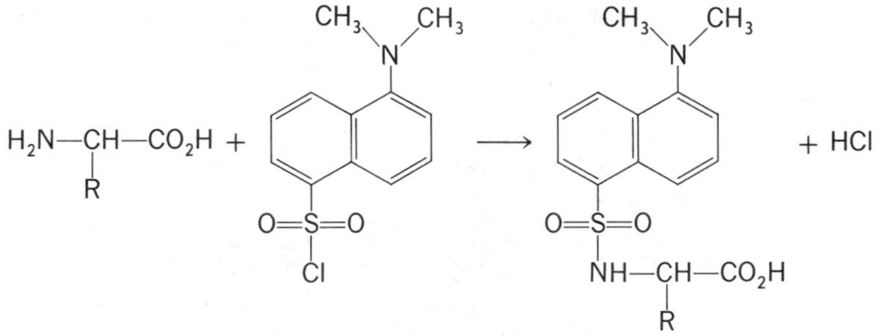

Dansyl chloride   Dansyl amino acid derivative

Both FDNB and dansyl chloride allow only the amino terminal amino acid to be determined. The well-known reaction of isothiocyanates with amines was ingeniously modified by Per Edman to degrade a polypeptide chain in cycles, releasing a derivative of the next amino terminal amino acid residue at each cycle without degrading the remaining polypeptide chain. Box 3.C describes an automated device that carries out these cycles under machine control with such efficiency and precision that the sequence of more than 60 amino acid residues from the amino terminus have been determined in a single analysis. Most proteins, of course, have more than 60 amino acid residues, and many such automated analyses determine less than 60 residues. To obtain the entire sequence of amino acids in a polypeptide chain, it is necessary to cut the chain at specific residues to produce specific fragments that are suitable for automated sequencing. Cutting at different sites with different reagents creates overlapping sets of fragments. From the amino acid sequences of two such overlapping sets of fragments, the investigator can deduce the entire amino acid sequence of the polypeptide chain. We describe here two of several procedures that have been developed for creating sets of large peptides from a polypeptide.

BOX 3.C   AUTOMATED DETERMINATION OF AMINO ACID SEQUENCES IN PROTEINS AND PEPTIDES

Edman derivatized the amino groups of proteins with phenylisothicyanate, known as the Edman reagent. Both α-amino groups and the ε-amino groups (of lysine residues) react with this reagent to form the phenylthiocarbamoyl derivatives. However, only the α-amino derivatives are capable of undergoing a cyclization reaction that allows repetitive removal and identification of amino acid residues from the amino terminus of a protein or peptide. The series of reactions is presented on the next page.

The first step is the coupling reaction. A nucleophilic attack of protein amino groups on phenylisothiocyanate generates the phenylthiocarbamoyl derivative. In anhydrous acid, the α-amino derivative participates in a cyclization reaction that cleaves the peptide bond and exposes the second amino acid of the polypeptide chain as the new amino terminal residue. The cyclic derivative is separated from the polypeptide and converted to the phenylthiohydantoin derivative of the amino acid by incubation in aqueous acid. The identity of this "PTH-amino acid" is determined by chromatography. The polypeptide, now one amino acid residue shorter, is subjected again to the coupling reaction, and the analysis is continued.

Critical to the cyclization step is the use of anhydrous acid. If water were present, acid catalyzed hydrolysis of peptide bonds would occur. The acid must be kept so dry that even asparagine and glutamine carboxamido groups remain. When properly carried out, the Edman degradation releases PTH-asparagine and PTH-glutamine residues at the expected cycles.

When it first was developed, the Edman procedure was done manually as an ordinary benchtop experiment. About 20 years ago, apparatuses and procedures for the automated determination of amino acid sequences were introduced. Many alternative approaches and refinements have appeared since that time, with development of methodologies for smaller scale, increasing cycle efficiencies ("repetitive yield", the fraction of molecules from which the amino terminal residue is released as the PTH derivative at each cycle) and the ability to analyze very small or very hydrophobic proteins that resisted analysis by the early automated procedures. An automated machine for determining the sequence of amino acids in proteins and peptides generally is referred to as a "sequenator" or "protein sequenator."

A sequenator has advantages beyond those of the labor saved and around-the-clock operation. Edman procedures have many steps, which makes them prone to errors. A sequenator may be programmed to carry out more than 50 operations per cycle, all controlled by reliable electronic circuitry. At several steps, oxygen and/or water must be rigorously excluded. This is accomplished in the sequenators by the use of a sealed reaction chamber that can be flushed with inert gas. Temperature control readily is accomplished in the automated apparatus.

In a recently developed "gas phase" sequenator, the sample is immobilized on a film of cationic polymer that has been deposited in the matrix of a glass fiber filter approximately 12 mm in diameter. The glass fiber disk is mounted in a chamber of 0.05 ml volume with connections that allow liquids and gases to be forced through the disk. Phenylisothiocyanate is introduced in heptane solution, in which the protein is insoluble. After evaporation of the heptane, the filter is exposed not to liquid base but to a vapor of trimethylamine and water. Thus, the coupling reaction occurs without danger of protein loss due to it dissolving in basic aqueous solution. Then, excess phenylisothiocyanate washes away in a stream of heptane. The cyclization reaction occurs as the disk is exposed to anhydrous trifluoroacetic acid vapors. The amino acid derivative flows into a 0.3 ml collection tube in a stream of organic solvent. The disk then is dried and is ready for the second cycle of coupling. These and other features allow, under optimum conditions, 25 cycles of analysis to be carried out on 0.1 nmole of protein with a repetitive yield more than 95%. Thus, amino acid sequence data can be obtained from only 5 μg of a protein of molecular weight 50,000.

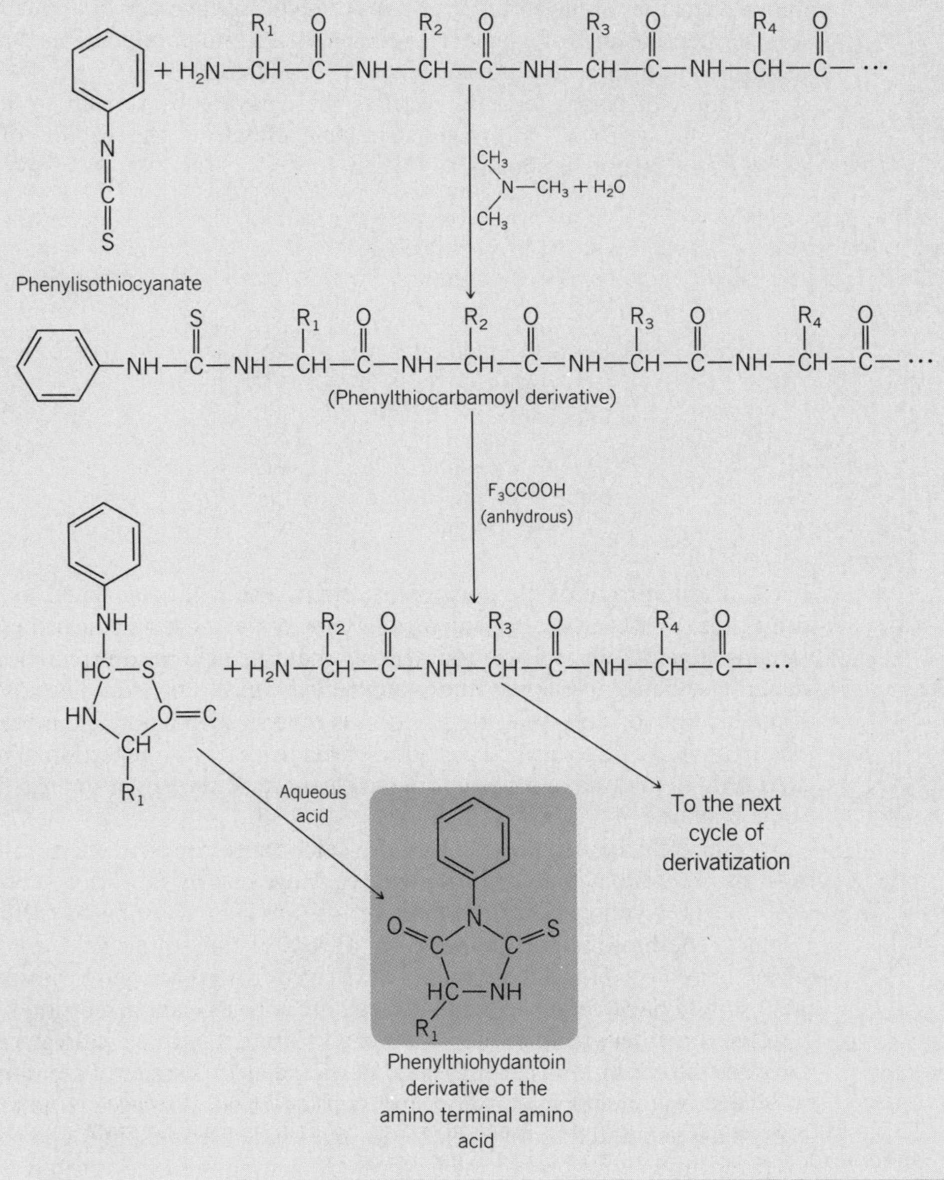

Phenylisothiocyanate

(Phenylthiocarbamoyl derivative)

F_3CCOOH
(anhydrous)

Aqueous acid

To the next cycle of derivatization

Phenylthiohydantoin derivative of the amino terminal amino acid

3.6.3 SPECIFIC CLEAVAGE OF POLYPEPTIDE CHAINS

The general name for an enzyme that catalyzes the hydrolysis of proteins is **proteinase.** The proteinase bovine pancreatic trypsin prefers unfolded (i.e., "denatured") polypeptide chains as substrates and will cleave such chains on the carboxyl side of lysine and arginine residues. A complete digestion of the substrate thus will give a set of peptides, "tryptic peptides," that have lysine or arginine carboxyl terminal residues, except for the peptide that is derived from the carboxyl end of the original polypeptide. Free lysine and arginine also may be released if lysine and/or arginine residues are adjacent at some position(s) in the sequence. The abundance of lysine and arginine residues in proteins is such that tryptic peptides will have, on the average, only about eight residues. This is too few to take advantage of the powers of automated sequencing. Cleavage at arginine residues alone would give a more favorable average of about 20 residues per peptide and, of course, also reduces the number of peptides that must be purified and submitted for automated sequencing.

Acylation of lysine residues restricts the cleavage by trypsin to arginine residues. One useful acylating agent is maleic anhydride, the reaction of which with a lysine residue is shown.

The α-amino group of the polypeptide chain also will be acylated by maleic anhydride. However, maleic anhydride is one of the most specific protein-derivatizing reagents, and its reaction with other amino acid residues can be held to virtually undetectable levels under controlled conditions. Another advantage of maleic anhydride is that the reaction is reversible in acid. The investigator may treat with acid a purified peptide derived from trypsin digestion of malelylated protein and make it susceptible to cleavage by trypsin at the regenerated lysine residues.

Specific cleavage of a polypeptide also can be accomplished chemically. The most effective and widely used chemical cleavage reagent is cyanogen bromide, $Br—C \equiv N$. Cyanogen bromide cleaves on the carboxyl side of methionine residues. Methionine accounts for 4% or less of the amino acid residues in proteins on the average. Thus, peptides derived from cyanogen bromide treatment should have, on the average, about 25 residues. Cyanogen bromide reacts with the thioether group of methionine, alkylating it with a cyano group. The reactive sulfonium ion that is formed participates in a series of reactions that involve simultaneous cyclization and peptide bond cleavage. A new amino terminus is generated that is suitable for automated sequencing. The new carboxyl terminus is a homoserine lactone. (A lactone is a cyclic carboxylic acid ester.) The overall reaction is as follows:

$$
\begin{array}{c}
CH_3 \\
| \\
S \\
| \\
CH_2 \\
| \\
CH_2 \quad O \qquad\quad R \quad\ O \\
| \quad\ \ \| \qquad\quad | \quad\ \| \\
-NH-CH-C-NH-CH-C- \ + \ BrC\equiv N \longrightarrow
\end{array}
$$

$$
\begin{array}{c}
CH_2 \\
CH_2 \quad O \qquad\qquad\quad R \quad\ O \\
| \qquad\quad \qquad\quad\ | \quad\ \| \\
-NH-CH-C \ \ \ + \ \overset{+}{H_3}N-CH-C- \ + \ H_3C-S-C\equiv N + HBr \\
\quad\quad\quad\ \ \backslash \\
\quad\quad\quad\quad O
\end{array}
$$

3.7 BIOLOGICALLY ACTIVE OLIGOPEPTIDES

An oligopeptide, already defined in Section 3.6, is simply a short polypeptide. It is composed of a small number of amino acids, sometimes arbitrarily stated as ten or fewer residues, joined by peptide bonds. Two amino acids joined by a peptide bond form a dipeptide; a peptide containing three amino acids is a tripeptide, and so on. Sometimes the term **peptide** is substituted for "oligopeptide." We discuss a chemical property of peptides and then the structures of some biologically active peptides.

Like an α-amino acid, a peptide has an α-amino and an α-carboxyl group. It is of interest to note the changes in the pK_a values for these groups that occur when amino acid residues are incorporated into an oligopeptide. Thus, in small peptides, the α-carboxyl group generally has a pK_a value in the range 3 to 4, indicating that it is less acidic than the α-carboxyl group of a free amino acid (pK_a range of 1.7–2.4). In a peptide, the α-carboxyl group no longer has a protonated α-amino group in its immediate vicinity, making it more difficult to remove a proton from the carboxyl group. Thus, the α-carboxyl group becomes a weaker acid. Similarly, the α-amino group of a peptide becomes a weaker base, with a pK_a value in the range 7.5 to 8.5. This is outside the range of 9 to 11 observed for the α-amino groups of free amino acids.

Peptides have roles as hormones, antibiotics, toxins, and metabolic intermediates. An example of a metabolically important peptide is glutathione. Glutathione is a tripeptide that is ubiquitous in nature and in mammals is required to prevent oxidative damage to red blood cells. It has a structure that deviates from that of a standard peptide and that we use to illustrate the procedure for naming a simple peptide (Structure 3.4). The chemical name for glutathione is γ-glutamylcysteinyl glycine. The suffix -yl signifies the amino acid residue whose carboxyl group is linked in peptide linkage to the amino group of the next amino acid in the peptide. In the case of peptides containing glutamic (or aspartic) acid, the carboxyl group involved in the peptide linkage must be identified. In glutathione, it is the γ-carboxyl that is bound in the peptide linkage. This obviously is an unusual situation, for when glutamic acid (or aspartic acid) occurs in proteins, it is the α-carboxyl that is bound in peptide linkage. When no prefix is given, the α linkage is understood.

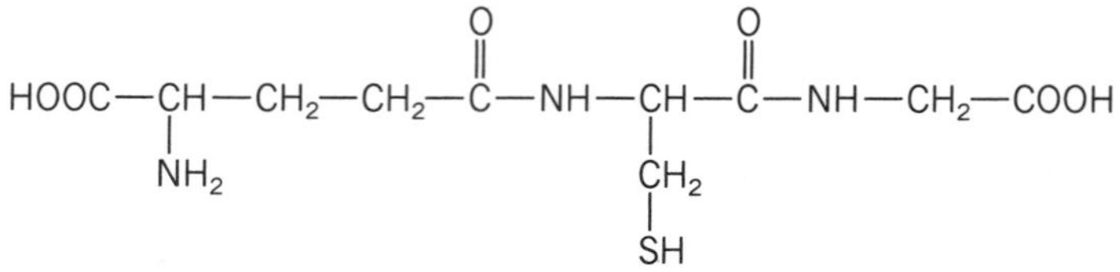

γ-Glutaminylcysteinylglycine
(Glutathione)

STRUCTURE 3.4

A very important group of naturally occurring peptides is that of the peptide hormones, which are responsible for intercellular and interorgan communication. Tens of groups of peptide hormones are recognized. Vasopressin and oxytocin are nonapeptides that are formed in the pituitary gland and can assume a cyclical structure by forming disulfide linkages between the amino-terminal cysteine and a cysteine in the interior of the peptide. Seven of the nine amino acids in the two peptides are identical. Nevertheless, the physiological effects are quite different. Oxytocin causes the contraction of smooth muscle; vasopressin causes a rise in blood pressure by constricting the peripheral blood vessels. The structures of these peptide hormones follow. The student may find it instructive to identify the amino terminus and trace the course of the peptide chains of these hormones. Note that the carboxyl terminal residue of these peptides is the amide of glycine. Carboxyl terminal amidation is almost unknown among proteins, but occurs frequently in naturally occurring peptides. Protection against proteinases of the carboxypeptidase type is a possible function. Insulin, produced by the pancreas, is a hormone consisting of two polypeptide chains containing a total of 51 amino acid residues. The synthesis of the peptide hormones discussed in this paragraph proceeds by proteolysis of polypeptide precursors, as is discussed in Sections 3.15 and 20.12.

Several antibiotics are peptides of comparatively simple structure and are synthesized by mechanisms that are distinct from those of protein synthesis. Gramicidin (Structure 3.3) and tyrocidin are examples of such compounds. The biosynthesis of gramicidin is discussed in Chapter 20. Penicillin, another antibiotic, contains the valine and cysteine residues, but these are not linked by peptide bonds. Rather, a strained four-membered ring and a sulfur-containing ring are found.

Synthetic peptides also may have biological activity. An especially simple example is the artificial sweetener Aspartame. Aspartame is L-aspartyl-L-phenylalanyl-methyl ester. It is approximately 200 times as sweet as sucrose and is considered to lack some of the unpleasant aftertaste associated with other synthetic sweeteners.

Peptides now can be synthesized for research purposes (Box 3.D). These find uses in studies on hormone activity, as substrates for enzymes and as "immun-

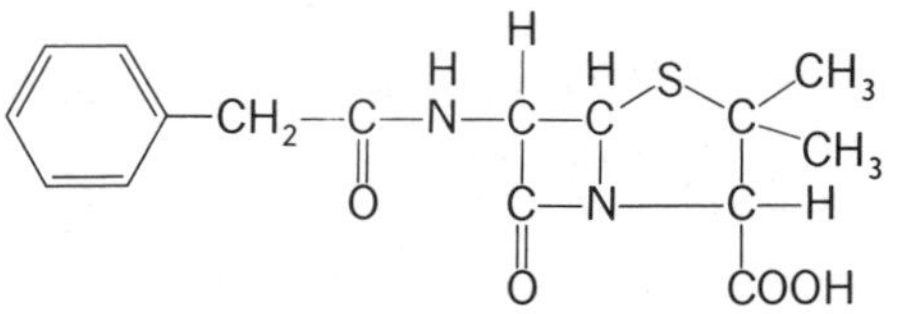

Penicillin G
(Benzyl penicillin)

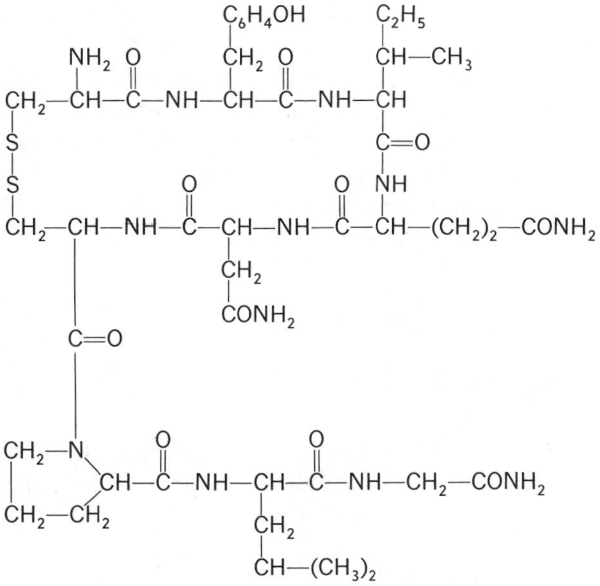

Oxytocin

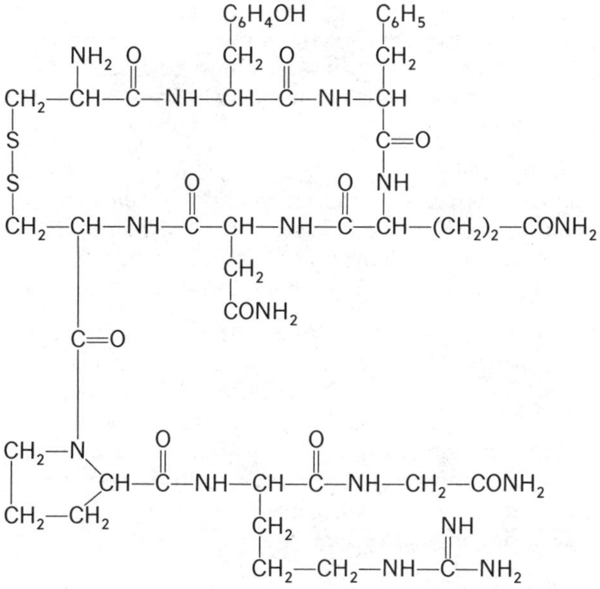

Vasopressin

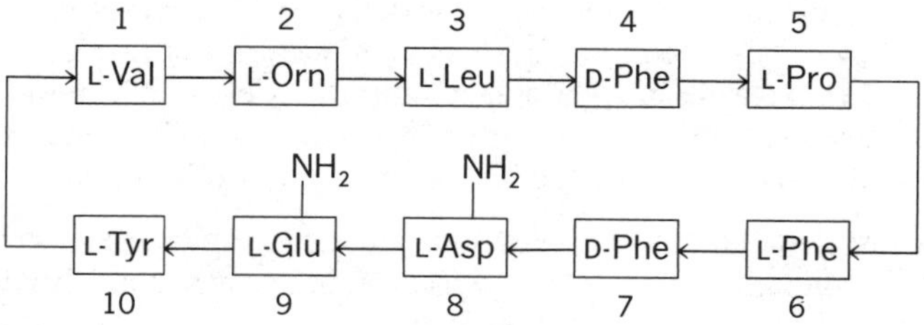

Tyrocidin

ogens," substances capable of eliciting antibodies when injected into animals (see Section 3.11 and Section 22.1). Sometimes, portions of the structure of a rare protein have been determined from analysis of its gene or microsequence analysis (Box 3.C), but available quantities of the protein do not allow antibodies to be induced. A synthetic peptide that corresponds to the carboxyl terminal region of the protein or hydrophilic stretches of the polypeptide chain often will induce the desired antibodies. Usually, the synthetic peptide is conjugated to a protein such as serum albumin, to increase the efficiency of antibody production.

3.8　THE FORMS THAT PROTEINS TAKE

Up to this point, we have discussed proteins primarily in terms of their covalent structure, the polypeptide chain. The essence of a protein's function, and the diversity of functions that proteins have, resides in their ability to bring amino acid residues that are distantly arrayed along the peptide chain into contact in

---

BOX 3.D　SYNTHESIS OF A POLYPEPTIDE ON A SOLID SUPPORT

The chemical synthesis of a polypeptide is a study in the use of protective groups. Protective groups must be used not only to block the reactive amino acid side chains, such as those of lysine, cysteine, and tyrosine, but also to direct the orderly addition of amino acid residues. Protective groups dictate that several chemical steps and purifications must be employed for each new amino acid residue introduced. In such a situation, great benefit derives from the use of a solid, insoluble support for the growing polypeptide chain.

A scheme for the chemical synthesis of a polypeptide chain appears at the end of this box. The strategy is to add a carboxyl-activated, amino-blocked amino acid residue to the amino group of a peptide or amino acid residue that is attached to a polymer by its carboxyl end. This core reaction, shown as the second reaction, in the central portion of the scheme, causes the polypeptide chain to grow, one residue per cycle, from its carboxyl terminal end to its amino terminal end.

The amino acid that will be the carboxyl terminal residue of the final polypeptide enters the reaction attached to the polymer by an ester bond. The next amino acid enters the reaction as its "t-BOC" amide derivative. The t-BOC group, abbreviating the chemical name "tertiary-butyl-oxycarbonyl-," is $(CH_3)_3C—O—(CO)—$. The t-BOC-amino acid reacts with dicyclohexylcarbodiimide (DCC) in a solution of dimethylformamide to give an active ester. The active ester reacts with the amino group of the polymer-bound amino acid or peptide, also in dimethylformamide, to give the very stable dicyclohexyl urea and the peptide with a new peptide bond. Activation and coupling occur as a single experimental step, simplifying the synthesis.

Incubation of the resin with a solution of HCl in anhydrous acetic acid removes the t-BOC group without hydrolyzing the ester bond to the polymer support. Subsequent purification of the intermediate peptide is accomplished simply by washing the polymer beads with the appropriate solvents. The polymer then is ready for incubation with the t-BOC derivative of the next amino acid that is to be added, in a formamide solution of DCC. Obviously, the protecting groups that are used for the R groups of the amino acids must be able to resist many cycles of exposure to the conditions used for coupling and subsequent removal of the t-BOC group. Another set of conditions is used to remove the peptide from the resin and all of the R group protecting groups.

The steps just described have been automated using valves and electronic circuitry to introduce solution after solution into a reaction vessel containing the polymer beads. Currently, most chemical peptide synthesis is by automated procedures.

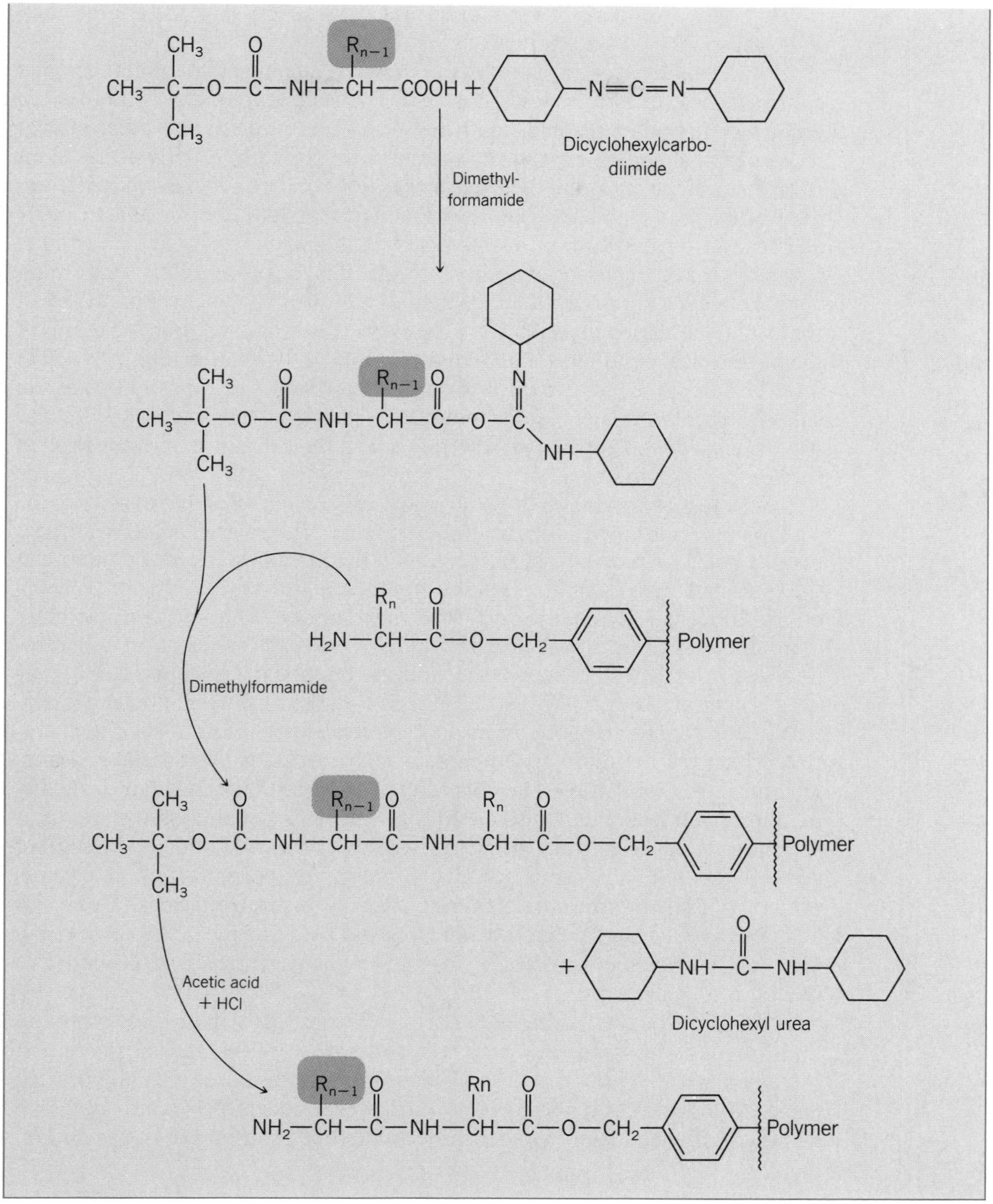

three-dimensional space. Proteins have not only a definite covalent structure, defined by the order of amino acids along the polypeptide chain, but also a definite three-dimensional structure. As we point out in Section 3.15, the three-dimensional structure also is controlled primarily by the order of amino acid residues. The stability of these three-dimensional structures is the result of noncovalent bonding (Sections 3.11 and 3.12). The positioning of reactive functional groups relative to one another in an environment of controlled polarity allows proteins to bind other molecules specifically, to become catalysts and to take on other functions.

The general form of most proteins is globular, meaning that they are roughly spherical. The technique known as x-ray crystallography gives information about the three-dimensional structures of globular proteins and other macromolecules that can be crystallized or otherwise placed in orderly arrays. This complex and laborious method has revealed the three-dimensional structures of several tens of globular proteins with sufficient resolution to be able to locate all of the nonproton atoms. From x-ray crystallography results, and other data, it appears that the interior of a globular protein is packed as tightly with amino acid residues as amino acids are packed in amino acid crystals. About 75% of the available interior space is within the surfaces at which nonbonded atoms of the amino acid residues are expected to contact. The remaining 25% of the interior volume is subdivided into spaces that, with few exceptions, are too small to accommodate a water molecule. The solvent in which the globular protein is dissolved appears to penetrate only a short distance from the protein surface in most instances.

Since most of the side chains of amino acids are nonpolar, the interior of the protein molecule is predominantly apolar. In fact, the polarity of an amino acid residue is a good predictor of its location in the three-dimensional structure of the molecule, with polar amino acids confined mainly to the exterior. When a polar amino acid side chain is located in the interior of a protein, the possibility of functional significance should be kept in mind. A surface location for hydrophobic amino acid side chains may indicate binding of lipids or other hydrophobic substances, possibly in a membrane or at the surface of another protein.

Not all proteins are globular. A major group of **fibrous proteins** is the keratins, which are the main constituents of hair, beaks, nails and claws, scales, horns, hooves, and wool. The collagens of skin, cartilage, and bone, already mentioned in Section 3.4, also are fibrous and the most abundant proteins in the mammalian body, amounting to about 6% of the mass and up to a third of the total protein. Collagens are glycoproteins composed largely of glycine, proline, hydroxyproline, and alanine. Collagens form a triple helix (Figure 3.4 and Section 3.10) and are cross-linked to give a rigid and inextensible material. Silk and the contractile proteins tropomyosin and paramyosin are other examples of fibrous proteins.

Some proteins are neither globular nor fibrous. Myosin, a muscle protein that functions in some noncontractile systems as well, is one of the largest proteins known, with more than 1700 amino acid residues in its polypeptide chain. Myosin is not a fibrous protein. However, it is far less compact than a globular protein and appears to assume an extended conformation in solution.

FIGURE 3.4   The collagen triple helix. Each chain forms an extended left-hand helix. The three chains are wound in a right-hand helix. The course of one of these is indicated by the dashed line.

The terminal amino and carboxyl groups of the polypeptide chain and the ionizable amino acid side chains all have the potential to contribute to the total charge of a protein in solution and to have a role in the functions of the protein. One of the consequences of the specific folding of a polypeptide chain is that the pK_a values for these functional groups in proteins frequently deviate significantly from the values observed for the free amino acids. The pK_a differences that occur, when the α-amino and/or α-carboxyl groups are those of peptides rather than free amino acids, was discussed in the previous section. Of even greater importance in determining the pK_a values of terminal and side chain groups, especially with regard to protein function, is the new environment that an amino acid residue finds itself in when it is incorporated into a folded protein molecule. In a hydrophobic environment in the interior of a protein, and even close to its surface, ionization occurs less readily, and the pK_a values are shifted from their free solution values. Also, several closely spaced ionizable groups will influence each other's pK_a values both in nonpolar and polar environments. The pK_a values of carboxyl groups (α, β, and γ) of proteins usually fall in the range 2 to 5.5. The ionizations of tyrosine and histidine residues in proteins have been subject to special scrutiny because they can be detected by spectroscopic techniques. Tyrosine hydroxyl groups have been observed to have pK_a values from 9 to 12 and histine imidazole groups values from 5 to 8.

The shift of pK_a values away from the values found for the free amino acids indicates that the environment in the interior of a protein can have a powerful effect on the reactivity of side chain functional groups. Specific amino acid residues in protein molecules participate in chemical reactions that free amino acids exhibit to only a limited or undetectable extent. For example, single serine and cysteine residues of certain proteinases form esters and thioesters, respectively, with the carboxyl ends of peptides during proteinase action, as indicated in Section 4.5. For free amino acids in solution, the equilibrium for such an esterification reaction lies far to the side of hydrolysis: free alcohol or sulfhydral and free carboxylic acid, indicating that the conditions within the protein molecule must be very different from the conditions in solution.

Histidine, with its lone electron pair in the ring nitrogen, often acts as an efficient acid–base catalyst, a property it shows only weakly as the free amino acid. Histidine also serves as a metal ligand in the iron-containing proteins hemoglobin and cytochrome c. As we described in Section 3.4, the N—1 nitrogen of the imidazole ring of histidine can be phosphorylated to form a high-energy N—P bond, as occurs in some transport proteins (see Section 8.12). Lysine is intimately involved in binding pyridoxal phosphate, lipoic acid, and biotin (see Chapter 5) and, like serine and histidine, is the component of the active site of certain enzymes, such as muscle aldolase. Other amino acids that are known to show unusual reactivities when they have a role in protein function are aspartic and glutamic acids, arginine, tyrosine and methionine, and the terminal amino and carboxyl residues.

3.9 THE IONIC CHARGE OF PROTEIN MOLECULES

Since a protein may have both positively and negatively charged, ionized amino acid residues, the net charge on a protein may be negative or positive. Or the protein may have no net charge, depending on the pK_a values of the groups involved and the pH of the solution. The pH at which the protein has no net electric charge is termed the **isoelectric point,** or pI. In solutions with pH values above the isoelectric point, the protein will have a net negative charge; at lesser pH values, it will be positively charged. More proteins have pI values below 7 than above 7, so that most proteins are negatively charged at neutral pH. A technique that takes advantage of the ionic charge of a protein molecule is **electrophoresis,** which is described in Box 3.E.

BOX 3.E   POLYACRYLAMIDE GEL ELECTROPHORESIS OF PROTEINS

Probably, biochemists analyze proteins more often by electrophoresis through a polyacrylamide gel matrix than any other single technique. The procedure is so common that the acronym PAGE, for "polyacrylamide gel electrophoresis" sometimes is used without prior definition (a practice we do not recommend!). PAGE requires simple equipment and relatively inexpensive supplies, but delivers remarkable resolution of complex mixtures. Primarily, it is an analytical method, but preparative PAGE is applied for purification of small samples, for example, for amino acid analysis (Box 3.B) and partial amino acid sequence determination (Box 3.C). PAGE is so widely applied and so often poorly understood that we take the space necessary to present some details of the method and its theory.

Electrophoresis is the migration of charged molecules in solution due to the influence of an externally applied electric field. In polyacrylamide gel electrophoresis, the molecules being analyzed migrate in an electric field that is imposed on the aqueous solution that is trapped in the matrix of the gel. Acrylamide and methylene-*bis*-acrylamide are the compounds from which a polyacrylamide gel is polymerized.

$$CH_2{=}CH{-}\underset{\substack{|\\NH_2}}{C}{=}O \qquad\qquad CH_2{=}CH{-}\underset{\substack{|\\NH{-}CH_2{-}NH}}{C}{=}O \quad O{=}\underset{\substack{|\\}}{C}{-}CH{=}CH_2$$

Acrylamide   Methylene-*bis*-acrylamide

Acrylamide and smaller concentrations of methylene-*bis*-acrylamide are dissolved in the buffered solution in which the electrophoresis is to occur. Compounds that decompose into free radicals initiate the polymerization reaction, which produces long polyacrylamide chains that are cross-linked by occasional introduction into the chain of methylene-*bis*-acrylamide residues in place of acrylamide residues.

Linear polyacrylamide chains, not cross-linked, will produce a viscous solution. Cross-linking by incorporation of methylene-*bis*-acrylamide residues causes all of the chains in the solution to become part of one or a very small number of molecules. These very large, branched molecules and the aqueous solution they trap become the clear, rubbery, polyacrylamide gel.

The gel in a gel electrophoresis experiment has two important functions: stabilizing the system against convective disturbances and providing pores, that is, passageways, through the gel. The flow of electric current during electrophoresis generates heat, which will be dissipated more rapidly at the boundaries than in the interior of a solution. The temperature gradients generated by uneven heat dissipation will cause convection of a solution that is not a gel. A gel virtually can eliminate convective mixing. In the second function of the gel, the pores retard the movement of large molecules more than they retard the movement of small molecules.

A number of gels are suitable for stabilized electrophoresis experiments, but only polyacrylamide and agarose (Box 6.C) currently find wide application. The pores in agarose gels generally are larger than those in polyacrylamide gels, and agarose gels are used for the electrophoresis of nucleic acids and the very largest proteins. Advantages of polyacrylamide gel include the hydrophilic matrix (due to the carboximido groups) and, for a given gel concentration, a distribution of pore sizes that appears to be uniform throughout the gel.

The arrangement of gel and electrode vessels for a typical gel electrophoresis experiment are shown in the diagram here. The gel is a fraction of a mm to a few mm thick and is polymerized in place between two glass plates. Thus, the gel is bounded on its front and back surfaces by glass plates, which are held a uniform distance apart by "spacers" that also define the edges of the gel. The glass plates and spacers are connected to the upper and lower reservoirs in such a way that the gel is in contact with the vessel buffers.

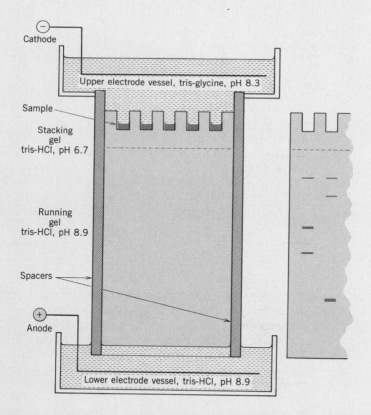

Each electrode, usually a platinum wire, facilitates an electrolysis reaction in the electrode vessel. Oxygen evolves at the anode (positive electrode), hydrogen at the cathode, when a direct current power supply is connected. The applied voltage usually is 50 to a few hundred V, depending on the gel and buffers used. The electrodes and electrode reactions serve to convert the flow of electrons from the power supply to a flow of ions in the vessels and gel. The heat generated in the gel by the current flow is rapidly radiated through the glass because of its large surface area and because the gel is thin.

The great resolving power of PAGE is due not only to the properties of the polyacrylamide gel, but also to a special arrangement of gels and buffers. As the figure indicates, two gels actually are used: a "running gel" under and in contact with a "stacking gel." The stacking gel has "wells," which are rectangular depressions a few mm wide cast in place by a form when the stacking gel is poured. The stacking gel is poured only shortly before electrophoresis to avoid excessive diffusion of buffers between the stacking and running gels.

The system shown is for anionic proteins with pI values below about 6. The three buffers of this system are the pH 8.9 running gel buffer, the pH 6.7 stacking gel buffer, and the upper electrode buffer. The samples that are to be analyzed are solutions of protein and sucrose or glycerol in stacking gel buffer. They are delivered with a fine pipet to the stacking gel wells. The sample solution displaces upper electrode buffer from the well and remains in the bottom of the well because its density is greater, due to the glycerol or sucrose.

After the samples are in the wells and the power supply is connected and voltage applied, the combination of two gels and three buffers causes a startling effect. As the anionic proteins migrate down into the stacking gel, they become *more* concentrated than they were in the sample solution. A sample protein applied as a 0.1 mg/ml solution several mm deep can be expected to concentrate to a few tens of mg/ml, which corresponds to a zone only about 0.02 mm thick in the vertical direction on the gel.

What causes the proteins to become more concentrated? The effect is attributed, in part, to the retardation of protein migration rate as protein molecules encounter the stacking gel. However, the more important effect is the result of the pH of the stacking gel and the compositions of the upper electrode buffer and the stacking gel buffer. We consider next the rationale for these buffers.

A system designed to resolve anionic proteins generally will have only one cation in the solutions, a cation that is a component of the buffering system. In our example, the cation is the protonated form of the alkyl amine "Tris." The name Tris is derived from "*tris*-hydroxymethylaminomethane," more systematically designated 2-amino-2-hydroxymethyl-propane-1,3-diol. The pK_a of Tris at room temperature is approximately 8.1, so it will effectively buffer the running gel. Buffering of the pH 6.7 stacking gel and sample solutions will be less effective (Section 1.14). The protonated form of Tris obviously will migrate upward toward the cathode in the system shown.

$$CH_2OH$$
$$|$$
$$HOCH_2—C—CH_2OH$$
$$|$$
$$NH_2$$

Tris

The system has two anions: chloride and glycinate (the carboxylate anion, unprotonated amine form of glycine, see Box 3.A). Chloride ions have a greater mobility (rate of migration toward the anode in a unit electrical field) than glycinate ions. The Tris-glycine upper electrode buffer (which, of course, contains protonated Tris cations and glycinate anions) serves as a generator of glycinate ions. The glycinate ions, in rapid equilibrium with zwitterionic glycine, migrate down into the stacking gel and eventually the running gel, following the chloride ions toward the anode. In the pH 6.7 stacking gel, most of the glycinate anions will become protonated to form glycine zwitterions. These have no net charge and thus will not migrate in the electrical field. The effective mobility of the glycine-glycinate mixture thus will be very small. In the stacking gel, the anionic proteins of the sample will have mobilities intermediate between those of chloride ions and the glycine-glycinate pair. Since the ionic current must be the same everywhere in the gel, the result is chloride ions followed by proteins followed by glycine-glycinate. The great difference in the mobilities of chloride and glycine-glycinate causes the proteins to be trapped in a very narrow zone.

A radical change occurs when the narrow protein zone and its trailing glycine-glycinate reach the pH 8.9 stacking gel, which for complex reasons actually increases to about pH 9.3. At this pH, a much greater proportion of glycinate anions is formed, as the student may verify using the Henderson–Hasselbalch equation (Section 1.11). The new glycine-glycinate pair has a mobility in excess of that of the anionic proteins, and the proteins, in fact, are being further retarded by the greater concentration of the polyacrylamide in running gel, compared to the lower concentration stacking gel. Thus, the proteins drop behind the interface between the chloride and glycine-glycinate pair and are released from the stacking condition. They then separate in the running gel, their mobilities determined by a combination of the negative charge on the protein and the protein dimensions. The larger the protein, of course, the greater the degree to which the polyacrylamide gel retards its mobility.

The experimenter obviously can alter the separations achieved in PAGE by adjustments of the polyacrylamide gel concentration and the running gel pH. When glycinate ions have reached the bottom of the gel, the power supply is disconnected and the gel is removed from the apparatus. Soaking the gel in a solution of protein-specific stain that does not bind to polyacrylamide reveals the location of protein zones. We represent a portion of the stained gel to the right of the gel apparatus in the figure. Locations of the zones are recorded by photography or measurements of distances from the interface between the stacking and running gels.

One of the characteristics that the investigator would like to determine for a newly discovered protein is its mass. When the entire amino acid sequence becomes known, the exact molecular weight of the protein can be determined by summing the weights of the amino acid residues. However, generally this information will be available only late in the characterization of the protein. It is surprising that the technique of polyacrylamide gel electrophoresis (Box 3.E), which depends on the charge of the protein molecule, should also provide information about the size of a protein molecule. In fact, because it is such an easy and inexpensive technique, gel electrophoresis commonly is used for estimating protein molecular weights, especially in the preliminary stages of the characterization of a protein. The methodology is described in Box 3.F. The reader should be aware, however, that the molecular weights estimated by gel electrophoresis are just that, estimates. They are based on the comparison of the mobilities of the proteins under study and other proteins of known molecular weight. Amino acid sequence determination and certain physical methods such as sedimentation and diffusion, which will not be discussed here, give more reliable measures of the molecular weight.

BOX 3.F   ESTIMATING PROTEIN MOLECULAR WEIGHTS BY POLYACRYLAMIDE GEL ELECTROPHORESIS

Box 3.E describes in outline how a polyacrylamide gel electrophoresis (PAGE) experiment is performed and how the *charge and size* of the protein molecule influence the distance that it migrates. Two modifications of the PAGE procedure allow the biochemist to extract information about the size of a protein molecule or polypeptide chain independently of its charge. Both procedures require reproducible measurements of the distances that a given zone of protein has traveled through the gel. For this reason, a "tracking dye" usually is added to the sample protein solution. The dye bromophenyl blue, for example, has a mobility that is less than that of chloride ions but greater than that of the glycine-glycinate pair at all pH values encountered in the gel described in Box 3.E. Hence, the dye will mark the boundary between chloride and glycine-glycinate. The relative mobility or R_m of any protein zone is defined as the ratio of two distances. Divide the distance that the protein zone has migrated from the interface between the stacking and running gels by the distance that the tracking dye has migrated from the same interface. R_m values vary from 0 to 1.

Graph A shows that the logarithm of the R_m for a given protein is linearly related to the gel concentration. The slope and intercept of the line are dependent on the size and charge of the protein and hence generally will be different for different proteins. In fact, for many proteins, the slope of such a line (Graph A) is proportional to the protein molecular weight (Graph B). This last result depends on the proteins having a similar conformation, since the gel concentration influences protein mobility according to protein dimensions rather than protein mass. Only if protein conformations are similar will the dimensions of the different protein molecules be similarly related to the protein molecular weights. Fortunately, this situation holds, at least to a first approximation, for most globular proteins (Section 3.8). By analyzing both properly selected "standard proteins" of known molecular weight and proteins of unknown molecular weights on the same set of gels, the investigator obtains data like that shown in Graphs A and B and can estimate the molecular weights of the unknowns.

From Graph B, we can conclude that the proteins 1 and 2 have a similar molecular weight. If the PAGE system is resolving anionic proteins, we conclude from Graph A that protein 1 is slightly more negatively charged than protein 2. This situation is characteristic of **isoproteins (isoenzymes** if the proteins are enzymes; see Section 4.12). Isoproteins are very similar to each other in size and structure but may differ in net charge. Often, this reflects different but closely related genes encoding the proteins, but differential modification of amino acid residues and certain other explanations also are possible. Note the relationship between proteins 3 and 4. Protein 4 behaves as if it is smaller but less highly negatively charged than protein 3. Analysis in an approximately 9% polyacrylamide gel might fail to resolve these two proteins, showing the importance of using several gel concentrations when assessing the purity of a protein preparation.

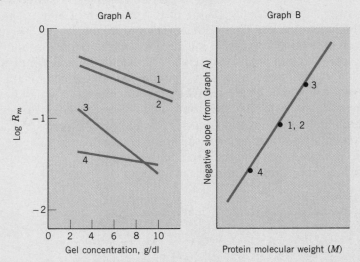

In the second modification of the PAGE procedure, often abbreviated "SDS-PAGE," sample proteins are heated in a solution of the detergent sodium dodecyl sulfate (SDS), and SDS is incorporated into the electrophoresis gel and electrode buffers. The SDS denatures and forms an elongated, noncovalently bonded complex with each protein molecule. The size of the complex apparently is controlled primarily by the length of the polypeptide chain that forms the core of the complex. The mass ratio of SDS to protein in these complexes is greater than one and usually varies little with the sequence of amino acids in the polypeptide chain. The result is that the negative charge of the complex is controlled primarily by the size of the polypeptide chain.

Because the charge contributions of the amino acid side chains usually are negligible in SDS-protein complexes, the charge to mass ratio of such complexes is almost independent of the size of the polypeptide chain. Free in solution, such complexes would have similar electrophoretic mobilities. In a gel, however, the large complexes are significantly more retarded in their mobilities than the small complexes, and the separations achieved in an SDS-PAGE experiment are strictly dependent on the size of the polypeptide chain for many proteins. Hence, a simple plot of R_m versus log M for a single gel concentration (Graph C) often is sufficient to give an estimate of the polypeptide chain molecular weight.

More polypeptide chain molecular weight have been estimated by SDS-PAGE than any other method. Although the results usually are accurate to a few thousands for "typical" soluble, globular proteins of molecular weight 15,000 or greater, small proteins, glycoproteins, and some other proteins migrate "abnormally," and the method, of course, gives erroneous estimates of molecular weight for these proteins. Also, SDS-PAGE cannot give an estimate for the molecular weight of a protein that has more than one polypeptide chain, unless the numbers of the chains in the protein are known.

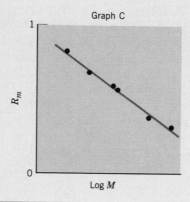

Graph C

3.10 THE FOLDING OF POLYPEPTIDE CHAINS

To help define a complicated macromolecule such as a protein in descriptive terms, four structural levels are considered.

1. *Primary structure.* The concept of the primary structure of a protein already was introduced in Section 3.3. It is defined as the linear sequence of amino acid residues in the polypeptide chain. Implied, of course, is the peptide linkage between each of the amino acids (Figure 3.1), but no other forces or bonds are indicated by the term "primary structure."

2. *Secondary structure.* This term refers to regular folding patterns of contiguous portions of the polypeptide chain. Examples include the α-helix and the β-pleated sheet, both of which are discussed below.

3. *Tertiary structure.* This term refers to the three-dimensional structure, especially the bonds between amino acid residues that are distant from each other in the polypeptide chain and the arrangements of secondary structure elements relative to one another. The term is applied to globular proteins, rarely to fibrous proteins.

The term **conformation** refers to the secondary and tertiary structure jointly, that is, the folding pattern of the polypeptide chain and all of contacts between amino acid side chains of that polypeptide chain.

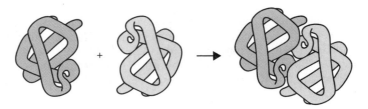

FIGURE 3.5   A protein dimer unit, illustrating the quaternary structure of a globular protein composed of two protomers.

4. *Quaternary structure.* This defines the structure resulting from interactions between separate polypeptide units of a protein containing more than one subunit. Thus, the enzyme phosphorylase *a* contains two identical subunits that alone are catalytically inactive, but when joined as a dimer, form the active enzyme as shown in Figure 3.5. This type of structure is called a **homogeneous quaternary structure;** if the subunits are dissimilar, a **heterogeneous quaternary structure** is obtained. Another term employed to describe the subunits of such a protein is **protomer,** and a protein made up of more than one protomer is an **oligomeric protein.** In specific terms, hemoglobin is an oligomeric protein having a heterogeneous quaternary structure, consisting of two identical α-chain protomers and two identical β-chain protomers (i.e., $\alpha_2\beta_2$). Quaternary structure is considered further in Section 3.14.

3.11   THE ELEMENTS OF SECONDARY STRUCTURE

The recognition of a particular secondary structure in a protein molecule generally is accomplished by consideration of the polypeptide backbone alone, without analysis of the amino acid side chains. This is possible because secondary structures are stabilized by hydrogen bonds (see Section 1.4) between peptide imide and carbonyl groups of the polypeptide backbone and, generally, not by bonds between side chains. Three classes of secondary structures generally are recognized: helices, sheets, and bends. These structures owe their existence not only to polypeptide chain hydrogen bonds, but also to steric limitations on the rotations of bonds in the polypeptide chain.

The peptide bond, which is an imide (substituted amide) bond, has a planar structure. The reason for this is that the electrons are delocalized in the amide linkage, giving the C—N bond considerable (estimated to be about 40%) double bond character, as shown by the resonance structures.

$$O=C-N \overset{H}{\underset{R}{\diagup}} \longleftrightarrow {}^{-}O \diagup C=N^{+} \overset{H}{\underset{R}{\diagup}}$$

Thus, the planar peptide bond can be represented as

The six atoms within the plane are related to each other by bond lengths and angles that vary little from amino acid residue to amino acid residue.

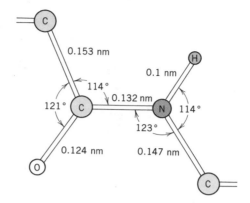

Only three of these bonds are part of the peptide chain per se: the α-carbon to carbonyl carbon bond, the C—N bond, and the imide nitrogen to α-carbon bond. Since the double bond character of the C—N bond limits rotation about it, only the first and the last allow rotation. The rotation angles ψ and ϕ, which establish the relative positions of any two successive amide planes along the polypeptide chain, are defined in Figure 3.6. The α-carbon atoms can be thought of as swivel centers for the adjacent amide planes.

There is a definite geometric consequence to fixing all of the ψ values of a chain to one value and all of the ϕ angles to another value: The chain will assume a helical configuration, or at the very least, one of two degenerate forms obtained by projecting a helix onto a plane: a circle or a zig-zag structure.

Helix   Circle   Zig-Zag

Thus, if steric hindrance to rotation should limit the values of ψ to a narrow range and ϕ to a narrow range, one of the above regular structures should be favored. Knowledge of the ϕ and ψ rotation values for a series of contiguous amino acid residues will completely define the secondary structure of the corresponding region of the polypeptide chain. Of the possible choices for a secondary structure, the circle is, of course, eliminated for polypeptide chains of indefinite length. The three most abundant regular secondary structures actually found in proteins are the α-helix, the parallel β-pleated sheet, and the antiparallel β-pleated sheet, the pleated sheets being constructed of polypeptide chains that are in a zig-zag conformation. The α-helix is represented in Figures 3.7 and 3.8, and the pleated sheet structures are depicted in 3.9 and 3.10.

3.11.1 THE α-HELIX

The protein α-helix is a right-handed helix, meaning that the chain rotates clockwise as one views down the helix axis at the polypeptide chain proceeding into the distance. The α-helix has 3.6 amino acid residues per turn and is stabilized by nearly straight hydrogen bonding between an imide group (—NH—) in the polypeptide chain and a carbonyl group (—CO—) at a posi-

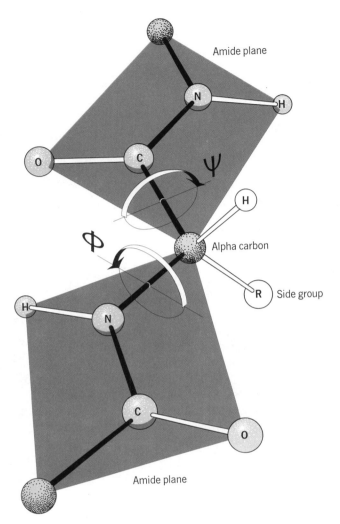

FIGURE 3.6   Allowed rotations about bonds in the polypeptide chain. Two rotational bond angles are defined: ψ for the α-carbon to a carbonyl carbon bond and ϕ for the imide nitrogen to α-carbon bond. (*Source:* Reprinted from R. E. Dickerson and I. Geis, *The Structure and Action of Proteins.* Menlo Park, Calif.: W. A. Benjamin, Inc., © 1969 by I. Geis.)

tion four residues away in the same chain (Figure 3.8). Because every —NH— and —CO— group can form a straight hydrogen bond in this manner, the α-helix is especially well stabilized. Under these conditions, the ϕ values range from 113 to 132° and the ψ values from 123 to 136°, ranges of angles that are particularly sterically favored according to both theoretical and experimental studies.

Although the α-helix is characterized by the locations of the atoms of the polypeptide chain, the amino acid side chains do influence the probability that a given sequence of amino acid residues will be found in an α-helix. Since the α-carbon atom is the swivel point for the chain, the R groups associated with the α-carbon atom sterically influence the adjacent amide planes. Glycine residues

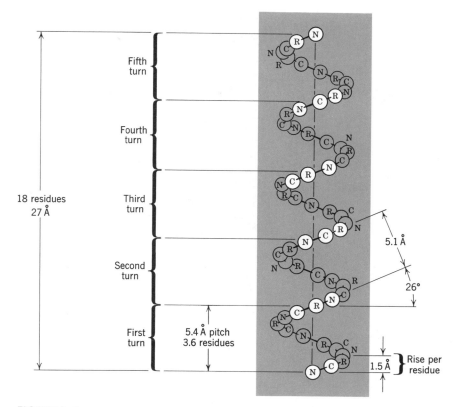

FIGURE 3.7 Representation of a polypeptide chain in an α-helical configuration. The shaded circles indicate atoms that are behind the plane of the page. The white circles represent atoms that are above the plane of the page. (*Source:* From L. Pauling and R. B. Corey, Proceeds of International Wool Textile Research Conference, B, 249, 1955, as redrawn in C. B. Anfinsen, *The Molecular Basis of Evolution*, New York: Wiley, 1959, p. 101.)

do *not* foster α-helix formation because the side chain, merely an H atom, is too small to constrain the ψ and φ angles to those appropriate for an α-helix. The polypeptide chain is too flexible at a glycine residue. In contrast, the ring structure of a proline residue so thoroughly constrains the values of ψ and φ that they cannot assume values necessary for the α-helix. Both glycine and proline residues are considered to be "helix breakers." Glutamic acid, leucine, methionine, phenylalanine, and certain other amino acid residues occur very frequently in α-helices and are considered to be α-helix-promoting residues. Other amino acid residues certainly influence α-helix formation, but the relationships apparently are more complex.

3.11.2 β-PLEATED SHEETS

Figure 3.9 shows all of the atoms of pleated sheets as if they were confined to a plane, in order to clearly represent the hydrogen bonding. In fact, the β-pleated sheet structures that have been observed in proteins and are predicted by theory do *not* have their atoms constrained exactly to the same plane. As is implied by the name, the sheets are pleated as shown in Figure 3.10. The atoms lie in a

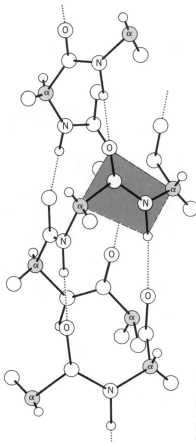

FIGURE 3.8 A right-handed α-helix. The hydrogen bonding between the backbone of each residue and the fourth residue further along the polypeptide chain is shown as parallel dashed lines. The α-carbons, peptide nitrogen atoms, and oxygen atoms are marked. One amide plane is shaded in.

folded plane and the side chains of succeeding residues along the polypeptide chain protrude alternately above and below the general plane of the structure. The β-pleated sheet structures, like the α-helix, allow the maximum amount of hydrogen bonding between polypeptide backbone imide and carbonyl groups, and the H bonds formed are straight enough to be stable. The β-pleated sheets seen in protein structures are not only pleated but also slightly twisted. This is predicted by a detailed consideration of the steric hinderances about bonds in the polypeptide chain.

3.11.3   OCCURRENCE OF SECONDARY STRUCTURES IN PROTEINS

The α-helix and β-pleated sheet structures are found in both globular and fibrous proteins. *The wide and frequent occurrence of the α-helix and β-pleated sheet structures appears to be due to the combination of sterically favored disposition of the polypeptide chain* and *favorable hydrogen bonding.* A helix that is less stable than an α-helix, known as the 3_{10}-helix, occurs in short segments in a few proteins. A few other types of helices are theoretically possible. Flat, untwisted β-pleated sheets have been observed but are rare, and no other type of regular sheet structure is known.

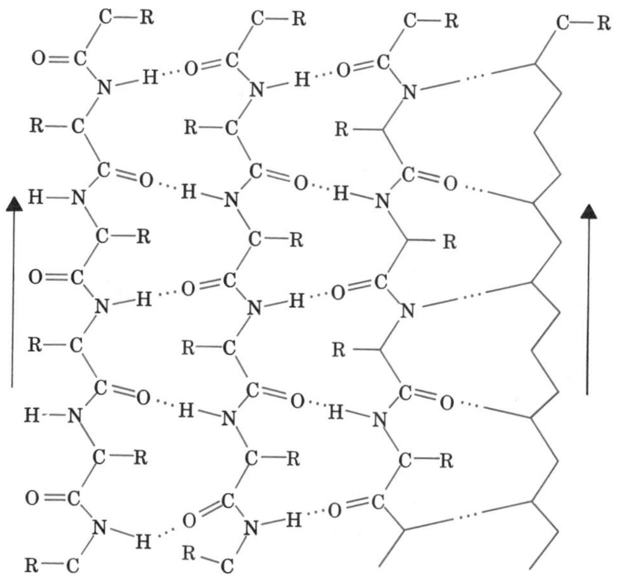

Parallel chain β-pleated sheet (stretched keratin)

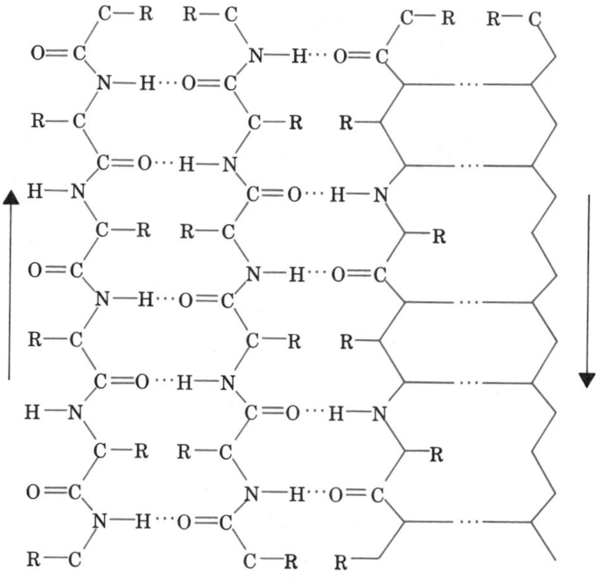

Antiparallel β-pleated sheet (silk)

FIGURE 3.9 The arrangement of polypeptide chains and hydrogen bonds in parallel chain and antiparallel chain β-pleated sheets.

α-Keratin and paramyosin are examples of fibrous proteins that have α-helix structures. Parallel β-pleated sheets and antiparallel β-pleated sheets form the structures of the fibrous proteins stretched β-keratin and silk, respectively. These fibrous proteins have very long stretches of polypeptide chain entirely folded into the named secondary structures. In globular proteins, the individual

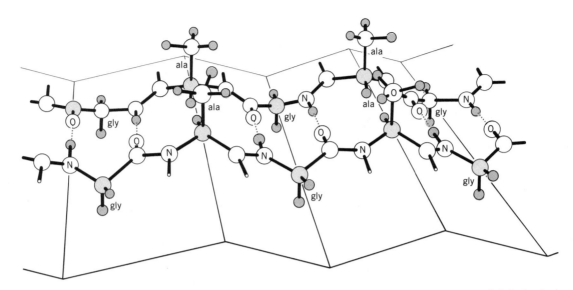

FIGURE 3.10 The arrangement of polypeptide chains in an antiparallel β-pleated sheet. The twist of the sheet is not shown.

α-helical and β-pleated sheet structures are smaller than those of fibrous proteins. Only a few α-helical segments of more than 20 residues have been detected in globular proteins, with an average length of about 11 residues, or three turns of the helix. Rarely do more than 10 contiguous amino acid residues participate in one strand of a β-pleated sheet structure of a globular protein. Nevertheless, these secondary structures contribute significantly to globular protein structure, as indicated by the α-helix content of several proteins in Figure 3.11. Examples of pleated sheet structures in globular proteins include a six-stranded antiparallel β-pleated sheet that represents a significant part of the structure of the proteinase bovine pancreatic α-chymotrypsin. A four-stranded parallel β-pleated sheet is found in the enzyme glutathione reductase.

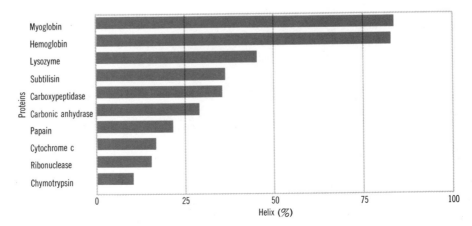

FIGURE 3.11 The percentage of amino acid residues that form parts of α-helices in several proteins.

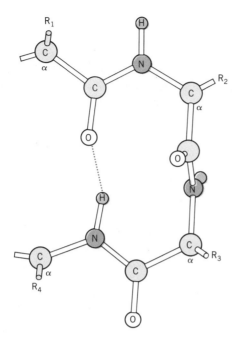

FIGURE 3.12 An example of a reversing turn, a secondary structure that has been observed in the structures of globular proteins.

The last element of secondary structure that we consider here is the **reversing turn** or **reversing bend.** A reversing turn causes the polypeptide chain to reverse its direction, as might occur at the surface of a protein. Figure 3.12 shows one type of reversing turn. A glycine residue is in the middle of this type of bend, as is required by steric constraints. Note that the reversing turn is closed by a hydrogen bond between a polypeptide imide nitrogen atom and carbonyl oxygen.

The fibrous protein collagen (Section 3.8 and Figure 3.4) forms a very extended left-hand helix (in contrast to the compact, right-handed α-helix). Three of these extended left-hand helices in parallel wind around each other to form a right-handed super-helix that is stabilized by interchain hydrogen bonds. The amino acid composition of collagen is rather unusual, being composed of 25% glycine and another 25% proline and hydroxyproline. Because of the high glycine and proline content, no α-helix occurs. The collagen super-helix is not expected in other proteins because it requires a glycine residue at every third position in the polypeptide chain in order to achieve stability. This amino acid sequence is characteristic only of collagen. A different type of super-helix, composed of coils of α-helices, is found in other fibrous proteins such as α-karatin and tropomyosin. The special structures of these fibrous proteins sometimes are referred to as "supersecondary structures" because they represent a level of structure that appears to be intermediate between secondary and tertiary structures.

3.12 TERTIARY STRUCTURE AND THE FORCES THAT MAINTAIN IT

We have pointed out the importance of steric effects and hydrogen bonding in maintaining the secondary structures of proteins. These factors also contribute to the stability of tertiary structure, the long-range interactions between amino acid residues in the three-dimensional structures of proteins. Hydrogen bonds, those of both secondary and tertiary structures, are stabilized in the nonpolar interior of the protein molecule because hydrogen bonds are ionic. The attraction between opposite electric charges increases as the dielectric constant of the medium is decreased. The disulfide bond, already described in Section 3.4, is considered to be part of the tertiary structure. Finally, tertiary structures also are supported by other ionic bonds and hydrophobic forces. These are presented diagramatically in Figure 3.13.

Ionic bonds result from the electrostatic association of ionized residues of opposite ionic charge, usually favored by the nonpolar interior environment of the protein.

Hydrophobic forces are of less obvious origin than the other forces discussed above. A hydrophobic force results in an energetically favorable association of two or more hydrophobic amino acid side chains in the interior of the protein. Transferring these hydrophobic side chains from the interior of the protein to the aqueous environment that surrounds the protein molecule is energetically not favored. Since the transfer of residues from the interior of a protein molecule to the surrounding solvent is exactly what will occur on denaturation (unfolding) of the protein molecule, the hydrophobic force stabilizes the protein. This stabilization effect is probably related to an unfavorable increase in order, that is, a decrease in entropy, which is expected to accompany a transition to the aqueous phase. Water is a highly associated solvent (Section 1.4), with a significant fraction of the molecules hydrogen-bonded in the tetrahedral

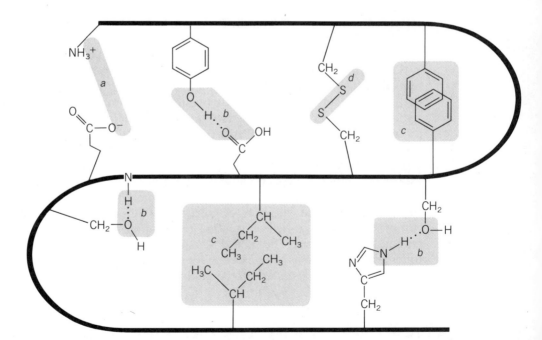

FIGURE 3.13   Disulfide bonds and some types of noncovalent bonds that stabilize tertiary protein structure. (*a*) Ionic bonds due to electrostatic interactions. (*b*) Hydrogen bonding between hetero-atoms of side chains and between side chain and polypeptide backbone. (*c*) Hydrophobic interactions between nonpolar side chains. (*d*) Disulfide bonds.

arrangement that is characteristic of ice. Introducing a nonpolar molecule into water increases the order in the arrangement of water molecules immediately surrounding the nonpolar molecule. The cagelike structure that forms will have a lowered entropy relative to the bulk of the water, giving a positive (i.e., unfavorable) free energy of solution even if the enthalpy is slightly negative. Hence, the nonpolar amino acid side chains tend to remain in the interior of the protein molecule, as favored by the hydrophobic force.

3.13 SUBSTRUCTURES OF GLOBULAR PROTEIN MOLECULES

Types of secondary structures and a general principle of globular protein structure, the largely nonpolar interior of the molecule, have been discussed in previous sections. Here, we present briefly some other aspects of the folding of the polypeptide chain in globular proteins.

1. The content of α-helix and β-pleated sheets varies widely from protein to protein. The muscle oxygen-binding protein myoglobin has a considerable α-helix content, whereas the proteinase α-chymotrypsin has primarily β-sheet structures. Many proteins have substantial amounts of both α-helices and β-pleated sheets. Examples include a number of enzymes that catalyze steps in the metabolism of phosphorylated sugars or organic acids, such as triose phosphate isomerase and lactate dehydrogenase. In these molecules, the polypeptide chain tends to fold in such a way that α-helices and β-pleated alternate along the chain. In several other groups of proteins, the α-helices and β-pleated sheets are segregated into different portions of the molecule.

2. Known polypeptide chain conformations are not knotted. That means that if one were to pick up the molecule by the amino terminus and the carboxyl terminus and stretch it out, the linear polypeptide chain would result. Also, the polypeptide chain is not rolled up like a ball of string. Rather, the polypeptide chain generally courses from one surface of the molecule to another, where it reverses its course.

3. Perhaps, the most useful general concept of globular protein structure is that of the **domain.** Protein molecules sometimes show roughly planar regions that have few intrusions of protein structure as indicated in Figure 3.13. These regions of nearly open space divide the protein molecule into domains. Domains are especially apparent in immunoglobulin molecules as indicated below and illustrated in Figure 3.19. In other protein molecules, domains are less obvious and more a matter of conjecture or definition. Statistical tests that analyze the density of atom populations around each amino acid residue in the three-dimensional structure sometimes detect domains not apparent by viewing the protein structure model. Often, the course of the polypeptide chain leaves one domain for a second and, after forming the second domain, returns to the first, making two polypeptide connections between the two domains. Most recognized domains have from 100 to 200 amino acid residues, though both larger and smaller domains have been identified.

Domains appear to have structural and evolutionary significance. An active center of an enzyme is the localized region of the molecule at which the enzyme binds and acts on its substrates. Active centers often are formed by at the

FIGURE 3.14   A view of the α-chymotrypsin molecule that shows the division between two domains and the active site. Only the course of the polypeptide chain is traced, as represented by the positions of the α-carbon atoms.

junction of two or more domains of the enzyme protein. Figure 3.14 shows a view of bovine α-chymotrypsin that reveals the separation between the two domains of that protein and its relationship to the active center of the enzyme.

Domains have been postulated to facilitate the evolution of proteins by providing "preformed" sequences of amino acids that will assume a stable conformation and/or provide a specific function. The enzyme pyruvate kinase has a domain that resembles very closely a domain of triose phosphate isomerase and another that could have been "borrowed" from any of several dehydrogenases. The implication is that segments of genes were duplicated and moved and joined and that the new genes so formed were retained if they provided a favorable adaptation for the organism.

The enzyme phosphofructokinase, which has a key role in the metabolism of carbohydrates, provides a dramatic example of likely gene duplication. The phosphofuctokinases of some bacteria and mammals are homogeneous tetramers, but the mammalian enzyme is about twice the size of the bacterial enzyme. The amino terminal half of the mammalian protomer, the carboxyl terminal half of the mammalian protomer, and the bacterial protomer all have homologous amino acid sequences. The two halves of the mammalian enzyme thus appear to be related "super domains." The bacterial and mammalian enzymes have the same number of active centers, four per tetramer. However, fructose bisphosphate activates the mammalian enzyme but not the bacterial enzyme. The latter activation sites may have evolved from the "extra" active centers that would have been present in an enzyme that resulted from duplication of the bacterial gene.

3.14 STRUCTURES OF SPECIFIC PROTEINS

We consider some special properties of five selected proteins—wool keratin, cytochrome *c*, hemoglobin, immunoglobulin G, and ribulose-*bis*-phosphate carboxylase—as an illustration of the diversity of structure and function exhibited by these macromolecules.

3.14.1 WOOL α-KERATIN

As we have indicated in Section 3.11, fibrous proteins use three arrangements of the polypeptide chain to serve the structural needs of tissues: α-helix, β-pleated sheet, and, for collagen, triple-helix structures. The wool fiber has been studied thoroughly by a number of physical techniques including x-ray diffraction and electron microscopy. The secondary structure of wool α-keratin is the right-handed α-helix. Three such α-helices form a left-handed coil, called a protofibril. The structure of the protofibril is presented diagrammatically in Figure 3.15.

The protofibril is stabilized by cross-linking disulfide bridges. The protofibrils are, in turn, coiled into higher-order, cablelike structures called microfibrils, which are about 8 nm across. Several hundred similar microfibrils are embedded in an amorphous protein matrix to form a macrofibril. A number of macrofibrils are formed in a cell and there are oriented in an elongated, filamentous parallel manner to form the complete wool fiber as the cell dies. Thus, a wool fiber consists of a very large number of polypeptide chains that are held together by hydrogen bonding and cross-linking disulfide bridges as well as an insoluble protein matrix. The process of giving hair a "permanent wave" includes the reduction of disulfide bonds and their reoxidation after the hair has been held in the desired form.

When α-keratin is exposed to moist heat and stretched, it is converted to a different conformational form—namely, β-keratin. The hydrogen bonds stabilizing the α-helical structure are broken under these conditions and an extended parallel β-pleated sheet conformation results. The ability of the same polypeptide chains to assume two very different conformations under different conditions has contributed to our understanding of both structures.

3.14.2 CYTOCHROME c

Cytochrome *c* is a small, abundant, stable, and colored protein found in the energy-producing machinery of all aerobic cells. These attributes have made it

FIGURE 3.15 Structure of a protofibril of wool α-keratin. Three α-helices are shown wrapped together in a left-hand super-helix, called the protofibril, that is stabilized by disulfide bonds.

one of the first and most well studied of globular proteins. The single iron atom in the cytochrome c molecule alternates between the FeII and FeIII oxidation states as it functions in the transport of electrons. We already have discussed the covalent bonding of the iron-containing heme group by thioether linkages to two cysteine residues of the protein. As depicted in Figure 3.16, the heme iron

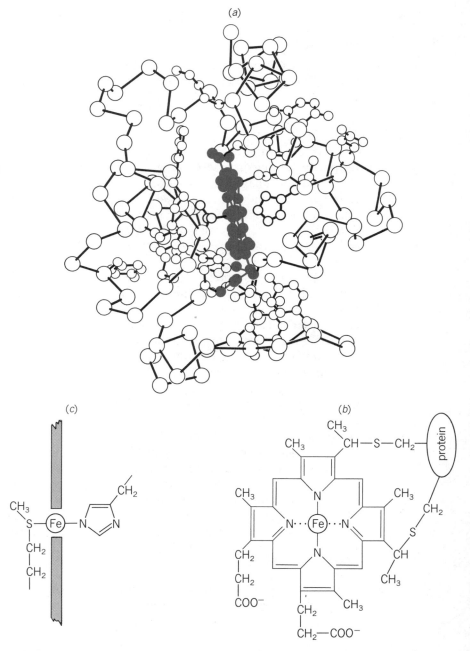

FIGURE 3.16 Cytochrome c and its heme group. (*a*) The general disposition of the polypeptide chain, showing the pocket in which the heme resides. (*Source:* Modified from T. Takano et al., *Journal of Biological Chemistry* Vol. 248 (1973), p. 5244); (*b*) structure of the cytochrome c heme group; (*c*) a view perpendicular to the plane of the heme group, showing bonding of heme iron to histidine and methionine residues.

also is bonded on opposite sides of the heme plane to a histidine nitrogen and methionine sulfur. Hydrophobic amino acid residues line the cavity in the cytochrome *c* structure in which the heme is located.

The heme is so central to the cytochrome *c* structure that its removal results in the unfolding of the polypeptide chain. If crystals of oxidised, FeIII, cytochrome *c* are reduced, the crystal shatters, showing that even a change in the oxidation state of the iron strongly influences conformation of the protein. The cytochrome *c* molecules of animals, plants, fungi, and bacteria have from 104 to 111 amino acid residues, 35 of which are found to be invariant from species to species. The electron-transporting ability of cytochrome *c* can be assayed *in vitro* by incubating it with the enzyme cytochrome oxidase. Cytochrome *c* and cytochrome oxidase from the same organism or distantly related organisms function equally well in this assay, indicating that each combination of invariant and variant amino acid residues fulfills the functional requirements of the cytochrome *c* oxidation reaction.

Because cytochrome *c* appears to have the same or very similar function in a very diverse set of organisms, and because of a number of other properties, the variations in cytochrome *c* amino acid sequence have been made the basis of a phylogenetic tree. In this tree, the most closely related organisms have the fewest differences in the amino acid sequences of their cytochrome *c*. Although such a tree generally parallels trees based on classical criteria such as morphology, members of some different mammalian genera have a greater similarity in their cytochrome *c* sequences than members of a single amphibian genus. Phylogenies based on amino acid sequences thus provide an additional measure of the relatedness of organisms. Amino acid sequences can be especially valuable in establishing distant relationships.

3.14.3   HEMOGLOBIN

The oxygen-carrying protein of red blood cells, hemoglobin, was one of the first proteins to be purified in large, hundred gram quantities because it is packaged already partially purified in red blood cells and is present in human blood at a concentration of about 150 g/l. Because of its abundance and some other favorable properties, biochemists and x-ray crystallographers have learned as much about this protein as any other. Hemoglobin, sometimes abbreviated Hb, accounts for about 98% of the oxygen-carrying capacity of blood, roughly 200 ml of gaseous oxygen per liter of blood. The protein, or "globin," portion of hemoglobin controls the affinity of the molecule for oxygen. This is illustrated by the hemoglobins of organisms that live under conditions of very low oxygen tension, such as parasitic intestinal worms. Such hemoglobins bind oxygen more than 1000-fold more tightly than mammalian hemoglobins, but all of these hemoglobins have the same heme group. The heme iron is bonded to a histidine residue on one side of the heme plane. At oxygenation, the iron is liganded to oxygen on the opposite side of the heme, in a pocket on the surface of the hemoglobin molecule. The heme iron remains in the FeII state during oxygenation and deoxygenation.

Figure 3.17 shows an oxygen-binding curve for normal human hemoglobin, hemoglobin A, the $\alpha_2\beta_2$ tetramer that binds four oxygen molecules. The oxygen concentration is expressed in Figure 3.17 as oxygen pressure, which would be approximately 0.21 atmospheres at sea level, but less on the average in the lungs and even less in muscle tissue. Also shown is the curve for the single polypeptide

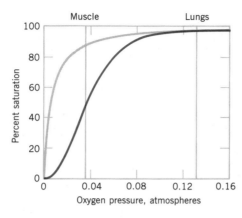

FIGURE 3.17 Comparison of oxygen binding by hemoglobin (colored curve) under conditions found in red blood cells, and by myoglobin. A value of 100% saturation with oxygen corresponds to one mole of oxygen bound ($Mb(O_2)$) and four moles of oxygen bound ($Hb(O_2)_4$) for myoglobin and hemoglobin, respectively.

chain cousin of hemoglobin, the muscle protein myoglobin, abbreviated Mb. A much lower concentration of oxygen is required to half-saturate (50%) myoglobin than hemoglobin. This indicates the greater affinity of myoglobin for oxygen, as is required if myoglobin is to bind oxygen at concentrations that cause hemoglobin to release it. Note that the oxygen-binding curves in Figure 3.17 show that a small decrease in oxygen tension below its average value in muscle would cause a much greater release of oxygen from hemoglobin than myoglobin, thus tending to maintain the rate of transfer of oxygen to myoglobin.

The shape of the myoglobin oxygen-binding curve indicates high affinity for oxygen and a simple saturation process characteristic of a single equilibrium constant for oxygen binding. The hemoglobin curve, in contrast, shows a reduced affinity for oxygen at low oxygen concentrations and a rapid change in affinity over the range of oxygen concentrations found in muscle. Obviously, several equilibrium constants are needed to explain the hemoglobin oxygen-binding curve. The differences between hemoglobin and myoglobin are subtle: The 141-amino acid residue α protomers and 146-amino acid residue β protomers of hemoglobin, and the 153-amino acid residue myoglobin molecule are similar in amino acid sequence and conformation. The key to the difference in behavior of hemoglobin and myoglobin is the tetrameric (four protomers) $\alpha_2\beta_2$ structure of hemoglobin, especially the interactions between protomers. This is indicated by the oxygen-binding curves for purified hemoglobin α or β protomers, which resemble that of myoglobin. Assembly into a tetrameric structure is, of itself, insufficient to explain the differences. An oxygen-binding curve for a tetramer of myoglobin in which the "protomers" did not interact would be expected to be the same as that of ordinary, single polypeptide chain myoglobin. A "communication" between hemoglobin protomers, depending on their state of oxygenation, both suppresses the affinity of deoxyhemoglobin for oxygen and causes the affinity of the unliganded protomers to increase dramatically, once one or more of the protomers of a given molecule have become liganded. This behavior is said to be **allosteric** and is characteristic also of several enzymes, as described in Section 4.12.

A consequence of allosteric binding of oxygen by hemoglobin is that the intermediate states of oxygenation, $Hb(O_2)$, $Hb(O_2)_2$, and $Hb(O_2)_3$, are poorly represented in solutions of partially oxygenated hemoglobin, which is mostly Hb and $Hb(O_2)_4$. The basis of this is a shift in the relative positions of the protomers, as diagrammed in Figure 3.18, which indicates strong contacts between α–β subunit pairs. The association of the α_1—β_1 pair with the α_2—β_2 pair is weaker than the association within these pairs. Thus, the latter association remains relatively rigid in the deoxyhemoglobin to oxyhemoglobin transition. The α_2—β_2 pair is shown in Figure 3.18 in the same position for deoxyhemoglobin and oxyhemoglobin, whereas the α_1—β_1 pair has been rotated counter-clockwise on oxygenation. The arrangement of protomers in oxygenated hemoglobin causes the heme groups to have a greater affinity for oxygen than is allowed by the arrangement of protomers in deoxyhemoglobin. It is important to remember that hemoglobin can assume either the deoxyhemoglobin conformation or the oxyhemoglobin conformation in any of the five states of oxygenation, from Hb to $Hb(O_2)_4$. However, the equilibrium is very much in favor of the deoxyhemoglobin form in the absence of oxygen. When one protomer becomes liganded with oxygen, the equilibrium is shifted in favor of the oxyhemoglobin arrangement of Figure 3.18, increasing the affinity of the remaining three protomer hemes for oxygen. Each successive binding of oxygen, to form $Hb(O_2)_2$, $Hb(O_2)_3$, and finally $Hb(O_2)_4$, further shifts the equilibrium in favor of the oxyhemoglobin arrangement of protomers.

How is the shift in equilibrium in favor of the oxyhemoglobin form accomplished? The heme iron lies slightly out of the plane of the porphyrin ring system, to the histidine side, in unliganded hemoglobin. On oxygenation, the iron and its histidine ligand move closer to the prophyrin plane. This results in a series of shifts in the polypeptide chain conformation that favors the oxyhemoglobin arrangement of protomers.

Oxygen is not the only ligand of hemoglobin that can shift the equilibrium between protomer arrangements. The phosphorylated carboxylic acid 2,3-*bis*-phosphoglycerate, chloride ions, protons, and carbon dioxide all are physiologically significant ligands of hemoglobin. They all bind more tightly to deoxyhemoglobin than to oxyhemoglobin, thus favoring the deoxyhemoglobin form. Some aspects of proton binding (the "Bohr effect") already have been discussed in Section 1.15.

The properties of naturally occurring mutants of human hemoglobin have contributed to our understanding of hemoglobin function. About 200 such mutations have been characterized with regard to amino acid sequence changes, which are single amino acid substitutions in most cases, and oxygen-binding. An example is hemoglobin Philly, in which β-chain residue Tyr_{35} (i.e., a tyrosine residue at position 35 from the amino terminus) is replaced by Phe_{35}. The loss of the tyrosine hydroxyl group reduces hydrogen bonding between protomers. Hemoglobin Philly has a myoglobinlike oxygen-binding curve and

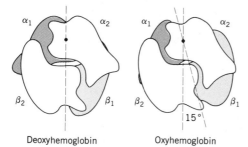

FIGURE 3.18 Shift in structure of hemoglobin on oxygenation. The tetrahedral arrangement of hemoglobin subunits is shown here with the α_1- and β_1-subunits, tightly associated with each other, below the α_2- and β_2-subunits, also tightly associated with each other. The close α_1—β_1 and α_2—β_2 contacts are not shown. Instead, a space between the protomers was drawn for the sake of clarity.

does not bind 2,3-*bis*-phosphoglycerate or exhibit a Bohr effect. It appears to be locked into the oxyhemoglobin protomer arrangement even in the absence of oxygen. Inositol hexaphosphate, a more highly negatively charged analogue of 2,3-*bis*-phosphoglycerate, does bind to hemoglobin Philly. This restores the normal oxygen-binding curve that is characteristic of hemoglobin A, perhaps by restoring interprotomer connections that were lost due to the phenylalanine-to-tyrosine substitution.

3.14.4 IMMUNOGLOBULIN G

The immunoglobulins are groups of structurally related proteins of vertebrates that include the antibody molecules of the immune system. Immunoglobulin G (IgG) molecules are serum proteins that combine with antigens: foreign proteins, polysaccharides, or other macromolecules that represent a potential threat to the organism by their presence in the body. The quaternary structure of an IgG molecule has four polypeptide chains: two "light chains" of about 215 amino acid residues and two "heavy chains" of fewer than 500 amino acid residues. These four polypeptide chains are linked by disulfide bridges and form 12 domains, as shown in Figure 3.19. The joint of the Y-shaped molecule provides flexibility so that the two combining sites, which are located at the end of each arm of the Y and bind the antigen, do not have to be a fixed distance apart.

Because there are two antigen-recognizing sites, networks of antibody–antigen chains will be formed when antibodies react with antigens that have more than one site recognized by the antibodies. Usually, such networks are

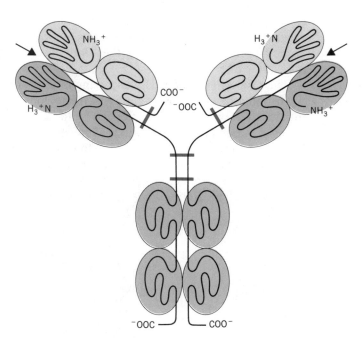

FIGURE 3.19 A diagram of the structure of immunoglobulin G molecules. Two light chains (L) and two heavy chains (H) are joined by disulfide bridges to form a roughly Y-shaped molecule.

FIGURE 3.20 Quaternary structure of ribulose-*bis*-phosphate carboxylase shown in two views. The molecule has eight small protomers, drawn as shaded spheres, and eight large protomers.

insoluble. The detection of a precipitate by any of several methods provides an assay of antibody activity. The reaction also can be a test for the presence of specific antigens, providing many types of assays that are useful in medicine, agriculture, and elsewhere (see Box 4.E). In the body, the insoluble, or soluble, antibody–antigen complexes that form when IgG molecules bind antigen molecules are recognized by other elements of the immune system to render them inactive and remove them.

Each combining site is formed at the interface of two amino terminal domains, one of a light chain and one of a heavy chain. Thus, each of the four polypeptide chains has variable regions called **variable domains** V_L and V_H, of about 100 amino acid residues that form the combining sites at the upper ends of the two arms of the Y. The amino acid sequence in the variable regions is different for each type of IgG molecule, according to the antigenic sites that it can recognize and bind. The remainder of the molecule shows only a few variations in different IgG molecules. Sites that combine with antigens are marked with arrows in Figure 3.19. Certain regions within the variable domains, corresponding in part to those residues that actually contact antigen molecules, are termed "hypervariable" because of the great variation seen with the antigen specificity of the IgG. The constant portion of the IgG molecule has covalently bound carbohydrate side chains, making IgG a glycoprotein. Immunoglobulins, each of which is specific for a very limited set of antigens, are synthesized by a very complex mechanism that is described, in part, in Chapter 22.

3.14.5 RIBULOSE-*bis*-PHOSPHATE CARBOXYLASE

We end this section with a note on the enzyme ribulose-*bis*-phosphate carboxylase. In plants, this protein catalyzes the reaction of ribulose-1,5-*bis*-phosphate with carbon dioxide to produce two molecules of 3-phosphoglycerate in the most important "carbon dioxide fixing" reaction. Ribulose-*bis*-phosphate carboxylase occurs in such high concentrations in plants that it undoubtedly is the most abundant protein in the biosphere. It has an interesting quaternary structure with eight small, 100-amino acid residue protomers and eight large, about 500-amino acid residue, protomers arranged as shown in Figure 3.20. The function of this important enzyme is discussed in Chapter 15.

3.15 THE PROCESSES BY WHICH PROTEINS TAKE THEIR SHAPE

The similar forms and function of proteins that have similar amino acid sequences, such as hemoglobin and myoglobin, imply a causal connection between amino acid sequence and protein conformation. One could imagine that the conformation of a protein is determined by its amino acid sequence if folding of the polypeptide chain is a spontaneous process, that is, if the folded protein has a lower energy than the unfolded protein, and a "kinetic path" exists for the unfolded polypeptide chain to reach the "native" conformation. Alternatively, the protein might be formed on or within a scaffolding of the cell that subsequently would be removed. Such a protein would not necessarily have a lower energy in its folded state than in an unfolded, completely denatured state. Actually, no choice between these alternatives is necessary. Both types of protein folding are known.

C. Anfinsen and colleagues demonstrated that the denatured, unfolded polypeptide chain of the ribonucleic acid degrading enzyme bovine pancreatic ribonuclease will spontaneously fold to produce the active enzyme. In one protocol, Figure 3.21, the purified, native enzyme was incubated in a concentrated solution of urea, with organic mercaptans (R—SH compounds). Concentrated urea solutions are better solvents than pure water for low polarity compounds. The stabilizing influence of hydrophobic forces in the interior of a globular protein thus is diminished or lost in a concentrated urea solution, which can dissolve the hydrophobic amino acid side chains. Ribonuclease has eight cysteine residues, which form four disulfide bonds in the native enzyme.

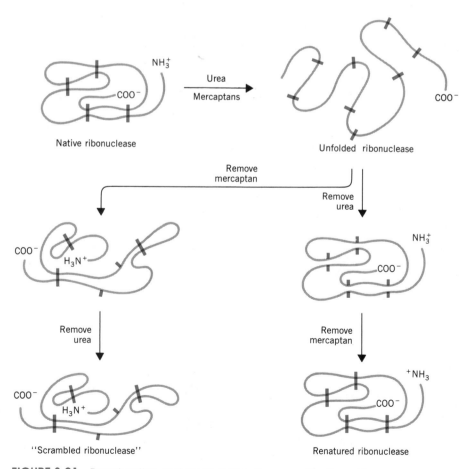

FIGURE 3.21   Denaturation and renaturation of pancreatic ribonuclease.

The combination of $8M$ urea solution and excess mercaptan such as mercaptoethanol denatures ribonuclease in the first step in Figure 3.21, giving eight cysteine residues in reduced form. The sample was divided into two portions in order to compare the effects of removing urea or mercaptan first. Removing the urea first allows the polypeptide chain to fold into nearly its proper conformation, but cysteine residues remain reduced. Oxidation of cysteine residues to cystine residues restores the native conformation. Oxidation of cysteine residues before removing urea causes random pairing of cysteine residues, giving "scrambled ribonuclease."

The denatured protein, when separated from urea, lacked any detectable enzyme activity. But air oxidation of the mercaptans allowed the disulfide bonds to form again over a period of a few minutes and, nearly simultaneously, the appearance of ribonuclease activity. This regenerated ribonuclease was indistinguishable from native ribonuclease in enzymic activity and physical characteristics, including the proper arrangement of cysteine residues in the four disulfide bonds. If disulfide bond formation were random, rather than being directed by the folding polypeptide chain, only 1% of the molecules would be expected to have all of the proper disulfide bridges. This and similar procedures have regenerated active proteins after denaturation with a variety of reagents. The four polypeptide chain IgG and even more complex protein-containing particles, including virus particles, have been renatured. However, not every protein is amenable to the denaturation–renaturation process.

Insulin is a peptide hormone with two polypeptide chains. The A chain of 21 amino acid residues is linked by two disulfide bonds to the 30 residue B chain, which has one internal disulfide bond. When it was subjected to denaturation and renaturation conditions, insulin was regenerated to only a few percent of its original activity, and most of the molecules had improper disulfide bonds. Why did insulin resist renaturation conditions that had been successful with far more complex proteins? The two polypeptide chains of insulin are, when first synthesized, parts of one chain. **Proinsulin,** represented in Figure 3.22, which itself is derived by specific proteolysis of a slightly larger precursor, has at its amino terminal end the B chain, at its carboxyl terminal end the A chain. Between

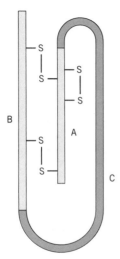

FIGURE 3.22 Structure of proinsulin. The A and B chains of the mature insulin molecule and the C chain of proinsulin are shown.

these is a sequence of 33-amino acid residues designated as the C chain, which lacks cysteine residues. In the islet cells of the pancreas, where insulin is synthesized, proteinases cleave the proinsulin to produce insulin. As expected, proinsulin readily can be renatured after denaturation under conditions similar to those previously described for pancreatic ribonuclease.

The C peptide of insulin can be regarded as a scaffolding for the proper folding of insulin. Removal of the C peptide from proinsulin gives insulin, but this native, biologically active insulin may not be folded in its lowest energy conformation because after denaturation, it renatures only to a very limited extent. Similarly denatured α-chymotrypsin, a three-chain enzyme (Figure 3.14; Section 4.13 and Table 4.3) renatures to only a limited extent, whereas its single chain precursor, α-chymotrypsinogen, refolds efficiently. Much more elaborate examples are known of scaffolding in the assembly of protein structures. The assembly of T4 bacteriophage (a virus that infects bacteria) requires not only the polypeptides that are found in the mature bacteriophage particles, but also the protein products of other T4 bacteriophage genes that act as structures or enzymically or both.

Since many proteins are able to fold properly from the unfolded polypeptide chain, biochemists have developed mathematical algorithms (computational rules) that attempt to predict protein conformation from amino acid sequence. These have successfully predicted portions of folded structure, but no complete three-dimensional structure of a protein molecule has been predicted by computation alone. The value of a fully successful prediction algorithm will be immense. The amount of effort required to determine the amino acid sequence of a polypeptide chain, especially if it is derived from nucleotide sequences (Chapter 6), is a very small fraction of the effort required to locate the atoms in crystals of the protein by x-ray diffraction. Predicting protein folding will require a greater theoretical knowledge than presently is available of not only the energetics of the folded protein but also the dynamics of the folding process. From the successful theory also should flow information about the dynamics of protein function, especially catalysis, the subject of the next chapter.

REFERENCES

1. *T. E. Creighton,* Proteins. *San Francisco, Calif.: Freeman, 1984.*
2. *A. Fersht,* Enzyme Structure and Mechanism. *San Francisco, Calif.: Freeman, 1977.*
3. *T. Palmer,* Understanding Enzymes. *New York: Wiley, 1985.*
4. *R. E. Dickerson and I. Geis,* Hemoglobin: Structure, Function, Evolution and Pathology. *Menlo Park, Calif.: Benjamin/Cummings, 1983.*

REVIEW PROBLEMS

1. A hormone was found to be a small, basic peptide. From the information next presented, write the most likely structure of the peptide.

   (a) A sample of the peptide was hydrolyzed in a 6*N* HCl solution that contained thioglycolic acid, as indicated in Box 3.B. Equimolar amounts of arginine, methionine, and tyrosine were found.

(b) A second sample of the peptide was treated with Sanger's reagent, purified and hydrolyzed in acid solution. *N,O-bis*-DNP-tyrosine and *O*-DNP-tyrosine were the only colored products recovered, and these were in approximately equal molar amounts.

(c) Digestion of the original peptide with trypsin released arginine and a new peptide.

(d) Treatment of the peptide hormone with cyanogen bromide released tyrosinyl-homoserine lactone and a basic peptide.

2. What is the expected net ionic charge at neutral pH for the peptide Tyr—Lys—Cys—Ala—Asp—His—Gly?

3. A protein was found to contain 0.29% by weight of tryptophan residues. Calculate the minimum molecular weight of the protein. The molecular weight of free tryptophan is 204.

4. What is meant by the "primary," "secondary," "tertiary," and "quaternary" structures of a protein?

5. Draw the structures of four or more possible chemical modifications of the peptide that is shown below, using known, naturally occurring modifications of amino acid residues in proteins to guide your choice of modifications.

Met—Lys—Tyr—Asn—Ser—Thr—Ser—Cys—His—Gly

C H A P T E R 4

ENZYMES

In the previous three chapters, we have presented the structures of many complex compounds, including sugars, amino acids, and the macromolecules composed of them. The succeeding three chapters similarly show the structures of other low and high molecular weight compounds. The syntheses of many of these compounds has been achieved in the laboratory (see, for example, Box 3.D), but the skills of the natural products chemists have been heavily taxed to accomplish these imitations of nature. Chemists have employed blocking groups and powerful reagents and often obtained only very limited yields in the synthesis of a biomolecule. The same syntheses are achieved by cells smoothly and routinely, principally through the actions of **enzymes,** the proteinaceous catalysts of the biosphere. Enzymes mediate not just the syntheses of biological compounds; they also catalyze reactions that supply the cell with energy, detoxify compounds, produce light, and so on. Enzymes are responsible for virtually all of the chemical reactions in cells in which covalent bonds are formed or broken. As the crucial mediators of biochemical reactions, enzymes are among the simplest but most elegant of biological machines and among the easiest to obtain in a pure and functional form. For these reasons, a considerable fraction of the efforts of biochemists has been devoted to the study of enzymes, and enzymes have been the subject of some of the most precise of biochemical measurements.

Enzymes also are of practical interest. The sales of enzymes for industrial and home use amount to hundreds of millions of dollars annually, worldwide. Industrial enzymes are used in the production of billions of dollars worth of chemicals and pharmaceuticals. It is reasonable to expect the industrial uses of enzymes to increase because modern techniques for gene manipulation (Chapter 22) make their production more efficient and less expensive. One of the oldest uses of an enzyme preparation, a very crude enzyme preparation in this case, is in cheese production. An extract prepared from the abomasum, or fourth stomach, of a suckling calf contains **chymosin,** formerly called rennin, an enzyme that cleaves a glycopeptide from one form of the milk protein casein. The remainder of the casein molecule, complexed with calcium, precipitates to form the curd in the initial step in cheese making. Chymosin accounts for about 10% of the use of enzymes in industrial and home products. Starch-digesting enzymes, in food processing, and protein-digesting enzymes, in applications such as food processing and laundry detergents, account for about three-quarters of enzyme use. Enzymes also find application as analytical reagents and in many other ways. Often, the enzyme is used in a solution. However, the application of "immobilized enzymes" (see Box 4.A) to industrial processes is growing.

BOX 4.A   IMMOBILIZED ENZYMES

In 1974, shortages caused a dramatic rise in the cost of table sugar, sucrose. Table sugar itself accounted for only a portion of the cane and beet sugar consumption, since sucrose is used to sweeten and thicken pastries, syrups, candies, and other foods. Corn syrup was and is a much less expensive source of sugars, primarily glucose. But glucose is not sufficiently sweet for the purposes of confectioners. The enzyme glucose isomerase catalyzes the conversion of glucose to fructose, which is about twice as sweet as glucose. Freeze-dried cells of certain Streptomyces (a family of fungi) species, which are rich in glucose isomerase, have been used in the commercial production of "high-fructose corn syrup." Stirring the dried cells with the proper dilution of corn syrup increases the fructose content. This process and the 1974 sucrose price rise stimulated a new application of a technology that had been developed in the 1950s: immobilized enzymes. Immobilized enzymes are enzymes that are connected by any of several means to a solid support, usually small beads. Probably, the immobilized enzyme that has converted the most moles of substrate to product under industrial conditions is glucose isomerase.

Enzymes bound to a solid support have a number of practical advantages. Immobilized enzymes easily can be removed from liquid suspension, often by settling alone, leaving virtually no enzyme to contaminate the product and allowing recovery of the enzyme in a reusable form. Frequently, immobilized enzymes are more stable to inactivation in an aqueous environment than the corresponding dissolved enzyme. In some industrial processes, the immobilized enzyme is reused several times over a period of months before it must be replaced. A disadvantage of immobilized enzymes is that they are unable to work on insoluble substrates and may have only low efficiency on polymeric substrates.

A frequent application of beads with bound enzyme is in a reactor column. The beads are packed in a cylinder with a porous bottom plate so that liquid but not beads can flow through the column, in a manner similar to that used in column chromatography (see Box 3.B) but on a much larger scale. Reactant, such as corn syrup solution, is passed through the reactor to convert it to product that appears in the effluent stream. Such a continuous process has obvious advantages in an industrial plant.

Several methods for immobilizing enzymes have been reported. When polyacrylamide and methylene-*bis*-acrylamide (Box 3.E) are polymerized in a solution of the enzyme, the enzyme becomes entrapped in the matrix. The polyacrylamide matrix allows low molecular weight substrates to reach the enzyme, but the enzyme remains in place. Of course, this approach will be successful only for an enzyme that is not significantly inactivated under the conditions of initiation and extension of the polyacrylamide chains. In other approaches, the enzyme may be adsorbed onto or chemically bonded to the surface of hard spheres or crystals of nearly chemically inert (under the enzyme reaction conditions) substances such as stainless steel, glass, or cellulose. The enzyme may be encapsulated in hollow spheres made of a substrate- and product-permeable material. Considerable effort has been expended into developing methods that anchor or otherwise hold the enzyme and yet allow it to remain active. One industrially important immobilizing process is the chemical coupling of glucose isomerase to glass beads.

Often, the enzyme itself is not immobilized. Rather, cells that contain the enzyme are immobilized. The whole cell approach reduces costs since the enzyme need not be purified, and often, the enzyme is protected by remaining in the cell. The new techniques of genetic engineering allow bacterial and fungal cells that contain abnormally high concentrations of specific enzymes to be produced.

4.1 GENERAL PROPERTIES OF ENZYMES

The existence and power of enzymes first were revealed in the nineteenth century when reactions that had been considered to occur only in the presence of cells were found also to be mediated by cell-free extracts. In 1860, Berthelot disrupted yeast cells by grinding them in a mortar with abrasive. This produced a cell-free extract. A fraction from this extract, precipitated by ethanol, catalyzed the conversion of sucrose to glucose plus fructose, a reaction that intact yeast cells carry out. Other hydrolytic reactions previously had been detected in other, even more crude, cell-free preparations such as gastric juice (hydrolysis of proteins) and saliva (hydrolysis of starch). In the early twentieth century, enzyme-catalyzed reactions already were the subject of mathematical analyses, but it was not until the 1920s that the protein nature of enzymes was unmistakable. This result was due to the crystallization of the urea-hydrolyzing enzyme urease by Sumner and several protein-hydrolyzing enzymes by Northrop. The enzyme preparations were demonstrated to have a constant ratio of enzyme activity to mass of protein through multiple crystallizations, giving strong evidence that the protein molecules themselves, rather than some contaminant, are the actual catalysts.

As in the example of urease, enzymes and classes of enzymes usually are named by appending the suffix -ase to the name or abbreviated name of a compound on which the enzyme acts. Enzymes that cause the cleavage of the polypeptide chain by reaction with water are referred to, as a class, as proteinases. Peptidases have a similar specificity, but facilitate the hydrolysis of peptides more effectively than the hydrolysis of the larger and compactly folded protein molecules. All of these enzymes are considered to be members of a larger grouping: the "hydrolases," enzymes that cleave bonds by the incorporation of a molecule of water. Although the suffix -ase now is widely used, many of the first discovered and most well-studied enzymes are not so named.

What compound(s) is the reactant(s) in a chemical reaction? And which is the product(s)? This is largely a matter of definition, of course. Similarly, in an enzyme-catalyzed reaction, the designation of substrate(s) [i.e., the reactant(s)] and the product(s) is arbitrary. However, investigators usually take into account such factors as the following:

1. Equilibrium (in favor of the products).
2. Or the direction in which the enzyme-catalyzed reaction can be assayed easily *in vitro* (substrate to product).
3. Or the presumed predominant reaction direction *in vivo* (net formation of the compounds that are designated as products; see also Section 4.4).

The latter two factors depend not only on the equilibrium constant but also the relative concentrations of substrates and products.

Enzymes increase the rates of chemical reactions. In some instances, the rate of a particular enzyme-catalyzed reaction is so small in the absence of the enzyme as to be undetectable against the background of competing reactions. Like all catalysts, an enzyme will function at a much smaller molar concentration than that of the reactant(s) on which it operates. Under these conditions, the enzyme does not alter the equilibrium of the reaction, and one mole of enzyme facilitates the conversion of many moles of reactant to product. We will consider this point again in developing a mathematical description of simple enzyme-catalyzed reactions in Section 4.3.

In comparison with most catalysts, enzymes are especially effective. They are efficient, catalyzing reactions at high rates at moderate temperatures and under otherwise mild conditions, making extremes of pH, pressure, and so on unnecessary. Enzymes also are particularly specific catalysts, often acting on only one form of an optically active compound even when presented with a mixture of chiral forms. Many enzymes also exhibit the ability to couple two chemical reactions that have very dissimilar reactants and products. This allows a reaction that is not energetically favored to be coupled to one that is energetically favored, so that a reasonable extent of conversion is achieved for the former reaction. Examples of coupled reactions are given in Section 4.9 and Chapter 9. Finally, many enzymes are controlled in their catalytic efficiency by substrate or product concentrations. They also may be controlled by metabolically relevant compounds that resemble neither substrates nor products of the enzyme. This property, which is of enormous importance for the regulation of metabolism, is termed "allosterism" and is discussed in Section 4.12.

4.2 HOW DO ENZYMES ACCELERATE REACTIONS?

Long ago, some investigators held romantic ideas about enzymes somehow being able to facilitate the conversion of substrate to product "by acting at a distance," operating outside the known laws of chemistry and physics. Although currently there is no enzymically catalyzed reaction that can be explained completely as a series of chemical reactions, step by microscopic step, it is axiomatic that the actions of enzymes can be explained in the terms of mechanistic organic chemistry and that an essential element in enzyme mechanisms is the binding of substrates (and products). As is described next, specific binding accounts for not only catalytic efficiency but also substrate selectivity.

Much now is known about the physical, chemical, and structural aspects of enzymes as protein molecules and the peculiar reactivities of specific amino acid residues within the enzyme molecule. Once it was believed that the identification and localization of the amino acid residues associated with a catalytic site would explain the catalytic activity of any enzyme. Now, biochemists realize that this approach, while essential, is insufficient. In recent years, enzyme chemists have designed ingenious reagents to probe and identify the active sites of enzymes. Highly sophisticated physical techniques such as nuclear magnetic resonance spectrometry and high-resolution x-ray crystallography have provided the enzyme chemist with critical information about protein structure that has enormous implications for understanding proteins as catalysts.

The efficiency of enzymic reactions clearly requires that the energy of activation for the enzymic reaction be significantly less than that for the corresponding nonenzymic reaction, as indicated in Figure 4.1.

This reduction in the energy of activation could be accomplished if the enzyme were to use the same mechanism as the nonenzymic reaction and simply facilitated it. However, it also is possible, in a given enzymic reaction, that the enzyme exploits a mechanism that is too energetically unfavorable to be important in an ordinary solution reaction. Finally, the enzyme may subdivide the reaction into a number of microscopic steps, each with a small energy of activation. That such subdivision does occur is shown by the recovery of "covalent intermediates," compounds formed by the reaction of an enzyme with a substrate prior to release of the last product of the reaction. Section 4.5 describes such an intermediate.

The binding of the substrate occurs near or at a region of the enzyme protein molecule that is designated as the **active site.** Structurally, the active site may be a crevice such as is found in the proteinases papain or chymotrypsin (Figure 3.14) or pancreatic ribonuclease A, or it may be a deep pit as in carbonic anhydrase, with the catalytically essential zinc atom at the bottom of the pit. Whatever the shape of the active site, the correct substrate apparently binds uniquely in the required orientation. The amino acid residues of the enzyme protein that form the active site usually are derived from nonadjacent segments

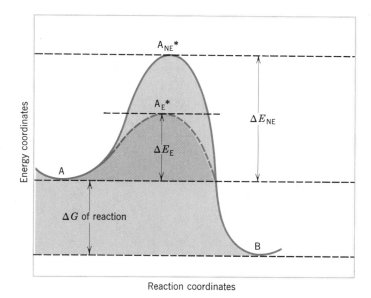

FIGURE 4.1   Energy barriers for both an enzymically catalyzed and the corresponding uncatalyzed reaction A → B. $A_{NE}*$ is the activated, transition-state complex for the nonenzymic reaction; A_E* is the corresponding complex for the enzyme-catalyzed reaction. ΔG is the difference between the greater energy of reactant A and the lesser energy of the product B. For the nonenzymic reaction, the symbol ΔE_{NE} is the difference between energies of the activated complex and the reactant. The corresponding quantity for the enzymic reaction is ΔE_E.

of the polypeptide chain. In pancreatic ribonuclease A, for example, the histidine residues at position 12 and at position 119 in the 124-amino acid residue chain both are part of the active site.

Current notions hold that the ionic, nucleophilic, and other reactive groups of the enzyme protein are properly positioned in the active site for catalysis. Within the active site, such properties as the dielectric constant, degree of solvation, and steric hindrance may be very different than in solution. In such an environment, reactive groups of both enzyme and substrate may have special reactivity. The precise shape of the active site allows portions of the substrates to be bound tightly, as is shown directly by x-ray crystallographic analyses of some complexes of enzyme and substrate analogue. The enzyme lysozyme, which is found in tears and eggwhite and many other biological fluids, catalyzes the hydrolysis of polysaccharide chains in the peptidoglycan layer of cell walls of certain bacteria. The calculated interaction energy of a portion of the active site of lysozyme with a glucopuranose ring, which involves noncovalent bonds only, is approximately -59 kJ/mole. This is comparable to the energy of an $N-H$ bond in an alkyl ammonium ion.

The already mentioned ability of enzymes to distinguish between two or more chiral forms of a substrate proves that at least three points of contact must occur between enzyme and substrate, in order to make a chiral distinction. Figure 4.2 indicates how the notion of "three-point attachment" will allow an enzyme to reject the wrong chiral form of the nonsubstrate. The groups a_1 and a_2 in Figure 4.2 may be the same or different, and in a chiral substrate, a_1, a_2, b, and d all will be different. The corresponding binding sites for three groups in the active site are designated by the same letters in double quotes. Three points of attachment will allow an enzyme to act in a chiral fashion even on a nonchiral substrate, indicated by $a_1 = a_2$ in Figure 4.2. Regardless of whether a_1 and a_2 are the same or different, there is only one possible fit of the substrate in the active site. It is clear that most substrates will have many more than three "points of attachment" to the enzyme in order to explain both the efficiency and the specificity of enzymically catalyzed reactions. These additional interactions between substrate and enzyme can be expected to increase the specificity, both chiral and other, of enzymes beyond the contributions possible with just three points.

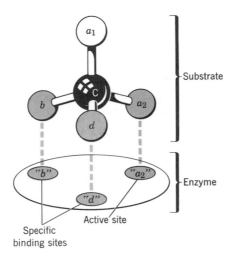

FIGURE 4.2 Diagrammatic representation of the positioning of a substrate to its active site in the enzyme protein structure.

An example of a chirally specific reaction is that catalyzed by L-amino acid oxidase, which attacks only the L-amino acids. The distinct D-amino acid oxidases act only on the D-amino acid isomers.

$$\text{L-Amino acid} \xrightarrow[\text{L-Amino acid oxidase}]{O_2} \alpha\text{-Keto acids} + NH_3 + H_2O_2$$

$$\text{D-Amino acid} \xrightarrow[\text{D-Amino acid oxidase}]{O_2} \alpha\text{-Keto acids} + NH_3 + H_2O_2$$

A few enzymes, such as the racemases, recognize both chiral forms of the substrate. The racemases catalyze the equilibration of the two optical forms. Thus, alanine racemase catalyzes the reaction.

$$\text{L-Alanine} \rightleftharpoons \text{D-Alanine}$$

Still other enzymes have specificities toward geometric or *cis-trans* isomers. Fumarase will readily add water across the double-bond system of the *trans* isomer fumaric acid, but is completely inactive toward the *cis* isomer maleic acid.

Enzymes commonly treat even two identical groups of a substrate differently. As an example of such a substrate, consider citrate, which is a symmetrical, nonchiral compound. The net effect of the action of the enzyme aconitase is to interchange the hydrolxyl group on the central carbon of citrate and a proton of one and only one of the two, otherwise identical, acetic acid groups. The product of the reversible reaction is the chiral compound isocitrate (isocitric acid).

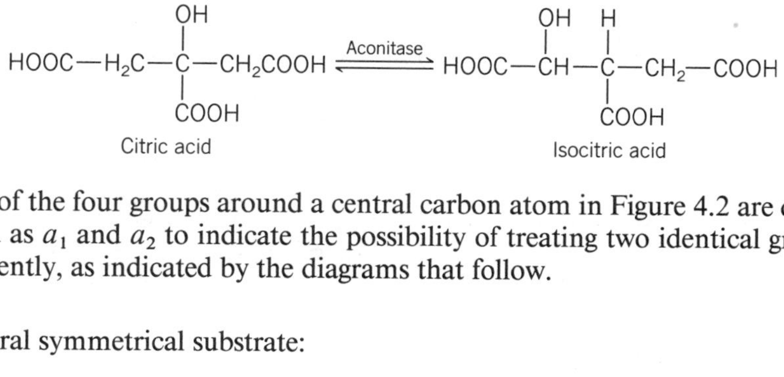

$$\text{HOOC}-H_2C-\underset{\underset{\text{COOH}}{|}}{\overset{\overset{\text{OH}}{|}}{C}}-CH_2COOH \underset{\text{Aconitase}}{\rightleftharpoons} \text{HOOC}-\overset{\overset{\text{OH}}{|}}{CH}-\underset{\underset{\text{COOH}}{|}}{\overset{\overset{\text{H}}{|}}{C}}-CH_2-COOH$$

Citric acid   Isocitric acid

Two of the four groups around a central carbon atom in Figure 4.2 are designated as a_1 and a_2 to indicate the possibility of treating two identical groups differently, as indicated by the diagrams that follow.

General symmetrical substrate:

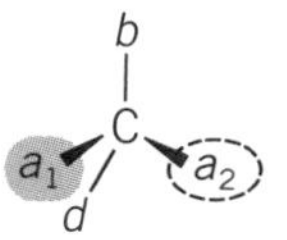

where $a_1 = a_2$

Specific symmetrical substrate:

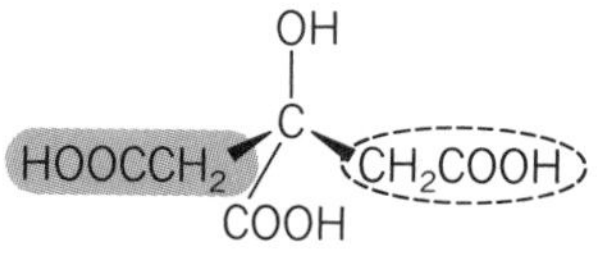

Citric acid

Specific chiral product:

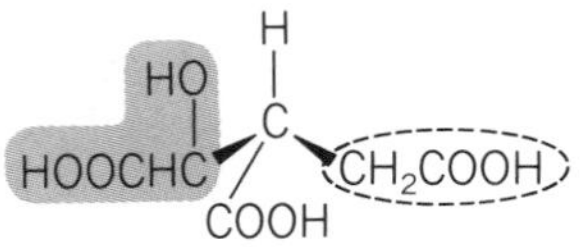

Isocitric acid

As is indicated in Figure 4.1, both the enzymically catalyzed reaction and the ordinary solution reaction must proceed by activation of the substrate (or enzyme-substrate pair) to form the activated "transition state complex," which has a very specific, difficult-to-achieve, high-energy conformation that readily can breakdown into the product(s). Increasing the concentration of transition state complexes is an essential element in catalysis. The active site is capable of bringing reactants together at an effective concentration and in favorable conformations and orientation. These conditions can be achieved in free solution, at least in theory, but only as rare events. In addition, the enzyme, by its affinity for different portions of the substrate(s), can apply mechanical and/or electrostatic strain to the substrate(s). By proper selection of the amino acid side chains and portions of the polypeptide backbone that form the wall of the active site, the polarity and the positions of reactive groups, such as acids and bases, presumably can be optimized for catalysis.

Comparisons of the energetics of forming a transition state complex in an enzymically catalyzed reaction and one that occurs free in solution show that the more favorable energetics of the former reaction is primarily due to a reduction in the entropic term. That is, the enzyme apparently is able to reduce the difference between the freedom of motion available to the substrate and that available to the transition state complex. The transition-state complex is by definition highly constrained with regard to translation, vibration, and rotation. The process of binding the substrate obviously is, of itself, constraining, thereby reducing the entropy difference. This phenomenon sometimes is referred to as "entropy trapping."

The importance of increasing the concentrations of reactants and reducing their freedom of motion is illustrated even by ordinary organic chemical reactions. Compare the hydrolysis of aspirin in water to the hydrolysis of an acetyl phenol with a group, X, situated ortho to the acetoxy group and selected to have the inductive effect of a carboxyl group. In the second reaction, acetate is added as a catalyst, to substitute for the ring carboxyl of aspirin that serves as an internal catalyst.

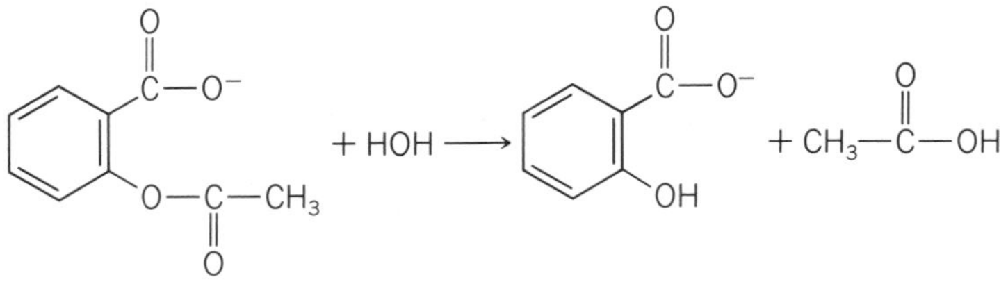

Aspirin

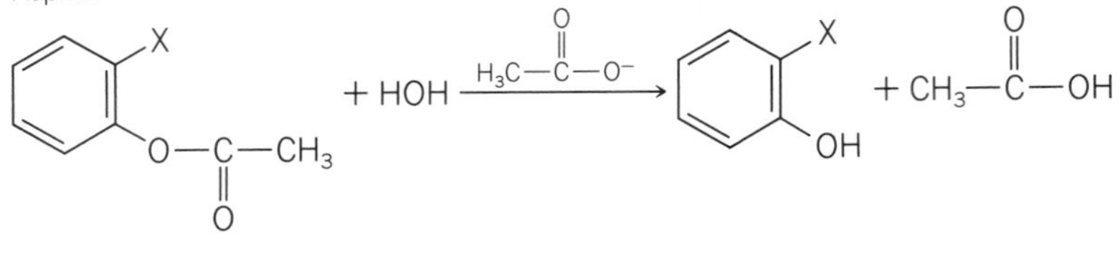

The first-order rate constant for the hydrolysis of aspirin has the units of $\sec^{-1}$, whereas the second-order rate constant for the hydrolysis of the acetylphenol derivative, as promoted by acetate anion, has the units $M^{-1}\sec^{-1}$. The ratio of the first-order rate constant to the second-order rate constant is $13M$! That is, the incorporation of the carboxyl group into the same molecule, at the ortho position, appears to be equivalent to increasing the acetate concentration, in the promoted hydrolysis of the acetylphenol derivative, to $13M$ at room temperature. However, position and orientation effects, as well as concentrations, can promote the reaction. A study of the temperature dependence shows that it is primarily the entropy term that is reduced in the more favorable aspirin reaction.

Linus Pauling stated one of the important concepts in enzyme catalysis in 1948.

I think that enzymes are molecules that are complementary in structure to the activated complexes of the reactions that they catalyze. The attraction of the enzyme molecule for the activated complex would thus lead to a decrease in its energy, and hence to a decrease in the energy of activation of the reaction and to an increase in the rate of the reaction

Several observations support Pauling's notion that enzymes should bind substrates tightly, but bind the transition state complex or some very closely related structure even more tightly. Section 4.5 reports one such observation.

The need for an active site of precise and controlled design undoubtedly is a factor that contributes to the evolution of enzymes as large molecules. Possibly, only a large molecule can have sufficient rigidity and sufficiently exclude solvent to provide the necessary environment for efficient catalysis. The large size presumably also is necessary to provide sites for the binding of molecules that control enzyme activity (see Section 4.12). Finally, the large size may be necessary to resist proteinases and other potentially damaging agents.

4.3   RATE LAW FOR A SIMPLE ENZYME-CATALYZED REACTION

Let us consider a reaction with one reactant, the substrate S, and one product, P.

$$S \rightleftharpoons P \tag{4.1}$$

If the reaction is catalyzed by an enzyme, the substrates and products and their equilibrium concentrations remain the same, which means that the enzyme, E, accelerates both the forward and back reactions.

$$S \overset{E}{\rightleftharpoons} P \tag{4.2}$$

It is by binding the substrate that the enzyme effects conversion to product, and this fact can be expressed as a minimum of two equilibria.

$$E + S \underset{k_{-1}}{\overset{k_1}{\rightleftharpoons}} ES \underset{k_{-2}}{\overset{k_2}{\rightleftharpoons}} E + P \tag{4.3}$$

Of course, many other intermediates are possible in such a reaction, due to interconversions of various forms of the enzyme and/or its complexes with substrate and product. For example, EP is expected to be an intermediate between ES and E + P. However, since one step generally is rate-limiting,

Equation 4.3, incomplete as it is, provides a useful description of the rate processes for many enzyme-catalyzed reactions.

The most common kinetic measurements of enzyme-catalyzed reactions are those of the steady increase in product concentration that occurs in a dilute solution of the enzyme, to which has been added an excess of substrate. Very shortly after E and S are mixed, the product begins to form at a constant rate. This constant increase in the concentration of product, $d[P]/dt$, is designated v_0, the "steady-state initial velocity." v_0 is considered to be an initial velocity because only a very small proportion of the substrate is consumed while the measurements are being made. So [S] for practical purposes is constant, and the v_0 that is measured corresponds to the initial value of [S].

The concentration of the product, [P], will be very small under the conditions just described. Thus, we can neglect the back reaction of E with P to form ES, and equation 4.3 becomes

$$E + S \underset{k_{-1}}{\overset{k_1}{\rightleftharpoons}} ES \overset{k_2}{\longrightarrow} E + P \tag{4.4}$$

Implicit in Equations 4.3 and 4.4 is the recycling of E. In these equations, the form of the enzyme that is released with P from ES has been designated with the same symbol, E, as the species that reacted with S. This identity is necessary if E is to be a catalyst. A catalyst, by definition, is not consumed by the reaction it catalyzes. The recycling of E is explicitly represented in Equation 4.5.

$$E + S \underset{k_{-1}}{\overset{k_1}{\rightleftharpoons}} ES \overset{k_2}{\longrightarrow} E + P \tag{4.5}$$

In the reaction described by Equation 4.5, there will be a very short "induction phase," usually only a fraction of a second in duration, before the steady-state initial velocity is achieved. Generally, the induction phase is not detected or is ignored in steady-state kinetics.

We define the total concentration of enzyme in all forms as E_0. Obviously, $E_0 = [E]$ before the substrate is added. After substrate is added, $E_0 = [E] + [ES]$. During the short induction phase, [ES] increases to its steady-state value and remains constant during the steady-state portion of the reaction, when v_0 is measured and is constant. Briggs and Haldane used the steady-state condition of constant [ES] and the initial velocity condition of constant [S] to develop equations that relate v_0 to kinetic constants and to the concentrations of enzyme and substrate, E_0 and [S]. Their results, outlined here, were published in 1925.

Because the product can be formed only by the breakdown of ES, governed by rate constant k_2, we can conclude that the velocity under initial rate conditions is

$$v_0 = k_2[ES]. \tag{4.6}$$

If we are able to evaluate the steady state [ES] in terms of E_0, [S], and kinetic constants, Equation 4.6 will give us the desired result of Briggs and Haldane.

Constant [ES] requires that the rates of ES formation and ES breakdown be equal. The rate of formation of ES is given by $k_1[E][S]$. Because ES can be lost in two reactions, breaking down into E + S or E + P, the rate of breakdown of ES is given by $k_{-1}[ES] + k_2[ES]$. Setting the rates of ES formation and ES breakdown equal to each other gives

$$k_1[E][S] = (k_2 + k_{-1})[ES]$$

Segregating the variables from the constants and combining the constants gives

$$\frac{[E][S]}{[ES]} = \frac{k_2 + k_{-1}}{k_1} \equiv K_m \qquad (4.7)$$

where K_m is designated as the Michaelis constant, after the developer of earlier forms of simple enzyme kinetics equation.

Since [ES] rather than [E] occurs in Equation 4.6, we substitute $(E_0 - [ES])$ for [E] in Equation 4.7 to obtain

$$\frac{(E_0 - [ES])[S]}{[ES]} = K_m$$

Solving this for [ES] gives

$$[ES] = \frac{E_0[S]}{[S] + K_m} \qquad (4.8)$$

Substituting this result in Equation 4.6 gives

$$v_0 = \frac{k_2 E_0[S]}{[S] + K_m} \qquad (4.9)$$

Equation 4.9 is one form of the Briggs–Haldane result.

4.4 ENZYMICALLY CATALYZED REACTIONS EXHIBIT SATURATION KINETICS

In some experiments, the concentration of the enzyme, E_0, may not be known, and it is convenient to substitute V_{max}, the maximum velocity, for the product $k_2 E_0$.

$$v_0 = \frac{V_{max}[S]}{[S] + K_m} \qquad (4.10)$$

The justification for this modified version of the Briggs–Haldane equation is that the maximum velocity will be achieved when essentially all of the enzyme is in the ES form, that is, when $[ES] \cong E_0$. According to Equation 4.10, representing the simple model of Equation 4.4, when [S] is equal to K_m, the denominator of 4.10 will be 2[S] and the numerator will be $[S]V_{max}$. This means the velocity v_0 will be equal to just half the maximum velocity, V_{max}, when $[S] = K_m$. Under this condition, $[ES] = 0.5E_0$, as indicated by Equation 4.8. Figure 4.3 shows the characteristic rectangular hyperbola that is obtained by graphing Equation 4.10.

As [S] is increased to several times the value of K_m, it approaches a **saturation** value, which is indicated by v_0 approaching V_{max}. Such a concentration of substrate, 20 or 100 times K_m, is said to be a **saturating concentration of substrate** because virtually all of the enzyme must be in the [ES] form. No further increase in the velocity of the reaction is possible because the steady-state

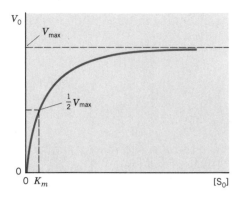

FIGURE 4.3 A graph of the initial, steady-state velocities of a series of reactions catalyzed by the same concentration of enzyme, varying the substrate concentration, (S), for each reaction. The reaction velocity approaches V_{max} at very large values of (S).

concentration of free E, [E], is vanishingly small. Please note the phrase "steady-state concentration." Even at saturation, the enzyme must, of course, constantly cycle between E and ES forms to accomplish catalysis, as indicated by Equation 4.5.

At saturating concentrations of substrate, then, equation 4.6 becomes

$$v_0 = k_2 E_0 \equiv V_{max} \qquad (4.11)$$

Thus, k_2 is the ratio V_{max}/E_0, which is the number of moles of substrate converted to product per unit time divided by the number of moles of enzyme. This is the **turnover number** for a single substrate, single product reaction. The turnover number for most enzymes is in the range of 1 to 1×10^4 per second, with only a few enzymes having turnover numbers of 10^5 per second or greater. Carbonic anhydrase, which catalyzes the hydration of CO_2 to form carbonic acid, has a turnover number of about 10^6 per second. As is discussed in Section 4.6, carbonic anhydrase is one of a small group of enzymes that achieve or approach the theoretical limits for efficiency of enzyme catalysis. The turnover number, which is simply k_2 for an enzymic reaction described by Equation 4.4, is a more complex derivative of kinetic constants for multisubstrate, multiproduct enzymes.

K_m, as defined by Equation 4.7 for a single-substrate reaction, is a ratio of rate constants. For most enzymes, which do not have very large turnover numbers, k_2 will be smaller than k_1 or k_{-1}. Under this condition, K_m will be approximately equal to $K_s \equiv k_{-1}/k_1$, the equilibrium constant for the dissociation of ES into E and S. Thus, small values of K_s indicate strong affinity of the enzyme for the substrate, that is, formation of the ES complex is thermodynamically favored. Typical K_m values for the natural substrates of enzymes are in the range of 10^{-2} to $10^{-7}M$. When [S] is only a small fraction of K_m, the rate of the enzymically catalyzed reaction will, of course, be very small, as indicated in Figure 4.3, and it will be approximately proportional to [S]. Thus, an enzyme-catalyzed reaction is expected to be first-order in substrate concentration at low [S] but zero-order for large [S], as indicated in Figure 4.3.

In metabolic pathways, several enzymes may participate in the synthesis of a final product, each enzyme acting on the product of the previous enzyme. In such a sequence of reactions, one reaction, the rate-limiting reaction, may control the rate of synthesis of the final product. The limitation may be due to the K_m or V_{max}, or both, of one reaction. Consider the sequence that follows:

$$A \xrightarrow[E_a]{} B \xrightarrow[E_b]{} C \xrightarrow[E_c]{} D$$

K_m: 10^{-2} M 10^{-4} M 10^{-2} M

If the concentration of A in the cell is 10^{-4} M, the enzyme E_a will be only fractionally saturated. E_b most likely will have a greater affinity for its substrate B because of the smaller K_m value. The result will be that as B is produced, it will be consumed rapidly by its conversion to C in a reaction catalyzed by enzyme E_b. Thus, the A $\rightarrow$ B reaction very likely will be limiting in the A $\rightarrow$ D series of reactions. A similar argument can be made for a series of reactions catalyzed by enzymes with different V_{max} values. However, be aware that K_m and V_{max} values can be influenced by pH, temperature, concentrations of ions, and other conditions that may be different in the cell and under the *in vitro* conditions used to determine K_m and V_{max}. Other tests are required to determine what reaction is rate-limiting *in vivo*.

Experimentally, the K_m can be estimated from a graph of v_0 versus [S], as in Figure 4.3, by estimating a value of V_{max} and then finding the value of [S] that corresponds to a v of $V_{max}/2$. For a simple, enzyme-catalyzed reaction, K_m is this value of [S]. However, V_{max} is an asymptotic value and therefore difficult to estimate accurately. Also, very large values of [S], for example, more than 20 times K_m, may not be attainable because of limited solubility or great expense of the substrate. The Lineweaver–Burk plot is one of several kinds of analyses that may be performed to estimate K_m and V_{max}. These analyses use all of the v versus [S] data rather than just those points that are close to V_{max} and K_m. The Lineweaver–Burk plot is based on a version of Equation 4.10 that has been modified by taking the reciprocal of both sides of the equation. The result is

$$\frac{1}{v} = \frac{K_m}{V_{max}}\left(\frac{1}{[S]}\right) + \frac{1}{V_{max}}$$

which has the form of a simple linear equation, $y = mx + b$, where $m = K_m/V_{max}$ and $b = 1/V_{max}$. Figure 4.4 shows the form of a Lineweaver-Burk plot, $1/v$ versus $1/[S]$. A weakness of analyses that use the Lineweaver-Burk plot is that the points for small values of [S] tend to most strongly influence results obtained. However, every analysis of enzyme kinetics data has its assumptions and its differential weighting of experimental points, and these must be consid-

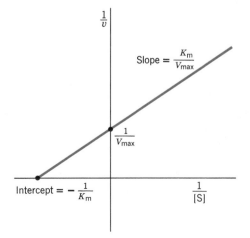

FIGURE 4.4 A Lineweaver–Burk plot, as predicted by Equation 4.11. The data points all will correspond to positive values of $1/v$ and $1/(S)$, of course, but frequently, the line is extrapolated to negative values of $1/(S)$ in order to provide an estimate of K_m.

ered in evaluating the results. In most analyses, it is important to use [S] values that are both significantly greater and significantly less than K_m. In addition, the investigator must keep in mind always that K_m and V_{max} values derived from Lineweaver-Burk plots and similar analyses will not be correct unless the experimental system and conditions correspond to the theory that underlies the analysis. For example, the rates must be true initial rates, no inhibitor must be present at an effective concentration, and so on.

Up to this point, we have emphasized studies on purified enzymes. However, new studies of an enzyme almost always begin with preparations that contain the enzyme of interest, other enzymes, and other kinds of materials. The influence of the contaminants must be considered. Fortunately, the specificity and high activity of many enzymes make it possible to determine some of the properties of the enzyme of interest even before it is purified. One of the most fundamental determinations that must be made is whether or not the reaction under study actually is enzyme-catalyzed. Since enzymes are proteins and proteins are subject to denaturation (Chapter 3), an enzyme-catalyzed reaction almost always will be sensitive to heat and other denaturing conditions. The determination of sensitivity to denaturation sometimes is referred to as the "boiled enzyme test."

A second test is based on the saturation behavior of enzyme-catalyzed reactions. At saturating [S], in the absence of inhibitors and at constant conditions of pH, temperature, and the like, the amount of product that is formed will depend linearly on the product of the time of incubation and the concentration of the impure enzyme. Thus, if the reaction is carried out at saturating [S] for various times and with various dilutions of crude enzyme solution, the quantity (time × concentration of crude enzyme) should be proportional to the amount of product formed if the reaction actually is enzyme-catalyzed. Nonenzymic reactions of S generally will not behave in this way. For example, the acid-catalyzed (nonenzymic) hydrolysis of S at constant pH will be independent of the dilution of the "enzyme" solution.

It also is possible to learn about the approximate size and shape of the active enzyme protein before it has been purified, providing that the enzyme is sufficiently stable and a sufficiently easy and accurate assay for the enzyme has been developed. Box 4.B indicates how the sedimentation rate of the enzyme in centrifugal fields may be estimated. The sedimentation rate depends, in part, on the molecular mass of the enzyme. Box 4.C describes a technique that reveals

---

BOX 4.B   SUCROSE DENSITY GRADIENT CENTRIFUGATION

Proteins and even complex assemblies of proteins that are found in cells form true solutions when they are isolated. That is, they produce uniform mixtures with aqueous solvent and will not settle out in the earth's gravitational field. However, very high-speed centrifuges, called ultracentrifuges, develop centrifugal fields that are hundreds of thousands times greater than the earth's gravitational field. If a solution of a protein is placed in a sealed tube in the rotor of such an ultracentrifuge, and the rotor is operated at speeds of tens of thousands of revolutions per minute for periods of hours, the protein may form a pellet at the periphery of the tube. At the very least, the protein solution will be more concentrated at the periphery of the tube. Thus, the ultracentrifuge can accomplish a gentle, nondenaturing concentration of a large protein or subcellular particle. However, concentration by ultracentrifugation usually is impractical for the smallest proteins.

The **sedimentation coefficient** is a characteristic constant of a molecule or particle, in the same sense that the molecular weight is a characteristic constant. The sedimentation coefficient is a function of the size, shape, and density of the molecule or particle. For a series of molecules of the same density and shape (e.g., spherical), the sedimentation increases with the mass of the molecule. A sphere is the most compact form that a solid of a given mass and density can take. Therefore, for molecules of the same density and mass, the sedimentation coefficient will decrease as the shape of the molecule deviates further and further from being spherical. Since almost all native enzymes are compact structures, as is necessary to maintain the active site and other structural features, they tend to have a shape that is nearly spherical. Since enzymes are composed principally of amino acids, different enzymes tend to have approximately the same density. For these reasons, if the sedimentation coefficients of several enzymes are compared, the order of their sedimentation coefficients corresponds closely to the order of their molecular weights. The technique of **sucrose density gradient centrifugation,** abbreviated SDGC, employs an ultracentrifuge to sediment large enzymes and other subcellular particles in a controlled fashion so that information about the molecule or particle can be obtained. If the enzyme or particle can be assayed with sufficient selectivity and sensitivity, even crude preparations can be analyzed to determine an approximate sedimentation coefficient and approximate molecular weight.

In SDGC, a "gradient" of sucrose concentration in buffered aqueous solution is prepared in a centrifuge tube, as indicated in (*a*) of the diagram. Gradients are constructed by any of several techniques that are not presented here. (*b*) shows how the sucrose concentration will vary in the tube as a function of distance for a "7 to 25% linear gradient." The sucrose concentration is 7% (= 70 mg/ml) at the meniscus and 25% at the bottom of the tube. Since sucrose solutions are more dense than water, the sucrose concentration gradient also is a solution density gradient. The sample of enzyme solution has no sucrose in it and a protein concentration that is much less than the least sucrose concentration of the gradient. The sample is pipetted as a thin layer on the gradient, as indicated by the colored area in (*a*) and (*b*). (*a*), (*b*), and (*c*) are arranged so that vertical distances on the diagram all correspond to the same distance along the long axis of the centrifuge tube.

When the centrifuge tube is exposed to a very strong centrifugal field that is parallel to its long axis, the enzyme will migrate in the centrifugal field toward the bottom of the tube. Because of the stabilizing influence of the density gradient, with its continuously increasing density toward the bottom of the tube, different species in the sample will migrate as zones. The rate of migration for each zone will be directly related to the sedimentation coefficient of the molecules in that zone. The separation of a sample that had two differently sedimenting species is shown in (*c*). The concentration contribution, and therefore the density contribution, of the proteins is small compared to that of the sucrose, which is indicated by the small sample zones that are superimposed on the stabilizing sucrose density gradient in (*c*).

The centrifugation is generally accomplished with a "swinging bucket rotor" as indicated in (*d*). The centrifuge tube with gradient and sample is placed in a bucket, shown in black, which is attached to the rotor in such a way that the tube is vertical [right-hand side of (*d*)] in the earth's gravitational field with the rotor at rest. In the centrifugal field of a spinning rotor, the bucket is horizontal, as shown on the left-hand side of (*d*). After centrifugation, the bucket returns to the vertical orientation, which avoids mixing the contents of the tube. The liquid from the tube is dripped from a hole punched in the bottom of the tube, to divide the gradient into fractions as indicated in (*e*). Fraction-by-fraction enzyme assays will reveal the position to which the enzyme migrated in the gradient during centrifugation, as indicated by the dashed line in (*c*). Protein analyses allow other protein zones to be located. Standards of known sedimentation coefficient and molecular weight can be provided. Usually, these are other enzymes applied to the same or identical gradients and centrifuged in the same rotor. The sedimentation coefficient and molecular weight of the enzyme under study can be estimated by interpolation within the range of values provided by the standards.

Density gradient centrifugation also may be carried out with other density gradient solutes in place of sucrose, such as glycerol.

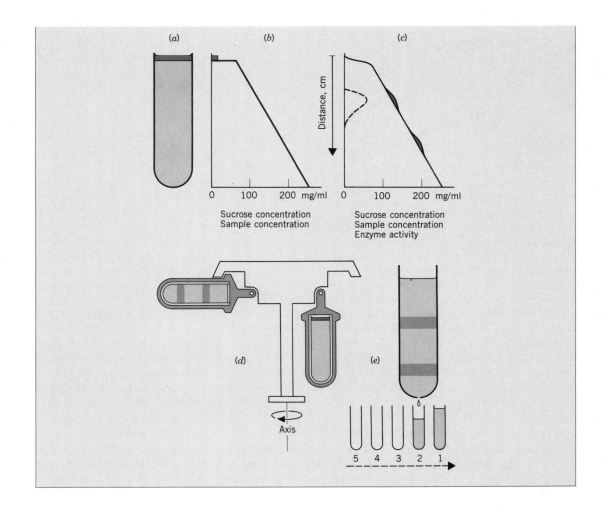

BOX 4.C   EXCLUSION CHROMATOGRAPHY

We already have discussed column chromatography in this text, for example, the use of ion exchange chromatography for the analysis of amino acids in Box 3.B. In column chromatography, we can define the total liquid volume in the column as V_t. V_t is simply the total volume of the resin bed in the column minus the volume of solid matrix of the resin. For column chromatographic procedures, such as ion exchange, that involve adsorption of solutes to the column resin, a solute will be eluted from the column only after a volume of chromatography solvent has passed through the column that exceeds, usually by several fold, V_t. In contrast, there is a type of chromatography that is called "gel permeation chromatography," "size exclusion chromatography," or simply, **exclusion chromatography.** The separations in this type of chromatography generally are achieved by passing a volume of just V_t through the column.

The resin beads of an exclusion chromatography column usually consist of a cross-linked polymer gel, such as polyacrylamide (Box 3.E), agarose (Box 6.B), dextran (polyglucose), or polystyrene (Box 3.B). However, porous glass beads and other materials also have been used. The critical characteristics of an exclusion chromatography resin are that it be porous and the pores be, on the average, of dimensions comparable to the dimensions of the molecules that are to be separated. The porous beads must be strong enough to resist compression in the flowing stream of chromatography solvent and, contrary to the requirements for adsorption chromatography, must *not* adsorb the substances that are to be resolved.

The void volume, V_0, is defined as the volume of liquid in the column that is external to the resin or other beads. A macromolecule that is too large to enter the pores of the beads will emerge from the column in an effluent volume of V_0 because only the volume of V_0 is available to such a large macromolecule. Very small molecules, such as water and salts dissolved in it, will emerge in a volume of V_t. Molecules of intermediate size will be eluted with a volume that is larger than V_0 but smaller than V_t. The algebraic representation of this notion is

$$V_e = V_0 + (V_t - V_0)K_D$$

where V_e is the elution volume of a particular macromolecule. K_D, the distribution coefficient of that macromolecule, varies between 0 and 1. K_D is the average fraction of the internal volume of the beads that is available to a particular molecule. Thus, if the molecule is so small that the entire internal volume is accessible to it, K_D will be 1 and $V_e = V_t$.

The figure presented here shows a chromatogram, which is the result of a chromatography experiment on the commercially available exclusion resin Sephadex G-75. Sephadex G-75 is composed of cross-linked dextran. The sample was an extract of pancreas gland, a rich source of hydrolytic enzymes. Enzyme assays for ribonuclease (RNase) and a proteinase were performed fraction by eluted fraction. The chromatogram shows that the two activities were nearly completely resolved. Since the RNase emerged first from the column, it has a smaller K_D value than the proteinase. Many other proteins also eluted from the column, but the specificity of the enzyme assays allows other proteins to be ignored.

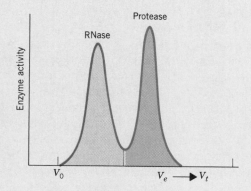

The K_D of a particular molecule depends on its dimensions, molecules with larger dimensions finding less access to the interior of the beads and therefore having a smaller K_D. For reasons stated in Box 4.B, enzymes tend to have similar conformations. Therefore, for enzymes, larger dimensions generally means greater molecular weight. The exact, theoretical relationship between molecular weight and K_D remains obscure. However, a plot of K_D versus the logarithm of the molecular weight tends to be linear for the K_D range of about 0.15 to 0.85. A comparison of the K_D values and molecular weights of standard proteins with the K_D of an enzyme, determined as in the illustration presented, allows an estimate of the molecular weight of the enzyme. An example of the type of calibration curve that might be prepared is shown next, for Sephadex G-150, a dextran resin with larger pores than those of Sephadex G-75. As is indicated in the figure, a K_D value of 0.4 corresponds to a molecular weight of 30,000 for this resin.

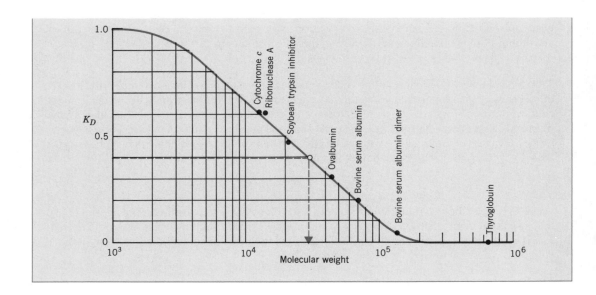

the approximate dimensions of the enzyme, again without necessarily having the enzyme in a pure state. A more complete understanding of the enzyme as protein and catalyst will require the clean, purified entity, which, in turn, requires methods for enzyme purification.

Purification of an enzyme essentially is the process of discarding contaminants, the most important of which are proteins. Thus, the specific activity often is used as a measure of how well a given purification step actually purifies. The specific activity is defined as the number of units of enzyme activity divided by the number of milligrams of protein in the same volume. This obviously requires that both the protein concentration and the number of units of enzyme activity per unit volume be determined. A unit of enzyme activity is defined in terms of the number of moles of substrate that are converted to product per unit of time. For example, an **international unit,** abbreviated **I.U.,** of enzyme activity corresponds to the amount of enzyme that causes the loss of 1 μmol of substrate per minute under specified conditions, usually a saturating concentration of substrate. Another unit of enzyme activity is the **katal,** abbreviated **kat,** defined as the amount of enzyme that causes the loss of 1 mol of substrate per second under specified conditions. Very often, other much smaller units of enzyme

TABLE 4.1   Important Terms in Enzymology

1. *Enzyme unit.* The amount of an enzyme that will catalyze the transformation of a specified amount of substrate into product under defined conditions.
2. *Specific activity.* The number of units of enzyme per unit mass of protein, usually milligrams of protein.
3. *Turnover number.* The number of moles of substrate converted to product per minute per mole of enzyme. Related quantities are the **catalytic center activity,** which is the turnover number per active site of the enzyme protein, for enzymes with more than one active site. The **molar catalytic activity** is the turnover number in the units of $\sec^{-1}$.

activity are defined because the sensitivity of current assays allows very much less than one micromole changes in the substrate amount to be determined. Table 4.1 defines some terms that are widely used in describing enzymes.

4.5 COVALENT INTERMEDIATES IN ENZYME CATALYSIS

Chymotrypsin is a proteinase and peptidase. With peptide and protein substrates, it catalyzes most efficiently the hydrolysis of peptide bonds in which the carboxyl group of aromatic amino acids participate. Thus, most of the new peptides formed in a chymotrypsin-catalyzed hydrolysis reaction have aromatic amino acid residues at the carboxyl end. Chymotrypsin also has esterase activity. An example of an ester substrate is the p-nitrophenyl ester of N-acetyltyrosine

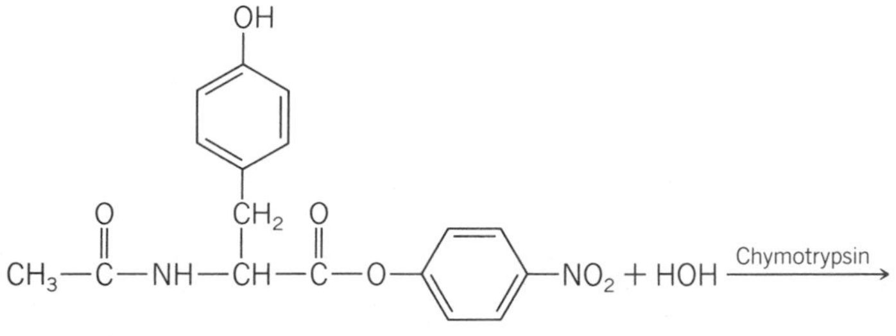

N-Acetyltyrosine-p-nitrophenyl ester

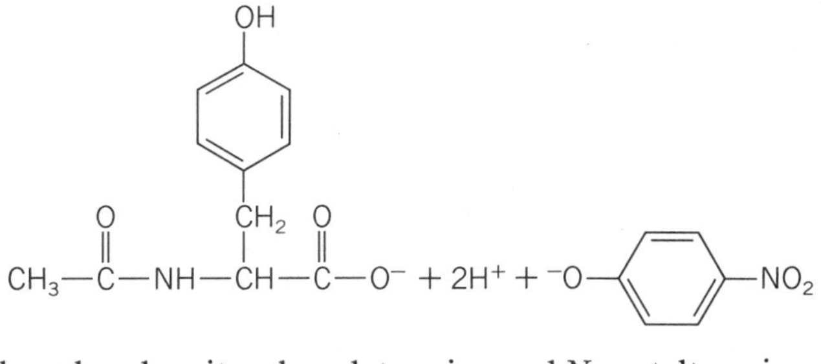

which is hydrolyzed to the colored p-nitrophenylate anion and N-acytyltyrosine by the action of chymotrypsin. The rate of the reaction can be monitored conveniently by the light absorbance due to the p-nitrophenylate anion. However, other esters, such as ethyl and methyl esters, also are substrates. Generally, the rate of hydrolysis of ester substrates by chymotrypsin is greater than the rate for a comparable amide substrate, as is expected from the relative rates of ordinary acid- and base-catalyzed hydrolysis of the two types of compounds.

The acyl group of the esters need not be aromatic for a measurable rate of reaction. The p-nitrophenyl ester of acetic acid is a substrate for chymotrypsin, although a poorer substrate than the esters of N-acetylated aromatic amino acids. Another ester, the p-nitrophenyl ester of trimethylacetic acid, reacts with chymotrypsin in an especially interesting way.

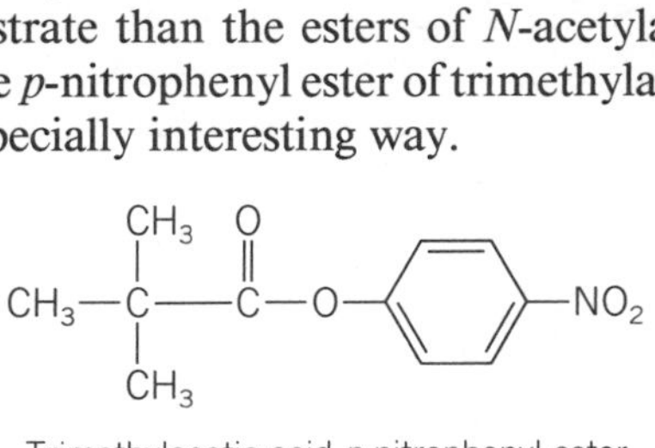

Trimethylacetic acid p-nitrophenyl ester

As in reactions of chymotrypsin with good *p*-nitrophenyl ester substrates, the *p*-nitrophenylate anion is released. However, it is released only in a 1 : 1 molar ratio to the amount of chymotrypsin, even when the ester was added in great molar excess. When the chymotrypsin was recovered from the reaction mixture, it failed to catalyze the hydrolysis of either ester or amide substrates. That is, it had been inactivated. Since the *p*-nitrophenylate anion but not trimethylacetate had been released in the reaction, acylation of the chymotrypsin was suspected. This suspicion was confirmed by treatment of the modified chymotrypsin with the strong nucleophile hydroxylamine, NH_2OH, that released the hydroxamic acid derivative of trimethyacetic acid.

$$CH_3-\underset{\underset{CH_3}{|}}{\overset{\overset{CH_3}{|}}{C}}-\overset{\overset{O}{\|}}{C}-NH-OH$$

Trimethylacetic acid hydroxamate

The treated chymotrypsin was again active as a proteinase and esterase, showing that whatever site had been acylated in the modified enzyme is vital to chymotrypsin function. Various chemical analyses revealed that the site of acylation is the serine residue at position 195 in the chymotrypsin polypeptide chain. X-ray crystallography of chymotrypsin that had been inactivated by reaction of the trimethylacetic acid *p*-nitrophenyl ester also showed that the trimethylacetic acid group is bound to serine 195, as an ester. This acylation is not just an artefact of a particular ester substrate, but rather is an essential step in all chymotrypsin catalyzed hydrolyses, as indicated by the scheme in Figure 4.5. Free chymotrypsin, (*a*) in Figure 4.5, is shown reacting with a substituted amide to form the enzyme-substrate complex, (*b*). Formation of the transition state complex, (*c*), leads to release of the amine portion of the substrate and generation of the acyl-enzyme intermediate, (*d*). Reaction of the acyl-enzyme with water generates the transition-state complex for the second segment of the reaction, (*e*). The enzyme-product complex (*f*) then dissociates to give the carboxylic acid and free chymotrypsin (*g*). The active serine, Ser—195 of chymotrypsin, is represented by —CH_2—OH. The proton-binding "pocket" shown in the upper left-hand quadrant of each chymotrypsin "icon" represents a protonatable histidine residue, His—57 that also is essential for chymotrypsin action. Other binding sites are shown for the carbonyl oxygen of the substrate carboxyl group and the aromatic or hydrophobic aliphatic side chain, R. The reactions shown are the same for an ester substrate except that R′—OH replaces R′—NH_2, and so on.

Apparently, when the trimethylacetic acid ester is the substrate, the acyl-intermediate (Figure 4.5*d*) that is formed is only very slowly hydrolyzed because of the abnormal R group presented by the trimethylacetic acid. Hydroxylamine is such a powerful nucleophile that it is able to displace the trimethylacetyl group from the serine 195 amino acid residue of chymotrypsin, restoring the enzyme to its original active form.

The scheme in Figure 4.5*c* and *e* shows a tetrahedral arrangement of bonds about the carbonyl carbon of the ester or amide substrate as a part of a structure that is or is like the transition-state complex. That a tetrahedral intermediate is

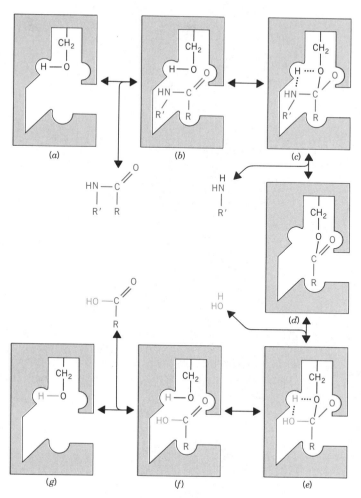

FIGURE 4.5 A covalent intermediate in the catalytic mechanism of chymotrypsin. Only a few of the documented steps in the chymotrypsin reaction are shown, and these are described in the text.

important in the chymotrypsin reaction is indicated by the ability of aldehydes such as

to strongly inhibit chymotrypsin. This aldehyde is an analogue of the chymotrypsin amide substrate, *N*-acetyl-prolyl-alanyl-prolyl-tyrosine amide.

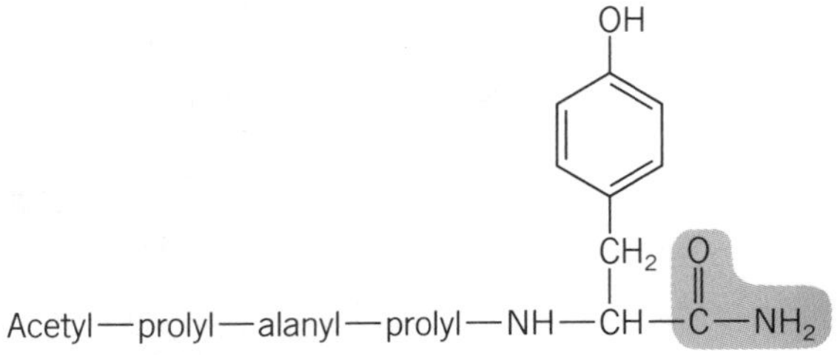

OH

CH$_2$

Acetyl—prolyl—alanyl—prolyl—NH—CH—C—NH$_2$

This is hydrolyzed to ammonia and the corresponding acid, *N*-acetyl-prolyl-alanyl-prolyl-tyrosine, by the catalytic action of chymotrypsin. The corresponding alcohol

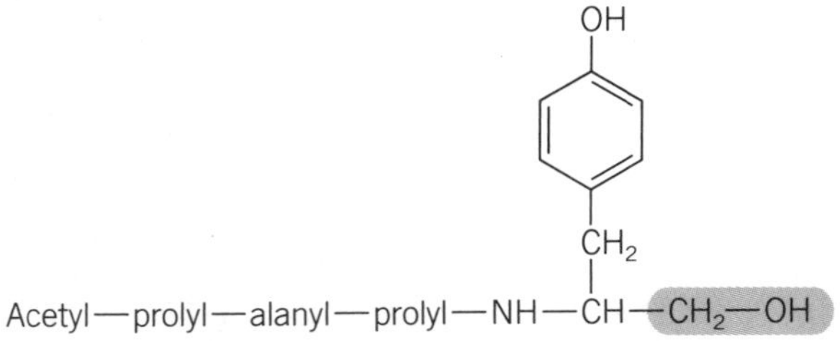

OH

CH$_2$

Acetyl—prolyl—alanyl—prolyl—NH—CH—CH$_2$—OH

is neither a substrate nor a good inhibitor of chymotrypsin. Presumably, the aldehyde *is* a good inhibitor because in its hydrated form, which is expected to be the dominant form in aqueous solution, it has the "carbonyl" carbon in the tetrahedral form. The structure of the hydrated aldehyde is

OH

CH$_2$ OH

Acetyl—prolyl—alanyl—prolyl—NH—CH—CH

OH

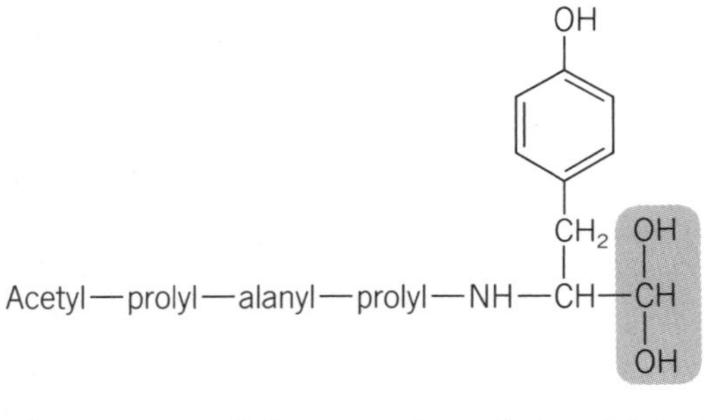

The postulated structure of the complex of the aldehyde inhibitor with chymotrypsin is presented diagramatically in Figure 4.6. A corollary of the notion of Pauling that is quoted in Section 4.2 is that an enzyme should bind an analogue of the activated transition-state complex more tightly than it binds a substrate. The strong inhibition of chymotrypsin by the hydrated aldehyde is in agreement with this corollary.

The corollary also is in agreement with the results from several enzyme systems in which series of homologous substrates of a given hydrolase are compared. The substrates may be chosen to have approximately the same rate constants for nonenzymic hydrolysis but different expected affinities for the hydrolase enzyme, based on substrate structure. Surprisingly, such substrates show similar K_m values, but the "high affinity" substrates typically are enzymically hydrolyzed more rapidly. The implication of these results is that the supposed increased binding of the "good" substrates did not influence the K_m values, the apparent affinity of enzyme for substrate. Rather, the enzyme was able to use the binding energy to distort the ES complex closer to the conforma-

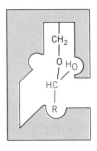

FIGURE 4.6 The postulated complex between a hydrated aldehyde inhibitor and chymotrypsin. Chymotrypsin is represented diagrammatically as in Figure 4.5. The complex is tetrahedral in its configuration about the carbonyl carbon. Thus, it is expected to resemble the transition-state complex, but without the amino or alcohol group of a usual chymotrypsin substrate.

tion of the activated complex. The resulting increase in the population of the activated complexes will, of course, accelerate the reaction, even as the value of k_2 remains approximately constant.

We have selected chymotrypsin for discussion in this section because it is an intensely studied enzyme and hence is a good example of how an enzyme may be studied using the experimental methods and concepts of organic and physical organic chemistry. Such methods have been applied to many enzyme systems.

4.6 REACTION RATES AND REACTION EQUILIBRIUM CONSTANTS

Catalytic amounts of enzyme do not alter the equilibrium constant of the catalyzed reaction. However, the catalytic constants for an enzyme-catalyzed reaction and the equilibrium constant for that reaction are interrelated. The interrelationship places limits on the velocities of enzyme-catalyzed reactions and the ranges of concentrations of substrates and products in cells.

J. B. S. Haldane derived a relationship between the constants of the Briggs–Haldane equation (4.10) and the reaction equilibrium constant for a single substrate, single product, reversible reaction. The derivation begins with an equation of the form of Equation 4.3. We do not present the derivation here but only the result, the Haldane relationship.

$$K_{eq} = \frac{V_{max(s)} K_{m(p)}}{V_{max(p)} K_{m(s)}} \qquad (4.12)$$

The kinetic constants for both the forward and the backward reactions are represented in the Haldane relationship. $K_{m(s)}$ is the Michaelis constant for the substrate. $K_{m(p)}$ is the Michaelis constant for the product. $V_{max(s)}$ is the maximum velocity in the forward direction, whereas $V_{max(p)}$ is the maximum velocity in the reverse direction, Equation 4.3. The Briggs–Haldane equations for the forward and reverse directions, respectively, are

$$v_{0(f)} = \frac{V_{max(s)}[S]}{[S] + K_{m(s)}}$$

and

$$v_{0(r)} = \frac{V_{max(p)}[S]}{[S] + K_{m(p)}}$$

where the subscripts f and r indicate the forward and reverse reactions, respectively. $K_{m(s)}$ is defined by Equation 4.7. $K_{m(p)}$ is defined by

$$K_{m(p)} = \frac{k_2 + k_{-1}}{k_{-2}}$$

Given the fact that the equilibrium constant is fixed, it is clear from the Haldane relationship and algebra that an increase in $V_{max(f)}$ will require either an increase in $V_{max(r)}$ or $K_{m(s)}$, or it will require a decrease in $K_{m(p)}$, or some combination of these. That is, if the enzyme is modified in its structure to increase the velocity of the forward reaction at a given substrate concentration, that change will have its consequences on other kinetic properties of the enzyme. Those consequences are that the velocity in the reverse direction will be increased, or the Michaelis constants for one or both directions will be altered, or there will be some combination of these changes. The effect will be that low substrate concentrations or high product concentrations will more adversely affect the net forward reaction than they would if the V_{max} in the forward direction were smaller.

Thus, the Haldane relationship must place limitations on the evolution of enzymes, and cells would not necessarily be given a selective advantage by the development of enzymes that can catalyze reactions at the maximum possible forward velocity. Most likely, enzymes are responsible not only for catalysis but also, through their K_m values, for maintaining substrates and products at specific concentrations. The Haldane relationship shows that an enzyme cannot be independently optimized for both K_m and V_{max} in both forward and reverse directions. More complex Haldane relationships have been derived for multiple-substrate, multiple-product reactions, with similar conclusions. So the discussion here can be generalized to complex enzyme-catalyzed reactions. In addition, the response of the enzyme to changes in substrate concentration may be steeper than predicted by Equation 4.10 if it is capable of allosteric activation by substrate. As is indicated in Section 4.12, such an enzyme should allow very close control over substrate concentration *in vivo*.

Certain enzymes do seem to have been optimized for catalysis. Examples are carbonic anhydrase, already mentioned in Section 4.4, triose-phosphate isomerase, and fumarase. The last two catalyze the interconversion of glyceraldehyde phosphate and dihydroxyacetone phosphate and the hydration of fumaric acid, respectively. The rates at which these enzymes catalyze their reactions correspond to the rates at which the respective substrates can diffuse through aqueous medium to the enzyme surface. Since there is no mechanism by which the enzyme can change the rate of diffusion, further increases in catalytic efficiency will not change the velocity of the reaction.

4.7 CLASSIFICATION OF ENZYMES

As we have already mentioned, one important characteristic of an enzyme is its substrate specificity; that is, because of the conformation of the complex protein molecule, the uniqueness of its active site, and the structural configuration of the substrate molecule, an enzyme often will select only a specific compound or a set of related compounds for attack.

An enzyme will usually exhibit **group specificity**; that is, a general group of

TABLE 4.2 Classification of Enzymes

1. *Oxidoreductases.* These enzymes catalyze reactions in which one substrate is oxidized, acting as a hydrogen donor, and another substrate is reduced. The oxidoreductases include the dehydrogenases, which convert single bonds to double bonds, and the oxidases, which use oxygen as the oxidant. Others in this class are the peroxidases, which use H_2O_2 as the oxidant, the hydroxylases, which introduce hydroxyl groups, and the oxygenases, which introduce molecular oxygen in place of a double bond in the substrate.

2. *Transferases.* Enzymes in this class transfer one-carbon groups (e.g., methyl), aldehydic or ketonic groups, or phosphoryl groups or amino groups, and so on from one substrate to another.

3. *Hydrolases.* Hydrolases catalyze the hydrolytic (water-adding) cleavage of C—O, C—N, C—C, P—O, and other single bonds. The class includes peptidases, esterases, glycosidases, phosphatases, and the like.

4. *Lyases.* These enzymes cleave bonds without the addition of water, but by elimination reactions to form double bonds or rings. Also in this class are enzymes that catalyze the reverse reaction, that is, adding groups across double bonds or to rings. Usually, C=C, C=O, or C=N bonds are acted on, and examples of enzymes in the class are decarboxylases, aldolases, and dehydratases.

5. *Isomerases.* Racemases, epimerases *cis-trans* isomerases, intramolecular oxidoreductases, mutases, and intramolecular transferases form this class, which alters the structure but not the atomic composition of substrates by moving a group from one position to another in one molecule.

6. *Ligases.* Enzymes in this class, also known as the synthetases, couple the hydrolysis of a pyrophosphate in ATP or other nucleoside triphosphate to a second reaction in which two molecules are joined. For example, an RNA ligase will form a new phosphodiester bond (Chapter 6) as it joins two RNA fragments, a reaction that also results in the hydrolysis of ATP.

compounds may serve as substrates. Thus, a series of aldohexoses may be phosphorylated by a kinase and ATP. If the enzyme will use only one substrate for phosphorylation (e.g., glucose but no other monosaccharide), it is said to have an **absolute group specificity.** Such an enzyme is referred to as a "glucokinase." Another kinase may have a **relative group specificity** if it catalyzes the phosphorylation of two or more aldohexoses. Such an enzyme might be referred to as a "hexokinase."

On a broader scale, enzymes may be classified according to the type of reaction that they catalyze. An inspection of Table 4.2, which classifies enzymes into six major groups, demonstrates their versatility.

4.8 ANALYTICAL AND PREPARATIVE USES OF ENZYMES

The specificity of enzymes, and their ability to catalyze reactions that otherwise would be immeasurably slow, allow quantitative analyses of individual substances in biological fluids such as blood and in other complex mixtures. Successful assays have been based on measurements of the rate of reaction, others on complete conversion of a substrate to a product that can be quantitated. For example, D-glucose is oxidized to the lactone of gluconic acid (Section 2.5.3) in the presence of oxygen and D-glucose oxidase according to the following equation:

$$\text{D-Glucose} + O_2 \xrightarrow{\text{D-Glucose oxidase}} H_2O_2 + \text{Gluconolactone} \qquad (4.13)$$

The extent of this reaction can be assessed in a number of ways. For example, if the enzyme peroxidase and a peroxidase substrate are included in the reaction mixture, the peroxidase substrate will be oxidized. Certain peroxidase substrates are complex organic compounds that will act as a chromogen, that is, a weakly or noncolored, color-producing substance.

$$\text{Chromogen} + H_2O_2 \xrightarrow{\text{Peroxidase}} H_2O + \tfrac{1}{2}O_2 + \text{Dye}$$

With the proper choice of conditions, including enzyme concentrations, oxygen concentration, and chromogen, the amount of dye produced will be determined by the amount of D-glucose present in the solution. The concentration of the dye is determined by spectrophotometry, which depends on the dye's ability to absorb light in the visible range.

The assay just described uses coupled enzyme reactions to produce an amount of an easily measured compound, the dye, that is directly proportional to the amount of the substance that is to be determined, D-glucose in this case. Because of the simplicity of this and many other clinically useful, enzyme-catalyzed reactions, they have been automated in many instances for very efficient and routine analyses of biological samples. In automated systems, computer-controlled machines transfer and mix solutions, make timed spectrophotometric, electrochemical, and other measurements, record results, and make the final calculations. Automated systems offer improved accuracy over the corresponding manual assays because of the consistency with which mixing, transfers, and the like are made in the automated system. In addition, the "stamina," efficiency, and rapidity of the machines allow many standards, for example, of D-glucose, to be tested, so that the reliability of the system is tested frequently.

Enzyme-catalyzed reactions even allow the continuous monitoring of some biologically important compounds. A "glucose electrode" has been developed, based on the D-glucose oxidase reaction (Equation 4.13). Box 4.D describes the construction of such an electrode, which is able to produce an electrical current that is related to the concentration of D-glucose in the solution in which it is immersed. One important use of such an electrode, should it be possible to manufacture it in a sufficiently small, reliable, and long-lived form, would be to

BOX 4.D   A GLUCOSE ELECTRODE

Combining the specificity of enzymes with the continuous monitoring capabilities of electrochemical electrodes can be the basis for powerful new instruments.

A Clark electrode allows the continuous measurement of the concentration of dissolved oxygen in a solution. The Clark electrode has within it two electrodes, one of platinum and one silver-coated with silver chloride. If 0.7 V, with the platinum electrode negative, is applied to this pair of electrodes and they are placed in contact with a solution that contains chloride ions, oxygen will be reduced at the platinum electrode, and silver will be oxidized to silver chloride at the silver–silver chloride electrode. At 0.7 V, protons and most other ions are not reduced. Teflon (polytetrafluoroethylene) is a plastic that is permeable to oxygen but not ions or water. Therefore, a Teflon membrane is used in the Clark electrode to separate a thin layer of a chloride salt solution from the oxygen-containing solution that is to be analyzed. If, in the diagram that follows, the barrier membrane and the layer that is labeled "GO + C" were removed, the structure would be that of a Clark electrode.

The efficient reduction of oxygen at the platinum electrode surface causes the oxygen concentration there to be zero. Nevertheless, only a negligible decrease in oxygen concentration occurs in the solution outside of the Teflon membrane. The Teflon membrane provides resistance to oxygen flow, so that very little of the oxygen in the solution outside the Teflon membrane actually is reduced. However, what is reduced causes a current to flow from the 0.7-V source, through the electrodes and solution. This small current is proportional to the rate of oxygen reduction. The rate of oxygen reduction is, in turn, proportional to the oxygen concentration outside the Teflon membrane. It is the difference between the oxygen concentration in the solution and at the electrode surface (zero) that establishes the gradient of oxygen concentration in the membrane and, hence, the rate at which oxygen flows through the membrane. Thus, a properly calibrated measurement of the current is, in effect, a measurement of the oxygen concentration in the solution.

The Clark electrode is incorporated into one form of "glucose electrode" by the addition of two layers to the Teflon membrane, as indicated in the figure. A gel that contains glucose oxidase and the enzyme catalase (GO + C) is applied directly to the Teflon membrane and is separated from the solution by a membrane that is permeable to low molecular weight substances such as glucose, water, and oxygen but not permeable to proteins and other substances that might foul the electrode. According to Equation 4.13, glucose and oxygen that enter the GO + C layer will react to form gluconolactone and hydrogen peroxide. Catalase catalyzes the conversion of hydrogen peroxide to water and oxygen. The net result is that one-half mole of oxygen is consumed per mole of glucose oxidized. The greater the concentration of glucose, the more effectively the oxygen will be removed. Hence, increasing concentrations of glucose decrease the current in the Clark electrode, which is measured by electronic circuitry (indicated by the meter face in the figure). If we use selected materials and the proper dimensions of membrane and enzyme gel layer, it is possible to construct a glucose electrode that responds efficiently to glucose concentrations that are in the normal range for human blood, 0.65 to 1 g/l. The glucose concentration can be estimated, even in solutions of varying oxygen concentration, by comparing the current produced in two similar but not identical assemblies like those shown in the figure; the second control assembly lacks glucose oxidase and serves as a control.

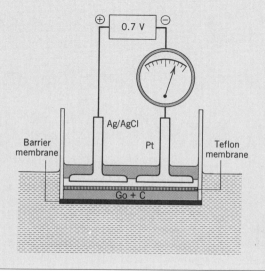

BOX 4.E   ENZYMES IN DETECTION SYSTEMS

The catalytic powers of enzymes allow their use in sensitive detection systems for either qualitative or quantitative assays of molecules that are neither substrates nor products. This is accomplished through the coupling of enzymes to antibodies (Figure 3.17) or other specific binding proteins. We present two examples of such detection systems here. Both use immobilized proteins, but for analytical applications rather than the preparative applications described in Box 4.A. In these systems, advantage is taken of the simple purification steps, by washing alone, that can be accomplished when molecules are bound to a solid substrate.

An enzyme-linked immunosorbent assay, or ELISA, can detect and quantitate nanogram amounts of a protein or other antigen in a complex mixture. There are several variations of the ELISA assay, only one of which is described here. The assay usually is carried out in wells of a few millimeters diameter that are molded into plastic trays, usually 96 wells per tray. Two such wells are shown diagrammatically here, already coated with IgG antibodies against the antigen of interest. (See Figure 3.19 and Section 3.14.4 for IgG structure and function.) The wells can be washed extensively without removing the antibody molecules, which have their antigen-binding sites still available, as indicated in (*a*) of the illustration. In (*b*), the wells are filled with known dilutions—for example, a series of two-fold dilutions, of a solution that contains an unknown amount of the antigen. The antigen is indicated by the stars. After a period of incubation to allow binding of antigen to immobilized IgG molecules, the wells are washed to remove other substances [dots in (*b*)] that were in the solutions of antigen, leaving only antigen-IgG complexes as shown in (*c*).

To the wells is added a solution of the same IgG that was attached to the well walls in (*a*), except that the IgG has been modified. The IgG molecules have been chemically coupled to an enzyme, indicated by "E." The IgG-enzyme conjugates are allowed to bind to determinants of the antigen that

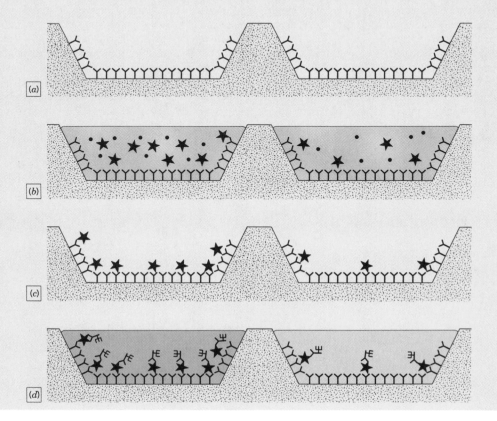

were not already occupied by their binding to the well-bound IgG molecules. This will occur only in those wells that received a sufficient amount of antigen in the first step, as indicated in (*b*). Unbound IgG-enzyme complex is removed by washing. In the last step (*d*), a solution of a chromogenic substrate is added. For example, if a phosphatase is the enzyme that was coupled to the IgG, *p*-nitrophenyl-phosphate is a suitable substrate. *p*-Nitrophenyl-phosphate is not colored, but on hydrolysis, it gives the colored *p*-nitrophenylate anion. The plate is incubated to allow the enzyme-catalyzed reaction to proceed. The reaction is quenched with a reagent that denatures the enzyme, and the extent of product formation is determined spectrophotometrically, usually with a device that measures light absorbance directly in the well of the ELISA plate.

When the relative amounts of the different reagents used in the ELISA assay have been adjusted properly, color development is proportional to the amount of antigen over a certain antigen concentration range. For example, in the last step, the substrate concentration must be saturating so that the rate of the reaction will be proportional to the amount of IgG-bound enzyme that is present and will be independent of [S]. The amplification that is achieved by the conversion of many moles of substrate to product by one mole of enzyme allows a sensitivity to be achieved with the ELISA assay that rivals the sensitivity of assays based on radioisotopes.

In the second type of enzymic-detection assay, proteins are separated by gel electrophoresis, under denaturing or nondenaturing conditions, as described in Box 3.F. After electrophoresis, the zones are transferred to a nitrocellulose sheet to give a "contact print" of zones from the gel. The transfer may be accomplished by electrophoresis perpendicular to the gel sheet or "blotting" as described for nucleic acid zones in Box 6.G. The process of blotting removes denaturants, so even proteins that were separated under denaturing conditions tend to refold to the extent that they often can be recognized by specific antibodies. (*a*) of the figure at the end of this box represents diagrammatically a nitrocellulose sheet in cross section with protein molecules from two different zones blotted to the surface of the nitrocellulose, although in reality, the proteins are bound within the sheet as well as on the surface. In (*b*) the sheet has been incubated with specific antibody. The IgG molecules, which are specific for the antigen of interest, bound to the zone 2 protein but not to the zone 1 protein.

A second, enzyme-linked antibody provides qualitative identification of the detected zone 2 protein, as is indicated in (*c*). For example, if the first antibody, which was used in (*b*), was rabbit IgG, the second, enzyme-linked antibody might be IgG from goats that had been injected with rabbit IgG.

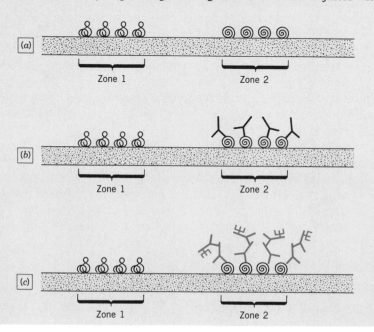

This enzyme-conjugated "goat antirabbit IgG" thus is a general agent that can be used with any of several rabbit antibodies, reducing the number of IgG-enzyme conjugates that must be prepared. To detect the protein zone that the rabbit IgG had recognized, the nitrocellulose sheet is soaked in a solution of a chromogenic substrate. The product of the reaction must be not only colored but also insoluble so that the colored zone does not spread but remains at the location of the protein zone that is being detected.

monitor continuously glucose concentration in the blood of diabetics (Type I diabetes, no insulin produced) and perhaps even to control insulin administration by an "artificial pancreas."

Since one enzyme molecule can convert many molecules of substrate the product, very small amounts of an enzyme can be detected, given a sufficiently sensitive test for the product. Currently, many sensitive qualitative and quantitative assays for biologically important substances rely on reagents that are chemically coupled complexes of enzyme and another protein, a protein which is capable of binding to the substance of interest. Two such analyses are described in Box 4.E. The chemical coupling of enzymes to glass beads and other insoluble matrices is the basis of both analytical and preparative procedures in the laboratory. Even industrial scale production can be accomplished with such **immobilized enzymes,** as is described in Box 4.A.

4.9   KINETICS OF MULTISUBSTRATE REACTIONS

Many of the most intensely studied enzymes, such as chymotrypsin, effectively have only one substrate. There is a second reactant, but it is water, which remains at a constant concentration in the reaction. However, among all enzymes single-substrate enzymes are a rarity. Here, we describe some of the mathematical formalisms that are used to analyze the kinetic behavior of multisubstrate enzymes. Kineticists recognize three general mechanisms for multisubstrate enzyme systems. Two of these mechanisms, termed **ordered** and **random,** require that all substrates must be added to the enzyme before any products can be released. In the third mechanism, called **ping pong,** one or more products is released from the enzyme before all the substates have reacted with the enzyme.

A short-hand notation simplifies the description of multisubstrate, enzyme-catalyzed reactions. All substrates, if not specifically stated, are designated as A, B, C, and D. All products are designated P, Q, R. Different forms of the enzyme are represented by E, F, G. E is the form of the enzyme that is free or at least most nearly free of any form of the substrates or products. The short-hand notation is applied to each of the following three general mechanisms.

4.9.1   ORDERED MECHANISM

In this general mechanism, there is a precise order in which substrates associate with the active sites of the enzyme. All of the products are released in a specific order and only after all of the substrates have reacted.

The reaction

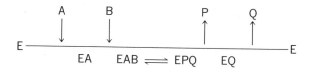

$$E + A \rightleftharpoons EA \xrightarrow{\quad B \quad} EAB \rightleftharpoons EPQ \xrightarrow{\quad P \quad} EQ \rightleftharpoons E + Q$$

shows A as the first substrate to react with the enzyme E. Only then can B react to form the complex EAB. As catalysis occurs, first P and then Q are released, and in that order. The short-hand notation of this reaction is

$$
\begin{array}{ccccccc}
& A & B & & P & Q & \\
& \downarrow & \downarrow & & \uparrow & \uparrow & \\
E & & & & & & E \\
& EA & EAB \rightleftharpoons EPQ & EQ & &
\end{array}
$$

This specific type of reaction is referred to as having an **ordered Bi, Bi mechanism.** The term **Bi** indicates two substrates or two products, so an ordered Bi, Bi mechanism requires two substrates and two products. **Uni** indicates a single substrate or a single product, and **Ter** three substrates or three products. A good example of an ordered Bi, Bi mechanism is that catalyzed by alcohol dehydrogenase. The two substrates are ethanol and NAD^+, and the two products are acetaldehyde and NADH. NAD^+ and NADH are nucleotide coenzymes that are described in Chapter 5. The proton that is produced in the reaction can be ignored if the reaction is carried out in a buffered solution, that is, constant H^+ concentration.

$$CH_3CH_2OH + NAD^+ \underset{\text{Alcohol dehydrogenase}}{\rightleftharpoons} CH_3CHO + NADH + H^+ \quad (4.14)$$

Experimental kinetic analyses of alcohol dehydrogenase revealed that the mechanism is ordered Bi, Bi, with the specific shorthand notation of

$$
\begin{array}{ccccc}
NAD^+ & CH_3CH_2OH & & CH_3CHO & NADH + H^+ \\
\downarrow & \downarrow & & \uparrow & \uparrow \\
E & & & & E \quad (4.15) \\
& E \cdot NAD^+ & (E \cdot NAD^+ \cdot CH_3CH_2OH \rightleftharpoons & & \\
& & ENADH \cdot CH_3CHO) & &
\end{array}
$$

4.9.2 RANDOM MECHANISM

When two substrates A and B add to an enzyme in random order and two products P and Q are released in a random order, such a sequence is designated as a random Bi, Bi mechanism.

Thus, a general reaction would be

$$
\begin{array}{ccc}
E + A \rightleftharpoons EA \searrow^{B} & & Q \diagup^{EP \rightleftharpoons E + P} \\
& EAB \rightleftharpoons EPQ & \\
E + B \rightleftharpoons EB \nearrow_{A} & & P \searrow^{EQ \rightleftharpoons E + Q}
\end{array}
\quad (4.16)
$$

The short-hand notation is

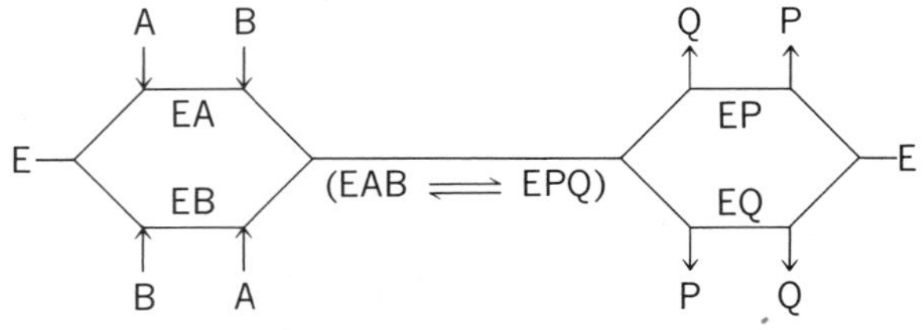

$$(4.17)$$

The random Bi, Bi mechanism describes the reaction that catalyzed by glycogen phosphorylase. In the equation here, inorganic phosphate is abbreviated Pi, and the product glycogen is, of course, one glucose unit smaller than the substrate glycogen.

$$\text{Glycogen} + \text{Pi} \underset{\text{glycogen phosphorylase}}{\rightleftharpoons} \text{glucose-1-phosphate} + \text{glycogen}$$

$$\text{Glycogen} + \text{Pi} \underset{\text{Phosphorylase}}{\rightleftharpoons} \text{Glucose-1-Phosphate} + \text{Glycogen}$$

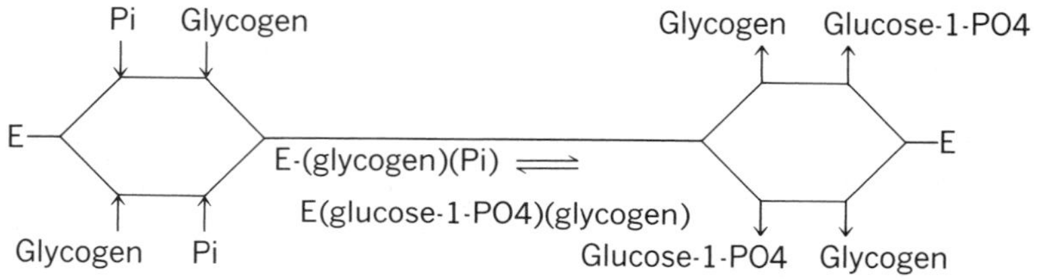

$$(4.18)$$

4.9.3   PING PONG MECHANISM

For two substrates and two products, the general mechanism is depicted as

$$\text{E} + \text{A} \rightleftharpoons \text{EA} \rightleftharpoons \text{EP} \longrightarrow \text{F} \longrightarrow \text{FB} \rightleftharpoons \text{EQ} \rightleftharpoons \text{EQ} + \text{Q} \quad (4.19)$$

In this sequence, the enzyme complexes formed are EA, FP, FB, and EQ with A being first converted to P and then B to Q. F designates a modified enzyme (i.e., X-enzyme where X might be a phosphorylated, carboxylated, or other functional group attached to the enzyme transiently). F combines with B, and subsequently, X is transferred to B to form the second product Q and regenerate the original form of the enzyme, E. The short-hand notation for a Bi, Bi ping pong mechanism is

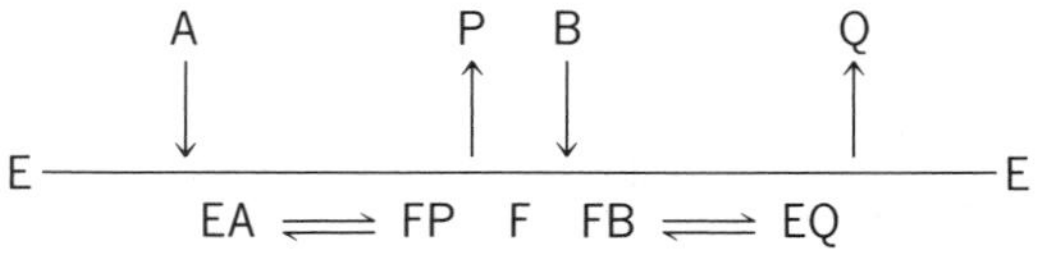

$$(4.20)$$

An example of a reaction of the ping pong type is that catalyzed by rat liver acetyl CoA carboxylase.

$$\text{Acetyl CoA} + \text{ATP} + \text{HCO}_3^- \longrightarrow \text{Malonyl CoA} + \text{ADP} + \text{Pi}$$

This is designated as a Bi, Bi, uni, uni, ping pong mechanism because first two substrates add to the enzyme, then two products are released, then another substrate adds and the final product is released.

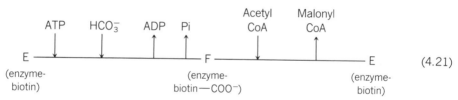

$$E \longrightarrow F \longrightarrow E \qquad (4.21)$$

(enzyme-biotin)   (enzyme-biotin—COO⁻)   (enzyme-biotin)

Acetyl CoA carboxylase catalyzes a coupled reaction of the type already mentioned in Section 4.1. That is, it mediates the energetically unfavorable formation of a carbon-carbon bond by coupling the reaction to the structurally unrelated but energetically favorable hydrolysis reaction of ATP to ADP and inorganic phosphate.

To determine the order of the additions of substrates and products in a multisubstrate, multiproduct enzyme system generally requires a variety of experiments, including detailed kinetic analysis of the reaction rates with all but one of the substrates and products set at fixed concentrations while one substrate or product is varied. The equilibrium constants for the binding of substrates and cofactors alone and in the presence of others, product inhibition kinetics (Section 4.11), and other measurements all can contribute to determining the type of mechanism. References to these methods and their application are given at the end of this chapter.

4.10 EFFECTS OF TEMPERATURE AND pH

A chemical transformation A → B, as illustrated by Figure 4.1, involves the random activation of molecules in the A population to a specific, high-energy conformation that is designated the transition state. Those molecules that are in the transition-state conformation will, at a relatively fixed frequency, undergo the transformation to product, B. Thus, the rate of the reaction will be proportional to the concentration of the transition-state species. The concentration of the transition-state species, in turn, depends on the amount of thermal energy required to produce the transition-state species of the reacting molecules. As is explained in Section 4.2, in an enzyme-catalyzed reaction, the activation energy is less than in the corresponding uncatalyzed reaction. The transition state level is more readily attained with a result that more molecules enter the transition state and form product B. The enzyme, of course, does not alter the ΔG for the reaction but only reduces the activation energy that molecule A must attain before it can undergo change.

The familiar Arrhenius equation relates the specific reaction rate constant, k, to temperature

$$\log k = \log A - E_A/2.3\ RT \qquad (4.22)$$

where A is a proportionality constant, E_A is the activation energy, R is the gas constant, and T the absolute temperature. The equation predicts that the rate of the reaction, be it enzymically catalyzed or not, will be increased with increasing temperature. However, since enzymes are proteins and any protein will be denatured if the temperature is raised sufficiently, enzyme-catalyzed reactions show an increase in rate with increasing temperature only within a relatively small and low temperature range. The combined effects of temperature on the enzyme-catalyzed reaction and the enzyme protein are indicated in Figure 4.7.

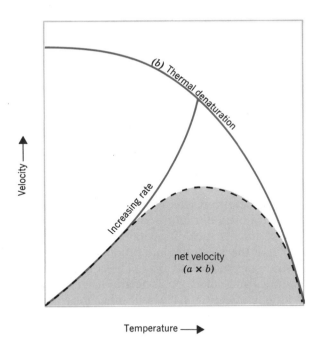

FIGURE 4.7   Effect of temperature on the reaction rate of an enzyme-catalyzed reaction. (*a*) represents the increasing rate due to increasingly populated transition state with additional thermal energy. (*b*) represents the effect of thermal denaturation on the fraction of the original enzyme molecules that remain active. The dashed curve represents the combination, *a* × *b*, the net velocity of the reaction.

The optimum temperature of the enzyme-catalyzed reaction will depend on several factors, including how long the enzyme is incubated at the test temperature before the substrate is added and the type of organism from which the enzyme was derived. Longer incubations before addition of substrate will allow more time for inactivation of the enzyme protein; enzymes from thermophillic organisms generally will be much more stable at elevated temperature than enzymes from other organisms.

Changes in pH will profoundly affect the degree of ionization of the amino, carboxyl, and other ionizable residues in a protein. Since ionizable amino acid residues may be present in the active site of the enzyme, and other ionizable residues may be responsible for maintaining the protein's conformation, it is not surprising that the pH of the solution may markedly affect enzyme activity.

Moreover, since many substrates are ionic in character (e.g., ATP, NAD^+, amino acids), the active site of an enzyme may require particular ionic species of the substrate for optimum activity. These effects are probably the main determinants of the shape of the curve that represents enzyme catalytic activity as a function of pH, an example of which appears in Figure 4.8. Usually, a bell-shaped curve is obtained. The plateau of the curve usually is small and the rates decrease rapidly with pH on either side of the maximum. The rate decreases represent changes in the state of ionization of groups of enzyme or the

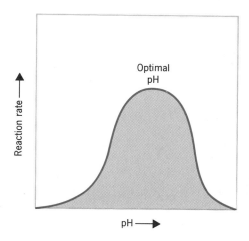

FIGURE 4.8 Typical effect of pH on the reaction rate for an enzyme-catalyzed reaction.

substrate, or both, that are critical, with regard to the state of ionization, to the enzyme-catalyzed reaction.

For a few enzymes, the plateau may be wider than is indicated in Figure 4.8, or the enzyme may show a less steep loss of activity with pH, possibly due to compensating changes in two or more ionizable groups. The pH optimum of an enzyme should be assessed early in any investigation of an enzyme so that assays may be performed at the optimum pH in well-buffered solutions. Clearly, the sensitivity of enzyms to pH changes is one of the reasons that maintenance of pH is critical to the survival of the cell. Obviously, then, it would be of great value in understanding the regulation of cellular metabolism if we had better knowledge of how pH is controlled or modified in the cellular geography.

4.11 ENZYME INHIBITORS

A great variety of naturally occurring and synthetic compounds have the ability to bind reversibly or irreversibly to specific enzymes and alter their activity. Enzyme inhibitors reduce or eliminate the catalytic activity of the enzyme. Included among such inhibitors are drugs, antibiotics, toxins, and antimetabolites, as well as many natural products of enzymic reactions. Two general classes of inhibitors are recognized, according to whether the inhibitory action is irreversible or reversible.

4.11.1 IRREVERSIBLE INHIBITORS

An irreversible inhibitor forms a covalent bond with a specific functional group, usually an amino acid side chain, which is, in some manner, associated with the catalytic activity of the enzyme. In addition, there are examples of enzyme inhibitors that covalently bind not directly at the active site, but at another site that physically blocks the active site or in some other way prevents enzyme function. An irreversible inhibitor cannot be released by dilution or dialysis; its effects cannot be reversed simply by increasing the concentration of substrate. The velocity of the reaction is reduced to an extent that corresponds to the fraction of enzyme molecules that have been inactivated. In terms of enzyme kinetics, the effect of the irreversible inhibitor is like that of the reversible noncompetitive inhibitors described next in Section 4.11.2.2.

An example of an irreversible inhibitor of chymotrypsin has the nonsystematic name tosyl-phenylalanyl-chloromethyl ketone, or **TPCK.**

$$CH_3 - \underset{}{\bigcirc} - SO_2 - NH - \underset{\underset{CH_2-\bigcirc}{|}}{CH} - CH_2 - Cl$$

TPCK

Even at very low concentrations, TPCK quantitatively inactivates chymotrypsin. TPCK is a chemically reactive analogue of a chymotrypsin substrate, with the usual carboxylic acid amide or ester replaced by the chloromethyl group. The great specificity of TPCK is indicated by its *inability* to inactivate the related proteinase trypsin. Trypsin, like chymotrypsin, is a serine proteinase but unlike chymotrypsin, it is specific for lysine and arginine residues at the sites in the polypeptide chain that are cut. Analyses of the TPCK-inactivated chymotrypsin show that the TPCK reacts with a single amino acid residue of chymotrypsin, the catalytically critical His — 57. The reaction product evidently is the result of a nucleophilic attack of histidine on the electropositive chloromethyl carbon of TPCK. As is indicated by Section 4.5 and in Figure 4.5, histidine 57 is located near serine 195 in the chymotrypsin active site and is known to participate in the chymotrypsin-catalyzed reaction by facilitating proton transfer. Thus, active site-specific, irreversible inhibitors may be used as tools to study the enzyme active site.

A particularly interesting type of irreversible inhibition is that exhibited by the so-called k_{cat} inhibitors. Such an inhibitor is said to be latent because it shows only weak chemical reactivity until it has been modified by the action of a specific enzyme. On binding to the active site and activation by the usual catalytic activity of the enzyme, the new, highly reactive inhibitor reacts chemically with the enzyme, leading to its irreversible inhibition. The enzyme literally commits suicide! These inhibitors have great potential as drugs and highly specific probes for enzyme active sites since they are not converted from the latent to the active form except by their specific target enzymes. An excellent example is the inhibition of D-3-hydroxyl decanoyl ACP dehydrase of (*E. coli*) by the latent inhibitor D-3-decynoyl-*N*-acetyl cystamine. In the reaction sequence here, —S—NAC represents the *N*-acetylcysteine moiety that forms a thioester bond to the 10-carbon, acetylenic fatty acid, 3-decynoic acid (see Chapter 7 for nomenclature), in the structure of the latent inhibitor. The action of D-3-hydroxyl decanoyl ACP dehydrase converts the acetylenic group to two double bonds. Nucleophilic attack by an active site histidine of the enzyme at one of the reactive double-bonds leads to irreversible inactivation of the enzyme. See Section 13.13.2 for the metabolic significance of D-3-hydroxyl decanoyl ACP dehydrase.

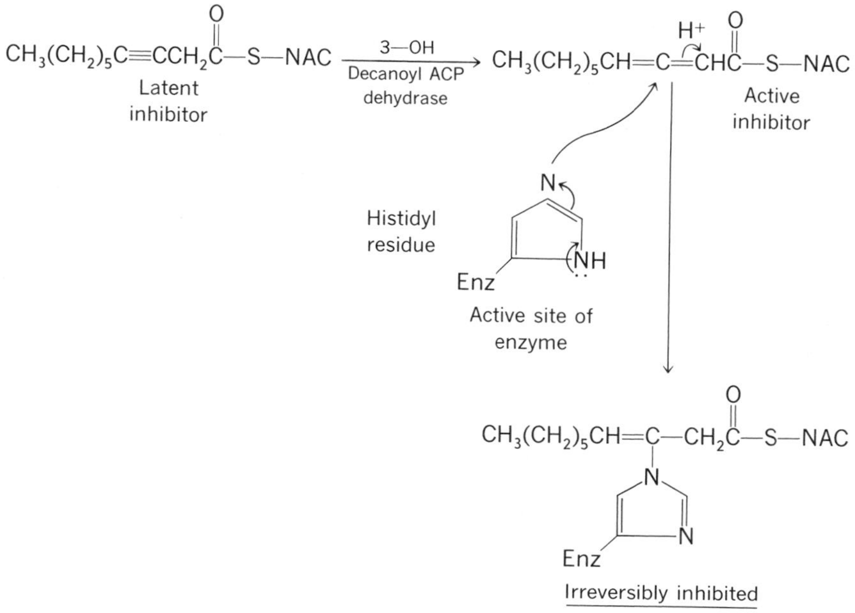

The reaction scheme shows:

$CH_3(CH_2)_5C\equiv CCH_2\overset{O}{\underset{\parallel}{C}}-S-NAC$ (Latent inhibitor) $\xrightarrow[\text{Decanoyl ACP}]{\text{3—OH}}$ dehydrase $CH_3(CH_2)_5CH=C\overset{H^+}{=}\overset{O}{\underset{\parallel}{C}}HC-S-NAC$ (Active inhibitor)

Histidyl residue, Enz, Active site of enzyme

$CH_3(CH_2)_5CH=C-CH_2\overset{O}{\underset{\parallel}{C}}-S-NAC$

Enz — Irreversibly inhibited

4.11.2 REVERSIBLE INHIBITORS

Reversible inhibitors have only a transient association with the enzyme. As the term implies, this type of inhibition involves an equilibrium between the enzyme and the inhibitor, the equilibrium dissociation constant (K_1) being a measure of the affinity of the inhibitor for the enzyme. Three distinct types of reversible inhibition are known, and they are recognized by the way they alter or fail to alter the dependence of reaction rate on substrate concentration.

4.11.2.1 Competitive Inhibition. In competitive inhibition, compounds that may or may not be structurally related to the natural substrate combine reversibly with the enzyme at or near the active site. In **feedback inhibition,** the product of a reaction that is one or more metabolic steps beyond the reaction catalyzed by the enzyme of interest acts as an inhibitor of the enzyme. Feedback inhibition is of great importance in metabolic control, and it frequently is of the competitive type. The assumption in competitive inhibition is that the inhibitor and the substrate compete for the same site, which leads to the series of the following reactions:

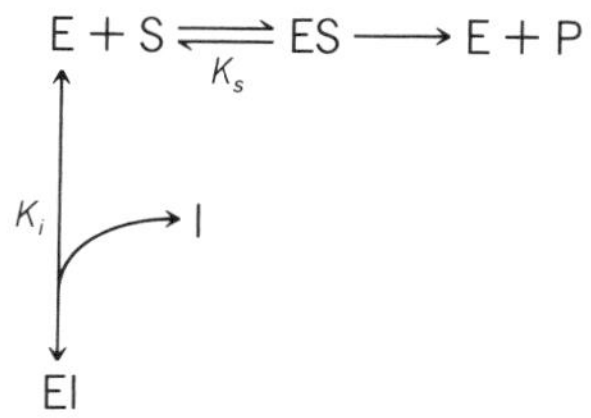

$$E + S \underset{K_s}{\rightleftharpoons} ES \longrightarrow E + P$$
$$K_i \uparrow \downarrow I$$
$$EI$$

(4.23)

In competitive inhibition, ES and EI complexes are formed, but EIS complexes never are produced. One can conclude that high concentrations of substrates will overcome the inhibition by causing the reaction sequence to be displaced to the right according to Equation 4.23. Table 4.3 summarizes the kinetic parame-

TABLE 4.3 Summary of Kinetic Expressions for Conversion of Substrate to Product under Various Types of Inhibition

TYPE OF INHIBITION	EQUATION	V_{max}	APPARENT K_m
None	$v = \dfrac{V_{max}[S]}{K_m + [S]}$	—	—
Competitive	$v = \dfrac{V_{max}[S]}{K_m\left(1 + \dfrac{I}{K_i}\right)} + [S]$	No change	Increased
Noncompetitive	$v = \dfrac{(V_{max}[S])(K_m + [S])}{1 + I/K_i}$	Decreased	No change
Uncompetitive	$v = \dfrac{V_{max}[S]}{K_m + S\left(1 + \dfrac{I}{K_i}\right)}$	Decreased	Decreased

ters of competitive inhibition and Figure 4.9 shows the typical kinetic curves observed in this type of inhibition. A classic example of competitive inhibition is the effect of malonic acid on succinic dehydrogenase, which readily oxidizes succinic acid to fumaric acid. If increasing concentrations of malonic acid, which closely resembles succinic acid in structure, are added, succinic dehydrogenase activity falls markedly. This inhibition can be reversed by increasing, in turn, the concentration of the substrate succinic acid.

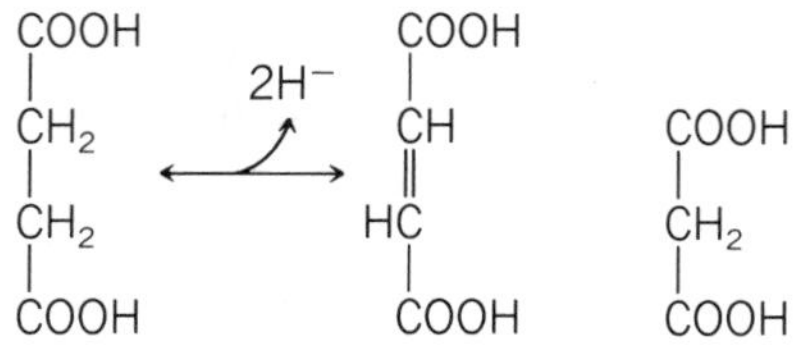

```
COOH              COOH
 |        2H⁻       |            COOH
CH₂       ↗        CH            |
 |      ⟶          ‖            CH₂
CH₂               HC             |
 |                 |            COOH
COOH              COOH

Succinic          Fumaric       Malonic
 acid              acid          acid
```

4.11.2.2 Noncompetitive Inhibition.

Compounds that reversibly bind with either the enzyme or the enzyme substrate complex are designated as noncompetitive inhibitors. The reactions are

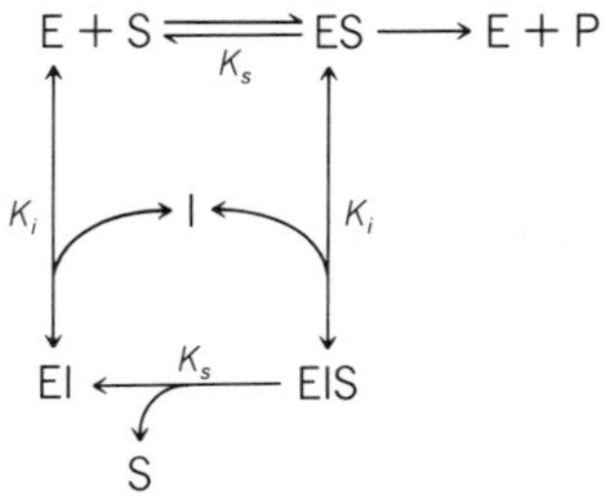

$$\begin{array}{ccccc}
E + S & \underset{K_s}{\rightleftharpoons} & ES & \longrightarrow & E + P \\
K_i \Big\downarrow & I & & \Big\downarrow K_i & \\
EI & \underset{K_s}{\longleftarrow} & EIS & & \\
& S & & &
\end{array}$$

(4.24)

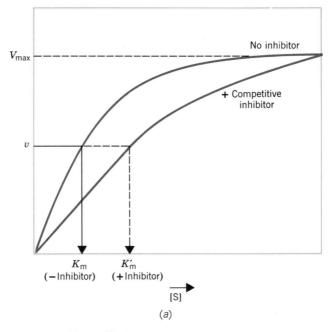

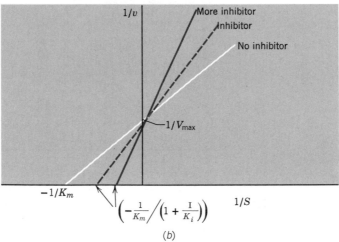

FIGURE 4.9 Effects of a competitive inhibitor on reaction rate as a function of substrate concentration. Note the unchanged value of V_{max} but increased K_m that are characteristic of competitive inhibition and are seen in both the direct plot (a) and the Lineweaver–Burk plot (b).

Noncompetitive inhibition therefore differs from competitive inhibition in that the inhibitor can combine with ES, and S can combine with EI to form in both instances EIS. This type of inhibition is not completely reversed by high substrate concentration since the cyclic sequence of Equation 4.24 will occur regardless of the substrate concentration. Since the inhibitor-binding site is not identical to the active site, nor does it modify the active site directly, the K_m is not altered. Refer to Table 4.3 for the equation defining noncompetitive inhibition. Figure 4.10 show typical direct and reciprocal plots for a reaction that

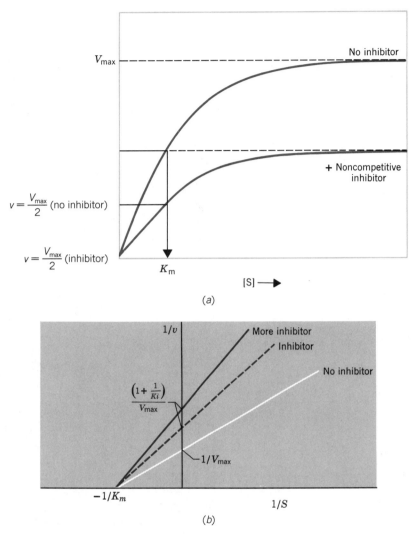

FIGURE 4.10   Effects of a noncompetitive inhibitor on reaction rate as a function of substrate concentration. Note the decreased value of V_{max} but unchanged K_m that are characteristic of noncompetitive inhibition and are seen in both the direct plot (a) and the Lineweaver–Burk plot (b).

shows noncompetitive inhibition. Equations that predict these results appear in Table 4.3.

4.11.2.3   Uncompetitive Inhibition. Compounds that reversibly combine with only the ES complex but not the free enzyme are called uncompetitive inhibitors. The inhibition is not overcome by high substrate concentrations; interestingly, the K_m', that is, the apparent K_m or $S_{0.5}$ in the presence of the inhibitor, is consistently smaller than the K_m value of the uninhibited reaction. This implies that S is more effectively bound to the enzyme in the presence of the inhibitor. Equation 4.25 describes systems that show uncompetitive inhibition.

$$E + S \underset{K_s}{\rightleftharpoons} ES \longrightarrow E + P$$

$$I \longleftarrow K_i$$

(4.25)

$$EIS$$

Comparison of this reaction to Reaction 4.24 would suggest that an element of uncompetitive inhibition is always a component of noncompetitive inhibition since in both cases EIS is formed. Figure 4.11 depicts typical curves for uncompetitive inhibition, and Table 4.3 gives the corresponding equation.

The types of reversible inhibition that have been observed are not limited to competitive, noncompetitive, and uncompetitive. Sometimes, inhibition appears to be of a mixed type, a combination of noncompetitive with one of the other two types of reversible inhibition. In some systems as the substrate concentration increases, the reaction rate at first increases and then decreases,

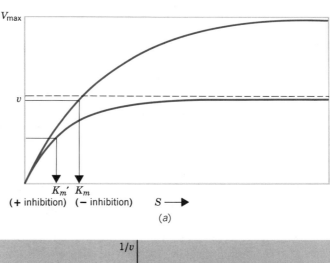

(a)

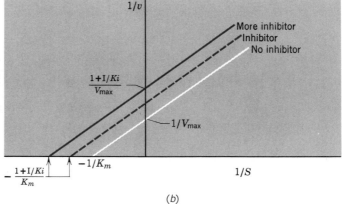

(b)

FIGURE 4.11 Effects of an uncompetitive inhibitor on reaction rate as a function of substrate concentration. Note the decrease in the apparent K_m that gives K_m' (also called $S_{0.5}$). The V_{max} also is decreased. These results are seen in both the direct plot (a) and the Lineweaver–Burk plot (b).

reflecting **substrate inhibition.** Usually, this is caused by substrate binding to the enzyme at a site that is not the active site, resulting in the activation of an inhibition mechanism of the **allosteric** type, as is discussed in Section 4.12.

Reversible inhibitors may be used in the efficient and specific purification of enzymes. In **affinity chromatography,** the inhibitor is covalently bound to a resin, usually with some organic molecule serving as a "linker" or "spacer" to reduce steric constraints on binding of the inhibitor by the enzyme. The modified resin is poured into a column (see Boxes 3.B and 4.A). When even a crude mixture such as a cell extract is passed over the column, the selectivity provided by the inhibitor often will give the resin the ability to bind only the enzyme of interest. Washing the column with a solution that is buffered at or near the pH optimum of the enzyme removes impurities but not the enzyme. Subsequent washing with a solution that is buffered at a pH that is not near the pH optimum usually will remove the enzyme in a highly purified form. Other examples of applications of affinity chromatography are the purification of hormone receptors with a chromatography resin of immobilized hormone and the purification of enzymes that act on DNA by use of insoluble complexes of DNA and cellulose.

4.12   ALLOSTERIC EFFECTS IN ENZYME-CATALYZED REACTIONS

The phenomenon of substrate inhibition just described is an example of homotropic negative cooperativity because a given molecule, the substrate, is able to inhibit the action of the enzyme on the same type of molecule, the substrate. Already discussed in Section 3.14 is the homotropic positive cooperativity of oxygen binding by hemoglobin. The binding of one oxygen molecule by deoxyhemoglobin results in an increased affinity of the hemoglobin in its binding of succeeding oxygens. Feedback inhibition of an enzyme by the end-product of a series of reactions, the metabolically important enzyme control phenomenon discussed briefly in Section 4.11, is an example of heterotropic negative cooperativity. Similarly, activation of an enzyme by a nonsubstrate and nonproduct would be an example of heterotropic positive cooperativity.

Allosteric inhibition is the result of the interaction of an enzyme molecule with a biologically significant **effector** molecule that inhibits the enzyme by acting at a site other than the active site. **Allosteric activation** is a similar phenomenon that results in the stimulation of reaction rate. Originally, the name "allosteric effector" was reserved for regulators that are unlike substrates and products (heterotropic inhibition or activation). However, substrates and/or products (homotropic inhibition or activation) are considered to be allosteric effectors when they act at sites other than the active site. Enzymes that are under allosteric control by substrate will deviate from the behavior predicted by the Briggs–Haldane equation and similar equations for multisubstrate and multiproduct reactions. That is, in initial velocity kinetics, the plot of v_0 versus [S] will not be hyperbolic. Activation by substrate, instead, will produce an S-shaped or "sigmoidal" curve. For this reason, the quantity K_m has little or no meaning for enzymes under allosteric control. Instead, the quantity $S_{0.5}$, already introduced in the preceding section, is used to specify the substrate concentration that will produce a reaction velocity that is half V_{max}.

The potential advantage of allosteric activation by a substrate is illustrated in Figure 4.12, which shows both a normal hyperbolic response curve as predicted by the Briggs–Haldane equation (4.9) and the sigmoidal type of curve that results from allosteric activation by substrate. The Briggs–Haldane equation predicts that the substrate concentration must be exactly 27 times greater when $v = 0.75 \, V_{max}$ than when $v = 0.1 \, V_{max}$, that is, that $[S]_{0.75}/[S]_{0.1} = 27$. Values of less than 27 imply positive homotropic cooperativity. Similar calculations for the sigmoidal curve of the figure give a value of 2.3 for $[S]_{0.75}/[S]_{0.1}$. Thus the hyperbolic curve requires more than a 10-fold greater change in substrate concentration than the sigmoidal curve, 27-fold versus 2.3-fold, for the same change in the rate of substrate consumption. The effect of the sigmoidal type of response is that the substrate will tend to be held within more narrow confines of concentration. That is, since the enzyme responds to a small increase in

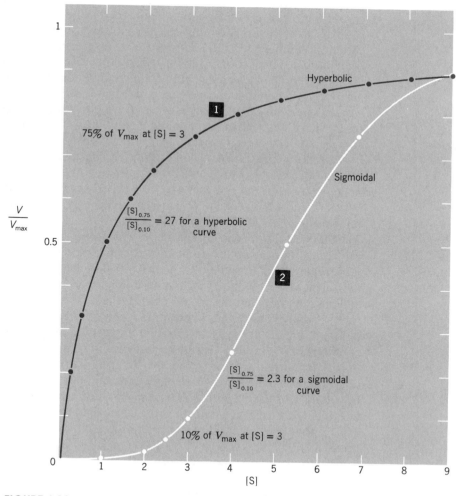

FIGURE 4.12 The effects of substrate concentration on the velocity of a reaction catalyzed by ① an enzyme that obeys the Briggs–Haldane equation and ② an allosteric enzyme that is activated by its own substrate and shows sigmoidal kinetics.

substrate concentration with a great increase in rate, it will tend to rapidly consume substrate and therefore lower its concentration. Section 4.6 discusses other aspects of the role of enzymes in maintaining relatively fixed concentrations of molecules in the cell.

Among the most interesting of allosteric phenomena are those that are heterotropic, in which the rate of an enzyme-catalyzed reaction is influenced by a molecule that is neither a substrate nor a product of the enzyme. At a given substrate concentration, the rate may be influenced by changing V_{max} or $S_{0.5}$, or both. Commonly, the V_{max} remains unchanged by the heterotropic allosteric effector. Allosteric activators often cause a sigmoidal curve to be transformed into a hyperbolic curve, for example, by shifting curve 2 to curve 1 in Figure 4.12. An allosteric inhibitor may cause the velocity-substrate curve to become more sigmoidal. The significance of allosteric effects for the control of metabolism is discussed in Chapter 18.

A few allosteric enzymes have only a single polypeptide chain. Presumably, the binding of the allosteric effector causes a small but significant change in the conformation of the polypeptide chain so that binding of the substrate or catalytic efficiency or both are altered. However, most of the enzymes that exhibit allosterism are oligomeric proteins. Oligomeric proteins already have been discussed in Section 3.14. Oligomeric enzymes may have distinct *regulatory* and *catalytic* sites on the same or different polypeptide chains. Often, they catalyze a reaction that under *in vivo* conditions is far from equilibrium and is at a branch point in a metabolic pathway. A reaction at a branch point is one that results in the synthesis of a compound that is the precursor of two or more other compounds that are synthesized by distinct pathways (see Chapter 18).

Models that are meant to explain allosteric behavior of oligomeric enzymes rely on interactions between the protein subunits as the basis for the nonhyperbolic behavior of the enzyme. The "concerted-symmetry" model of Monod, Wyman, and Changeux (MWC model) assumes that the oligomeric enzyme can exist in either of two forms, T or R, that differ in their enzymic properties. The MWC model for enzyme allosteric activation is illustrated in Figure 4.13. It is identical in form to the model for oxygen binding by hemoglobin, Section 3.14, Figure 3.16. Presumably, the two forms of an allosteric enzyme, like the two forms of hemoglobin, may assume different conformations within subunits and different orientations of one subunit relative to another. Each of these two forms has all of its subunits in the same conformation, but that conformation is different for each of the two forms. In homotropic allosteric activation, according to the MWC model, a substrate molecule may bind only to the more active of the two forms of the enzyme, the R form. Thus, increasing the substrate concentration increases the rate of reaction not only by increasing the fraction of E that is in the ES form (Equation 4.5), but also by shifting more of the enzyme into a more active form with all four subunits in the R conformation. The result is a sigmoidal curve for reaction velocity versus substrate concentration.

The "sequential" model of Koshland, Nemethy, and Filmer (KNF model, based on earlier suggestions of Adair and Pauling, and the "induced fit" model of Koshland) is similar to the MWC model, but it allows the two conformations of polypeptide chains to coexist within one oligomeric enzyme as indicated in Figure 4.13c. The KNF model is more complex than the MWC model and can be made very general and applicable to most allosteric enzymes by providing

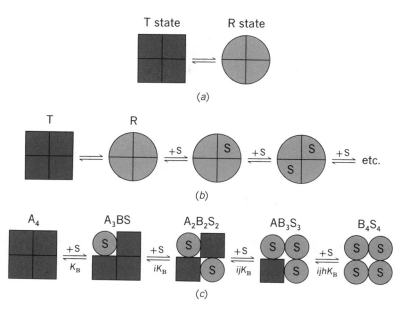

FIGURE 4.13 Comparison of MWC and KNF for allosteric control of an enzyme by its substrate. As described in the text, (a) and (b) illustrate the MWC model for a tetrameric enzyme and for that enzyme in the presence of substrate, respectively. In the sequential KNF model (c), the transition from nonsubstrate-binding to substrate-binding protomers is not concerted.

for modulated interactions between subunits, as dictated by the geometry of the oligomers. Each consecutive binding of S alters the affinity for S, from K_m to iK_m to ijK_m to $ijhK_m$ for species A_4, A_3BS, $A_2B_2S_2$ and AB_3S_3, respectively, in Figure 4.13c.

Even when the MWC or KNF model, or any other model, fits the data derived from a given system, that does not prove that the model is correct. Establishing the correctness of a model requires more direct analyses. A classical example of a test for interaction between subunits of an oligomeric enzyme was performed on *E. coli* aspartate carbamoyltransferase (ACTase). The reaction that is catalyzed by ACTase is

$$
\underset{\text{Aspartic acid}}{\overset{\text{COO}^-}{\underset{\overset{|}{\underset{\text{H}_3\overset{+}{\text{N}}-\text{CH}-\text{COO}^-}{\overset{|}{\text{CH}_2}}}}}} +
\underset{\text{Carbamoyl phosphate}}{\overset{\text{O}}{\underset{\overset{||}{\underset{\text{NH}_2}{\overset{|}{\text{C}-\text{OPO}_3^-}}}}}}
\overset{\text{ACTase}}{\rightleftharpoons}
\underset{\substack{\text{N-Carbamoyl} \\ \text{aspartate}}}{\overset{\text{COO}^-}{\underset{\overset{|}{\underset{\overset{||}{\text{NH}_2}}{\overset{\overset{|}{\text{CH}_2}}{\text{C}-\text{NH}-\text{CH}-\text{COO}^-}}}}{}}}
\quad + \text{Pi}
$$

Cytidine triphosphate (CTP; see Section 6.2 for structure) acts as a feedback inhibitor of ACTase. Gerhart and Schachman showed that the two different types of subunits in ACTase could be dissociated. The **catalytic subunit,** *c,* has the ACTase enzymic activity (see also Section 4.13) and exhibits hyperbolic kinetics and is not inhibited by CTP. The other subunit, the **regulatory subunit,** is designated *r*. Combining *r* and *c* subunits restored allosteric control by CTP. The complete enzyme consists of six subunits of each type.

4.12.1 ISOENZYMES

An **isoenzyme,** sometimes termed "isozyme," is one of two or more forms of an enzyme that exist in the same organism and catalyze the same reaction. The most thoroughly studied isozyme system is that of lactic dehydrogenase, LDH, which catalyzes the oxidation of lactic acid to pyruvic acid. Through starch gel electrophoresis, a technique that is very similar to polyacrylamide gel electrophoresis (Box 3.E), five different forms of LDH have been detected in the tissues of vertebrate organisms under nondenaturing conditions. There are two predominant forms of LDH isozymes, the M form, which is characteristic of skeletal muscle, and the H form, which is characteristic of heart muscle. These are the products of two related genes, each of which encodes a polypeptide chain of molecular weight 35,000. An active LDH enzyme molecule consists of four of these polypeptide chains, giving a molecular weight of 140,000 for the oligomeric protein.

The pure M tetramer, M_4, and the pure H tetramer, H_4, have been isolated. If conditions that are mildly denaturing are used, the tetramer can be dissociated and, when the denaturing agent is removed, renatured. When the M and H pure tetrameric isozymes are mixed and subject to mild denaturation and renaturation, five species were formed: M_4, M_3H, M_2H_2, MH_3, and H_4. These results show the relatedness of the M and H forms and are consistent with an arrangement of subunits in a tetramer that places each polypeptide chain in an equivalent position. The three heterogeneous and two homogeneous M and H isozymes of LDH are observed in tissue extracts, as well as after assembly *in vitro*. A few hundred systems of isozymes are known, and in many instances, the different forms of an enzyme have been shown to have different kinetic and allosteric properties. The differing properties presumably are advantageous to the function of different tissues in which the specific isozymes occur.

Very specific **activity stains** have been developed for several isozymes. These allow all of the enzyme zones that contain a particular type of activity, for example, LDH activity, to be located after gel electrophoresis. The activity stain distinguishes the enzyme of interest from other activities that may have been resolved in the gel, even when very crude mixtures such as cell extracts have been analyzed. This technology has been widely applied in genetic studies because large populations of individuals of a given species can be analyzed for a given isozyme or set of isozymes with relatively little effort. Most of the enzymes of glycolysis, which are described in Chapter 10, have either two or four identical polypeptide chains, and isozymes of these enzymes frequently have been observed. Box 3.F gives information about the interpretation of the electrophoresis patterns that are characteristic of isozymes, and Box 4.E describes a related technique for locating specific zones after gel electrophoresis.

4.13 ENZYME PRECURSORS AND ENZYME ASSOCIATIONS

The concept of the proprotein was introduced in Section 3.15 with the example of proinsulin. Like proinsulin, certain enzymes are synthesized in a form, the **proenzyme,** that is larger than the mature enzyme. The proenzyme subse-

TABLE 4.4 Conversion of Zymogens to Active Enzymes

ZYMOGEN	ACTIVATING AGENT	ACTIVE ENZYME		INACTIVE PEPTIDE
Pepsinogen	$\xrightarrow{\text{H}^+ \text{ or Pepsin}}$	Pepsin	+	Fragments
Trypsinogen	$\xrightarrow{\text{Enterokinase or Trypsin}}$	Trypsin	+	Hexapeptide
Chymotrypsinogen A	$\xrightarrow{\text{Trypsin + Chymotrypsin}}$	α-Chymotrypsin	+	Amino acid residues
Procarboxypeptidase A	$\xrightarrow{\text{Trypsin}}$	Carboxypeptidase A	+	Fragments
Proelastase	$\xrightarrow{\text{Trypsin}}$	Elastase	+	Fragments

quently is cut at specific sites by the action of one or more proteinases to generate the active enzyme. The proenzyme form of a proteinase is referred to as a **zymogen.** Table 4.4 lists some proteinases that are synthesized as zymogens. These proteinases act outside the cell but, of course, must be synthesized in cells. Synthesis of the proteinase in the inactive, zymogen form protects the cell against the potentially damaging effects of the proteinase. It also allows the enzyme polypeptide chains to accumulate to high concentrations and to be rapidly activated to the catalytic form when the proteinase is needed.

Many active enzymes require, in addition to one or more polypeptide chains, one or more **cofactors** to be catalytically active. Cofactors may be roughly categorized into (a) prosthetic groups, (b) coenzymes, and (3) metal ions. Prosthetic groups are tightly bound. Examples are the porphyrin moiety of peroxidase and the flavin-adenine dinucleotide of succinic dehydrogenase. Coenzymes are less tightly bound and usually can be removed by the process of dialysis. To carry out **dialysis,** the enzyme is placed in a sac that is permeable to the coenzyme but not the enzyme protein, and the sac is immersed in a very large volume of stirred buffer so that any coenzyme that dissociates from the enzyme protein will become so diluted by the large volume that the protein will not be able to bind it again. An enzyme with its coenzyme, or prosthetic group, removed is designated an **apoenzyme.** Adding the correct, concentrated coenzyme to the apoenzyme will, in many instances, restore enzyme activity.

Metal ions that are necessary for enzymic activity may be either loosely or tightly bound, depending on the enzyme. All of the tightly bound ions are divalent or trivalent. Usually, they participate in the catalytic mechanism, for example, as oxidizing or reducing agents, but some are necessary for maintaining the proper conformation of the enzyme protein. The properties of several coenzymes and cofactors are considered in greater detail in Chapter 5.

We have discussed in Section 4.12 examples of enzymes that have two or four identical polypeptide chains. ACTase was an example of an enzyme with two kinds of subunits: one catalytic and the other regulatory. Recently, H. Schachman and his colleagues found that the functional active site of ACTase is not present in an individual catalytic subunit. Rather, the active site is formed at the junction between catalytic subunits. Thus only associated catalytic subunits show ACTase activity.

The enzyme lactose synthetase from mammary gland, like ACTase, is an example of an enzyme that has the activity of a catalytic subunit modified by a noncatalytic polypeptide chain. The precursors of lactose in mammary gland are glucose and an activated form of galactose, uridine diphosphate galactose, which is abbreviated UDP-galactose. The structure and metabolic significance of this compound are discussed in Section 10.6.2. The reaction catalyzed by lactose synthetase is

$$\text{UDP-Galactose} + \text{Glucose} \rightleftharpoons \text{UDP} + \text{Lactose} \tag{4.26}$$

The enzyme has two kinds of protein subunits, neither of which will catalyze the above reaction. However, the catalytic subunit of lactose synthetase does catalyze another, related reaction, which is

$$\text{UDP-Galactose} + \begin{array}{c}\text{N-Acetyl} \\ \text{glucoseamine}\end{array} \rightleftharpoons \text{UDP} + \begin{array}{c}\text{N-Acetyl} \\ \text{lactoseamine}\end{array} \tag{4.27}$$

Addition of the apparently noncatalytic subunit inhibits Reaction 4.27 and facilitates Reaction 4.26. The noncatalytic protein was found to be α-lactalbumin, a milk protein that is found in mammary gland but not in other tissues. Apparently, the catalytic subunit of lactose synthetase has two principal functions, lactose synthesis in mammary gland and N-acetyl lactoseamine synthesis in other tissues.

Tryptophan synthetase also is composed of two types of polypeptide chains. The reaction catalyzed by the complete tetrameric enzyme, $\alpha_2\beta_2$, is

$$\begin{array}{c}\text{Indole} \\ \text{glycerophosphate}\end{array} + \text{L-Serine} \xrightarrow{\alpha_2\beta_2} \text{L-Tryptophan} + \begin{array}{c}\text{Glyceraldehyde-} \\ \text{3-phosphate}\end{array}$$

Each type of subunit also is catalytically active, but cannot catalyze the entire tryptophan synthetase reaction. The two "half reactions" that are catalyzed by separated subunits are

$$\begin{array}{c}\text{Indole} \\ \text{glycerophosphate}\end{array} + \xrightleftharpoons{\alpha} \text{Indole} + \begin{array}{c}\text{Glyceraldehyde-} \\ \text{3-phosphate}\end{array}$$

and

$$\text{Indole} + \text{L-Serine} \xrightarrow{\beta_2} \text{L-Tryptophan}$$

The complete $\alpha_2\beta_2$ complex has a turnover number [Reaction (1)] that is 30 to 100 times as great as that observed for reactions (2) or (3) as catalyzed by the subunits. The reaction intermediate indole appears to remain bound to the active tetrameric $\alpha_2\beta_2$ enzyme and is not released into the medium.

Retention of reaction intermediates also is a characteristic of certain **multienzyme** complexes. One of the most elaborate of these is the fatty acid synthetase complex of animal and yeast cells. The functions of synthetase complex are described in greater detail in Section 13.9. For our purposes here, it is sufficient to note that one molecule of acetyl-coenzyme A and seven molecules of malonyl-coenzyme A are combined in a series of reactions that extend the acetyl group in two carbon units derived from malonate, with the release of CO_2, to generate the 16-carbon palmitic acid. Each 2-carbon addition requires a series of five reactions, each catalyzed by a different enzyme in the complex. Other activities also are required for palmitic acid synthesis and are part of the complex. The growing fatty acid chain remains bound to the complex. Presumably, this allows efficient synthesis and "channels" the intermediate length fatty acids to remain in the synthetase pathway so that they are not drained off to other reactions. In yeast, the fatty acid synthetase complex is one of the most elaborate subcellular particles found in the cell.

REFERENCES

1. *T. Palmer,* Understanding Enzymes, *2nd ed. New York: Wiley, 1985.*
2. *G. G. Hammes,* Enzyme Catalysis and Regulation. *New York: Academic Press, 1982.*
3. *A. Fersht,* Enzyme Structure and Mechanism. *San Francisco, Calif.: W. H. Freeman and Co., 1977.*
4. *C. Walsh,* Enzymic Reaction Mechanisms. *San Francisco, Calif.: W. H. Freeman and Co., 1979.*
5. *I. H. Segel,* Enzyme Kinetics. *New York: Wiley, 1975.*
6. *International Union of Biochemistry,* Enzyme Nomenclature. *New York: Academic Press, 1984.*

REVIEW PROBLEMS

1. Some enzyme-catalyzed reactions of highly ionic substrates are nearly independent of the ionic strength of the solution, which is in strong contrast to the ordinary reactions of ionic compounds in aqueous solution. What is a reasonable explanation for this observation?

2. Consider the two constants K_s and K_m of the same enzyme. Are the two quantities likely to be more nearly equal for an enzyme that is nearly catalytically perfect or for an enzyme that has a small turnover number?

3. Why do almost all enzyme-catalyzed reactions show a pH optimum?

4. A single-substrate enzyme catalyzes a reaction that has a reaction rate dependence on substrate concentration that is predicted by the Briggs–Haldane equation, giving a K_m of 1 mM. If the enzyme is assayed at [S] = 0.001 mM, the observed velocity is 1 mM/min. What is the predicted velocity for [S] = 0.002 mM?

5. A certain fixed concentration of an inhibitor caused a 90% reduction in the velocity of a single-substrate, single-product, enzyme-catalyzed reaction at saturating concentration of substrate. In the presence of the inhibitor, the less-than-saturating substrate concentration at which the velocity of the reaction was reduced to half of its maximum value was the same substrate concentration that caused the uninhibited reaction to proceed at half of V_{max}. Is it possible or impossible for the inhibitor to be acting as a competitive inhibitor? What effect should the inhibitor have on the enzyme-catalyzed reverse reaction?

6. What are the likely biochemical and physiological significances of a velocity versus [S] curve that is sigmoidal rather than hyperbolic?

CHAPTER 5

VITAMINS AND COENZYMES

Vitamins are organic compounds required in small amounts in the diets of animals in order to ensure healthy growth and reproduction. Vitamins are essential nutrients because the animal cannot synthesize such compounds in amounts adequate for its daily needs. The absence of a vitamin in the diet, or its poor absorption from the digestive tract, usually produces a disease with characteristic symptoms. The detailed description of the deficiency symptoms and the amounts required for alleviation of these symptoms is more properly the subject of nutrition. Therefore, many fascinating stories can be told of how man came to realize that he required more than the three major groups of foodstuffs: carbohydrates, lipids, and proteins.

5.1 VITAMINS—A CLASSIFICATION

Vitamins may be classified according to their solubility into two groups: the water-soluble (B) vitamins and the fat-soluble vitamins. Many of the water-soluble vitamins are components of larger coenzyme molecules. When they are encountered as essential nutrients in an animal diet, they are converted (biosynthesized) into the coenzyme molecule where they function in "essential" metabolic reactions. The same coenzymes with few exceptions occur in plants where they are also assembled from the smaller vitamin molecules. Plants, however, have the ability to synthesize the vitamin (from CO_2, NH_3, and H_2S) and, in fact, serve as excellent sources of these dietary essentials. Table 5.1 lists the water soluble vitamins required by humans, their related coenzyme, and the enzymatic function that they perform.

TABLE 5.1   Water-Soluble Vitamins, Effects of Deficiency, Related Coenzyme, and Chemical Reactions Catalyzed

VITAMIN	DEFICIENCY SYMPTOMS	COENZYME	REACTION CATALYZED
B_1, thiamin	Beri-beri, polyneuritis	Thiamin pyrophosphate	Cleavage or formation of carbon-carbon bonds adjacent to carbonyl carbon atoms
B_2, riboflavin	Dermatitis, impaired growth and reproduction	Flavin mononucleotide (FMN); flavin adenine dinucleotide (FAD)	Oxidation–reductions including hydride anion transfer, one electron transfer
Nicotinic acid	Pellagra, dermatitis, black tongue (in dogs)	Nicotinamide adenine dinucleotide (NAD^+); nicotinamide adenine dinucleotide phosphate ($NADP^+$)	Oxidation–reductions involving hydride anion transfer
Pantothenic acid	Neurological problems	Coenzyme A	Transfer of acyl groups
B_6, pyridoxine	Neurological disorders, dermatitis	Pyridoxal phosphate	Transamination, decarboxylation, racemization
Lipoic acid	None demonstrated	Lipoamide	Oxidation–reduction
Biotin	None demonstrated in absence of avidin; with avidin, dermatitis	Biotin carboxyl-carrier-protein	Carboxylation, transcarboxylation
Folic acid	Anemia	Tetrahydrofolic acid	Transfer of one-carbon units
Vitamin B_{12}	Pernicious anemia	Coenzyme B_{12}	Rearrangements
Ascorbic acid	Scurvy	—	Hydroxylation

5.2   THIAMIN

5.2.1   STRUCTURE

Thiamin, or vitamin B_1, has the following structure:

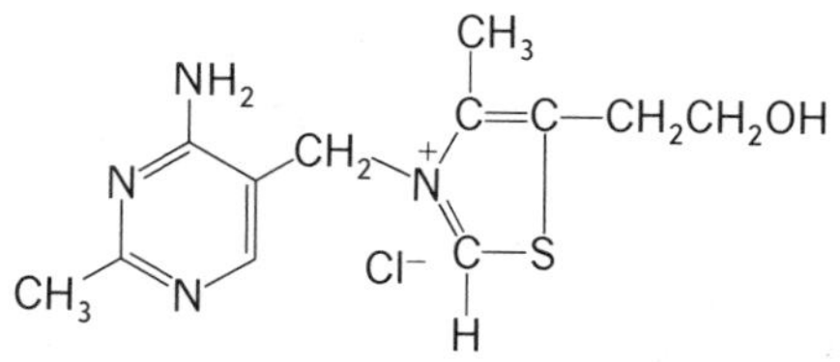

The study of this compound and the role it played in correcting the nutritional disease *beri-beri*, prevalent in the Far East, gave rise to the name **vitamin.** Casimir Funk in 1912 described the compound as a "vital amine" and it became known as Vitamin B. When other vitamins with similar chemical properties (water soluble, nitrogen-containing) were described, it was renamed B_1 to distinguish it from B_2, B_3, and B_6.

5.2.2 OCCURRENCE

Thiamin occurs in the outer coats of the seeds of many plants including the cereal grains. Thus, unpolished rice and foods made of whole wheat are good sources of the vitamin. In animal tissues and yeast, it occurs primarily as the coenzyme thiamin pyrophosphate or cocarboxylase.

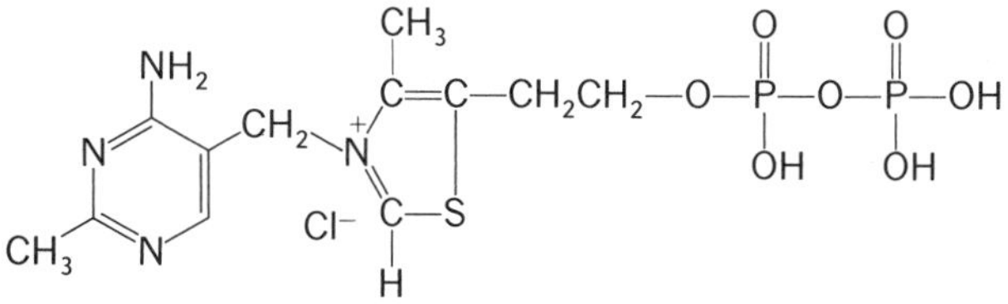

Thiamin pyrophosphate
(cocarboxylase)

Animals, other than ruminants whose bacteria can provide the vitamin, require thiamin in their diet. A deficiency of this vitamin in man produces the classic disease known as "beri-beri." In dry beri-beri, muscular weakness and loss of weight, neuritis, and evidence of involvement of the central nervous system are the symptoms. Wet beri-beri leads to edema and impaired cardiac function. In experimental animals, thiamin deficiency leads to early signs of impairment of brain function.

Although beri-beri has long been known in areas of the world where polished rice is the chief source of calories, thiamin is one of the vitamins that may be inadequately supplied in American diets. The recommended allowance is 0.5 mg/day per 1000 calories and many adults take in less. Since the vitamin is water-soluble and cannot be stored, an adequate supply can and should be acquired by eating seeds (beans, peas, corn) or products made of whole wheat flour. As with other water-soluble vitamins, overcooking may leach out and/or destroy the thiamin originally present in the food source.

5.2.3 BIOCHEMICAL FUNCTION

Thiamin pyrophosphate participates as a coenzyme in α-keto acid dehydrogenases (Sections 12.2 and 18.4.2), pyruvic decarboxylase (Section 10.4.7), transketolase (Section 11.4.6), and phosphoketolase, an enzyme concerned with the metabolism of pentoses in certain bacteria. For example,

$$\text{D-Xylulose-5-}P + P_i \xrightarrow[\text{Cocarboxylase}]{\text{Phosphoketolase}} \text{Acetyl-}P + \text{Glyceraldehyde-}P \qquad (5.1)$$

It should be noted that yeasts can decarboxylate pyruvic acid because they contain thiamin pyrophosphate (cocarboxylase) *and* the apoenzyme (decarboxylase). Animal cells contain thiamin pyrophosphate when the thiamin supply is adequate, but they lack the apoenzyme, the decarboxylase. That is why decarboxylation in these cells is carried out as an **oxidative** decarboxylation as illustrated in Section 12.2.

In all these reactions, the common site of action is C-2 of the thiazole ring. The hydrogen atom at this position tends to dissociate as a proton to form a

carbanion. The carbanion

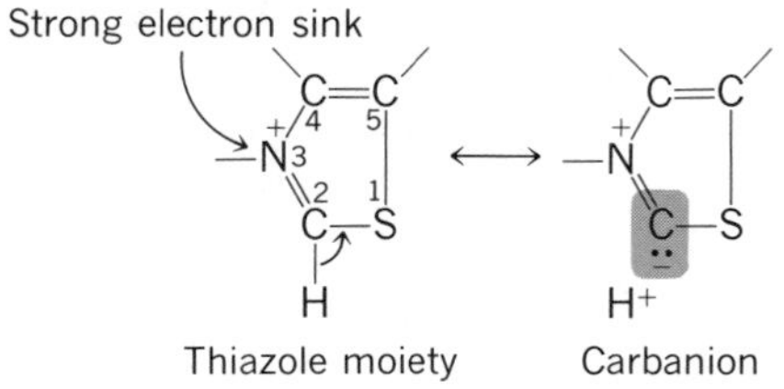

Thiazole moiety   Carbanion

participates in the decarboxylation of pyruvic acid as shown in the next diagram. The adduct formed undergoes decarboxylation after the appropriate rearrangement of electrons and acetaldehyde dissociates with the regeneration of the carbanion.

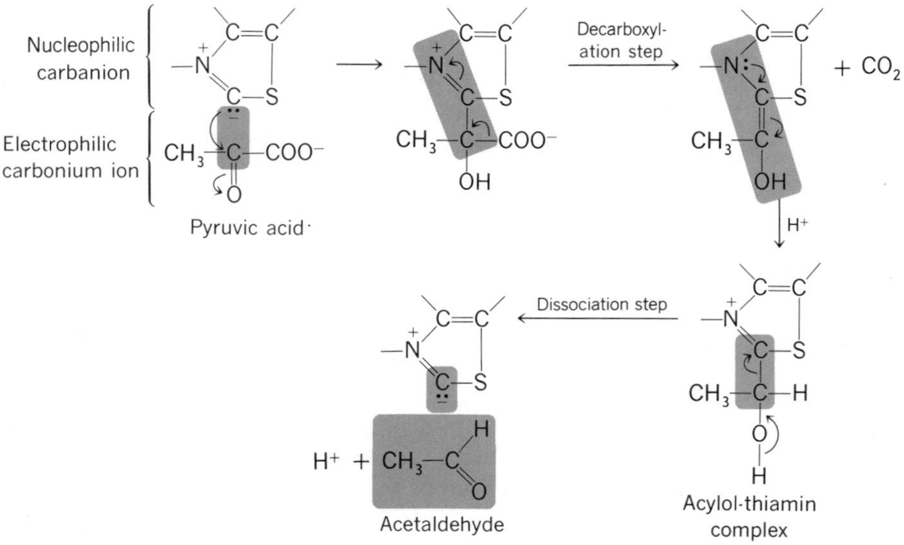

The detailed mechanisms for the other enzymatic reactions involving thiamin pyrophosphate are discussed under the headings just indicated.

5.3 RIBOFLAVIN

5.3.1 STRUCTURE

Riboflavin (vitamin B_2) consists of the sugar alcohol D-ribitol attached to 7,8-dimethyl-isoalloxazine.

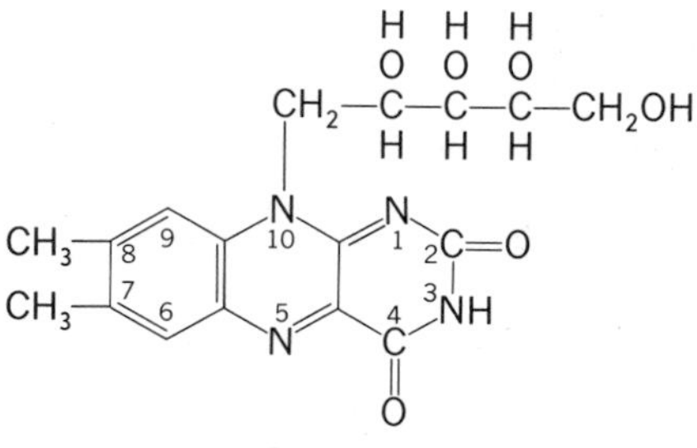

Riboflavin

The vitamin occurs as a component of the two flavin coenzymes, flavin mononucleotide (FMN) and flavin adenine dinucleotide (FAD). Although the flavin coenzymes were given the names mono- and dinucleotide, the names are not accurate chemically speaking, since the compound attached to the flavin moiety is the sugar alcohol ribitol and not the aldose sugar ribose, and the isoalloxazine ring is not a purine or pyrimidine

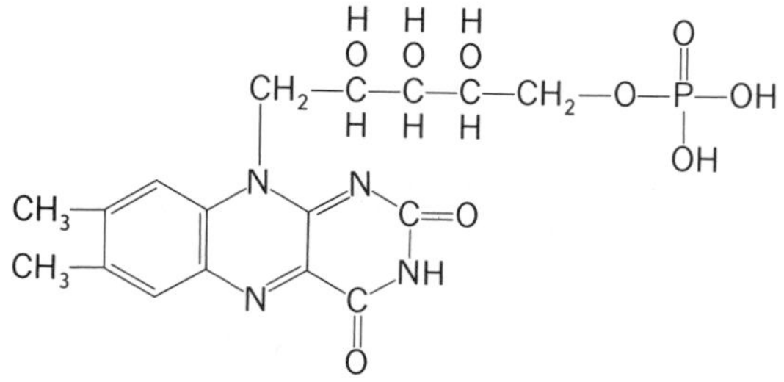

Flavin mononucleotide (FMN)
(Riboflavin monophosphate)

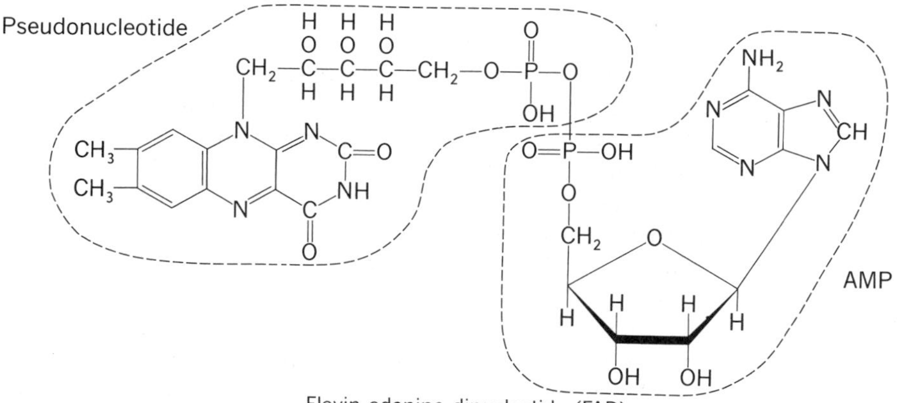

Flavin adenine dinucleotide (FAD)

derivative. Thus, these compounds are better described as pseudonucleotides.

5.3.2 OCCURRENCE

Riboflavin is synthesized by green plants, many bacteria, and fungi but not by any animals. Since it is available in animal tissue in the form of the flavin coenzymes (see preceding section), an animal can obtain the vitamin by eating tissues such as liver that contain a high concentration. The primary source, however, is plant material, although commercial production by yeasts and certain microorganisms is practiced.

The symptoms of riboflavin deficiency are difficult to observe in man. Signs such as a dark red tongue, dermatitis, and cheilosis similar to those of niacin deficiency have been observed. In rats, in whom an experimental deficiency can be produced, growth is impaired, changes occur in the lens of the eye leading to blindness (cataract), nerve degeneration is observed, and there is impaired reproduction. With riboflavin deficiency, however, there is no clear-cut evidence of impaired oxidative activity that is expected considering the known biochemical role of its flavin coenzymes.

5.3.3 BIOCHEMICAL FUNCTION

Riboflavin functions as a coenzyme because of its ability to undergo oxidation – reduction reactions. On reduction, the yellow color disappears since the reduced flavin is colorless. On exposure to air, the yellow color of the oxidized form reappears. As indicated in the diagram, the overall reaction consists of the addition of two hydrogen atoms in a 1,4 addition to form the reduced, colorless flavin.

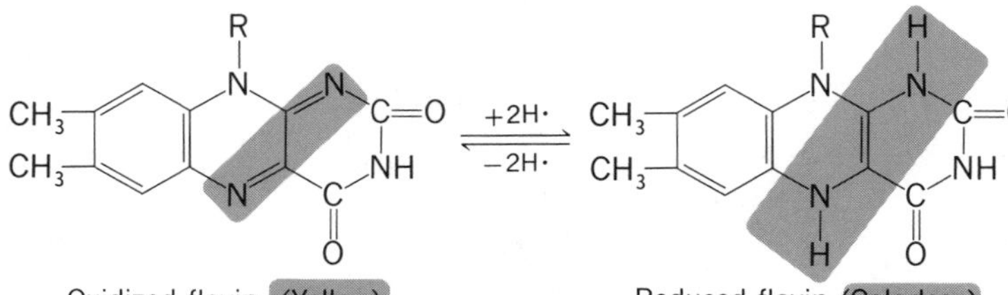

Oxidized flavin (Yellow)   Reduced flavin (Colorless)
R = Remainder of FMN or FAD molecule

Although the ability of FMN and FAD to function as coenzymes for a group of proteins known as **flavoproteins** has been known for decades, the mechanisms by which they function are still under investigation. It is clear, however, that more than one mechanism is involved. For example, the flavoprotein dihydroorotate dehydrogenase involved in UMP synthesis catalyzes the oxidation of dihydroorotic acid by nicotinamide adenine dinucleotide (NAD^+) (Section 17.14.1) by transferring a hydride anion from the dihydroorotic acid to NAD^+. The enzyme succinic dehydrogenase that catalyzes the oxidation of succinate to fumarate and contains FAD as a covalently linked prosthetic group also probably functions by removing hydride anions from succinic acid.

Other flavoproteins function by the successive transfer of electrons one at a time; in such reactions, semi-quinone free radicals and their anions are intermediates. These radicals, which are resonance-stabilized,

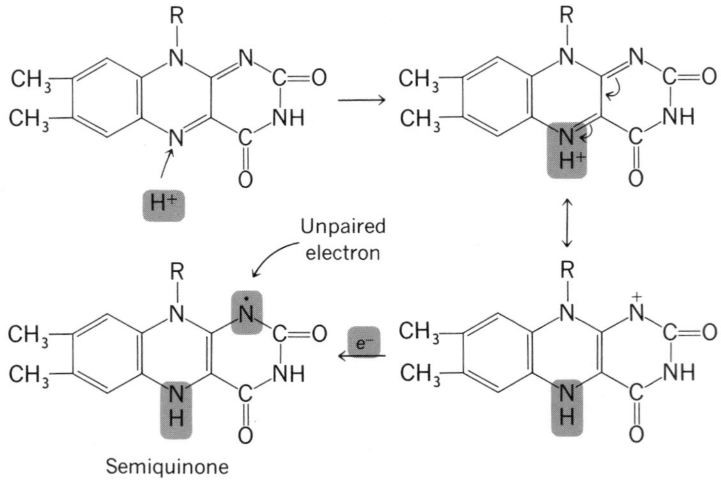

Semiquinone

are able to function as intermediaries between two-electron reducing agents such as NADH and one-electron oxidants such as the ferric ion (Fe^{3+}) found in the mitochondrial electron transport chain. Some flavoproteins contain metals such as molybdenum (Mo^{6+}) and iron (Fe^{3+}) and are known as **metalloflavoproteins.** These metals usually participate by being alternately reduced and oxidized. The iron in iron flavoproteins is frequently of the nonheme iron type (NHI) found in the iron-sulfur proteins known as ferredoxins (Section 14.2.3).

Because of the diverse nature of flavoproteins, this group of catalysts can function as dehydrogenases, oxidases, hydroxylases, and oxidative-decarboxylases. As dehydrogenases, they catalyze the transfer of electrons to acceptors other than O_2, frequently components of the electron transport chain. Flavoprotein oxidases catalyze the transfer of two electrons to O_2 and form H_2O_2. Flavoprotein hydroxylases catalyze the introduction of one or more oxygen atoms into the substrate being oxidized; an example is xanthine oxidase that catalyzes the conversion of hypoxanthine to xanthine, and the conversion of xanthine to uric acid.

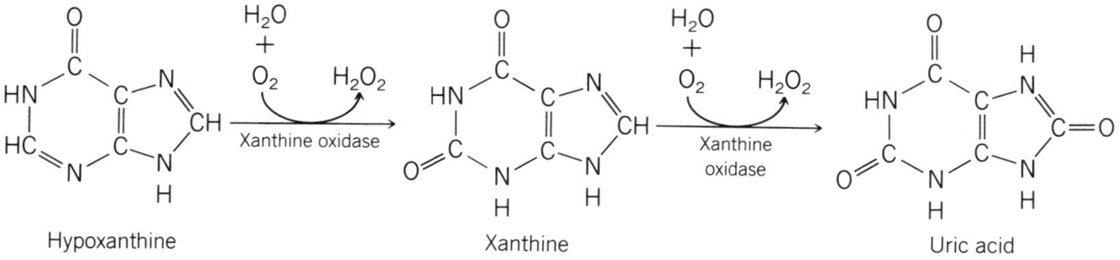

Hypoxanthine   Xanthine   Uric acid

The enzyme known as tryptophan oxidase-decarboxylase or tryptophan monooxygenase catalyzes the conversion of tryptophan to indole acetamide.

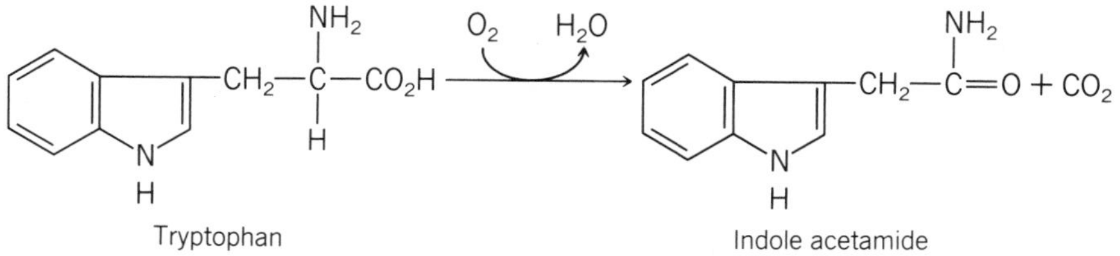

Tryptophan   Indole acetamide

Table 5.2 lists some reactions catalyzed by flavoproteins.

In contrast to the nicotinamide nucleotide dehydrogenases (Section 5.4), the flavin coenzymes tend to be very firmly associated with the protein (apoenzyme) component of the enzyme and are usually carried along during purification of the flavoprotein. In fact, the flavin coenzymes are usually only separated from the apoenzyme after treatment with acid in the cold or even by boiling. The separation by acid in the cold is usually reversible and the addition of FAD or FMN to the apoenzyme restores its catalytic activity. In succinic dehydrogenase, the coenzyme is *covalently* linked to the protein through the methyl group in the isoalloxazine ring.

TABLE 5.2 Some Reactions Catalyzed by Flavoproteins and Metallo-flavoproteins

ENZYME	ELECTRON DONOR	PRODUCT	COENZYME AND OTHER COMPONENTS	ELECTRON ACCEPTOR
D-Amino acid oxidase	D-Amino acids	α-Ketoacids + NH_3	2FAD	$O_2 \longrightarrow H_2O_2$
L-Amino acid oxidase (liver)	L-Amino acid	α-Ketoacids + NH_3	2FAD	$O_2 \longrightarrow H_2O_2$
L-Amino acid oxidase (kidney)	L-Amino acid	α-Ketoacids + NH_3	2FMN	$O_2 \longrightarrow H_2O_2$
L(+)-Lactate dehydrogenase (yeast)	Lactate	Pyruvate	1FMN; 1 heme (cyt b_5)	Respiratory chain
Glycolic acid oxidase	Glycolate	Glyoxylate	FMN	$O_2 \longrightarrow H_2O_2$
NAD$^+$-cytochrome c reductase	NADH	NAD$^+$	2FAD, 2Mo, NHI*	Cytochrome c_{ox}; respiratory chain
NAD$^+$-cytochrome b_5 reductase	NADH	NAD$^+$	FAD; Fe	Cytochrome b_5
Aldehyde oxidase (liver)	Aldehydes	Carboxylic acids	FAD; Fe, Mo	Respiratory chain
α-Glycerol phosphate dehydrogenase	sn-Glycerol-3 phosphate	Dihydroxy-acetone phosphate	FAD; Fe	Respiratory chain
Succinic dehydrogenase	Succinate	Fumarate	FAD; Fe, NHI*	Respiratory chain
Acyl CoA (C_6-C_{12}) dehydrogenase	Acyl CoA	Enoyl CoA	FAD	Electron-transferring flavoprotein
Nitrate reductase	NADPH	NADP$^+$	FAD; Mo, Fe	Nitrate
Nitrite reductase	NADPH	NADP$^+$	FAD; Mo, Fe	Nitrite
Xanthine oxidase	Xanthine	Uric acid	FAD; Mo, Fe	O_2
Lipoyl dehydrogenase	Reduced lipoic acid	Oxidized lipoic acid	2FAD	NAD$^+$
Dihydroorotate dehydrogenase	Dihydroorotic acid	Orotic acid	2FMN; 2FAD, 4Fe	

* NHI, non-heme iron protein.

5.4 NICOTINAMIDE; NICOTINIC ACID

5.4.1 STRUCTURE

The vitamin known as niacin or vitamin B_3 is nicotinic acid. Another form of the vitamin is the amide, nicotinamide or niacinamide.

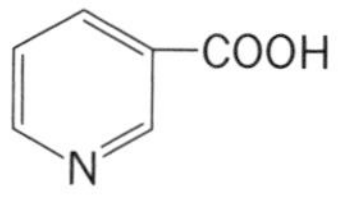

Nicotinic acid 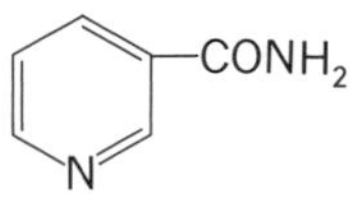   Nicotinamide

5.4.2 OCCURRENCE

Niacin is widely distributed in plant and animal tissues; meat products are an excellent source of the vitamin. The coenzyme forms of the vitamin are the **nicotinamide nucleotide** coenzymes; namely, nicotinamide adenine dinucleotide (NAD^+) and nicotinamide adenine dinucleotide phosphate ($NADP^+$).

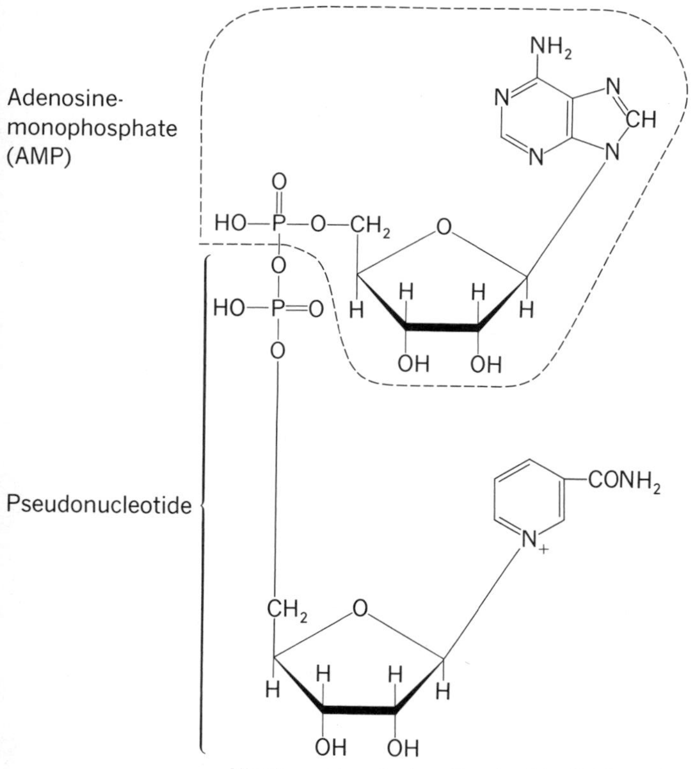

Nicotinamide adenine dinucleotide (NAD^+)
Diphosphopyridine nucleotide (DPN^+)
or coenzyme I

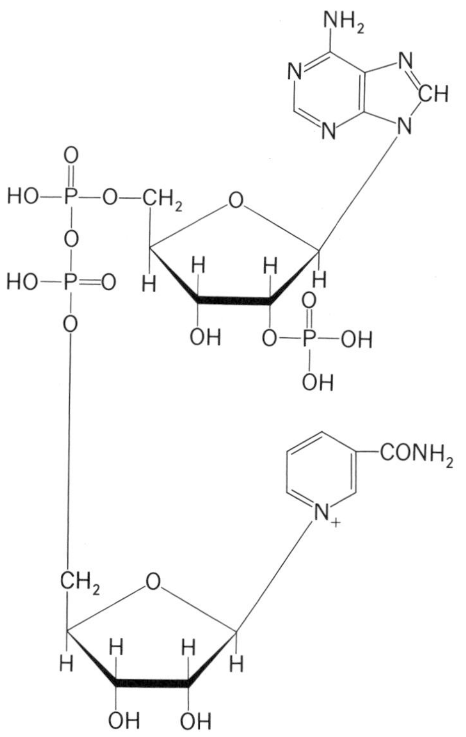

Nicotinamide adenine dinucleotide phosphate ($NADP^+$)
Triphosphopyridine nucleotide (TPN^+)
or coenzyme II

The compounds are composed of a nucleotide (AMP) and a pseudonucleotide, since nicotinamide is not a purine or pyrimidine. Nicotinamide is a pyridine derivative, and for this reason, NAD^+ and $NADP^+$ are sometimes referred to as pyridine nucleotides.

Although the structure and physiological role of these coenzymes were fairly evident by 1935, nicotinic acid was not recognized as a vitamin until 1937, when Elvehjem at the University of Wisconsin established its essential nature. A deficiency of niacin causes pellagra in man and black tongue in dogs. The symptoms of pellagra are dermatitis, especially of skin areas exposed to light; a sore, dark-colored tongue; an inability to digest and assimilate food; and intestinal hemorrhaging. Since niacin gives rise to NAD^+ and $NADP^+$, one might expect that certain essential oxidation–reduction reactions would be affected in niacin deficiency. However, no serious inhibition of such processes has ever been observed.

Nicotinic acid is unusual in that, although a vitamin, man can synthesize it in small quantities from the amino acid tryptophan. Thus, if the dietary source of tryptophan is adequate, some of the daily requirements of 20 mg by the adult human might be met in this way. Since only about one milligram of nicotinic acid can be formed from 50 to 60 mg of tryptophan, there must be an external supply of the vitamin provided.

5.4.3   BIOCHEMICAL FUNCTION

The nicotinamide nucleotides serve as coenzymes for enzymes known as dehydrogenases that catalyze oxidation-reduction reactions. Indeed, these coenzymes are better described as cosubstrates since it is easy to demonstrate their stoichiometric participation in the reaction catalyzed by the dehydrogenase.

As an example, consider the reaction catalyzed by alcohol dehydrogenase, an enzyme widely distributed in nature. The enzyme catalyzes the reversible oxidation of one mole of ethanol by one mole of NAD^+ and forms one mole each of acetaldelyde, NADH and H^+.

$$CH_3CH_2OH + NAD^+ \rightleftharpoons CH_3CHO + NADH + H^+ \tag{5.2}$$

The apparent equilibrium constant of this reaction may be written

$$K_{app} = \frac{[CH_3CHO][NADH]}{[CH_3CH_2OH][NAD^+]} \tag{5.3}$$

When determined experimentally, K_{app} was approximately 10^{-4} at pH 7.0 and 10^{-2} at pH 9.0. The equilibrium constant is therefore obviously related to the pH; this is because a H^+ is a product of Equation 5.2 when alcohol is oxidized. Clearly, the reaction from left to right will be favored by a low H^+ concentration or high pH; on the other hand, the equilibrium would be displaced to the left at high H^+ concentration or low pH.

In order to understand the production of an equivalent of H^+ ion in this reaction, consider the reduction of NAD^+ (or $NADP^+$) more carefully. An examination of the reactions catalyzed by nicotinamide nucleotide dehydrogenases shows that the equivalent of two hydrogen atoms are usually removed from the substrate. In the oxidation of ethanol to acetaldehyde, this occurs by transfer of a hydride ion (a hydrogen atom with an additional electron, H^-) and the release of a proton, H^+.

The oxidized and reduced forms of NAD^+ ($NADP^+$) have the formulas as shown where R equals the remainder of the coenzyme molecule.

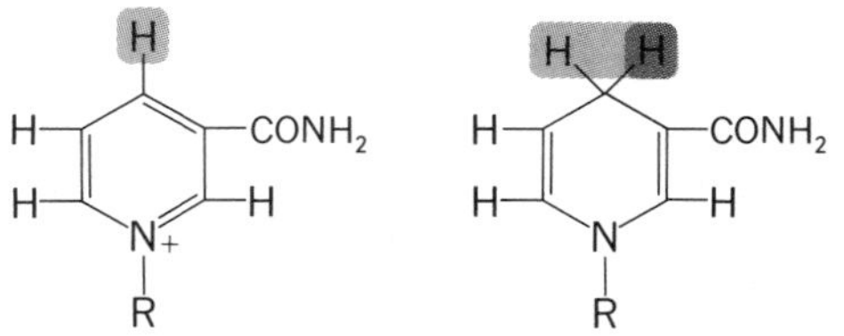

Oxidized
NAD+ or NADP+

Reduced
NADH or NADPH

The NADH (NADPH) is produced by the addition of a hydride anion to the oxidized nucleotide at position 4, where the added hydrogen is known to enter the ring. This can be more readily pictured if we write a resonance form of oxidized NAD$^+$ in which the carbon at position 4 possesses the positive charge usually placed on the nitrogen atom. The proton, required to balance the reaction when a hydride anion is removed from the substrate, is released in solution.

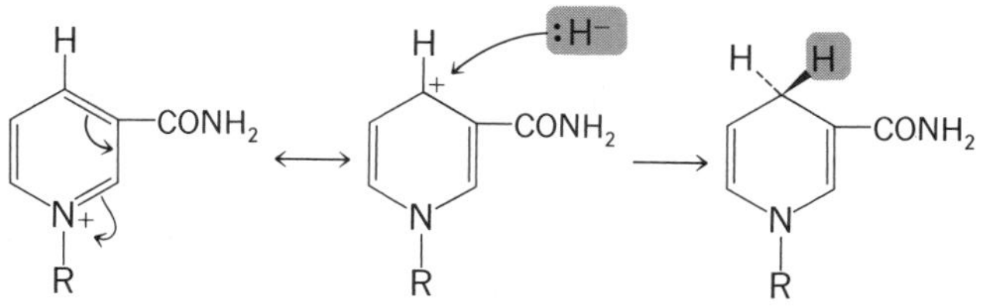

Note that the two hydrogens at position 4 in the reduced NAD$^+$ or NADP$^+$ project out from the planar pyridine ring. The hydride anion that adds to the oxidized coenzyme could add as just shown so that it projects to the front of the ring. It could also add from the rear to form the structure that follows.

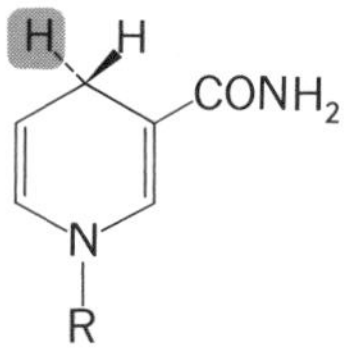

TABLE 5.3 Some Reactions Catalyzed by Nicotinamide Nucleotide Enzymes

ENZYME	SUBSTRATE	PRODUCT	COENZYME
Alcohol dehydro-genase	Ethanol	Acetaldehyde	NAD$^+$
Isocitric dehydro-genase	Isocitrate	α-Ketoglutarate + CO$_2$	NAD$^+$, NADP$^+$
Glycerolphosphate dehydrogenase	sn-Glycerol-3-phosphate	Dihydroxyacetone phosphate	NAD$^+$
Lactic dehydro-genase	Lactate	Pyruvate	NAD$^+$
Malic enzyme	L-Malate	Pyruvate + CO$_2$	NADP$^+$
Glyceraldehyde-3-phosphate dehydrogenase	Glyceraldehyde-3-phosphate + H$_3$PO$_4$	1,3-Diphosphogly-ceric acid	NAD$^+$
Glucose-6-phos-phate dehydro-genase	Glucose-6-phos-phate	6-Phosphogluconic acid	NADP$^+$
Glutamic dehydro-genase	L-Glutamic acid	α-Ketoglutarate + NH$_3$	NAD$^+$, NADP$^+$
Acetaldehyde dehydrogenase	Acetaldehyde	Acetic acid	NAD$^+$

The dehydrogenases that utilize NAD^+ and $NADP^+$ show great specificity with regard to the side of the pyridine ring that the hydride ion approaches. Those in which the added hydrogen projects toward the reader when the ring is shown above are known as A-type dehydrogenases. These include the alcohol dehydrogenase of yeast and the lactic dehydrogenase of heart muscle. Examples of B-type dehydrogenases are liver glucose dehydrogenase (NAD^+) and yeast glucose-6-phosphate dehydrogenase ($NADP^+$).

The nicotinamide nucleotide enzymes exhibit several general modes of action. The dehydrogenases that require NAD^+ and $NADP^+$ catalyze the oxidation of alcohols (primary and secondary), aldehydes, α- and β-hydroxy carboxylic acids, and α-amino acids (Table 5.3). These reactions are frequently readily reversible. In other instances, the value of the equilibrium constant may determine that under physiological conditions the reaction will proceed in only one direction. The reaction, however, may result in either the reduction or oxidation of the nicotinamide nucleotide. Because of this, the nicotinamide nucleotides may readily accept electrons directly from a reduced substrate and donate them directly to an oxidized substrate in a coupled reaction. Thus, the reduction of acetaldehyde to ethanol by yeast (in the presence of alcohol dehydrogenase) is linked to the oxidation of glyceraldehyde-3-phosphate (in the presence of glyceraldehyde-3-phosphate dehydrogenase). A similar coupled reaction occurs with pyruvate to lactate in animal tissues.

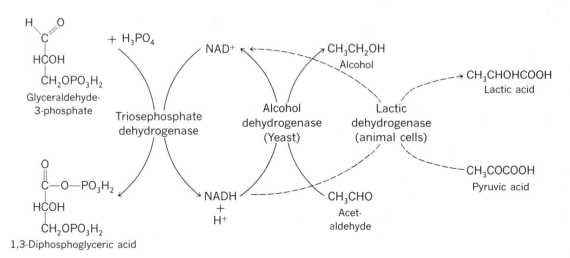

The nicotinamide nucleotides and their dehydrogenases have been a favorite subject for the study of the kinetics and the mechanisms of enzyme action. Several of the dehydrogenases are available in the form of highly purified, crystalline proteins. In addition, there is a convenient method for distinguishing the reduced nicotinamide nucleotide from its oxidized form. The method is based on the observation by Warburg that the free forms of the reduced coenzymes strongly absorb light at 340 nm while the oxidized coenzymes do not. (When the NADH or NADPH is bound to the protein component of the dehydrogenase, the maximum absorption is about 335 nm.)

The absorption spectra of the oxidized and reduced nicotinamide nucleotides are shown in Figure 5.1; the molar absorbancy a_m for the two coenzymes is

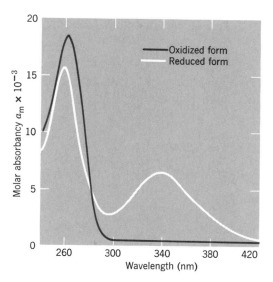

FIGURE 5.1 Absorption spectra of the oxidized and reduced nicotinamide nucleotides.

identical. By measuring the change in the absorption of light at 340 nm during the course of a reaction, it is possible to follow the reduction or oxidation of the coenzyme. An example of such measurements is given in Figure 5.2, in which the reduction of NAD^+ in the presence of ethanol and alcohol dehydrogenase is shown.

In Figure 5.2, the absorbancy at 340 nm is plotted as a function of time. After equilibrium is obtained and no further reduction of NAD^+ occurs, acetaldehyde is added. In adding a product of the reaction, the equilibrium of Reaction 5.2 is displaced to the left and some of the reduced NADH is reoxidized, as indicated by a decrease in the absorption of light at 340 nm. If additional alcohol is added now, the equilibrium is again adjusted, this time from left to right, and NAD^+ reduction results, as shown by the increase in light absorption at 340 nm.

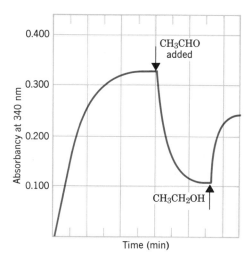

FIGURE 5.2 The reduction and reoxidation of NAD^+ in the presence of ethanol, acetaldehyde, and alcohol dehydrogenase.

5.5 VITAMIN B$_6$ GROUP

5.5.1 STRUCTURE

Three compounds belong to the group of vitamins known as B$_6$. They are **pyridoxal, pyridoxine,** and **pyridoxamine.**

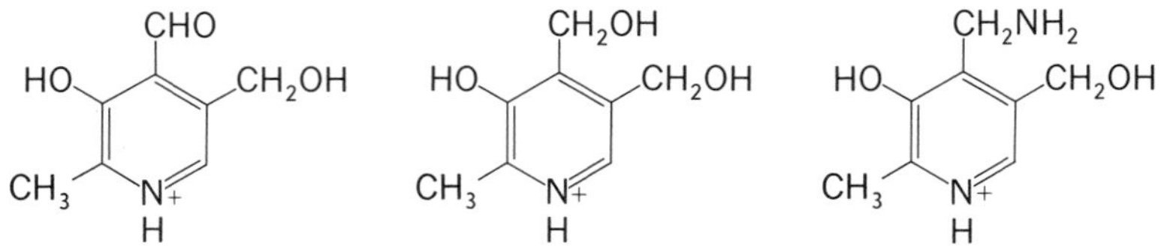

Pyridoxal

Pyridoxine or pyridoxol

Pyridoxamine

5.5.2 OCCURRENCE

The three forms of vitamin B$_6$ are widely distributed in animal and plant sources; cereal grains are especially rich sources of the vitamin. Pyridoxal and pyridoxamine also occur in nature as their phosphate derivatives that are the coenzyme forms of the vitamin.

Ingested pyridoxine is converted in the liver to pyridoxal phosphate by the following sequence:

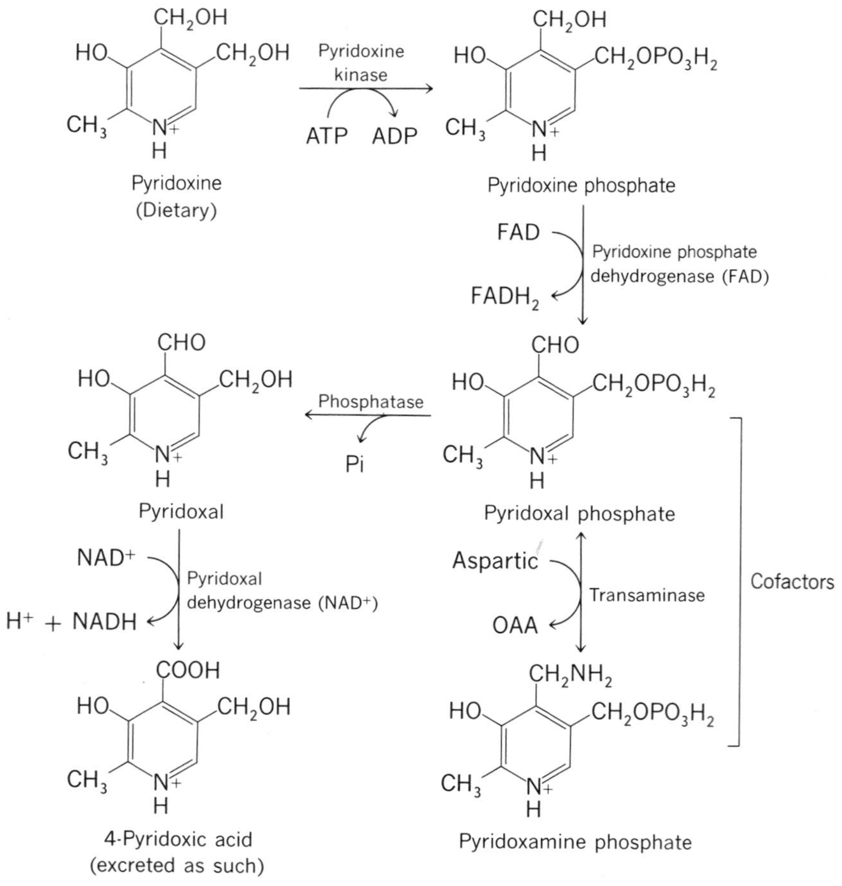

Pyridoxine (Dietary)

Pyridoxine kinase

ATP   ADP

Pyridoxine phosphate

FAD

Pyridoxine phosphate dehydrogenase (FAD)

FADH$_2$

Pyridoxal

Phosphatase

Pi

Pyridoxal phosphate

NAD$^+$

Pyridoxal dehydrogenase (NAD$^+$)

H$^+$ + NADH

Aspartic

Transaminase

OAA

Cofactors

4-Pyridoxic acid (excreted as such)

Pyridoxamine phosphate

Approximately 90% of pyridoxine administered to man is rapidly converted to 4-pyridoxic acid and excreted as such.

All three forms of the vitamin are effective in preventing vitamin B_6 deficiency symptoms that, in rats, occur initially as a severe dermatitis. Extreme deficiency in animals causes convulsions similar to those of epilepsy and indicates a profound disturbance in the central nervous system. The different forms of vitamin B_6 also serve as growth factors for many bacteria.

5.5.3 BIOCHEMICAL FUNCTION

Pyridoxal phosphate is a versatile coenzyme that participates in the catalysis of several important reactions of amino acid metabolism known as transamination, decarboxylation, and racemization.

Transamination:

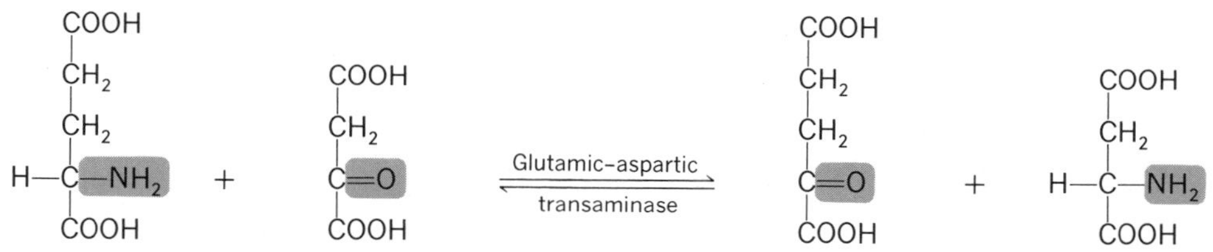

Glutamic acid (Donor amino acid) + Oxalacetic acid (Acceptor keto acid) ⇌ [Glutamic–aspartic transaminase] α-Ketoglutaric acid (Product keto acid) + Aspartic acid (Product amino acid)

Decarboxylation:

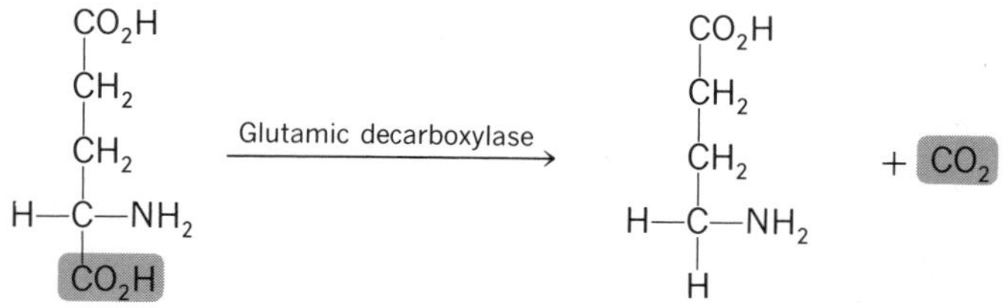

L-Glutamic acid → [Glutamic decarboxylase] γ-Amino butyric acid + CO_2

Racemization:

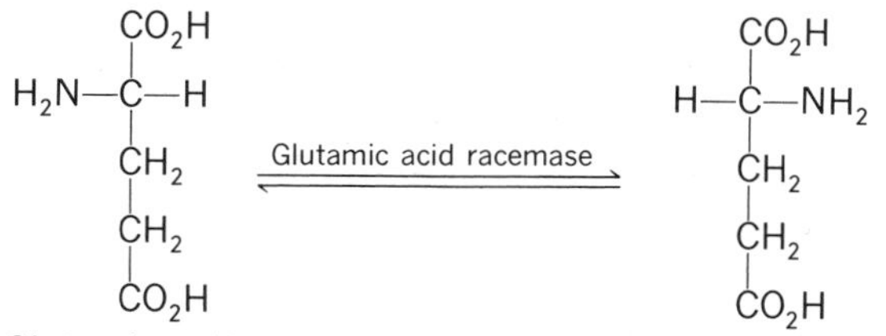

L-Glutamic acid ⇌ [Glutamic acid racemase] D-Glutamic acid

Each reaction is catalyzed by a different, specific apoenzyme, but in each case, pyridoxal phosphate functions as the coenzyme. The mechanisms for these reactions are as follows:

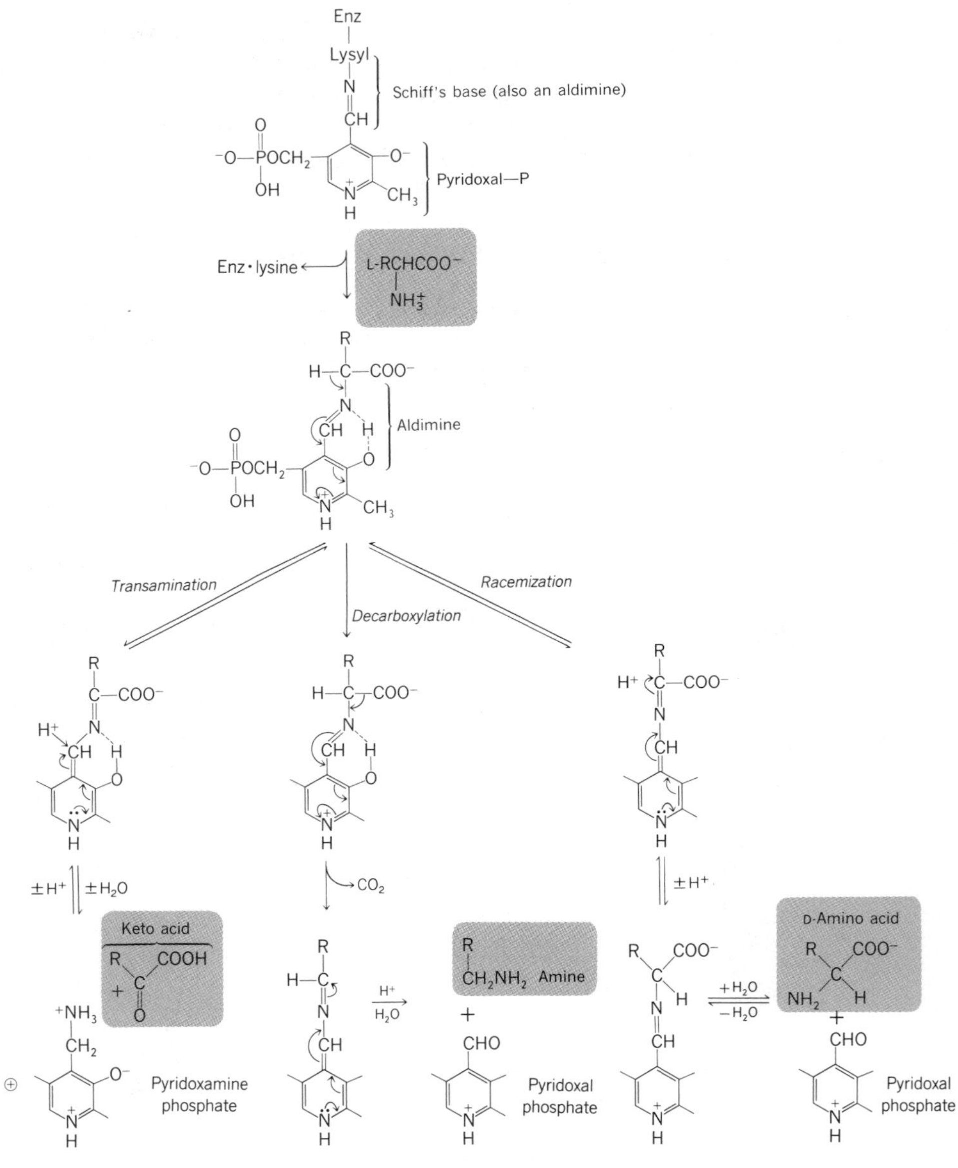

Enz
|
Lysyl
N — Schiff's base (also an aldimine)
‖
CH

$-O-\overset{O}{\underset{OH}{P}}OCH_2$ ... O^- — Pyridoxal—P
$\overset{+}{N}$ ... CH_3
H

Enz·lysine ←   L-RCHCOO⁻
                    |
                   NH₃⁺

R
|
H—C—COO⁻
N — Aldimine
‖
CH   H
|
$-O-\overset{O}{\underset{OH}{P}}OCH_2$ ... O
N ... CH_3
H

Transamination   *Racemization*

Decarboxylation

R
|
C—COO⁻
H⁺ N
‖
CH   H
... O
N
H

±H⁺ ‖ ±H₂O

Keto acid
R   COOH
 \ /
  C
  ‖
  O
+NH₃
CH₂
... O⁻   Pyridoxamine
$\overset{+}{N}$ phosphate
H

R
|
H—C—COO⁻
N
‖
CH   H
... O
$\overset{+}{N}$
H

→ CO₂

R
|
H—C
N
‖
CH
$\overset{+}{N}$
H

$\xrightarrow[H_2O]{H^+}$

R
|
CH₂NH₂ Amine
+
CHO   Pyridoxal
$\overset{+}{N}$ phosphate
H

R
|
H⁺ ⟩C—COO⁻
N
‖
CH
$\overset{+}{N}$
H

±H⁺

R   COO⁻
 \ /
  C
  |
  H
N
‖
CH
$\overset{+}{N}$
H

$\xrightleftharpoons[-H_2O]{+H_2O}$

D-Amino acid
R   COO⁻
 \ /
  C
 / \
NH₂ H
+
CHO   Pyridoxal
$\overset{+}{N}$ phosphate
H

There is now good evidence to support the concept that pyridoxal phosphate is loosely bound as a Schiff's base to the ε-amino group of a lysyl residue in all enzymes involving pyridoxal phosphate. Chemical reduction with sodium borohydride reduces the Schiff's base to a secondary amine and binds pyridoxal phosphate irreversibly to the protein.

Approximately 60 specific reactions of amino acids involving pyridoxal phosphate have been discovered, one of which is the interconversion of serine and glycine. Of unusual interest is the fact that pyridoxal phosphate is found bound to lysine in animal and plant phosphorylases. If the coenzyme is removed from the protein, phosphorylase activity disappears but can be restored by adding pyridoxal phosphate. The precise role of pyridoxal phosphate in this system is unknown.

5.6 LIPOIC ACID

5.6.1 STRUCTURE

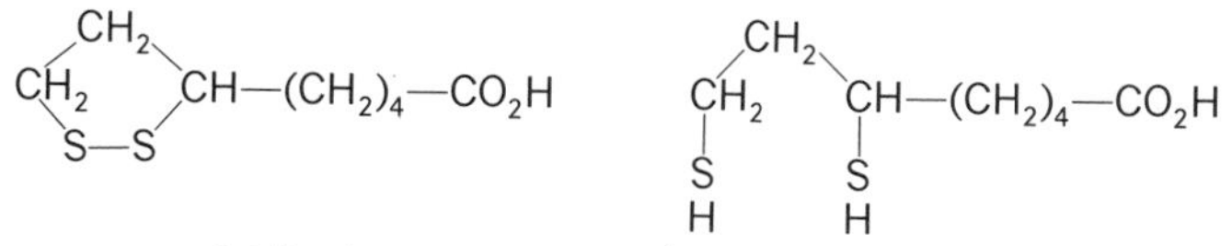

Oxidized   Reduced

Lipoic acid

5.6.2 OCCURRENCE

Lipoic acid was discovered when it was observed as a growth factor for certain bacteria and protozoa. It can therefore be termed a vitamin, an essential nutrient, for those organisms. There is no evidence of a requirement by man who presumably can synthesize it in the amount required. Liver and yeast are rich sources of lipoic acid, but considering the role that it plays as a cofactor, the compound must occur widely in nature. Lipoic acid exists in both oxidized and reduced forms due to the ability of the disulfide linkage to undergo reduction. Bound to protein, lipoic acid is released by acidic, basic, or proteolytic hydrolysis. Careful hydrolysis of lipoyl-protein complexes discloses that lipoate is covalently bound to lysine as ε-N-lipoyl-L-lysine. This structure, which has a striking resemblance to biocytin (ε-N-biotinyl-L-lysine) isolated as a hydrolysis product of biotin-protein complexes, indicates that in lipoyl enzymes, the lipoic acid is bonded to lysyl residues of the protein.

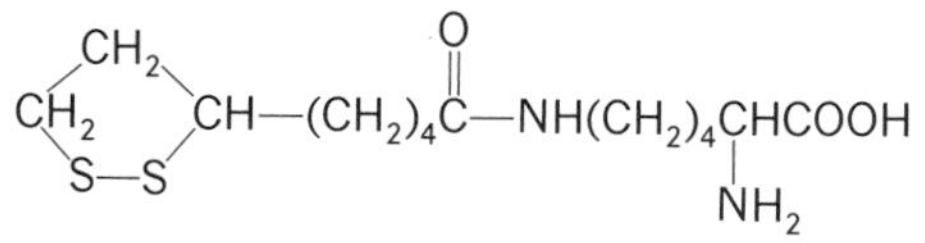

ε-N-Lipoyl-L-lysine

5.6.3 BIOCHEMICAL FUNCTION

Lipoic acid is a cofactor of the multienzyme complexes **pyruvic dehydrogenase** and **α-ketoglutaric dehydrogenase** (Sections 12.2 and 18.4.2). In these complexes, the lipoyl-containing enzymes catalyze the generation and transfer of acyl groups and, in the process, undergo reduction followed by reoxidation. In the initial step involving the lipoic acid moiety, an acylol-thiamin complex (Section 5.2.3) reacts with the oxidized lipoic residue to form an additional complex that subsequently rearranges to form the free thiamin residue and the acyl-lipoic acid complex. It is in this reaction that the acylol moiety is oxidized to an acyl group and the oxidized lipoic reduced.

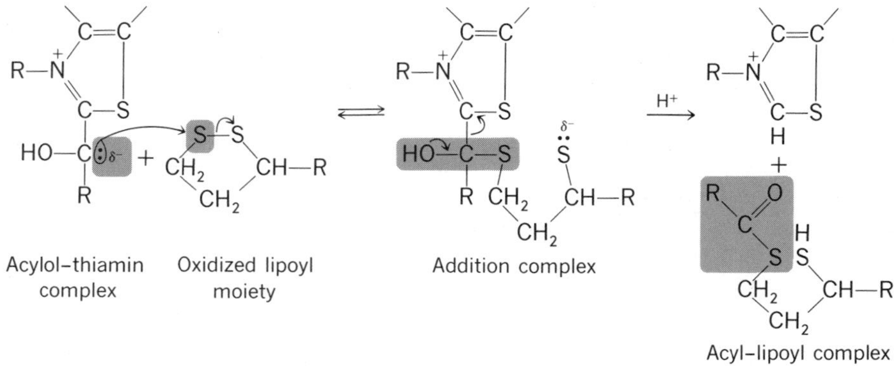

Acylol–thiamin complex | Oxidized lipoyl moiety | Addition complex | Acyl–lipoyl complex

Next, the acyl group is transferred from the acyl-lipoic acid complex to coenzyme A to form acyl CoA and the reduced lipoyl moiety.

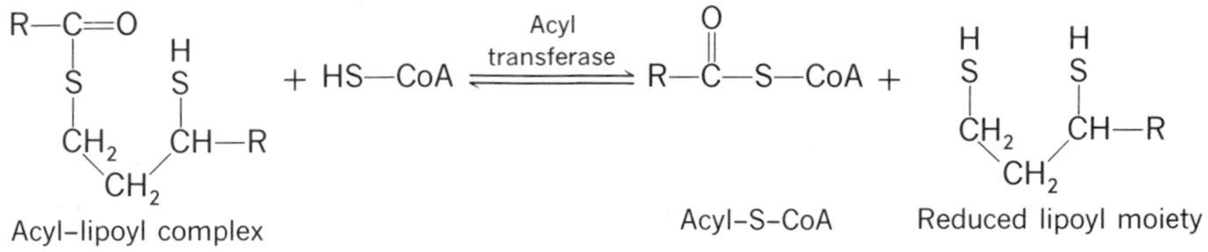

Acyl-lipoyl complex | Acyl–S–CoA | Reduced lipoyl moiety

Finally, the reduced lipoic acid moiety is oxidized by a FAD-containing enzyme to regenerate the oxidized lipoyl moiety and allow the process to be repeated.

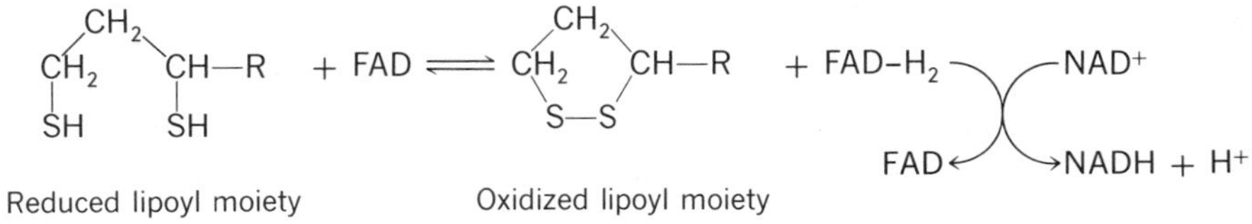

Reduced lipoyl moiety | Oxidized lipoyl moiety

These enzymes will be considered in more depth in Section 12.2.

5.7 BIOTIN

5.7.1 STRUCTURE

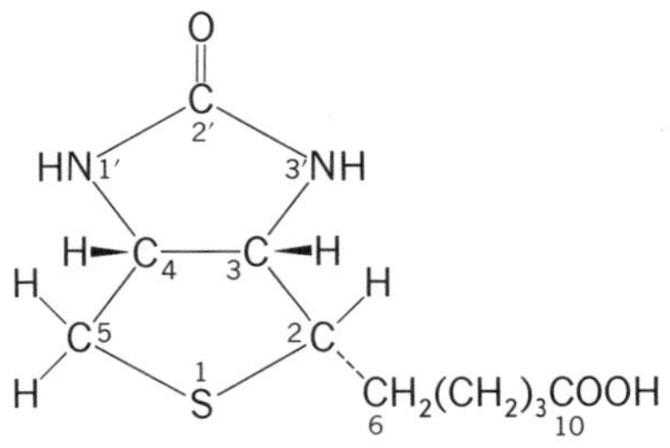

5.7.2 OCCURRENCE

The essential nature of biotin was established by its ability to serve as a growth factor for yeast and certain bacteria. In animals, the biotin requirement is met by the intestinal bacteria that can synthesize the vitamin. Known as the "anti-egg white injury factor," a nutritional deficiency of this vitamin may be induced in animals by feeding them large amounts of avian egg white. Egg white contains a basic protein known as avidin that has a remarkably high affinity for biotin or its simple derivatives. At 25°C, the binding constant is about 10^{15}. Avidin is therefore an extremely effective inhibitor of biotin-requiring systems and is employed by the biochemist to test for possible reactions in which biotin may participate.

Biotin is widely distributed in nature with yeast and liver as excellent sources. The vitamin occurs mainly in combined forms bound to protein through the ε-N-lysine moiety. Biocytin, ε-N-biotinyl-L-lysine, has been isolated as a hydrolysis product from biotin-containing proteins.

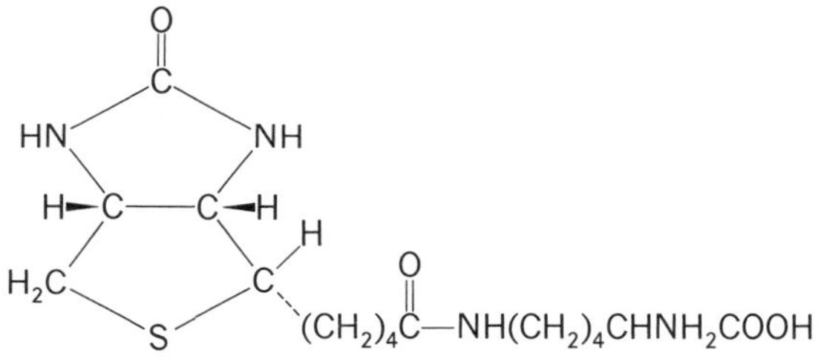

Biocytin

Because of their linkage with proteins through covalent peptide bonds, neither biotin nor lipoic acid is dissociated by dialysis, a technique commonly used to remove readily dissociable groups such as the nicotinamide nucleotides. As a result, no enzymes have been described that can be reactivated by the simple expedient of adding biotin to the apoenzyme.

5.7.3 BIOCHEMICAL FUNCTION

Biotin, bound to its specific enzyme protein, is intimately associated wih carboxylation reactions. The overall reaction catalyzed by biotin-dependent carboxylases can be divided into two discrete steps.

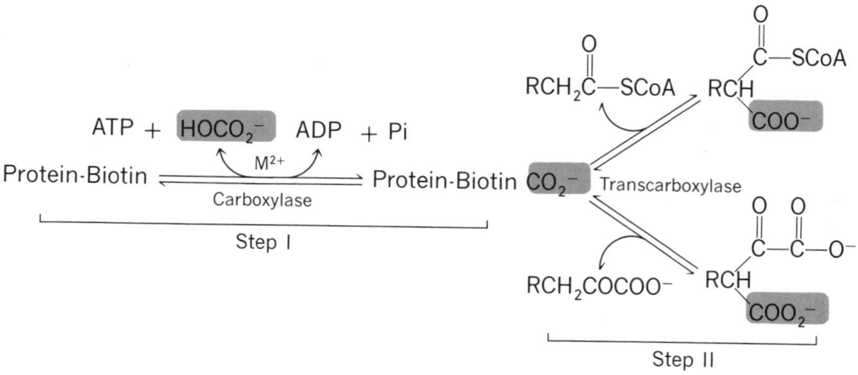

The first step involves the formation of carboxyl biotinyl enzyme; the second step involves carboxyl transfer to an appropriate acceptor substrate depending on the specific transcarboxylase that is involved. Pyruvic carboxylase is an example of an enzyme employing an α-keto acid as an acceptor (Section 10.7.2), while acetyl CoA carboxylase (see the following diagram) and propionyl CoA carboxylase (Section 13.7) are examples of an acyl CoA serving as the specific acceptor.

The mechanism of the conversion of acetyl CoA to malonyl CoA in *E. coli* has been intensively studied; the results show clearly the following sequence in which three proteins participate: (1) biotin carboxylase, (2) biotin carboxyl carrier protein itself (BCCP), and (3) acetyl CoA, malonyl CoA transcarboxylase.

(a) $ATP + HCO_3^- + BCCP \xrightleftharpoons[\text{Biotin carboxylase}]{\text{Mg}^{++}} ADP + Pi + BCCP-CO_2^-$

(b) $BCCP-CO_2^- + \text{Acetyl CoA} \xrightleftharpoons[\text{Trans-carboxylase}]{} BCCP + \text{Malonyl CoA}$

The chemical reactions involved in these steps are believed to be

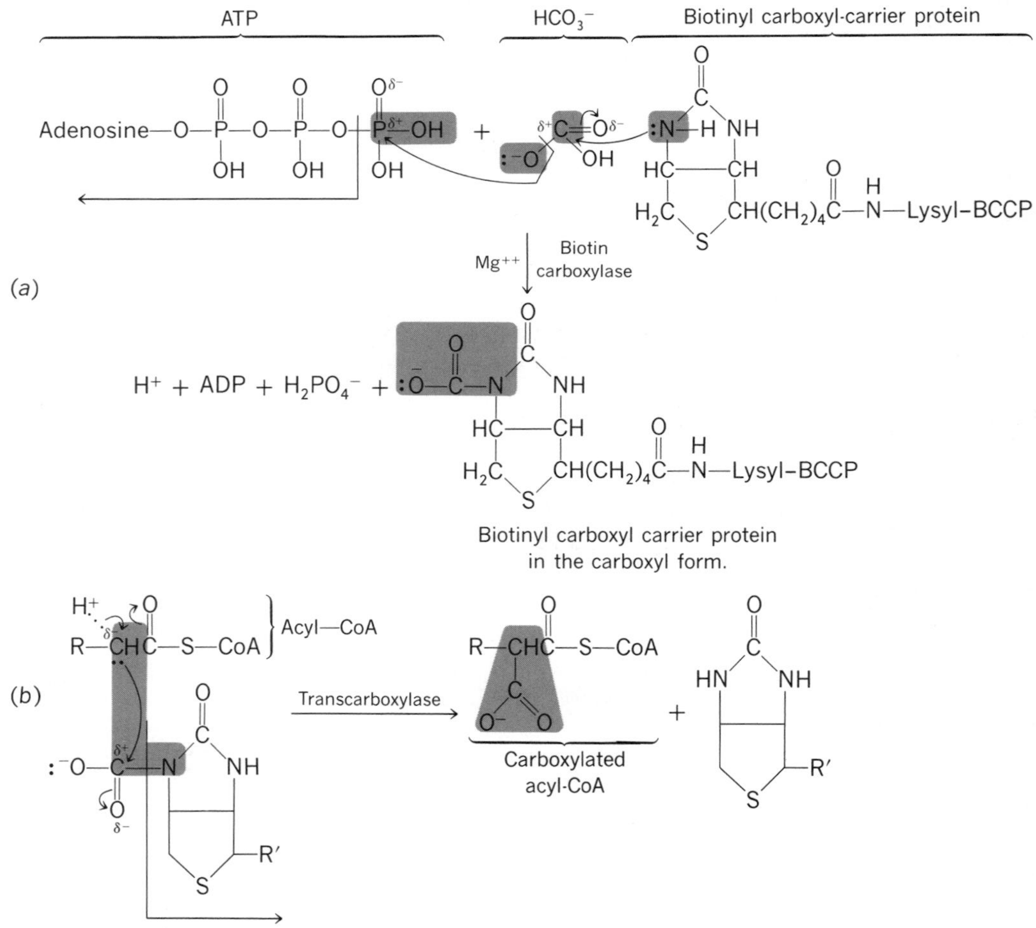

(a)

Biotinyl carboxyl carrier protein
in the carboxyl form.

(b)

Playing a key role in these steps is BCCP, a dimeric protein with a molecular weight of 44,000, containing two moles of biotin/mole of dimer linked to the polypeptide chain via two lysyl bridges. Biotin carboxylase is a dimer with a molecular weight of 98,000 and two similar subunits of 49,000 each, whereas transcarboxylase is a tetramer of 130,000 molecular weight with subunits of 30,000 and 35,000 each.

In animals, the enzyme is a protein with a molecular weight of 460,000 and consisting of two identical subunits of 230,000 each, with one biotin per subunit. The enzyme appears to exist in two forms, an active and an inactive form and these forms are interconverted by a phosphorylation/dephosphorylation cycle (see Section 18.4 for a more detailed discussion of this type of regulation). The active form that does not require citrate for further activation contains four to five moles of phosphate per mole of subunit; the inactive form contains up to eight to nine moles of phosphate per mole of subunit and becomes active only in the presence of citrate. The inactive (but citrate-dependent) form is converted back to the active (citrate-independent) form by a protein phosphatase.

Animal acetyl CoA carboxylase is an unusual protein because the addition of citrate to the protein causes massive polymerization. Earlier data suggested that citrate aggregation is correlated with activation of the enzyme, but recent data clearly show that activation of the enzyme occurs *prior* to polymerization. Polymerization of the carboxylase therefore is a reflection of the physical properties of the enzyme but plays no role in the activation processes.

Furthermore, the rates of synthesis and degradation of acetyl CoA carboxylase increase or decrease, respectively, the levels of the enzyme in the cell depending on the nutritional state of the animal. That is to say, a well-fed animal will have high levels of carboxylase protein, whereas a starved animal will have low levels.

In summary, the animal, bacterial, yeast, and plant enzymes catalyze the identical reaction. However, unlike the bacterial enzyme, the other systems involve a single polypeptide with multifunctions. In the bacterial system, three separate proteins make up the acetyl CoA carboxylase; in the animal system, each function is found in certain areas or domains of the polypeptide. The enzyme protein is thus like a pearl necklace; each pearl represents a domain or activity site and the string is the polypeptide. In the animal system, we therefore have a BCCP domain, a biotin carboxylase domain, a transcarboxylase domain, a citrate-binding domain, and a phosphorylation site. All these domains must be positioned precisely in the polymerized oligomer for the expression of full activity by the enzyme.

5.8 FOLIC ACID

5.8.1 STRUCTURE

2-Amino-4-hydroxy-6-methyl pteridine   *p*-Aminobenzoic acid (PABA) moiety   Glutamic acid

Folic acid [Pteroyl-L-glutamic acid (Pte glu), F]

5.8.2 OCCURRENCE

Folic acid and its derivatives, which are chiefly the tri- and hepta-glutamyl peptides, are widespread in nature. The vitamin cures nutritional anemia in chicks and serves as a specific growth factor in a number of microorganisms. Since extremely small amounts are needed by experimental animals, it is very difficult to produce folic acid deficiencies. Intestinal bacteria provide the small amounts necessary for growth. Derivatives of folic acid play an important but unknown role in the formation of normal erythrocytes.

5.8.3 BIOCHEMICAL FUNCTION

Although folic acid is the vitamin, its reduction products are the actual coenzyme forms. An enzyme, L-folate reductase, reduces folic acid to dihydrofolic acid (H_2F); this compound is

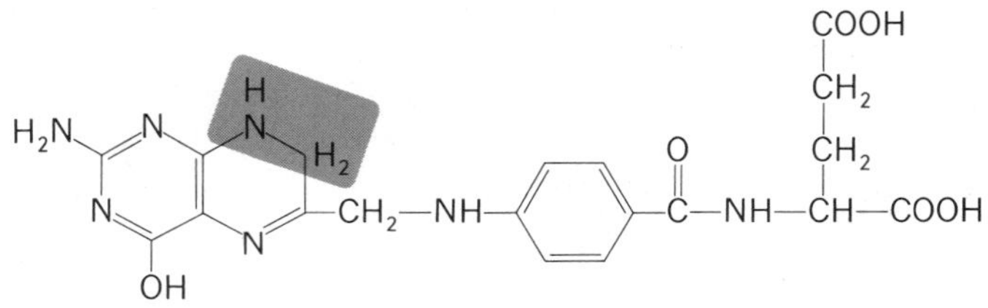

Dihydrofolic acid (H_2F)

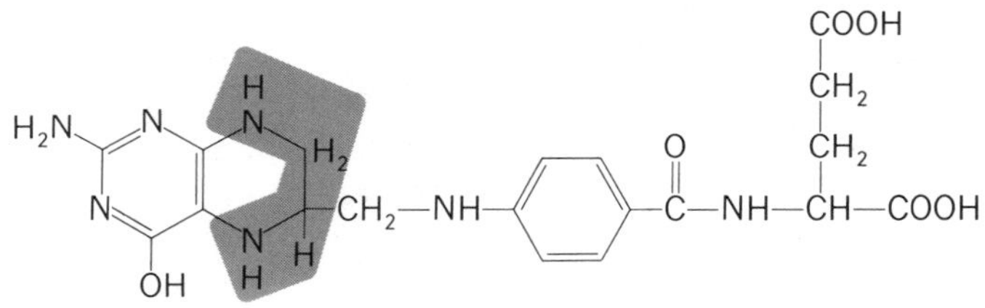

Tetrahydrofolic acid (H_4F)

reduced, in turn, by dihydrofolic reductase to tetrahydrofolic acid (H_4F). The reducing agent in both reactions is NADPH.

$$F + NADPH + H^+ \xrightarrow{\text{Folate reductase}} H_2F + NADP^+$$

$$H_2F + NADPH + H^+ \xrightarrow{\text{H}_2\text{F Reductase}} H_4F + NADP^+$$

The central role of tetrahydrofolic acid, H_4F, is that of a carrier for a one-carbon unit at the oxidation level of formate (or formaldehyde). The formate unit is used in the biosynthesis of pyrimidines, purines, serine, and glycine. The chemistry of this formate unit is complex, but involves initially the activation of formic acid.

$$H_4F + ATP + HCOOH \xrightarrow[\text{synthetase}]{\text{10-Formyl H}_4\text{F}} N^{10}\text{-formyl-}H_4F + ADP + P_i$$

N^{10}-Formyl-H_4F undergoes ring closure to $N^{5,10}$-methenyl-H_4F (as shown in the following diagram),

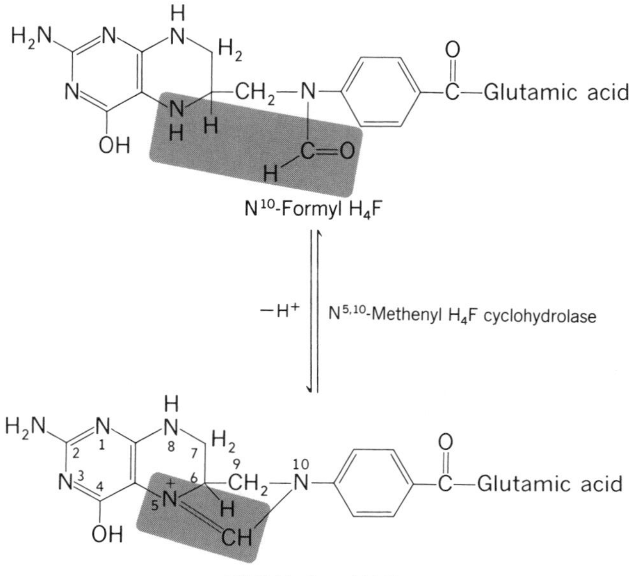

N^{10}-Formyl H_4F

$-H^+$ | $N^{5,10}$-Methenyl H_4F cyclohydrolase

$N^{5,10}$-Methenyl H_4F

$N^{5,10}$-Methenyl-H_4F is then reduced by NADPH in the presence of a specific dehydrogenase to form $N^{5,10}$-methylene-H_4F, the derivative of H_4F that functions as a coenzyme.

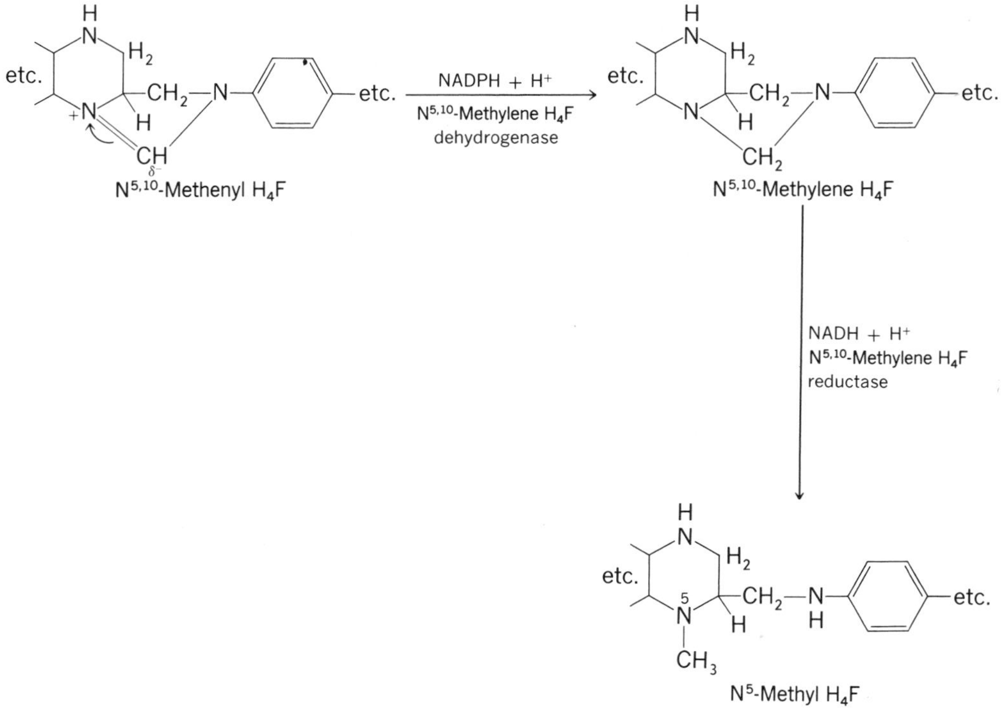

$N^{5,10}$-Methenyl H_4F

NADPH + H$^+$
$N^{5,10}$-Methylene H_4F
dehydrogenase

$N^{5,10}$-Methylene H_4F

NADH + H$^+$
$N^{5,10}$-Methylene H_4F
reductase

N^5-Methyl H_4F

While tetrahydrofolic acid and its C_1 derivatives participate in a considerable number of reactions, we shall only consider those that are summarized next.

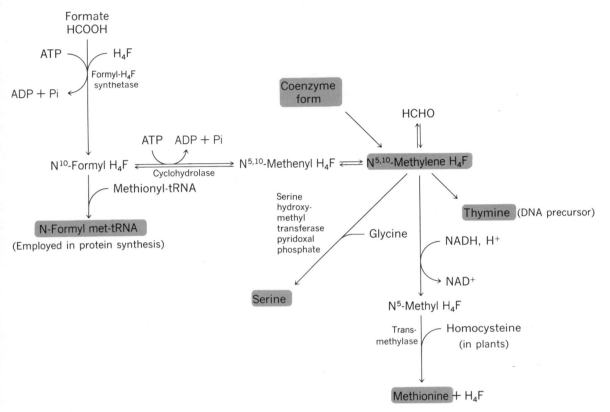

5.8.3.1 Serine-Glycine Interconversion.

As shown on page 189, $N^{5,10}$-methylene-H_4F, in the presence of pyridoxal phosphate, serine hydroxymethyl transferase, and glycine, forms serine. Of interest is the observation that not one but two derivatives of important vitamins, folic acid and pyridoxal phosphate, are required as cofactors for the utilization of formate and glycine to form serine; that is an excellent example of the intermeshing of vitamins in the tissue economy.

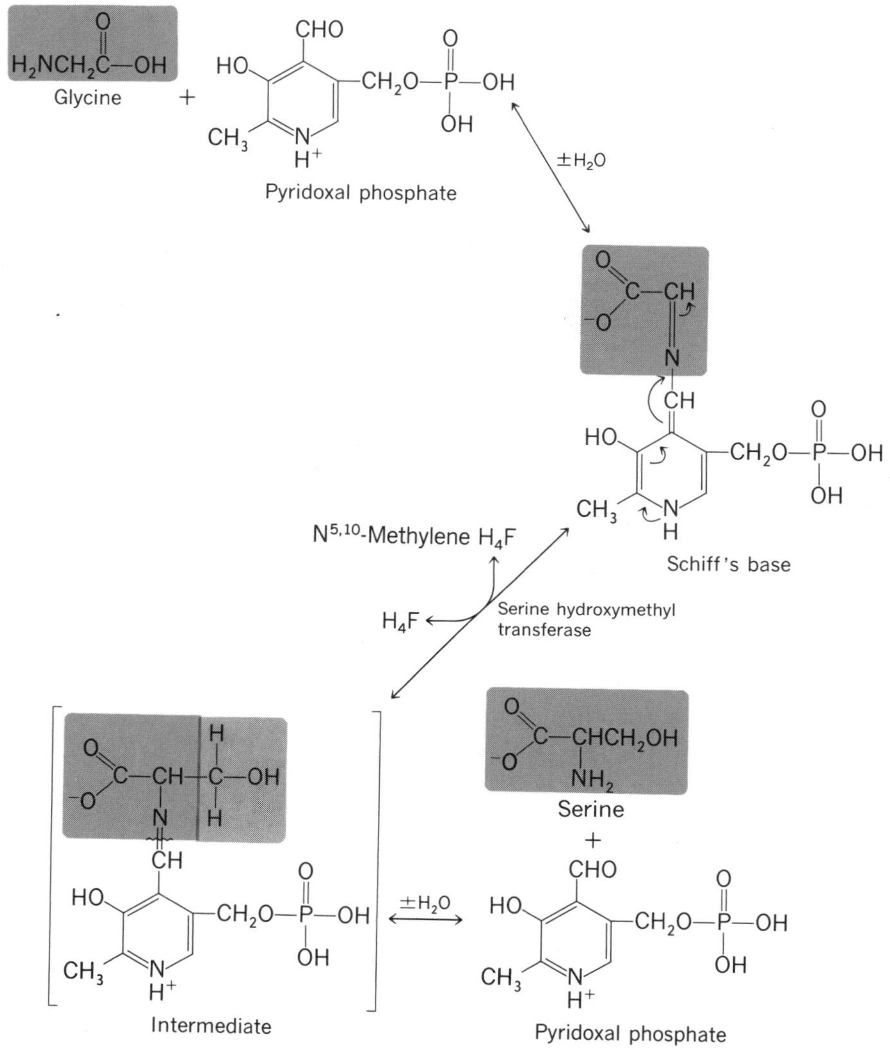

5.8.3.2 Thymidine-5′-phosphate Biosynthesis. This sequence is of critical importance since the pyrimidine base, thymine, is one of four bases found in all DNAs. As shown on page 190, the series of reactions revolve around the regeneration of $N^{5,10}$-methylene-H_4F and the transfer of the C_1 group to deoxyuridine 5′-PO_4 by the enzyme thymidylate synthetase. Cobamid coenzyme (Section 5.9) is necessary for these reactions but its function is as yet unknown.

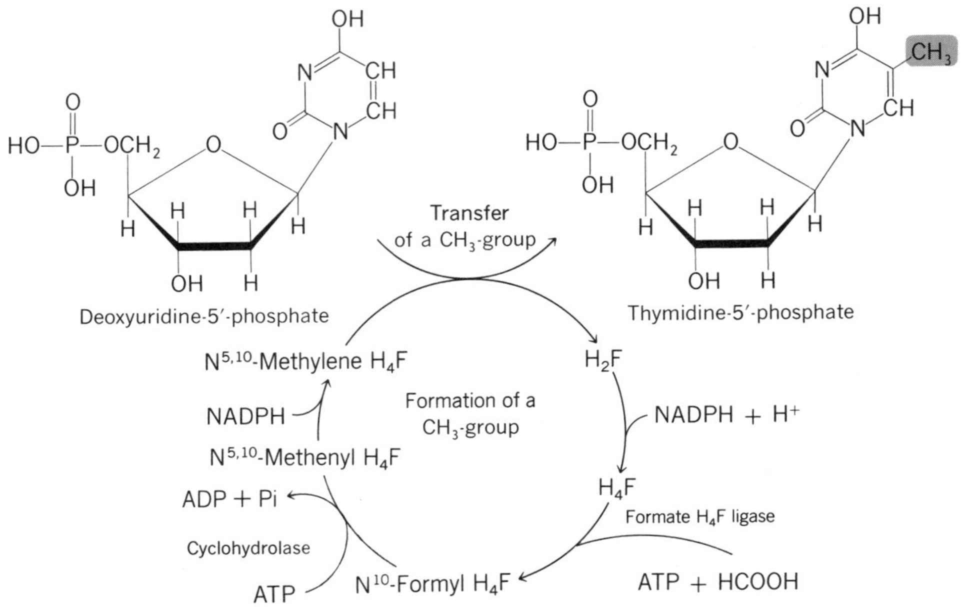

Deoxyuridine-5'-phosphate

Thymidine-5'-phosphate

Transfer of a CH_3-group

$N^{5,10}$-Methylene H_4F

H_2F

Formation of a CH_3-group

NADPH

NADPH + H^+

$N^{5,10}$-Methenyl H_4F

ADP + Pi

H_4F

Formate H_4F ligase

Cyclohydrolase

ATP

N^{10}-Formyl H_4F

ATP + HCOOH

5.8.3.3 Biosynthesis of Methionine.

Two important systems are known that synthesize methionine.

Sequence (*a*): Homocysteine + N^5-Methyl-H_4F $\xrightarrow[\substack{N^5\text{-}CH_3\text{-}H_4F\\ \text{Homocysteine}\\ \text{transmethylase}}]{Mg^{++}}$ Methionine + H_4F

Sequence (*b*): Cobalamin-Enz (Inactive)

S-Adenosyl methionine   S-Adenosyl homocysteine

$FADH_2$ FAD

Methyl cobalamin-Enz (Active)

H_4F

Cobalamin-Enz (Active)

Homocysteine

N^5-Methyl-H_4F

Methionine

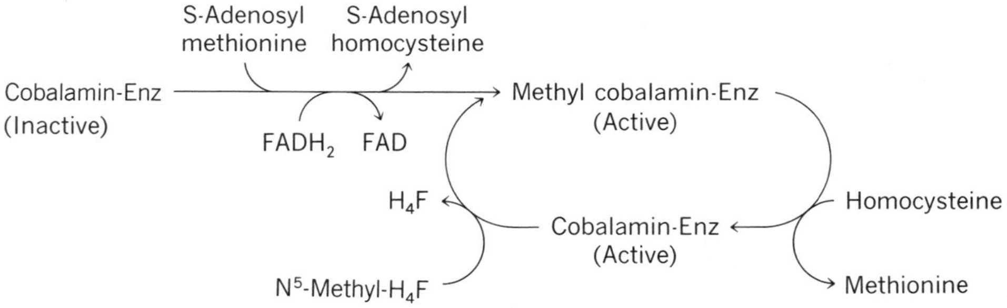

Many bacteria and all plants synthesize methionine by sequence (*a*); this sequence does not occur in mammalian tissues. Because no cobalamin enzymes occur in plants, sequence (*a*) is the only pathway for the synthesis of methyl groups in all plants.

Sequence (*b*) is present in many bacteria and occurs exclusively in mammalian tissues. As illustrated in the previous diagram, the first step involves the reduction and methylation of the inactive cobalamin enzyme, cobalamin-N^5-methyl H_4F homocysteine methyltransferase, to the active form with *S*-adenosyl methionine (Section 17.5.2) as the methylating reagent. The methylated active enzyme now transfers its methyl group to homocysteine, the acceptor, to form methionine and regenerate the active cobalamin enzyme that now is methylated by N^5-Methyl-H_4F. Thus, *S*-adenosyl methionine initiates the reaction, whereas N^5-Methyl-H_4F serves as the methyl donor for the synthesis of methionine.

5.9 VITAMIN B_{12}

5.9.1 STRUCTURE

Vitamin B_{12} as it is isolated from liver is a cyanocobalamin whose structure is shown here.

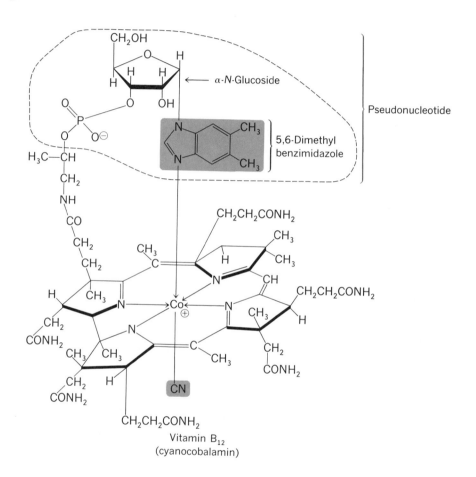

Vitamin B_{12}
(cyanocobalamin)

Vitamin B_{12} may also be isolated with anions other than cyanide; for example, hydroxyl, nitrite, chloride, or sulphate. Still other vitamin B_{12}-like compounds in which the 5,6-dimethylbenzimidazole moiety is replaced by other nitrogenous bases have been isolated from bacteria. In pseudo-vitamin B_{12}, the nitrogenous base is adenine; in another form of the vitamin, the base is 5-hydroxy-benzimidazole.

5.9.2 OCCURRENCE

Vitamin B_{12}, which has been found only in animals and microorganisms and not in plants, occurs as part of a coenzyme known as coenzyme B_{12}, which has the following structure. In the coenzyme, the position occupied by cyanide in vitamin B_{12}

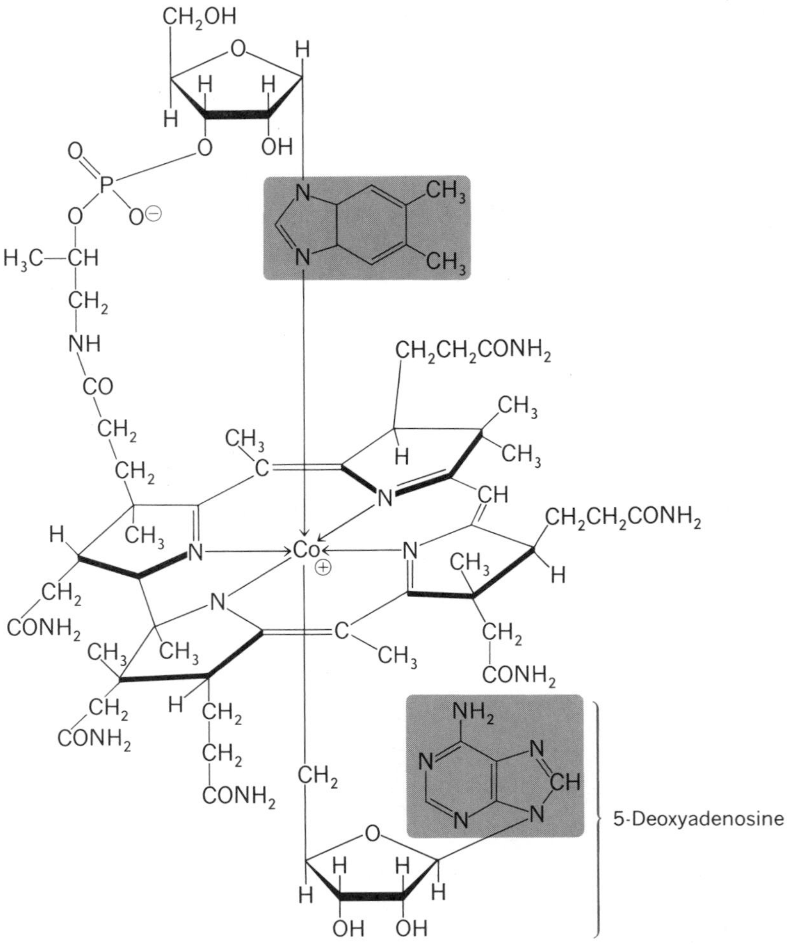

Coenzyme B$_{12}$;
cobamide coenzyme

is bonded directly to the 5′-carbon atom of the ribose of adenosine. This peculiar organometallo bonding is interesting as the methylene group is a reactive center in this coenzyme.

The coenzyme is relatively unstable, and in the presence of light or cyanide is decomposed, respectively, to the hydroxycobalamin or cyanocobalamin form of the vitamin. Hence, the very distinct possibility exists that vitamin B$_{12}$ occurs in nature chiefly as coenzyme B$_{12}$.

Pseudo-vitamin B$_{12}$ occurs with adenine rather than 5,6-dimethylbenzimid-azole as the base attached to ribose; there also exists a coenzyme form of pseudo-vitamin B$_{12}$. A coenzyme form of the vitamin that contains 5-hydroxy-benzimidazole also occurs.

Vitamin B$_{12}$ was first recognized as an agent (extrinsic factor) useful in the prevention and treatment of pernicious anemia. The intrinsic factor, a muco-polysaccharide from gastric mucosa cells, forms a complex with the extrinsic factor that is absorbed from the ileum. If the intrinsic factor is not present, vitamin B$_{12}$ is not absorbed; vitamin B$_{12}$ is also a growth factor for several bacteria and a protozoan, *Euglena.*

5.9.3 BIOCHEMICAL FUNCTION

The coenzyme is synthesized from vitamin B_{12} by a specific B_{12} coenzyme synthetase.

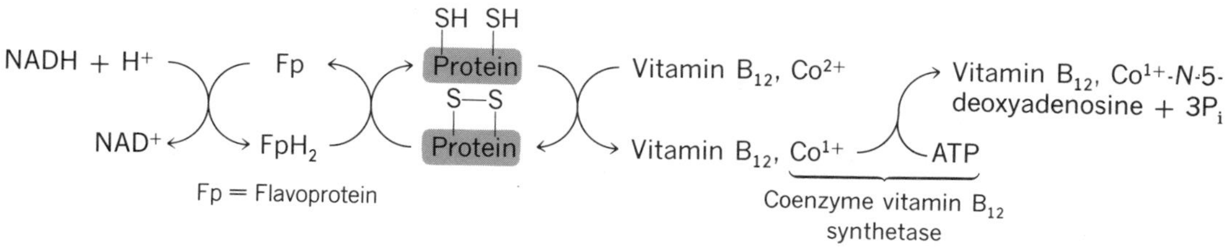

Fp = Flavoprotein

Coenzyme vitamin B_{12} synthetase

The reducing system is complex in that it involves a NADH-flavoprotein-di-sulfide (S—S) protein system. The reductant, NADH, transfers its electrons via a flavoprotein to the specific disulfide (S—S) protein to form a dithiol (SH, SH) protein that converts vitamin $B_{12}(Co^{2+})$ to vitamin $B_{12}(Co^{1+})$. This reduced form then becomes the substrate for the alkylation reaction with ATP.

The B_{12} coenzyme participates in approximately 11 distinct biochemical reactions as well as reactions by which CH_3-vitamin B_{12}-enzyme complex is either reduced to methane or carboxylated by CO_2 to form acetate. Of all these reactions, only that catalyzed by methylmalonyl CoA mutase occurs in animal tissue: All 11 reactions have been discovered and described in bacterial systems. No vitamin B_{12} coenzyme-linked reactions have been observed in higher plants. Of interest, persons with pernicious anaemia excrete methyl malonate in their urine. Obviously, the presence of this acid in significant amounts in urine serves as a diagnostic test for this disease.

Vitamin B_{12} coenzyme reactions can be grouped into four general systems. Specific examples of these four general reactions are the following.

1. *General:* Carbon-carbon bond cleavage.
   Specific: L-Methylmalonyl CoA mutase, which uses 5'-deoxy-adenosyl-cobalamin as a coenzyme (Section 13.7)

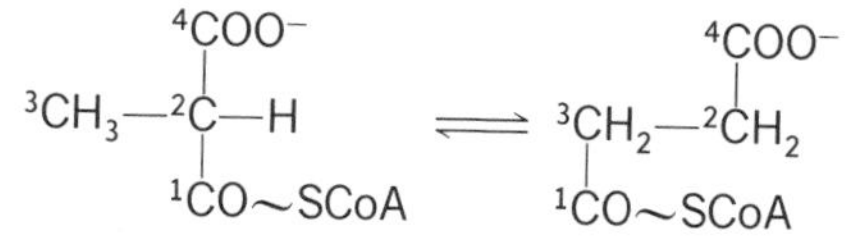

L-Methylmalonyl CoA   Succinyl CoA

2. *General:* Carbon-oxygen bond cleavage.
   Specific: Diol dehydrase. This type of reaction occurs in bacteria. The enzymatic mechanism is very complicated.

$$CH_3CHCH_2OH \longrightarrow CH_3CH_2CHO + H_2O$$
$$\overset{|}{OH}$$

3. *General:* Carbon-nitrogen bond cleavage.
Specific: D-α-Lysine mutase.

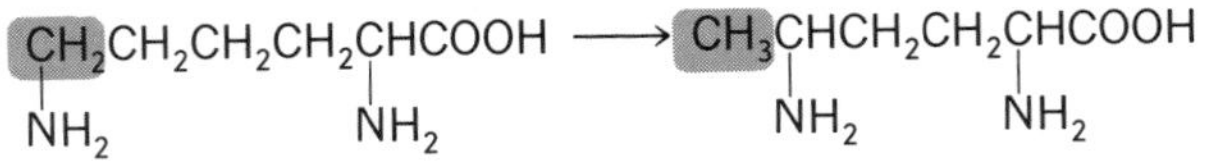

$$CH_2CH_2CH_2CH_2CHCOOH \longrightarrow CH_3CHCH_2CH_2CHCOOH$$
$$\quad\quad NH_2 \qquad\qquad\quad NH_2 \qquad\qquad NH_2 \qquad\quad NH_2$$

4. Methyl activation.

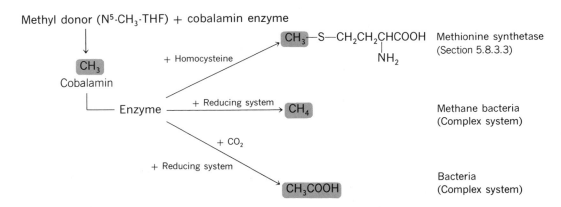

5.10 **PANTOTHENIC ACID**

5.10.1   STRUCTURE

Pantothenic acid is required by animals as well as microorganisms. It was first detected because of its ability to stimulate the growth of yeast. Like biotin, pantothenic acid is produced by intestinal bacteria and it is therefore difficult to demonstrate a nutritional deficiency.

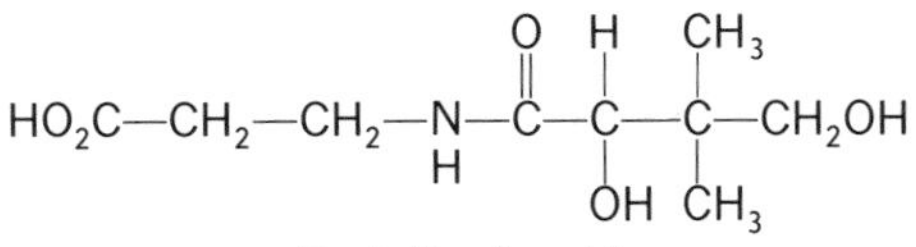

$$HO_2C—CH_2—CH_2—N—\overset{O}{\overset{\|}{C}}—\overset{H}{\underset{OH}{C}}—\overset{CH_3}{\underset{CH_3}{C}}—CH_2OH$$

Pantothenic acid

5.10.2   OCCURRENCE

The vitamin occurs in nature as a component of coenzyme A and acyl-carrier protein (ACP). Coenzyme A shown on p. 195 was discovered and named because it is required for the enzymic acetylation of aromatic amines. It is the coenzyme for acetylation. Coenzyme A was isolated and its structure determined in the late 1940s by F. Lipmann. The complete chemical synthesis of the coenzyme was described by Khorana in 1959.

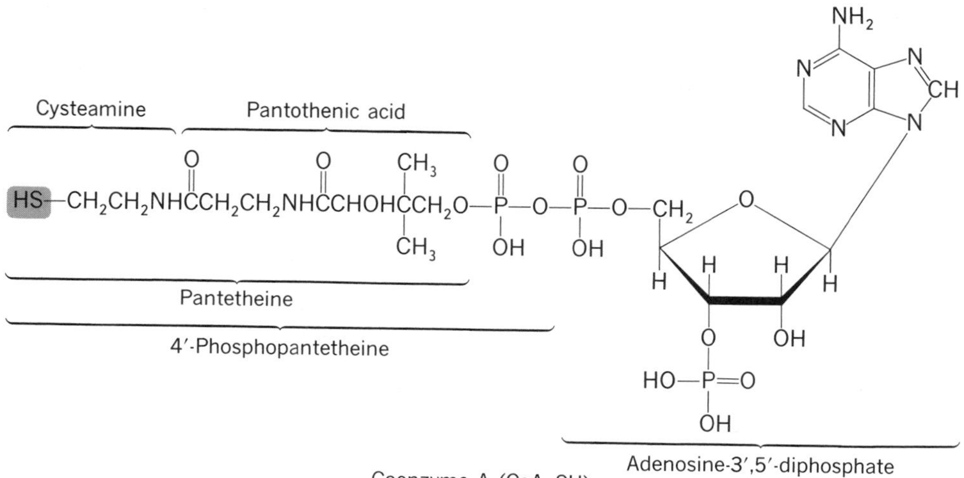

Coenzyme A (CoA–SH)

5.10.3 BIOCHEMICAL FUNCTION

Thioesters formed from coenzyme A and carboxylic acids possess unique properties that account for the role the coenzyme plays in biochemistry. These properties are best understood when compared with certain properties of oxygen esters. It is possible to write a resonance form of an oxygen ester in which

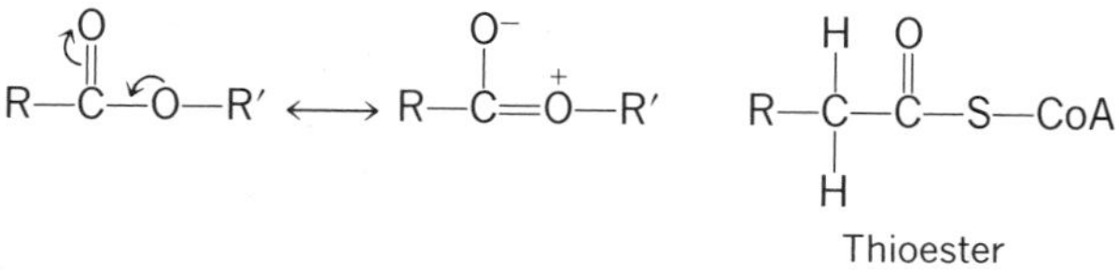

Thioester

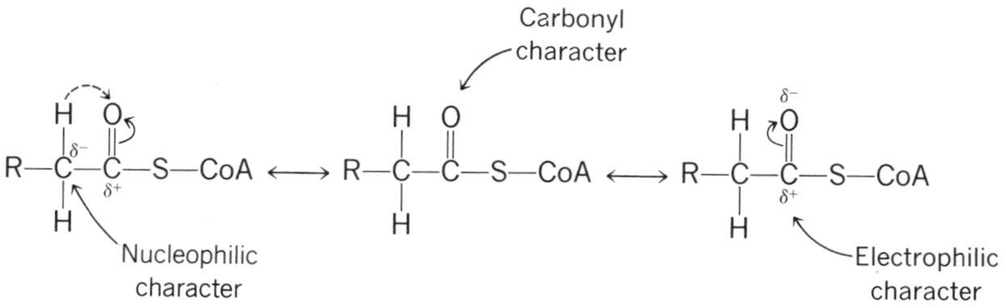

the ester oxygen atom contains a positive charge and is double-bonded to the carboxylic acid carbon. Sulfur, however, does not readily release its electrons for double bond formation, and thioesters therefore do not exhibit the resonance forms written for the oxygen ester. Instead, thioesters exhibit considerable carbonyl character in which a fractional positive charge may be represented on the carboxyl carbon; the carboxyl oxygen therefore exhibits a partial negative charge. With the fractional positive charge on the carboxyl carbon, the hydrogen atom on the adjacent α-carbon will tend to dissociate as a proton leaving a fractional negative charge on that α-carbon. These two possibilities are responsible for the electrophilic character of the carboxyl carbon atom in thioesters as well as the nucleophilic character of the α-carbon atom. Moreover, the inability of thioesters to possess the resonance forms written above for ordinary oxygen esters explains their significantly greater instability and the higher $\Delta G'$ of hydrolysis exhibited by these compounds.

Nucleophiles such as amines, ammonia, water, thiol compounds, and phosphoric acid can attack the electrophilic site and displace the —S—CoA group. Electrophiles such as CO_2, acyl CoA, or the CO_2BCCP complex (Section 5.7.3) can, in turn, attack the nucleophilic site.

Nucleophile (δ^-)

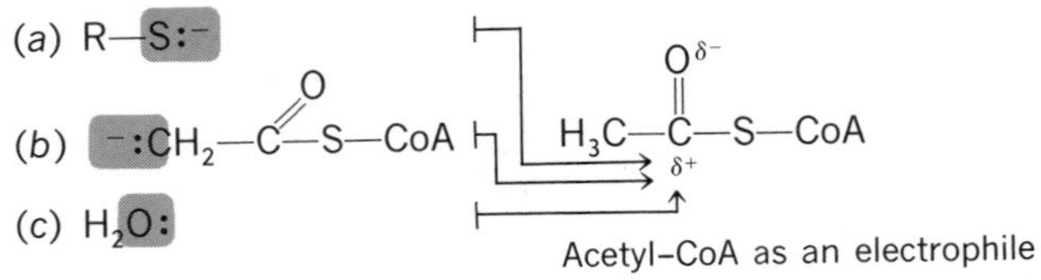

(a) R—S:⁻

(b) ⁻:CH_2—C(=O)—S—CoA

(c) H_2O:

H_3C—C(=O, δ^-)—S—CoA, δ^+

Acetyl–CoA as an electrophile

Electrophile (δ^+)

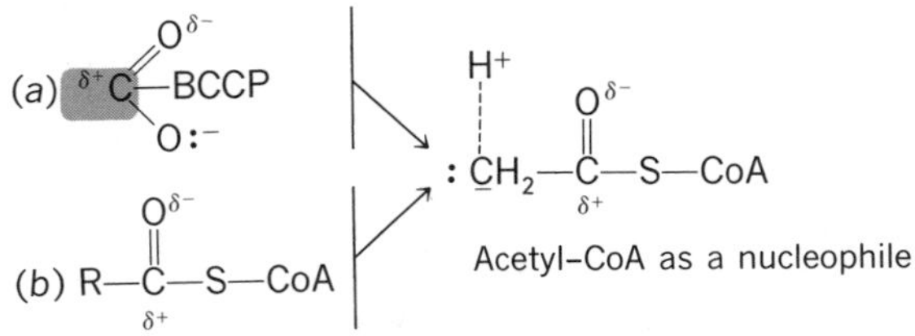

(a) δ^+C(=O, δ^-)(O:⁻)—BCCP

(b) R—C(=O, δ^-)(δ^+)—S—CoA

H⁺
: CH_2—C(=O, δ^-)(δ^+)—S—CoA

Acetyl–CoA as a nucleophile

Throughout the text, numerous examples are given of the reactivities of thioesters of coenzyme A. Most if not all of these reactions can be explained on the basis of the dual reactivity of the ester. The student should attempt to gather the many coenzyme A reactions and explain the mechanisms to his or her own satisfaction. Several examples and further discussion will be found in Chapters 12 and 13.

An interesting heat-stable protein, of low molecular weight and called **acyl carrier protein** (ACP), plays an important role in the biosynthesis of fatty acids (see Section 13.9). A distinctive feature of this protein is the 4'-phosphoryl-pantetheine moiety that is covalently bonded to the hydroxyl group of a serine residue in the protein. Having the pantetheine structure, the molecule can serve as an acyl carrier in a manner analogous to coenzyme A through thioester formation with its sulfhydryl group. Soluble ACPs occur in plant and bacterial tissues, but in animal tissues, part of the ACP molecule is tightly bound to the fatty acid synthetase polypeptide (see Figure 13.13).

The E. coli ACP as well as a number of plant ACPs has been carefully studied. Its molecular weight is 8700 and it has 77 amino acid residues. Its complete sequence with Ser* as the site for 4'-phosphopantetheine is

$\overset{1}{NH_2}$-Ser-Thr-Ile-Glu-Glu-Arg-Val-Lys-Lys-$\overset{10}{Ile}$-Ile-Gly-Glu-

Gln–Leu–Gly–Val–Lys–Gln–$\overset{20}{Glu}$–Glu–Val–Thr–Asp–Asn–Ala–Ser–

Phe–Val–$\overset{30}{Glu}$–Asp–Leu–Gly–Ala–Asp–$\overset{36}{\overset{*}{Ser}}$–Leu–Asp–Thr–$\overset{40}{Val}$–Glu–

Leu–Val–Met–Ala–Leu–Glu–Glu–Glu–$\overset{50}{Phe}$–Asp–Thr–Glu–Ile–Pro–

Asp–Glu–Glu–Ala–$\overset{60}{Glu}$–Lys–Ile–Thr–Thr–Val–Gln –Ala–Ala–Ile–

$\overset{70}{Asp}$-Tyr-Ile-Asn-Gly-His-Gln-$\overset{77}{Ala}$-COOH

The bonding of 4'-phosphopantetheine to the protein component of ACP is depicted as

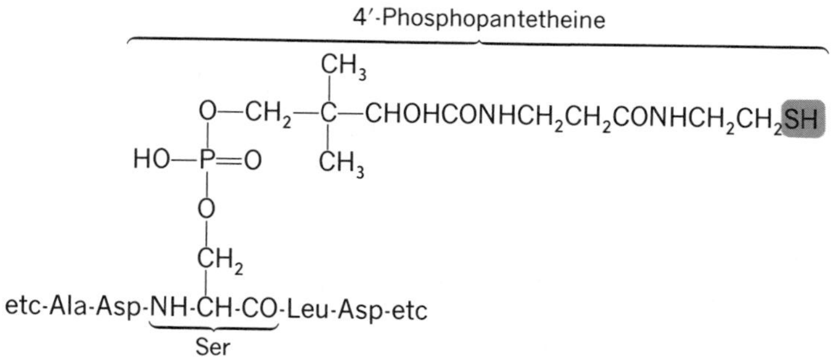

4'-Phosphopantetheine

$$
\begin{array}{c}
\hspace{3em} CH_3 \\
O—CH_2—\overset{|}{\underset{|}{C}}—CHOHCONHCH_2CH_2CONHCH_2CH_2SH \\
HO—P{=}O \hspace{1em} CH_3 \\
\overset{|}{O} \\
\overset{|}{C}H_2 \\
etc\text{-}Ala\text{-}Asp\text{-}NH\text{-}CH\text{-}CO\text{-}Leu\text{-}Asp\text{-}etc \\
Ser
\end{array}
$$

E. coli ACP has been chemically synthesized by the Merrifield procedure (Box 3.D). Plant ACPs are quite similar to the *E. coli* ACP in terms of physical and biochemical properties. The function of ACP will be discussed in greater detail in Chapter 13.

5.11 ASCORBIC ACID (VITAMIN C)

The vitamin-coenzyme relationships that have been described so far are those of vitamins that are soluble in water. Those vitamins lacking a known coenzyme function, to be described now, include only one additional water-soluble vitamin, namely ascorbic acid. While no coenzyme relationship is established, a significant amount of information regarding the physiological role of these compounds is available in most cases.

5.11.1 STRUCTURE

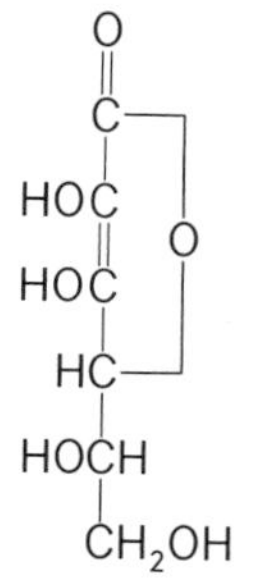

$$
\begin{array}{c}
O \\
\parallel \\
C— \\
HOC \hspace{1em} \\
\parallel \hspace{1em} O \\
HOC \\
HC— \\
HOCH \\
CH_2OH
\end{array}
$$

L-Ascorbic acid

5.11.2 OCCURRENCE

Plants and many animals can synthesize ascorbic acid from D-glucose. Important exceptions to this statement include man and other primates and the guinea pig. The enzyme that is missing in the species that are unable to produce the ascorbic acid is L-gulonoxidase, which converts L-gulonolactone to 3-keto-L-gulonolactone.

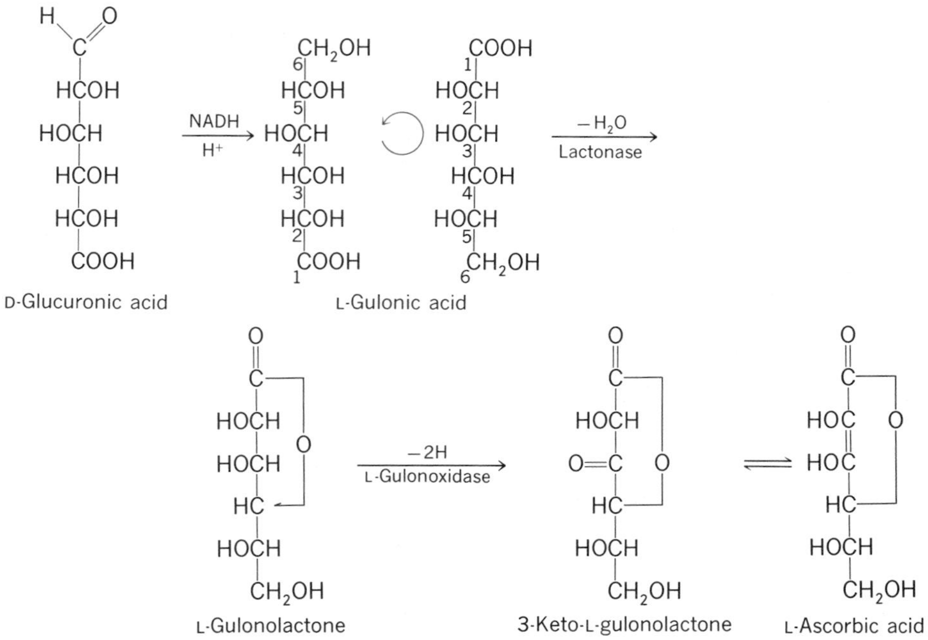

D-Glucuronic acid   L-Gulonic acid

L-Gulonolactone   3-Keto-L-gulonolactone   L-Ascorbic acid

5.11.3   BIOCHEMICAL FUNCTION

The absence of ascorbic acid in the human diet gives rise to scurvy, a disease characterized by edema, subcutaneous hemorrhages, anemia, and pathological changes in the teeth and gums. The disease was known to the ancients, especially among sailors, who often traveled for extended periods of time without sources of fresh fruits and vegetables that were known to prevent scurvy. A primary characteristic of scurvy is a change in connective tissue. In ascorbic acid deficiency, the mucopolysaccharides of the cell ground substance are abnormal in character, and there are significant changes in the nature of the collagen fibrils that are formed. The presence of ascorbic acid is required for the formation of normal collagen in experimental animals. At the enzyme level, ascorbic acid is involved in the conversion of proline to hydroxyproline, an amino acid found in relatively high concentrations in collagen.

The biochemical role that ascorbic acid plays is undoubtedly related to it being a good reducing agent. Its oxidized form, dehydroascorbic acid, is capable of being reduced again by various reductants including glutathione (GSH), and the two forms of ascorbate constitute a reversible oxidation – reduction system. In the case of collagen formation, ascorbic acid can function as the external reductant that is required in the conversion of proline to hydroxyproline. Ascorbic acid can function as an external reductant in the hydroxylation of *p*-hydroxyphenylpyruvic acid to homogentisic acid in the liver and in the conversion of dopamine to norepinephrine that occurs in the adrenals. Moreover, guinea pigs that are maintained on an ascorbic-acid-deficient diet will excrete *p*-hydroxyphenylpyruvic acid in their urine. Thus, it appears that the biochemical role of ascorbic acid is related to its involvement in hydroxylation reactions in the cell. It is interesting in this connection that the highest concentrations of ascorbate in animal tissues are found in the adrenals.

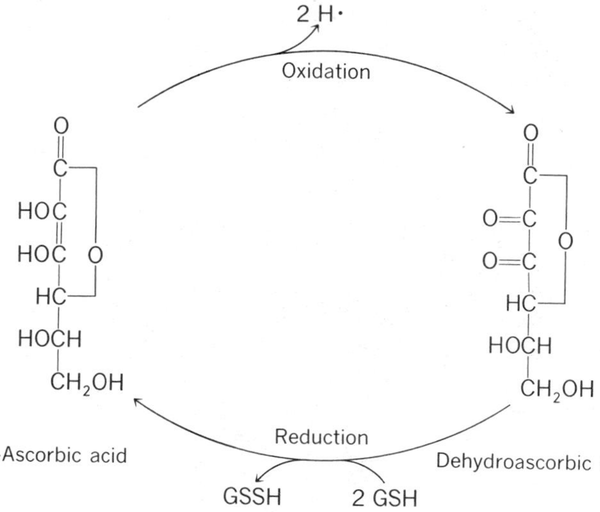
2 H·
Oxidation
HOC
HOC
HC
HOCH
CH_2OH
O=C
O=C
Reduction
GSSH
2 GSH

L-Ascorbic acid

Dehydroascorbic acid

5.12 FAT-SOLUBLE VITAMINS

Vitamins A, D, E, and K are known as the fat-soluble vitamins. They are all derived from isoprene units and are therefore classified as isoprenoid compounds. In sharp contrast to the water-soluble B vitamins and vitamin C, these fat-soluble vitamins can be stored in the body, especially the liver. Since they can be stored until used, excessive intake of these vitamins can produce toxic symptoms. Vitamin K deficiencies in healthy humans are rare since the intestinal bacteria produce the daily amounts required. Only if the bacterial synthesis is inhibited, as during antibiotic treatment, might it be necessary to supplement the intake of vitamin K. Man produces vitamin D with the aid of sunlight from 7-dehydrocholesterol normally found in his skin. Vitamin E deficiency has never been observed in humans; this vitamin is ubiquitous in plant foods normally encountered in the human diet. Vitamin A is therefore the only fat-soluble vitamin that might be deficient in the diet. About 1 mg/day is required by the adult human and this can easily be obtained by eating yellow plant foods (carrots, pumpkin, squash) containing carotenoids (α, β, and γ carotenes).

5.13 VITAMIN A GROUP

5.13.1 STRUCTURE

Vitamin A_1 or **retinol** and its aldehyde derivative, **retinal,** have the following structures:

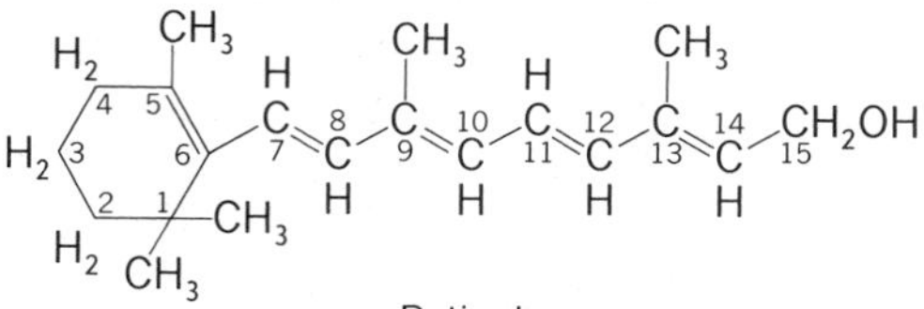
CH_3
CH_2OH

Retinol
(Vitamin A_1)

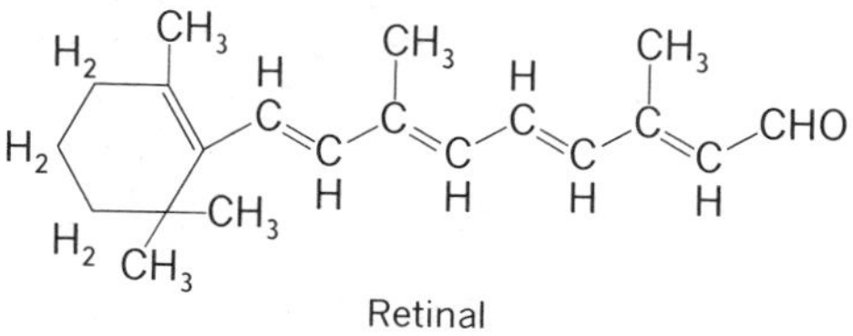
CH_3
CHO

Retinal
(Vitamin A_1-aldehyde)

These compounds are formed from their parent substance β-carotene, which is called a provitamin.

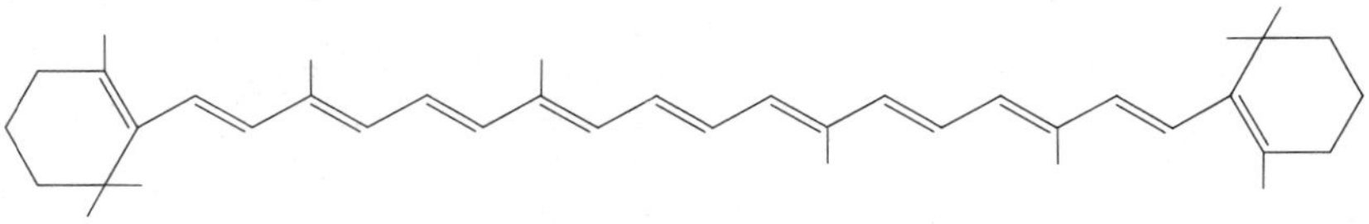

β-Carotene

An oxygenase located in the intestinal mucosa cleaves the β-carotene, yielding two moles of Vitamin A_1 aldehyde or retinal, which is then reduced to retinol by alcohol dehydrogenase.

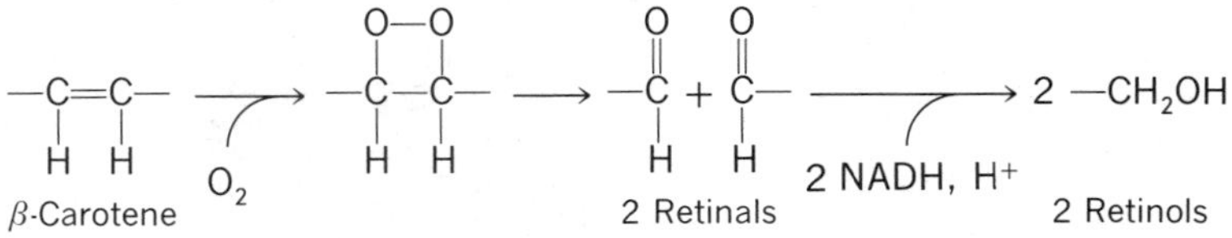

The all-*trans* configuration of the double bonds in the carotene is retained in the retinal and retinol that are formed.

5.13.2 OCCURRENCE

β-Carotene together with α- and γ-carotene and cryptoxanthine are synthesized by higher plants, but not by animals. Thus, green, leafy vegetables are good sources of provitamins for retinol. Because of their hydrophobic character, the carotenes are also found in milk, animal fat deposits, and liver where they are stored by the animal. Liver oil from fresh-water fish contains 3-dehydroretinol (vitamin A_2).

The classic symptoms of severe retinol deficiency are the keratinization processes that occur in epithelial cells; in the eyes, this process gives rise to **xeropthalmia.** An early sign of retinol deficiency in man and experimental animals is night blindness (poor ability to discriminate the intensity of light). Retarded growth and skeletal abnormalities are also observed when immature animals receive an inadequate supply of the vitamin.

While an adequate supply of retinol is required for the proper health of animals, excesses of retinol can be injurious. This happens because animals are incapable of excreting excess quantities of retinol (and other fat-soluble vitamins) and, instead, store the vitamin in fatty tissues and organs. Excesses then are harmful. Retinol toxicity has been observed in extreme cases, in which excessive amounts (e.g., 500,000 units per day) have been ingested over a considerable period of time. Some symptoms of retinol excess are bone fragility, nausea, weakness, and dermatitis.

The role (to be described next) of vitamin A_1 in the visual process is clearly related to night blindness associated with an inadequate supply of the vitamin. However, there is no clear explanation of the manner in which retinol exerts the other physiological effects just described.

5.13.3 BIOCHEMICAL FUNCTION

Retinol (vitamin A_1) and its aldehyde, retinal, are reactants in chemical changes that occur during the visual process in the rods of the eye. The retina of the human eyes, and most animals' eyes, contain two types of light receptor cells: rods and cones. The rods are used for seeing at low intensities of light (scotopic vision; shades of grey), whereas the color vision (photopic vision) is located in the cones. Only rod vision will be discussed briefly here. Retinol is oxidized in the rods by a specific retinol dehydrogenase to all-*trans* retinal and then converted by a retinal isomerase to 11-*cis*-retinal. In the dark, this aldehyde couples with opsin, forming the light-sensitive rhodopsin. When light strikes rhodopsin, 11-*cis*-retinal is isomerized to all-*trans*-retinal, which cannot form hydrophobic bonds to the opsin-phospholipid complex. The all-*trans*-retinal then forms a Schiff's base with a lysine group in the opsin, and this is finally hydrolyzed to all-*trans*-retinal and the opsin-phospholipid complex. These reactions are illustrated in the following diagram.

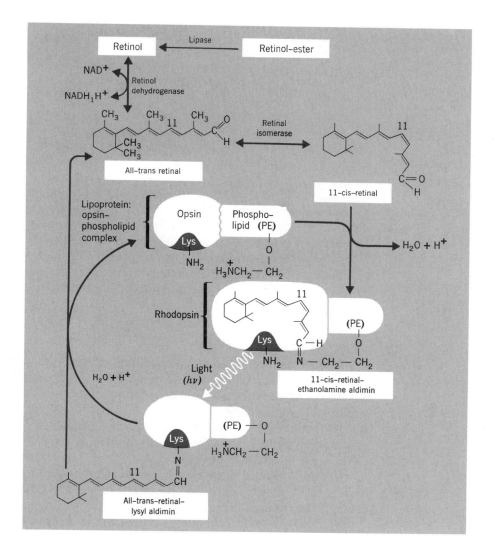

Of primary concern is the problem of how the action of light on rhodopsin results in a nervous excitation leading to vision. This mechanism is still unclear, but it appears that changes in rhodopsin conformation, ion permeabilities, and cyclic AMP may be factors of importance in the process.

5.14 VITAMIN D GROUP

5.14.1 STRUCTURE

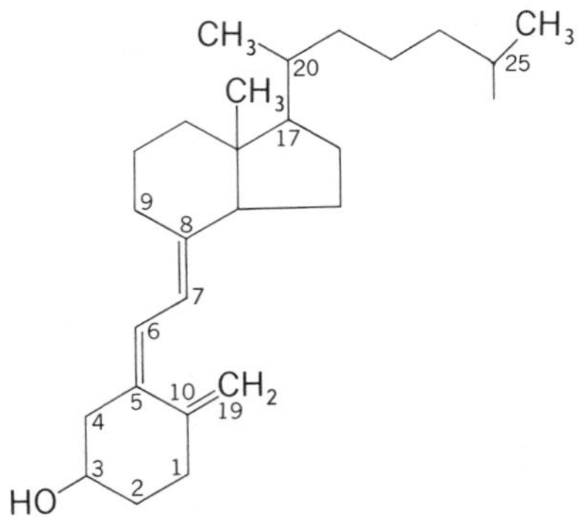

Vitamin D$_3$
Cholecalciferol

5.14.2 OCCURRENCE

Several compounds are known to be effective in preventing rickets; all are derived by irradiation of different forms of provitamin D; thus, vitamin D$_2$ (calciferol) is produced commercially by the irradiation of the plant steroid, ergosterol. In animal tissues, 7-dehydrocholesterol, which occurs naturally in the epidermal layers, can be converted by ultraviolet irradiation to vitamin D$_3$. The latter vitamin is also present in fish oil.

5.14.3 BIOCHEMICAL FUNCTION

Vitamin D$_3$ when given to rachitic animals increases the permeability of the intestinal mucosal cells to calcium ion, apparently by changing the character of the plasma membrane to calcium permeation. Recently, it has been shown that vitamin D$_3$ induces the appearance of a specific calcium-binding protein (CaBP) in the intestinal mucosa of a number of animals. This protein has been isolated and purified; it has a molecular weight of 24,000 and binds one atom of calcium per molecule of protein.

Vitamin D behaves more like a hormone than as the cofactor of an enzyme. That is, its effect is in controlling the production of a specific calcium-binding protein rather than influencing directly the activity of a specific enzyme.

Vitamin D$_3$ is not the active form of the vitamin. Instead, vitamin D$_3$ undergoes two hydroxylation reactions, the first in the microsomal fraction of liver, intestinal mucosa, and kidney, and the second in the kidney, before it is transported in its modified form to the target tissue. These reactions may be summarized as follows:

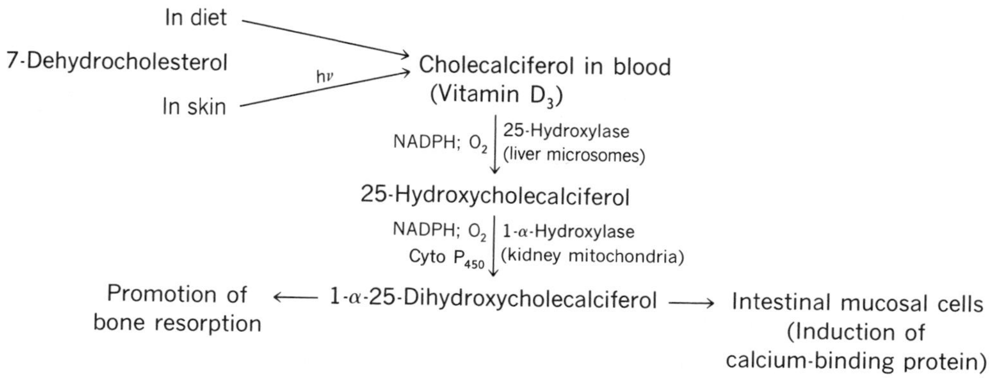

In diet ⟍

7-Dehydrocholesterol $\xrightarrow{h\nu}$ Cholecalciferol in blood
(Vitamin D_3)

In skin ⟋

NADPH; O_2 | 25-Hydroxylase
(liver microsomes)

25-Hydroxycholecalciferol

NADPH; O_2 | 1-α-Hydroxylase
Cyto P_{450} (kidney mitochondria)

Promotion of ⟵ 1-α-25-Dihydroxycholecalciferol ⟶ Intestinal mucosal cells
bone resorption   (Induction of
                                                          calcium-binding protein)

The structure of 1-α-25-dihydroxycholecalciferol, the active compound eventually formed from vitamin D_3, is also given.

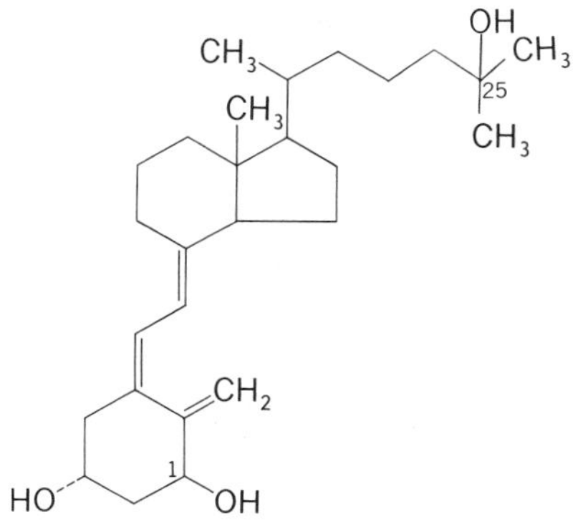

1-α-25-Dihydroxycholecalciferol

In target cells, for example, the intestinal mucosa cells, 1-α-25-$(OH)_2$-D_3 is coupled in the cytosol to a special receptor protein. This complex is transported to the nucleus where it binds to DNA and stimulates RNA polymerase II. The result is synthesis (transcription) of mRNA coding for a specific calcium-binding protein, CaBP. The mRNA is transported to the ribosomes for synthesis (translation) of CaBP.

5.15 VITAMIN E GROUP

5.15.1 STRUCTURE

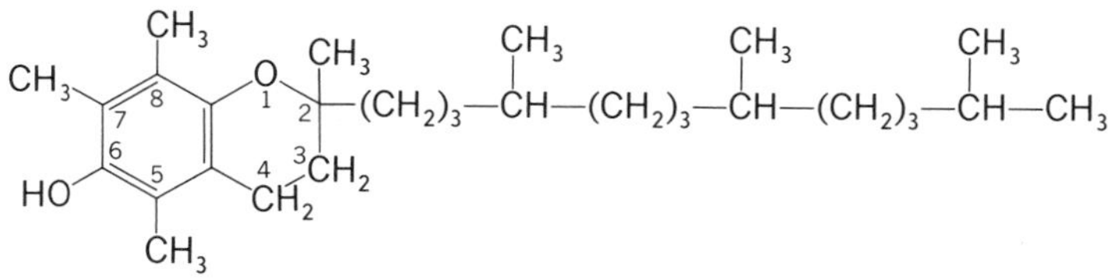

α-Tocopherol

5.15.2 OCCURRENCE

Tocopherols occur in plant oils in varying amounts. Apparently, the most widespread and biologically active form of the tocopherols is α-tocopherol,

5,7,8-trimethyl-tocol. Other tocopherols are β-(5,8-dimethyl), γ-(7,8-dimethyl), and δ-(8-methyl)-tocol. Besides the tocol series, another series, the toco-trienols, also occurs in nature although it is less widespread. The only difference between the two series of vitamins is that the trienols have a long side chain consisting of three isoprene entities instead of the fully hydrogenated side chain in the tocol series. Large amounts are found in wheat germ oil and corn oil, for example. Tocopherols are also found in animal body fat. There is some evidence that all α-tocopherol in heart muscle is localized in the mitochondria.

5.15.3 BIOCHEMICAL FUNCTION

Characteristic symptoms of avitaminosis E vary with the animal species. In mature female rats, reproductive failure occurs. Fetuses die during pregnancy and are absorbed from the uterus. In male rats, germinal tissue degenerates. With rabbits and guinea pigs, acute muscular dystrophy results; in chickens, vascular abnormalities occur. In humans, no well-defined syndrome of vitamin E deficiency has been detected.

The most prominent effect that has been demonstrated for tocopherol *in vitro* is a strong antioxidant activity. It has been suggested that the biochemical activity of tocopherol is related to its capacity to protect sensitive mitochondrial systems from irreversible inhibition by lipid peroxides. Thus, in mitochondria prepared from tocopherol-deficient animals, there is a profound deterioration of mitochondrial activity because of hematin-catalyzed peroxidation of highly unsaturated fatty acids normally present in these particles.

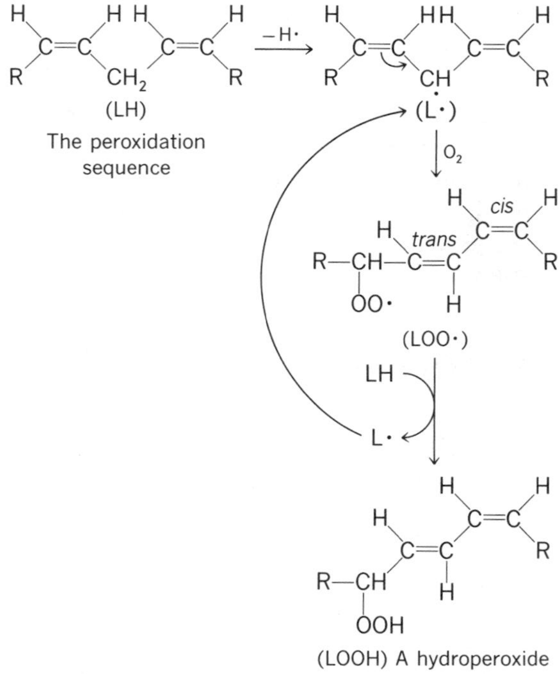

The peroxidation sequence

α-Tocopherol functions as a chain breaker by participating in the following reactions:

$$LOO\cdot + \alpha TH \longrightarrow LOOH + \alpha T\cdot$$
$$L\cdot + \alpha TH \longrightarrow LH + \alpha T\cdot$$

Chemically, α-tocopherol may undergo the following sequence of reactions leading to the formation of α-tocopherol quinone:

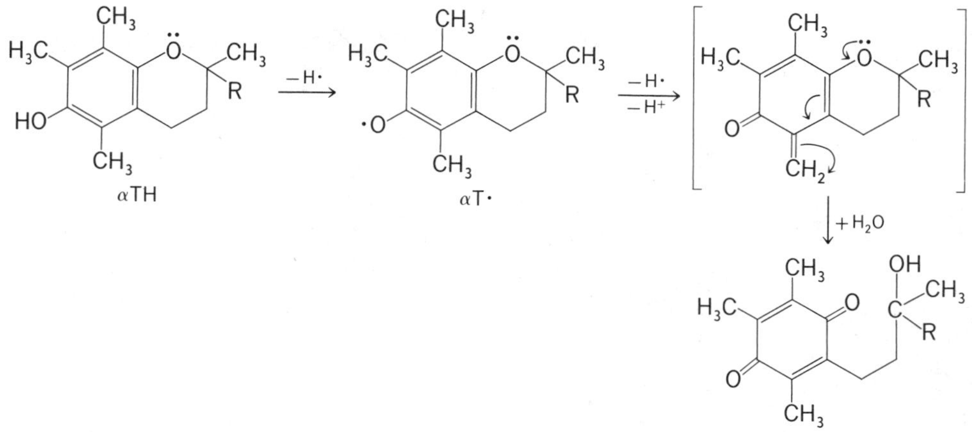

α-Tocopherol quinone

Thus, α-tocopherol acts as a breaker of free radical chain reactions and thereby inhibits the destructive peroxidation of, for example, polyunsaturated fatty acids that are always associated with membrane lipids. However, for several years, nutritionists have observed a striking similarity between the nutritional effects of α-tocopherol and very small amounts of dietary selenium (0.05 parts per million per day).

It is now known that selenium is an essential component of the enzyme glutathione peroxidase that scavenges toxic hydroperoxy compounds in tissues by the following reaction:

$$2\,GSH \quad + \quad ROOH \quad \xrightarrow[\text{peroxidase}]{\text{Glutathione}} \quad G\!-\!S\!-\!S\!-\!G + ROH + H_2O$$

Glutathione   Alkyl hydroperoxide   Oxidized glutathione

The body therefore has two lines of defense against toxic hydroperoxides: (1) α-tocopherol, which prevents in part the formation of these compounds, and (2) glutathione peroxidase, which converts toxic hydroperoxides (ingested or formed endogenously) to harmless primary and secondary alcohols.

5.16 VITAMIN K GROUP

5.16.1 STRUCTURE

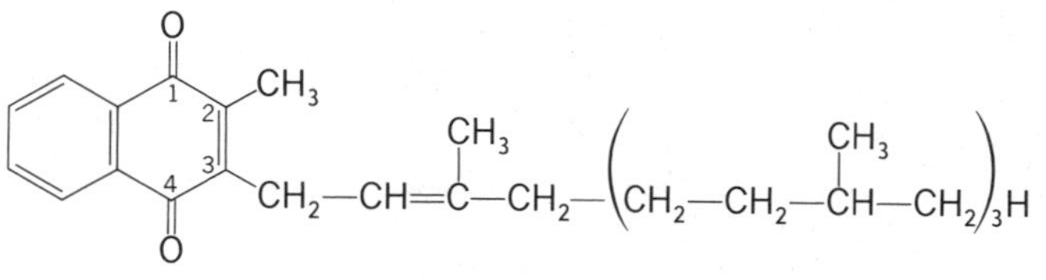

Vitamin K_1 (phytyl-menaquinone)

Vitamin K_1 was first isolated from alfalfa and has the phytyl side chain consisting of four isoprene units, three of which are fully reduced. In the vitamin K_2 series, six to nine isoprene units occur in the side chain.

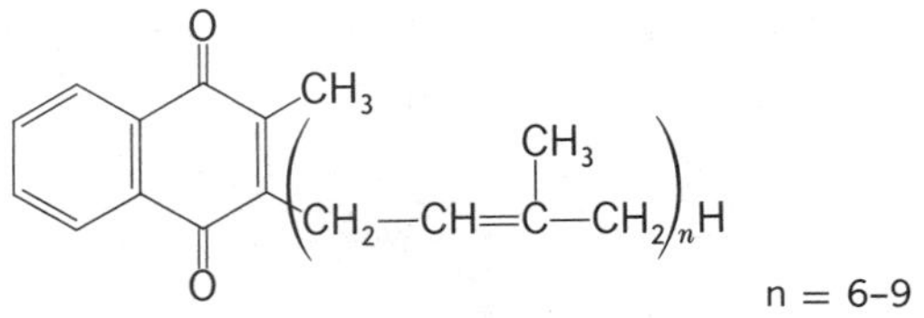

n = 6–9

Vitamin K_2 series

Vitamins K_2 are isolated from bacteria and purified fish meal. However, vitamins K_2 are also known in which one of the isoprene units is hydrogenated, for example, from *Mycobacterium phlei*. Menadione, menaquinone, or 2-methyl-1,4-naphthoquinone has the same quinone or ring moiety and exhibits the same vitamin activity as vitamin K_1 on a molar basis, possibly because it is readily converted to vitamin K_1.

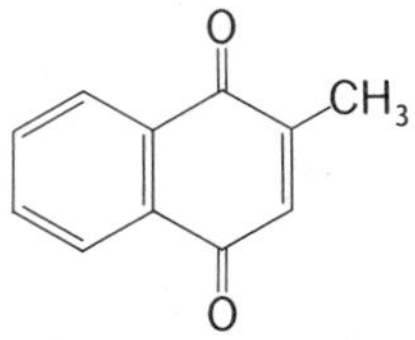

Menadione; menaquinone
(2-methyl-1,4-naphthoquinone)

5.16.2 OCCURRENCE

Vitamin K_1 was first isolated from a plant source and plant foods remain a good source of the vitamin. Vitamins of the K_2 series are formed by bacteria, notably those in the intestine. Thus, a deficiency of Vitamin K is hard to demonstrate in healthy animals. A deficiency may occur in man under conditions in which those bacteria are destroyed or their growth inhibited. Thus, when antibiotics are administered, particularly over an extended period, vitamin K levels may be lowered to the point where blood clotting time (see the next section) is dangerously prolonged. Biliary obstruction or other conditions in which decreased intestinal absorption of lipids exists also can give rise to vitamin K deficiency.

5.16.3 BIOCHEMICAL FUNCTION

A deficiency of Vitamin K results in decreased clotting of the blood. The biochemical explanation for this observation has been recently resolved. Near the amino terminal end of the protein prothrombin, a cluster of glutamic acid residues are present. Vitamin K plays a direct role in the post-translational modification of these residues to γ-carboxyl-glutamyl residues. As a result of this modification, prothrombin can now bind Ca^{2+} ions that are a requirement for the conversion of prothrombin to thrombin. The Vitamin K-dependent γ-glutamyl carboxylase catalyzes the reaction on the next page.

Whereas normal prothrombin has a number of γ-carboxyl glutamyl residues, in Vitamin K-deficient animals only glutamyl residues are present and thus the prothrombin is defective.

The latter stages of the blood-clotting process have been known for many

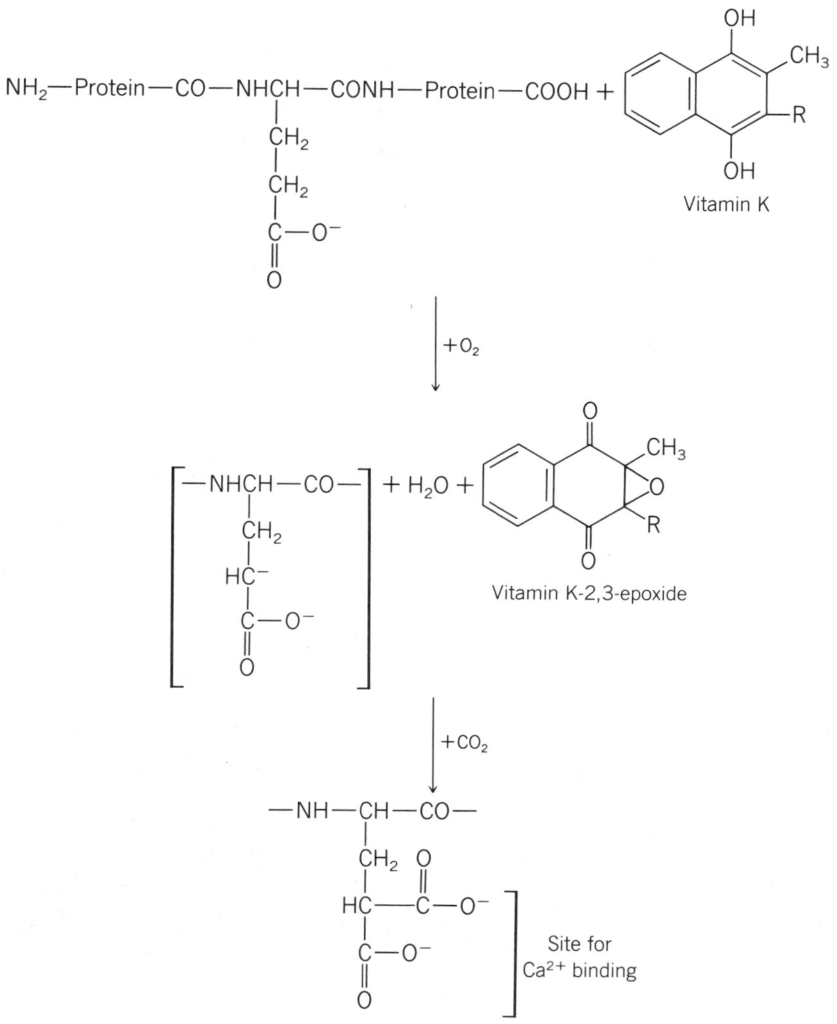

$$NH_2—Protein—CO—NHCH—CONH—Protein—COOH +$$

with side chain:
$$CH_2$$
$$CH_2$$
$$C—O^-$$
$$\|$$
$$O$$

Vitamin K

$$\downarrow +O_2$$

$$\left[—NHCH—CO— \right] + H_2O +$$

with side chain:
$$CH_2$$
$$HC^-$$
$$C—O^-$$
$$\|$$
$$O$$

Vitamin K-2,3-epoxide

$$\downarrow +CO_2$$

$$—NH—CH—CO—$$

with side chain:
$$CH_2 \quad O$$
$$\qquad \|$$
$$HC—C—O^-$$
$$C—O^-$$
$$\|$$
$$O$$

Site for
Ca^{2+} binding

years and may be stated as follows: Prothrombin, a plasma proenzyme or zymogen, is converted into thrombin, a proteolytic enzyme, by the combined action of several factors (see the diagram on p. 208). Thrombin, in turn, converts fibrinogen into fibrin, the protein from which clots are made. Recent chemical studies show that fibrinogen is a dimer (MW = 330,000) consisting of three polypeptide chains designated as α, β, and γ. When thrombin acts on fibrinogen, a total of four peptide bonds are cleaved and two small polypeptides (MW = 9000) are released. This modified fibrinogen molecule is now known as fibrin and is transferred from the form of a "soft" clot to that known as a "hard" clot by Ca^{++} ions and still other proteins.

Research has shown that blood clotting is a much more complicated process than just represented. In particular, the process is one involving a "cascade" phenomenon in which an active factor is produced from an inactive form and it, in turn, activates the conversion of a subsequent inactive form to an active one. The process may be represented as shown next in a simplified form.

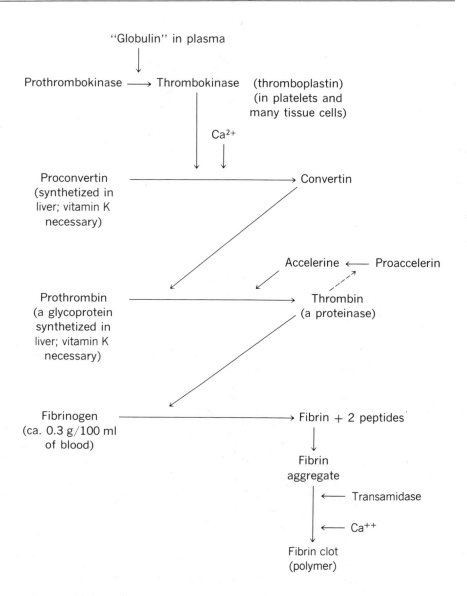

5.17 METALS IN BIOCHEMISTRY

The compounds that compose the living cell are constructed primarily from the six elements C, H, O, N, P, and S. Many organisms also contain relatively large amounts of Na, K, Ca, Mg, Fe, and Cl. Thus, a 70-kg man will contain over 1 kg of Ca and 600 g of P in his skeleton. His soft tissues will contain about 200 g K, 40 g Na, and 250 g Cl, and his blood will contain approximately 20 g Na, 24 g K, and 20 g Cl. His body will also contain approximately 20 g Mg and 4 g Fe, about half of which will be present in his blood.

In the case of animals, these elements must be acquired in their diet in amounts sufficient to meet the needs including amounts that are excreted. Thus, an adult man needs to take in 10 to 20 g of NaCl per day, primarily to replace the NaCl that is excreted in his urine or sweat.

Higher plants also must acquire these elements. The absorption is almost exclusively (except for CO_2 and O_2) through their root system. The quantities required are large relative to certain other materials and, for this reason, have been termed macronutrients.

Most organisms will also need Mn, Co, Mo, Cu, and Zn (they contain many more elements), all of which are known as components of certain enzymes. These elements, together with boron and iron (but not Co), have been termed micronutrients because they are required in relatively low amounts by higher plants.

Animals must also acquire these micronutrients in their diet, which they do primarily by eating plant materials. Ni, Al, Sn, Se, V, F, and Br have been also detected but, with the exception of Se (Section 5.15.3), little is known of the role of these elements, if any, in the living organism. Some general remarks concerning the roles of micronutrients, as well as Mg, Ca, and Fe, in enzyme catalysis follow.

5.17.1   METALS IN ENZYMES

Approximately one-third of the known enzymes have metals as part of their structure, require that metals be added for activity, or are further activated by metals. In the first of these cases, metals have been built into the structure of the enzyme molecule and cannot be removed without destroying that structure. Such enzymes include the metalloflavoproteins (Section 5.3.3), the cytochromes (Section 14.2.5), and the nonheme iron proteins, the ferredoxins (Section 14.2.3). In other instances, metals react reversibly with proteins to form metal-protein complexes that constitute the active catalyst. In many instances, the complex represents a specific, catalytically active, conformation of the protein; the role of the metal appears to be one of stabilizing that conformation.

An important example of the role of metals in regulating cellular activity is found in the interplay between Ca^{+2} and the calcium-binding protein called *calmodulin* (CAM). While Ca^{+2} and calmodulin alone are ineffective, the complex of Ca^{+2} and calmodulin (CAM·Ca^{+2}) has been observed to convert less active forms of several enzymes (e.g., adenylate cyclase, synthase·phosphorylase kinase (SPK), Ca^{+2}-dependent protein kinase, Ca^{+2}-dependent ATPase) to more active forms. This may be depicted as follows:

$$CAM + Ca^{+2} \rightleftharpoons CAM \cdot Ca^{+2}$$
$$CAM \cdot Ca^{+2} + Enz_{less\ active} \rightleftharpoons ENZ \cdot CAM \cdot Ca^{+2}_{more\ active}$$

Metals resemble protons (H^+) in that they are electrophiles that are capable of accepting an electron pair to form a chemical bond. In doing so, metals may act as general acids to react with anionic and neutral ligands. While their larger size relative to that of protons is a disadvantage, this is compensated for by their ability to react with more than one ligand. In general, metal ions react with two, four, or six ligands. If they do so with two, the complex is linear.

$$X-M-X$$

If four ligands react, the metal may set in the center of a square (planar) or a tetrahedron (tetrahedral).

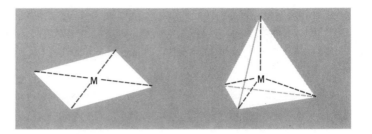

When six ligands react, the metal sits in the center of a octahedron.

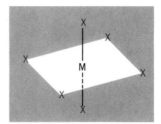

Amino acids either free or in their peptide linkages in proteins possess several groups capable of complex formation with metal ions. The carboxyl and amino groups of an amino acid can bind a metal as shown.

Obviously, the free amino and carboxyl groups in a protein can do the same, perhaps bending the protein into a specific, active conformation. The sulfhydryl group of cysteine and the imidazole ring of histidine are other important ligands moieties in metal-protein complexes. In the case of histidine, its residue is known to be at the site that binds the Fe atom in the cytochromes. The carbonyl group ($>$C$=$O) in the peptide bond can also be bonded to metal ions. In most of these examples, the metal is acting as an electron sink for the electron pair and can form a relatively stable complex.

One other important example of an excellent metal ligand is the phosphate ester found in ATP, ADP, sugar phosphates, the nucleic acids, and phosphorylated substrates. Metallic cations, especially Mg^{++}, can react with the oxygen atom of phosphate esters to neutralize their charge and subsequently catalyze reactions by acting as general acid catalysts. Indeed, almost all enzymes that catalyze reactions involving phosphate groups require Mg^{++} (or Mn^{++}) as a cofactor. Specific examples of metal ion involvement in enzymes will be discussed as they are encountered in subsequent chapters.

REFERENCES

1. *C. T. Walsh,* Enzymatic Reaction Mechanisms. *San Francisco, Calif.: W. H. Freeman and Co., 1977.*

2. *G. Popjak, "Stereospecificity of Enzymic Reactions." In* The Enzymes, *Vol. 2, P. Boyer, ed. New York: Academic Press, 1970.*

3. *D. B. McCormick and L. D. Wright, "Vitamins and Coenzymes." In the series* Methods in Enzymology, *Vols. 62, 66, and 67. New York: Academic Press, 1979, 1980.*

REVIEW PROBLEMS

1. Describe the biological roles of the following fat-soluble vitamins.
   (a) Vitamin A   (c) Vitamin E
   (b) Vitamin D   (d) Vitamin K

2. Explain how vitamin K could act as an intermediate in an electron transport system. Use structural formulas.

3. Draw the structures of the following water-soluble vitamins and the coenzymes they are converted to *in vivo.*
   (a) Niacin   (c) Riboflavin
   (b) Thiamin   (d) Pyridoxal

4. Identify the "business end" of the coenzyme molecules (i.e., the part of the structure that is most involved in an enzyme-catalyzed reaction) listed in Problem 3. Write a partial reaction, showing clearly how the coenzyme functions.

5. Outline the experimental procedure you would follow in order to determine whether or not an experimental animal requires ascorbic acid (vitamin C) as a vitamin.

NUCLEIC ACIDS AND THEIR COMPONENTS

Nucleic acids have a variety of roles in organisms. The low molecular weight components of nucleic acids are widely utilized in metabolic reactions. In complexes with proteins, certain high molecular weight nucleic acids provide structure; a few even are catalytic. However, the unique and premiere service of nucleic acids is as repositories and transmitters of genetic information. They make it possible for cells to function according to specific patterns and to give rise to new cells that either function similarly or develop new functions, according to plans encoded in the nucleic acid.

Genetic information is encoded in a nucleic acid molecule in a particularly general and simple fashion. Four different units make up each informational nucleic acid molecule. They are four letters in a code. The nucleic acid molecule is linear (not branching), and the units are like letters on a printed page or digital magnetic signals on a computer tape. Given the proper machinery, the order of the units can be interpreted, just as magnetic signals can be interpreted. The cell interprets the information in many nucleic acid molecules as sequences of amino acids in protein and peptide molecules. The synthesis of proteins, of definite amino acid sequences in definite, controlled amounts apparently constitutes most of what we see as the expression of the heredity of an organism, that is, its phenotype.

The "interpretation" of the encoded information, be it a protein molecule or data or even music, resembles not at all the physical characteristics of the recording medium. That is, the encoding is truly symbolic. How the machinery of the cell interprets its "tapes" is the principal theme of Part III of this text. In this chapter, we consider the covalent structures of low and high molecular weight forms of nucleic acids. As is the case with proteins, noncovalent bonds make essential contributions to nucleic acid function. Noncovalent bonding is especially important in the transfer of genetic information during nucleic acid and protein synthesis.

6.1   COMPONENTS AND ORGANIZATION OF NUCLEIC ACIDS

The monomeric unit of a nucleic acid molecule is a **nucleotide residue.** The nucleotide, in turn, has three readily recognized constituents:

A sugar.
A purine or pyrimidine base.
An esterified phosphoric acid moiety.

The sugar is either ribose or deoxyribose, dividing the nucleic acids into two major groups. **Ribonucleic acid (RNA)** has as its sugar D-ribose in the furanose configuration, whereas **deoxyribonucleic acid (DNA),** as its name implies, has D-2-deoxyribose, also in the furanose configuration.

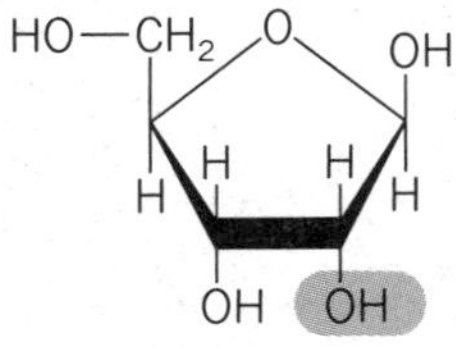

β-D-Ribofuranose 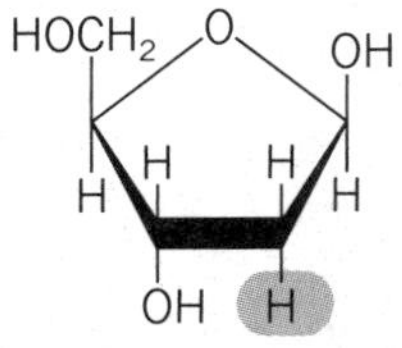   β-D-2-Deoxyribofuranose

The systematic names of the purine and pyrimidine bases are derived from the parent compounds, purine and pyrimidine. Adenine (6-aminopurine) and guanine (2-amino-6-oxopurine) are the two common purine bases of both DNA and RNA.

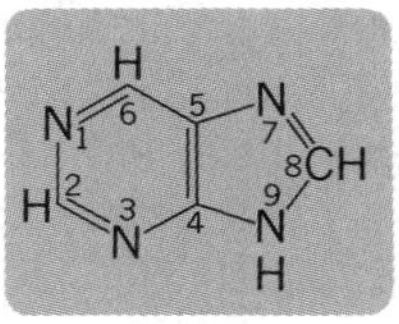

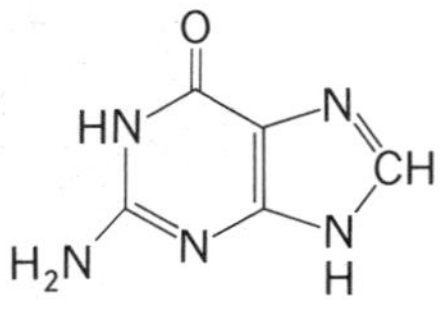

Purine 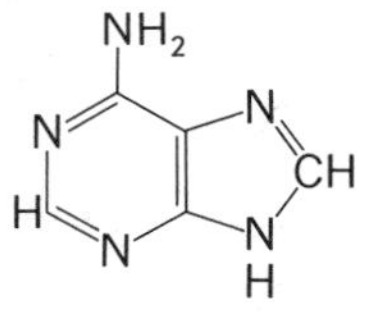   Adenine   Guanine
                   (6-Aminopurine)   (2-Amino-6-oxopurine)

The common pyrimidine bases of RNA are cytosine (2-oxo-4-aminopyrimidine) and uracil (2,4-*bis*-oxopyrimidine), whereas DNA has cytosine and thymine as the abundant pyrimidines. Many other purine and pyrimidine bases are found in specific nucleic acid molecules. Most are rare. One not-so-rare pyrimidine is the 5-methylcytosine that is 7% of the bases in some plant DNAs and up to 3% of the bases in other eucaryotic DNAs. *N*-6-Methyladenine occurs in lesser amounts in eucaryotic DNAs, and 5-methylcytosine and both *N*-6-methyladenine bases are present in some procaryotic DNAs. Other, more unusual bases also are known, and some of these appear here.

A purine or pyrimidine base linked to D-ribofuranose forms a **ribonucleoside.** The *N*-β-glycosidic bond connects carbon 1 of the ribose to N-9 of a purine base or N-1 of a pyrimidine base. To distinguish the base and sugar ring numbering systems, the carbon atoms of the sugar are designated with primes: C-1′, C-2′, through C-5′. The *N*-glycosidic bond may be considered to have been formed by removal of a molecule of water derived from the hemiacetal hydroxyl of the sugar and the *N*-bound proton of the base. The **deoxynucleosides** have similar structures. Both kinds of **nucleosides** have the β-configuration,

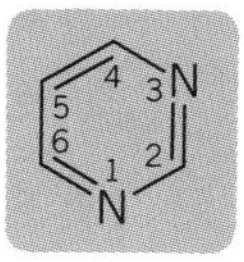

Pyrimidine 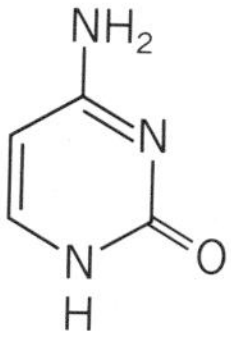   Cytosine 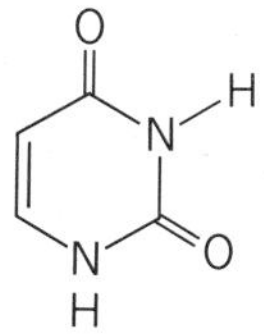   Uracil

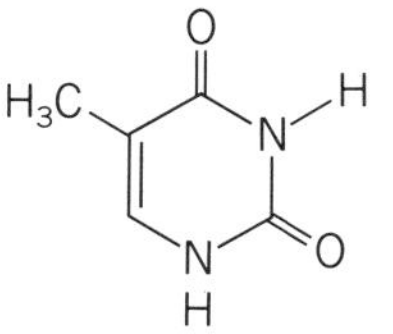

Thymine 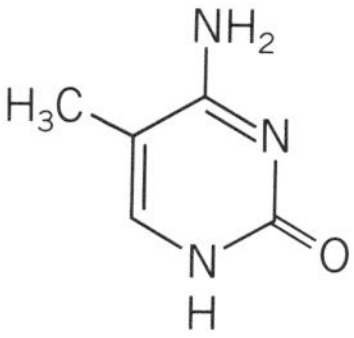   5-Methylcytosine

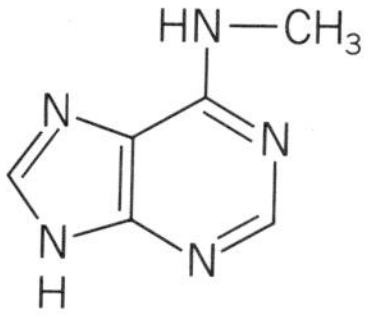

N-6-Methyladenine

defined as having the base and sugar C-5′ on the same side of the furanose ring of the D sugar.

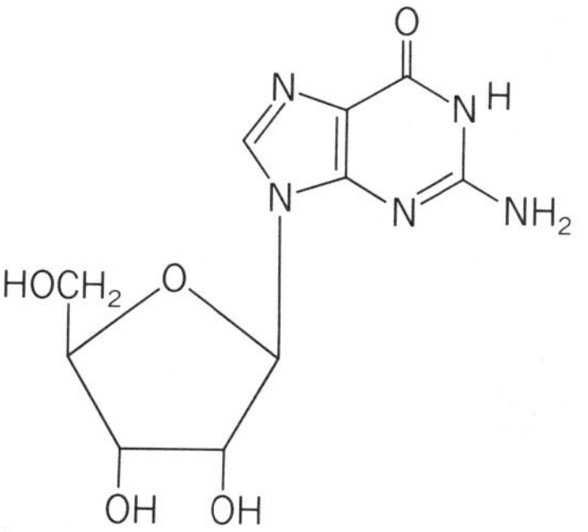

Guanosine
(9-β-D-Ribofurnosyl
guanine)

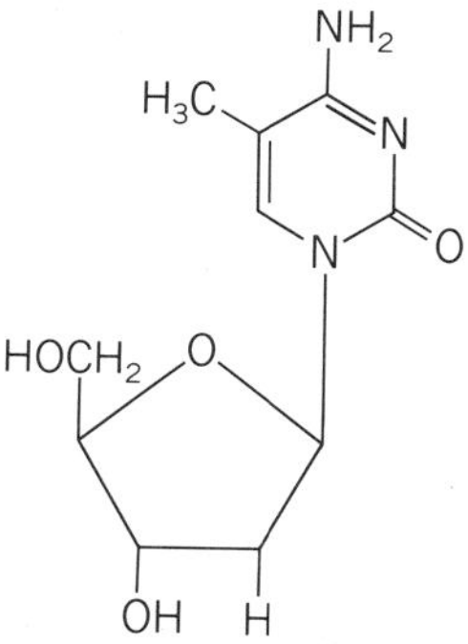

Deoxyribosyl-5-methylcytidine
(1-β-D-2-Deoxyribofuranosyl-
5-methylcytosine)

By convention, 1-β-D-ribofuranosylthymine is referred to as ribothymidine.

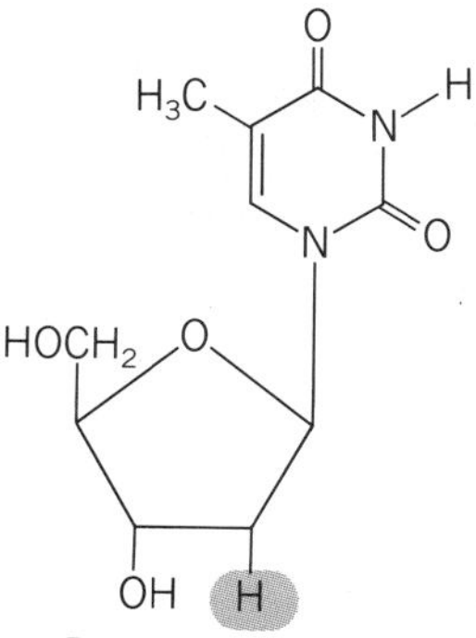

Deoxyribothymidine

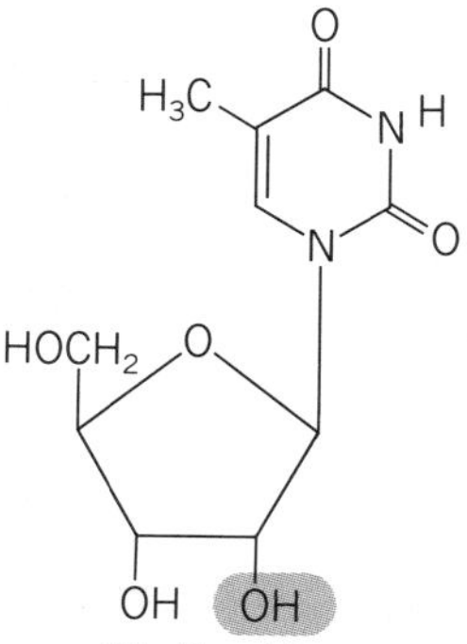

Ribothymidine

TABLE 6.1   Names of Nucleosides

BASE	RIBONUCLEOSIDE	SYMBOL	DEOXYRIBONUCLEOSIDE	SYMBOL
Adenine	Adenosine	A	2'-deoxyribonucleoside or deoxyadenosine	dA
Guanine	Guanosine	G	2'-deoxyribosylguanine or deoxyguanosine	dG
Cytosine	Cytidine	C	2'deoxyribosycytosine or deoxycytidine	dC
Uracil	Uridine	U	2'-deoxyribosyluracil or deoxyuridine	dU
Thymine	Ribothymidine	rT	2'-deoxyribosylthymine or deoxythymidine	dT

Table 6.1 gives the names of some purine and pyrimidine bases and their corresponding ribonucleosides and deoxybribonucleosides. Note that the suffix -osine is not consistently used in this accepted nomenclature and that ribothymidine and deoxyuridine are rare.

A **nucleotide** is a nucleoside phosphate, a nucleoside esterified to phosphoric acid. One or more of the three hydroxyls, 2', 3', or 5', may be esterified. However, the most common nucleotides are the nucleoside-5'-phosphates. The standard abbreviations for nucleoside- and deoxynucleoside-5'-phosphates are given in Table 6.2.

Alternative names for AMP are adenylic acid or adenylate; alternative names for dCMP are deoxycytidylic acid and deoxycytidylate, and so on. The 2'- and 3'-monophosphates of adenosine sometimes are referred to as 2'-adenylate and 3'-adenylate. However, for these and other nucleotides, abbreviations of Table 6.2 are reserved for the nucleoside 5'-phosphates. Another designation for 5'-adenylate acid is pA, and 3'-adenylate has the abbreviation Ap. Deoxyadenosine-5'-phosphate is pdA.

TABLE 6.2   Abbreviations for Nucleoside-5'-phosphates

BASE	RIBONUCLEOTIDE	DEOXYRIBNONUCLEOTIDE
Adenine	AMP	dAMP
Guanine	GMP	dGMP
Cytosine	CMP	dCMP
Uracil	UMP	dUMP
Thymine	rTMP	dTMP

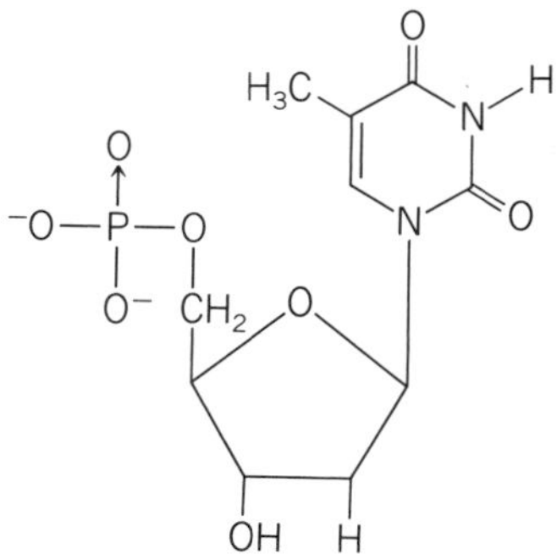

Thymidine-5′-phosphate
(dTMP)

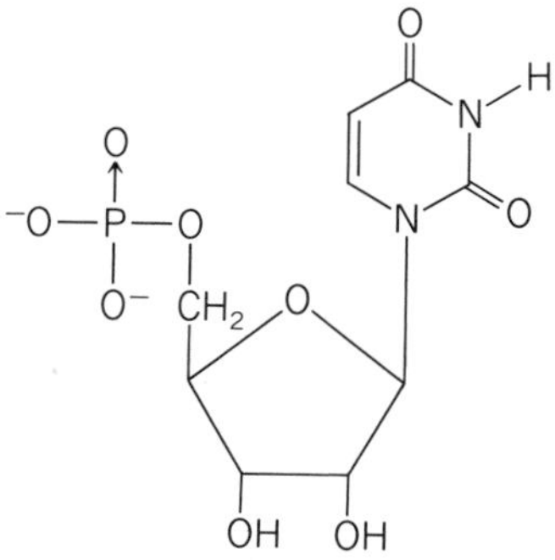

Uridine-3′-phosphate

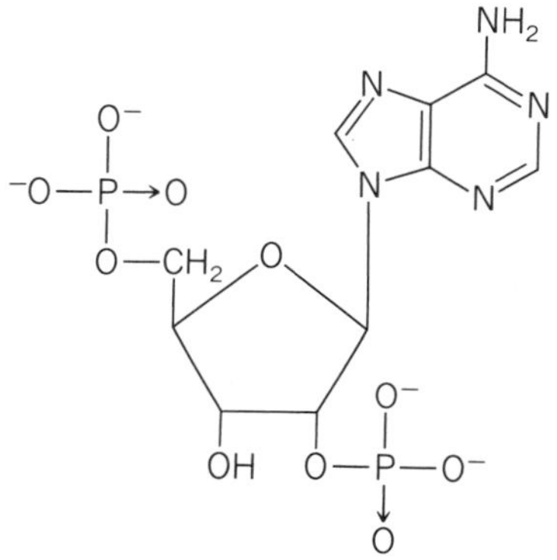

Adenosine-2′,5′-*bis*-phosphate

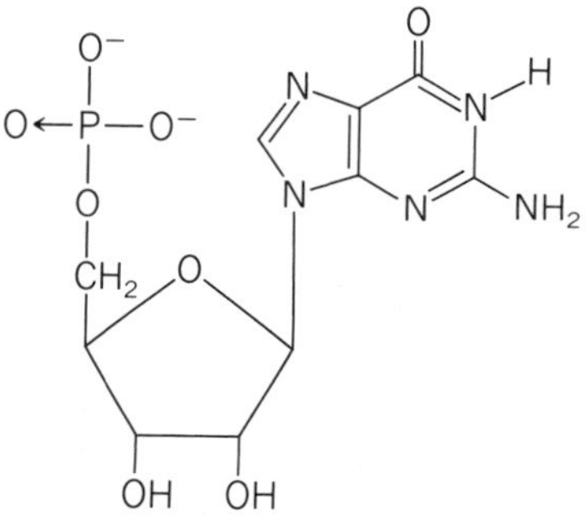

Guanosine-5′-phosphate
(GMP)

6.2 NUCLEOSIDE DI- AND TRIPHOSPHATES; CYCLIC NUCLEOTIDES

As was just indicated, the nomenclature "*bis*-phosphate" refers to a nucleoside that has two of its hydroxyl groups singly phosphorylated. The 5′-hydroxyl group of nucleosides also may be esterified to phosphoric anhydrides: pyrophosphate and triphosphate. These result in nucleoside diphosphates (not *bis*-phosphates) and nucleoside triphosphates, respectively.

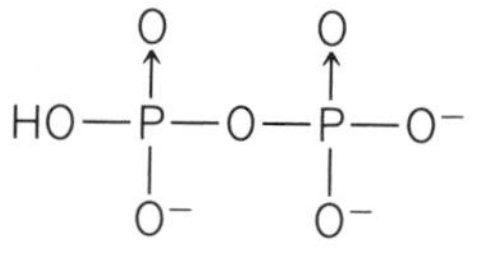

Pyrophosphate

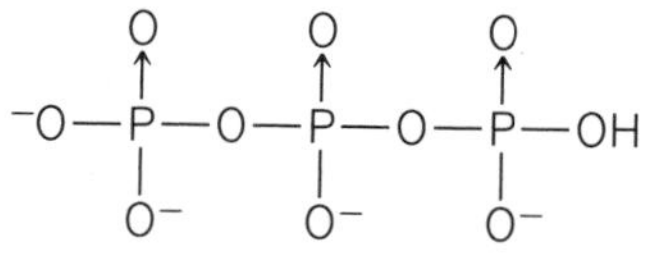

Triphosphate

Adenosine triphosphate serves as energy currency in the cell, and it is by far the most important compound in that regard, as is discussed in Chapter 9. Phosphates of ATP are designated α, β, and γ as indicated.

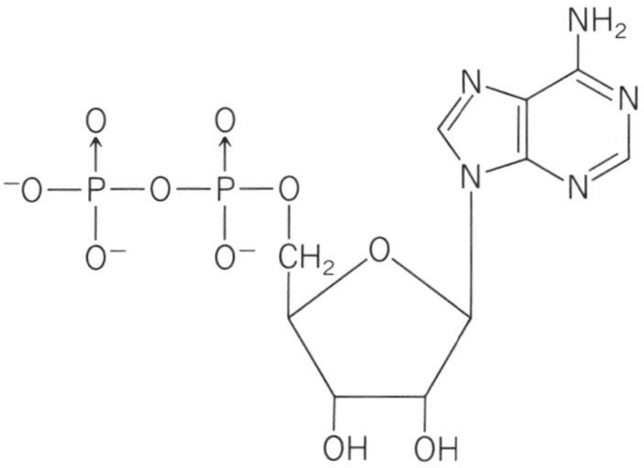

Adenosine diphosphate

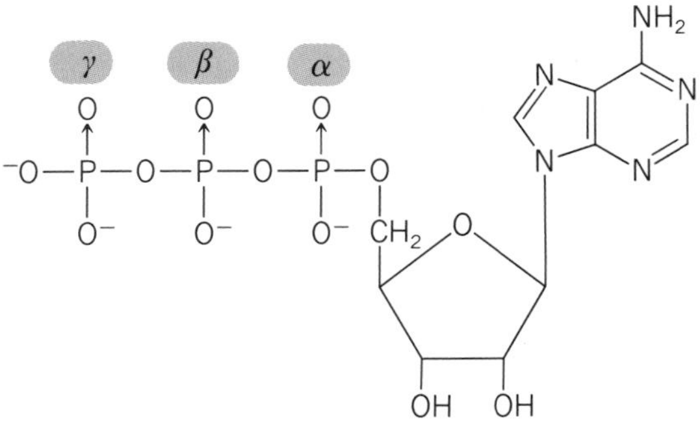

Adenosine triphosphate

Hydrolysis of either of the phosphoric anhydride bonds is thermodynamically a very favored reaction at neutral pH, releasing about more than twice as much energy as the hydrolysis reaction of an ordinary phosphoric acid ester, such as hydrolysis of AMP to adenosine and inorganic phosphate. The two phosphoric anhydride hydrolyzing reactions are

$$ATP^{4-} + H_2O \rightleftharpoons ADP^{3-} + HPO_4^{2-} + H^+$$
$$ATP^{4-} + H_2O \rightleftharpoons AMP^{2-} + HP_2O_7^{3-} + H^+$$

In the human body, about 2 kg of ATP is hydrolyzed per day in the two reactions presented, and a similar amount is synthesized. Most ATP hydrolysis is coupled to other reactions, to utilize the energy that was stored in ATP. The result is that ATP hydrolysis is used to "drive" other biochemical reactions that would otherwise be thermodynamically unfavorable. The relative contribution of all factors to the favorable energy of hydrolysis of ATP is unknown. However, the high negative charge density of ATP relative to its hydrolysis products and the resonance stabilization of the products are important contributors (see also Chapter 9). Nevertheless, ATP is stable in aqueous solution at neutral pH in the absence of enzymes that use it as substrate. This is an important characteristic in its role as energy currency. The triphosphates of all of the common nucleosides and deoxynucleosides also are important: They are the precursors for RNA and DNA synthesis, respectively.

Two kinds of cyclic phosphodiesters of nucleosides are well studied. The 3′ : 5′-cyclic phosphodiesters of adenosine and guanosine are important inter-

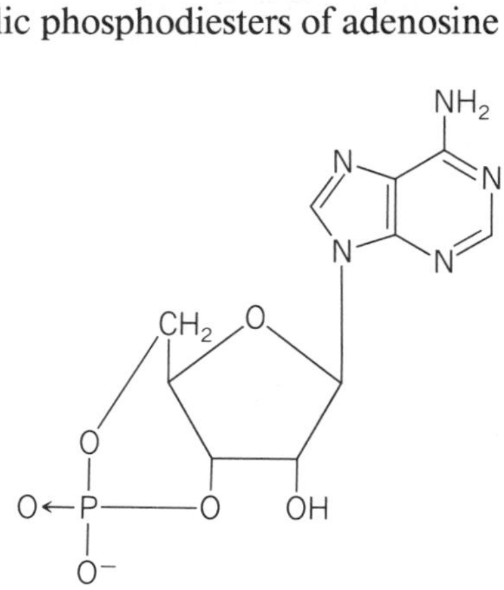

Adenosine 3′ : 5′-cyclic
phosphate

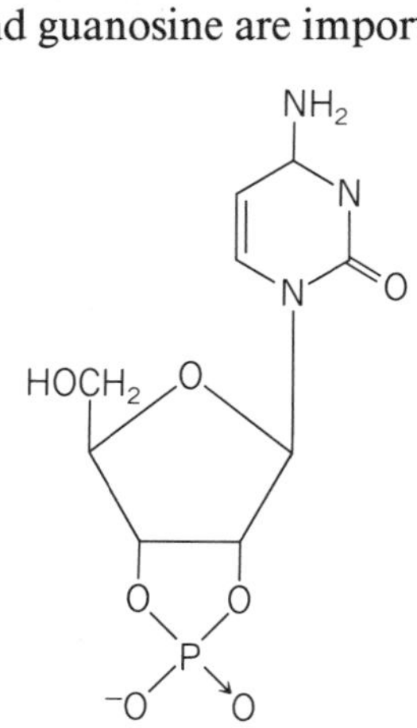

Cytidine 2′ : 3′-cyclic
phosphate

mediates in the action of some hormones. The 2′ : 3′-cyclic phosphodiesters of nucleosides are intermediates in the hydrolysis of RNA (see Section 6.4).

6.3 COVALENT STRUCTURE OF RNA AND DNA

The backbone of a nucleic acid molecule is a chain of alternating sugar and phosphate residues resulting from the formation of **3′ → 5′ phosphodiester bonds** between nucleotide residues. The term "nucleotide residue," rather than "nucleotide," is appropriate because the structure is no longer strictly that of a nucleotide. Water can be considered to have been split out in the process of forming the covalent bond, so a nucleotide residue has one fewer oxygen atom and two fewer hydrogen atoms than the corresponding nucleotide. Another term for 3′ → 5′ polymerized nucleotide residues is "polynucleotide." Often, **polynucleotide** implies a synthetic polymer of undetermined nucleotide sequence or of a single type of residue, such as polyadenylic acid [abbreviated poly(A)]. **Nucleic acid** implies a polynucleotide of defined size and nucleotide sequence, especially as isolated from a natural source. An **oligonucleotide** is, of course, a molecule composed of only a few nucleotide residues.

Figure 6.1 shows the structure of polynucleotide chains as represented by two oligonucleotides. Because nucleic acids are linear polymers with only one kind of internucleotide bond, the covalent structure is unambiguously represented by any of several symbolic forms. In one of these symbolic forms, the sugar of the nucleoside residue is represented by a vertical line, with C-1′ at the top and C-5′ at the bottom.

or
pGpApUpC
or
pG – A – U – C

or
pdGpdApdTpdCp
or
pdG – dA – dT – dCp

The nucleoside residue itself is a line with the single letter abbreviation of the nucleoside written above it. Alternatively, the letter alone may represent the nucleoside. A lowercase "p" or a hyphen denotes a phosphodiester bond. A terminal **phosphoryl group** (esterified phosphoric acid) also is indicated by p.

It is obvious from the structure in Figure 6.1 that the polynucleotide chain is not symmetric. It has a definite direction along the chain. The chain end that has a free 3′—OH or 3′-phosphoryl group or some group other than an ordinary phosphodiester bond is designated the 3′ end. The characteristic of the 5′ end is, of course, that the terminal residue does not have the 5′ group involved in an ordinary phosphodiester bond.

Abbreviated polynucleotide and oligonucleotide structures are *always* to be written with the 5′ end on the left and the 3′ end on the right, unless the structure is labeled otherwise. For economy in representing long sequences, the

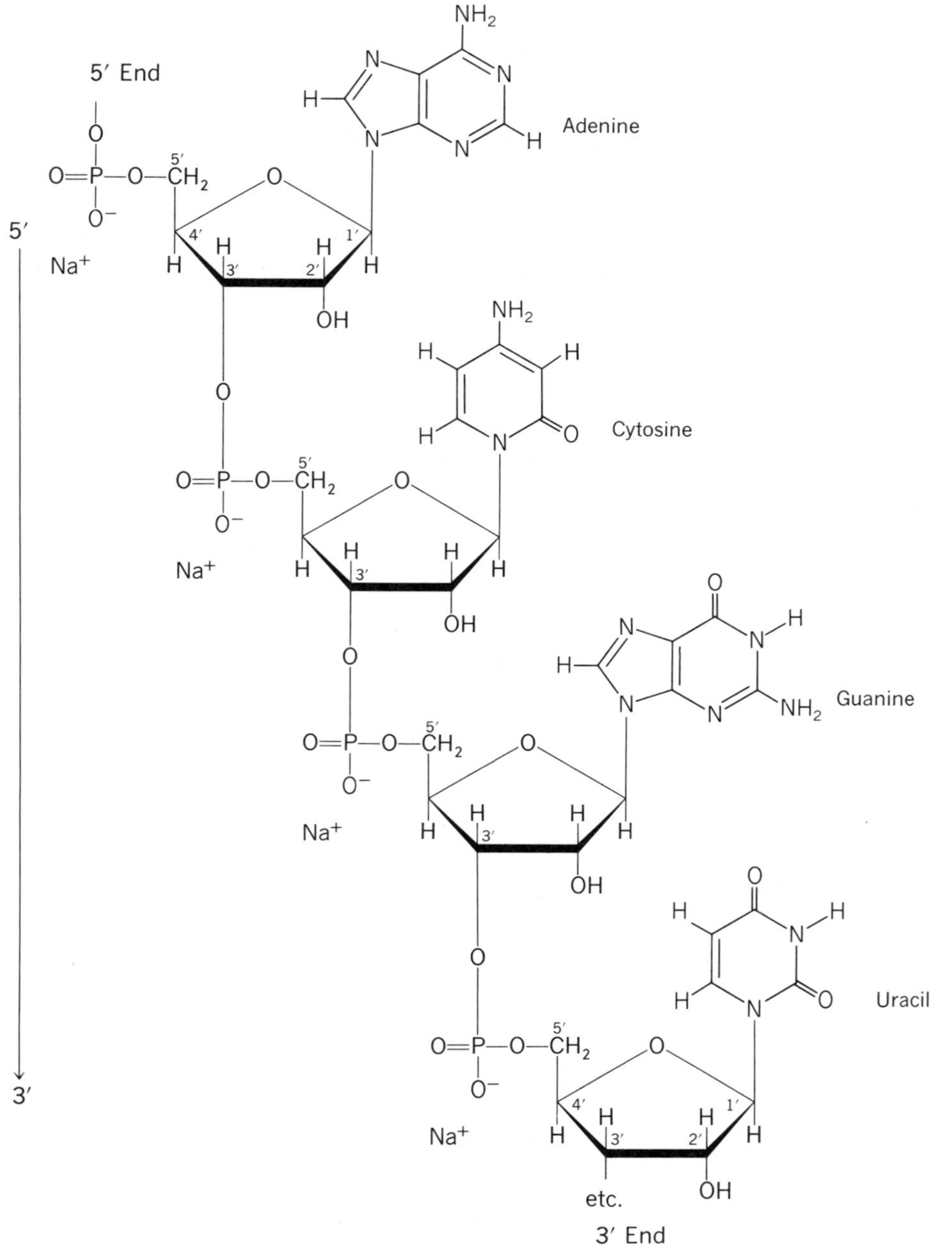

FIGURE 6.1 Structure of (*a*) a tetraribonucleotide and (*b*) a tetradeoxyribonucleotide showing the linear structure formed by 3′ → 5′ phosphodiester bonds between ribonucleotide and deoxyribonucleotide residues. The sodium ionic form is represented, though the counterions of a nucleic acid are readily exchanged with other cations in the medium.

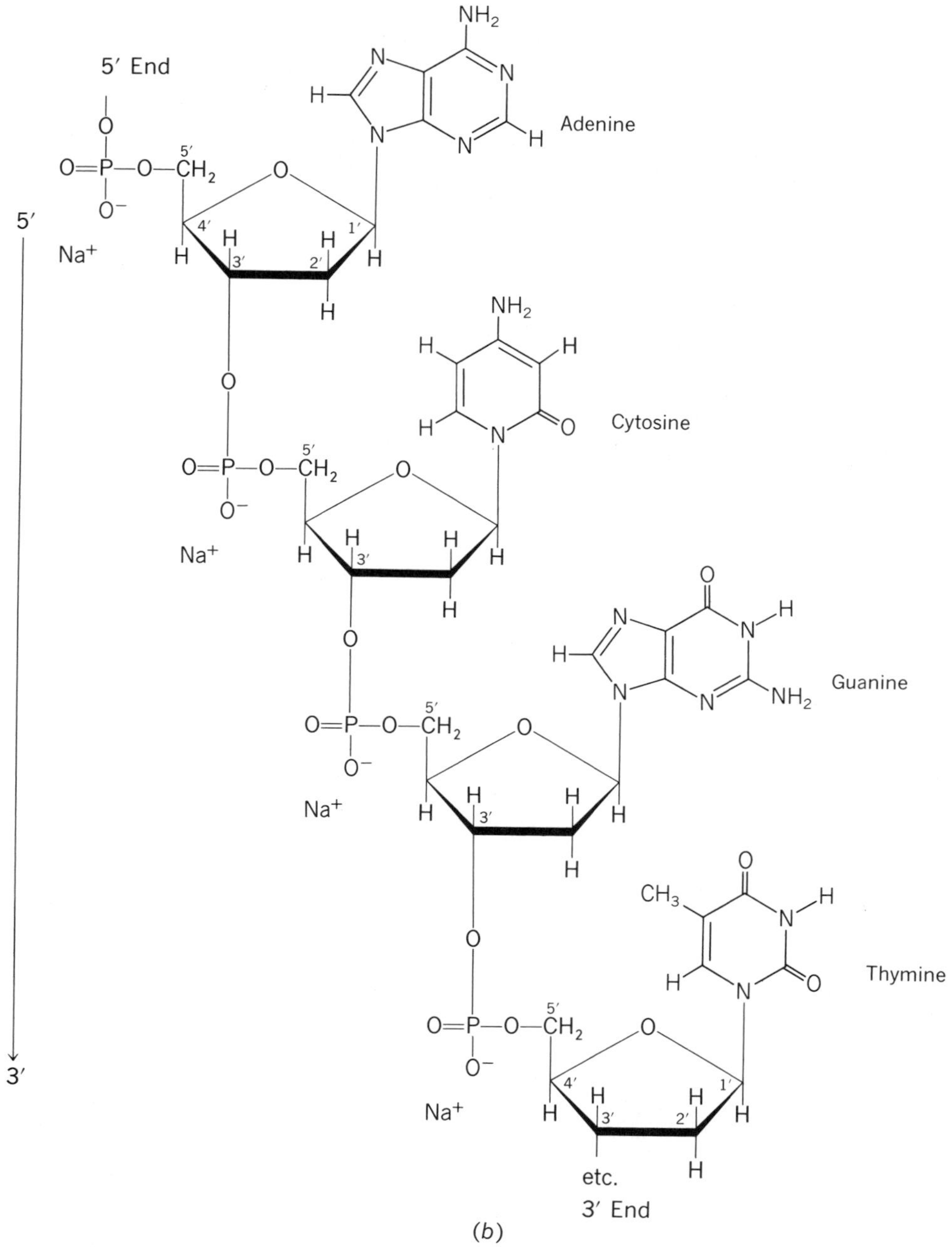

FIGURE 6.1 (continued)

phosphodiester bonds are not represented, though terminal groups usually are. A single "d" at the 5′ end indicates a sequence that consists entirely of deoxyribonucleotide residues. Examples are HO—UACACAUCUCAACUCp for an oligoribonucleotide and pdCATGTCCCATG—OH for an oligodeoxyribonucleotide. The convention also is useful for representing mononucleotides. AMP is pA and deoxycytidine-3′-phosphate is dCp. Nucleoside-2′-phosphates are not represented by this convention.

A single RNA molecule may have more than 10,000 nucleotide residues in it. DNA molecules may be far larger, as discussed in Section 6.10. That the DNA molecules may be so much larger than RNA molecules is partly a consequence of the greater stability of DNA, a topic of the next section. Gel electrophoresis is by far the most widely used technique for estimating the sizes of nucleic acid molecules. The charge on nucleic acid molecules is a topic of Box 6.A, and gel electrophoresis of nucleic acids is described in Box 6.B.

BOX 6.A   TAUTOMERIC FORMS AND PROTON REACTIONS OF NUCLEOSIDES AND NUCLEOTIDES

The structural formulas in this chapter present the nucleoside bases in the tautomeric form that appears to predominate according to spectroscopic analyses: the keto ($>C=O$) and amino ($-NH-$) form. These are favored over the enol ($\geqslant C-OH$) and imino ($=N-$) form, which is shown here for thymidine (structures on the right and left) and for guanosine (structure on the right).

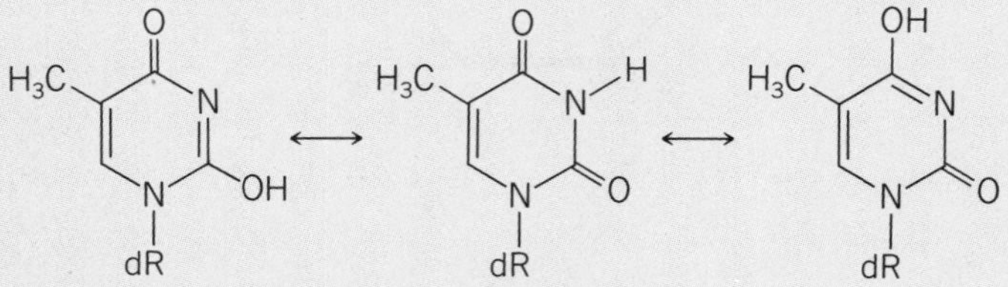

Thymidine

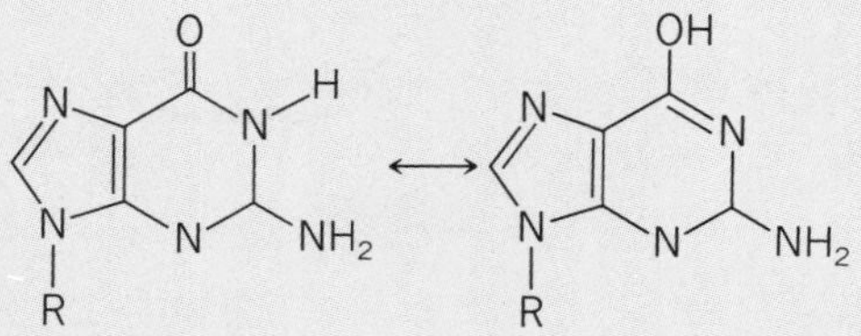

Guanosine

The keto and amino form fits the base-pairing scheme of Watson and Crick for double-stranded DNA (Section 6.5), whereas the enol and imino forms do not. Although the energy of the enol and imino form is greater than that of the keto and amino form, the energy difference is not great, so the enol and amino form may have biological significance. The tautomeric rearrangements require **intramolecular** proton transfers, and therefore, the tautomeric equilibria should be pH-independent.

However, nucleosides, nucleotides, and nucleotide residues have important protonic reactions. The important proton association and dissociation reactions of the nucleoside bases are shown near the end of this Box.

Like uridine, thymidine dissociates a proton with a pK_a of approximately 9.2. In the acid dissociation reactions of guanosine, thymidine, and uridine, the negative charge would be expected to spread from the ring nitrogen to the keto oxygen. The pK_a values of the bases are approximately 0.3 units greater in the polynucleotide than the nucleoside because of the negative charges contributed by the phosphodiester bonds, which are next described. That is, each protonated group becomes a weaker acid.

The proton dissociation reactions of phosphoric acid and phosphoric acid esters have been described in Section 1.13. Of the three proton dissociations of phosphoric acid, only one remains in a phosphodiester, and it has a pK_a of less than 1. In a polynucleotide chain, the most easily protonated base is

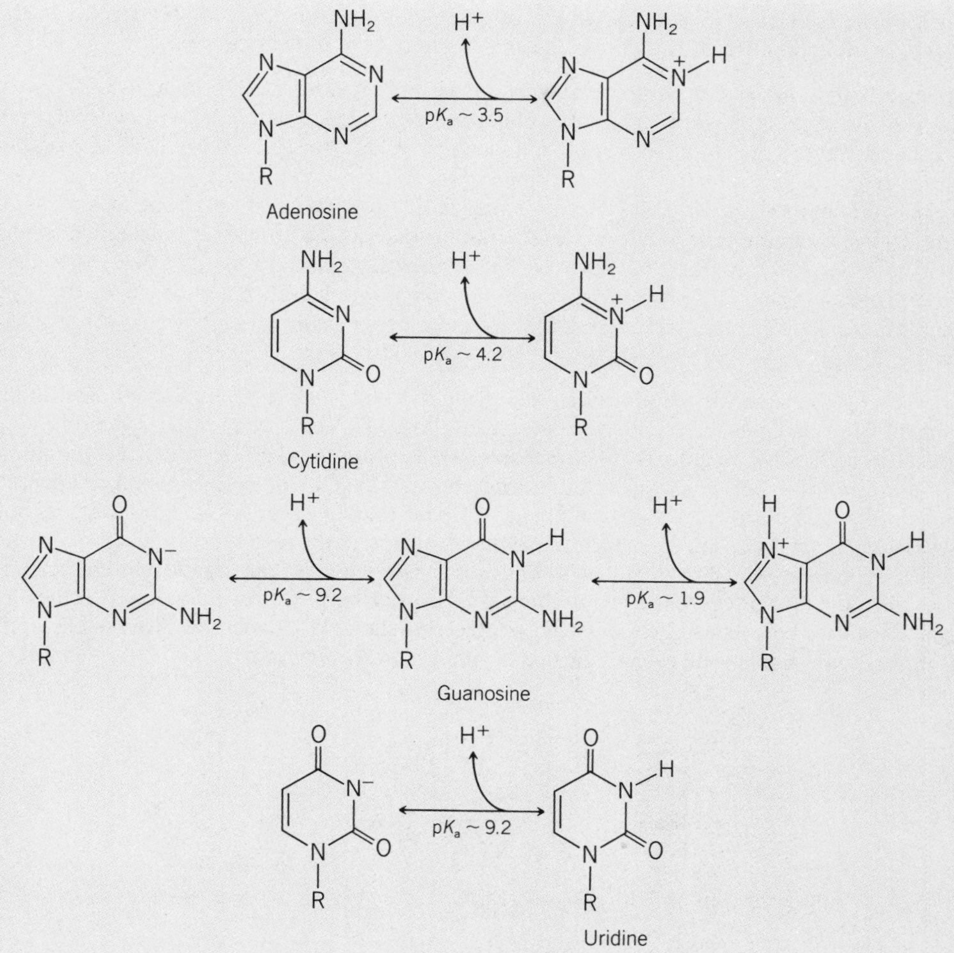

Adenosine

Cytidine

Guanosine

Uridine

cytosine, with a pK_a of approximately 4.5. Dissociation of a proton from a guanine, uracil, or thymine base will be half-complete only at a pH of about 9.5. Therefore, at neutral pH, there will be a net charge of -1 per nucleotide residue, from the dissociated phosphodiester bond. Ionizations of the bases make virtually no contribution at neutral pH.

BOX 6.B   GEL ELECTROPHORESIS OF NUCLEIC ACIDS

Both the size and the charge of a protein, a nucleic acid, or any other ionized molecule affect its electrophoretic mobility in a gel. Increasing the charge will, of course, increase the mobility if other factors are held constant. We described in Box 3.E how a gel retards the movement of different protein molecules so that, usually, the larger the molecule, the more its movement is retarded. Box 3.F described two methods by which the influences of (1) the ionic charge of a protein and (2) the resistance of the protein to transport through the gel were resolved; these methods are varying the polyacrylamide gel concentration and using the detergent SDS to put proteins in a denatured form in which the detergent charge overwhelms the protein charge. These approaches allow the molecular weight of the protein to be estimated without any exact information about its ionic charge. The size of a nucleic acid molecule, in contrast to the requirements for proteins, may be derived from its electrophoretic mobility in a *single* kind of gel and *without* the use of detergent.

That such simple molecular weight estimates are possible is a direct consequence of the simple relationship that holds between the number of nucleotide residues in a nucleic acid molecule and its charge at neutral pH: one negative charge per nucleotide residue (see Box 6.A). In RNA, for example, the mass of a ribonucleotide residue, neglecting the necessary counterion (e.g., sodium), varies from 304 for cytidylate to 344 for guanylate. This means that for RNAs of only roughly similar base compositions, the molecular weight is closely proportional to the number of nucleotide residues. Thus, the charge-to-mass ratio for a set of RNA molecules, or for a set of DNA molecules, is approximately constant, and the order of zone movement in gel electrophoresis will be the reverse order of molecular weights, provided that the molecules under examination have similar conformations.

One group of nucleic acid molecules that have reliably similar conformations are linear, double-stranded DNAs; for example, those created by cutting large DNA molecules with restriction endonucleases (Box 6.D). DNA fragments of 600 or fewer base pairs may be analyzed on polyacrylamide gels of concentration 4 to 20% in an apparatus like that shown in Box 3.E or in a horizontal gel apparatus, often called a "submarine gel" because the gel is under reservoir buffer. No discontinuous buffer system (Box 3.E) is needed or used; the same buffer is present in the gel and reservoirs. Nevertheless, very sharp zones are seen and great resolution is achieved because of the sample concentration that occurs when the movement of the highly charged DNA molecules is drastically slowed as they meet the gel. The photo here (courtesy of BioRad Laboratories) shows DNA zones resolved by electrophoresis through polyacrylamide gel and located by staining with silver ions.

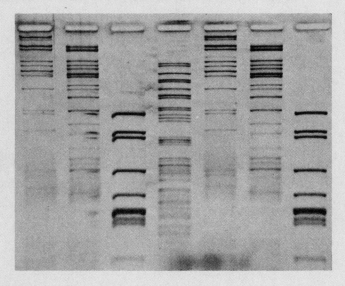

Some of the zones in the photo have less than a nanogram of DNA. DNA also can be detected, more rapidly but with slightly reduced sensitivity, by staining with ethidium bromide. Ethidium bromide binds to DNA in a process termed **intercalation,** insertion between base pairs. In that low-polarity environment, the ethidium bromide is highly fluorescent, and DNA zones are detected by visible light photography of the gel illuminated by ultraviolet light.

Large DNA fragments fail to penetrate even the most dilute polyacrylamide gels that are practical to prepare. However, the larger pores of agarose gels resolve larger DNA molecules. Agarose is the component of bacteriological agar-agar that is responsible for the strength of an agar-agar gel. Agarose gels are not formed by polymerization. Rather, noncovalent bonding between polysaccharide strands is responsible for gel formation. The graph here (*Source:* T. Maniatis, E. F. Fritsch, and J.

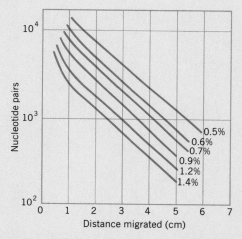

Ethidium bromide   Glyoxal

Sambrook, *Molecular Cloning,* Cold Spring Harbor Laboratory, New York, 1982, with permission) shows the dependence of the distance migrated during electrophoresis on the number of base pairs in the molecule and the agarose gel concentration. DNA zones in agarose gel most often are detected by ethidium bromide staining.

The size range of double-stranded, linear DNA molecules that can be resolved on polyacrylamide and agarose gels is from 10 to 20,000 base pairs, although the useful range of any given gel is only a factor of a few tens. Special electrophoresis techniques allow even larger DNA molecules to be resolved.

The sizes of single-stranded RNA and DNA molecules also are estimated from electrophoretic mobilities in gels. However, reliable estimates require that the molecules of unknown size and the appropriate RNA or DNA molecular weight standards be placed in similar conformations. This is accomplished by incorporating concentrated solutions of urea or formamide into the gel and performing electrophoresis at an elevated temperature, or by derivatizing the single-stranded nucleic acid with an aldehyde such as formaldehyde or glyoxal. The aldehydes form derivatives with nucleic acid bases and prevent hydrogen bonding between bases.

6.4 ANALYSIS OF STABILITY AND NUCLEOTIDE SEQUENCES OF NUCLEIC ACIDS

Perhaps the most fundamental analysis of the covalent structure (as distinguished from the conformation) of a nucleic acid is the **base ratios.** This is the relative number of moles of each kind of nucleotide residue in the nucleic acid. Analysis of base ratios requires a method for resolving four kinds of nucleotides or nucleosides or bases and a method for cleanly generating them by hydrolysis of the nucleic acid without a serious side reaction. If RNA is exposed to 0.3 *M*

NaOH for 18 hours at 37°C, it is quantitatively converted to nucleoside-2'-phosphates and nucleoside-3'-phosphates, abbreviated collectively as nucleoside-(2',3')-phosphates. As indicated by Figure 6.2, the mechanism of this reaction has the 2'-alkoxide (R—O⁻) group that is formed in strongly alkaline solution attacking the phosphorus atom. Most of the intermediate cyclic 2':3'-phosphodiester bonds survive under conditions that cleave most of the phosphodiester bonds, so nucleoside-2':3'-cyclic phosphates are the first nucleotides seen in the course of the reaction. Note that this part of the reaction

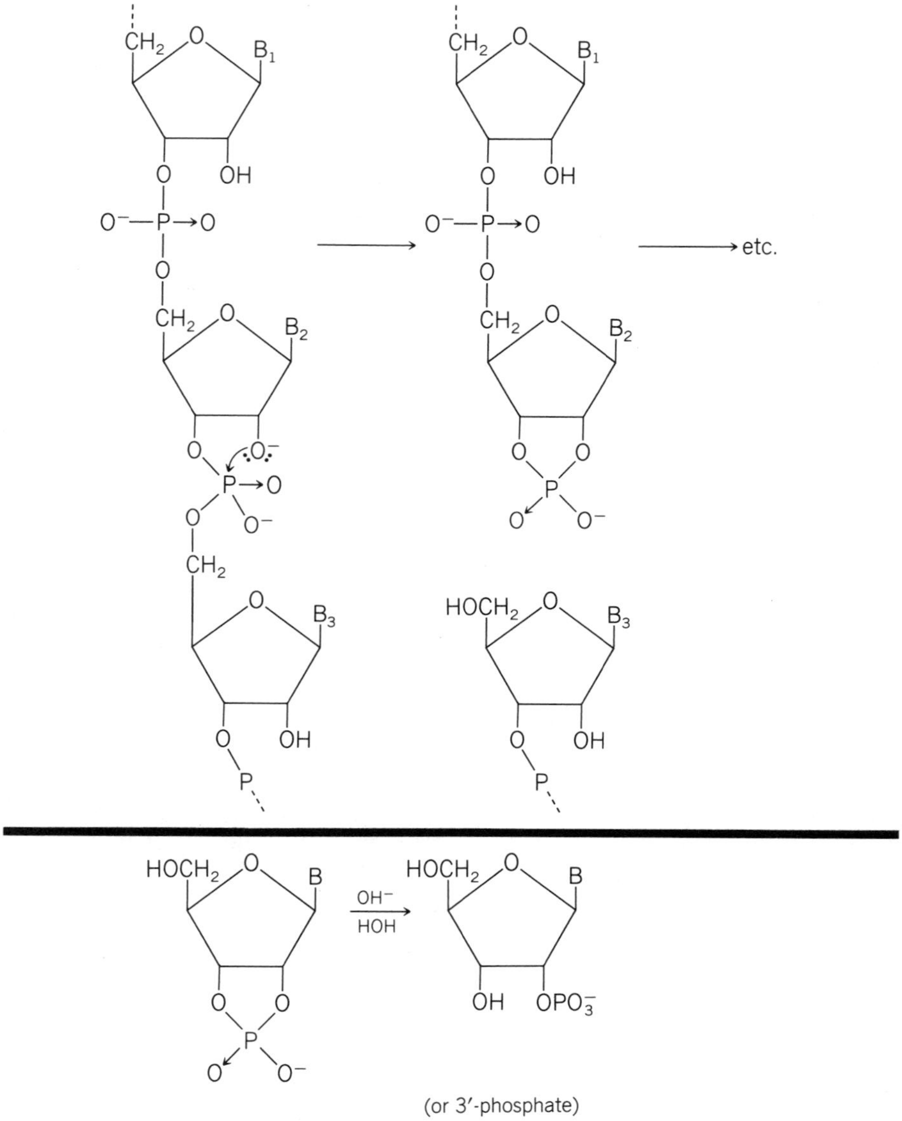

(or 3'-phosphate)

FIGURE 6.2   Mechanism of hydrolysis of RNA by base. The cleavage of a chain phosphodiester bond gives a new 5'—OH group and a 2':3'-cyclic phosphodiester bond, respectively, at the ends of the two polynucleotide fragments that are generated. The base-catalyzed opening of the cyclic phosphodiester bonds is a slower reaction. Hence at an intermediate stage in the series of hydrolysis reactions, cyclic nucleoside-2':3'-phosphates can be recovered in good yield. The final product is a mixture of nucleoside-(2',3')-monophosphates.

constitutes a polynucleotide chain cleavage (phosphoryl group migration) without hydrolysis.

Nucleotides with $2':3'$-cyclic phosphodiester groups are abbreviated A > p, G > p, and so on to indicate the cyclic phosphodiester. During continued incubation with base, hydroxyl ion attack hydrolyzes the cyclic phosphodiester bonds to give the two kinds of phosphomonoesters in roughly equal yields. If the hydrolysis conditions were significantly less vigorous than those just indicated, subsequent base ratio analysis would be thwarted by a too complex mixture of products, both cyclic diesters and monoesters being present. Significantly more vigorous conditions would cause some deamination of cytidylate to give uridylate, also invalidating the analysis.

Phosphodiester bonds in RNA also are hydrolyzed in acidic solution, by a mechanism that also involves the $2'$-hydroxyl group but that is different from the mechanism that describes hydrolysis in base. Side reactions prevent full recovery of the nucleotide bases and the reaction is not useful for accurate base ratio analysis. Because of the $2'$-hydroxyl group, RNA is less stable to hydrolysis than is DNA, at all pH values.

As was just indicated, the lack of a $2'$-OH group makes DNA much more stable in dilute base than RNA. In acidic solution, DNA is depurinated, that is, the N-glycosidic bonds to guanine and adenine are hydrolyzed and the free bases are released. More vigorous conditions of acid-catalyzed hydrolysis convert DNA to phosphoric acid, deoxyribose degradation products, free bases, and other products. For this reason, the degradation of DNA to deoxyribonucleotides for the purposes of base ratio analysis usually is accomplished with enzymes. The complete degradation of RNA to nucleotides (pN's or Np's) also can be accomplished by enzyme-catalyzed hydrolysis. We describe the actions of some of these nucleases in the next three paragraphs.

Enzymes that catalyze the hydrolysis of DNA are called deoxyribonucleases, abbreviated DNases, whereas those that catalyze RNA hydrolysis are ribonucleases or RNases. Nucleases catalyze the hydrolysis of either RNA or DNA, though "nuclease" also is a generic term for either an RNase or a DNase. Phosphodiesterases attack DNA or RNA and certain other phosphodiesters as well. Nucleases and many RNases show little or no specificity with regard to the nucleotide residue at which they cleave the polynucleotide chain. For example, incubation with ribonuclease T_2 degrades RNA to nucleoside-3'-phosphates. Incubation with nuclease P1 degrades RNA (or DNA) to nucleoside-5'-phosphates. However, ribonuclease T_1 cleaves only on the 3' side of guanylate residues. The products of the digestion are oligonucleotides with the 3' end . . . Gp and no internal guanylate residues.

The cyclic phosphodiester is an intermediate in the reaction catalyzed by ribonucleases that leave 3'-phosphoryl groups as the final product. This implicates the 2'-hydroxyl in the hydrolysis mechanism, just as in hydroxyl-ion-catalyzed hydrolysis. Enzymes such as spleen phosphodiesterase cleave either DNA or RNA to nucleoside-3'-phosphates and thus do not use the 2'-hydroxyl group in their reaction mechanism. This result was among the first pieces of evidence favoring $3' \rightarrow 5'$ rather than $2' \rightarrow 5'$ phosphodiester bonds in RNA. Among the preferred methods for degrading DNA to deoxyribonucleotides is digestion by pancreatic DNase and nuclease P1. The former cleaves DNA to oligodeoxyribonucleotides with 5'-phosphoryl groups and the latter completes the hydrolysis to dpN's. The products of either of these digestions can be analyzed by chromatography to estimate base ratios.

The key to the information content of a nucleic acid molecule is, of course, not the base ratios but the exact sequence of nucleotide residues in it. Several methods have been developed for determining the sequences of nucleotide residues in RNA or in DNA. The methods for DNA are more efficient and reliable than those for RNA. Box 6.C describes the two most widely used methods for determining the nucleotide sequences of DNA molecules.

BOX 6.C   DETERMINING THE SEQUENCES OF NUCLEOTIDES IN DNA

The task of determining the nucleotide sequence of a DNA molecule seems at first consideration to be formidable, if not impossible, because even the smallest DNAs from viruses have more than 2000 nucleotide residues. For years, biochemists despaired of ever having the ability to complete such a sequence. Two developments so revolutionized DNA analysis that now the sum of nucleotide residues in complete and fragmentary DNA sequences numbers in the few millions, and single laboratories are capable of determining a few thousand nucleotide residues in a week or two under ideal conditions. These developments are (1) the technology to specifically cut long DNAs into fragments (Boxes 6.D and 6.E) that more readily can be manipulated than the full-length DNA and (2) methods for rapidly determining the sequences of the DNA fragments, the topic here.

Early approaches to DNA and RNA nucleotide sequence determination were similar to those that had been applied to proteins: Obtain the sequence of small fragments, then put these in order using the nucleotide sequences of overlapping fragments. Unlike proteins, a nucleic acid usually has only four types of monomeric units, which means many very similar oligonucleotides will be generated when the molecule is fragmented. Sequence isomers such as dpCAGATT and dpCAAGTT are virtually impossible to separate. Clearly, an entirely different approach was needed. In fact, two related approaches were developed by A. M. Maxam and W. Gilbert, and by F. Sanger. Both methods are referred to as "ladder sequencing" because of the appearance of the electrophoresis gels from which nucleotide sequences are "read."

The tack taken by Maxam and Gilbert is first to introduce radioactivity into one end only of a DNA molecule. After limited cleavage of the DNA by four chemical reactions, the reaction products are resolved by polyacrylamide gel electrophoresis (Box 6.B). The gel is placed next to photographic film so that the locations of radioactive zones are revealed by the autoradiographic exposure when the film is developed. From this autoradiogram, long stretches, up to 600 residues, of the nucleotide sequence of the DNA molecule are determined by inspection according to the principles outlined next.

Usually, the terminal radioactive group required by the Maxam and Gilbert method takes the form of a ^{32}P-labeled moiety. Methods for labeling either the 5' or the 3' end are available. For example, the enzyme polynucleotide kinase catalyzes the transfer of a phosphoryl group from the γ position of ATP (Section 6.2) to a 5'-hydroxyl group of either DNA or RNA. If the polynucleotide that is to be labeled already bears a phosphoryl group, it is removed by prior incubation with another enzyme, a phosphomonoesterase. With $[\gamma^{32}P]$ATP as substrate, the reaction produces a 5' "end-labeled" polynucleotide.

$$^{-}O-^{32}\overset{\overset{\displaystyle O^-}{|}}{\underset{\underset{\displaystyle O}{\downarrow}}{P}}-O-\overset{\overset{\displaystyle O^-}{|}}{\underset{\underset{\displaystyle O}{\downarrow}}{P}}-O-\overset{\overset{\displaystyle O^-}{|}}{\underset{\underset{\displaystyle O}{\downarrow}}{P}}-O-Adenosine + HO-NpNpNp \ldots \longrightarrow$$

$$ADP + {}^{-}O-^{32}\overset{\overset{\displaystyle O^-}{|}}{\underset{\underset{\displaystyle O}{\downarrow}}{P}}-O-NpNpNp \ldots$$

If the DNA fragment is double-stranded, the reaction will introduce radioactivity into each 5′ end. Prior to the sequencing reactions, the strands are separated by gel electrophoresis under special conditions, or the DNA molecule is cut specifically at one site and the two partial molecules are separated.

One of the Maxam and Gilbert reactions calls for incubation of the DNA in dimethylsulfate at 20°C, pH 8, for 10 minutes. This introduces a methyl group at the 7 position of guanylate residues, but the reaction is under such mild conditions that methylation at other sites is negligible. The conditions also are adjusted so that only one guanylate in 20 or 30, or even 60 or 150, is derivatized, according to how long a sequence is to be determined in one experiment.

7-Methyldeoxyguanosine

The methylation of guanylate destroys the aromatic character of the guanine imidazole ring system, which makes the *N*-glycosidic linkage very labile. When the *N*-glycosidic bond is broken, the 1′ carbon becomes a potential aldehyde. The DNA, separated from the dimethylsulfate and reaction products, is incubated for about 30 minutes at 90°C in alkaline solution. A β-elimination reaction occurs under these conditions, essentially removing the deoxyribose residue from the polynucleotide chain and producing two new fragments, each with a phosphorylated end.

Of the two new fragments, only one is labeled. Because so few guanylate residues were methylated, on the average, any two cleavage sites in this "G reaction" will be 100 or more nucleotide residues apart. Since detection of separated reaction products is by autoradiography after gel electrophoresis, which detects only end-labeled reaction products, a distribution of oligodeoxyribonucleotides will be displayed. Each oligodeoxyribonucleotide represents, by its size, the number of nucleotide residues from the 5′ end of the original DNA molecule to a guanylate residue. Consider the reaction products from a specific DNA fragment, cleaved by removing dG residues as described above.

*pdCGATATAAACGCTGGTCCATGGTTTAA

⇓

*pdCp + pATATAAACGCTGGTCCATGGTTTAA
*pdCGATATAAACp + pCTGGTCCATGGTTTAA
*pdCGATATAAACGCTp + pGTCCATGGTTTAA
*pdCGATATAAACGCTG + pTCCATGGTTTAA
*pdCGATATAAACGCTGGTCCAT + pGTTTAA
*pdCGATATAAACGCTGGTCCATGp + pTTTAA
*pdCGATATAAACGCTGGTCCATGGTTTAA

The symbol *p indicates the radioactive 5′ phosphoryl group, and the phosphodiester bonds are not explicitly represented. The last structure is, of course, the unreacted starting material, to remind us that the Maxam and Gilbert reactions are designed to be partial reactions, which means that some of the initial amount of DNA fragment will remain undamaged after the reactions are complete.

In the example G reaction, **radioactive** zones are expected to correspond to pdNp, pd(Np)$_{10}$, pd(Np)$_{13}$, pd(Np)$_{14}$, pd(Np)$_{20}$, pd(Np)$_{21}$, and the original molecule. Seeing these zones, the investigator concludes that there are deoxyguanylate residues located at positions 2, 11, 14, 15, 21, and 22 but at no other positions with the possible exception of position 1.

Unfortunately, not all of the Maxam and Gilbert reactions remove a single kind of deoxynucleotide residue from the DNA molecule. Frequently used are a T + C reaction, which removes deoxycytidine and deoxythymidine residues with about equal efficiency, a C reaction, and an A > G reaction. The last gives strong zones that correspond to cleavage at deoxyadenylate residues and weaker zones that correspond to cleavage at deoxygualylate residues. In theory, the sequence of all of the deoxyribonucleotide residues except that at the 5′ end can be deduced from the positions of zones formed by electrophoresis of the products of the four reactions: G, A > G, C, and T + C. In practice, the fragments of five and fewer residues usually cannot be resolved and must be determined by other methods. The next drawing of this Box gives the appearance of the gel under the assumption that even the smallest fragments were resolved. The drawing also has a "complete ladder" that displays the products of reactions at all deoxyribonucleotide residues, although such a lane is rarely necessary. The sequence read from the drawing is pdNGATATAAACGCTGGTCCATGGTTTAA. The student should write out the structures of oligodeoxyribonucleotides that are expected in some of the specific zones of the gel.

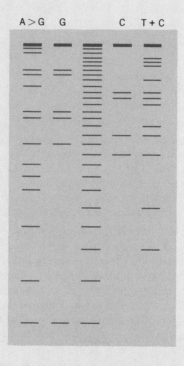

The method of F. Sanger, sometimes referred to as "dideoxyribonucleotide-sequencing," gives a similar, but simpler, pattern of zones after electrophoresis of reaction products, but the DNA fragments in those zones are produced in an entirely different manner. They are synthesized enzymically, *in vitro*, by transcription of a single-stranded DNA template, the origin of which is beyond the scope of this presentation. First, the single-stranded DNA is hybridized to a specific, synthetic (Box 6.F) oligodeoxyribonucleotide that is complementary to the single-stranded DNA, in the sense that two strands of B–DNA are complementary.

The synthetic oligodeoxyribonucleotide, called a primer, is designed to hybridize (Section 6.6) to a region of the single-stranded DNA that is to the 3′ side of the part of the polynucleotide chain that is of interest. Usually, the single-stranded DNA is hundreds or thousands of nucleotides long. We show here a small portion of such a DNA molecule hybridized to a primer.

3′—HO—TGACCGGCAGCAAAATG—5′

5′ . . . CGATATAAACGCTGGTCCATGGTTTAACGTCGTGACTGGCCGTCGTTTTACACTTGAAA . . . 3′

As is described in Chapter 19, when such a complex is incubated with a DNA-dependent DNA polymerase and deoxyribonucleotide triphosphates (Section 6.2), the deoxynucleotide residues are added to the 3′ end of the primer sequentially and under the control of the single-stranded DNA template. The result will be the synthesis of a **transcript.**

. . . GCTATATTTGCGACCAGGTACCAAATTGCAGCACTGACCGGCAGCAAAATG—5′
5′ . . . CGATATAAACGCTGGTCCATGGTTTAACGTCGTGACTGGCCGTCGTTTTAC A CTTGAAA . . . 3′

If to such a reaction is added a deoxyribonucleoside triphosphate with a modified sugar residue, such as dideoxyguanosine triphosphate, the chain will be terminated whenever a dideoxyguanylate residue is incorporated in place of a deoxyguanylate residue because the former lacks a 3′-hydroxyl group to participate in the next phosphodiester bond.

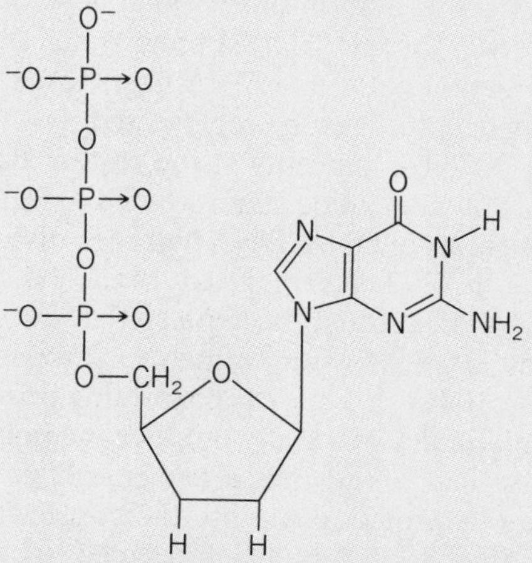

Dideoxyguanosine triphosphate

Thus, some of the transcripts will be prematurely terminated, but only at the sites of insertion of deoxyguanylate residues, that is, sites of deoxycytidylate residues in the single-stranded DNA template. By balancing the ratio of dideoxyguanosine triphosphate to deoxyguanosine triphosphate in the reaction mixture, the researcher can control the average length of the prematurely terminated transcripts. This corresponds to adjusting the chemical cleavage conditions in the Maxam and Gilbert reaction to achieve the proper degree of partial reaction.

In the Sanger dideoxyribonucleotide-terminating procedure, four separate transcription reactions are initiated simultaneously, one each with the dideoxynucleoside triphosphate derivative of deoxyadenosine, deoxycytidine, deoxyguanosine, and deoxythymidine. All reaction mixtures have all four deoxynucleoside triphosphates as well, one of them labeled with radioactive phosphate in order to make the transcripts detectable after electrophoresis by autoradiography. The products are analyzed and interpreted as in the Maxam and Gilbert method, except that it is the sequence of the transcript that is determined. Both the Maxam and Gilbert and the Sanger methods continue to find wide application, because each has its advantages. The Sanger method generally requires less effort, whereas the Maxam and Gilbert method is less sensitive to effects of specific sequences, such as those that are rich in deoxyguanylate and deoxycytidylate residues.

6.5 NUCLEIC ACID DOUBLE HELICES

By 1950, E. Chargaff had analyzed the base ratios of DNAs from several animal and bacterial sources. Although the base ratios of different DNAs varied widely, certain regularities were apparent. Most fundamental was the molar equivalence of deoxyadenlyate to deoxythymidylate and of deoxycytidylate to deoxy-guanylate. The base ratios could be expressed simply as the % GC. That is, a DNA with 40% GC content has deoxyadenylate, deoxycytidylate, deoxyguany-late, and deoxythymidylate in the ratio of 30:20:20:30 to satisfy Chargaff's rule of molar equivalence.

Another indication of the regular structure of DNA came from x-ray diffrac-tion experiments, notably in studies by R. Franklin and M. Wilkins. When x rays pass through ordered collections of DNA molecules and subsequently intercept a sheet of photographic film, the diffracted x rays produce a regular pattern. Although the pattern is not a picture of the DNA molecule, it does show that the DNA molecule has a definite structure and can indicate whether a given model for the conformation of the polynucleotide chain in DNA is possibly valid or definitely not valid. The simplicity and certain other characteristics of the x-ray diffraction pattern of DNA fibers revealed to x-ray crystallographers that the DNA structure can be represented by a simple model that is helical.

In the early 1950s, concepts of the regular base composition and regular folding of the polynucleotide chain seemed to be in conflict with the presumed biological functions of DNA. DNA had been discovered as a major constituent of salmon sperm. O. T. Avery, C. M. MacLeod, and M. McCarty had geneti-cally changed (transformed) bacteria so that they elaborated a polysaccharide coat that they formerly were incapable of making. Avery and his colleagues showed in the 1940s that the "transforming principle" was DNA from other strains of bacteria that normally produce the polysaccharide coat. A. Hershey and M. Chase had found that a bacteriophage (bacterial virus) injected its DNA, but not its protein coat, into the host bacterium to begin an infection. Thus, the nonliving DNA molecule seemed to be important to the genetics of animals, bacteria, and viruses. It seemed reasonable to some that DNA would be found to be the molecule of which genes are made. How could gene(s) for a polysaccharide coat and genes that direct the production of new bacteriphage particles be made from a molecule that seemed from x-ray crystallography and other physical and chemical studies to be so similar regardless of its biological source?

J. Watson and F. Crick reconciled the properties of uniform structure and diverse genetic function when they conceived their now well-known model for DNA structure, the most influential model of a biological molecule ever con-structed. Figure 6.3 provides two representations of the model. Two polynu-cleotide chains are intertwined to form the familiar double-helix with the more polar, hydrophilic ("water-loving") deoxyribose and phosphate residues on the outside. The less polar base pairs form the interior of the cylindrical molecule. In this model, the base pairs are roughly perpendicular to the long axis of the molecule and are stacked to give roughly the appearance of a spiral staircase. The bases of one base pair lie in the same plane. Thus, a base pair has approxi-mately the same thickness as a base: 0.34 nm. One turn of the helix has ten base pairs, so the pitch (length of one full turn) of the helix is 3.4 nm. Note that the two strands of a helix are oriented in an antiparallel rather than parallel fashion. That is, the 5'-to-3' direction of the two chains is opposite.

The base pairs are the conceptual, as well as the structural, core of the model. If we ignore for the moment steric restraints imposed by the polynucleotide chain, there are tens of ways that the nucleic acid bases could be juxtaposed or

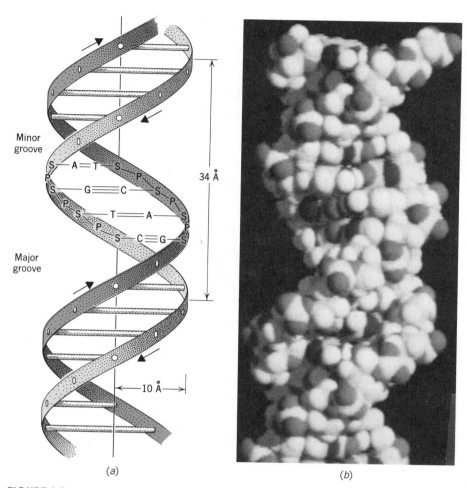

FIGURE 6.3 Two representations of the B conformation of double-stranded DNA. In (*a*), the course and opposite polarities of the two strands are shown, with the alternating sugar (deoxyribose) and phosphate residues indicated by S—P—S—P—S . . . Also shown are the G : C and A : T base pairs that form the core of the molecule. (*b*) A space-filling model of B–DNA, with the course of the two polynucleotide chains marked. (Courtesy of Dr. John E. Johnson.)

stacked in pairs. As indicated in Figure 6.4, crucial features of the particular pairing proposed by Watson and Crick are

1. The C-1′ atoms of the deoxyribose residues are separated by the same distance in A : T and C : G base pairs.
2. The angle formed between a line connecting the two C-1′ atoms of a base pair and an *N*-glycosidic bond is the same for all four nucleotide residues.
3. Strong, straight hydrogen bonds establish specific pairing between dA and dT residues and between dC and dG residues.

The proposed base pairs at once satisfied genetic and structural requirements, including Chargaff's rules. Similar dimensions of the two types of base pairs in two orientations allow a regular structure, but place no constraint on the se-

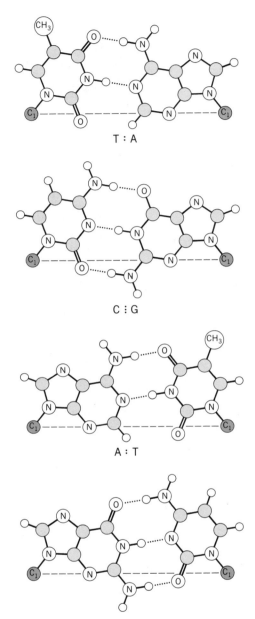

T : A

C : G

A : T

G : C

FIGURE 6.4 The four possible Watson and Crick base pairs, representing the two possible orientations of a G : C and an A : T base pair within a double helix.

quence of nucleotide residues in one chain. Thus, any gene can be encoded in a sufficiently long strand of DNA, and the locations of deoxyribose and phosphate residues will be the same. However, the nucleotide sequence in one strand dictates the nucleotide sequence in the other strand according to the Watson and Crick base pairing and the antiparallel structure. This is illustrated by the following structure, written in a shorthand notation in which the strand with the 5′ to 3′ orientation is uppermost:

5′ . . . pdGATCCGCG . . . . 3′
3′ . . . dCTAGGCGCp. . . . 5′

The structure described in Figure 6.3 captures the essential features of ordinary DNA, that is, DNA with all four nucleotide residues well represented. It is designated the "Watson–Crick B helix" to distinguish it from structures for DNA of unusual base composition or for DNA in water-poor environments. The correctness of the B structure was indicated not only by its agreement with Chargaff's rules and the genetic requirements for DNA function, but also because it correctly predicted the DNA x-ray scattering pattern, at least to a first approximation. It was sterochemically satisfactory, meaning that a space-filling molecular model (Figure 6.3b) could be built from correctly sized model atoms, and the model was consistent with the known structures of mononucleotides. In the years since the publication of the B structure in 1953, modifications have been introduced to bring the structure into closer agreement with higher quality x-ray diffraction data.

DNA dissolved in buffered aqueous salt solutions behaves as if it is a Watson–Crick B helix except that it appears to have 10.5 rather than an even 10 base pairs per turn of the double helix. The diameter of the molecule, approximately 2 nm, and the length per base pair, approximately 0.34 nm, are as predicted. These dimensions anticipate one of the most startling properties of DNA, the extraordinary length-to-diameter ratio. For example, the genome of the bacterium *Escherichia coli* is a single, circular DNA molecule of approximately 4.3 million base pairs (4300 kilobase pairs or 4300 kb). The chromosomes of higher organisms also appear to be composed of single DNA molecules. At 0.34 nm per base pair, the contour length of *E. coli* DNA is approximately 1.5 mm, giving a length-to-diameter ratio of 730,000. Because the DNA molecule is so long, the slightest disturbance of the solution can produce local regions that are flowing in different directions. This produces a stretching force on a long DNA molecule and may break it. Such "shearing" forces normally break DNA molecules of the size of *E. coli* DNA into thousands of pieces, and it is these pieces rather than the intact molecules that most often are studied in the laboratory.

Note that although the linear DNA double helix is composed of two intertwined, covalent, linear molecules, the convention is to refer to the two molecules together as the "DNA molecule." The two covalent units of this "noncovalent" molecule obviously are held together by hydrogen bonds. The structure also is stabilized by the forces between the bases stacked one on another to form the "spiral staircase" in the interior of the molecule. Destabilizing the structure are the electrostatic repulsive forces between the negatively charged phosphodiester groups (see Box 6.A) and thermal motion. The DNA molecule is not static but exhibits a random, dynamic, and localized separation of bases and rapid reforming of the hydrogen-bonded base pairs. Raising the temperature increases thermal motion and shifts the equilibrium in favor of separation of bases; at a high enough temperature, the two strands of a linear DNA fragment unwind from each other and separate. This process is termed "melting." Extremes of pH also will melt DNA. DNA denaturation is a reversible process that is described in Section 6.6.

Some double-stranded DNAs are circular. The DNA will be circular whether one or both strands actually is a circular chain of nucleotide residues connected by phosphodiester bonds. If both strands are circular, the two strands are intertwined and cannot be separated so long as both strands are intact. Figure 6.5 shows diagrammatically how the two covalent, circular strands can be

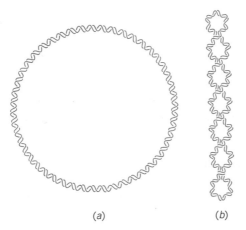

(a)   *(b)*

FIGURE 6.5   Superhelical configuration of covalently closed circular DNA. (*a*) Represents an imaginary circular DNA of only 500 base pairs. If we assume 10 base pairs per turn of the helix, the circular DNA will have 50 turns. (*b*) Shows the superhelical form expected for a 500 base-pair circular DNA that has only 44 turns of the double helix rather than 50.

inseparable even though they are not covalently linked to each other. The Watson and Crick model predicts, and the linear form of a double-stranded DNA defines, the number of turns that the B form DNA should have in its lowest energy form. The number of turns is approximately one per 10 base pairs. An important consequence of "covalently closed" circular DNA is the possibility that the actual number of turns of the double helix, imposed by the two covalent circular strands, will be different than the number of turns in the corresponding linear DNA. An exact match of the number of turns of the helix and the number of turns anticipated by the Watson and Crick model for B DNA will give a circular molecule without supercoils. In contrast, the supercoiled form of a circular, covalently-closed, double-stranded DNA molecule occurs when the actual number of turns in the helix is either greater or less than the number of turns that would occur if the molecule were linear. The conflict between (1) the actual number of turns in the DNA strands and (2) the number of turns that the low-energy form of B–DNA would assume if not constrained by the covalent circle results in supercoiling of the DNA. The spontaneous process of supercoiling allows the DNA to have the proper number of turns that the covalent structure of the covalently closed, circular DNA has prevented.

The phenomenon of supercoiling easily can be demonstrated with a rope. Lay the rope flat on a table and tape the ends together so as to form a covalent circle. The configuration of the rope then corresponds to Figure 6.5*a*. Now, remove the tape. Hold one end of the rope and twist the other end of the rope one, two, three, and so on complete rotations. Holding the rope in this tense form, tape the ends again. Notice that the rope assumes a configuration like that shown in Figure 6.5*b*.

Restriction endonucleases are among the most useful tools for analyzing double-stranded DNA. We present information on these very important enzymes in Box 6.D and Box 6.E, which describe the origin and properties of restriction endonucleases and some of their applications in the study and manipulation of DNA sequences.

Double-stranded DNA is not the only form of double-stranded nucleic acid. Double-stranded RNA is found in a few viruses and in small quantities in some fungi and plants. Double-stranded RNA is forced, because of steric constraints induced by the 2′ hydroxyl group, into a conformation that is called the A form. In the A form, the base pairs are tilted relative to the helix axis, and certain other

BOX 6.D   RESTRICTION ENDONUCLEASES AND THEIR USES

Restriction endonucleases are DNA-hydrolyzing enzymes that are highly specific with regard to the nucleotide sequence at which they will cut DNA. They are essential to the *in vitro* construction of chimeric DNA sequences, which are DNA sequences that are derived from two or more sources and combined into one molecule. The result is *in vitro* recombination, the core of **recombinant DNA technology** (see Chapter 22). Restriction endonucleases were discovered because of the surprising way in which the infectivity of certain bacteriophages, that is, viruses of bacteria, varied on transfer from one bacterial host to another. For example, a preparation of the DNA bacteriophage λ was recovered after growth on *E. coli* strain C. The bacteriophage was 1000 to 10,000 times more infectious to *E. coli* strain C than it was to *E. coli* strain K. Bacteriophage λ derived from *E. coli* strain K was highly infectious to either strains C or K, but would revert to poor infectivity for strain K after it had replicated in C. Thus, bacteriophage λ can be restricted from efficient infection of strain K or be modified on replication in K to show high infectivity on that host. Many other such **restriction–modification** systems have been discovered.

In several systems, the basis of restriction and modification has been traced to a pair of enzymes: (1) a restriction endonuclease that can destroy foreign DNA when it enters the bacterial cell and (2) a DNA-modifying enzyme that covalently modifies the DNA of the cell that forms the restriction endonuclease, for example, by methylation of bases. This modification prevents the restriction endonuclease from cutting the cell's DNA. In the bacteriophage λ example, the K strain host has a restriction–modification system. Those few bacteriophage DNA molecules that escape restriction become modified, and their progeny become modified, so that they are able to replicate in *E. coli* strain K.

The demonstration in an organism of an endonuclease that is specific for foreign DNA is taken as prima facie evidence for a restriction–modification system. Research on restriction endonucleases illustrates an important principle of research: how difficult it is to determine whether a particular investigation is or is not of practical or even fundamental value. Bacteriophage restriction once was thought to be of no practical and little fundamental interest, but the study of restriction led to the discovery of one of the most important tools of modern biochemistry, restriction endonucleases, and their application to new industrial processes.

Two general classes, or types, of restriction endonucleases are recognized. Those of type II are useful in preparing *in vitro* recombinants (Chapter 22) because they catalyze the cutting of a DNA molecule within a specific nucleotide sequence. The table here gives examples of restriction endonucleases and shows how they are named according to the genus and species of the source organism and the order of their discovery. For example, "EcoRI" is the first restriction endonuclease to have been identified from *E. coli* strain R; HaeIII is the third restriction endonuclease to have been discovered in *Haemo*-philus *ae*gyptius. For all of the examples in the table except MnlI, the cutting site in only one strand is presented. Thus, for EcoRI, the new ends created by cutting are

```
5'- . . . GAATTC . . . -3'          5'- . . . G-OH        pAATTCC . . . -3'
                            ────→                      +
3'- . . . CTTAAG . . . -5'          3'- . . . CTTAAp      HO-G . . .  -5'
```

This reaction shows the 5'-phosphoryl and 3'-OH that are characteristic of the new ends and the symmetry of the cut site. The slash in G/AATTC, for example, shows the phosphodiester bond that is cleaved.

Note that the staggered cut produced by restriction endonucleases such as EcoRI and PstI gives "sticky ends," which have a weak potential to form molecular hybrids by base pairing. This property assists in connecting DNA fragments together when forming *in vitro* recombinants. However, even "blunt ends" such as those produced by the action of SmaI or HaeIII can be joined (Chapter 22). Most restriction endonucleases recognize symmetrical sites, that is, sites that are the same after a 180° rotation, as in the EcoRI example. Others, such as AccI, have some sites that are symmetrical and others that are not. HinfI is an example of an enzyme that recognizes nearly, but not perfectly,

Some Type II Restriction Endonucleases

DESIGNATION	SOURCE	CLEAVAGE SITE(S)
AccI	*Acinetobacter calcoaceticus*	GT/AGAC GT/ATAC GT/CGAC GT/CTAC
HaeIII	*Haemophilus aeqptius*	GG/CC
EcoRI	*Escherichia coli* strain R	G/AATTC
HinfI	*Haemophilus influenzae* strain Rf	G/AATC G/ACTC G/AGTC G/ATTC
MnlI	*Moraxella nonliquifaciens*	5′—CCTCNNNNNNN/NNNN 3′—GGAGNNNNNNN/NNNN
PstI	*Providencia stuarti*	CTGCA/G
SmaI	*Serratia marsescens*	CCC/GGG

symmetrical sites. In contrast, the site cut by MnlI is unsymmetric and has no requirement for the identities of the base pairs present at the seven locations to the left of the cut site, indicated by N's. Restriction endonucleases generally are double-stranded DNA-specific; a few that recognize GC-rich sequences, such as HaeIII, also cut single-stranded DNA. Experiments indicate that the cutting of single-stranded DNAs is due to transient or stable associations of regions from different parts of the polynucleotide chain into a double-stranded conformation.

The property that distinguishes restriction nucleases from other deoxyribonucleases is their ability to cleave only at specific sequences of base pairs rather than randomly. The number of sites in any DNA molecule that are cleaved by a restriction endonuclease depends, on the average, on the number of bases recognized by the restriction endonuclease. EcoRI, PstI, and SmaI all have sites composed of six base pairs. If, for example, a DNA had a 50% GC base composition, the probability of having any given deoxynucleotide residue at a given site is 0.25. Any given dinucleotide sequence has a probability of $0.25 \times 0.25 = 0.0625$. The probability of any six-base sequence is then $(1/4)^6 = 1/4096$. For enzymes such as HaeIII and Mnl1, which recognize a sequence of four nucleotide residues, and HinfI, which recognizes a sequence of five residues but places no limitation on the central residue, the probability of locating the recognized sequence is 1/256. The AccI site has six residues but allows either of two nucleotide residues at either of the two central locations in that site, which gives a probability of 1/1024. Nucleotides sequences in DNA are, of course, not random, so the calculated probabilities provide only a rough guide as to the number of sites to expect. That is, in a sequence of about 4000 residues, one EcoRI site would be expected on the average, but some 4000-residue sequences have several EcoRI sites and others none. The use of restriction endonucleases in physically mapping nucleotide sequences is described in Box 6.E.

BOX 6.E   MAPPING DNA SEQUENCES WITH RESTRICTION ENDONUCLEASES

Restriction endonucleases are valuable not only for preparing DNA fragments to be assembled into new sequence arrangements, but also for more mundane requirements such as producing fragments for nucleotide sequence analysis (Box 6.C) and comparing very long stretches of DNA for similarities that can be noted from arrangements of restriction endonuclease sites. "Restriction mapping" is the assignment of the restriction sites of different enzymes along the polynucleotide chain. Restriction mapping can be accomplished without determining the nucleotide sequence of the DNA because the sites of restriction cuts can be deduced from the apparent sizes of double-stranded DNA fragments as estimated from agarose gel, or occasionally polyacrylamide gel, electrophoresis (Box 6.B).

The plasmid pBR322 is a double-stranded covalent DNA circle of 4362 base pairs. If pBR322 is treated with the restriction endonucleases AccI, EcoRI, and PstI (see Box 6.D) singly and in pairs, the linear fragments observed after agarose gel electrophoresis have the sizes indicated in base pairs (bp).

AccI	1595 bp and 2767 bp
EcoRI	4362 bp
PstI	4362 bp
AccI + EcoRI	651 bp, 1595 bp, and 2116 bp
AccI + PstI	1366 bp, 1401 bp, and 1595 bp
EcoRI + PstI	750 bp and 3612 bp

From the results of the digestion of pBR322 with AccI + EcoRI, it is obvious that the single EcoRI site is in the larger of the two AccI-generated DNA fragments, because the smaller, 1595 bp fragment survives the digestion with EcoRI. Similarly, the single PstI site is in the larger AccI fragment. The EcoRI and PstI sites are 750 bp apart according to the results of the double digestion. Thus, the map of the larger AccI-generated fragment must be

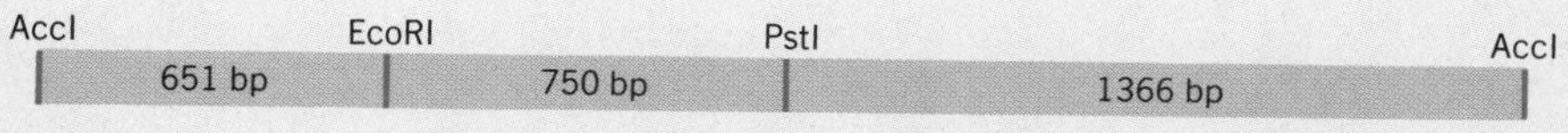

This map, combined with the remaining 1595 bp AccI-generated fragment, will give two primitive, circular **restriction endonuclease maps** of plasmid pBR322. One is chosen by convention to be the basis of a nucleotide residue numbering system. The following diagram shows a more complete restriction endonuclease map of pBR322.

Often, one restriction endonuclease, let us call it *X*, will have several sites within a fragment that is generated by a second restriction endonuclease, *Y*, as shown in the map of pBR322 by the four TaqI sites within AccI fragment A. Locating the several sites of restriction endonuclease *X* unambiguously generally will require approaches beyond those given in the example here, such as partial digestions with restriction endonuclease *X* to give fragments that still have one or more sites of *X*. For example, partial digestion of pBR322 with TaqI might yield fragments corresponding to D—E_1 and to E_1—E_2, and so on. The pBR322 map also illustrates how restriction sites for two enzymes can coincide. The nucleotide sequence beginning at base 649, CGTCGACC, has a site for AccI [GT(A/C)(G/T)AC], for SalI [GTCGAC], and for Taq I [TCGA]. In contrast, the sequence at the second AccI site, beginning with base 2244, is TGTATACT. This sequence does not contain a SalI or a TaqI site.

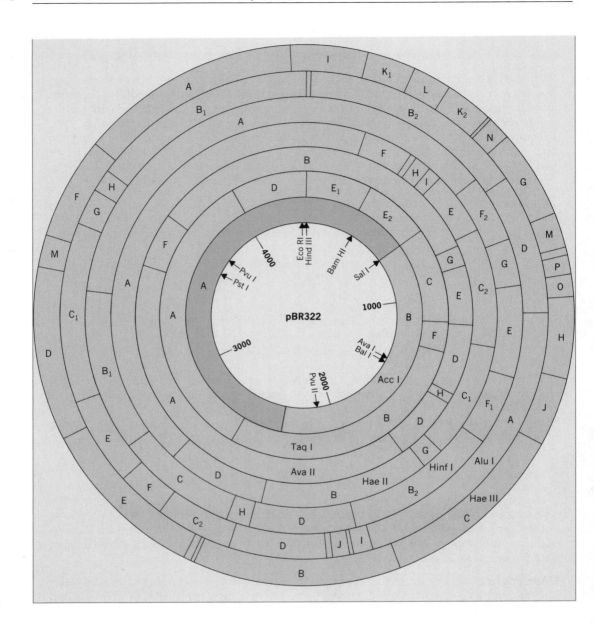

changes also distinguish this double helix from that of B–DNA. Artificially-formed DNA–RNA hybrids, with one strand of each type of polynucleotide, also assume the A conformation, as does double-stranded DNA in alcohol-water solutions.

6.6   DENATURATION AND RENATURATION OF NUCLEIC ACIDS

The melting of a DNA molecule increases its absorption of ultraviolet light. The explanation for this observation is that electronic interactions between stacked bases reduce their ability to absorb light of wavelength approximately 260 nm.

A much smaller proportion of bases are in the stacked configuration in the single, melted strands than in the double helix and the absorbance approaches the greater, more "hyperchromic" value that is characteristic of individual mononucleotides. Figure 6.6a shows how the absorbance increases with increasing temperature, giving sigmoidal curves. Because G:C base pairs are more stable than A:T base pairs, the temperature at the midpoint of the transition, T_m, is greater for DNAs with higher %GC content. In 0.15 M NaCl, 0.015 M sodium citrate, pH 7, the relationship between % GC content and T_m is

$$\%GC = 2.44(T_m - 69.3)$$

In solutions of lower salt concentration, the negative charges on the phosphodiester groups repel each other more effectively, and the T_m is decreased for any DNA. "Melting curves" for DNAs in a solution of lower ionic strength are shown in Figure 6.6b.

As was indicated in the previous section, the base pairs of double-stranded RNA are tilted at an angle of 70° to the long axis of the molecule, rather than the approximately 90° of the B–DNA helix. The tilting results in a more compact molecule. For this reason, double-stranded RNA is more stable than double-stranded DNA and exhibits a greater T_m value for the same %GC content.

One of the most important aspects of the Watson and Crick model for DNA is its genetic implications. The pairing of two strands, facilitated by their complementary sequences, indicates that an isolated strand could guide the synthesis of a new strand that, after synthesis, would be complementary to the first strand. Nucleic acid molecules are, in fact, replicated and transcribed into new

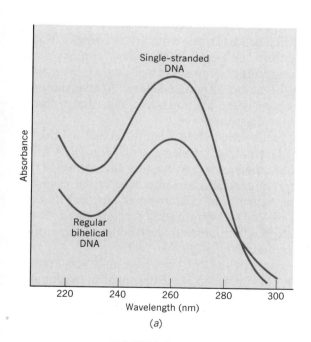

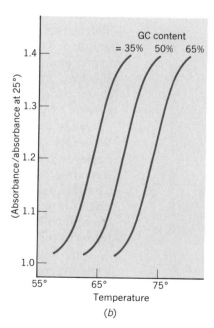

(a)

(b)

FIGURE 6.6 DNA melting curves detected by changes in absorbance of ultraviolet light. (a) Spectra of double-stranded DNA and the corresponding melted, single-stranded DNA. (b) Three DNA samples in separate solutions of buffered 0.015M sodium chloride were heated to various temperatures, and the absorbance at 260 nm was measured. The values of the T_m are 64°C, 69°C, and 76°C, respectively, for the three DNAs.

nucleic acid molecules under the guidance of a template strand, following the rules of Watson and Crick base pairing, as described in Chapter 19. Sometimes, the new strand remains associated with the template in a double helix, as in the replication of double-stranded DNA. In other instances, such as RNA synthesis under the direction of a double-stranded DNA template, the strand being synthesized apparently remains associated with its template only briefly and in the region at which the chain is growing. Strands of DNA or of RNA, or of RNA and DNA, that are complementary to each other will spontaneously form a double helix in solution under the appropriate conditions. This ability of two molecules to recognize each other without the intervention of any external agent, such as an enzyme, has provided biochemists with an extremely important analytical tool. Relationships between very large and complicated nucleic acid molecules can be discerned, relationships that would not be apparent by other analyses.

Molecular hybridization, between two single strands of nucleic acid to form a double helical product, is a very specific reaction that requires exactly complementary or, at least, very nearly exactly complementary strands as reactants. The rate and extent of the reaction depend on the concentration of the two kinds of reactant single strands and the period of time of the reaction. Detailed analysis of these reaction variables indicates that the slowest step in the reaction, the rate-limiting step, is associated with the proper encounter of two strands so that complementary portions are aligned. Formation of the rest of the helix is very rapid. The temperature and the salt concentration and other conditions of the solvent are usually adjusted so that the temperature of the molecular hybridization reaction is about 25°C below the temperature at which the helix would melt to yield single strands. At this temperature, reactions between single strands give both perfectly matched and not quite perfectly matched helices. Perfectly matched helices are more stable, and because single strands and helices are in equilibrium, the most perfect helices eventually come to predominate. At lower temperatures, the imperfect helices are more stable, and they prevent the efficient formation of perfect helices. At temperatures closer to the melting point, the extent of the reaction is less, hence the compromise at 25°C below the melting temperature.

One of the most important discoveries derived from molecular hybridization experiments is that some nucleotide sequences, especially in DNA from eucaryotes, are repeated hundreds, thousands, and even a million times in the genome of the organism. Other sequences occur only once in the genome. Consider a DNA that has some sequence of unknown size repeated in it an unknown number of times. A molecular hybridization reaction can define a quantity called the complexity, N. N is the number of nucleotide residues in one repeat of the repeating sequence, in one strand.

The complexity N can be illustrated with DNAs that have no or very limited repeating sequences. *E. coli* DNA has about 4.3×10^6 base pairs in its genome, which if they all were unique sequences would correspond to a complexity of 4.3×10^6. Bacteriophage T7 DNA has about 73,000 base pairs, for an N of 73,000. One microgram of any DNA would have about 3×10^{-9} moles of nucleotide residues. However, the number of moles of DNA depends on the size of the molecule. One microgram of T7 DNA corresponds to 3.9×10^{-14} moles, whereas 1 μg of *E. coli* DNA is only 3.4×10^{-16} moles. If 0.5 μg of each DNA are combined, 3×10^{-9} moles of nucleotide residues are present in the resulting 1 μg. T7 and *E. coli* DNAs have no extensive nucleotide sequences in common. If the DNA mixture is dissolved in buffered salt solution and melted and then incubated under hybridization conditions, complementary T7 DNA strands will hybridize to each other and complementary *E. coli* DNA strands will hybridize to each other.

However, T7 DNA strands, because of their very much greater *molar* concentration, will "find" each other more rapidly and hybridize more rapidly than the *E. coli* DNA strands. In general, increasing the concentration of a DNA will increase the rate at which it hybridizes, so that the time required for half of the DNA of one type to hybridize will be reduced in proportion to the increase in total DNA concentration. A useful quantity is the product of the initial concentration of melted DNA, C_0, in moles of nucleotide residues per liter, multiplied by the time required for half of the DNA to hybridize, the $t_{1/2}$. This quantity, $C_0 t_{1/2}$, will be about 115 times as great for *E. coli* DNA as for T7 DNA. The calculation is 3.9×10^{-14} moles divided by 3.4×10^{-16} moles = 115. The important conclusion is that N is proportional to the $C_0 t_{1/2}$ value, at which the helix formation for a given DNA in a reaction mixture is half-complete.

DNA molecular hybridization experiments are almost always performed with fragments rather than the intact molecules. The DNA analyzed in Figure 6.7 was in fragments of approximately 400 base pairs. If different DNA samples are all treated to give fragments of the same size, the relationship between N and $C_0 t_{1/2}$ for the DNAs will be the same, and the N values will correspond to the number of nucleotide residues in the original, unbroken DNAs. The $C_0 t_{1/2}$ value for *E. coli* is about 4 *M*-sec. Box 6.F describes some applications of DNA molecular hybridization.

Note that for covalently closed, circular DNA melting is possible. However, because the two strands are intertwined, hybridization or renaturation is almost a unimolecular process. The result is that such DNA renatures exceedingly rapidly compared to the linear DNAs shown in Figure 6.7. DNA – RNA hybrid molecules also may be produced by incubation under renaturation conditions.

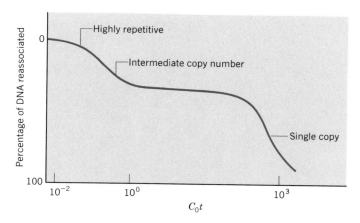

FIGURE 6.7   Hybridization of DNA from a eucaryote showing the wide variation in product of concentration and time of hybridization required to display all of the classes of DNA that are present. The value of $C_0 t_{1/2}$ for single-copy DNA in the example is less than $10^3 M$-sec.

BOX 6.F   MOLECULAR HYBRIDIZATION ON A SOLID SUPPORT

Through the use of restriction endonucleases (Box 6.D) and gel electrophoresis (Box 6.B), it is possible to resolve and display specific portions of very large DNA molecules and even to learn the order of the restriction endonuclease-generated fragments within the large DNA molecule (Box 6.E). Individual DNA fragments, of course, can be isolated and their nucleotide sequence determined (Box 6.C). Often, the analysis of such DNA fragments could be advanced more rapidly if the nucleotide sequence relationships of individual DNA fragments to other nucleic acids could be determined easily. For example, if a particular messenger RNA were in hand, identifying which DNA fragments from a restriction endonuclease digestion contained the same nucleotide sequences could lead to the isolation of the corresponding gene. This could be done by cutting the corresponding DNA zones from a gel, zone by zone, and hybridizing each to the RNA in solution (Section 6.6). However, E. Southern has introduced a more direct, less laborious, generally more sensitive method that also has greater resolution. His method often is referred to as "Southern blotting."

After the DNA fragments have been resolved by electrophoresis, usually in agarose gel, the gel is immersed in base to denature the DNA fragments, that is, separate their strands, *in situ*. After rapid neutralization, the gel is laid on a stack of buffer-soaked filter paper or a sponge. Paper made of nitrocellulose or some other DNA-binding material is laid on the gel, and on this sheet is placed a stack of dry filter paper in an arrangement shown here.

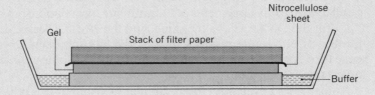

Capillary action draws the buffer through gel and binding paper and into the filter paper. DNA fragments bind and do not pass into the top filter paper stack. The result is essentially a "contact print" of the gel onto the binding paper. Baking at 80° permanently fixes the DNA to the cellulose nitrate or other support. The sheet with attached DNA fragments is placed in a plastic pouch of the type used to freeze foods. Radioactive DNA or RNA is added to the pouch in a solution that is suitable for a hybridization reaction. The pouch is sealed and incubated to allow hybridization.

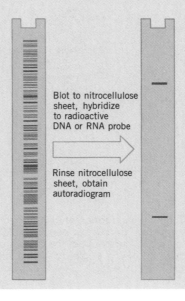

Molecular hybridization occurs only to those portions of DNA molecules that have sufficient nucleotide sequence homology to the radioactive DNA or RNA, which is referred to as a **probe** because of its ability to locate specific sequences in the mass of unrelated DNA sequences present on the sheet. A Southern blotting experiment is outlined on the previous page.

When the nucleic acid that is transferred to the binding sheet is RNA rather than DNA, the technique is referred as "Northern blotting." The technology has been extended to proteins, which similarly can be "contact-printed" onto nitrocellulose sheet. In this case, the detection is by binding of antibody molecules, and the technique, in laboratory jargon, is termed "Western blotting."

They are less stable than either DNA–DNA or RNA–RNA hybrids except when certain solvents are used to make DNA–DNA hybrids less stable than DNA–RNA hybrids.

6.7 CLASSES OF NUCLEIC ACIDS

Nucleic acid function is a major theme of Part III of this book. As a prologue to that discussion, we indicate here the major types of nucleic acids. All organisms thus far characterized have nucleic acid as their genetic material. In all cells, the genetic material is DNA, whereas a given virus may possess any of four types of nucleic acid as the genomic material: single-stranded or double-stranded DNA or RNA. Virus nucleic acids are discussed in Chapter 22.

DNA is found in only a limited variety of forms in both procaryotic (bacteria, blue-green algae) and eucaryotic (fungi, plants, animals) cells. The complexity of the *E. coli* chromosome, as assessed by molecular hybridization analyses (Section 6.6), and its physical size, as assessed by electron microscopy and other techniques, are consistent. That is, it seems that the procaryotic chromosome is a single, large, circular, double-stranded DNA molecule. The contour length of such a DNA is in excess of 1 mm, whereas the longest dimension of the bacterial cell is about 0.001 mm. Obviously, the bacterial chromosome must be extensively folded in the cell. The DNA is supercoiled (see Section 6.5) and complexed with protein and RNA, all of which apparently contribute to a compact conformation of the DNA in the bacterial cell.

In eucaryotes, the inheritance of genetic characters is linked, certain characters generally being inherited together. These linkage groups correspond to chromosomes, correlating a physical entity that can be seen even with the light microscope, the metaphase chromosome, with the genetic properties of the organism. At least some chromosomes of eucaryotes contain a single, very large DNA molecule. We leave the discussion of the size and folding of DNA in the chromosomes of eucaryotes to Section 6.10.

Bacteria may acquire extrachromosomal DNA elements, called **plasmids.** Plasmids are double-stranded circles of about 4 to more than 200 kb in size (see also Box 6.E). Plasmids replicate independently of the bacterial chromosome. This semiautonomous existence is possible because the plasmid encodes some of the genetic information necessary for its replication, and usually, the bacterial cell continues to maintain and support the plasmid because the plasmid contributes some function that gives the bacterium a selective advantage. The classical example of such an adaptation is resistance to an antibiotic. "Hospital strains" of bacteria that are able to replicate even in the presence of several antibiotics have acquired one, and usually several, drug resistance plasmids. Plasmids generally control the number of copies of the plasmid that are present per cell, which may be only 1 to 3 or 30, or more. Usually, the larger plasmids are present in fewer copies. Other examples of plasmid-encoded functions are (1) formation of the F-pilus or conjugation tube, a structure that facilitates the transfer of plasmids from one bacterial cell to another; (2) production of toxins that inhibit or kill related but nonplasmid bearing bacteria; and (3) the ability to metabolize particular compounds for the nutrition of the bacterial cell. As is

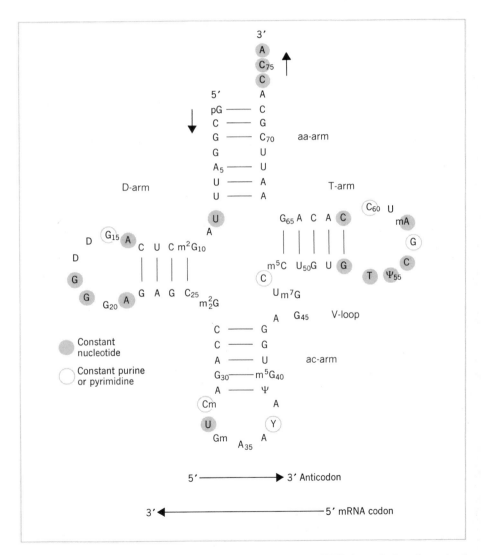

FIGURE 6.8   Nucleotide sequence of yeast phenylalanine tRNA shown in the clover-leaf configuration. Shaded circles are constant in all tRNAs and open circles indicate positions that are occupied constantly by either purines or pyrimidines. Abbreviations: A, adenosine; T, thymidine; G, guanosine; C, cytidine; U, uridine; D, dihydrouridine; ψ, pseudo-uridine; Y, a purine nucleoside; m, methyl; and m_2, dimethyl. The lines indicate hydrogen bonding. Four arms and one loop are defined by this structure.

described in Chapter 22, plasmids are important tools in recombinant DNA technology.

Some of the organelles of eucaryotes, the mitochondria and the chloroplasts, as well as a few other, less widely distributed organelles, have a semiautonomous existence because they have their own DNA. The mitochondrial DNA of plants, for example, present a complex picture. From the number and mobility of zones produced on electrophoresis of restriction endonuclease-generated DNA fragments (Box 6.D), from 200,000 to 2,500,000 base pairs of DNA account for all of the nucleotide sequences present in the mitochondria from various plant species. However, the mitochondrial DNA of a given species is not of uniform size and may be a mixture of circular and linear forms. If for maize mitochondrial DNA, we assign a value of 1.0 to the amount of DNA sequences identified as unique by analysis of restriction endonuclease-generated fragments, the actual mitochondrial DNA molecules are circles of approximately 0.4, 0.8, 1.0, 1.4, 1.6, and 2.0 times this size. Plant mitochondria also often have small plasmid-like DNAs and even double-stranded RNA.

A typical cell may contain ten times as much RNA as DNA. Three classes account for most of the RNA that is to be found in the cells of either procaryotes or eucaryotes: (1) ribosomal RNA, or rRNA, that is the most abundant; (2) transfer RNA, or tRNA; and (3) messenger RNA, abbreviated mRNA. All are synthesized by enzymes that copy their nucleotide sequences from a DNA template in a process called transcription (Chapter 19), and all participate in protein synthesis (Chapter 20), but in different ways: rRNA in ribosomes (Section 6.10), the particle on which protein synthesis takes place; mRNA (Chapter 19) as the intermediary that transfers genetic information from DNA to the ribosome where it is interpreted as amino acid sequences; tRNA (Section 6.8) as part of the interpretive system, carrying amino acids to the site of protein synthesis and specifying what amino acid is to be incorporated according to the information in the mRNA. Other RNAs occur in smaller amounts and participate in DNA synthesis and in the cutting and splicing of RNA sequences, among other functions.

6.8 FOLDING OF SINGLE-STRANDED NUCLEIC ACIDS

The regular bihelical structure of most DNAs contrasts with the structure of most RNA molecules. Nevertheless, those RNA molecules that have been studied in detail appear to possess a definite structure. Only one method can give a detailed and complete picture of the three-dimensional arrangement of atoms in a nucleic acid molecule: x-ray crystallography. Nuclear magnetic resonance spectroscopy gives less complete but nevertheless detailed information about nucleic acid conformation. However, both of these techniques are of little use, currently, for molecules of more than 100 nucleotide residues.

Like proteins, nucleic acids can be considered to have several levels of structure. The primary structure is, of course, the sequence of nucleotide residues. Secondary structure is considered to be the **regular helices** that may form by base pairing between close or distant portions of the polynucleotide chain. Tertiary structure is represented by base-stacking and base-pairing interactions that bring together distant portions of the polynucleotide chain and do not result in regular helical structures. Regular helices have base pairs that are mainly of the Watson and Crick type, although G : U base pairing occurs and apparently has an energy only slightly less than that of an A : U pair. Tertiary interactions often result from more unusual types of base pairing.

Complete three-dimensional structures are available only for a few tRNA molecules. Figure 6.8 shows the nucleotide sequence of yeast phenylanine tRNA arranged so as to give a pattern of base pairing that is extensive and consistent with most of the known tRNA sequences. That is, Figure 6.8 gives a secondary structure for yeast phenylalanine tRNA. Note the G : U base pair in the helical stem that bonds the 3′ region to the 5′ region of the structure. tRNAs have unusual nucleotide residues in addition to the four standard residues, as was indicated in Section 6.1.

Note the contrast between Figures 6.8 and 6.9, which shows the three-dimensional structure of the same tRNA derived from x-ray crystallography. The same base pairing is present, but the conformation of the molecule and the tertiary interactions could not be anticipated from the secondary structure.

Very little is known about the three-dimensional structures of larger nucleic acid molecules. Methods that have been applied to large RNA molecules include (1) measurements of RNA cleavage by or reaction with nucleases and chemical reagents that are sensitive to the secondary structure of the molecule,

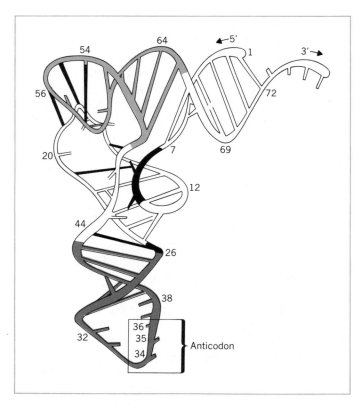

FIGURE 6.9 Schematic model of yeast phenylalanine tRNA derived from an electron density map of the crystalline tRNA. See Figure 6.8 for definitions of arms and loops in the secondary structure. (*Source:* Courtesy of S. N. Kim.)

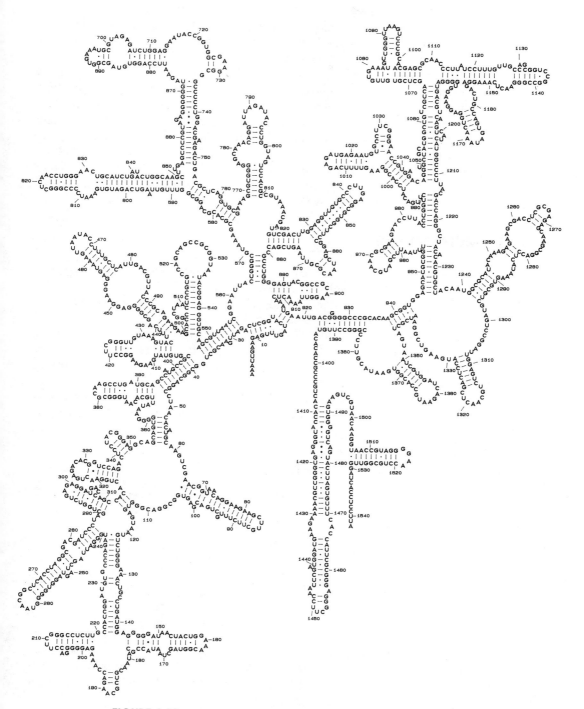

FIGURE 6.10 Secondary structure model for *E. coli* 16S ribosomal RNA. The proposed folding for the 1542 nucleotide residue RNA is derived from a variety of chemical and biological data. (*Source:* H. F. Noller and C. R. Woese, Science *212*:403 (1981), with permission.)

(2) reaction with cross-linking agents, and (3) observations of phylogenetic relationships. These methods even have the potential to give limited information about tertiary interactions. In making phylogenetic comparisons, examples of the same functional type of RNA from several different organisms are selected. If the molecules have very similar sequences, then the few sequence differences may give clues to the RNA folding pattern. For example, in a comparison of two RNAs, an A residue at a particular position in one molecule may be substituted by a C residue in the other. If also, at a second position in the sequence, a U residue is substituted by a G residue, one may suspect that the first position and the second position in each of the molecules are associated by Watson and Crick base pairing. A secondary structure model for an *E. coli* ribosomal RNA is shown in Figure 6.10.

6.9   MUTAGENESIS

Mutagenesis is the induction of a mutant, meaning an organism or nucleic acid molecule of altered genetic character. Mutants result from alterations in the nucleotide sequence. Mutants can be induced in many ways. In the laboratory, they most often have been induced by radiation or chemical treatment of the organism or its nucleic acid, or by supplying mutagenic agents to the organism or a system in which the nucleic acid is replicating. More recently, developments in the synthesis of oligonucleotides (Box 6.G) allow precisely controlled mutagenesis by programmed synthesis of altered sequences.

An example of a mutagenic reagent is bisulfite. Bisulfite reacts specifically with cytidylate residues in RNA and converts them to uridylate residues.

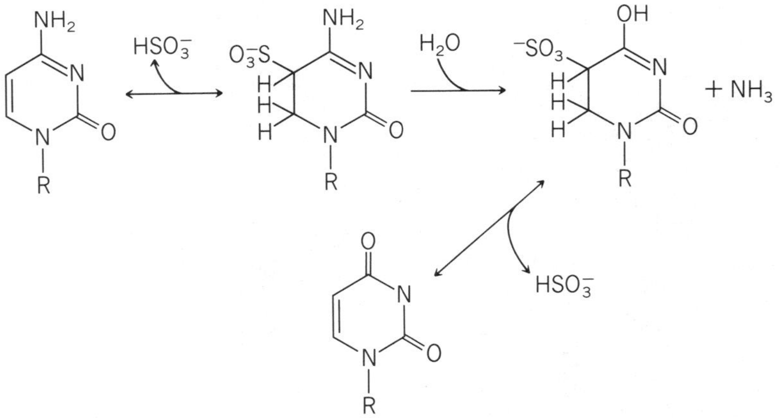

Two examples of *in vivo* mutagens are 5-bromouracil and 2-aminopurine. Bromouracil, because of its altered tautomeric equilibrium compared to that of thymine, can base-pair with either A or G residues. 2-Aminopurine can base-pair with cytosine or thymine/uracil.

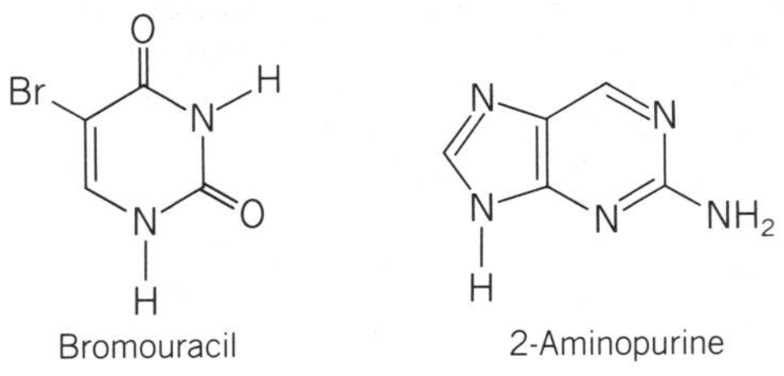

Bromouracil   2-Aminopurine

BOX 6.G   CHEMICAL SYNTHESIS OF OLIGODEOXYRIBONUCLEOTIDES

The combination of sophisticated organic chemistry and recombinant DNA methodologies gives the biochemist the opportunity to produce polynucleotides of almost any desired nucleotide sequence, to create nucleic acids that are not available from natural sources. The chemical side of this alliance most often takes the form of an automated oligodeoxyribonucleotide synthesizer that takes advantage of solid support technology similar to that in use for oligopeptide synthesis (Box 3.D). Chemically modified nucleotide residues are added sequentially to the polynucleotide chain, which is bound by its 3' end to a solid support of polymeric or crystalline beads. Because of the solid support, purification of the polynucleotide intermediates between synthetic steps is simply a matter of washing the resin beads. Commercially available synthesizers can produce oligodeoxyribonucleotides of 50 or even 100 residues in micromolar amounts.

Several systems for synthesizing phosphodiester bonds between deoxynucleoside residues have been developed. One of the most widely used approaches, **phosphoramidite chemistry,** does not use phosphate esters as the starting material for synthesis. The phosphorus atom in a phosphate ester is in the $+5$ oxidation state, whereas a **phosphite** ester is in the $+3$ oxidation state. Both are shown here in the fully protonated state.

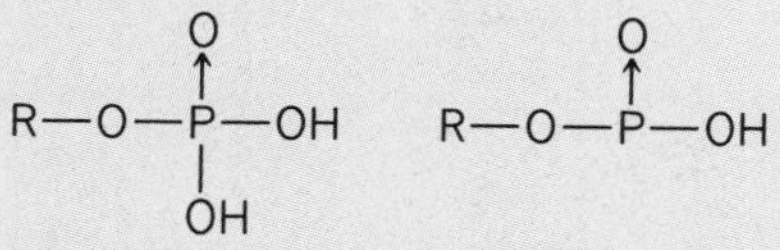

Phosphate ester   Phosphite ester

A phosphoramidite is a phosphite amide. The structure of a deoxycytidine phosphoramidite derivative that is used in oligodeoxyribonucleotide synthesis is given here.

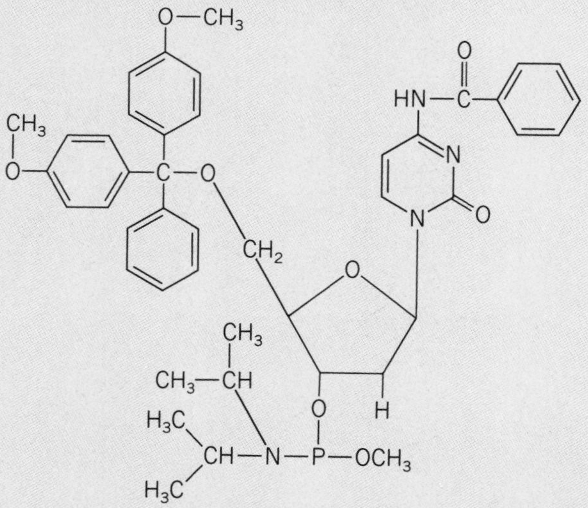

Reactive groups of the deoxynucleoside phosphoramidite derivatives must, of course, be blocked chemically in order to prevent undesirable side reactions during the coupling that will lead to the formation of new polynucleotide phosphodiester bonds. The cytosine base of the deoxycytidine phosphoramidite derivative is blocked by a benzoyl group, and the 5'-hydroxyl group of this and other deoxynucleoside phosphoramidite reactants are blocked with a dimethoxy-trityl group in an ether linkage that is especially labile in acid.

The reaction series at the end of this box shows the first steps in the synthesis of an oligodeoxyribonucleotide. The 3' most nucleoside residue of the final product enters the reaction coupled by its 3'-hydroxyl group to the solid support. The base of this nucleoside, as well as those of the nucleotide residues subsequently added, except thymine, require appropriate blocks. The 5'-hydroxyl group of the deoxynucleoside is able to displace the protonated *bis*-isopropyl amido group of the phosphoramidite in the coupling reaction. In the second step, a stable phosphotriester bond is formed by iodine oxidation of the phosphorus atom from its $+3$ to its $+5$ state. Mild acid removes the dimethoxytrityl group, but not the methoxy group of the phosphotriester or the base-blocking groups. Thus, the support-bound product of the reaction series has, at this stage, a free 5'-hydroxyl group and is ready to react with the deoxynucleoside phosphoramidite of residue corresponding to base B_{n-2}. Continuation of these cycles causes the polynucleotide chain to grow in the 3' to 5' direction. At the end of the series of reactions, the phosphate-methyl groups and the blocks of the deoxyribonucleoside bases are removed by incubation in alkaline organic solvent, and the polynucleotide is removed from the resin by treatment with ammonium hydroxide. After purification of the product, the 5'-dimethoxytrityl group is removed to give the oligodeoxyribonucleotide.

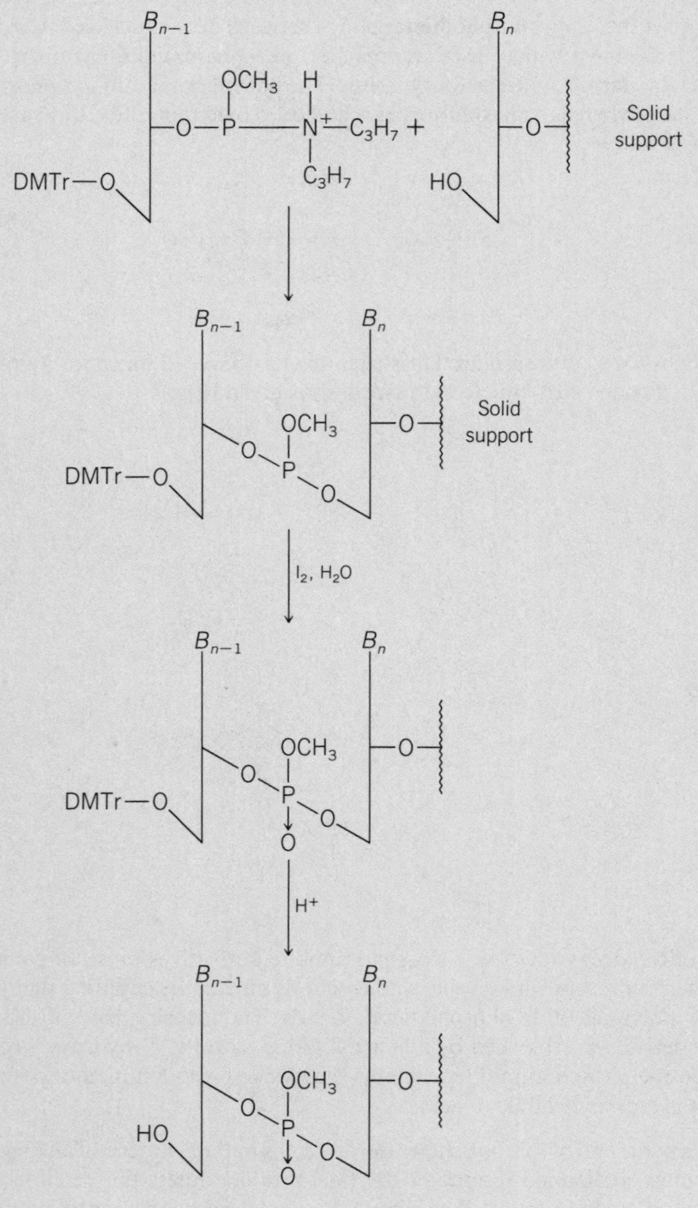

Most mutations are disadvantageous to an organism and its progeny. Mutagenesis also is associated with carcinogenesis. Therefore, mutagens, as useful as they are as tools in biochemistry and genetics, must be handled with care and properly contained.

6.10 NUCLEOPROTEINS

When cells are disrupted, most of the DNA and RNA present in them is released as complexes with protein, rather than free nucleic acids. These **nucleoprotein complexes** or **nucleoproteins,** although stable to the cell disruption treatments, generally have no covalent connections between nucleic acid and protein. Biochemists also refer to **ribonucleoproteins** and **deoxyribonucleoproteins.** Examples are chromatin, the deoxyribonucleoprotein that accounts for most of the mass of eucaryotic chromosomes, and ribosomes, the ribonucleoprotein particles on which peptide bonds are formed in protein biosynthesis. Virus particles also either are nucleoproteins or contain them. Certain virus particles are the most stable nucleoproteins known, presumably because of their need to survive in extracellular environments. Because of the complexity of nucleoproteins, their study requires not only the techniques of protein and nucleic acid biochemistry but other techniques as well. We consider, as an example of how nucleoproteins may be studied, some of the experiments that have partially defined chromatin structure.

6.10.1 CHROMATIN

Very sophisticated physical analyses have revealed that at least some eucaryotic chromosomes have a single DNA molecule with a molecular weight of larger than 10^{10}. Thus for a multichromosomal organism, several DNA molecules with contour lengths of more than 10^7 μm must be packed in the approximately 10-μm-diameter nucleus. The DNA not only must be packaged; it must be packaged in such a way that it can be replicated and serve as a template for RNA synthesis (Chapter 19) without exceeding the bounds of the nucleus. Chromatin, which may be defined operationally as the deoxyribonucleoprotein that is released when nuclei are disrupted, is about two-thirds protein and one-third DNA, with smaller amounts of RNA and other compounds.

The protein portion of chromatin is roughly half histone proteins, which are basic proteins that bind tightly to DNA and neutralize about one-fifth of the ionic charge of the DNA phosphodiester groups. Histones fall into five groups: H1, H2A, H2B, H3, and H4. Compared to the other histones, with molecular weights of 11,000 to 14,500, H1 is larger, with a molecular weight of 21,000. H1 also has a much greater ratio of lysine to arginine residues than is characteristic of other histones. H1 varies from species to species and tissue to tissue, whereas H3 and H4 are highly conserved in all tissues of plants and animals and even in yeasts. H2A and H2B show an intermediate degree of conservation. The remaining "nonhistone" proteins have many more members, all occurring at much lower concentrations than the histone proteins. Among the nonhistone proteins are candidate regulators and catalysts of gene expression, whereas histone proteins probably act primarily to control the folding of the DNA double helix.

Treating purified chromatin with low concentrations of a nuclease, such as micrococcal nuclease, has a dramatic effect on the structure of the double-stranded DNA that was isolated from the partially digested chromatin. Agarose gel electrophoresis from nuclease-treated chromatin produces a "ladder" of zones that at first glance resembles the ladders produced in DNA-sequencing experiments (Box 6.C). However, the "rungs" of this ladder are separated not by differences of one nucleotide residue, but by increments of about 200 base pairs. As shown in Figure 6.11, zones corresponding to 200, 400, 600, 800, and so on base pairs are seen.

Different nucleases produce the same result, indicating that it is the conformation and availability of the DNA rather than the specificity of the nuclease that determines the pattern of chromatin degradation. Graded, increasingly vigorous treatments of chromatin with nucleases produce changes in the pattern for DNA isolated from the treated chromatin. Most of the DNA is shifted to the 200 base-pair zone and, with more severe nuclease treatments, to even more rapidly migrating zones. Chromatin that has its DNA degraded to form the 200 base-pair ladder also has a simple chemical composition. For each 200 base pairs of DNA, the treated chromatin has one histone H1 molecule and two molecules each of the H2A, H2B, H3, and H4 histones. This corresponds to the composition of the chromatin particle that is the source of the 200 base-pair DNA fragment, the **nucleosome.**

Further treatment of nucleosomes with nuclease removes an additional 30 to 35 base pairs from the DNA, but does not change the composition of histone proteins. The DNA removed is considered to be part of a "linker," a segment of the DNA strand that is not tightly bound to histone proteins and apparently connects segments of about 165 base pairs that *are* protected by histone protein. The residual deoxynucleoprotein, with one molecule of H1 and two molecules each of the other four histone proteins, is designated the **chromatosome.**

Even more vigorous nuclease treatment of the chromatosome reduces the

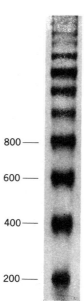

800 —

600 —

400 —

200 —

FIGURE 6.11   Analysis of cutting of double-stranded DNA in chromatin by mild nuclease treatment. Treated chromatin was exposed to a two-phase phenol-water system to remove proteins before electrophoresis of the DNA through agarose gel.

DNA content to about 147 base pairs and releases H1. The result is the **nucleosome core.** Chromatin fibers, chromatosomes, and nucleosome cores have been the objects of many kinds of analyses, including differential extractions to determine which intermolecular bonds resist which solvents, chemical cross linking of proteins to determine which proteins are neighbors in the deoxynucleoprotein structure, and x-ray and neutron diffraction to learn the relative positions of DNA and protein molecules and even to trace the DNA and protein chains. This great diversity of information has revealed the general outline of the structure of chromatin and its nuclease degradation products.

The nucleosome cores have a surprising structure. The DNA is *not* buried inside of the nuclesome structure, but occurs as two loops of a helix. This is, in fact, a superhelix, a left-handed helix of right-handed helical B–DNA formed around a protein center, as shown in Figure 6.12. The result is a disklike structure. The inherent stiffness of a B–DNA helix would require 150 or more base pairs of DNA to form a circle without excessive distortion of the B–DNA structure. Therefore, it is not surprising that the 100 base-pair loops of the nucleosome are distorted. They have "kinks" in the B–DNA structure, presumably imposed by the interaction of DNA and histone proteins. The heart of the nucleosome is a tetramer of two H3 and two H4 histones. H2A-H2B dimers are located on the "faces" of the disks. At the next level of complexity, the chromatosome, histone H1 appears to "seal off" the two turns of DNA superhelix. The chromatin itself has another level of helical structure, a helix of nucleosomes. In this last helix, the structure of which is not fully understood, the long axis of the H1 protein cylinders (Figure 6.12) probably lie perpendicular to the new helix axis. Even higher orders of folding also contribute to the ability of the nucleus to contain the highly elongated structure of chromosomal DNA.

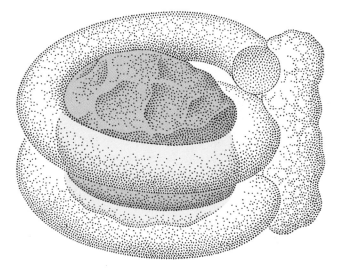

FIGURE 6.12 Arrangement of DNA and protein in a nucleosome, the repeating unit of chromatin. Two turns of DNA superhelix, represented in the illustration as kinked tubing, surround a protein core (dark color) as described in the text. The cylinder-like H1 protein is shown at the right.

6.10.2   RIBOSOMES

The ribosome is the one the most abundant of subcellular organelles. It is present in most of the cells of every cellular organism. Because of their abundance and relatively easy purification, ribosomes, especially those of microorganisms and rat liver, have been studied since the 1950s. Ribosomes are recovered from *E. coli* cells by disrupting the cells in buffered 5 mM $MgCl_2$ solution and subjecting the extract to high-speed centrifugation, using techniques similar to those described in Box 4.B. Up to 50% of the dry weight of the soluble portion of bacterial cells is ribosomes, the bulk of which is a particle that is designated the **70S ribosome.** Given that the particle is isolated by centrifugation, it is not surprising that it is named for the rate at which it sediments in a centrifugal field, 70 Svedberg units (see Box 4.B). The 70S ribosome is the structure of the bacterial cell on which peptide bonds are formed during protein synthesis (Chapter 20). Although it is many times more complex than a protein or tRNA molecule, the ribosome nevertheless has a regular structure that successfully has been analyzed by chemical, genetical, and physical methods. It is composed almost entirely of RNA and protein, with small amounts of polyamines such as spermidine, $H_2N—(CH_2)_3—NH—(CH_2)_4—NH_2$, and metal ions such as magnesium.

One of the first insights into that structure came when the 70S ribosome was exposed to buffered 0.5 mM $MgCl_2$, a concentration of the divalent metal ion that is one-tenth that usually used for isolation of the 70S ribosome. The 70S ribosome dissociated into the 50S and the 30S ribosomal subunits.

$$70S \text{ Ribosome} \rightleftharpoons 50S \text{ Ribosomal} + 30S \text{ Ribosomal}$$
$$\text{subunit} \qquad \text{subunit}$$

That is, reduced magnesium ion concentration shifted the equilibrium to the right.

Because of their function in protein synthesis, ribosomes are, of course, present in almost all of the cells of eucaryotes. One of the few exceptions is the mature mammalian red blood cell, which loses its complement of ribosomes as well as its nucleus during its development. The ribosomes of eucaryotes are somewhat larger than those of procaryotes and are designated **80S ribosomes.** 80S ribosomes also at least partially dissociate when the magnesium ion concentration is reduced, into 60S and 40S ribosomes. In Table 6.3, we compare some of the properties of ribosomes from procaryotic and eucaryotic cells. Most procaryotic ribosomes have very similar properties; more variation is seen among the ribosomes of eucaryotes. Two important organelles of eucaryotes, the mitochondrion and the chloroplast, have ribosomes. However, these organelle ribosomes resemble the ribosomes of bacteria more closely than the cytoplasmic ribosomes of eucaryotes, possibly reflecting a procaryotic evolutionary origin of mitochondria and chloroplasts.

Bacterial ribosomes have been the most thoroughly studied. Although cytoplasmic ribosomes of eucaryotes are larger and more complex than those of bacteria, including an additional type of RNA, the 5.8S RNA, both types of ribosomes have a similar organization of RNA and protein. Many analogies of structure and function have been made between the proteins from each type of ribosome. We will discuss here only the 30S subunit of the *E. coli* ribosome.

TABLE 6.3   Ribosomes and Their Components

SOURCE	PROCARYOTE 70S	EUCARYOTE (CYTOPLASM) 80S
Particle weight	2.7×10^6	4.3×10^6
Mass fraction of RNA	~57%	~59%
Large subunit	50S	60S
RNAs	23S	26S–28S
	(2904 nt, *E. coli*)	(~5100 nt, rat liver)
	5S (120 nt)	5S (118 nt)
		5.8S (150 nt)
Proteins		
Number	32	~50
Molecular weight range	5381–24,599	12,000–42,000
Small subunit	30S	40S
RNAs	16S	18S
	(1542 nt, *E. coli*)	(1789 nt, yeast)
Proteins		
Number	21	~30
Molecular weight range	8369–26,613 for S2 through S21; S1, 61,159	11,000–42,000

Note: nt = Nucleotide residues; the proteins of the 30S subunit of E. coli *ribosomes are designated S1 through S21.*

The methods for studying nucleoprotein structure that have been applied to chromatin, especially chemical cross linking (of protein to protein and protein to RNA), also have been applied to ribosomes. In addition, our knowledge of ribosomes has been enhanced by (1) investigations of mutants with altered or even missing ribosomal proteins, (2) electron microscope observations of the sites on the surface of the ribosome to which antibodies (Figure 3.19) against specific ribosomal proteins bind, and (3) the ability to assemble ribosomal subunits from purified ribosomal RNA and ribosomal proteins. In contrast to the structure of the nucleosome, the nucleic acid and protein of the ribosome appear to comingle throughout the particle, with the RNA slightly more centrally located, on the average, than the proteins.

The 30S ribosomal subunit of *E. coli* is a slightly elongated particle with dimensions approximately $11 \times 11 \times 23$ nm. As is indicated by Figure 6.13, the particle has a constriction or neck. Many of the important functions of the 30S subunit that have been shown to be necessary for protein synthesis (Chapter 20) have been associated with proteins located at one region of this constriction, on the left-hand side of the drawing in Figure 6.13. The locations of many of the 21 proteins of the 30S subunit are known, and some are shown in Figure 6.13. Protein S1 is unusual not only for its large size (Table 6.3) but also because it is less tightly bound to the 30S particle than the other proteins, and it can dissociate and reassociate with the particle.

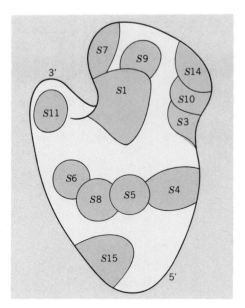

FIGURE 6.13 Arrangement of some of the proteins of the 30S ribosomal subunit of the bacterium *E. coli*. The particle is about 23 nm in its longest dimension. The locations of some of the 21 proteins of the 30S subunit, $S1$, $S3$, and so on, are shown, as well as the 3' and 5' ends of the 16S ribosomal RNA. (*Source:* Adapted from H. G. Wittmann, *Annual Reviews of Biochemistry,* Vol. 52 (1982), pp. 35–65.)

6.10.3   TOBACCO MOSAIC VIRUS

Virus particles, the extracellular infectious form of the virus (see Chapter 22), are termed **virions.** The virions of tobacco mosaic virus (TMV), the first plant virus to be intensively studied, are far more regular in structure than ribosomes. The particle weight is about 40×10^6 and consists of 95% protein and 5% RNA. The latter is a single molecule of about 6400 nucleotide residues. The protein, called TMV coat protein, is of one kind, with a molecular weight of 17,500. The virions of TMV are rigid rods of helical structure about 300 nm long and 18 nm in diameter, with a central cavity or pore that runs the length of the rod. The rod is of uniform structure over its length. The appearance of a portion of the TMV rod is shown in Figure 6.14, which shows about 20% of the full-length TMV rod. The structure of the rod is known in detail from x-ray diffraction studies of the rods and isolated TMV coat protein. There are, on the average, three nucleotide

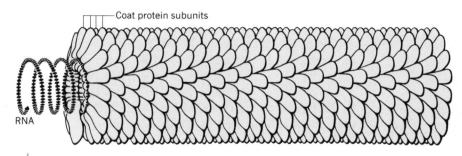

FIGURE 6.14 The structure of tobacco mosaic virus particles. The right-hand helix of RNA and protein is shown, with each coat protein molecule having the appearance of a wooden shoe and the RNA represented as a string of beads, one bead per nucleotide residue. (*Source:* Modified from P. J. G. Butler, *Journal of General Virology,* Vol. 65 (1984), pp. 253–279, with permission.)

residues of the RNA associated with each protein subunit. The helical course of the RNA takes it between the helically arrayed protein subunits as can be seen on the left side of Figure 6.14.

H. Fraenkel-Conrat and co-workers discovered in the 1950s how to isolate TMV coat protein and TMV RNA. They mixed TMV RNA and TMV coat protein in neutral pH phosphate buffer and found that infectious particles formed *spontaneously* over a period of a few hours. This was the first demonstration of such a spontaneous biological assembly reaction. The particles that formed had the properties of authentic TMV virions from plants, as assessed by a number of tests. Much later, the reaction conditions were modified to accomplish assembly in minutes. The assembly reaction has been studied in considerable detail. The protein that initially reacts with TMV RNA is not the dispersed coat protein but possibly helical aggregates of about 30 to 40 molecules. Contrary to initial expectation, the protein aggregate does not bind to either the 3' or 5' end of TMV RNA, but to a specific sequence that is about 15% of the distance from the 3' to 5' end, near residue 5500.

```
5' . . . GA    G          G            GAAGA
        GACG  AGGG—CCCAU  GAACUUA—CA       A
        CUGU  UCCC    GGUA—CUUGAGU  GU      G
3' . . . AG    A    U   A              A   UGCU
             G     A
           U A G
             |
      residue 5500
```

Figure 6.15 is a diagram of the initial complex between TMV RNA and the coat protein aggregate. An interesting consequence of this initiation mechanisms is that the 5' end of the RNA must thread through the central pore of the TMV rod as more coat protein is bound. The RNA-protein complex in Figure 6.15*b* and *c* is helical, and the helical conformation of the RNA is shown in the sketch, below the diagram of the initiation complex. Note that the portion of the RNA that is connected to the 5'-end is threaded through the central pore of the ribonucleoprotein complex. Thus the RNA gradually threads through the pore, so that the end of the RNA that is labeled as 5' moves upward in the diagram of Figure 6.15. Eventually, the end of the virus rod that protects the 5'-end of the RNA will be the end most distant from the observer in (c), after the RNA has been completely threaded through the pore.

The TMV system illustrates several principles. By use of many copies of a single type of coat protein, the virus produces a large and stable structure and protects the RNA in the harsh extracellular environment through the investment of only one gene, the coat protein gene. The coat protein is able not only to protect the RNA but also to recognize a particular sequence of nucleotide residues in the RNA in order to initiate assembly, which is facilitated by a complex change in conformation of the protein. The RNA also controls the assembly reaction since it must fold in such a way that there is one and only one origin of assembly. Two origins would not be compatible with the threading of the TMV RNA through the central pore of the ribonucleoprotein. The reaction is spontaneous, requiring no input of energy or facilitation by an enzyme, which shows that the free RNA and free coat protein are at a higher level of energy than the assembled TMV rod, and they can facilitate their own assembly reactions.

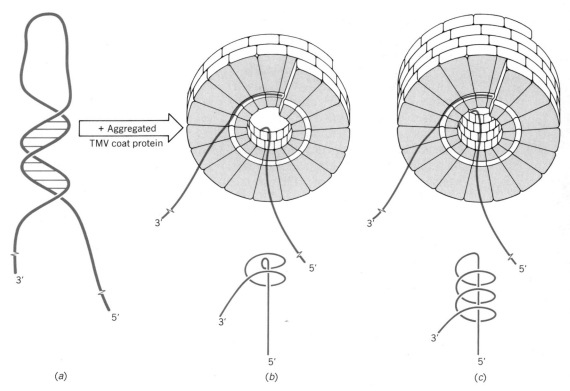

FIGURE 6.15 Early steps in the assembly of the tobacco mosaic virus particle. A portion of the TMV RNA molecule is shown in (a), corresponding to a region that is about 1100 nucleotide residues from the 3' end. This region reacts with aggregated TMV coat protein to initiate the assembly of the virus particle, as indicated in (b). As assembly continues, more coat protein adds to the complex on the face that is away from the observer in (b), giving the result seen in (c).

6.11 CATALYTIC AND OTHER ACTIVITIES OF RNA

The great majority of biological catalysts are proteins, as described in Chapter 4. Ribosomal RNA molecules contribute not only to the structure but also to the functioning of ribosomes in protein synthesis, as described in Chapter 20, and other RNA molecules show specific chemical reactivities in association with proteins. A few RNA molecules show such reactivities in the absence of protein. The enzyme ribonuclease P, of bacteria, catalyzes the cleavage of an RNA molecule that contains the nucleotide sequences of one or more tRNA molecules and additional sequences as well. The products are tRNAs and RNA fragments that contain the additional sequences.

Ribonuclease P is composed of a polypeptide chain of molecular weight 17,000 and an RNA molecule of 400 nucleotide residues. Both the RNA and the protein are required for ribonuclease P activity under physiological conditions. However, if the magnesium ion concentration is increased to more than 100mM, which is several times physiological concentrations, the RNA portion, RNase P RNA, catalyzes the ribonuclease P reaction using authentic precursor tRNA molecules as substrates. That is, a phosphodiester bond is cleaved, and it is the same phosphodiester bond that is cleaved by the complete ribonuclease P. The mechanism of this reaction is not understood. However, it is clear that no

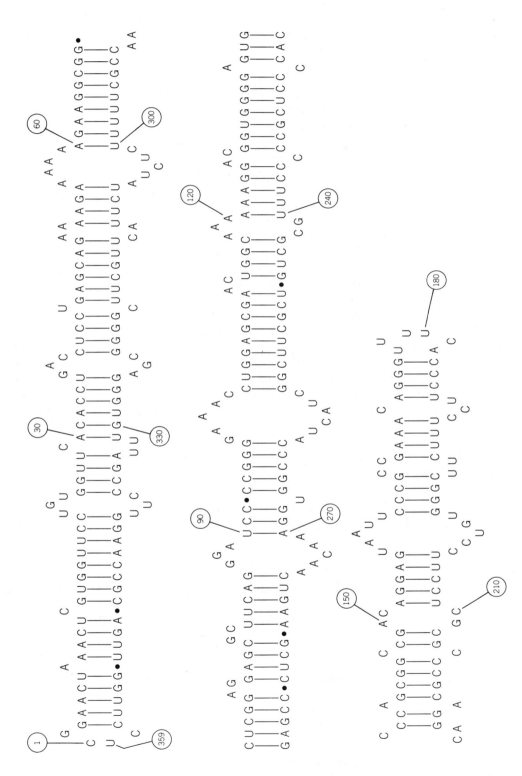

FIGURE 6.16   Nucleotide sequence and likely secondary structure of potato spindle tuber viroid. The RNA loop of 359 residues is displayed on three lines for clarity. Vertical lines show base pairing and residue numbering is indicated at intervals of 30.

protein is required for the high magnesium reaction and the RNase P RNA acts as a true catalyst, converting many moles of substrate to product without the RNase P RNA being changed. The reaction exhibits a hyperbolic dependence of rate on substrate concentration and a pH optimum. Thus, it is reasonable to assume that the 17,000 molecular weight polypeptide chain of native ribonuclease P functions principally to stabilize an active conformation of RNase P RNA, rather than to catalyze the reaction that is characteristic of the enzyme.

A "self-splicing" reaction of RNA also has been studied extensively. The 26S ribosomal RNA of the protozoan *Tetrahymena* is formed as a precursor that has a 413 nucleotide sequence inserted in what otherwise would be the mature 26S rRNA. The precursor undergoes a reaction with guanosine that results in the 413 residue "intervening sequence" (see Section 19.17) being excised and the ends of the 26S RNA being joined to produce intact 26S rRNA. No protein is required for this spontaneous, *in vitro* reaction.

Clearly, RNAs facilitate far fewer biochemical reactions than are facilitated by proteins, the enzymes. However, we can speculate that such may not have been the case at all stages of the evolution of life on our planet. If we must choose one type of molecule as *the* primordial molecule of life, from among those biological macromolecules known today, that molecule would be RNA. RNA is the one macromolecule that can be both informational (Chapter 19) and catalytic.

Viroids are the smallest infectious agents that have been characterized. They are covalent RNA circles, usually of 400 or fewer nucleotide residues, not protected by coat protein. They infect plants, replicating in the infected cells and producing copies of the viroid by a mechanism that does not involve DNA but rather the transcription of complementary RNA strands (see Section 19.1). Figure 6.16 shows the nucleotide sequence and likely secondary structure of the first viroid to be studied extensively, potato spindle tuber viroid (PSTV). PSTV is able to infect potato, tomato, and a few other plants, to replicate and cause disease. How all of this is accomplished by a structure of only 359 nucleotide residues remains a mystery.

REFERENCES

1. R. L. P. Adams, R. H. Burdon, A. M. Campbell, D. P. Leader, and R. M. S. Smellie, The Biochemistry of the Nucleic Acids, *9th ed. London: Chapman and Hall, 1981.*

2. V. A. Bloomfield, D. M. Crothers, and I. Tinoco, Physical Chemistry of Nucleic Acids. *New York: Harper and Row, 1974.*

3. R. W. Old and S. B. Primrose. Principles of Gene Manipulation: An Introduction to Genetic Engineering. *Los Angeles, Calif.: University of California Press, 1981.*

REVIEW PROBLEMS

1. Draw the structures of
    (a) 5-hydroxymethylcytosine
    (b) 5-*O*-methylguanosine
    (c) *N*-1-methyladenosine
    What sugar moiety, if any, does each of these compounds have?

2. Estimate the ionic charge at pH 8 for
   (a) Adenosine
   (b) AMP
   (c) ADP
   (d) Adenosine-2′ : 3′-cyclic phosphate
   (e) pdApdC

3. Which is more resistant to degradation at pH 11, DNA or RNA? Which is more resistant to degradation at pH 3?

4. Compare an A : T and a C : G base pair in double-stranded DNA with regard to
   (a) Thickness in the direction perpendicular to the plane of the bases
   (b) Stability of the base pair at pH 12.5
   (c) Number of hydrogen bonds
   (d) Distance from one C-1′ atom to the other C-1′ atom

5. A sample of double-stranded DNA was found to have thymidylate as 19% of the nucleotide residues. What should be the %GC of this DNA? How does the melting temperature of a double-stranded DNA change with base composition?

6. Give the formula (nucleotide sequence) of an oligodeoxyribonucleotide that should form a perfectly base-paired complex with CpCpApUpApUpCpC.

7. What are two forces or bonds that contribute strongly to the stability of specifically folded DNA and RNA structures?

8. Compare double-stranded DNA and double-stranded RNA for the following properties:
   (a) Structure of the monomeric units
   (b) Arrangement of base pairs relative to the helix axis
   (c) Melting temperature

9. What types of RNA molecules make up the bulk of RNA in a typical cell?

10. What information has contributed to our understanding of the folding of "single-stranded" RNA molecules?

11. Name three nucleoprotein structures. What types of forces and bonds are likely to stabilize such structures.

LIPIDS

Lipids are characterized by their sparing solubility in water and considerable solubility in organic solvents, physical properties that reflect the hydrophobic nature of their structures. A rather heterogenous group of compounds, lipids may be traditionally classified as (1) acyl glycerols, (2) waxes, (3) phospholipids, (4) sphingolipids, (5) glycolipids, (6) alkyl glyceryl ethers, and (7) terpenoid lipids, including carotenoids and steroids. All classes are widely distributed in nature.

7.1 FATTY ACIDS

The principal component associated with most lipids is the monocarboxylic acid that has even number of carbon atoms (4 to 30) in a straight chain. Fatty acids of animal origin are quite simple in structure; that is, their fatty acids are straight chained (mostly 16 to 22 carbons) and may have from 0 to 6 double bonds. Somewhat more varied, bacterial fatty acids may be saturated, monoenoic, branched chain, or may even contain a cyclopropane ring (lactobacillic acid). Plant fatty acids are considerably more varied and may have acetylenic bonds, epoxy, hydroxy, and keto groups or cyclopropene and cyclopentene rings. See Table 7.1 for typical examples of fatty acids and their structures.

7.1.1 REACTIVITIES

The chemical reactivities of fatty acids reflect the reactivity of the carboxyl group, other functional groups, and the degree of unsaturation in the hydrocarbon chain. Because free fatty acids are toxic, they occur as such only to a very limited extent in the cell. These acids are therefore found bound as oxygen esters in complex lipids (i.e., triacyl glycerols, glycolipids, and phospholipids).

Ester bonds are susceptible to both acid and base hydrolysis. Acid hydrolysis differs from base hydrolysis in that the former is reversible whereas the

TABLE 7.1   Structures of Common Fatty Acids

ACID	STRUCTURE	MELTING POINT (°C)
Saturated fatty acids		16
Acetic acid	CH_3COOH	−22
Propionic acid	CH_3CH_2COOH	−7.9
Butyric acid	$CH_3(CH_2)_2COOH$	−3.4
Caproic acid	$CH_3(CH_2)_4COOH$	32
Decanoic acid	$CH_3(CH_2)_8COOH$	44
Lauric acid	$CH_3(CH_2)_{10}COOH$	54
Myristic acid	$CH_3(CH_2)_{12}COOH$	63
Palmitic acid	$CH_3(CH_2)_{14}COOH$	70
Stearic acid	$CH_3(CH_2)_{16}COOH$	75
Arachidic acid	$CH_3(CH_2)_{18}COOH$	80
Behenic acid	$CH_3(CH_2)_{20}COOH$	84
Lignoceric acid	$CH_3(CH_2)_{22}COOH$	
Monoenoic fatty acids		
Oleic acid	$CH_3(CH_2)_7\overset{cis}{CH=CH}(CH_2)_7COOH$	13
cis-Vaccenic acid	$CH_3(CH_2)_5\overset{cis}{CH=CH}(CH_2)_9COOH$	44
Dienoic fatty acid		
Linoleic acid	$CH_3(CH_2)_4(\overset{cis}{CH=CH}CH_2)_2(CH_2)_6COOH$	−5
Trienoic fatty acids		
α-Linolenic acid	$CH_3CH_2(\overset{cis}{CH=CH}CH_2)_3(CH_2)_6COOH$	−10
γ-Linolenic acid	$CH_3(CH_2)_4(\overset{cis}{CH=CH}CH_2)_3(CH_2)_3COOH$	—
Tetraenoic fatty acid		
Arachidonic acid	$CH_3(CH_2)_4(\overset{cis}{CH=CH}CH_2)_4(CH_2)_2COOH$	−50
Unusual fatty acids		
α-Elaeostearic acid	$CH_3(CH_2)_3\overset{trans}{CH=CH}\overset{trans}{CH=CH}\overset{cis}{CH=CH}(CH_2)_7COOH$ (conjugated)	48
Tariric acid	$CH_3(CH_2)_{10}C≡C(CH_2)_4COOH$	51
Isanic acid	$CH_2=CH(CH_2)_4C≡C-C≡C(CH_2)_7COOH$	39
Lactobacillic acid	$CH_3(CH_2)_5\overset{\underset{\textstyle CH_2}{\diagup\diagdown}}{CH-CH}(CH_2)_9COOH$	28
Vernolic acid	$CH_3(CH_2)_4CH\overset{cis}{\underset{O}{\diagup\diagdown}}CHCH_2CH=CH(CH_2)_7COOH$	—
Prostaglandin (PGE$_2$)	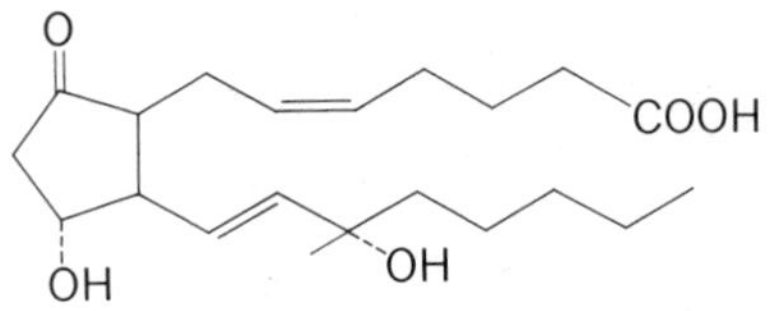	—

BOX 7.A   ANALYSES OF LIPIDS: GAS-LIQUID CHROMATOGRAPHY

Any volatile compound(s) may be injected into a column containing a liquid absorbant supported on an inert solid. The basis for the separation of the components of the volatile mixture is the difference in the partition coefficients of the components as they are carried through the column by an inert gas such as helium. The actual apparatus is quite simple, as may be seen in Scheme 7.1. The sample is introduced at A. The carrier gas transports the injected volatile material into the column, where the components partition into the liquid absorbent (which coats inert particles) and separate; eventually, a fraction passes through a suitable detecting device, which sends signals to a recorder, which in turn converts the signals into a sequence of peaks. Two detecting devices (of many) will be briefly described to give the student a grasp of the technique.

The **thermal conductivity cell** is a detecting device based on the principle that heat is conducted away from a hot wire by a gas passing over it. Two fine coils of wire with a high temperature coefficient of resistance are placed in two parts of the metal block (C^1 and C). Suitable electrical resistors are inserted in the circuit of C^1 and C to form a Wheatstone bridge circuit. When current is passed through the bridge, the wires C^1 and C are heated. Final equilibrium temperature of the wires depends on the thermal conductivity of the gas passing over the wire coil. If the gas is the same, the wires will have the same temperature and the same resistance, and therefore the bridge is balanced; if an effluent gas now passes through C^1 while only the carrier gas passes through C, the wire temperature will differ; the resistance in turn will be changed, and the bridge becomes unbalanced. The extent of unbalance is measured with a recording potentiometer as indicated.

The second type of detector device is a **hydrogen flame ionization** detector. It has extreme sensitivity, a wide linear response, and is insensitive to water. In theory, when organic material is burned in a hydrogen flame, electrons and ions are produced. The negative ions and electrons move in a high-voltage field to an anode and produce a very small current, which is changed to a measurable current by appropriate circuitry. The electrical current is directly proportional to the amount of material burned.

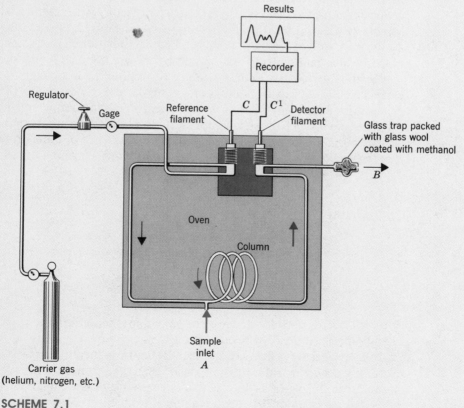

SCHEME 7.1

latter is irreversible. The last step in base hydrolysis is irreversible because in the presence of excess base, the acid exists as the fully dissociated anion that has no tendency to react with alcohols. In acid hydrolysis, however, the system is essentially reversible in all its steps and reaches an equilibrium rather than going to completion. Thus, strong bases, not strong acids, are used for the hydrolysis of ester bonds in simple and complex lipids (also called saponification).

Ester
linkages

$$\alpha \quad CH_2OCOR^1$$
$$\beta \quad R^2COOCH \xrightarrow[\substack{\text{In alkali, called} \\ \text{saponification}}]{\text{3 OH}^-} HOCH \quad + R^1COO^- + R^2COO^- + R^3COO^-$$
$$\alpha' \quad CH_2OCOR^3 \qquad\qquad CH_2OH$$

Triacyl glycerol   Glycerol   Fatty acids

Free fatty acids undergo dissociation in water.

$$RCOOH \rightleftharpoons RCOO^- + H^+$$

$$K_a = \frac{[H^+][RCOO^-]}{[RCOOH]}$$

Since $pK_a = -\log K_a$, the acid strength is determined by the dissociation of the acid. Thus, the pK_a of most fatty acids is about 4.76 to 5.0. Stronger acids have lower pK_a values and weaker acids have higher pK_a values. The effective concentration of an acid is also an important factor. Since acetic acid is very soluble in water, its acid properties are readily measured. On the other hand, stearic acid, with its long, hydrophobic, hydrocarbon side chain, is highly insoluble in water; consequently, its acid properties are not readily measurable.

Because fatty acids are composed of a hydrophobic component, the hydrocarbon chain, and a hydrophilic component, the carboxylate group, these molecules are termed **amphipathic** with the hydrophobic component preferring interaction with each other, whereas the hydrophilic component interacts with the surrounding aqueous environment. A consequence of these properties is that fatty acids tend to associate in a defined manner, forming micelles.

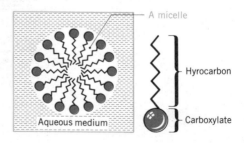

We shall discuss the importance of amphipathic compounds when we consider membrane structures (see Section 8.2.1).

Other properties of fatty acids reflect the nature of their hydrocarbon chains. Naturally occurring saturated fatty acids that have from one to eight carbon atoms are liquid, whereas those with more carbon atoms are solids. Stearic acid has a melting point of 70°C, but, with the introduction of one double bond, as in oleic acid, the melting point drops to 14°C, and the addition of more double bonds further lowers the melting point. When a double bond is in the hydrocarbon chain of a fatty acid, geometric isomerism occurs. Most unsaturated fatty acids are found as the less stable *cis* isomers rather than as the more

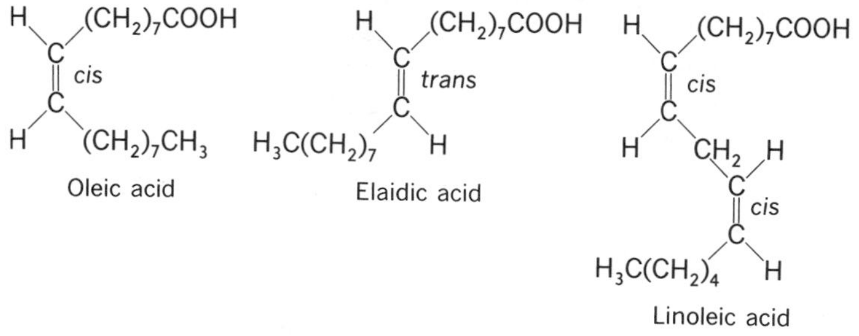

Oleic acid

Elaidic acid

Linoleic acid

stable *trans* isomers. Structurally, the hydrocarbon chain of a saturated fatty acid has a zigzag configuration, as indicated in Structure 7.1, with the carbon-carbon bond forming a 109° bond angle.

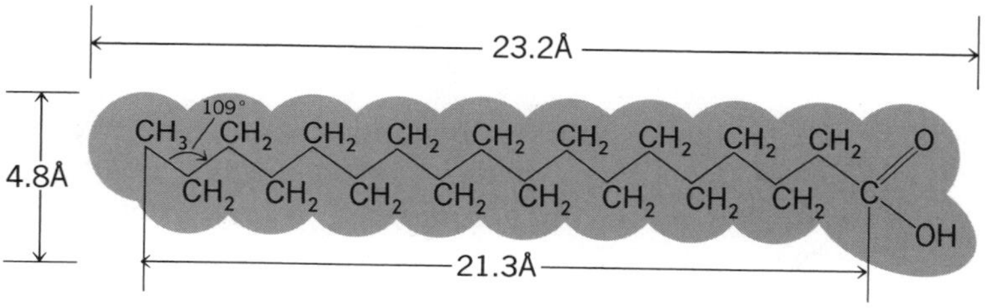

STRUCTURE 7.1

When a *cis* 9,10 double bond is introduced, as in oleic acid, the combination of the *cis* configuration and the sigma and pi bonds of the double bond produce the bent molecule indicated in Structure 7.2.

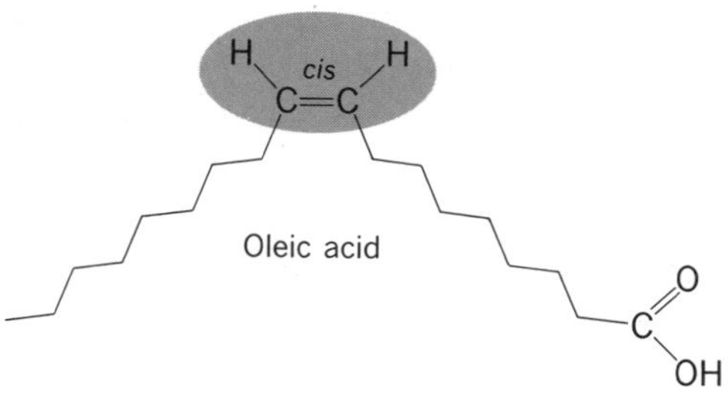

Oleic acid

STRUCTURE 7.2

Linoleic acid, with two double bonds in the hydrocarbon chain, has its alkene chain even more severely bent. Therefore,

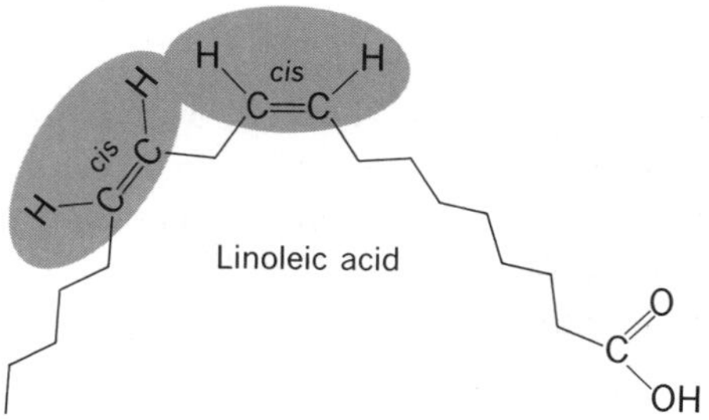

Linoleic acid

STRUCTURE 7.3

when we examine compounds containing double bonds in hydrocarbon chains, we should picture these not as straight chains, occupying a minimum of space, but as large, bulky groups that are considerably bent if they are unsaturated. It is of interest that membranes in animal and plant cells are rich in polyunsaturated fatty acids. On the other hand, bacteria do not contain polyunsaturated fatty acids. Their principal unsaturated fatty acid is the monoenoic acid, *cis*-vaccenic acid.

$$\overset{\text{cis}}{CH_3-(CH_2)_5-CH=CH(CH_2)_9-COOH}$$

In addition to geometric isomerism, another structural aspect involving double bonds in naturally occurring fatty acids is the **nonconjugated double-bond system** of the polyunsaturated fatty acids. Linoleic acid is an example of the nonconjugated type, in which the double bonds are interrupted by a methylene group. This arrangement is called a pentadiene structure.

$$-CH_2-CH=CH-CH_2-CH=CH-CH_2-$$
Nonconjugated double bond system

However, an industrially important polyunsaturated fatty acid, α-elaeostearic acid, the principal acid in tung oil, is isomeric with α-linolenic acid, but differs from it by having a conjugated triene system. Its structure, in contrast to α-linolenic acid, is

$$\overset{\text{trans}\quad\text{trans}\quad\text{cis}}{CH_3(CH_2)_3CH=CHCH=CHCH=CH(CH_2)_7COOH}$$

and it illustrates the **conjugated double-bond system.**

$$-CH_2-CH=CH-CH=CH-CH=CH-CH_2-$$
Conjugated double bond system

These two types of multidouble-bond systems exhibit important differences in chemical reactivity. The nonconjugated or 1,4-pentadiene system has a methylene group flanked by double bonds on both sides. The methylene group may be attacked by a reagent containing a heavy metal such as iron or copper to form a free radical leading to a series of reactions with molecular oxygen.

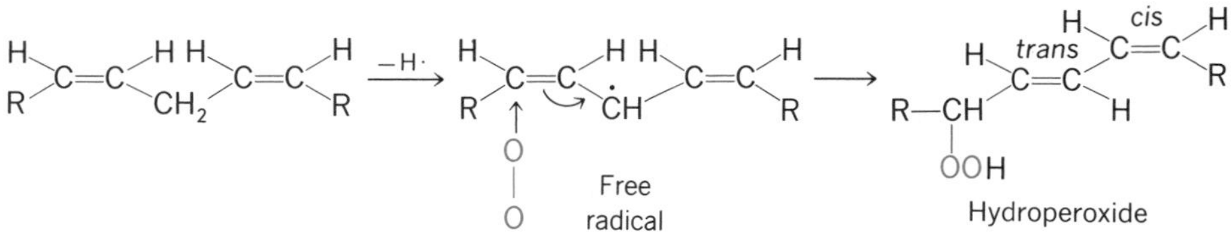

The conjugated double-bond systems are much more reactive because of considerable delocalization of **pi** electrons. Fatty acids with these systems undergo extensive polymerization, a valuable property used by the paint industry. Both retinol and the carotenes are excellent examples of important conjugative systems in biomolecules (see Section 8.13.1). These conjugative bond systems play an important role in the visual processes of the retina (see Section 5.13). We will indicate additional examples elsewhere in this text.

7.2 NOMENCLATURE

Lipid biochemists employ a useful shorthand notation to describe fatty acids. The general rule is to write first the number of carbon atoms, then the number of double bonds, and finally indicate the position of the first carbon of the double bond, counting from the carboxyl carbon. Thus, palmitic acid, a saturated C_{16} acid, is written as 16:0, oleic acid as 18:1(9), and arachidonic acid as 20:4(5,8,11,14). The *cis* configuration is assumed to be the only geometric isomer present. If the *trans* configuration occurs in the structure, it is so stated, that is, 18:3(6t,9t,12c). Polyunsaturated fatty acids also have their double-bond positions defined with respect to the terminal methyl group. Thus, linoleic acid can be represented either as 18:2 (9,12) or 18:2 (*n*-6) where *n* equals the number of carbon atoms in the molecule; the first double bond begins at six carbons from the methyl end of the fatty acid and the second would follow the *cis* nonconjugative rule in relation to the first double bond. Thus, 20:2(*n*-6) would be

$$CH_3CH_2CH_2CH_2CH_2CH=CHCH_2CH=CHCH_2CH_2CH_2CH_2CH_2CH_2CH_2CH_2COOH$$

or 20:2(11,14).

With reference to the nomenclature of the phospholipids, if either carbon 1 or 3 of glycerol is esterified by a fatty acid or phosphoric acid, carbon 2 becomes an asymmetric center, yielding antipodal forms. Thus, students as well as biochemists are often confused by the fact that L-3-glycerophosphate (I) is equivalent to D-1-glycerophosphate (II). To simplify this problem, the

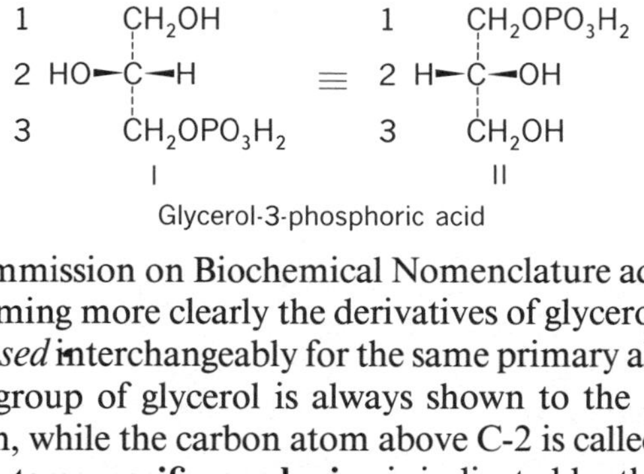

$$
\begin{array}{llcll}
1 & CH_2OH & & 1 & CH_2OPO_3H_2 \\
2 & HO\!-\!C\!-\!H & \equiv & 2 & H\!-\!C\!-\!OH \\
3 & CH_2OPO_3H_2 & & 3 & CH_2OH \\
& I & & & II
\end{array}
$$

Glycerol-3-phosphoric acid

IUPAC–IUB commission on Biochemical Nomenclature adopted the following system for naming more clearly the derivatives of glycerol. The numbers 1 and 3 *cannot be used* interchangeably for the same primary alcohol group. The second hydroxy group of glycerol is always shown to the left of C-2 in the Fischer projection, while the carbon atom above C-2 is called C-1 and the one below, C-3. This **stereospecific numbering** is indicated by the prefix *sn* before the stem name of the compound. Glycerol is thus labeled

$$
\begin{array}{ll}
CH_2OH & 1 \\
HO\!-\!C\!-\!H & 2 \longleftarrow \text{Stereospecific numbering (}sn\text{)} \\
CH_2OH & 3
\end{array}
$$

Clearly, compound I, now called *sn*-glycerol-3-phosphoric acid, is the optical antipode of *sn*-glycerol-1-phosphoric acid (III).

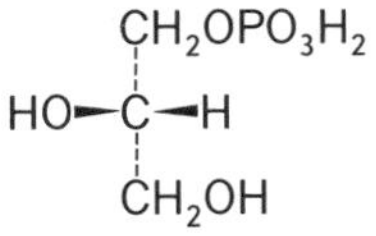

$$
\begin{array}{l}
CH_2OPO_3H_2 \\
HO\!-\!C\!-\!H \\
CH_2OH \\
III
\end{array}
$$

A mixture of both would be called *rac*-glycerol phosphoric acid.

The sterochemistry of a phosphatidyl choline would be defined by the term 3-sn-phosphatidyl choline. Keeping the definition of sn in mind, we simply write the structure as

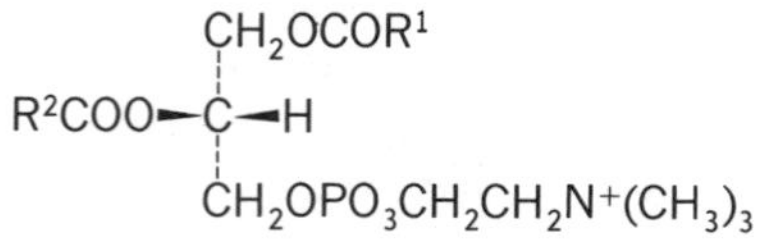

$$CH_2OCOR^1$$
$$R^2COO-C-H$$
$$CH_2OPO_3CH_2CH_2N^+(CH_3)_3$$

7.3   ACYL GLYCEROLS

The most widespread acyl glycerol is triacyl glycerol, also called triglyceride or neutral lipid. The general structure of a triacyl glycerol is

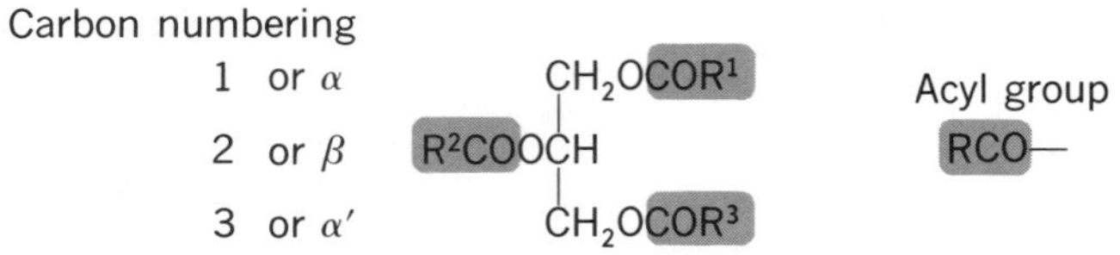

Carbon numbering

1 or α	CH_2OCOR^1	Acyl group
2 or β	R^2COOCH	$RCO-$
3 or α'	CH_2OCOR^3	

Triacyl glycerol

Diacyl glycerols and monoacyl glycerols do not occur in appreciable amounts in nature but are important intermediates in a number of biosynthetic reactions (see Chapter 13 for further details). Their structures are

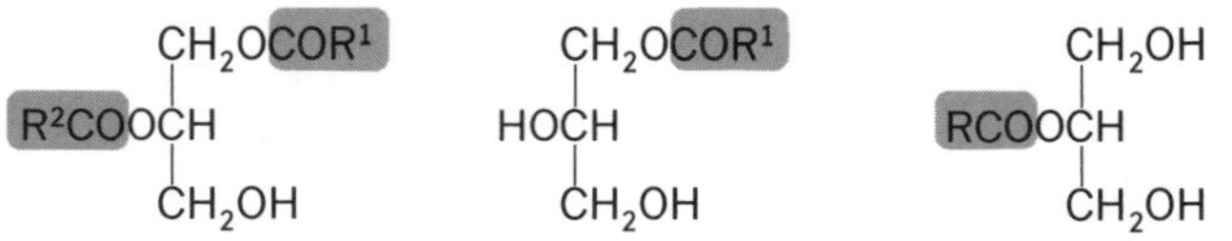

$$CH_2OCOR^1 \qquad CH_2OCOR^1 \qquad CH_2OH$$
$$R^2COOCH \qquad HOCH \qquad RCOOCH$$
$$CH_2OH \qquad CH_2OH \qquad CH_2OH$$

1,2,-Diacyl glycerol   1-Monoacyl glycerol   2-Monoacyl glycerol

Triacyl glycerols exist in the solid or liquid form, depending on the nature of the constituent fatty acids. Most plant triacyl glycerols have low melting points and are liquids at room temperature because they contain a large proportion of unsaturated fatty acids such as oleic, linoleic, or linolenic acids. In contrast, animal triacyl glycerols contain a higher proportion of saturated fatty acids, such as palmitic and stearic acids, resulting in higher melting points, and thus at room temperature, they are semisolid or solid. Table 7.1 lists some of the naturally occurring fatty acids, their structures, and their melting points.

BOX 7.B   THIN-LAYER CHROMATOGRAPHY

Thin-layer chromatography is adsorption chromatography performed on flat layers of adsorbent materials supported on glass plates. A thin uniform film of silica gel containing a binding medium, such as calcium sulfate, is spread onto a glass plate. The thin layer is allowed to dry at room temperature and then is activated by heating in an oven between 100 and 250°C, depending on the degree of activation desired. The activated plate is then placed flat on the laboratory bench and samples spotted carefully on the surface of the thin layer. Material ranging from 0.05 to 50 mg or more are readily spotted with micropipettes. After the solvent has evaporated, the plates are placed vertically in a glass tank that contains a suitable solvent. Within 5 to 30 minutes, a separation is produced by the solvent rising through the thin layer, differentially carrying the components of the spots from the origin, depending on adsorption of the components on the silica gel or the partition between the mobile solvent and the water held by the silica gel. The plate is removed from the solvent tank, permitted to dry briefly and then, depending on the type of compounds on the gel, the spots are detected by spraying the plate with a variety of reagents or dyes. Moreover, the thin inorganic layer of adsorbent can be used with reagents of a more corrosive nature. The possibility of using high-temperature techniques such as carbonization, in conjunction with a spray of concentrated sulfuric acid, offers a universal means of detection of great sensitivity. Thus, the speed, efficiency, and sensitivity of thin-layer chromatography has made this technique one of the most powerful available to the lipid biochemist.

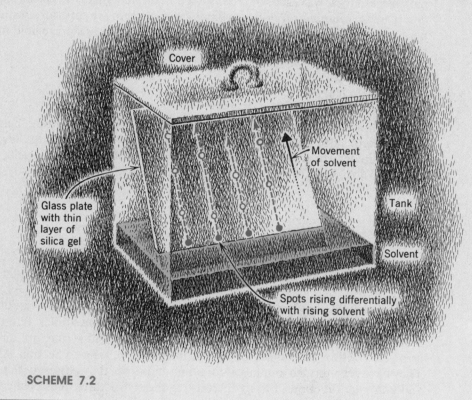

SCHEME 7.2

7.4 WAXES

Equally widespread are the waxes that serve as protective coatings on fruits and leaves, or that are secreted by insects (e.g., beeswax). In general, waxes are a complicated mixture of long-chain alkanes, with an odd number of carbon atoms ranging from C_{25} to C_{35}; and oxygenated derivatives such as secondary alcohols and ketones. Being highly insoluble in water and having no double bonds in their hydrocarbon chains, waxes are chemically inert. They serve admirably on leaf surfaces to protect plants from water loss and abrasive damage. Waxes play an important role in providing a water barrier for insects, birds, and animals such as sheep. This property has been dramatically demonstrated in recent years. When extensive oil spills have occurred in the ocean, detergents have frequently been used to solubilize the oil. Under these conditions, marine birds have great difficulty in maintaining their buoyancy, since the waxy layers covering their feathers were removed by both the oil and detergent. Another important type includes the esters of long-chain fatty acids and long-chain primary alcohols.

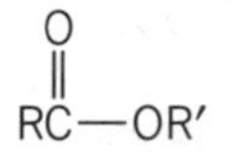

$$\underset{\text{Oxygen ester}}{RC-OR'} \quad \overset{O}{\parallel}$$

RC—OR'   Where R has C_{17}-C_{29} carbon atoms and R'
Oxygen ester C_{18}-C_{30} carbon atoms

These wax esters are of considerable commercial importance because they act as superior machine lubricants. For many years, sperm whales were the principal source of these wax esters but recently a unique plant that grows primarily in desert areas, *Simmondsia chinensis* or jojoba, may serve as a superior substitute because it synthesizes large amounts of oxygen wax esters as storage lipid in its seeds.

7.5 PHOSPHOLIPIDS

Phospholipids are so named because they contain a phosphorus atom. In addition, glycerol, fatty acids, and a nitrogenous base are key components. Several phospholipids, which are considered as derivatives of phosphatidic acid, are listed in Table 7.2. The structure of phosphatidic acid is

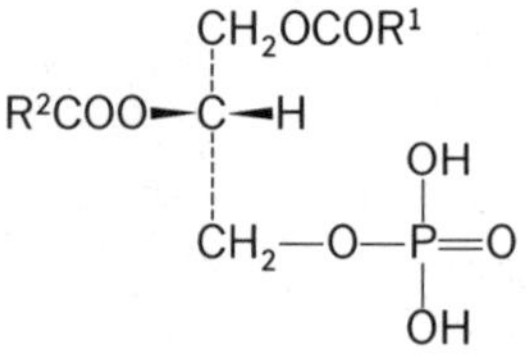

$$\begin{array}{c} CH_2OCOR^1 \\ | \\ R^2COO - C - H \\ | \quad\quad OH \\ | \quad\quad | \\ CH_2 - O - P = O \\ | \\ OH \end{array}$$

3-*sn*-Phosphatidic acid

Phospholipids are widespread in bacteria, animal, and plant tissues, and their general structures, regardless of their sources, are quite similar. Phospholipids, namely, phosphatidyl ethanolamine, choline, and serine, are always associated with membranes (for further information, refer to Section 8.2.1). They have been termed amphipathic compounds since they possess both polar and nonpolar functions, thereby permitting these structures to associate with both hydrophilic (polar) and hydrophobic (nonpolar) environments.

7.6 SPHINGOLIPIDS

Sphingolipids include an important group of compounds closely associated with animal membranes, in particular, nerve tissue. The central compound is called 4-sphingenine (formerly sphingosine). A variety of components can be attached to this structure to give important derivatives. 4-Sphingenine is formed

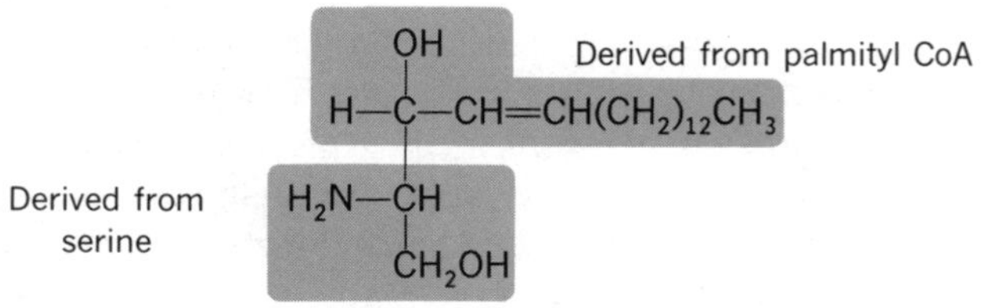

$$\begin{array}{c} OH \\ | \\ H - C - CH = CH(CH_2)_{12}CH_3 \\ | \\ H_2N - CH \\ | \\ CH_2OH \end{array}$$

Derived from palmityl CoA

Derived from serine

4-Sphingenine

TABLE 7.2 Some Amphipathic Lipids

PHOSPHOLIPID	USUAL FATTY ACID (NONPOLAR COMPONENT)	BASE (POLAR COMPONENT)
3-*sn*-Phosphatidyl choline (lecithin) 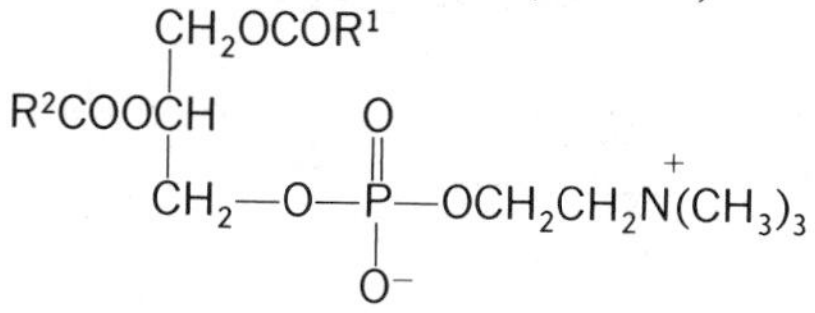 CH_2OCOR^1 R^2COOCH $CH_2-O-\overset{O}{\underset{O^-}{\overset{\|}{P}}}-OCH_2CH_2\overset{+}{N}(CH_3)_3$	Stearic or palmitic (R^1) polyunsaturated (R^2)	Choline
3-*sn*-Phosphatidyl aminoethanol (cephalin) 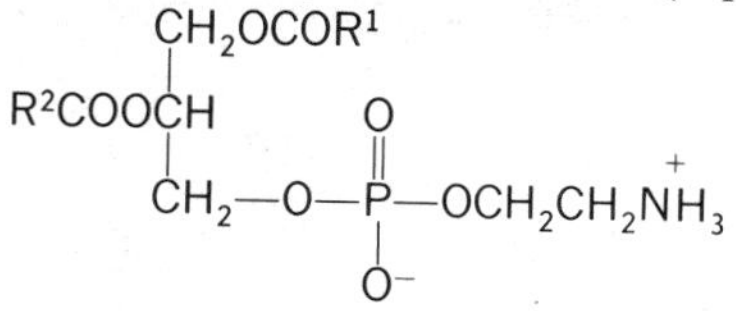 CH_2OCOR^1 R^2COOCH $CH_2-O-\overset{O}{\underset{O^-}{\overset{\|}{P}}}-OCH_2CH_2\overset{+}{N}H_3$	Stearic or palmitic (R^1) polyunsaturated (R^2)	Aminoethanol
3-*sn*-Phosphatidyl serine 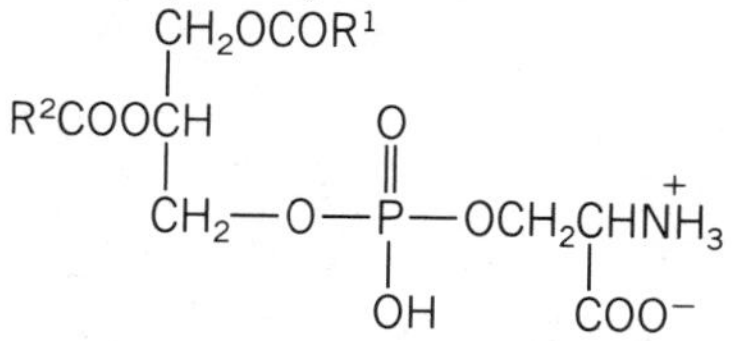 CH_2OCOR^1 R^2COOCH $CH_2-O-\overset{O}{\underset{OH}{\overset{\|}{P}}}-OCH_2\underset{COO^-}{\overset{+}{CHNH_3}}$	Stearic or palmitic (R^1) polyunsaturated (R^2)	Serine
3-*sn*-Phosphital aminoethanol (plasmalgen) 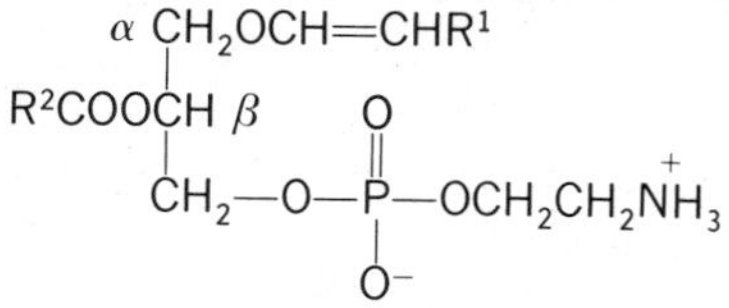 $\alpha\ CH_2OCH=CHR^1$ $R^2COOCH\ \beta$ $CH_2-O-\overset{O}{\underset{O^-}{\overset{\|}{P}}}-OCH_2CH_2\overset{+}{N}H_3$	Unsaturated ether (α) Linoleic (β)	Aminoethanol
1-Alkyl phospholipid (α-glyceryl ether) 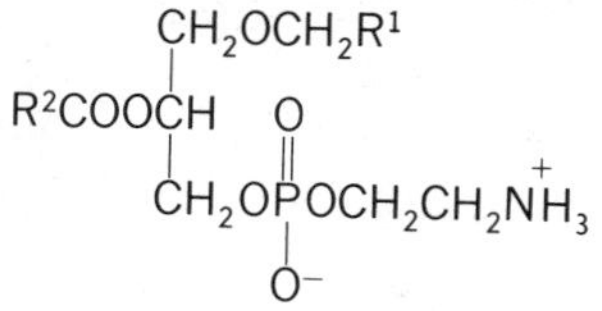 $CH_2OCH_2R^1$ R^2COOCH $CH_2O\overset{O}{\overset{\|}{P}}OCH_2CH_2\overset{+}{N}H_3$ O^-	R^2 probably an unsaturated fatty acid	Aminoethanol
3-*sn*-Phosphatidyl inositol 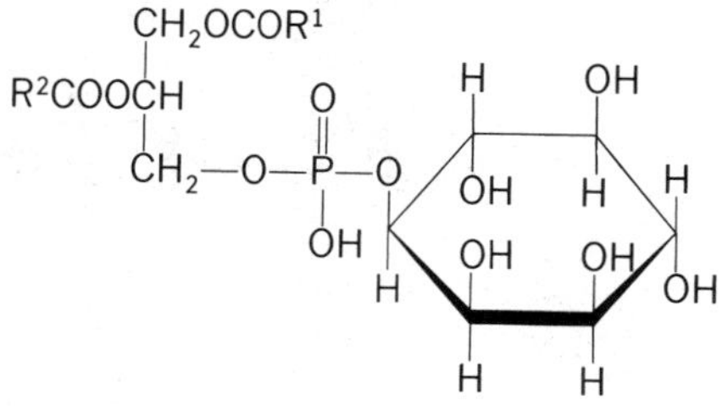 CH_2OCOR^1 R^2COOCH $CH_2-O-\overset{O}{\underset{OH}{\overset{\|}{P}}}-O$ (inositol ring)	Palmitic (R^1) Arachidonic (R^2)	Myoinositol replaces base
3-*sn*-Phosphatidyl glycerol 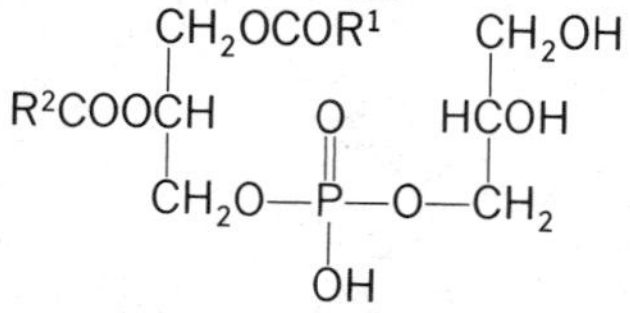 $CH_2OCOR^1 \quad CH_2OH$ $R^2COOCH \quad HCOH$ $CH_2O-\overset{O}{\underset{OH}{\overset{\|}{P}}}-O-CH_2$	Polyunsaturated fatty acid (R^1, R^2)	Glycerol replaces base

from a rather complex series of reactions involving palmityl-CoA and serine. The fully reduced compound is called sphinganine (formerly dihydrosphingosine). Some important derivatives are presented here.

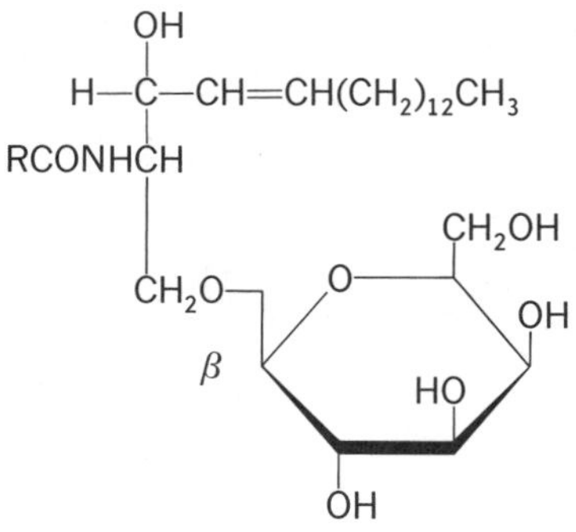

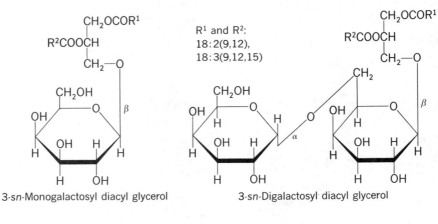

Cerebroside

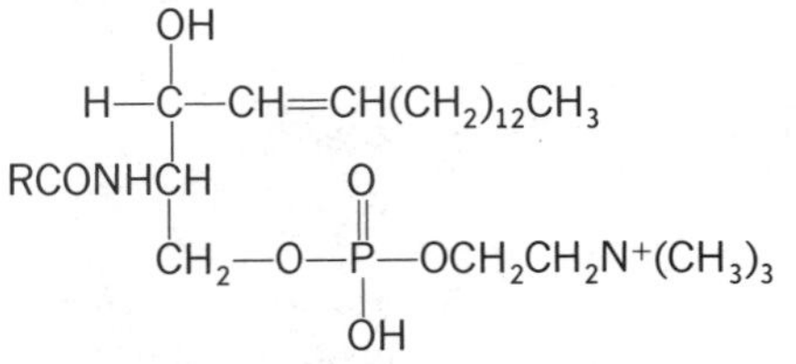

Sphingomyelin

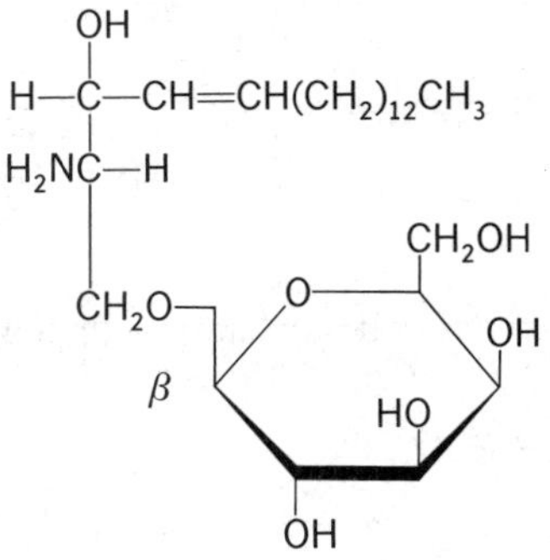

Psychosine

7.7 GLYCOLIPIDS

Another group of compounds are collected in the class of glycolipids because they are primarily amphipathic carbohydrate-glyceride derivatives and do not contain phosphate. These include the galactolipids and the sulfolipids, found primarily in chloroplast membranes. Their structures are

3-sn-Monogalactosyl diacyl glycerol

R^1 and R^2:
18:2(9,12),
18:3(9,12,15)

3-sn-Digalactosyl diacyl glycerol

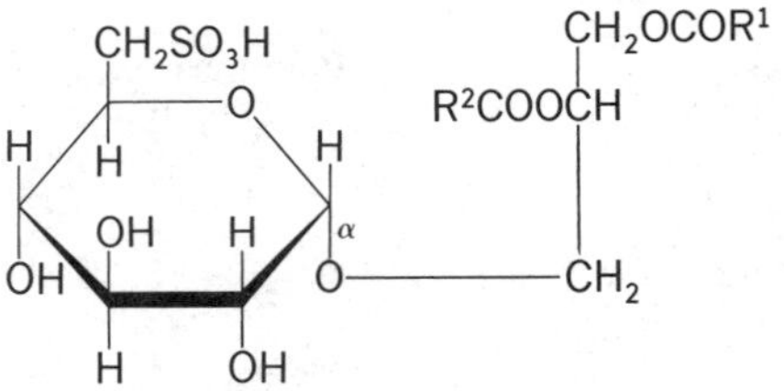

3-sn-Sulfonyl-6-deoxyglucosyl diacyl glycerol

7.8 GLYCERYL ETHERS

An interesting group, the glyceryl ethers, are found in varying portions in marine organisms and other animal species. As indicated below, Position 1 has either a saturated or unsaturated alkyl ether; Position 2 usually contains an acyl moiety; and Position 3 may have another acyl component or a phosphoryl choline group:

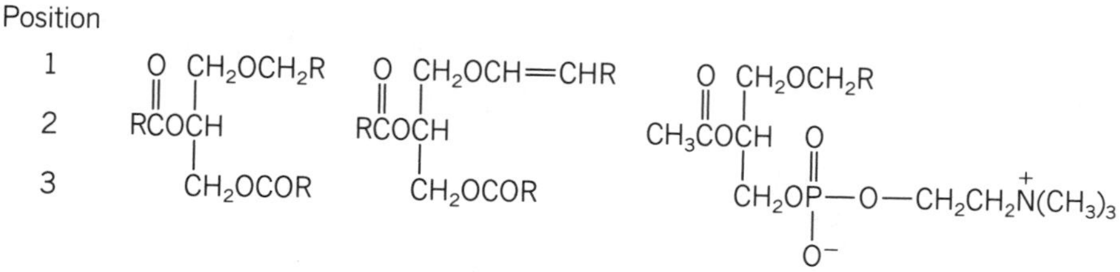

Position

1-Alkyl-2,3-
diacyl-*sn*-glycerol

1-Alkenyl-2,3-
diacyl-*sn*-glycerol

1-Alkyl,2-acetyl,
3-phosphoryl choline-
sn-glycerol

The function of these unique lipids is not clear, but recently the 1-alkyl-2-acetyl-*sn*-glycerol-3-phosphoryl choline has been found to be a platelet aggregating factor that exerts its effect at 10^{-10} M concentration. It follows, therefore, that at such a low concentration, this complex lipid does not participate as a component of membrane structure but functions at a site that must relate to a cascade or amplification response leading to an important physiological effect.

7.9 TERPENOIDS AND STEROLS

The terpenoids are a very large and important group of compounds that are actually made up of a simple repeating unit, the isoprenoid unit; this unit, by ingenious condensations, gives rise to such compounds as rubber, carotenoids, steroids, and many modified terpenes. Isoprene, which does not occur in nature, has as its actual biologically active counterpart isopentenyl pyrophosphate, which is formed by a series of enzymically catalyzed steps from mevalonic acid. Isopentenyl pyrophosphate undergoes further reactions to form squalene, which, in turn, can condense with itself to form cholesterol. Another typical terpenoid product is β-carotene, which is cleaved in the intestinal mucosa cells to form retinol. The various structural relationships are indicated in the diagrams here. Note the repeating isoprenoid unit in all these compounds. We shall discuss some of the biosynthetic reactions in Chapter 13.

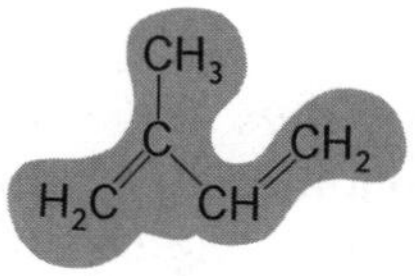

"Isoprenoid unit"

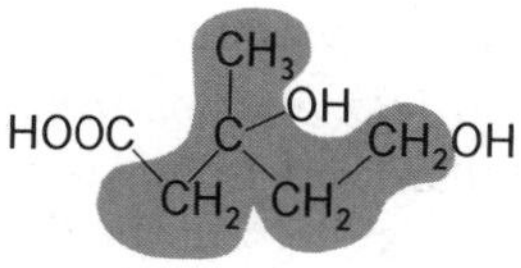

Mevalonic acid

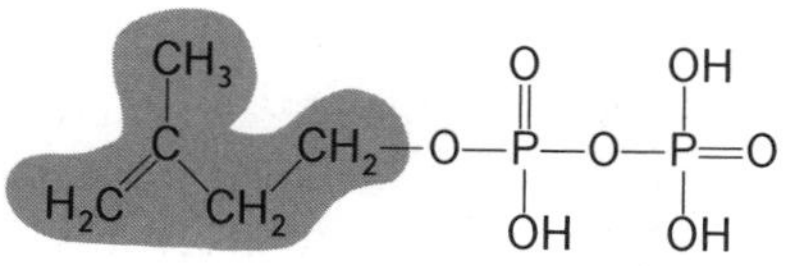

Isopentenyl pyrophosphate

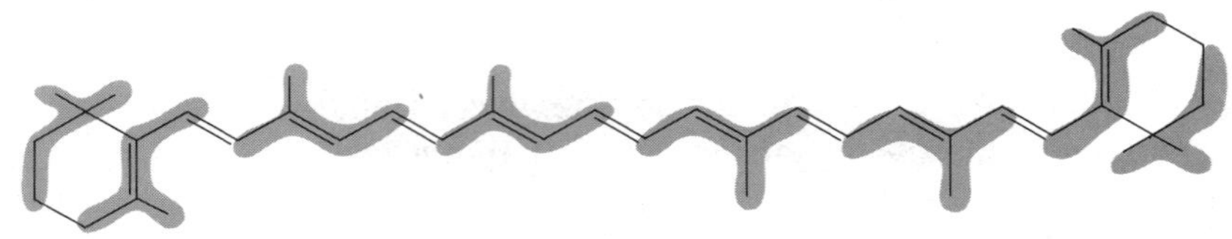

β-Carotene

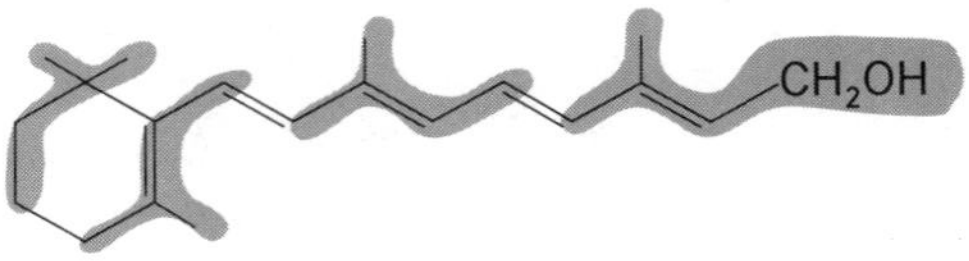

Retinol
(Vitamin A₁)

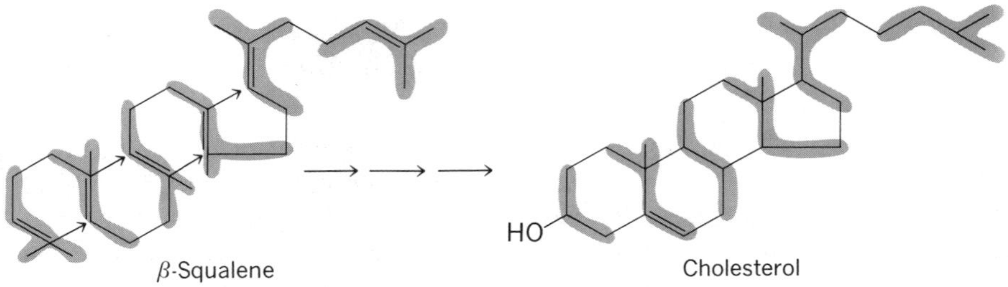

β-Squalene   → → →   Cholesterol

7.10   FUNCTION OF LIPIDS

In recent years, it has become apparent that lipids play extremely important roles in the normal function of a cell. Not only do lipids serve as highly reduced storage forms of energy, but they also play an intimate role in the structure of cell membranes and the organelles found in the cell. Later in Chapter 8, we shall discuss these aspects in some detail. In Chapter 13, we will review further functional aspects of lipids.

Lipids participate directly or indirectly in metabolic activities such as the following.

1.   *Major sources of energy in animals, insects, birds, and high lipid-containing seeds* (see Section 13.1).

2.   *Activators of enzymes.* Three microsomal enzymes—namely, glucose-6-phosphatase, stearoyl CoA desaturase, monooxygenases, and D-β-hydroxybutyric dehydrogenase (a mitochondrial enzyme)—require phosphatidyl choline micelles for activation. Many other enzymes can be cited that require lipid micelles for maximal activation.

3.   *Components of the electron-transport system in the inner membrane of mitochondria* are buried in a milieu of phospholipids. The same applies to the photophosphorylation system in the thylakoid membranes of all chloroplasts in green plants. The major lipids in these membranes are mono- and di-galactosyl diglycerides.

4.   Arachidonic acid, 20:4 (5,8,11,14), is the specific precursor for all prostaglandins and leukotrienes, compounds that function in a number of specific animal cells at incredibly low concentrations, that is, 10^{-12} M. Arachidonic

acid is bound as an acyl moiety to the 2-position of a number of phospholipids and thus is inactive as a substrate. When released by the action of phospholipase A_2, the free arachidonic acid is converted by a cyclooxygenase and other enzymes to form, among many products, two very important prostaglandin derivatives, prostacyclin I_2 and thromboxane A_2. The former is synthesized principally in the arterial walls of blood vessels and is the most active physiological vasodialator so far discovered. The latter is synthesized in blood platelets and is the most potent vasoconstrictor so far known. Leukotrienes have recently been shown to be responsible for broncheostriction in asthmatic attacks. The structures are

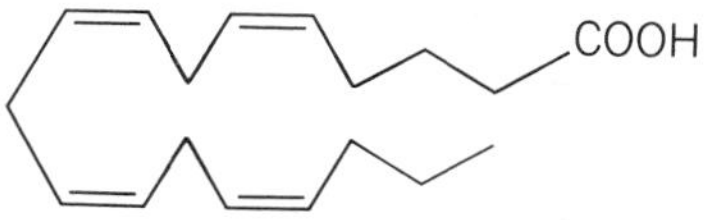

Arachidonic acid

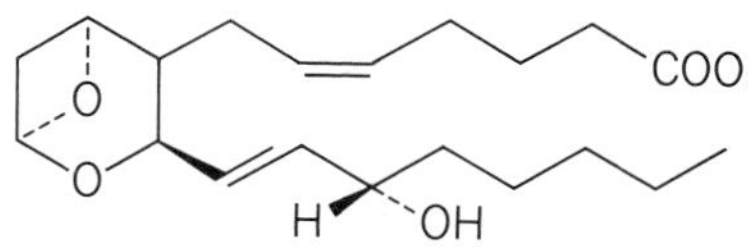

Thromboxane A_2

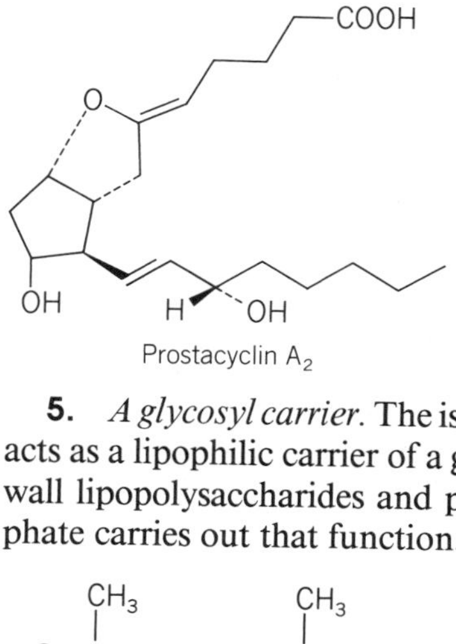

Prostacyclin A_2

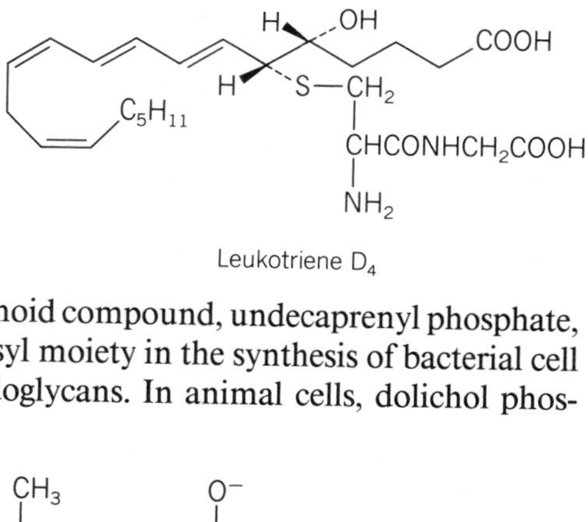

Leukotriene D_4

5. *A glycosyl carrier.* The isoprenoid compound, undecaprenyl phosphate, acts as a lipophilic carrier of a glycosyl moiety in the synthesis of bacterial cell wall lipopolysaccharides and peptidoglycans. In animal cells, dolichol phosphate carries out that function.

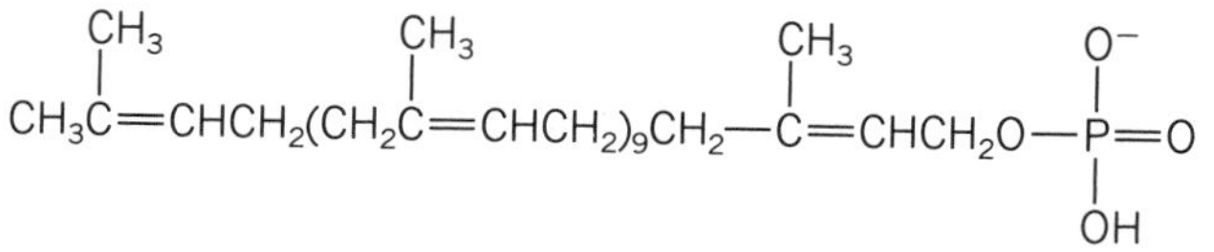

Undecaprenyl phosphate

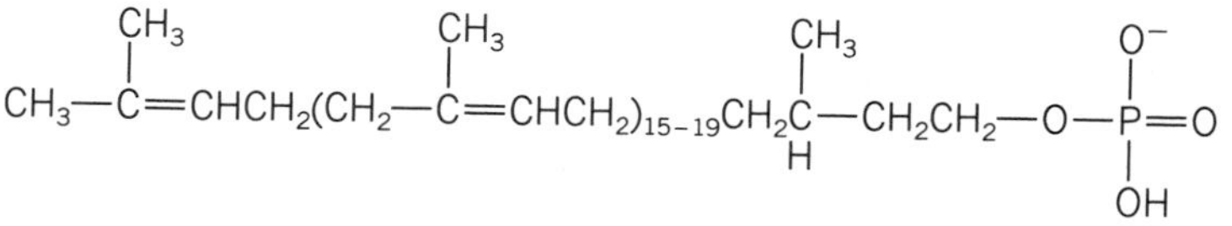

Dolichol phosphate

6. *A substrate in the indirect decarboxylation of serine to ethanolamine* is phosphatidyl serine. It is decarboxylated by a specific decarboxylase to phosphatidyl ethanolamine. The direct decarboxylation of serine to ethanolamine has never been demonstrated.

$$\text{Phosphatidyl serine} \longrightarrow \text{Phosphatidyl ethanolamine} + CO_2$$

7. Phosphatidyl choline with oleic acid in the 2 position is the specific substrate for the Δ^{12} desaturase in plants that converts oleic acid to linoleic (see Chapter 13).

$$\text{Oleyl-phosphatidyl choline} \xrightarrow[\text{NADH}_2]{O_2} \text{Linoleyl-phosphatidyl choline}$$

8. Phosphatidyl-inositol triphosphate serves as a key precursor in the formation of a second messenger (see Section 18.5.2).

7.11   LIPOPROTEINS

Lipids are not transported in the free form in circulating blood plasma, but move as chylomicrons, very low-density lipoproteins, or free fatty acid-albumin complexes. In addition, lipoproteins occur as components of membranes. We shall discuss the role of lipoproteins briefly in Sections 13.4 and 13.5.

Lipoproteins are classes of biomolecules in which the lipid components consist of triacyl glycerol, phospholipid, and cholesterol (or its esters) in remarkably consistent proportions within each class of lipoproteins (see Table 7.3). The protein components, in turn, have a relatively high proportion of nonpolar amino acid residues that can participate in the binding of lipids. Studies have clearly excluded covalent and ionic bonds as being involved in the tight binding of the lipid to specific apoproteins. The principal binding force is the hydrophobic interaction between apoproteins and lipids. As we have already mentioned, hydrophobic interaction is the tendency of hydrocarbon components to associate with each other in an aqueous environment. An example of hydrophobic bonding between a lipid and a protein is a retinol-to-retinol binding protein or a sterol-to-sterol carrier protein.

Lipoproteins are also found in the membranes of mitochondria, endoplasmic reticuli, and nuclei. The electron transport system in mitochondria appears to contain large amounts of lipoproteins. Lamellar lipoprotein systems occur in the myelin sheath of nerves, photoreceptive structures, chloroplasts, and the membranes of bacteria.

TABLE 7.3   Composition of Some Lipoproteins

SOURCE	LIPOPROTEIN	PARTICLE MASS (DALTONS)	PROTEIN	PHOSPHOLIPID	CHOLESTEROL (FREE + ESTER)	TRIAC GLYCE
Blood serum	Chylomicron	$10^9 - 10^{10}$	2	3–6	2–5	80–9
	Very low-density	$5 - 100 \times 10^6$	5–10	15–20	10–25	40–8
	Low-density	2×10^6	25	20	45	10
	High-density	0.25×10^6	40–50	30	20	1–5
Egg yolk	β-Lipovitellin	4×10^5	78	12	1	9
Milk	Low-density	4×10^6	13	52	0	35

7.12   COMPARATIVE DISTRIBUTION OF LIPIDS

With the advent of modern lipid techniques, much work has been directed toward an elucidation of the nature of lipids in a wide number of organisms. In general, procaryotic cells and eucaryotic cells (those without and with membrane-enclosed organelles, respectively) differ remarkably in their lipid composition. A brief survey of these differences will now be presented.

7.12.1 PROCARYOTIC CELLS

In general, a bacterial cell has over 95% of its total lipid complement associated with its cell membrane; the remaining 5% is distributed between its cytoplasm and the cell wall. Bacterial cells are distinctive because of the complete absence of sterols in their cells; such cells are unable to synthesize the steroid ring structure although they are capable of forming extended linear isoprenoid polymers. With the exception of the mycobacteria, triacyl glycerols are missing in bacteria, and with the exception of *Bacilli,* which do contain some $16:2(5,10)$ and $16:2(7,10)$ polyunsaturated fatty acids, bacteria do not have the capacity to synthesize the conventional nonconjugated polyunsaturated fatty acids. Thus, bacteria are somewhat limited in their capacity for the synthesis of a broad spectrum of fatty acids and produce only saturated, monoenoic, cyclopropane, or branched-chain fatty acids. Indeed, a number of species such as the *Mycoplasma* and mutants of *E. coli* have even lost the capacity for monoenoic fatty acid synthesis and require for growth an external source of these fatty acids.

Equally unique is the complete absence of fatty acids in the primitive procaryotic organisms identified as *Archaebacteria.* These organisms include methanogenic bacteria (methane producing under an anaerobic environment), thermophilic bacteria (that live in environments of elevated temperatures), and extreme halophilic bacteria (that live in lakes with a high salt concentration). Instead of having as membrane lipids the usual phospholipids, these organisms contain glyceryl ethers with a terpenoid derivative, a phytanyl moiety, replacing the usual fatty acid derivative. A typical glyceryl ether would be 2,3-diphytanyl-*sn*-glycerol-1-phosphoryl-1′-*sn*-glycerol.

Why these primitive procaryotic organisms have these unique membrane lipids is a provocative question in terms of evolution.

7.12.1 EUCARYOTIC CELLS

7.12.1.1 Plants. In general, the seeds of higher plants have a rather fixed composition of fatty acids that are phenotypic expressions of their genotypes. The exotic fatty acids are normally found as triacyl glycerols in the mature seed and are rarely found in such tissue organelles as the chloroplast, or mitochondria. Throughout the higher plant kingdom, chloroplasts possess a remarkably constant pattern of fatty acids and complex lipids. In particular, the polyunsaturated fatty acid *a*-linolenic is always found associated with four highly polar, complex lipids that are unique to photosynthetic tissue: monogalactosyl-diacyl glycerol, digalactosyl-diacyl glycerol, sulfoquinovosyl-diacyl glycerol, and phosphatidyl glycerol. These lipids are closely associated with the lamellar membranes of chloroplasts. Higher plants synthesize a wide range of polyunsaturated fatty acids, the most important, in terms of human nutrition, being linoleic acid (see the following).

7.12.1.2 Animals.

The lipids of animal cells are equally complex and their composition is characteristic of a particular cell. Thus, a nerve cell is rich in sphingolipids, glyceryl ethers, and plasmalogens as well as phospholipids; an adipose cell, on the other hand, consists essentially of droplets of triacyl glycerols. There is one rather remarkable feature that is unique to cells of both lower and higher forms of animal life; namely, the inability to synthesize linoleic acid [18:2 (9,12)]. In general, eucaryotic cells readily synthesize oleyl CoA *de novo* from stearoyl COA by an aerobic mechanism in which a *cis*-9,10 position is introduced (counting from the carboxyl carbon). (See Section 13.13 for a discussion of the mechanism of this reaction.) However, animal cells completely lack the enzyme responsible for the further desaturation of oleic acid to linoleic acid, although this specific desaturase is widespread in plant tissues. Moreover, animal cells introduce further *cis* double bonds into the hydrocarbon chain only toward the carboxyl end, whereas plant cells always introduce additional double bonds toward the methyl end.

In animal cells:

$$18:2(9,12) \xrightarrow{-2H} 18:3(6,9,12) \xrightarrow{+C_2} 20:3(8,11,14) \xrightarrow{-2H} 20:4(5,8,11,14)$$

| Linoleic | γ-Linolenic | Homo-γ-linolenic | Arachidonic |

Linoleic
(from diet)

In plant cells:

$$18:1(9) \xrightarrow{-2H} 18:2(9,12) \xrightarrow{-2H} 18:3(9,12,15)$$

Oleic   Linoleic   α-Linolenic

It is for this reason that linoleic acid is an essential fatty acid in the diet of man because it alone serves as the precursor of arachidonic acid that, in turn, converts into the very important prostaglandins family and other important physiologically active compounds.

REFERENCES

1. *M. I. Gurr and A. T. James,* Lipid Biochemistry: An Introduction, *2nd ed. London: Chapman and Hall, 1975.*

2. *J. L. Harwood and N. J. Russell,* Lipids in Plants and Microbes. *London: George Allen and Unevin, 1984.*

3. *S. P. Colowick and N. O. Kaplan,* Methods in Enzymology. *New York: Academic Press (numerous annual volumes).*

REVIEW PROBLEMS

1. Given a mixture of acetic acid, oleic acid, and trioleyl glycerol in water, propose a procedure for the separation of these compounds from each other.

2. Write the structural formulas for the following acids:
   (a) 14:3(7,10,13)
   (b) 12:1(3 *trans*)
   (c) 10-CH$_3$—18:0
   (d) 18:2(6,9)
   (e) 12-hydroxy 18:1(9)
   (f) 20:4(5,8,11,14)

3. Which of the following compounds would be soluble, partly soluble, or insoluble in water?

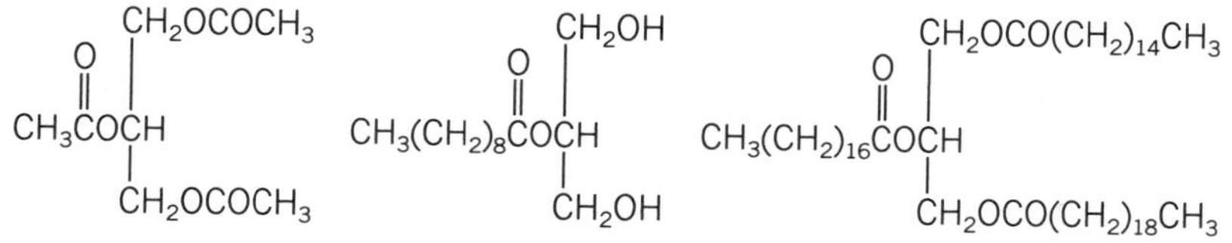

$$\begin{array}{ccc}
& \text{CH}_2\text{OCOCH}_3 & \text{CH}_2\text{OH} & \text{CH}_2\text{OCO(CH}_2)_{14}\text{CH}_3 \\
\text{O} & | & \text{O} \quad | & \text{O} \quad | \\
\| & | & \| \quad | & \| \quad | \\
\text{CH}_3\text{COCH} & & \text{CH}_3(\text{CH}_2)_8\text{COCH} \quad \text{CH}_3(\text{CH}_2)_{16}\text{COCH} \\
| & & | \\
\text{CH}_2\text{OCOCH}_3 & & \text{CH}_2\text{OH} \quad \text{CH}_2\text{OCO(CH}_2)_{18}\text{CH}_3
\end{array}$$

4. Write the structure for dioleylphosphatidyl choline.

5. What distinguishes the following compounds from each other:
   (a) A sphingomyelin?
   (b) A cerebroside?
   (c) A monogalactosyldiacyl glyceride?

6. Give a specific example of the *indirect* decarboxylation of an amino acid.

7. Cite at least three distinctive differences between *procaryotic* and *eucaryotic* cells in terms of their lipids.

PART 2

ENERGY METABOLISM AND BIOSYNTHESIS OF SMALL MOLECULES

C H A P T E R 8

THE CELL—
ITS BIOCHEMICAL
ORGANIZATION

In the first part of this text, we defined the physical and chemical principles operating in biological systems and described the chemistry of small and large biomolecules. In Part 2, which starts with Chapter 8, we will apply this knowledge to the metabolic processes involved in the synthesis and breakdown of carbohydrates, lipids, amino acids, and the common bases found in nucleic acids. We also will describe such important processes as respiration, oxidative phosphorylation, photosynthesis, and nitrogen fixation. Then, in Chapter 18, we will describe some current thoughts regarding the regulation of these many metabolic processes. All of these events occur in the basic structural unit found in nature, namely the cell. In this chapter, we shall define the cell both in structural and biochemical terms.

In general, two types of cells exist in nature: the procaryotic and the eucaryotic cell. By definition, the procaryotic cell has a minimum of internal organization. It possesses no membrane-bound organelle components, its genetic material is not enclosed by a nuclear membrane, nor is its DNA complexed with histones. Indeed, histones are not found in this cell. Its sexual reproduction involves neither mitosis nor meiosis. Its respiratory system is closely associated with its plasma membrane. Typical procaryotic cells include all bacteria and the cyanobacteria. All other cells are of the eucaryotic type.

A eucaryotic cell has a considerable degree of internal structure with a large number of distinctive membrane-enclosed organelles. For example, the nucleus is the site for informational components collectively called chromatin. Reproduction involves both mitosis and meiosis; the respiratory site is the mitochondrion; and in plant cells, the site of the conversion of radiant energy to chemical energy is the highly structured chloroplast.

In this chapter, we shall attempt to define the components of cells in terms of their structure and function. We hasten to add that the descriptions will refer to cells in general; to attempt to define the many specialized cells found in the animal and plant kingdom would necessarily blur the basic similarities and dissimilarities that we are attempting to emphasize. Figure 8.1 compares the general procaryotic and eucaryotic cells, illustrating diagramatically the

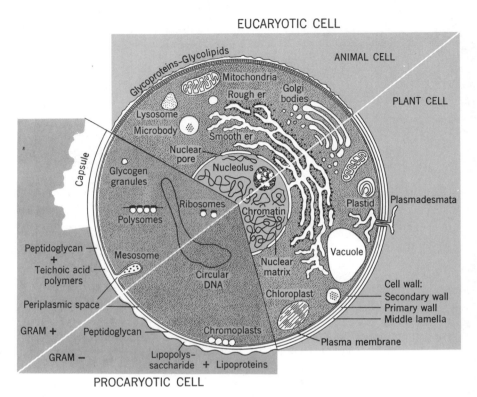

FIGURE 8.1 Schematic comparison of procaryotic and eucaryotic cells. This general diagram, of course, does not imply similar sizes and shapes for all cells.

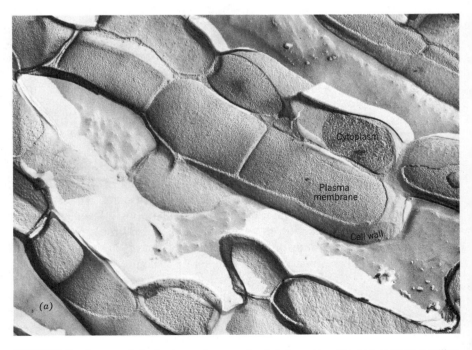

FIGURE 8.2 Freeze-etch electron micrographs. (a) A procaryotic cell, a gram-positive *Bacillus lichenformis* magnified 51,000 ×. (b) A procaryotic cell, a gram-negative *Nitro-*

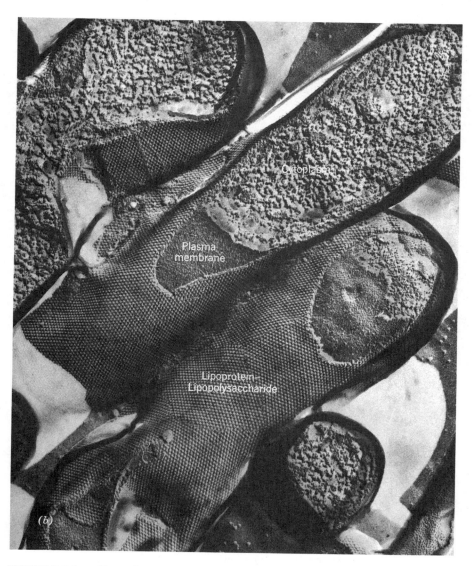

(b)

FIGURE 8.2 (continued)

somonas species magnified 237,000 ×. (The photos in (a) and (b) reproduced with permission of C. Rensen, Woods Hole Oceanographic Institution.) (c) A eucaryotic cell, Schizosaccharomyces pombe magnified 24,000 ×: CW, cell wall; Er, endoplasmic reticulum; G, Golgi apparatus; L, lipid body; Mi, mitochondria; N, nucleus; PL, plasmalemma; and V, vacuole. (The photo in (c) reproduced with permission of F. Kopp and K. Mühlethaler, Swiss Federal Institute of Technology.)

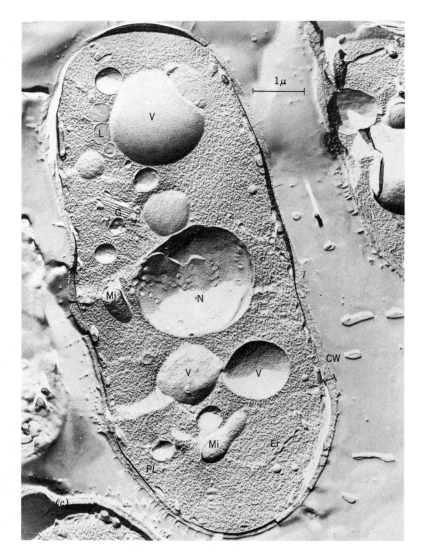

FIGURE 8.2 (continued)

differences as well as the similarities of these cells. Figures 8.2a, b, and c show freeze-etch electron micrographs of procaryotic gram-positive (G+) and gram-negative (G−) cells, and of a eucaryotic yeast cell. Figure 8.3 illustrates a general procedure for the isolation of organelle structure from eucaryotic cells.

8.1   CELL WALLS

The cell walls in procaryotic cells and most eucaryotic cells such as algae, fungi, and plants confer shape and rigidity to the cell itself. Without cell walls, cells would be spherical in shape and extremely fragile to slight osmotic changes of the external environment. This conclusion can be demonstrated by the conversion of a gram-positive bacterium with its thick cell wall to a protoplast devoid of cell wall but possessing its plasma membrane and its cytosolic components.

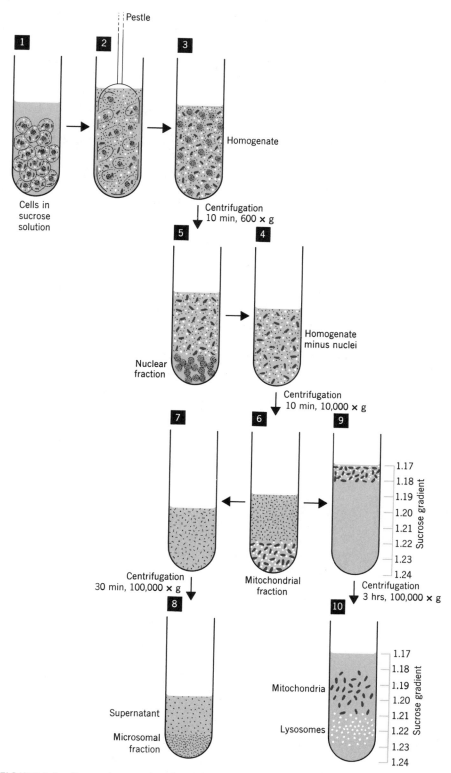

FIGURE 8.3 General procedure for obtaining cell organelles from eucaryotic organisms by gently disrupting the cells (2), differentially centrifuging (4–7), and then centrifuging through a gradient medium (9,10). (*Source:* Christian deDuve, ''The Lysosome.'' Copyright © May 1963, Scientific American, Inc., all rights reserved.)

Protoplasts can be readily prepared by exposing a suspension of gram-positive bacteria to the enzyme lysozome in an osmotically stabilized medium. Lysozyme hydrolyzes the peptidoglycan components of the wall, weakening the wall and allowing the plasma membrane-enclosed protoplast to escape into the isotonic environment. There, the protoplast can undergo normal replication and growth. However, if the environment is altered by the addition of water to form a hypotonic medium, immediate rupture occurs. In gram-negative bacteria, lysozyme breaks the peptidoglycan skeleton of the wall, forming a spheroplast that usually still has wall material attached to the cell.

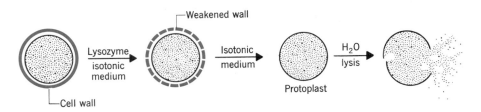

One exception to this requirement for cell walls is the procaryotic genus *Mycoplasma,* all members of which are devoid of cell walls. These organisms have adjusted to survival in the hostile osmotic environment by being parasitic, that is, living in plant and animal host cells where the osmotic environment is carefully maintained. These organisms have also included steroids, obtained from the host cell, in their plasma membranes. Since steroids (Section 7.9) are complex planar ring systems that allow stacking and interaction with other lipid components of the plasma membrane, considerable stability is conferred on the mycoplasma. Similar situations occur in the animal; an example is the erythrocyte that has no rigid cell wall. It remains structurally stable in the circulating osmotically favorable blood plasma but lyses instantly when transferred to water.

8.1.1   PROCARYOTIC CELL WALLS

As already observed in Figure 8.1, cell walls of procaryotic cells are rather complicated and markedly different from those in eucaryotic cells. Bacteria are roughly divided into gram-positive [G(+)] and gram negative [G(−)] cells based on the differential staining by a crystal violet-iodine reagent. In general, gram-positive cells have thick walls, up to 80% of which are composed of a meshlike macropolymer called a peptidoglycan. The chemical nature of this macropolymer has already been described (Section 2.8). Variations in the structure and composition of this peptidoglycan occur in a number of bacteria.

Superimposed on the peptidoglycan are the teichoic acid polymers that consist of repeating units of either glycerol or ribitol connected by internal phosphate diesters. D-Alanine is usually attached through an ester linkage to the polyhydroxyl alcohol. The teichoic acid polymers are probably intimately involved in the cell's antigenicity and susceptibility to phage infection. They definitely confer a strong negative charge on the surface of the cell wall because of their high content of ionized phosphate groups. The teichoic acids apparently are located in the region extending from the exterior of the plasma membrane

Ribitol teichoic acid from *Bacillus subtilis*

to the outer regions of the peptidoglycans. Gram-positive cell walls are also characterized by the absence of any significant lipid.

Of considerable interest is the fact that the enzyme lysozyme found in tears and saliva, bacteria, and plants readily hydrolyzes peptidoglycans at the β-1,4 linkage of *N*-acetyl muramic acid (Section 2.8) with the resulting weakening of the cell wall and subsequent rupture of the cell. Certain antibiotics such as penicillin specifically inhibit the synthesis of new cell walls in growing cells, and this leads to lysis and death to the cell. Since eucaryotic cells have entirely different cell walls or membranes when compared to procaryotic cells, penicillin has no effect on animal cells. Hence, this specificity leads to its great value in the treatment of infectious diseases caused by procaryotic cells, in particular, gram-positive organisms.

As suggested in Figure 8.1, gram-negative organisms have a somewhat more complex cell wall structure. Although little if any teichoic acid is found in these organisms and a thin strand of peptidoglycan similar in structure to those found in gram-positive cell walls is sandwiched between the cell membrane and the outer envelope, the major component of these organisms is a giant macropolymer called a lipopolysaccharide. Very complex in structure, its details are well defined, particularly in the lipopolysaccharide of Enterobacteriaceae. A generalized structure is presented in Figure 8.4. We do not, however, propose to discuss the details of its structure nor its biosynthesis here, because of the complexity of the subject. When lipopolysaccharides are released into the blood stream of an animal, they are very toxic, causing fever, hemorrhagic shock, and other tissue damage. They are, therefore, called endotoxins. The considerable

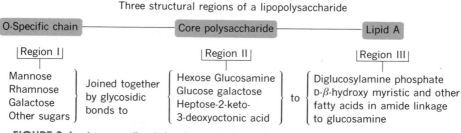

Three structural regions of a lipopolysaccharide

O-Specific chain		Core polysaccharide		Lipid A
|Region I|		|Region II|		|Region III|
Mannose Rhamnose Galactose Other sugars	Joined together by glycosidic bonds to	Hexose Glucosamine Glucose galactose Heptose-2-keto- 3-deoxyoctonic acid	to	Diglucosylamine phosphate D-β-hydroxy myristic and other fatty acids in amide linkage to glucosamine

FIGURE 8.4 A generalized structure of a lipopolysaccharide.

knowledge concerning this heteropolysaccharide can be found within the references at the end of this chapter.

The student should not consider that the bacterial cell wall is covered with a sheet of complex macromolecules. If this was the case, the organism would have some difficulty in obtaining metabolites for growth. The cell surface is actually punctured by a large number of pores through which biochemical compounds flow, but which prevent entry of very large molecules such as proteins or nucleic acids. One could think of a bacterial cell wall as a giant molecular sieve allowing small-molecular-weight compounds to pass through to the plasma membrane but retaining macromolecules. At the plasma membrane, a variety of transport mechanisms become operative (see Section 8.12).

8.1.2   PLANT CELL WALLS

In mature plant cells, the cell wall is composed of three distinct parts: the intercellular substance or middle lamella, the primary wall, and the secondary wall. The middle lamella is composed primarily of pectin polymers and may also be lignified. The primary wall consists of cellulose, hemicellulose (xylans, mannans, galactans, glucans, etc.), and pectins, as well as lignin. The secondary wall, which is laid down last, contains mostly cellulose, with smaller amounts of hemicellulose and lignin (Section 2.7.2).

A large number of openings called pits occur in various arrays and shapes in the secondary wall. Connecting adjacent cells are threadlike structures called **plasmodesmata,** which penetrate through the pits and the primary wall and middle lamella to the neighboring cell. It is thought that the endoplasmic reticulum of the cell extends through the plasmodesmata into the neighboring cell, thereby permitting a flow of metabolites and hormones from one cell to the next.

Much work has been expended to understand the structural role of cellulose in the plant cell walls. There is considerable agreement that cellulose forms microfibrils consisting of about 2000 cellulose molecules in cross section. These are arranged in orderly three-dimensional lattices around the cell, particularly in the secondary cell wall, to give great strength as well as plasticity to the wall.

There is now reasonably good evidence that the Golgi apparatus in the cytosol participates in the formation of the middle lamella and the adjacent primary cell walls as a plant cell divides during mitosis. This organelle, rich in enzymes for phospholipid and polysaccharide synthesis, releases small vesicles that line up and fuse in a linear fashion to form first a gellike matrix, which then develops into the middle lamella with the deposition of hemicelluloses and pectins.

It should be noted that a plant cell wall has associated with it a significant number of hydrolases including invertase, phosphatases, nucleases, and peroxidases. The significance of these hydrolases in the cell wall is not clear.

8.1.3   ANIMAL CELL SURFACES

The generalized animal cell has no rigid cell wall. However, (see Figure 8.1) associated with its plasma membrane are binding proteins, glycolipids, glycoproteins, enzymes, hormone receptor sites, and antigens that confer on the cell

TABLE 8.1 Chemical Composition of Plasma and Organelle Membranes

MEMBRANE	PROTEIN %	LIPID %	CARBOHYDRATE %	PROTEIN/LIPID RATIO
Plasma membranes				
Mouse live cell	44	53	3	0.85
Human erythrocyte	49	43	8	1.1
Amoeba	54	42	4	1.3
Gram-positive bacteria	70	20	10	3.0
Mycoplasma	59	40	1	1.6
Organelle membranes				
Mitochondrial outer membrane (liver)	51	47	2	1.1
Mitochondrial inner membrane (liver)	76	23	1	3.2
Chloroplast lamellae (spinach)	67	28	5	2.3
Nuclear membranes (rat liver)	61	36	3	1.6

surface the unique properties characteristic of a given cell. Thus, the plasma membrane of an animal cell is not smooth in appearance but rather has a "fuzz" over its surface. More will be discussed later in this chapter on the functions of the animal cell surface.

8.2 PLASMA MEMBRANES

The semipermeable barrier between the internal and external environment of the cell is called the plasma membrane. By means of a limiting membrane, the cell organizes its internal environment for specific purposes and expends energy to maintain this environment despite changes constantly occurring externally. Since the cell may also be a component of a larger unit in multicellular organisms, intercellular coordinations or interactions are necessary.

Chemical analyses of a large number of cell membranes have consistently revealed the presence of proteins and an array of complex polar lipids (Table 8.1). Indeed, of the total phospholipid found in the bacterial cell, over 95% is associated with its plasma membrane. Most if not all the carbohydrate components are covalently associated with glycoproteins and glycolipids.

8.2.1 MEMBRANE LIPIDS

Membrane lipids comprise the matrix that give form and structure to membranes and in which membrane proteins are imbedded. All membranes contain amphipathic lipids (see Section 7.1.1) that include phospholipids and glycolipids (Table 8.2).

TABLE 8.2 Lipids of Procaryotic and Eucaryotic Plasma and Organelle Membranes

MEMBRANE	LIPID (% OF MEMBRANE COMPONENTS)	LIPID COMPOSITION (% OF TOTAL LIPID)				
		PHOSPHO-GLYCERIDES	GLYCOSYL DIGLYCERIDES	SPHINGO-LIPIDS	STEROID	OTHER[a]
Plasma membranes						
B. subtilis	18	74	16	0	0	10
Erythrocytes (human)	29	37	0	21	23	19
Liver (rat)	40	45	0	10	20	25
Organelle membranes						
Spinach chloroplast lamella	52	10	74	0	1	15[b]
Endoplasmic reticulum	25	72	0	14	9	5
Mitochondria (liver)	26	—	—	—	—	—
Outer membrane	—	96[c]	0	1	3	—
Inner membrane	—	97[d]	0	2	1	—

[a] Pigments or specialized lipids.
[b] Carotenoids, chlorophyll, and quinones.
[c] 3% of which is cardiolipin.
[d] 21% of which is cardiolipin.

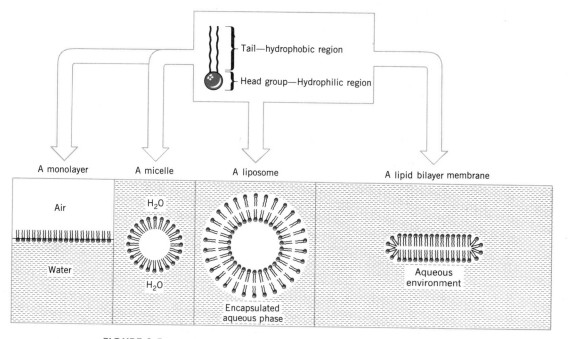

These lipids are characterized by having both hydrophobic (lipophilic) and hydrophilic (lipophobic) functions.

Why are these properties so important? Because of the physical properties of membrane lipids, these lipids are essentially insoluble in an aqueous system. When added to water, monomers of these lipids tend to arrange themselves as a monolayer in the interphase of an air–water system with the hydrophilic or head group penetrating into the water phase and the hydrophobic group or tail group protruding into the air phase (see Figure 8.5). When dispersed in water by sonication or other means, micelles are formed with structures of variable size. Micelles are highly stable and water soluble. In their newly ordered arrangement, the hydrophobic functions of the amphipathic lipids, namely the hydro-

FIGURE 8.5   Various structures of amphipathic lipids in an aqueous environment.

carbon chains, are arranged internally to exclude water and thus are held together by hydrophobic interaction forces. The hydrophilic functions, namely the phosphoryl base moieties of phospholipids, are in turn highly attracted to the aqueous environment. More complicated structures are liposomes and lipid bilayers.

Membranes undergo a physical phase transition from a flexible fluidlike liquid crystalline state to a solid gel structure as a function of temperature. The temperatures at which the phase transition occurs are dependent on the composition of the amphipathic lipids. Thus, lipids with more unsaturated fatty acids have lower transition temperatures than those with more saturated fatty acids; longer chain lengths have higher transition temperatures than shorter chain lengths; *cis* unsaturated fatty acids have lower transition temperatures than *trans* unsaturated fatty acids.

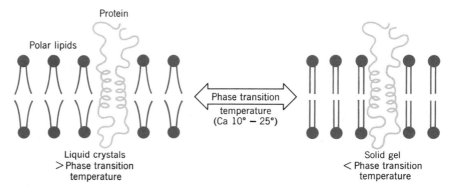

What is the significance of a thermal phase transition in membranes? Obviously, homeothermic animals, by controlling their internal temperature, do not expose their membrane systems to marked changes in temperature. However, poikilothermic organisms—these include a large array of cold-blooded vertebrates such as fish as well as plants and lower organisms—are exposed to marked shifts in temperature. Undoubtedly, all membrane-bound enzymes, transport processes, receptor sites, and the like associated with membranes are surrounded by a lipid milieu; therefore, their activities would be markedly altered by the physical state of the membrane lipids, which, in turn, is in part a reflection of the surrounding temperature. Thus, it can be shown that lipids of liver mitochondria from homeothermic animals have a higher proportion of saturated fatty acids than those obtained from poikilothermic organisms. These results correlate well with the presence or absence of thermal phase changes in the membrane lipids of these animals, respectively.

8.2.2   MEMBRANE PROTEINS

Superimposed onto bilayer lipid structures are the membrane proteins. These proteins are surrounded by phospholipids that are not as mobile as the bulk lipid of the membrane and are called **boundary lipids.** In general, there are two groups of proteins associated with membranes. One group, called peripheral or **extrinsic proteins,** is not imbedded in the hydrophobic core but is weakly bound by ionic bonds and divalent cation salt bridges to the lipid polar head group and/or intrinsic proteins and can be displaced by hypotonic solutions containing chelating reagents, mild detergents, or sonication. Examples include cytochrome c, which is loosely associated with the outer face of the inner membrane of mitochondria, and α-lactalbumin, which is loosely associated with the plasma membrane of mammary gland cells. Still another type of association is exemplified by cytochrome b_5, an electron carrier protein located in the cytoplasmic side of the endoplasmic reticulum of eucaryotic cells and intimately involved in the process of fatty acid desaturation. This protein has an exposed functional porphyrin system at one end of its polypeptide involved in electron transport, while the other end of the protein is tied onto the membrane lipids by virtue of its hydrophobic amino acid sequence. In addition, periplasmic binding proteins are classified as peripheral proteins of the outer face of the plasma membrane of bacteria.

The second class of proteins, called integral or **intrinsic proteins,** is tightly bound to the lipid bilayer by both ionic and hydrophobic forces. These penetrate deeply or transverse the membrane and can only be released by employing detergents or organic solvents. Because of their high hydrophobicity, these proteins tend to aggregate in aqueous solutions when the disrupting reagent such as a detergent is removed.

The concept of two classes of proteins buried in a lipid bilayer was proposed originally in 1972 by Singer and Nicolson. The current fluid-mosaic model is depicted in Figure 8.6. Intrinsic proteins include a large number of functional proteins that participate as transport carriers, drug and hormone receptor sites, antigens, and a large number of membrane-bound enzymes. For example, cytochrome P_{450} is classified as an integral protein of the endoplasmic reticulum of eucaryotic cells as is the NAD-cytochrome b_5-reductase, which is tightly coupled to the hemeprotein, and cytochrome oxidase, which is imbedded in the inner membrane of the mitochondrion. Other examples will be mentioned elsewhere in the text.

The fluid mosaic model illustrates the position of extrinsic and intrinsic as well as the amphipathic lipids in membranes and is quite compatible with most of the available experimental data. The model is easily adapted to plasma membranes and membranes of different organelles of eucaryotic cells and serves well as a working model in attempting to understand the function of membrane-localized enzymes.

A final comment that can be drawn from the current picture of plasma and organelle membrane structures is that membranes are asymmetric. That is, because of the two classes of proteins, extrinsic and intrinsic, the outer and inner faces of membranes may have markedly different physical, structural, and biochemical properties. As we will see, for example, the inner membrane of mitochondria is markedly asymmetrical for a number of reasons involving the transport of ions, the electron transport chain, and oxidative phosphorylation.

Moreover, the distribution of lipids in membranes is also asymmetric. For example, the outer monolayer of the erythrocyte is made up of phosphatidyl choline and sphingomyelin, while the inner layer consists of phosphatidyl ethanolamine, phosphatidyl serine, and phosphatidyl inositol. Finally, membrane lipids diffuse laterally very rapidly but obviously do not move from the outer to the inner monolayer and conversely.

Furthermore, with many membranes, short, branched chains of sugars are found associated with both external or cytoplasmic-oriented extrinsic proteins and the cytoplasmic external portions of the intrinsic proteins. These oligosaccharide chains are not, however, associated with the proteins that make up the

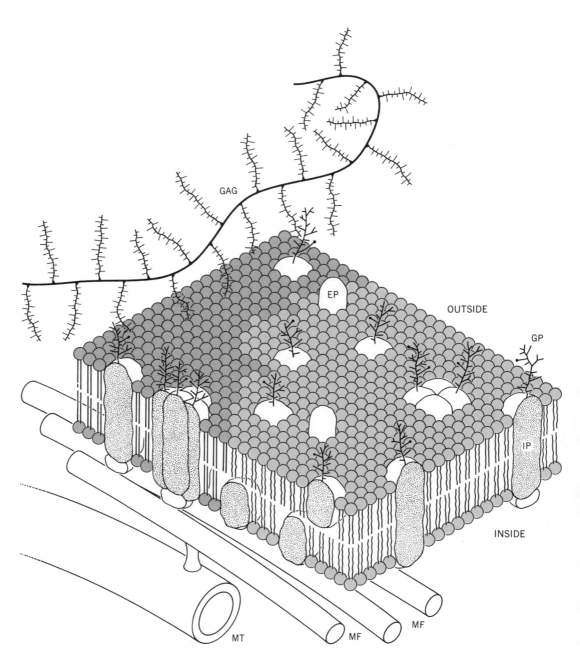

FIGURE 8.6   The fluid mosaic model of Singer and Nicolson. GAG, glycosaminoglycans; GP, glycoproteins; EP, extrinsic proteins; IP, intrinsic proteins; MF, microfilaments. Reproduced with permission from G. W. Nicholson in *Membranes and Transport*, Vol. II, p. 482 (A. N. Martonosi, ed.) Plenum (1982).

inner surface of the plasma membrane. The function of these sugar residues may include the specification of receptor sites for a number of hormones that are glycoproteins, serve to define a cell type specificity such as human blood types, and possibly define the adhesion of like cells to each other in a tissue.

8.3   NUCLEUS

Although in procaryotic cells, no nucleus per se is observed, a fibrillar area can be detected on the interior side of the plasma membrane that is associated with an extremely involuted double-stranded circle of DNA. It has been estimated that in a single bacterial cell 2 μm long, its DNA, if stretched out as a single fiber, would be over 1000 μm long, 500 times the length of its own cell body. Although histones are absent in these cells, histonelike proteins have been recently observed. Also, high concentrations of polyamines such as spermidine, cadaverin, and putrescine have been detected in the bacterial cell, and these compounds may participate in neutralizing negative charges on procaryotic DNA (Section 6.10).

In eucaryotic cells, the nucleus is a large dense body surrounded by a double membrane with numerous pores that permit passage of the products of nuclear biosynthesis into the surrounding cytoplasm. Internally, the nucleus contains chromatin or expanded chromosomes composed of DNA fibers closely associated with histones (Chapters 6 and 19). During nuclear division, the chromosomes contract and become clearly visible in the light microscope as the DNA chains undergo their programmed changes. In addition, the nucleoplasm contains enzymes such as DNA polymerases and RNA polymerase(s) (Chapter 19) for mRNA and tRNA synthesis; and, surprisingly, the enzymes of the glycolytic sequence, citric acid cycle, and the pentose phosphate pathway (Chapters 10–12) have been detected in the nucleoplasm. One to three spherical structures called the nucleolus are closely associated with the inner nuclear envelope and are presumably the sites of rRNA biosynthesis (Chapter 19). This dense suborganelle is nonmembraneous and contains RNA polymerase, RNAase, NADP pyrophosphorylase, ATPase, and *S*-adenosylmethionine-RNA-methyltransferase, but no DNA polymerase. Ribosomal RNAs are separately synthesized in the nucleolus and are then transported to the cytoplasm as discrete units to be assembled in the cytoplasm to form polysomes. We shall discuss the function of these nucleic acids in Chapters 19 and 20.

8.4   ENDOPLASMIC RETICULUM

The number of membrane-bound channels and vesicles called the endoplasmic reticulum is missing in procaryotic cells. However, this system is present in all eucaryotic cells (Figures 8.2*c* and 8.7).

Varying in size, shape, and amount, the endoplasmic reticulum extends from the cell membrane, coats the nucleus, surrounds the mitochondria, and appears to connect directly to the Golgi apparatus. There are two kinds of endoplasmic reticulum—the rough-surfaced type known as **ergastoplasm,** which has ribosomes associated with it externally, and the smooth type, which lacks ribosomes. When cells are disrupted by homogenization and fractionated by differential centrifugation, the pellet, which contains that material which sediments on centrifugation at 100,000 $\times$ *g* for 30 minutes, is called the microsomal fraction (microsomes) and contains small vesicles and fragments derived chiefly from the endoplasmic reticulum. A number of important enzymes are

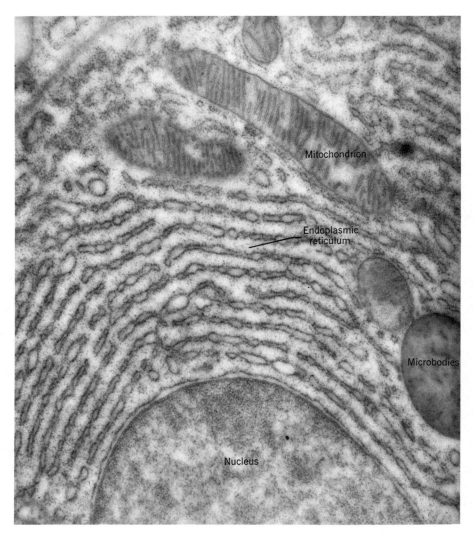

FIGURE 8.7   Electron micrograph of a cross section of an exocrine cell of the guinea pig pancreas showing the common organelle structures of this eucaryotic cell. (*Source:* Courtesy of G. E. Palade, Yale University Medical School.)

associated with the endoplasmic reticulum of mammalian liver cells. These include the enzymes responsible for the synthesis of sterols, triacylglycerols, and phospholipids; the detoxification of drugs by modification through methylation, hydroxylation, and so on, the desaturation and elongation of fatty acids; and the hydrolysis of glucose-6-phosphate. As a word of caution, however, a number of these activities are only associated with microsomes of liver cells and frequently are missing in microsomes from other eucaryotic tissues. Although cytochromes characteristic of mitochondria are absent, both cytochrome b_5, which serves as a limited electron carrier system in the desaturation of fatty acids, and cytochrome P_{450}, which participates in hydroxylation reactions in animal and plant cells, reside in the endoplasmic reticulum.

8.4.1 RIBOSOMES

The structure of ribosomes was discussed in Section 6.10.2. In procaryotic cells, ribosomes are grouped in clusters 10 to 20 nm in diameter, held together by mRNA to form polysomes. Because of the intense synthesis of proteins by growing bacterial cells, their cell matrix (sap) contains very many of these clusters.

Table 6.3 in Chapter 6 summarizes the information concerning both procaryotic and eucaryotic ribosomes (see also Chapters 19 and 20). In eucaryotic organisms, ribosomes are associated closely with the endoplasmic reticulum, thereby forming rough surface endoplasmic reticulum. Protein synthesis occurs on the endoplasmic reticulum. Newly formed proteins are secreted into the vesicular system and then transferred to Golgi bodies to be used there in the formation of lysosomes and other microbodies (see Chapter 20 for an illustration of the role of this system in insulin biosynthesis, as well as Sections 8.8, 8.9, 8.10, and 8.13 of this chapter).

As indicated in Section 6.10.2, the ribosomes found in mitochondria and chloroplasts closely resemble the bacterial ribosomes in size as well as sensitivity to protein inhibitors such as chloramphenicol. We shall discuss in considerable detail the functions of ribosomes in protein synthesis in Chapter 20.

8.5 MITOCHONDRIA

Since the nineteenth century, microscopists have observed in all eucaryotic cells small, rod-shaped particles 2 to 3 μm long, which were called mitochondria. In 1948, A. L. Lehninger showed that in the animal cell, the mitochondrion was the sole site for oxidative phosphorylation, the tricarboxylic acid cycle, and fatty acid oxidation. Because of the importance of these systems in the total economy of the cell, much research has been expended in defining the structure and function of these bodies that are found universally in eucaryotic cells but are totally missing in procaryotic cells.

Although procaryotic cells have no mitochondrial bodies, their plasma membrane appears to be the site of electron transport and oxidative phosphorylation. Thus, all the cytochrome pigments and a number of dehydrogenases associated with the tricarboxylic acid cycle, namely succinic, malic, and α-ketoglutaric dehydrogenase, are localized in the bacterial plasma membrane. In addition, enzymes involved in phospholipid biosynthesis and cell wall biosynthesis are also found in or on this membrane structure.

All mitochondria consist of a double membrane system. An outer membrane is separated from but envelops an inner membrane, which by invagination extends into the matrix of the organelle as cristae (see Figures 8.7 and 8.8). Considerable evidence suggests that all the enzymes of the electron transport system, namely the flavoproteins, succinic dehydrogenase, cytochromes b, c, c_1, a, and a_3, are buried in the inner membrane. In addition, the inner surface of the inner membrane has projecting into the matrix a cluster of knobs called inner-membrane particles. These structures (85 A diameter) include the coupling factor F_1 that has ATPase activity, a molecular weight of 280,000, and a hydrophobic channel called F_0 that transverses the inner membrane and con-

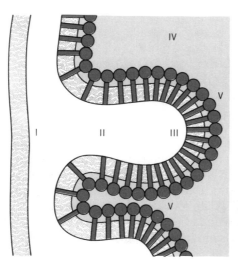

FIGURE 8.8   Cross section of a mitochondria showing: I, outer membrane; II, intermembrane space; III, inner membrane; IV, matrix; and V, inner membrane particles.

nects to the F_1 factor (see Chapter 14 for more details). This ATPase participates in the final step of oxidative phosphorylation by catalyzing the following reaction:

$$ADP + Pi \rightleftharpoons ATP + H_2O$$

The outer membrane is fully permeable to a large number of compounds with molecular weights up to 10,000, whereas the inner membrane of mitochondria possesses limited permeability. The outer membrane has a density of 1.13, and the inner membrane has a density of 1.21. These physical properties are exploited in separating these two membranes for further studies. The outer membrane has about three times more phospholipid than the inner membrane; hence, its slightly lower density. Phosphatidyl inositol is found exclusively in the outer membrane, whereas cardiolipin occurs almost exclusively in the inner membrane. These phospholipids therefore become useful markers for the identification of mitochondrial membrane fractions. Of the total protein in mitochondria, 4% is associated with the outer membrane, 21% with the inner membrane, and 67% with the matrix.

In the matrix, it has been calculated that the protein concentration may be as high as 60% (on a weight basis), with water and metabolite making up the remaining 40%.

Much effort has been expended in localizing the large number of enzymes associated with mitochondria. Table 8.3 lists the location of a number of enzymes in liver mitochondria and also identifies the so-called marker enzymes —that is, enzymes that, because of their exclusive location at a specific site in an organelle, are employed by biochemists to identify an experimentally obtained fraction as associated with that site. Thus, for example, a cell fraction with a high activity for succinic dehydrogenase but no activity for monoamino oxidase would allow the biochemist to conclude that the fraction was enriched with inner-membrane fragments of mitochondria with no contamination of outer-membrane fragments.

The matrix enzymes are present in high concentrations. It has been estimated that the level of important enzymes in the matrix may be as high as

TABLE 8.3 Localization of Some Liver Mitochondrial Enzymes

OUTER MEMBRANE	INTERMEMBRANE SPACE	INNER MEMBRANE	MATRIX
Rotenone–insensitive NADH–Cyt-b_5-reductase	Adenylate kinase	Cytochrome b, c, c_1, a, a_3	Malic dehydrogenase
	Nucleoside diphosphokinase	β-Hydroxybutyrate dehydrogenase	Isocitric dehydrogenase
Monoamine oxidase		Ferrochelatase	Glutamic dehydrogenase
Kynurenine hydroxylase		δ-Amino levulinic synthetase	Glutamic-aspartic transaminase
ATP-dependent fatty acyl CoA synthetase		Carnitine palmityl transferase	Citrate synthase
Glycerophosphate acyl transferase		—	Aconitase
			Fumarase
Lysophosphatidate acyl transferase		Fatty acid elongation enzymes (10)	Pyruvic carboxylase
Lysolecithin acyl transferase		Respiratory chain-linked phosphorylation enzymes	Protein synthesis enzymes
Phosphocholine transferase		Succinic dehydrogenase	Fatty acyl CoA dehydrogenase
Phosphatidate phosphatase		Cytochrome a_3 oxidase	Nucleic acid polymerases
Nucleoside diphosphokinase		Mitochondrial DNA polymerase	ATP-dependent fatty acyl CoA synthetase
Fatty acid elongating system C_{14}-C_{16}		—	GTP-dependent fatty acyl CoA synthetase

Shaded boxes indicate marker enzymes.

10^{-6} M. Since it has also been calculated that a number of substrate concentrations are approximately 10^{-6} M, the kinetic data obtained *in vitro* when enzymes are measured at 10^{-9} M and substrates at 10^{-6} M may be very different from that actually occurring in the mitochondrial matrix. Moreover with such a high concentration of soluble proteins, the matrix is probably a gel with a physical environment markedly different from that used in determining enzyme kinetics—that is, dilute, aqueous, buffered solutions with very low protein concentrations.

In addition to a large number of soluble enzymes, the mitochondrial matrix contains mitochondrial DNA, a circular double-stranded molecule somewhat smaller but very similar in shape to bacterial DNA. In bacteria, mitochondria, and chloroplasts, the DNA is histone-free and bound to membranes. Each mitochondrion has from two to six DNA circles amounting to about 0.2 to 1 μg of DNA/mg mitochondrial protein. This amount of DNA can code for about 70 polypeptide chains of 17,000 mol wt. Since DNA polymerase is found in the inner membrane and the matrix, presumably mitochondrial DNA is independently synthesized in the mitochondria. Replication is semiconservative (see Chapter 19). Curiously, mitochondrial DNA in animal cells has about 15×10^3 base pairs, but in plants the mitochondrial DNA is much larger and variable with about 100×10^3 base pairs. An explanation for these differences is not known. Of further interest is the observation that 70S ribosome particles are in the mitochondrial matrix as well as tRNA, mRNA, and protein-synthesizing enzymes that catalyze limited protein synthesis. A DNA-dependent RNA polymerase has also been detected in the matrix. This complete machinery for synthesizing proteins is concerned with the formation of about 10% of mitochondrial proteins. The remaining 90% is encoded by nuclear genes, translated by cytoplasmic polysomes and translocated into the various components of the mitochondrion (the translocation process is discussed in Section 8.13 of this chapter).

A number of biochemists have submitted the provocative speculation that both mitochondria and chloroplasts resemble very closely procaryotic cells with respect to size, distribution of respiratory enzymes, and the striking similarity of their DNA and RNA components. Perhaps, both mitochondria and chloroplasts originated from procaryotic endosymbionts that, over a long evolutionary period, were gradually integrated into their host.

8.6   CHLOROPLASTS

All eucaryotic organisms with photosynthetic capabilities have chlorophyll-containing organelles called chloroplasts. Only the structure of higher plant chloroplasts will be considered here, although lower plants possess these organelles in varying size, shape, and number. In a green leaf, a single palisade cell contains approximately 40 chloroplasts. Each chloroplast is about 5 to 10 μm in diameter and about 2 to 3 μm thick. About 50% of the dry weight of the chloroplast is protein, 40% lipid, and the remainder water-soluble small molecules. The lipid fraction consists of about 23% chlorophyll ($a + b$), 5% carotenoids, 5% plastoquinone, 11% phospholipid, 15% digalactosyl diglyceride, 36% monogalactosyl diglyceride, and 5% sulfolipid.

Chloroplasts have an envelope consisting of an outer and inner membrane. Internally are found a large number of closely packed membraneous structures called lamellae or thylakoids that contain the chlorophyll of the organelle. In one kind of chloroplast (C_3 and C_4 mesophyll cells), the lamellae are arranged as closely packed disks or stacks called grana (Figure 8.9a) that are interconnected to each other by intergrana or stroma lamellae. The grana stacks are the sites of oxygen evolution and photosynthetic phosphorylation. In other chloroplasts (C_4 bundle sheath cells), the chlorophyll-containing lamellae are not arranged in stacks but rather extend the length of the organelle (Figure 8.9b). The definition and characteristics of C_3 and C_4 plants are discussed in detail in Chapter 15.

The matrix embedding the lamellae is called the stroma and is the site of the carbon photosynthetic enzymes involved in CO_2 fixation, ribosomes, nucleic acid-synthesizing enzymes, and fatty acid synthesizing-enzymes (see Chapter 15 for more detail).

Chloroplasts are extremely fragile structures; only a brief exposure to distilled water will result in a bursting of the outer envelopes, loss of stroma protein, and marked changes in the appearance of the lamellar systems.

Chloroplasts contain circular chloroplast DNA. Ribosomes are of the 70S species and are very similar to those observed in mitochondria and bacteria. A DNA-dependent RNA polymerase also occurs in intact chloroplasts.

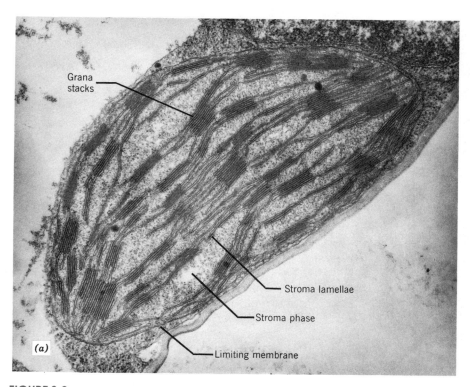

FIGURE 8.9 (a) Chloroplast of sugar cane mesophyll cell showing the grana stacks and the stroma lamellae (magnified 45,900 X). (b) Chloroplast of sugar cane bundle sheath cell showing the extended stroma lamellar system with an absence of grana stacks (magnified 35,500 X). White bodies are starch granules. (Source: With permission of W. M. Laetsch, University of California, Berkeley.)

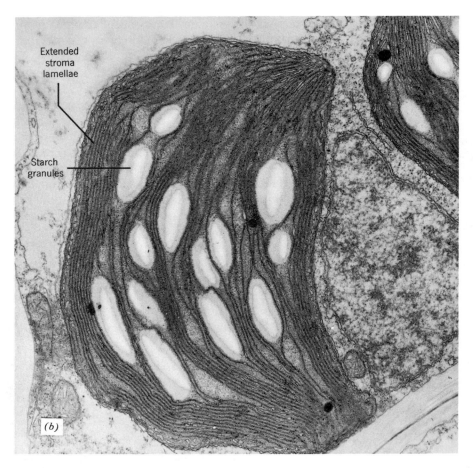

Extended
stroma
lamellae

Starch
granules

(b)

FIGURE 8.9 (continued)

As with mitochondrial protein synthesis, chloroplasts synthesize only 10 to 20% of their total proteins; the remaining proteins are nuclear encoded, translated in the cytoplasm, and translocated into the various sites of the chloroplast.

In photosynthesizing procaryotic cells such as *Rodospirillum rubrum,* small particles, about 60 nm in diameter, are attached to the inner surface of the cell membrane. These particles are called chromatopores; they have no limiting membrane and possess all the bacteriochlorophyll. They are therefore sites of bacterial photosynthesis. In the procaryotic cyanobacteria, no discrete chloroplasts are visible, but the photosynthetic lamellar membrane occupies most of the cell. These procaryotic cells therefore seem to be more advanced than the chromatophore-containing bacteria but less developed than eucaryotes such as the green algae that have membrane-limiting chloroplasts.

8.7   LYSOSOMES

In 1955, the Belgian biochemist de Duve discovered and described for the first time a new organelle, the lysosome. Found in all animal cells except erythrocytes in varying numbers and types, the lysosome, in general, is a rather large organelle consisting of a unit membrane enclosing a matrix containing about 30 to 40 hydrolytic enzymes. These enzymes are characterized by having acid pH optima; acid phosphatase is used as a marker enzyme for this organelle.

ENZYME GROUPS IN LYSOSOMES

Ribonucleases	Cathepsins (proteinase)
Deoxyribonucleases	Acid glycosidases
Acid phosphatases	Sulfatases
Lipases	Phospholipases

Collectively, the lysosomal enzymes act on a number of biopolymers. Thus, the proteases have a wide capacity for the hydrolysis of proteins, the acid nucleases for RNA and DNA, and the acid glycosidases for polysaccharides. A family of acid phosphatases also are present. The median value for the pH optima of these enzymes is around pH 5. Obviously, the lysosomal matrix must be acidic for the enzyms to be reactive. It is attractive to consider the lysosomal membrane that has a high specific activity for NADH dehydrogenase as serving as a hydrogen ion pump. All the enzymes, other than the esterases and the NADH dehydrogenase, are present as soluble proteins in the matrix of the lysosome.

In autophagic processes, cellular organelles such as mitochondria and the endoplasmic reticulum undergo digestion within the lysosome. The enzymes are active at postmortem autolysis. In the death of a cell, lysosomal bodies disintegrate, releasing hydrolytic enzymes into the cytoplasm with the result that the cell undergoes autolysis. There is good evidence that in the metamorphosis of tadpoles to frogs, the regression of the tadpole's tail is accomplished by the lysosomal digestion of the tail cells. Bacteria are digested by white blood cells by engulfment and lysosomal action. The acrosome, located at the head of the sperm, is a specialized lysosome and is probably involved in some manner in the penetration of the ovum by the sperm. Finally, a number of hereditary diseases involving the abnormal accumulation of complex lipids or polysaccharides in cells of the afflicted individual have now been traced to the absence of key acid hydrolases in the lysosomes of these individuals.

8.8 GOLGI APPARATUS (DICTYOSOMES)

Each eucaryotic cell contains a unique stack of smooth-surfaced compartments (or cisternae) that make up the Golgi complex. Recent evidence strongly suggests that this complex serves as a unique sorting device that receives newly synthesized proteins [all containing signal (or transit) peptides] from the endoplasmic reticuli in the **proximal** (*n cis*) compartment. There, posttranslational modifications begin to take place such as lipid additions or phosphorylations. Further changes occur as the proteins proceed through the central *n* **medial** compartment and then to the **distal** (*n trans*) compartment where terminal sugars are added and final sorting takes place. The fully modified proteins now leave the Golgi apparatus to be delivered either to lysosomes, plasma membrane, or are secreted such as insulin in the highly specialized pancreatic Isle of Langerhans cells (see Chapter 20.12). Interestingly, those proteins with no signal or transit peptide regions are rejected by the Golgi apparatus and presumably remain cytoplasmic proteins.

8.9   COATED VESICLES AND ENDOSOMES

All eucaryotic cells have transient structures that are involved in the transport of macromolecules from the exterior of the cell to its interior. These macromolecules (or ligands) include light density lipoproteins, iron-complexed transferin hormones, immunoglobins, and even viruses. The structures have been identified as **coated vesicles,** endosomes (also called receptosomes) and pinosomes.

Approximately 2% of the exterior surface of plasma membrane of animal cells are covered with characteristic areas called coated pits. In addition, cell surfaces are rich in receptor proteins that combine with the wide variety of macromolecules, also called ligands. Once the ligands combine with the receptors, the ligand-bound receptors move laterally into coated pits. There, the loaded pits are rapidly pinched off and are internalized as **coated vesicles.** Coated vesicles can be readily detected because they have a very characteristic bristle coat on their outer surface. Approximately 100 nm in diameter, vesicles are covered with an unusual protein, clathrin, with a molecular weight of 185,000 that is distributed in a polyhedral lattice arrangement.

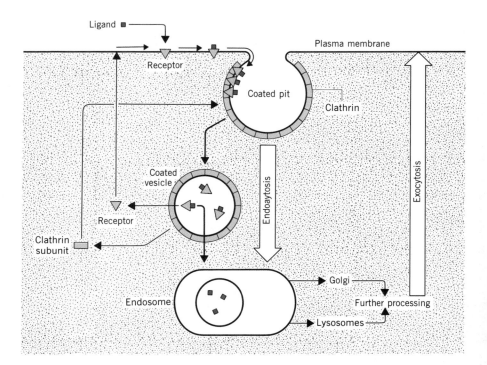

FIGURE 8.10   The endocytic transport of macromolecules such as proteins and viruses from outside the cell to its interior by the coated vesicle-endosome system.

Near the periphery of the cell's interior, another structure called an **endosome** (also named a receptosome) can be observed. These bodies have a heterogenous morphology with a diameter of 0.3 to 1 μm, do not contain hydrolytic enzymes, are less dense than lysosomes, and have an interior pH of 5.0.

There is good evidence that the internalized coated vesicles fuse with the endosomes and discharge their ligands into the interior of the endosomes (where the low pH disrupts the ligand-receptor linkage) with a simultaneous release of clathrin, ligand-free receptors, and membrane fragments, most of which recycle back to the plasma membrane to replenish the population of receptors and coated pits. The ligand-containing endosomes now move, apparently by microtubule guidance, to the interior of the cell where they may fuse with lysosomes or become associated with vesicles derived from the Golgi bodies. The system is depicted in Figure 8.10.

Still another structure called a pinosome can be observed in the animal cell. These structures are involved in nonreceptor specific endocytosis and also fuse with lysosomes. They have a clear interior matrix and are quite distinct from coated vesicles and endosomes, but their function is not too well defined.

Much research still remains to answer a number of puzzling questions concerning the biochemistry of the coated vesicles-endosome-Golgi body system, especially with regard to specificity, energy requirements for fusion, recycling and internalization of important proteins of a wide variety, as well as viruses, and so on. A very important physiological function of endocytosis is the control of cholesterol levels in humans by the LDL (light density lipoprotein) receptor pathway. This pathway is discussed in Section 13.16.

8.10 MICROBODIES

This term covers a number of single membrane-enclosed cytoplasmic organelles that have been found in both animal and plant cells and that are characterized by containing H_2O_2-producing oxidases and catalase. The particles are approximately 0.5 μm in diameter. These subcellular respiratory organelles have no energy-coupled electron transport systems and are probably formed by budding from smooth endoplasmic reticuli.

In animal cells, microbodies are known as peroxisomes. In the liver cell, mitochondria are about two to five times more numerous than peroxisomes, with lysosomes being about one and a half times less numerous than peroxisomes.

Recently, it has been shown that liver peroxisomes have an unusually active β-oxidation system capable of β-oxidizing long-chain fatty acids (C_{16}, C_{18}). Shorter chain fatty acids are not oxidized by peroxisomes but are rapidly utilized by mitochondria. The β-oxidation enzymes of peroxisomes are rather unique in that the first step of β-oxidation is catalyzed by a flavoprotein, an acyl CoA oxidase.

$$\text{acyl CoA} + O_2 \longrightarrow \alpha, \beta\text{-unsaturated acyl CoA} + H_2O_2$$

The product of the oxidase is converted to β-OH-acyl CoA by an enoyl hydratase and is further oxidized by β-OH-acyl CoA dehydrogenase as discussed in Section 13.5.1. Interestingly, these last two enzymes are associated with a single bifunctional protein having these two activities in sharp contrast to the mitochondrial enzymes that are separate proteins. The thiolase is identical to that found in mitochondria. The role of catalase in peroxisomes can now be explained because the hydrogen peroxide generated by the acyl CoA oxidase must be rapidly removed.

The peroxisomal β-oxidation system is inducible (modified by diet), whereas the mitochondrial system is constitutive (not modified by diet). Thus, the peroxisomal system responds to diets as well as hypolipidemic drugs such as clofibrate. Which β-oxidation system contributes to the total picture of fatty acids degradation in the cell is difficult to ascertain, but certainly, both together make possible the smooth breakdown of fatty acids in the liver cell. Note that in peroxisomes, there is no coupling of the energy released by β-oxidation to oxidative phosphorylation, whereas in mitochondria, the electrons derived from β-oxidation flow into the oxidative phosphorylation system (see Chapter 13).

In plants, peroxisomes play important roles in both catabolic and anabolic pathways. In seeds rich in lipids such as castor bean and soyabeans, microbodies called glyoxysomes are the sites for the breakdown of fatty acids to succinate (see Chapter 12 for a description of the glyoxylate bypass). Glyoxysomes contain citrate synthase, catalase, transaminases, uricase, and β-oxidation enzymes whose biochemistry is identical to that found in liver peroxisomes—namely, an acyl CoA oxidase, the bifunctional protein with two activities, that is, enoyl CoA hydratase and β-hydroxy-acyl CoA dehydrogenase and a thiolase. This organelle, which is present only during a short period in the germination of the lipid-rich seed and is absent in lipid-poor seed such as the pea, is an example of a highly specialized organelle, beautifully designed to convert fatty acids to C_4 acids that can then be further converted to sucrose (see Chapter 12 for a discussion of the conversion of succinate to sucrose). Because carbohydrate-rich seeds do not draw on storage lipids for carbon skeletons or energy, this organelle is appropriately not present in these seeds. Since the glyoxysome contains, in addition to the glyoxylate bypass enzymes, catalase and a H_2O_2-generating glycolic oxidase, it could be considered a highly specialized peroxisome. It should again be emphasized that plant mitochondria lack all of the β-oxidation enzymes.

In leaf tissues, peroxisomes serve as sites of photorespiration in the leaf cell. This process involves the oxidation of glycolic acid (a product of photosynthetic CO_2 fixation) to CO_2 and H_2O_2. They have a single membrane and granular matrix without lamellae that house catalase; glycolic oxidase; isozymes of malate dehydrogenase; NADP isocitrate dehydrogenase; and the transaminases, glyoxylate : glutamate, hydroxypyruvate : serine, and oxaloacetate : glutamate. They are approximately 1 μm in width, and they number from a few to one-seventh as many as mitochondria in the leaf cell (again, see Chapter 15 for a discussion of photorespiration).

In summary, the microbodies discussed in this section have in common very high catalase activity and one or more H_2O_2-generating oxidases spatially separated from other important sites of metabolism. They possess no respiratory chain system or energy-conserving system such as occurs in mitochondria. While many hypotheses can be raised concerning their function, future experimentation should allow an assessment of their importance in the total economy of the cell.

8.11 CYTOSKELETON

For many years, biochemists have considered the cytosol a compartment containing soluble enzymes, metabolites, and salts in an aqueous but gellike environment. Much evidence now supports the idea that this compartment contains actually a complex network of fine structures called microtubules, microfilaments, and microtrabaculae.

Microtubules are long, unbranched, slender cylindrical structures with an average diameter of about 25 nm. The structures are made primarily by the self-assembly of the heterodimer, tubulin (molecular weight 50,000). Colchicine inhibits the assembly process by combining with tubulin and is thus employed in investigations pertaining to the functions of this structure. One very important function of microtubules is their role in the assembly and disassembly of the spindle structures during mitosis. They also provide internal structure to the cell and since they seem to associate with the inner face of plasma membranes, they may be involved in transmembrane signals.

Microfilaments are considerably more slender cylinders made up of the contractile protein, actin, and linked to the inner face of the plasma membrane; these structures may be involved in the generation of forces for internal cell motion.

Microtrabaculae appear to be very fragile tubes that form a transient network in the cytosol. Whether or not soluble enzymes are clustered or associated with these structures to form unstable multienzyme complexes remains for future research to determine.

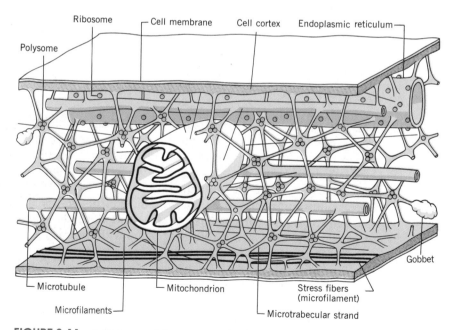

FIGURE 8.11 A diagram of the proposed cytoskeleton structure of a cell showing the various components of the extensive network in the cytosol. Reproduced with permission from "The Ground Substance of the Living Cell" by K. R. Porter and J. B. Tucker, p. 59. Copyright © March, 1981 by Scientific American, Inc. All rights reserved.

In summary, the student should think of the cytosolic compartment as a viscous gel in which various linear filaments, cylinders, and tubes form a complex network, the interaction of which, with all the other organelles in the cell, will provide worthy challenges for biochemists to explore and define. Figure 8.11 summarizes the interrelation of the cytoskeleton structures in the cytosol.

8.12   TRANSPORT PROCESSES

An essential role of biomembranes is to allow movement of all compounds necessary for the normal function of a cell across the membranal barrier. These compounds include a vast array of sugars, amino acids, steroids, fatty acids, anions, and cations to mention a few; these compounds must enter or leave the cell in an orderly manner. Only recently has much progress been made in understanding the biochemistry of transport processes.

There are at least four general mechanisms by which metabolites (solutes) can pass through biomembranes.

8.12.1   PASSIVE DIFFUSION

A few metabolites of low molecular weight are presumed to move or diffuse across the membrane. The rate of flow is directly proportional to the concentration gradient across the membrane. When the concentration gradient ceases to exist, no further flow occurs. Active transport inhibitors such as cyanide or azide do not affect the process. Although passive diffusion was believed to be an important transport mechanism in cells, current information suggests that this process is very limited. Water would be an example of a simple compound passing through a membrane by passive diffusion, but essentially all metabolites move through membranes by more sophisticated and hence better regulated processes.

8.12.2   FACILITATED DIFFUSION

This type of diffusion is somewhat similar to simple diffusion in that a concentration gradient is required and the process does not involve an expenditure of energy. However, it differs in several important respects from passive diffusion. First, the membrane contains a large number of specific components called carriers that facilitate the individual transport processes; that is, speed up the rate of diffusion much more than is predicted from simple diffusion. Second, the diffusion is mechanism-stereospecific. And, third, the rate of penetration of the metabolite approaches a limiting value with increasing concentration on one side of the membrane. The rate of the process is also temperature-sensitive. The kinetics mimic simple Michaelis–Menten enzyme kinetics (Chapter 4), that is, the system can be saturated. Thus, K_m and V_{max} values are easily measured and characterize the carrier systems.

The general mechanism can be explained by a specific carrier molecule present in the membrane, which forms a specific complex with the metabolite to be transported at the outer area of the membrane. The complex then, by diffusion, rotation, oscillation, or some other motion, can translocate the metabolite to the inward area of the membrane, where it could dissociate to discharge the metabolite. A suggested mechanism is depicted in Figure 8.12.

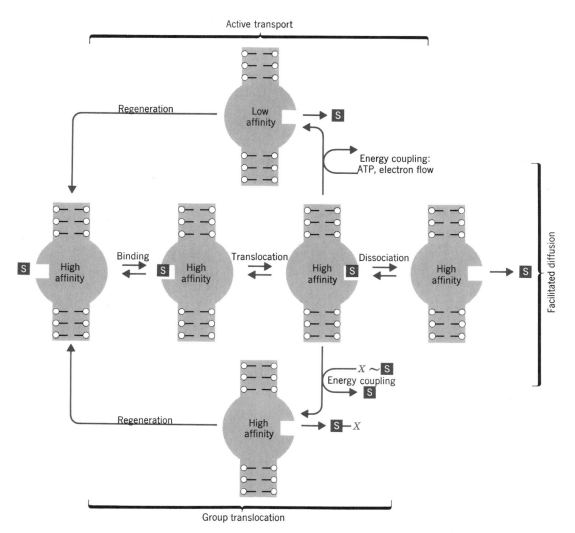

FIGURE 8.12   Different general models for facilitated diffusion, group translocation, and active transport.

Recent evidence has revealed over a 100 small molecular weight proteins localized in the outer surface of plasma membranes in gram-negative organisms that fulfill the requirements of facilitated diffusion. In the transport of sulfate by *Salmonella typhimurium,* a protein with a molecular weight of 34,000 has been isolated. It contains no lipid, carbohydrate, phosphorus, or SH groups. One sulfate molecule is specifically bound per molecule of protein, and the binding is very strong. Binding is, however, reversible and requires no ATP. The protein exhibits no enzymic activity. Under conditions of a low sulfate source, the organism produces about 10^4 molecules of the binding protein per bacterium. Osmotic shock treatment of cells results in considerable loss of the carrier protein, suggesting that the protein is near or on the cell membrane surface (periplasmic space). An increasing number of specific membrane transport proteins have now been isolated and crystallized. These include carrier proteins for glucose, galactose, arabinose, leucine, phenylalanine, arginine, histidine, tyrosine, phosphate, Ca^{2+}, Na^+, and K^+. They all have small molecular weights ranging from 9000 to 40,000 and occur as monomers.

8.12.3   GROUP TRANSLOCATION

A transport mechanism has been proposed for the transport of sugars across bacterial membranes; according to this mechanism, a specific sugar outside the cell is transported across the membrane and released inside the cell as a phosphorylated derivative. Active transport (Section 8.13.4) is achieved since the sugar phosphate cannot escape back through the membrane. This is diagrammed in Figure 8.12.

8.12.4   ACTIVE TRANSPORT

This process, one of the most important properties of all plasma membranes, is very similar to facilitated diffusion with the critical exception that the metabolite or solute moves across the membrane against a concentration gradient and this requires energy input. Use of an inhibitor such as azide or iodacetate that markedly decreases the production of energy in the cell greatly inhibits active transport. Neither passive diffusion or facilitated diffusion would be affected by the use of these inhibitors.

Most models for this mechanism postulate that the outside solute combines with a carrier, and then the carrier-solute complex is modified in the lipophilic membrane by energy input in such a way that the carrier's affinity for the solute is lowered, the solute is released into the interior of the cell, the high-affinity conformer of the carrier is regenerated, and the cycle is repeated. These ideas are depicted in Figure 8.12.

Recent evidence clearly indicates that the energy input required for active transport involves two distinctly different systems: (1) a non-ATP-utilizing respiration-linked process and (2) direct utilization of ATP.

8.12.4.1   Respiratory-Linked Active Transport.   These systems are coupled to the oxidation of a suitable substrate such as D-lactate, L-malate, or NADH that is catalyzed by a flavin-linked membrane-bound dehydrogenase. Electrons derived from the substrate are transferred to oxygen by means of a membrane-bound respiratory chain that is coupled to active transport within a segment of the respiratory chain between the primary dehydrogenase and cytochrome b_1 (see Figure 8.12). The generation or hydrolysis of ATP is not involved. The role of specific binding proteins in this process is not clear at present. Under anaerobic conditions, a reducible substrate can serve in place of oxygen as a terminal electron acceptor. A large number of metabolites are now known to be transported into bacteria by this type of system. Thus, in E. coli with D-lactate as the source of electrons, the following metabolites move into the interior of the cell: α-galactosides; galactose; arabinose; glucuronic acid; hexose phosphates; amino acids; hydroxy, keto, and dicarboxylic acids; and nucleosides.

8.12.4.2   ATP-Dependent Transport.   All plasma membranes have one enzyme activity in common, namely an ATPase that is activated by Mg^{2+}, K^+, and Na$^+$. The overall reaction hydrolyzes ATP in a stepwise fashion involving (1) a Na$^+$-dependent phosphorylation of the enzyme at the inside face of the plasma membrane and (2) a K$^+$-dependent hydrolysis of the phosphoenzyme on the outside face. During this process, K$^+$ is transported into the cell and Na$^+$ is moved out of the cell. The enzyme-phosphate bond has been identified with an aspartyl-β-phosphate residue.

$$\beta\;\overset{\displaystyle O}{\underset{\displaystyle \underset{\displaystyle \underset{\displaystyle |}{CH_2}}{|}}{C}}\!\!\sim\!\!O\!\!-\!\!\overset{\displaystyle O^-}{\underset{\displaystyle O^-}{P}}\!\!=\!\!O$$

$$-Pro-NH-\underset{}{CH}-CO-Lys-$$

·aspartyl·β·phosphate

This unique membrane-bound NaK ATPase has now been solubilized and has a mol wt of about 250,000 and two subunits: one with a mol wt of 84,000 to 100,000 and the other, a glycoprotein, with a mol wt of 55,000. The larger unit is phosphorylated by ATP.

Why should this NaK ATPase system be of such importance to the cell? Practically all aerobic cells have a high, relatively constant, intracellular K$^+$ concentration and a low Na$^+$ concentration, no matter what the exterior concentration of Na$^+$ or K$^+$ may be. Furthermore, a high internal concentration of K$^+$ is required for the maintenance of pyruvic kinase activity in glycolysis and optimal protein synthesis. Gradients of Na$^+$ + K$^+$ across plasma membranes must be maintained to allow transmembrane potential differences and nerve impulses in nerve cells. Of considerable importance, amino acids and sugars are transported into animal cell via a Na$^+$-cotransport system; that is, glucose or amino acids move into a cell only with a simultaneously comovement of Na$^+$. The facilitated diffusion of these metabolites into the cell can be successful only if the accumulating intracellular Na$^+$ can be pumped out. The activity of the NaK ATPase fits this requirement.

8.13 PROTEIN TRANSFER ACROSS MEMBRANES

We have examined the transport of macromolecules from outside the cell to the interior of the cell (see Section 8.9) and the transport of small solutes such as substrates, anions, and cations across plasma and organelle membranes (see Section 8.12). Although Chapter 20 will discuss in detail the mechanisms of protein synthesis, it is appropriate to complete this chapter with a brief discussion of the translocation of proteins newly synthesized in the cell into its different membranes or organelle compartments.

In recent years, there has been an explosion of knowledge concerning the transport of newly synthesized proteins in the cytosol to specific organelles in the cell. Although mitochondria and chloroplasts have the machinery for protein synthesis, this capability is limited to about 10 to 20% of all the proteins that make up the structural and functional proteins of these organelles. Obviously, the remaining 70 to 80% proteins must be encoded by nuclear DNA and translated by cytosolic mRNAs in conjunction with free or endoplasmic reticulum-bound polysomes. Mechanisms must then exist in cells for the rapid and specific transfer of newly synthesized proteins to target organelles and their membranes.

A number of years ago, it was observed that when a cell-free protein-synthesizing system was employed with the appropriate mRNA, the newly synthesized protein was in many cases significantly larger than when microsomes were included in the mixture. These proteins were therefore called preproteins or nascent proteins in contrast to the mature final products. An explanation for these observations became apparent as investigators sought answers to this puzzle.

A few definitions are now in order. **Translocation** involves the transport of an entire polypeptide chain across one or two membranes in a unidirectional manner from the site of synthesis to its final location. Translocation may be **cotranslational;** that is, the synthesis of the peptide is coupled to its transport across a membrane. Translocation may also be **posttranslational;** that is, the completed polypeptide has been released from its polysomal complex and moves vectorically across a membrane to another site where it is inserted into the appropriate membrane locus or modified and then inserted. A third mode is the simplest of them all; that is, the free polypeptide is directly inserted into a membrane without any further modification. This type of polypeptide usually has a COOH terminal hydrophobic tail called the **insertion sequence** that allows the polypeptide to anchor itself onto a membrane bilayer. Most cotranslational mechanisms involve the synthesis of a polypeptide with a transit sequence (also called a signal peptide) of about 15 to 25 amino acid residues at the NH_2 terminus of the polypeptide. The signal sequence of the preprotein moves vectorically through a membrane and is then clipped by a specific peptidase, thereby converting the preprotein to the final mature protein.

8.13.1   COTRANSLATIONAL TRANSLOCATION

As summarized in Figure 8.13a, a number of proteins are synthesized by rough endoplasmic-reticuli. The synthesis of the first stretch of amino acids called the transit sequence allows a specific attachment to a ribonucleoprotein complex called a signal recognition protein or SRP unit consisting of a complex of a small ribonucleic acid and six associated polypeptides with a collective molecular weight of 250,000. The ribosome (with its transit peptide-SRP complex) is then transferred to a "docking" protein (molecular weight of 70,000) on the ER membrane, and the completion of the synthesis of the preprotein begins. As the synthesis of the preprotein continues, it penetrates into and through the membrane via a tunnel or pore, into the lumen of the ER. When the transit sequence completes its penetration into the lumen, a transit peptidase in the lumen clips off the transit sequence. The polypeptide chain continues to extend into the lumen and there may undergo modification such as glycosylation or insertion of a cofactor and is eventually stabilized by attaining its final conformation. The lumenal protein can then be either secreted via appropriate paths or channeled into newly formed lysosomes near or at the site of the Golgi structure. As a rule, only preproteins with their transit sequences are cotranslationally transferred across membranes. Examples of proteins translocated by cotranslational translocation would include many serum proteins, hormones, milk proteins, egg proteins, and interferons.

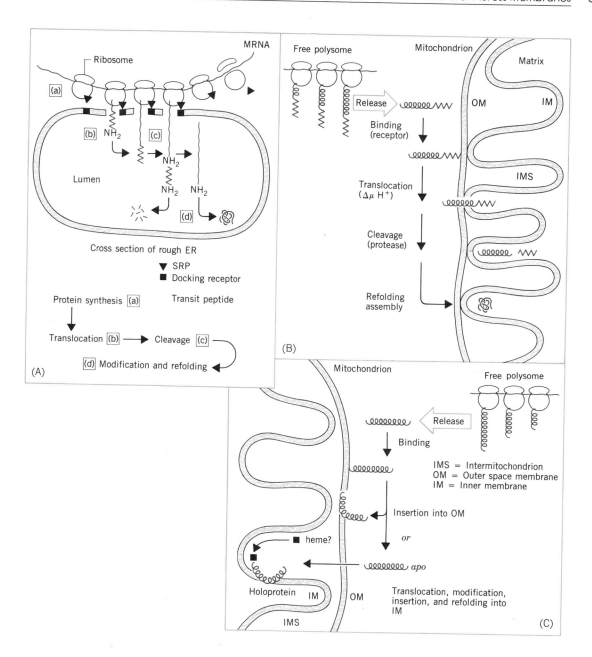

FIGURE 8.13 Various mechanisms for the transport of newly synthesized proteins through membranes. (A) Cotranslational translocation; (B) energy-driven posttranslational translocation; and (C) posttranslational translocation. Based in part on a diagram in "How Are Proteins Imported into Mitochondria" by G. Schatz and R. A. Butow in *Cell 32*, 316–318 (1983). Copyright © 1983 by MIT Press.

8.13.2 SIMPLE POSTTRANSLATIONAL TRANSLOCATION

A number of important ER membrane proteins include cytochrome P_{450}, NADH cytochrome P_{450} reductase, cytochrome b_5, and NADH cytochrome b_5 reductase. Cytochrome P_{450} is synthesized in the rough ER as a fully mature

TABLE 8.4 Brief Summary of Principal Biochemical Functions of Cell Organelles

Plasma membranes
Recognition sites
Translocator proteins
Receptors
Coated pits

Nucleus
Synthesis and storage of DNA
Polymerases for synthesis of DNA and RNA

Mitochondria
Synthesis of ATP
Electron transport chain
Tricarboxylic acid cycle
β-Oxidation system (animals only)
Ketone body synthesis/utilization
Limited protein synthesis

Chloroplasts
Photooxidation of $H_2O \equiv O_2$ formation
Photophosphorylation $\equiv$ ATP synthesis
CO_2 Fixation and NADPH generation
Glycolate formation $\equiv$ photorespiration
Fatty acid biosynthesis
$NO_2 \longrightarrow NH_3$ Reduction
$SO_4 \longrightarrow SH$ Reduction
Limited protein synthesis

Microbodies
Animal peroxisomes
   Purine oxidation
   Amino transferases
   β-Oxidation of long-chain fatty acids

Plant peroxisomes
   Photorespiration enzymes
   Limited β-oxidation
Glyoxysomes (plants)
   β-Oxidation enzymes
   Glyoxylate cycle enzymes

Lysosomes
Hydrolytic enzymes with acid pH optima

Golgi apparatus
Processing and sorting of proteins for transport to (1) secretory granules, (2) lysosomes, and (3) plasma membranes

Coated vesicles-endosomes
Translocation of external macromolecules into the internal area of cell by endocytosis

Endoplasmic reticulum
Protein synthesis
Phospholipid synthesis
Triglyceride synthesis
Sterol synthesis
Desaturation enzymes—cytochrome b_5
Hydroxylation rx—cytochrome P_{450}
Fatty acid elongation synthesis

Cytosol
Synthesis of proteins
Glycolysis
Gluconeogenesis
Pentose phosphate system
Fatty acid biosynthesis (animal)
Many more reactions

protein. Its amino-terminal sequence is very rich in hydrophobic amino acid residues and presumably serves as the insertion sequence into the membrane bilayer. No cleavage of the insertion sequence occurs. Post translational modification, that is, the insertion of the iron porphyrin into the apoprotein, presumably occurs also on the membrane surface.

Cytochrome b_5 is an intrinsic protein. It consists of a domain ($\sim 11,000$ daltons) at the NH_2 terminus where the iron porphyrin is located and a hydrophobic COOH terminus (5000 daltons) that is imbedded in the hydrophobic core of the lipid bilayer. Of interest is the fact that its insertion sequence is at the COOH terminus and it is synthesized by free polysomes. No cleavage occurs since the *in vitro* synthesized cytochrome b_5 is identical to that found embedded in the ER membrane. In both cases, there is no requirement for the SRP and docking proteins that were required in cotranslational translocation.

8.13.3   POSTTRANSLATIONAL TRANSLOCATION INTO MITOCHONDRIA AND CHLOROPLASTS

As has already been stated previously, the large majority of proteins that make up these organelles are synthesized in the cytoplasm and translocated into the interior of these organelles.

There are, in general, two types of proteins that are imported into mitochondria. The first is a preprotein synthesized by free polysomes and released with an amino terminus extension serving as a transit peptide. These extension lengths are in the order of 50 to 70 amino acid residues, in contrast to the much smaller transit peptide of secretory preproteins. These extensions also are more hydrophilic than the mature protein, presumably to facilitate the transport of the hydrophobic polypeptide through the cytosol into the mitochondrion. These preproteins then move through the rather porous outer membrane of the mitochondrion and then tunnel through the inner membrane. This last translocation requires an electrochemical potential across the membrane to permit the movement of the preprotein through the membrane. The mechanism for this requirement has not been completely resolved. On translocation, the NH_2 terminal extension is cleaved by a metal-requiring protease located in the matrix that only attacks the preprotein but not the mature protein. The clipped polypeptide is then further modified by the addition of such cofactors as thioctic acid, biotin, flavin, or iron porphyrins and then inserted into the appropriate inner membrane or matrix. Over 60 mitochondrial proteins have now been examined as to their translocation mode, and about 42 of these enter the mitochondrion as preproteins.

The second mode involves the synthesis of the mature protein from free polysomes and their insertion either directly (proteins of the outer membrane) or internalization of the apoprotein into the matrix, addition of the suitable cofactor, and then direct insertion into the membrane. Cytochrome c is an excellent example of this mode. No energized membrane is required in these translocations and the newly synthesized apo-cytochrome c is the polypeptide identical to the mature cytochrome c.

In chloroplasts, the translocation modes are very similar to those described for mitochondria. The small subunit of ribulose 1,5-*bis*-phosphate carboxylase, for example, is synthesized by free polysomes in the cytosol with the usual NH_2 terminal transit sequence, posttranslationally translocated into the stroma phase of the chloroplast and then clipped by an endopeptidase to form the mature subunit. Interestingly, uptake is an ATP-dependent process. These mechanisms are summarized in Figure 8.13B and C.

In summary, as knowledge concerning the cell and its organelles continues to grow, the complexity of the cell is matched by the beautiful, orderly processes that allow the cell to grow and divide and maintain itself. To define these processes with precision is a challenge that should excite all students of biochemistry for many years in the future. Table 8.4 summarizes the principal biochemical functions of the various components of the cell.

REFERENCES

1. *N. E. Tolbert, ed.,* The Cell, *Vol. 1. In* Biochemistry of Plants, *P. K. Stumpf and E. E. Conn, eds., New York. Academic Press, 1980.*

2. Trends in Biochemical Sciences, *Vol. 9, No. 4 (1984), New York: Elsevier Science Publishers.*

3. *E. E. Snell, ed.,* Annual Review of Biochemistry, *Palo Alto, Calif.: Annual Reviews, Inc. (Multiple annual volumes).*

REVIEW PROBLEMS

1. What is the common chemical composition of all plasma membranes? Relate this to the fluid mosaic model for plasma membranes. What would you predict concerning the chemical and physical characteristics of these components?

2. Write out the structure of a triacylglycerol and phosphatidylethanolamine and compare the structures in terms of hydrophobic and hydrophilic properties (if any).

3. Consider the roles of the following organelles in a eucaryotic cell:
    (a) Nucleus   **(c)** Mitochondrion
    (b) Endoplasmic reticulum   **(d)** Cytosol
    In the procaryotic cell, where do these functions occur?

4. Define and give an example of
    (a) Cotranslational translocation   **(c)** Transit peptide
    (b) Posttranslational translocation   **(d)** Insertion sequence

5. LDL (light density lipoprotein) is readily translocated from the exterior of a liver cell to its interior. Describe a possible mechanism for this transport.

INTRODUCTION TO METABOLISM AND BIOCHEMICAL ENERGETICS

One of the major reasons for studying biochemistry is to understand how living organisms utilize the chemical energy in their environment to carry out their biochemical activities. This requires an understanding of the simpler principles of physical chemistry and thermodynamics as they apply to living organisms. Also required is an appreciation of the so-called "energy-rich" compounds that permit the living organism to trap and subsequently utilize the chemical energy contained in the food materials it consumes.

The sun is the ultimate source of energy for all life on the planet Earth. That energy, as sunlight, is trapped by photosynthetic organisms and used to convert CO_2 into the organism's cellular material composed mainly of proteins, carbohydrates, and lipids, but also smaller amounts of nucleic acids, vitamins, coenzymes, and other compounds. Some of these products of photosynthesis (carbohydrates and lipids) are, in turn, utilized by nonphotosynthetic organisms, mainly animals, as a source of energy for growth, development, and reproduction. Other essential compounds that cannot be synthesized by animals (certain amino acids, fatty acids, and vitamins) are also provided by the photosynthetic organisms (mainly higher plants) when they are consumed by animals as food.

9.1 A DEFINITION OF METABOLISM

The chemical changes that occur when animals consume plant or animal tissues as food, or when plants carry out photosynthesis, are known as metabolism. **Intermediary metabolism** constitutes the sum of chemical reactions that the cell's constituents undergo. In the intact cell, both synthetic (anabolic) and degradative (catabolic) processes go on simultaneously, and energy released from the degradation of some compounds may be utilized in the synthesis of other cellular components. Thus, the concept of an **energy cycle** has developed in biochemistry in which fuel molecules, representing a source of potential chemical energy, are degraded through known enzymatic reactions to produce a few different energy-rich compounds.

Playing a key role in this energy cycle is the ATP–ADP system. ADP (adenosine diphosphate) is able to accept a phosphate group from other energy-rich compounds produced during metabolism and thereby be converted into ATP (adenosine triphosphate). The ATP, in turn, can be utilized to drive many biosynthetic reactions and, in addition, serve as a primary source of energy for specific physiological activities such as movement, work, secretion, absorption, and conduction. In doing so, it is generally converted back to ADP.

To appreciate the energy changes of the ATP–ADP system as well as other energy reactions in biochemistry, it is necessary to define and understand a few fundamental terms of **thermodynamics,** a science that relates the energy changes that occur in chemical and physical processes.

9.2   THE CONCEPT OF FREE ENERGY

One thermodynamic concept particularly useful to biochemists is **free energy** (G). We may speak of the **free-energy content** of a substance A, but this quantity cannot be measured experimentally. If A is converted to B in a chemical reaction, however,

$$A \rightleftharpoons B \tag{9.1}$$

it is possible to speak of the **change in free energy** (ΔG). This is the *maximum amount of energy made available* as A is converted to B. If the free-energy content of the product B (G_B) is less than the free-energy content of the reactant A (G_A), the ΔG will be a negative quantity. That is,

$$\Delta G = G_B - G_A$$
$$= \text{Negative quantity when } G_A > G_B$$

For ΔG to be negative means that the reaction occurs with a decrease in free energy. Similarly, if B is converted back to A, the reaction will involve an increase in free energy, that is, ΔG will be positive. Experience has shown that reactions which occur spontaneously do so with a *decrease* in free energy ($-\Delta G$). On the other hand, if the ΔG for a reaction is known to be positive, that reaction will occur only if energy is supplied to the system in some manner to drive the reaction. Reactions having a negative ΔG are termed **exergonic;** those that have a positive ΔG are called **endergonic.**

Experience has also shown that although the ΔG for a given process is negative, this fact has no relationship whatever to the rate at which the reaction proceeds. For example, glucose can be oxidized by O_2 to CO_2 and H_2O according to Equation 9.2.

$$C_6H_{12}O_6 + 6\,O_2 \longrightarrow 6\,CO_2 + 6\,H_2O \tag{9.2}$$

The ΔG for this reaction is a very large negative quantity, approximately $-686,000$ cal/mole of glucose. The large $-\Delta G$ has no relationship to the rate of the reaction, however. Oxidation of glucose may occur in a matter of a few seconds in the presence of a catalyst in a bomb calorimeter. Reaction 9.2 also goes on in most living organisms at rates varying from minutes to several hours. Glucose can nevertheless be kept in a bottle on the shelf for years in the presence of air without undergoing oxidation.

The factor that determines the *rate* at which a reaction proceeds is the *activation energy* for that process. Chemical theory postulates that reaction 9.1 will proceed by way of an intermediate or activated complex (e.g., A*). For A to proceed to B, A must pass through the complex A*, and energy must be expended on A to convert it to A*. If little energy is required, the reaction is said to have a low activation energy and the reaction will proceed readily. If the energy required is large, little perceptible conversion of A to B will occur and it will be necessary to provide sufficient energy to overcome the barrier to the reaction. The role of catalysts, including enzymes, is to lower the activation energy and allow the reaction to proceed (see Section 4.2).

The free-energy change of a reaction can be related to other thermodynamic properties of A and B by the expression

$$\Delta G = \Delta H - T\Delta S \tag{9.3}$$

In this expression, ΔH is the **change in heat content** that occurs as Reaction 9.1 proceeds at constant pressure; T is the absolute temperature at which the reaction occurs; and ΔS is the change in **entropy,** a term that expresses the degree of randomness or disorder in a system. The absolute heat H and entropy S contents of substances A and B are difficult to measure, but it is possible to measure the changes in these quantities as they are interconverted in Reaction 9.1. The ΔH for a reaction may be measured in a calorimeter, a device for measuring quantitatively the heat produced at constant pressure. To describe the measurement of ΔS and the absolute entropy content of chemical substances is beyond the scope of this book. However, it follows from Equation 9.3 that, as the entropy of the products increases over that of the reactants, the $T\Delta S$ term will become more positive and the ΔG will become more negative.

9.3 DETERMINATION OF ΔG

For Reaction 9.1, it is possible to derive the expression

$$\Delta G = \Delta G° + RT \ln \frac{[B]}{[A]} \tag{9.4}$$

where $\Delta G°$ is the **standard change in free energy,** soon to be defined; R is the universal gas constant; T is the absolute temperature; and [B] and [A] are the concentrations of A and B in moles per liter. Precisely, [B] and [A] should be replaced by the activities of A and B, a_A and a_B, respectively. As with pH, however, this correction is not usually made, because the activity coefficients are seldom known for the concentrations of compounds existing in the cell.

From Equation 9.4, the ΔG for a reaction is a function of the concentrations of reactant and product as well as the standard free-energy change $\Delta G°$. It is possible to evaluate $\Delta G°$ if we consider the ΔG at equilibrium. At equilibrium, there is no net conversion of A to B, and hence the change in free energy ΔG is 0. Similarly, the ratio of [B] to [A] is the ratio at equilibrium or the equilibrium constant K_{eq}. Substituting these quantities in Equation 9.4, we get

$$0 = \Delta G° + RT \ln K_{eq}$$
$$\Delta G° = -RT \ln K_{eq} \tag{9.5}$$

When the constants are evaluated ($R = 1.987$ cal/mole-degree; $25°C = 298°T$; and $\ln x = 2.303 \log_{10} x$), the equation becomes (at $25°C$)

$$\Delta G° = -(1.987)(298)(2.303) \log_{10} K_{eq}$$
$$= -1363 \log_{10} K_{eq} \tag{9.6}$$

This equation relating the $\Delta G°$ to K_{eq} is an extremely useful way to determine the $\Delta G°$ for a specific reaction. [Another way discussed in Section 9.6 relates $\Delta G°$ to a difference in oxidation–reduction potential $(\Delta E_0')$]. If the concentration of both reactants and products at equilibrium can be measured, the K_{eq} and, in turn, the $\Delta G°$ of the reaction can be calculated. Of course, if the K_{eq} is extremely large or extremely small, this method of measuring $\Delta G°$ is of little value, because the equilibrium concentration of the reactants and products, respectively, will be too small to measure. The $\Delta G°$ for each of a series of K_{eq} ranging from 0.001 to 10^3 is calculated in Table 9.1.

From inspection of Table 9.1, it is clear that reactions that have a K_{eq} greater than 1 proceed with a decrease in free energy. Thus, for Reaction 9.1, if the $K_{eq} = 1000$ (that is, if [B]/[A] is 1000), the tendency is for the reaction to proceed in the direction of the formation of B. If we start with 1001 parts of A, equilibrium will be reached only when 1000 parts (or 99.9%) of A have been converted to B. If Reaction 9.1 has a K_{eq} of 10^{-3} (that is, if [B]/[A] = 0.001), equilibrium will be attained when only 1 part or 0.1% of A has been converted to B.

It is also possible to evaluate $\Delta G°$ for the situation in which both the reactants and products are present at unit concentrations. When [A] = [B] = 1 M, Equation 9.4 becomes

$$\Delta G = \Delta G° + RT \ln \frac{1}{1}$$

$$= \Delta G°$$

Thus, $\Delta G°$ may be defined as the change in free energy when reactants and products are present in unit concentration, or more broadly, in their "standard state." The standard state for solutes in solution is unit molarity; for gases, 1 atm; for solvents such as water, unit activity. If water is a reactant or a product of a reaction, its concentration in the standard state is taken as unity in the expression for the ΔG (Equation 9.4). If a gas is either formed or produced, its standard state concentration is taken as 1 atm. If a hydrogen ion is produced or utilized in a reaction, its concentration will be taken at 1 M or pH = 0.

Since in the cell, few if any reactions occur at pH 0 but rather at pH 7.0, the standard free-energy change $\Delta G°$ is usually corrected for the difference in pH.

TABLE 9.1   Relation between K_{eq} and $\Delta G°$

K_{eq}	$\log_{10} K_{eq}$	$\Delta G° = -1363 \log_{10} K_{eq}$ (cal)
0.001	−3	4089
0.01	−2	2726
0.1	−1	1363
1.0	0	0
10	1	−1363
100	2	−2726
1000	3	−4089

Conversely, the equilibrium of a reaction may be measured at some pH other than 0. The standard free-energy change $\Delta G°$ at any pH other than 0 is designated as $\Delta G'$, and the pH for a given $\Delta G'$ should be indicated. Of course, if a proton is neither formed nor utilized in the reaction, $\Delta G'$ will be independent of pH and $\Delta G°$ will equal $\Delta G'$.

An example will demonstrate the use of these terms. In the presence of the enzyme phosphoglucomutase, glucose-1-phosphate is converted to glucose-6-phosphate. If we start with 0.020 M glucose-1-phosphate at 25 °C, it is observed that the concentration of this compound decreases to 0.001 M while the concentration of glucose-6-phosphate increases to 0.019 M. The K_{eq} of the reaction is 0.019 divided by 0.001, or 19. Therefore,

$$\begin{aligned}
\Delta G° &= -RT \ln K_{eq} \\
&= -1363 \log_{10} K_{eq} \\
&= -1363 \log_{10} 19 \\
&= (-1363)(1.28) \\
&= -1745 \text{ cal}
\end{aligned}$$

The $\Delta G°$ for this reaction will be independent of pH, since acid is neither produced nor used up in the reaction. This amount of free-energy decrease (-1745 cal) will occur when one mole of glucose-1-phosphate is converted to 1 mole of glucose-6-phosphate under such conditions that the *concentration of each compound is maintained at 1 M,* a situation quite different from the experimental situation just described for measuring the K_{eq}. Indeed, these conditions of **unit molarity** are difficult to maintain either in the test tube or the cell. It should be pointed out, however, that the concentration of a particular substance (e.g., glucose-6-phosphate) may frequently be maintained relatively constant at some concentration over a time interval, since it may be produced in one reaction while it is being used up in another. This condition of **steady-state** equilibrium undoubtedly exists in many biological systems and requires that thermodynamics be applied to the steady-state condition rather than to the equilibrium condition for which thermodynamics was first developed. A second complication is that the thermodynamic quantities discussed in this chapter apply only to reactions occurring in homogeneous systems, whereas much metabolism occurs in heterogeneous systems involving more than one phase. As a result, most of the values reported in the literature cannot be considered more than 10% accurate. Nevertheless, the concept of the standard free-energy change has found many fruitful applications in intermediary metabolism.

9.4 ENERGY-RICH COMPOUNDS

In all living forms, one compound repeatedly functions as a common reactant linking endergonic processes to others that are exergonic. This compound, adenosine triphosphate (ATP), is one of a group of "energy-rich" or "high-energy" compounds whose structure will now be considered. They are called "energy-rich" or "high-energy" compounds because they exhibit a large decrease in free energy when they undergo hydrolytic reactions. They are in general unstable to acid, alkali, and heat. In subsequent chapters, their biosynthesis and utilization will be described in detail.

9.4.1   PYROPHOSPHATE COMPOUNDS

Let us now consider the structure of ATP and its partner ADP in more detail. At pH 7.0 in aqueous solution, ATP and ADP are anions bearing a net charge of approximately -4 and -3, respectively. This results

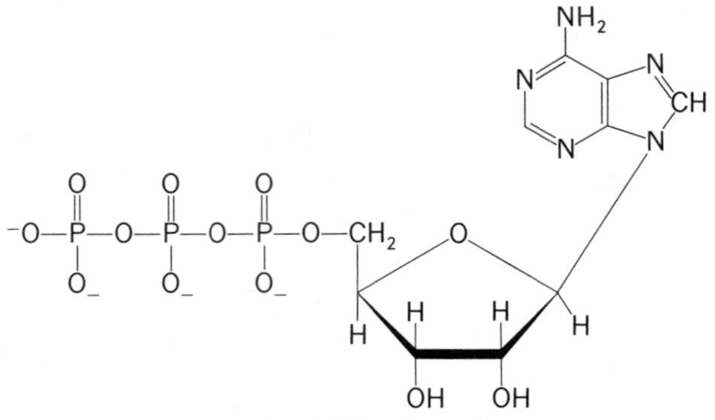

Adenosine triphosphate (ATP)

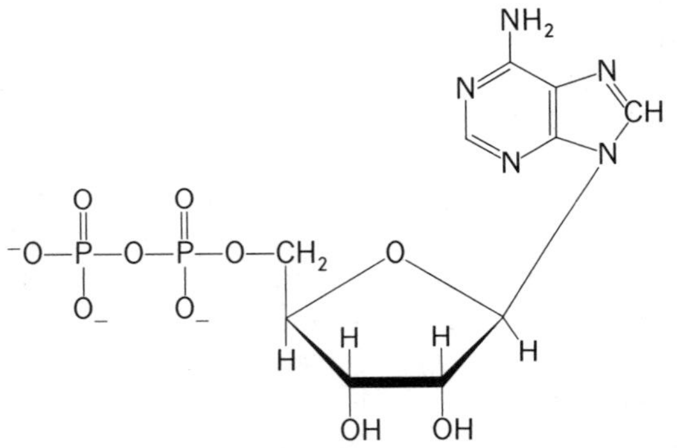

Adenosine diphosphate (ADP)

from the fact that the two dissociable protons on the interior phosphates of ATP (and the one interior phosphate of ADP) are primary hydrogens with pK_a's in the range of 2 to 3. The terminal phosphate of ATP (and ADP) have both a primary hydrogen with pK_a of 2 to 3 and a secondary hydrogen with pK_a of 6.5. Therefore, at pH 7.0, the primary hydrogen will be completely ionized and the secondary will be about 75% dissociated. In the cell, however, in which a relatively high concentration of Mg^{2+} exists, both ATP and ADP will be complexed with this cation in a one-to-one ratio to form divalent and monovalent complexes, respectively.

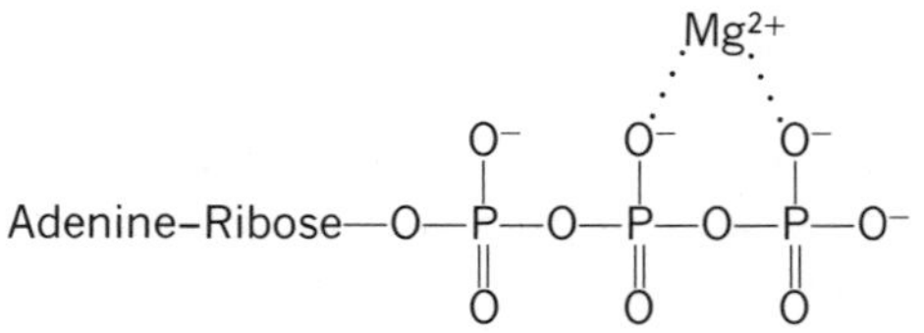

[ATP–Mg]$^{2-}$ complex 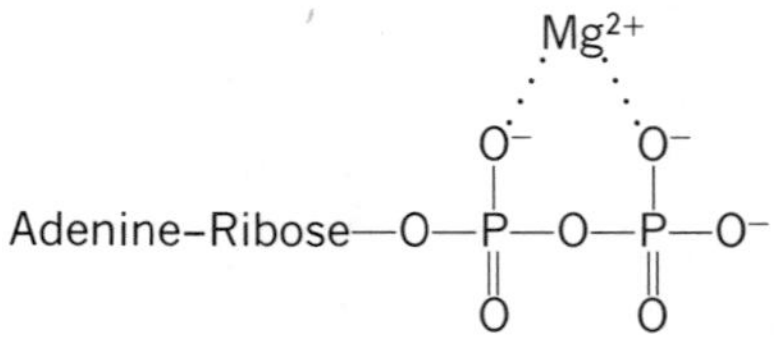   [ADP–Mg]$^-$ complex

It is informative to compare the $\Delta G'$ of hydrolysis of ATP with that of other phosphate compounds. The hydrolysis of the terminal phosphate of ATP, which is called orthophosphate cleavage, may be written as in Reaction 9.7.

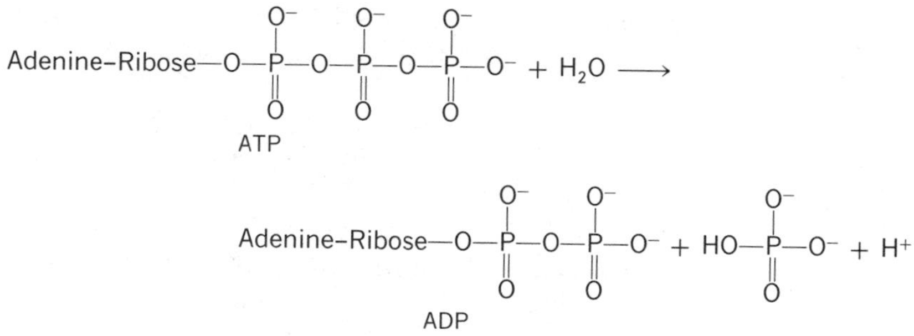

$$\text{Adenine–Ribose–O–}\overset{\overset{\displaystyle O^-}{|}}{\underset{\underset{\displaystyle O}{\|}}{P}}\text{–O–}\overset{\overset{\displaystyle O^-}{|}}{\underset{\underset{\displaystyle O}{\|}}{P}}\text{–O–}\overset{\overset{\displaystyle O^-}{|}}{\underset{\underset{\displaystyle O}{\|}}{P}}\text{–O}^- + H_2O \longrightarrow$$

ATP

$$\text{Adenine–Ribose–O–}\overset{\overset{\displaystyle O^-}{|}}{\underset{\underset{\displaystyle O}{\|}}{P}}\text{–O–}\overset{\overset{\displaystyle O^-}{|}}{\underset{\underset{\displaystyle O}{\|}}{P}}\text{–O}^- + \text{HO–}\overset{\overset{\displaystyle O^-}{|}}{\underset{\underset{\displaystyle O}{\|}}{P}}\text{–O}^- + H^+ \quad (9.7)$$

ADP

$$\Delta G' = -7300 \text{ cal (pH 7.0)}$$

The $\Delta G'$ at pH 7 has been estimated to be -7300 cal/mole. This is in contrast to the hydrolysis of glucose-6-phosphate, which results in a much smaller decrease in free energy.

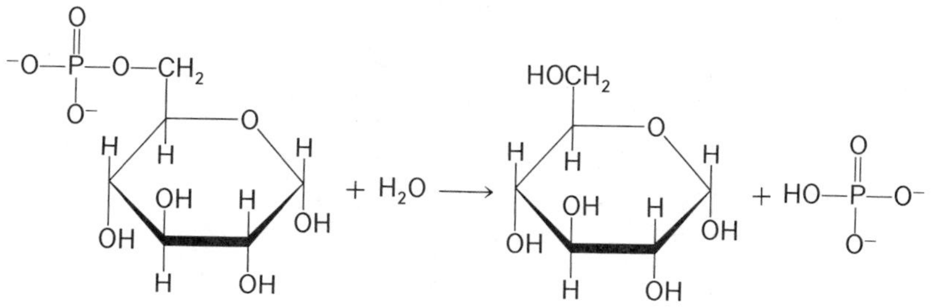

$$\Delta G' = -3300 \text{ cal (pH 7.0)} \quad (9.8)$$

We may properly ask why this large difference in the free energy of hydrolysis exists. On examining the several types of energy-rich compounds encountered in intermediary metabolism, we note several factors that are important but not all of which apply to every energy-rich compound. Regardless of the specific factors involved, it will be seen that the large decrease in free energy occurs during hydrolysis because the products are significantly *more stable* than the reactants. Important factors contributing to this large decrease in free energy are

1. Bond strain in the reactant caused by electrostatic repulsion (page 330).
2. Stabilization of the products by ionization (page 332).
3. Stabilization of the products by isomerization (page 333).
4. Stabilization of the products by resonance (pages 334–335).

In the case of ATP, the structure of importance in determining its character as an energy-rich compound is the pyrophosphate moiety that, at pH 7.0, is fully ionized.

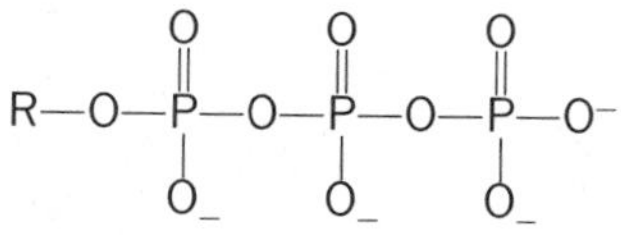

$$\text{R–O–}\overset{\overset{\displaystyle O}{\|}}{\underset{\underset{\displaystyle O_-}{|}}{P}}\text{–O–}\overset{\overset{\displaystyle O}{\|}}{\underset{\underset{\displaystyle O_-}{|}}{P}}\text{–O–}\overset{\overset{\displaystyle O}{\|}}{\underset{\underset{\displaystyle O_-}{|}}{P}}\text{–O}^-$$

There will be a tendency for the electrons in the P=O bond of the phosphates to be drawn closer to the **electronegative** oxygen atom, thereby producing a **partial negative charge** (δ^-) on that atom. This is compensated by a **partial**

positive charge (δ^+) on the phosphorus atom, resulting in a **polarization** of the phosphorus-oxygen bonding that may be indicated as

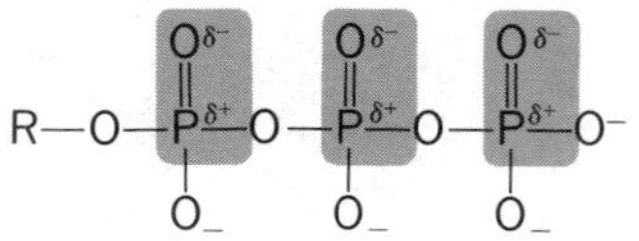

The existence of residual positive charges of this nature on adjacent phosphorus atoms in the pyrophosphate structures of ATP (and ADP) means that these molecules must contain sufficient internal energy to overcome the electrostatic repulsion between the adjacent like charges. When the pyrophosphate structure is cleaved, as on hydrolysis, this energy will be released and contribute to the total negative ΔG of the reaction. Although the P=O bond in glucose-6-phosphate can also be considered to have polar character, there is no adjacent phosphorus atom with a δ^+ charge.

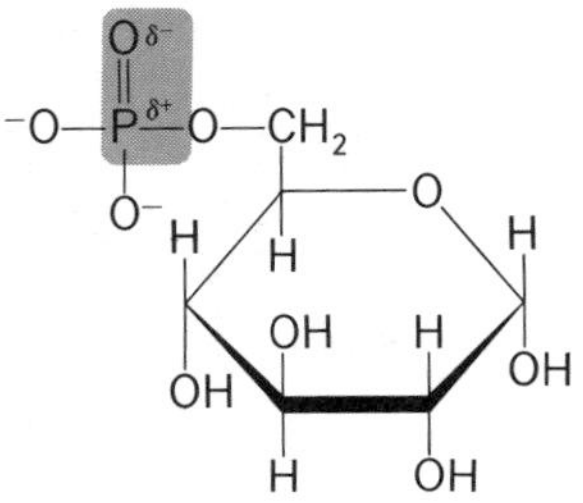

The argument for instability due to charge repulsion does not exist with this compound, and the ΔG of hydrolysis will be less for this reason.

Obviously, the same factor applies in the hydrolysis of ADP to AMP and

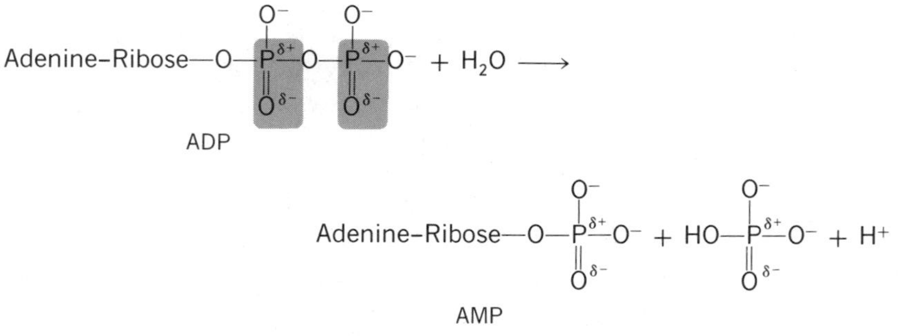

ADP

AMP

$$\Delta G' = -7300 \text{ cal (pH 7.0)}$$

inorganic phosphate where the observed $\Delta G'$ on hydrolysis at pH 7 is -7300 cal/mole. On the other hand, the hydrolysis of AMP to adenosine and H_3PO_4 is less ($\Delta G' = -2200$, pH 7) for lack of the same reason.

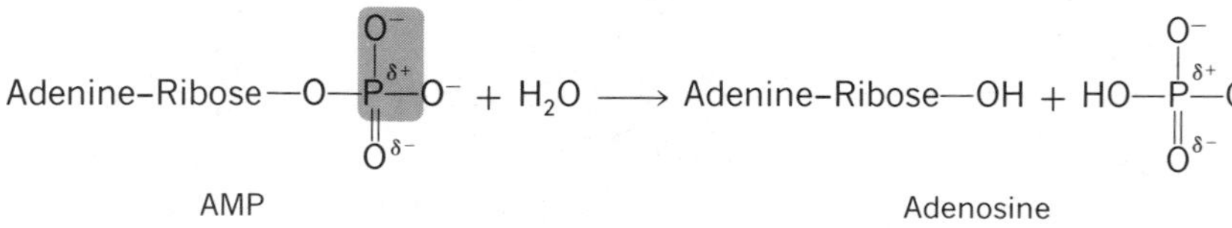

AMP   Adenosine

$$\Delta G' = -2200 \text{ cal (pH 7.0)}$$

Although ATP is converted to ADP in many reactions of intermediary metabolism, there are a number of important reactions in which the interior pyrophosphate bond of ATP is cleaved to yield AMP and inorganic pyrophosphate.

$$\text{Adenine–Ribose–O–} \overset{\overset{\displaystyle O^-}{|}}{\underset{\underset{\displaystyle O}{\|}}{P}} \text{–O–} \overset{\overset{\displaystyle O^-}{|}}{\underset{\underset{\displaystyle O}{\|}}{P}} \text{–O–} \overset{\overset{\displaystyle O^-}{|}}{\underset{\underset{\displaystyle O}{\|}}{P}} \text{–O}^- + H_2O \longrightarrow$$

ATP

$$\text{Adenine–Ribose–O–} \overset{\overset{\displaystyle O^-}{|}}{\underset{\underset{\displaystyle O}{\|}}{P}} \text{–O}^- + {}^-\text{O–} \overset{\overset{\displaystyle O^-}{|}}{\underset{\underset{\displaystyle O}{\|}}{P}} \text{–O–} \overset{\overset{\displaystyle O^-}{|}}{\underset{\underset{\displaystyle O}{\|}}{P}} \text{–O}^- + 2\,H^+ \qquad (9.11)$$

AMP

$$\Delta G' = -8600 \text{ cal (pH 7.0)}$$

This type of cleavage is known as the **pyrophosphate cleavage** and is in contrast to the **orthophosphate cleavage** in which ADP is formed (Reaction 9.7).

The ATP–ADP system is functional in nature because ADP, having been formed from ATP, can be rephosphorylated in energy-yielding reactions and be converted back to ATP. It is critical, therefore, that AMP and the pyrophosphate formed in the pyrophosphate cleavage be converted back to ATP. This is accomplished by two reactions catalyzed by enzymes widely distributed in nature. The first of these reactions, catalyzed by a pyrophosphatase, is the hydrolysis of pyrophosphate to yield two moles of inorganic phosphate.

$$ {}^-\text{O–} \overset{\overset{\displaystyle O^-}{|}}{\underset{\underset{\displaystyle O}{\|}}{P}} \text{–O–} \overset{\overset{\displaystyle O^-}{|}}{\underset{\underset{\displaystyle O}{\|}}{P}} \text{–O}^- + H_2O \longrightarrow 2\,\text{HO–} \overset{\overset{\displaystyle O^-}{|}}{\underset{\underset{\displaystyle O}{\|}}{P}} \text{–O}^- \qquad (9.12)$$

Pyrophosphate

$$\Delta G' = -8000 \text{ cal (pH 7.0)}$$

The second reaction is one in which ATP and AMP react to form two moles of ADP, which in turn can be further phosphorylated in several different energy-yielding reactions to regenerate ATP.

$$\text{Adenosine–Ribose–O–} \overset{\overset{\displaystyle O^-}{|}}{\underset{\underset{\displaystyle O}{\|}}{P}} \text{–O–} \overset{\overset{\displaystyle O^-}{|}}{\underset{\underset{\displaystyle O}{\|}}{P}} \text{–O–} \overset{\overset{\displaystyle O^-}{|}}{\underset{\underset{\displaystyle O}{\|}}{P}} \text{–O}^- + \text{Adenosine–Ribose–O–} \overset{\overset{\displaystyle O^-}{|}}{\underset{\underset{\displaystyle O}{\|}}{P}} \text{–O}^- \rightleftharpoons$$

ATP   AMP

$$\text{Adenosine–Ribose–O–} \overset{\overset{\displaystyle O^-}{|}}{\underset{\underset{\displaystyle O}{\|}}{P}} \text{–O–} \overset{\overset{\displaystyle O^-}{|}}{\underset{\underset{\displaystyle O}{\|}}{P}} \text{–O}^- + \text{Adenosine–Ribose–O–} \overset{\overset{\displaystyle O^-}{|}}{\underset{\underset{\displaystyle O}{\|}}{P}} \text{–O–} \overset{\overset{\displaystyle O^-}{|}}{\underset{\underset{\displaystyle O}{\|}}{P}} \text{–O}^- \quad (9.13)$$

ADP   ADP

The $\Delta G'$ for this reaction is approximately 0 because the K_{eq} is approximately 1.0. Examination of the means by which ADP can be converted back to ATP introduces two other energy-rich phosphate compounds, 1,3-diphosphoglyceric acid and phosphoenolpyruvic acid. Both of these are encountered during the conversion of glucose to pyruvic acid (see Chapter 10) and both have standard free energies of hydrolysis more negative than that of ATP.

9.4.2   ACYL PHOSPHATES

1,3-Diphosphoglyceric acid is an example of an acyl phosphate; its standard free energy of hydrolysis is -11.8 kcal/mole.

$$
\begin{array}{c}
\underset{\text{HCOH}}{\overset{\text{O}}{\overset{\|}{\text{C}}}}\text{—O—}\underset{\text{O}}{\overset{\text{OH}}{\overset{\|}{\text{P}}}}\text{—OH} \\
\text{HCOH} \\
\text{CH}_2\text{OPO}_3\text{H}_2
\end{array}
\;+\; \text{H}_2\text{O} \longrightarrow
\begin{array}{c}
\underset{\text{HCOH}}{\overset{\text{O}}{\overset{\|}{\text{C}}}}\text{—OH} \\
\text{HCOH} \\
\text{CH}_2\text{OPO}_3\text{H}_2
\end{array}
\;+\;
\text{HO—}\underset{\text{O}}{\overset{\text{OH}}{\overset{\|}{\text{P}}}}\text{—OH} \quad (9.14)
$$

1,3-Diphosphoglyceric acid   3-Phosphoglyceric acid

$\Delta G' = -11{,}800$ cal (pH 7.0)

Bond strain in the acyl phosphate is a significant factor contributing to the large negative standard free energy of hydrolysis of this class of compounds. The C=O bond of the acyl phosphate group may be considered also to have considerable polar character because of the tendency for the electrons in the double bond to be drawn closer to the electronegative oxygen. Energy is required to overcome the repulsion between the partial positive charges on the carbon and phosphorus atoms, such energy being released on hydrolysis of the acyl phosphate.

The relative tendencies of reactants and products to ionize at a particular pH have an important influence on the ΔG of a reaction. This factor may also be seen in the case of 1,3-diphosphoglyceric acid. In Reaction 9.14, the ionization of the reactants and products has not been indicated in the formulas. At pH 7, the reaction is more accurately represented as

$$
\begin{array}{c}
\underset{\text{HCOH}}{\overset{\text{O}}{\overset{\|}{\text{C}}}}\text{—O—}\underset{\text{O}}{\overset{\text{O}^-}{\overset{\|}{\text{P}}}}\text{—O}^- \\
\text{HCOH} \\
\text{CH}_2\text{OPO}_3^{2-}
\end{array}
\;+\; \text{H}_2\text{O} \longrightarrow
\begin{array}{c}
\underset{\text{HCOH}}{\overset{\text{O}}{\overset{\|}{\text{C}}}}\text{—O}^- \\
\text{HCOH} \\
\text{CH}_2\text{OPO}_3^{-2}
\end{array}
\;+\;
\text{HO—}\underset{\text{O}}{\overset{\text{O}^-}{\overset{\|}{\text{P}}}}\text{—O}^- + \text{H}^+ \quad (9.15)
$$

1,3-Diphosphoglyceric acid   3-Phosphoglyceric acid

where the primary and secondary hydrogen ions are ionized, while the tertiary hydrogen (on the inorganic phosphate) is not. The carboxylic acid group ($pK = 3.7$) formed on hydrolysis will also be extensively ionized. The effect of this ionization is to reduce the concentration of the actual hydrolysis product (the unionized acid) to a low level.

It should be stressed that the extent to which ionization is a factor in the $\Delta G'$ of the reaction (i.e., the extent to which products are stabilized in a reaction) will be dependent on the *difference* in the pK_a of the newly formed ionizable group and the pH at which the reaction occurs. It may be shown that the contribution of a new group with a pK_a of 1 unit *less* than the pH of the medium is -1363 cal/mole. If Reaction 9.15 were to occur at an acid pH (something less than 3) where the newly formed 3-phosphoglyceric acid is not significantly ionized, the ionization factor would contribute little to the $\Delta G'$ of hydrolysis of 1,3-diphosphoglyceric acid.

9.4.3 ENOLIC PHOSPHATE

The second compound encountered during the conversion of glucose to pyruvate that provides for the regeneration of ATP from ADP is phosphoenolpyruvic acid (PEP). The free-energy change on hydrolysis of this energy-rich **enolic phosphate** is $-14,800$ cal at pH 7.0.

$$
\begin{array}{c}
CO_2^- \quad O \\
| \quad\quad \| \\
C-O-P-O^- +H_2O \\
| \quad\quad | \\
CH_2 \quad O^-
\end{array}
\xrightarrow[\Delta G = -6800]{}
\begin{array}{c}
O \\
\| \\
HO-P-O^- \\
| \\
O^-
\end{array}
+
\begin{array}{c}
CO_2^- \\
| \\
C-OH \\
\| \\
CH_2
\end{array}
\xrightarrow[\Delta G = -8000]{\text{Tautomerization}}
\begin{array}{c}
CO_2^- \\
| \\
C=O \\
| \\
CH_3
\end{array}
\quad (9.16)
$$

Phosphoenol
pyruvate
Pyruvate
(unstable
enol form)
Pyruvate
(stable keto)

One can appreciate the large negative ΔG observed on hydrolysis of this compound if one recognizes that the inherently unstable enolic form of pyruvic acid is stabilized in PEP by the phosphate ester group. On hydrolysis, the unstable enol may be thought of as being formed, but it will instantly isomerize to the much more stable keto structure. It is estimated that the tautomerization occurs with a decrease in $\Delta G'$ of about 8000 cal/mole, therefore bringing the total $\Delta G'$ to $-14,800$ cal/mole. This tautomerization is of major importance in making PEP one of the most "energy-rich" phosphate compounds of biological importance.

9.4.4 THIOL ESTERS

A third type of energy-rich compound that can, in turn, be utilized to generate ATP from ADP (see Section 12.3.5) is the thioester, acetyl coenzyme A.
The $\Delta G'$ of hydrolysis of this compound is approximately -7500 cal.

$$
\begin{array}{c}
O \\
\| \\
CH_3-C-S-CoA + H_2O
\end{array}
\longrightarrow
\begin{array}{c}
O \\
\| \\
CH_3-C-O^-
\end{array}
+ \quad CoA-SH \quad + \quad H^+
$$

Acetyl—CoA
Coenzyme A

$$\Delta G' = -7500 \text{ cal (pH 7.0)}$$

An explanation for this larger $\Delta G'$ of hydrolysis is given in Section 5.10.3, where the unique properties of thioesters are discussed in detail.

9.4.5 GUANIDINIUM PHOSPHATES

A fourth type of energy-rich compound that plays an important role in energy transfer and storage is the guanidinium phosphate. This type of structure is found in phosphocreatine and phosphoarginine

$$
\begin{array}{c}
O^- \quad\quad CH_3 \\
| \quad H \quad | \\
^-O-P-N-C-N-CH_2-COO^- \\
\| \quad\quad \| \\
O \quad\quad {}^+NH_2
\end{array}
\quad\quad
\begin{array}{c}
O^- \\
| \quad H \quad H \\
^-O-P-N-C-N-(CH_2)_3-CH-COO^- \\
\| \quad\quad \| \quad\quad\quad\quad | \\
O \quad\quad {}^+NH_2 \quad\quad\quad\quad NH_3^+
\end{array}
$$

Phosphocreatine
Phosphoarginine

in muscles of vertebrates and invertebrates, respectively. These compounds are also known as phosphagens. The phosphagens are formed by the phosphorylation of creatine or arginine with ATP in the presence of the appropriate enzyme.

$$\text{Phosphocreatine} + \text{ADP} \xrightleftharpoons{\text{Creatine kinase}} \text{Creatine} + \text{ATP} \qquad (9.17)$$
$$\Delta G' = -3000 \text{ cal (pH 7.0)}$$

Since, however, the standard free-energy change on hydrolysis of these compounds is more negative by about -3000 than that of ATP, the equilibrium actually favors ATP formation. Phosphagens carry out their physiological role by furnishing a place to store energy-rich phosphate. When the concentration of ATP is high, Reaction 9.17 proceeds from right to left and phosphate is stored as energy-rich phosphocreatine. Then, when the level of ATP is depleted, Reaction 9.17 proceeds from left to right, and ATP concentration is increased.

The guanidinium phosphates, represented by phosphocreatine, are not inherently less stable because of bond strain as in the case of ATP and ADP. There are no obvious ionization or tautomerization processes that account for greater stability of the products over their reactants as in the case of the acyl and enolic phosphates.

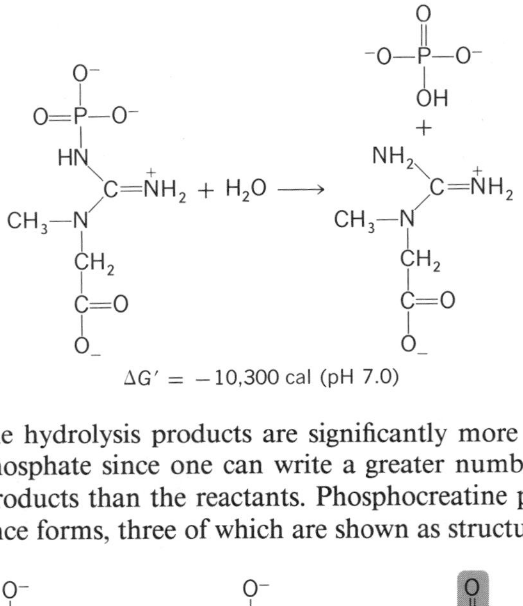

$$\Delta G' = -10,300 \text{ cal (pH 7.0)} \qquad (9.18)$$

Nevertheless, the hydrolysis products are significantly more stable than the guanidinium phosphate since one can write a greater number of **resonance forms** for the products than the reactants. Phosphocreatine possesses twelve possible resonance forms, three of which are shown as structures I to III.

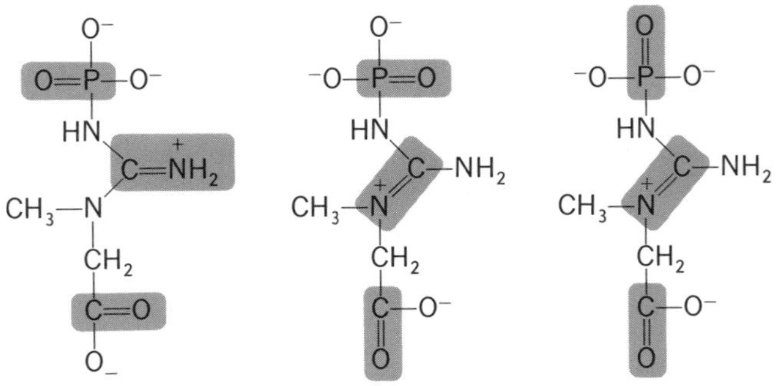

When, however, creatine lacks its phosphate group, one can write an increased number of resonance isomers that include structure IV, in which a positive charge is placed on the nitrogen atom formerly linked to the phosphate group. Since, in phosphocreatine, there is no *oxygen* atom between the P atom of the phosphate group and the ureido nitrogen, the

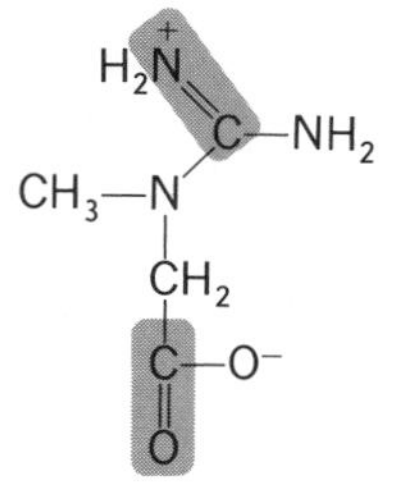

IV

partial positive charge on phosphorus would prevent a similar charge on an adjacent atom.

The five types of compounds just discussed may be contrasted with such compounds as glucose-6-phosphate or *sn*-glycerol-3-phosphate, which are phosphoric acid esters of organic alcohols and have relatively small values for the $\Delta G'$ of hydrolysis. When all these compounds are listed in Table 9.2, one can see that there is no sharp division between "energy-rich" and "energy-poor" compounds, and that several compounds including ATP occupy intermediate positions in the table. For that matter, the unique ability of ATP to participate in so many different reactions involving energy transfer may be ascribed to its truly intermediate position between the acyl and enolic phosphates that are generated in the breakdown of fuel molecules and the numerous acceptor molecules that are phosphorylated in the course of their metabolism.

While the discussion on pyrophosphate compounds in Section 9.4.1 dealt only with ATP and ADP, it should be noted that GTP, GDP, CTP, CDP, UTP, UDP as well as *d*ATP, *d*GTP, *d*TTP, and *d*CTP are also energy-rich com-

TABLE 9.2 Standard Free Energy of Hydrolysis of Some Important Metabolites

	$\Delta G'$ AT pH 7.0 (cal/mole)
phosphoenolpyruvate	$-14,800$
cyclic-AMP	$-12,000$
1,3-diphosphoglycerate	$-11,800$
phosphocreatine	$-10,300$
acetyl phosphate	$-10,100$
S-adenosylmethionine	$-10,000$
pyrophosphate	$-8,000$
acetyl CoA	$-7,500$
ATP to ADP and Pi	$-7,300$
ATP to AMP and pyrophosphate	$-8,600$
ADP to AMP and Pi	$-7,300$
AMP to adenosine and Pi	$-2,200$
UDP-glucose to UDP and glucose	$-8,000$
glucose-1-phosphate	$-5,000$
fructose-6-phosphate	$-3,800$
glucose-6-phosphate	$-3,300$
sn-glycerol-3-phosphate	$-2,200$

pounds. Moreover, there is specificity in the biological roles that these compounds play. Thus, UTP is primarily used for the biosynthesis of polysaccharides; GTP is employed in protein synthesis; and CTP is utilized in lipid synthesis. These three together with ATP are involved in RNA synthesis, while dATP, dGTP, dCTP, and dTTP are used in DNA synthesis. Although cyclic AMP (cAMP) exhibits a large decrease in free energy on hydrolysis, due to its unstable phosphodiester ring, it is not known to function by virtue of its "energy-rich" nature. Rather it is an allosteric effector and second messenger (Sections 10.8.1 and 18.5.1).

In the past, it has been common practice in biochemistry to refer to high-energy and low-energy phosphate bonds. Lipmann introduced the symbol $\sim$ ph to indicate a high-energy phosphate structure. This practice has resulted in the tendency to think of the energy as concentrated in the single chemical bond. This is erroneous, because as the previous discussion has stressed, the free-energy change ΔG depends on the structure of the compound hydrolyzed and the products of hydrolysis. Moreover, the ΔG refers specifically to the chemical reaction involved, namely, the **hydrolysis** of the compound.

9.5   COUPLING OF REACTIONS

In the cell, the energy made available in an exergonic reaction is frequently utilized to drive a related endergonic reaction and thereby it is made to do work. This is accomplished by coupling reactions that have **common intermediates.** A specific example can best illustrate this important principle.

During the conversion of glucose to lactic acid (or alcohol), the phosphorylated triose D-glyceraldehyde-3-phosphate is oxidized to 3-phosphoglyceric acid (Section 10.4). This reaction may be represented as the removal of two hydrogen atoms from the hydrated form of the aldehyde.

$$
\begin{array}{c}
\text{H}\diagdown\!\!\diagup\text{O} \\
\text{C} \\
\text{HCOH} \\
\text{CH}_2\text{OPO}_3\text{H}_2 \\
\text{D-Glyceraldehyde-3-phosphate}
\end{array}
\;+\; \text{H}_2\text{O} \longrightarrow
\left[
\begin{array}{c}
\text{OH} \\
\text{H}-\text{C}-\text{OH} \\
\text{HCOH} \\
\text{CH}_2\text{OPO}_3\text{H}_2
\end{array}
\right]
\longrightarrow
\begin{array}{c}
\text{O} \\
\text{C}-\text{OH} \\
\text{HCOH} \\
\text{CH}_2\text{OPO}_3\text{H}_2 \\
\text{3-Phosphoglycerate}
\end{array}
\;+\; 2\,\text{H}\cdot \quad (9.19)
$$

$$\Delta G' = -12,000 \text{ cal}$$

The carboxyl group of the newly formed acid (and the phosphate groups as well) would be ionized at pH 7.0. However, this ionization is intentionally not represented in Reactions 9.19 through 9.23 in order to avoid confusion with the release of a proton that occurs on oxidation of the aldehyde group in Reactions 9.21 and 9.23. The $\Delta G'$ for Reaction 9.19 is approximately $-12,000$ cal, indicating that the reaction is not readily reversible. However, living cells have evolved an elegant mechanism for coupling Reaction 9.19 to the generation of ATP, a process that, as we have seen, has a $\Delta G'$ of about 7300 cal/mole at 37°C.

$$\text{ADP} + \text{H}_3\text{PO}_4 \longrightarrow \text{ATP} + \text{H}_2\text{O}$$
$$\Delta G' = +7300 \text{ cal (pH 7.0)} \tag{9.20}$$

This is done through the participation of the common intermediate, 1,3-diphosphoglyceric acid, an acyl phosphate, whose formation would represent the expenditure of 11,800 cal (see Reaction 9.14).

The actual reaction in which 1,3-diphosphoglyceric acid is formed is a combined oxidation–reduction and phosphorylation reaction.

$$
\begin{array}{c}
\text{H} \quad \text{O} \\
\diagdown \diagup \\
\text{C} \\
| \\
\text{HCOH} \\
| \\
\text{CH}_2\text{OPO}_3\text{H}_2
\end{array}
\quad + \quad \text{NAD}^+ + \text{H}_3\text{PO}_4 \quad \longrightarrow \quad
\begin{array}{c}
\text{O} \\
\| \\
\text{C} - \text{OPO}_3\text{H}_2 \\
| \\
\text{HCOH} \\
| \\
\text{CH}_2\text{OPO}_3\text{H}_2
\end{array}
\quad + \quad \text{NADH} + \text{H}^+ \quad (9.21)
$$

D-Glyceraldehyde-3-phosphate   1,3-Diphosphoglyceric acid

$$\Delta G' = 1500 \text{ cal}$$

The acyl phosphate, in a subsequent reaction, is then utilized to convert ADP to ATP.

$$
\begin{array}{c}
\text{O} \\
\| \\
\text{C} - \text{OPO}_3\text{H}_2 \\
| \\
\text{HCOH} \\
| \\
\text{CH}_2\text{OPO}_3\text{H}_2
\end{array}
\quad + \text{ADP} \quad \longrightarrow \quad
\begin{array}{c}
\text{O} \\
\| \\
\text{C} - \text{OH} \\
| \\
\text{HCOH} \\
| \\
\text{CH}_2\text{OPO}_3\text{H}_2
\end{array}
\quad + \quad \text{ATP} \quad (9.22)
$$

$$\Delta G' = -4500 \text{ cal}$$

And the sum of the two reactions when coupled may be written as

$$
\begin{array}{c}
\text{H} \quad \text{O} \\
\diagdown \diagup \\
\text{C} \\
| \\
\text{HCOH} \\
| \\
\text{CH}_2\text{OPO}_3\text{H}_2
\end{array}
\quad + \quad \text{NAD}^+ + \text{H}_3\text{PO}_4 + \text{ADP} \quad \longrightarrow \quad
\begin{array}{c}
\text{O} \\
\| \\
\text{C} - \text{OH} \\
| \\
\text{HCOH} \\
| \\
\text{CH}_2\text{OPO}_3\text{H}_2
\end{array}
\quad + \quad \text{NADH} + \text{H}^+ + \text{ATP} \quad (9.23)
$$

D-Glyceraldehyde-3-phosphate   3-Phosphoglyceric acid

$$\Delta G' = -3000 \text{ cal}$$

Moreover, the $\Delta G'$ for Reaction 9.23 may be calculated by adding the $\Delta G'$ for Reaction 9.21 and Reaction 9.22; this amounts to -3000 cal. Note that Reaction 9.23 states, in effect, that a significant amount of energy made available in the oxidation of an aldehyde to a carboxylic acid has been utilized to drive the formation of ATP rather than simply being lost to the environment as heat. Moreover, in doing so, the cell has available an overall process that it is able to utilize in converting 3-phosphoglyceric acid back to glyceraldehyde 3-phosphate since the overall $\Delta G'$ is not as large as that for driving Reaction 9.19 from right to left.

Subsequent chapters contain many examples of coupled reactions in which a common intermediate plays a key role in conserving the total energy of the system.

9.6 ΔG AND OXIDATION–REDUCTION

The ΔG of a reaction that involves an oxidation–reduction process may be related to the difference in oxidation–reduction potentials (ΔE_0) of the reactants. A detailed discussion of electromotive force is beyond the scope of this book, but some appreciation of the energetics of oxidation–reduction reactions and the term **reduction potential** is necessary.

A reducing agent may be defined as a substance that tends to give up an electron and be oxidized.

$$Fe^{2+} \xrightarrow{\text{Oxidized}} Fe^{3+} + 1 \text{ electron}$$

Similarly, Fe^{3+} is an oxidizing agent because it can accept electrons and be reduced.

$$Fe^{3+} + 1 \text{ electron} \longrightarrow Fe^{2+}$$

Other substances such as H^+ or organic compounds such as acetaldehyde can serve as oxidizing agents and be reduced.

$$H^+ + 1 \text{ electron} \longrightarrow \tfrac{1}{2} H_2$$

$$CH_3\!-\!C\!\!\begin{smallmatrix}H\\ \\O\end{smallmatrix} + 2 H^+ + 2 \text{ electrons} \longrightarrow CH_3\!-\!\overset{H}{\underset{H}{C}}\!-\!OH$$

These reactions in which electrons are indicated as being consumed (or produced), but in which we have not indicated the donor (or acceptor), are called **half-reactions.** Clearly, the tendency or potentiality for each of these agents to accept or furnish electrons will be due to the specific properties of that compound, and hence, it is necessary to have some standard for comparison. That standard is H_2, which has been arbitrarily given the **reduction potential, E_0,** of 0.000 V at pH 0 for the half-reaction

$$H^+ + 1 e^- \longrightarrow \tfrac{1}{2} H_2 \tag{9.24}$$

Since a proton is consumed in Reaction 9.24, the potential of this half-reaction will vary with pH, and at pH 7.0, the reduction potential E_0' of Reaction 9.24 may be calculated to be -0.420 V. With this as a standard, it is possible to determine the reduction potential of any other compound capable of oxidation–reduction with reference to hydrogen. A list of such potentials, which includes several coenzymes and substrates to be discussed in subsequent chapters, is found in Table 9.3. Note that these potentials are for the reactions written as **reductions.** When any two of the half-reactions in Table 9.3 are coupled, the one with the *more* positive reduction potential will go as written (i.e., as a reduction), driving the half-reaction with the *less* positive reduction potential backward (i.e., as an oxidation). Qualitatively, one may observe that those compounds with the more positive reduction potentials (e.g., O_2 or Fe^{3+}) are good **oxidizing agents,** while those with the more negative reduction potentials are **reducing agents** (e.g., H_2 or NADH).

It is possible to derive the expression $\Delta G' = -n\mathscr{F}\Delta E_0'$, where n is the number of electrons transferred in an oxidation–reduction reaction, $\mathscr{F}$ is Faraday's constant (23,063 cal/V equivalent), and $\Delta E_0'$ is the difference in the reduction potential between the oxidizing and reducing agents. That is,

$$\Delta E_0' = [E_0' \text{ of half-reaction containing oxidizing agent}]$$
$$- [E_0' \text{ of half-reaction containing reducing agent}]$$

For example, consider the overall reaction resulting from coupling the two half-reactions involving acetaldehyde and NAD^+.

$$\text{Acetaldehyde} + 2 H^+ + 2 e^- \longrightarrow \text{Ethanol} \tag{9.25}$$

$$NADH + H^+ \longrightarrow NAD^+ + 2 H^+ + 2 e^- \tag{9.26}$$

Half-reaction 9.25 will go as a reduction because it has the higher reduction potential. Half-reaction 9.26 then will go as an oxidation *in the opposite direc-*

TABLE 9.3 Reduction Potentials of Some Oxidation–Reduction Half-Reactions of Biological Importance

HALF-REACTION (WRITTEN AS A REDUCTION)	E_0' AT pH 7.0 (V)
$\frac{1}{2} O_2 + 2 H^+ + 2 e^- \longrightarrow H_2O$	0.82
$Fe^{3+} + 1 e^- \longrightarrow Fe^{2+}$	0.77
Cytochrome a–$Fe^{3+} + 1 e^- \longrightarrow$ Cytochrome a–Fe^{2+}	0.29
Cytochrome c–$Fe^{3+} + 1 e^- \longrightarrow$ Cytochrome c–Fe^{2+}	0.25
Ubiquinone $+ 2 H^+ + 2 e^- \longrightarrow$ Ubihydroquinone	0.10
Dehydroascorbic acid $+ 2 H^+ + 2 e^- \longrightarrow$ Ascorbic acid	0.06
Oxidized glutathione $+ 2 H^+ + 2 e^- \longrightarrow$ 2 Reduced glutathione	0.04
Fumarate $+ 2 H^+ + 2 e^- \longrightarrow$ Succinate	0.03
Cytochrome b–$Fe^{3+} + 1 e^- \longrightarrow$ Cytochrome b–Fe^{2+}	−0.04
Oxalacetate $+ 2 H^+ + 2 e^- \longrightarrow$ Malate	−0.10
Yellow enzyme $+ 2 H^+ + 2 e^- \longrightarrow$ Reduced yellow enzyme	−0.12
Acetaldehyde $+ 2 H^+ + 2 e^- \longrightarrow$ Ethanol	−0.16
Pyruvate $+ 2 H^+ + 2 e^- \longrightarrow$ Lactate	−0.19
Riboflavin $+ 2 H^+ + 2 e^- \longrightarrow$ Riboflavin–H_2	−0.20
1,3-Diphosphoglyceric acid $+ 2 H^+ + 2 e^- \longrightarrow$ Glyceraldehyde-3-phosphate $+$ Pi	−0.29
$NAD^+ + 2 H^+ + 2 e^- \longrightarrow NADH + H^+$	−0.32
Acetyl CoA $+ 2 H^+ + 2 e^- \longrightarrow$ Acetaldehyde $+$ CoA–SH	−0.41
$H^+ + 1 e^- \longrightarrow \frac{1}{2} H_2$	−0.42
Ferredoxin–$Fe^{3+} + 1 e^- \longrightarrow$ Ferredoxin–Fe^{2+}	−0.43
Acetate $+ 2 H^+ + 2 e^- \longrightarrow$ Acetaldehyde $+ H_2O$	−0.47

tion from which it is given in Table 9.3. The overall reaction is

$$\text{Acetaldehyde} + \text{NADH} + \text{H}^+ \longrightarrow \text{NAD}^+ + \text{Ethanol} \qquad (9.27)$$

The $\Delta E_0'$ for Reaction 9.27 will be $-0.16 - (-0.32)$ or 0.16 V and the $\Delta G'$ for Reaction 9.27 will be

$$\Delta G' = (-2)(23,063)(0.16)$$
$$= -7400 \text{ cal}$$

Because this figure is a large negative quantity, the reaction is feasible thermodynamically. Whether the reaction will occur at a detectable rate is not indicated by the information at hand.

In a similar manner, the $\Delta G'$ may be calculated for the oxidation of NADH by molecular O_2, a common reaction in living tissues.

$$\text{NADH} + \text{H}^+ + \frac{1}{2} O_2 \longrightarrow \text{NAD}^+ + H_2O \qquad (9.28)$$

In this reaction, $n = 2$, $\Delta E_0' = 0.82 - (-0.32)$ or 1.14 V, and

$$\Delta G' = -n\mathcal{F}\Delta E_0'$$
$$= (-2)(23,063)(1.14)$$
$$= -52,600 \text{ cal}$$

Although the $\Delta G'$ is a large negative quantity, this has no bearing on whether NADH is rapidly oxidized. As a matter of fact, NADH is stable in the presence of O_2 and will react only in the presence of appropriate enzymes.

The *standard* reduction potential (E_0), in analogy with the standard free-energy change ($\Delta G°$), implies some specific condition or state of the reactants in an oxidation–reduction reaction. Just as $\Delta G°$ specifies that the reactants in a hydrolytic reaction, for example, are all present in their standard state (for solutes 1 M), the term E_0 specifies that the ratio of the oxidant to reductant in an oxidation–reduction reaction is unity. Therefore, just as the ΔG for a reaction in which the reactants are not present at 1 M can be related to $\Delta G°$ (Equation 9.4), the E for an oxidation–reduction reaction in which the oxidized form (oxidant) and reduced form (reductant) are not present in a 1:1 ratio can be related to E_0 by the Nernst equation.

$$E = E_0 + \frac{2.303RT}{n\mathcal{F}} \log \frac{[\text{Oxidant}]}{[\text{Reductant}]}$$

From this, it may be calculated that the E will be 0.030 V more positive than E_0 (therefore more oxidizing) if the ratio of the oxidant to reductant is 10:1 and 0.060 V more positive if that ratio is 100:1. Since there is no reason that this ratio should be 1:1 in biological systems, it is clear that the actual reduction potential (E) can vary significantly from the standard reduction potential (E_0).

This is but a brief discussion of some energy relationships encountered in biochemistry. Several references follow that can be consulted for greater detail.

REFERENCES

1. *I. H. Segel*, Biochemical Calculations, *2nd ed. New York: Wiley, 1976.*
2. *L. L. Ingraham and A. H. Pardee, "Free Energy and Entropy in Metabolism." In* Metabolic Pathways, *D. M. Greenberg, ed., 3rd ed., Vol. 1. New York: Academic Press, 1967.*
3. *A. L. Lehninger,* Bioenergetics, *2nd ed. Menlo Park, Calif.: Benjamin, 1971.*
4. *E. Racker,* A New Look at Mechanisms in Bioenergetics. *New York: Academic Press, 1976.*
5. *H. M. Kalckar,* Biological Phosphorylations, Development of Concepts. *Englewood Cliffs, N.J.: Prentice-Hall, 1969.*

REVIEW PROBLEMS

1. From your knowledge of energy-rich phosphate compounds, assign approximate values for the $\Delta G'$ of the following reactions:

$$CH_3-\overset{\underset{\parallel}{O}}{C}-O-CH_3 + C_2H_5OH = CH_3-\overset{\underset{\parallel}{O}}{C}-O-C_2H_5 + CH_3OH$$

$$CH_3-\overset{\underset{\parallel}{O}}{C}-O-PO_3H_2 + ADP = CH_3COOH + ATP$$

$$CH_3-\overset{\underset{\parallel}{O}}{C}-O-PO_3H_2 + C_2H_5OH = CH_3-\overset{\underset{\parallel}{O}}{C}-O-C_2H_5 + H_3PO_4$$

$$CH_3-\overset{\underset{\parallel}{O}}{C}-O-PO_3H_2 + H_2O = CH_3-\overset{\underset{\parallel}{O}}{C}-OH + H_3PO_4$$

2. The $\Delta G'$ for Equation 9.21 at pH 7.0 (Section 9.5) is given as $+1500$ cal/mole. *In vivo* the following concentrations are observed: (D-glyceraldehyde-3-phosphate) $= 10^{-4}$ M; (1,3 diphosphoglyceric acid) $= 10^{-5}$ M; and (inorganic phosphate), Pi $= 0.01$ M. What must the ratio of $NAD^+/NADH$ be in order for the reaction to proceed spontaneously from left to right?

3. The enzyme nucleoside diphosphate kinase catalyzes the following reaction:

$$GDP + ATP \rightleftharpoons GTP + ADP$$

Assuming the changes in free energy on hydrolysis of ATP (to ADP and H_3PO_4) and GTP (to GDP and H_3PO_4) are equal, calculate the concentration of the reactants and products at equilibrium, starting with 4 mM GDP and 4 mM ATP.

4. The $\Delta G'$ of hydrolysis of acetylphosphate to acetate and H_3PO_4 is $-10,000$ cal/mole (at pH 7.0). The $\Delta G'$ of hydrolysis of ATP to ADP and H_3PO_4 is -7300 cal/mole (at pH 7.0). Calculate the $\Delta G'$ and K_{eq} of the following reaction at pH 7.0 (assume the temperature is 25°C).

$$CH_3\underset{\underset{O}{\|}}{C}-OPO_3H_2 + ADP \rightleftharpoons CH_3COOH + ATP$$

CHAPTER 10

CARBOHYDRATES I: GLYCOLYSIS

The reader will now be introduced to the subject of intermediary metabolism as it pertains to carbohydrates. The reactions of alcoholic fermentation (or glycolysis) are described together with the energy relationships involved. Similarly, the process by which living organisms can reverse the glycolytic sequence and synthesize carbohydrates from simpler molecules such as lactic acid is described. Glycogen breakdown and entry of the sugar residues into the glycolytic sequence are addressed as well as the biosynthesis of polysaccharides. Mechanisms for regulating these processes as they occur in the intact cell are discussed.

10.1 CARBOHYDRATES AS A SOURCE OF ENERGY

Carbohydrates are a major source of energy for living organisms. In our food, the chief source of carbohydrate is starch, the polysaccharide produced by plants, especially the cereal crops, during photosynthesis. Plant starch is also the chief source of energy for animals, both domestic and wild. Relatively large amounts of starch can be stored in plant cells in times of abundant supply, to be used later when there is a demand for energy production. Cellulose, also produced by plants in large amounts as a structural carbohydrate, cannot be utilized by man because he lacks the necessary hydrolytic enzymes in his intestinal tract for hydrolysis.

In animals, glycogen occurs in most tissues, but about two-thirds of the body glycogen is found in skeletal muscles where it serves as a readily available source of energy for muscular contraction during vigorous exercise. Most of the remaining glycogen is found in the liver where it serves to maintain the blood glucose concentration. Simple sugars such as sucrose, glucose, fructose, mannose, and galactose are also encountered in nature and are utilized by living forms as food. Since glucose is the compound formed from both starch and glycogen on metabolism, our discussion of carbohydrate metabolism commences with this monosaccharide.

Glucose is used both by aerobic and anaerobic organisms. In the initial stages, the pathway (glycolysis) is the same in the two types of organisms. The glycolytic pathway supplies energy to anaerobic organisms by breaking down glucose to smaller entities without net consumption of oxygen and the end products are either excreted or accumulated in the cell. Aerobic organisms have evolved from anaerobic organisms and have retained the glycolytic pathway, but being able to utilize oxygen, the aerobic organisms have developed mechanisms that can complete the catabolism of end products forming CO_2 and H_2O. Thus, the glycolytic pathway accomplishes an incomplete breakdown of glucose and provides relatively small amounts of energy for the cell, whereas aerobic organisms, by the complete catabolism of glucose, gain much more energy. It can be argued, however, that the incomplete combustion of glucose was actually advantageous to evolving life forms since, in this way, they could produce molecules needed in their cellular structure and for their diverse functions.

10.2 GLYCOLYSIS AND ALCOHOLIC FERMENTATION

The sequence of reactions by which glucose is degraded anaerobically is called the **glycolytic sequence**. Strictly speaking, this refers to the production of two moles of lactic acid from one mole of glucose (Figure 10.1). Monosaccharides other than glucose can be broken down by the glycolytic sequence, provided they can be converted into an intermediate in that sequence. Energy is released in the form of ATP as the monosaccharide is degraded and several important metabolites are produced for use elsewhere in intermediary metabolism. All organisms, with the exception of blue-green algae, possess the ability to degrade glucose by this glycolytic process, forming two moles of pyruvic acid. Those cells and tissues that actually convert pyruvic acid to lactic acid as a major end product are much more limited. Notable examples are the skeletal (white) muscle of animals, lactic acid bacteria (that produce sour milk and sauerkraut), and some plant tissues (potato tubers). Skeletal muscle, with its poor oxygen supply and relatively few mitochondria but high concentration of glycolytic enzymes, is ideally designed for carrying out glycolysis; heart muscle, well supplied with oxygen and mitochondria, will convert only small quantities of pyruvic acid to lactic acid. As will be seen, most tissues that have an adequate supply of O_2 utilize the pyruvic acid directly by oxidizing it to acetyl CoA in the aerobic phase of carbohydrate metabolism (Chapter 12).

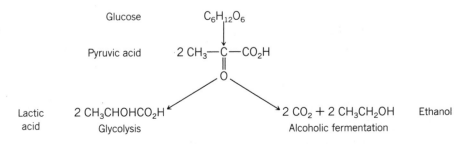

FIGURE 10.1   The process of glycolysis and alcoholic fermentation.

In alcoholic fermentation, two moles of CO_2 and ethanol are produced from one mole of glucose (Figure 10.1). This process, which occurs principally in yeasts and some other microorganisms, is identical to glycolysis in that two moles of pyruvic acid are produced from one of glucose. However, in alcoholic fermentation, pyruvic acid is converted to ethanol and CO_2 instead of lactic acid.

Note that both glycolysis and alcoholic fermentation proceed without the participation of molecular oxygen, even though oxidation has occurred in both processes. As evidence that oxidation has occurred, observe that some of the carbon atoms (the —COOH of lactic acid and CO_2) are more oxidized than they are in the glucose molecule, while others (the CH_3 groups of lactic acid and ethanol) are more reduced.

The sequence of reactions of glycolysis and alcoholic fermentation as it exists today was developed by the earliest workers in enzymology. In 1897, the Buchners in Germany obtained a cell-free extract of yeast that fermented sugars to CO_2 and ethanol. Shortly thereafter, the work of Harden and Young in England implicated phosphorylated derivatives of the sugars in alcoholic fermentation. Today, the glycolytic sequence is recognized as being composed of the reactions diagrammed inside the front cover of this text. A list of the pioneers in the field who were the architects of this scheme includes Embden, Meyerhof, Robison, Neuberg, the Cori's, Lipmann, Parnas, and Warburg. Their studies on the enzymatic aspects of glycolysis served as models for later workers examining the metabolism of lipids, amino acids, nucleic acids, and proteins, as well as respiration and photosynthesis. Many of the biochemical principles established by workers in the field of glycolysis apply equally well in other areas of intermediary metabolism; efforts will be made to identify these common principles.

10.3 GLYCOLYSIS: THE PREPARATIVE PHASE

Ten reactions are involved in the conversion of glucose to lactic acid. These may conveniently be divided into two groups. The first four reactions comprise a preparative phase in which glucose is phosphorylated and converted to D-glyceraldehyde-3-phosphate. This phase requires an expenditure of energy that is subsequently released when the glyceraldehyde-3-phosphate is oxidatively degraded to pyruvic acid.

10.3.1 HEXOKINASE AND GLUCOKINASE

Glucose is introduced into the glycolytic sequence by phosphorylation with ATP to yield glucose-6-phosphate.

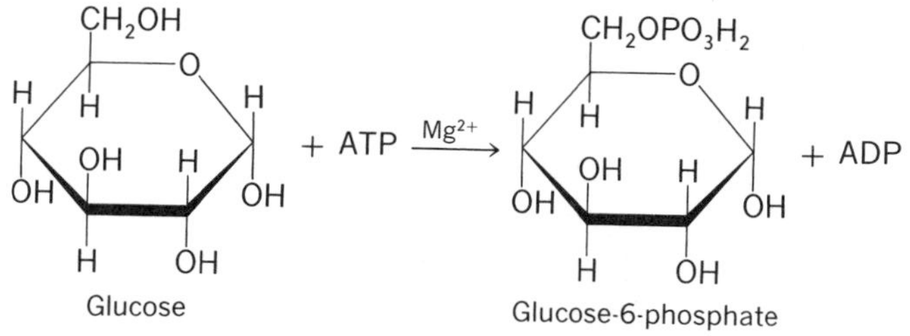

$$\Delta G' = -4000 \text{ cal (pH 7.0)}$$

This reaction is catalyzed by two different enzymes, one called **hexokinase** and one named **glucokinase**. Hexokinase is found in all cells that utilize glucose; several forms are found in animal tissues and the one in yeast was first discovered by Meyerhof in 1927. The yeast enzyme is a dimer consisting of two polypeptide chains of MW 55,000, each containing an active catalytic site. As the name implies, hexokinase will catalyze the transfer of phosphate from ATP to the primary alcohol of a number of hexoses, for example, glucose, fructose, and mannose forming the 6-phosphate. The animal hexokinases, which are monomers (MW = 100,000), exhibit product inhibition by glucose-6-phosphate that acts by binding at a site different from the catalytic site. Glucokinase is found almost exclusively in the liver where it plays an important role in removing excess glucose from the blood stream and storing it as liver glycogen.

Liver contains a fructokinase that produces fructose-1-phosphate rather than the 6-ester, and many animal tissues contain a galactokinase that produces galactose-1-phosphate. The metabolism of these esters and that of mannose-6-phosphate are discussed later in this chapter (see Sections 10.6.1 and 10.6.2).

It is informative to consider the energy changes that occur in these reactions. When glucose is phosphorylated to form glucose-6-phosphate, a compound having a low-energy phosphate ester grouping has been produced. The free energy of hydrolysis of this compound is about -3300 cal.

Glucose-6-phosphate   Glucose

$$\Delta G' = -3300 \text{ cal (pH 7.0)}$$

The phosphate group attached to the sugar was obtained from the terminal phosphate group of ATP. As we have seen, the free energy of hydrolysis of the latter compound is about -7300 cal.

$$ATP + H_2O \longrightarrow ADP + H_3PO_4$$
$$\Delta G' = -7300 \text{ cal (pH 7.0)}$$

In the hexokinase reaction, a high-energy bond of ATP was utilized and a low-energy structure (that of glucose-6-phosphate) was formed. In the terminology of the biochemist, the reaction catalyzed by hexokinase has resulted in the formation of a low-energy phosphate compound by the expenditure of an energy-rich phosphate structure. Normally, the loss of a high-energy bond by hydrolysis would result in the liberation of the -7300 cal as heat if changes in entropy are neglected. In the hexokinase reaction, part of that energy (-3300 cal) is conserved in the formation of the low-energy structure, and the remainder (-4000 cal) is liberated as heat, again neglecting entropy changes. Thus, we may estimate the free-energy change for the hexokinase reaction to be -4000 cal; that is, the reaction is strongly exergonic and the equilibrium of the reaction is far to the right. Experimental confirmation of this conclusion exists in that an equilibrium constant of 2×10^3 at pH 7.0 corresponding to a $\Delta G'$ of -4500 cal has been measured.

It is also informative to consider the possibility of reversing the hexokinase reaction (Equation 10.1). Given the K_{eq} of 2000, we can calculate with Equation 9.4 (Chapter 9) that one needs only a ratio of 200:1 of ADP to ATP to synthesize ATP and glucose if the ratio of glucose-6-phosphate to glucose is 10:1. In addition, the hexokinase reaction is demonstrably reversible in the test tube; but in the cell, the reaction never goes from right to left. Other factors clearly determine that the phosphorylation or glucose by ATP is a unidirectional process in the intact organism.

One of these is the affinity of the enzyme for its four substrates. The K_m's of liver hexokinases are as follows: glucose, 1×10^{-5} M; ATP, 1×10^{-4} M; glucose-6-phosphate, 0.08 M; and ADP, 3×10^{-3} M. Thus, if each substrate were present at 10^{-3} M, the enzyme would be saturated with glucose and nearly saturated with ATP, but would not yet be even half-saturated with ADP or glucose-6-phosphate. Since the enzyme rate is dependent on the degree of complex formation between the catalyst and its substrates (see Section 4.4), the phosphorylation of glucose will be greatly favored at 10^{-3} M when all substrates are present at 10^{-3} M. In addition, glucose-6-phosphate strongly inhibits liver hexokinase; that is, the enzyme exhibits end product inhibition. Clearly, the enzyme will cease to function as soon as any significant quantity of glucose-6-phosphate is produced, and it will remain inactive until the level of glucose-6-phosphate decreases as a result of its being used up in other reactions.

10.3.2 PHOSPHOHEXOISOMERASES

The next reaction in glycolysis is the isomerization of glucose-6-phosphate catalyzed by phosphoglucoisomerase.

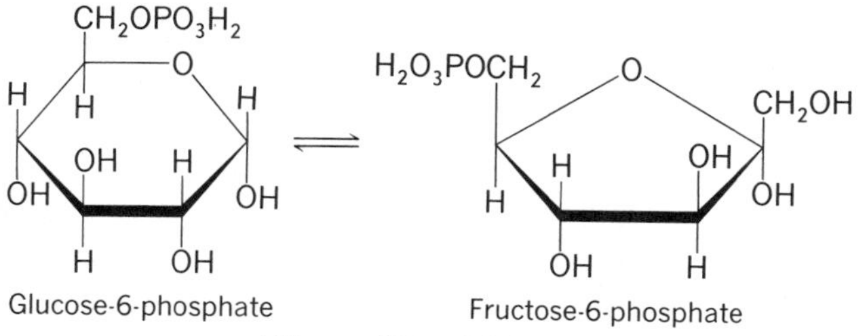

Glucose-6-phosphate   Fructose-6-phosphate

$$\Delta G' = +400 \text{ cal (pH 7.0)}$$

(10.2)

The K_{eq} of the reaction written from left to right is approximately 0.5. The human skeleton enzyme (MW 130,000) is a dimer of identical subunits that requires Mg^{2+}. An isomerase that catalyzes the conversion of mannose-6-phosphate to fructose-6-phosphate has been isolated from rabbit muscle. Although the three sugars glucose, fructose, and mannose are readily interconverted in dilute alkali (the Lobry de Bruyn–von Ekenstein transformation), the two isomerases are highly specific for fructose-6-phosphate and the corresponding hexose-6-phosphate for which they are named. Although Equation 10.2 has been written with the pyranose and furanose structures, the actual isomerization involves the open-chain form of the sugars, and an enediol (Section 2.5.2) is believed to be an intermediate.

10.3.3   PHOSPHOFRUCTOKINASE

This enzyme catalyzes the phosphorylation of fructose-6-phosphate by ATP. The enzyme requires Mg^{2+} and is specific for fructose-6-phosphate (Reaction 10.3). As with hexokinase, the high-energy bond of ATP is utilized to synthesize the low-energy ester-phosphate bond of fructose-1,6-bisphosphate. Using the arguments presented in the former case, we may expect that this reaction also should proceed with a large decrease in free energy and therefore should not be freely reversible. The $\Delta G'$ is -3400 cal/mole.

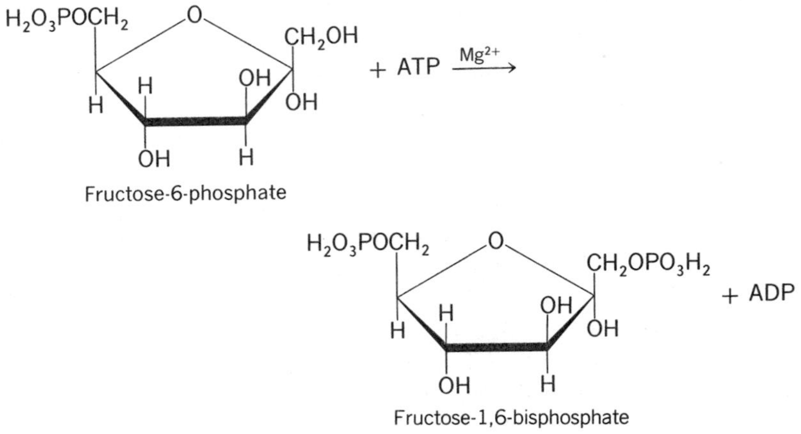

Fructose-6-phosphate

$$+ \text{ ATP} \xrightarrow{\ Mg^{2+}\ }$$

Fructose-1,6-bisphosphate

$$+ \text{ ADP} \qquad (10.3)$$

$$\Delta G' = -3400 \text{ cal (pH 7.0)}$$

Muscle phosphofructokinase (MW 360,000) is a tetramer of identical subunits that dissociates into inactive dimers.

Phosphofructokinase is an important site for metabolic regulation because its activity may be either increased or decreased by a number of common metabolites. Such effects are of the **allosteric type** (Section 4.12) in that they are the result of an interaction between the metabolite and the protein catalyst at a site other than the site where catalysis occurs. Thus, excess ATP and citric acid inhibit phosphofructokinase; that is, they are negative effectors. On the other hand, fructose 2,6-bisphosphate AMP, ADP, and fructose-6-phosphate stimulate the enzyme. The role of this enzyme and its companion fructose-1,6-bisphosphate phosphatase will be discussed later (Section 10.8).

Higher plants contain a phosphofructokinase that utilizes inorganic pyrophosphate as a phosphorylating agent instead of ATP.

$$\text{Fructose-6-phosphate} + \text{PPi} \rightleftharpoons \text{Fructose-1,6-bisphosphate} + \text{Pi} \qquad (10.3a)$$

This reaction, in contrast to that catalyzed by the ATP-dependent kinase, is reversible. The plant enzyme, which is more abundant in many plant tissues than the ATP-dependent enzyme, is also modulated by fructose 2,6-bisphosphate in such a way that glycolysis is promoted.

10.3.4   ALDOLASE

The last of the four reactions directly involved in the preparative phase of glycolysis is catalyzed by the enzyme aldolase. This reaction involves the cleavage of fructose-1,6-bisphosphate between $C-3$ and $C-4$ to form two triose (sugar) phosphates.

$$\begin{array}{c} CH_2OPO_3H_2 \\ | \\ C=O \\ | \\ HOCH \\ | \\ HCOH \\ | \\ HCOH \\ | \\ CH_2OPO_3H_2 \\ \text{Fructose-1,6-bisphosphate} \end{array} \rightleftharpoons \begin{array}{c} CH_2OPO_3H_2 \\ | \\ C=O \\ | \\ CH_2OH \\ + \\ \\ \\ HCOH \\ | \\ CH_2OPO_3H_2 \end{array}$$

Dihydroxy acetone phosphate

Glyceraldehyde-3-phosphate

(10.4)

$$\Delta G' = +5500 \text{ cal (pH 7.0)}$$

The K_{eq} for Reaction 10.4 from left to right is 10^{-4} M; this corresponds to a $\Delta G'$ of $+5500$ cal. Such values for the K_{eq} or $\Delta G'$ would appear to indicate that the reaction does not proceed from left to right. However, in a reaction of this sort, in which one reactant gives rise to two products, the equilibrium is strongly influenced by the concentration of the compounds involved. One can easily show by a simple calculation that, as the initial concentration of fructose-1,6-bisphosphate is lowered, a progressively larger amount of it will be converted to the triose. Thus, at an initial concentration of 0.1 M fructose bisphosphate, about 97% of it will remain when equilibrium is attained. However, at 10^{-4} M fructose bisphosphate initially, only 40% will remain at equilibrium.

10.3.5 TRIOSE PHOSPHATE ISOMERASE

The production of D-glyceraldehyde-3-phosphate in the aldolase reaction technically completes the preparative phase of glycolysis. The second or **energy-yielding** phase involves the oxidation of glyceraldehyde-3-phosphate, a reaction that is examined in detail in the next section. Note, however, that only half of the glucose molecule has been converted to D-glyceraldehyde-3-phosphate by Reactions 10.1 to 10.4. If cells were unable to convert dihydroxy acetone phosphate to glyceraldehyde-3-phosphate, half of the glucose molecule would accumulate in the cell as the ketose phosphate or be disposed of by other reactions. During evolution, this problem has been solved by the cell acquiring the enzyme **triose phosphate isomerase** that catalyzes the interconversion of these two trioses and permits the subsequent metabolism of all six carbon atoms of glucose.

$$\begin{array}{c} H \quad O \\ \diagdown C \diagup \\ | \\ HCOH \\ | \\ CH_2OPO_3H_2 \\ \text{Glyceraldehyde-3-phosphate} \end{array} \rightleftharpoons \begin{array}{c} CH_2OH \\ | \\ C=O \\ | \\ CH_2OPO_3H_2 \\ \text{Dihydroxy acetone phosphate} \end{array}$$

(10.4a)

$$\Delta G' = -1800 \text{ cal (pH 7.0)}$$

10.4 GLYCOLYSIS: THE ENERGY-YIELDING PHASE

10.4.1 GLYCERALDEHYDE-3-PHOSPHATE DEHYDROGENASE

This reaction, which is the first in the energy-yielding or second phase of glycolysis, is also the first reaction in the glycolytic sequence to involve oxidation–reduction. It is also the first reaction in which a high-energy phosphate compound has been formed where none previously existed.

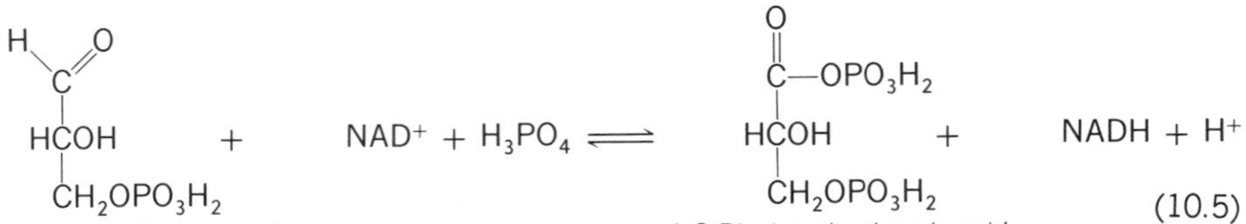

$$\begin{array}{ccc}
\text{Glyceraldehyde-3-phosphate} + \text{NAD}^+ + \text{H}_3\text{PO}_4 & \rightleftharpoons & \text{1,3-Bisphosphoglyceric acid} + \text{NADH} + \text{H}^+
\end{array} \tag{10.5}$$

$$\Delta G' = +1500 \text{ cal (pH 7.0)}$$

As a result of the oxidation of an aldehyde group to the level of a carboxylic acid, much of the energy that presumably would have been released in the form of heat has been conserved in the formation of the acyl phosphate group of 1,3-bisphosphoglyceric acid. The oxidizing agent involved is NAD^+. The energetics of this reaction together with that of Reaction 10.6 (next presented), which results in the formation of ATP, have been described in more detail in Chapter 9.

The $\Delta G'$ for Equation 10.5 is about $+1500$ cal. This corresponds to a K_{eq} of 0.08 and means that the reaction is readily reversible. This is to be expected, since the cell has modified the strongly exergonic oxidation of an aldehyde to a carboxylic acid into a reaction in which much of that energy is conserved as an acyl phosphate.

The rabbit muscle enzyme (MW 145,000) consists of four identical subunits, each of which contains two binding sites, one each for glyceraldehyde-3-phosphate and NAD^+. The catalytic site that binds the aldehydic substrate contains a thiol group (of cysteine) that plays an essential role in the reaction mechanism. Initially, the substrate and the thiol group combine to make a thiohemiacetal.

$$R-CHO + HS-Enz \rightleftharpoons R-\underset{\underset{OH}{|}}{\overset{\overset{H}{|}}{C}}-S-Enz$$

NAD^+ is then reduced and a thioester intermediate is formed.

$$R-\underset{\underset{OH}{|}}{\overset{\overset{H}{|}}{C}}-S-Enz + NAD^+ \rightleftharpoons R-\underset{\underset{O}{\|}}{C}-S-Enz + NADH + H^+$$

The thioester then reacts with inorganic phosphate to regenerate the free thiol group on the protein and 1,3-bisphosphoglyceric acid.

$$R-\underset{\underset{O}{\|}}{C}-S-Enz + H_3PO_4 \rightleftharpoons R-\underset{\underset{O}{\|}}{C}-OPO_3H_2 + Enz-SH$$

The sum of these three partial reactions is Reaction 10.5

10.4.2   PHOSPHOGLYCERYL KINASE

This reaction accomplishes the transfer of the phosphate from the acyl phosphate formed in the preceding reaction to ADP to form ATP. The name of the enzyme is derived from the reverse reaction, in which a high-energy phosphate is transferred from ATP to 3-phosphoglyceric acid.

$$\begin{array}{c} \underset{\|}{\overset{O}{C}}-OPO_3H_2 \\ HCOH \\ CH_2OPO_3H_2 \end{array} \quad + \quad ADP \quad \underset{}{\overset{Mg^{2+}}{\rightleftharpoons}} \quad \begin{array}{c} CO_2H \\ HCOH \\ CH_2OPO_3H_2 \end{array} \quad + \quad ATP \quad (10.6)$$

1,3-Diphosphoglyceric acid \qquad 3-Phosphoglyceric acid

$$\Delta G' = -4500 \text{ cal (pH 7.0)}$$

In this reaction, the acyl phosphate group ($\Delta G'$ of hydrolysis of $-11,800$ cal) has been utilized to drive the phosphorylation of ADP and make ATP ($\Delta G'$ of hydrolysis of -7300 cal). From these considerations alone, one would predict that the $\Delta G'$ for Reaction 10.6 would be -4500 cal, corresponding to a K_{eq} of 2×10^3. A value of 3.1×10^3 has been reported in the literature. Although, the thermodynamics favor ATP formation, Reaction 10.6 will be reversed if the ratio of ATP:ADP is 10:1 and the ratio of 3-phosphoglycerate to 1,3-bisphosphoglycerate exceeds 200:1. In contrast to Reactions 10.1 and 10.3, in which other factors (product inhibition, allosteric modifiers) determine that those reactions proceed in only one direction, Reaction 1.6 catalyzed by phosphoglyceryl kinase *does* proceed from right to left when glycolysis is reversed in the cell. This reversal of Reaction 10.6 is undoubtedly aided by the positive $\Delta G'$ for Reaction 10.5 described in the preceding section. By combining Reactions 10.5 and 10.6, we may write

$$\begin{array}{c} \underset{H}{\overset{O}{\diagdown C \diagup}} \\ HCOH \\ CH_2OPO_3H_2 \end{array} + NAD^+ + ADP + H_3PO_4 \overset{Mg^{2+}}{\rightleftharpoons} \begin{array}{c} CO_2H \\ HCOH \\ CH_2OPO_3H_2 \end{array} + ATP + NADH + H^+$$

D-Glyceraldehyde-3-phosphate \qquad 3-Phosphoglyceric acid

$$\Delta G' = -3000 \text{ cal}$$

and by adding the $\Delta G'$ for Reactions 10.5 and 10.6, we obtain an overall $\Delta G'$ of -3000 cal for the combined process.

10.4.3 PHOSPHOGLYCERYL MUTASE

Phosphoglyceryl mutase catalyzes the interconversion of the two phosphoglyceric acids. The equilibrium constant for the reaction from left to right is 0.17.

$$\begin{array}{c} CO_2H \\ HCOH \\ CH_2OPO_3H_2 \end{array} \rightleftharpoons \begin{array}{c} CO_2H \\ HCOPO_3H_2 \\ CH_2OH \end{array} \quad (10.7)$$

3-Phosphoglyceric acid \qquad 2-Phosphoglyceric acid

$$\Delta G' = +1050 \text{ cal (pH 7.0)}$$

This enzyme contains a serine whose hydroxyl group is capable of accepting a phosphate from 2,3-bisphosphoglyceric acid to form a **phosphoenzyme.**

$$E\text{—OH} + 2,3\text{-Bisphosphoglycerate} \underset{\diagdown}{\overset{\diagup}{\rightleftharpoons}} \begin{array}{l} E\text{—OP} + 2\text{-Phosphoglycerate} \\ \\ E\text{—OP} + 3\text{-Phosphoglycerate} \end{array}$$

Depending on which phosphate group is donated, either 2-phosphoglyceric acid or 3-phosphoglyceric acid is formed. When the concentration of 3-phosphoglycerate is high, the dephosphorylated form of the enzyme and 2,3-bis-phosphoglycerate are formed and then react to produce 2-phosphoglycerate and the phosphoenzyme.

10.4.4   ENOLASE

The next reaction in the degradation of glucose involves the dehydration of 2-phosphoglyceric acid to produce phosphoenol pyruvic acid, a compound with a high-energy enolic phosphate group.

$$
\begin{array}{c}
CO_2H \\
| \\
HCOPO_3H_2 \\
\| \\
CH_2OH
\end{array}
\quad \overset{Mg^{2+}}{\rightleftharpoons} \quad
\begin{array}{c}
CO_2H \\
| \\
C{-}OPO_3H_2 + H_2O \\
\| \\
CH_2
\end{array}
\qquad (10.8)
$$

2-Phosphoglyceric acid   Phosphoenol pyruvic acid

$$\Delta G' = -650 \text{ cal (pH 7.0)}$$

The equilibrium constant for this reaction is 3. The standard free-energy change is small; the $\Delta G' = -650$ cal and the reaction is freely reversible. It is interesting that by this simple process of dehydration, an energy-rich enolic phosphate ($\Delta G'$ of hydrolysis is $-14,800$ cal) is formed.

Earlier in this chapter (Reaction 10.5), the production of an energy-rich acyl phosphate was coupled to an oxidation–reduction reaction, and it was possible to rationalize the synthesis of the energy-rich structure as a consequence of that oxidation. In Reaction 10.8, a different chemical reaction, namely dehydration, is involved and it is more difficult to comprehend the synthesis of the enolic phosphate, especially when it is formed in a reaction involving a minimal $\Delta G'$. The difficulty lies primarily in the biochemist's definition of "energy-rich," and another way of looking at 2-phosphoglyceric acid and phosphoenol pyruvic acid is required. Although one compound is energy-poor and the other is energy-rich (if we consider the $\Delta G'$ of hydrolysis of these compounds), about the *same* amount of energy would be produced if they were oxidized to CO_2, H_2O, and H_3PO_4. The key to this puzzle is found in the fact that the dehydration catalyzed by enolase results in a rearrangement of electrons in the two molecules so that a significantly larger amount of the total *potential* energy of the compound is released on hydrolysis. The explanation as to *why* phosphoenol pyruvic acid releases an unusually large amount of energy on hydrolysis was discussed earlier (Section 9.4.3).

Enolase requires Mg^{2+} for activity. In the presence of Mg^{2+} and phosphate, fluoride ions strongly inhibit the enzyme. This effect is related to the formation of a magnesium-fluorophosphate complex that is only slightly dissociated and thereby effectively removes Mg^{2+} from the reaction mixture.

10.4.5   PYRUVIC KINASE

Pyruvic kinase catalyzes the transfer of phosphate from phosphoenol pyruvic acid to ADP to produce ATP and pyruvic acid.

$$
\begin{array}{c}
CO_2H \\
| \\
C{-}OPO_3H_2 \\
\| \\
CH_2
\end{array}
\; + \; ADP \quad \overset{Mg^{2+},\, K^+}{\rightleftharpoons} \quad
\begin{array}{c}
CO_2H \\
| \\
C{=}O \\
| \\
CH_3
\end{array}
\; + \; ATP
\qquad (10.9)
$$

Phosphoenol pyruvic acid   Pyruvic acid

$$\Delta G' = -6100 \text{ cal}$$

Because of the large $\Delta G'$ of hydrolysis of phosphoenol pyruvate ($-14,800$ cal), one would expect that the $\Delta G'$ for Reaction 10.9 would be approximately -7500 cal. The literature contains values for the K_{eq} as high as 3×10^4, corresponding to $\Delta G'$ of -6100 cal. Indeed, the equilibrium is so far to the right that is is difficult to obtain an accurate value for the K_{eq} (and therefore the $\Delta G'$) of the reaction. Therefore, this reaction, like those catalyzed by phosphofructokinase (Reaction 10.3) and hexokinase (Reaction 10.1), is physiologically irreversible in the cell and offers the potential for regulation of glycolysis.

Pyruvic kinase has been purified from numerous sources and shown to be a tetramer (MW 165,000 to 240,000, depending on source) containing four identical subunits. It requires both Mg^{2+} and K^+ for activity. It is an allosteric enzyme subject to regulation by several effectors. Its activity is high when glucose needs to be converted to pyruvate (or lactate) and low when gluconeogenesis is occurring. ATP, alanine, and acetyl CoA are negative effectors, while AMP and fructose 2,6-bisphosphate are positive effectors.

Pyruvic kinase also exists as a phosphoprotein whose catalytic activity is even more strongly inhibited by ATP and alanine. The phosphorylated form of the enzyme is produced from the dephosphorylated form by the action of ATP in the presence of a protein kinase. The phosphorylated form is converted back to the dephosphoform by hydrolysis catalyzed by a protein phosphatase.

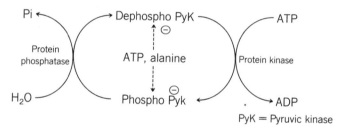

PyK = Pyruvic kinase

These phosphorylation–dephosphorylation cycles and their implications in terms of metabolic regulation will be further discussed in Sections 18.4 and 18.5. The regulation of pyruvic kinase is coordinated with that of phosphofructokinase and is described in greater detail later (Section 10.8).

10.4.6 LACTIC DEHYDROGENASE

The last reaction of glycolysis results in the production of L(+)-lactic acid when pyruvic acid is reduced by NADH. Note that because of the production of NADH in Reaction 10.5 and its

$$
\begin{array}{ccc}
CO_2H & & CO_2H \\
| & & | \\
C{=}O & + \text{ NADH } + H^+ \rightleftharpoons & HOCH & + \text{ NAD}^+ \qquad (10.10)\\
| & & | \\
CH_3 & & CH_3 \\
\text{Pyruvic acid} & & \text{L(+)-Lactic acid}
\end{array}
$$

$$\Delta G' = -6000 \text{ cal (pH 7.0)}$$

utilization in this reaction, there is no accumulation of NADH in a tissue carrying out glycolysis. It should be stressed that the supply of NAD^+ in any cell is limited. Therefore, it is conceivable that Reaction 10.5 could not occur when all the NAD^+ in the cell becomes reduced (glycolysis) and the breakdown of glucose could not proceed. This situation is avoided by the reoxidation of

NADH catalyzed by lactic dehydrogenase and, in effect, the oxidation of 3-phosphoglyceraldehyde is coupled to the reduction of pyruvate.

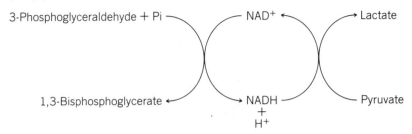

The equilibrium for the reaction catalyzed by lactic dehydrogenase is far to the right at pH 7.0 ($K_{eq} = 2.5 \times 10^4$); however, as pointed out in Section 5.4.3, the K_{eq} of reactions involving the nicotinamide coenzymes is greatly dependent on pH, and one can measure pyruvate formation under certain conditions; that is, one can reverse Reaction 10.10 by operating at a higher pH of 8 or 9.

Lactate, produced in Reaction 10.10, is the end of the glycolytic sequence as it occurs in muscle. This hydroxyacid, in contrast to pyruvate, is not convertible to other products. For it to be metabolized, it must be reoxidized to pyruvate and then converted to numerous other products as pyruvate is catabolized, or, by the process of gluconeogenesis (Section 10.7.2), resynthesized into glucose (or glycogen).

Lactic dehydrogenase (MW 140,000) is a tetrameric protein that exists in two forms found in heart (H_4) and skeletal muscle (M_4). The heart enzyme is inhibited by high (10^{-3}) concentration of pyruvate, but is active at lower values. On the other hand, the muscle enzyme (M_4) does not achieve its maximum velocity until the pyruvate concentration is 3×10^{-3} M and maintains its activity in much higher concentrations of pyruvate. Kaplan has pointed out that these properties are consistent with the tasks that the two different tissues have to perform. The heart requires a steady supply of energy that can best be achieved by converting glucose to pyruvate and then oxidizing the pyruvate to CO_2 and H_2O via the Krebs cycle (Chapter 12). This process derives the maximum amount of energy from the glucose molecule but requires an adequate supply of oxygen. In skeletal muscle, there can be sudden demands for energy at very low oxygen concentrations. This energy can be supplied by the reactions in glycolysis in which ATP is generated (Reactions 10.6 and 10.9) but in which O_2 is not involved. Such demands would, of course, require relatively large amounts of pyruvate to be formed and, in turn, reduced to lactic acid.

In support of this thesis, tissues such as the heart that are continually contracting are found to have H_4-lactic dehydrogenase, while the flight muscles of chicken and pheasants that make sporadic short flights have predominantly M_4-lactic dehydrogenase. On the other hand, the storm petrel, which is capable of sustained flight, has predominantly the H_4 type of enzyme in its flight muscles. These examples represent the extremes characteristic of highly aerobic (H_4) or anaerobic (M_4) processes; other tissues intermediate between these two extremes are known that contain the isozymes HM_3, MH_3, and M_2H_2.

10.4.7   PYRUVIC DECARBOXYLASE

At the same time that the series of reactions constituting the glycolytic sequence were being studied in muscle, alcoholic fermentation was being examined in yeast extracts. Fortunately for students in the biological sciences, the sequence of reactions that converts glucose to alcohol and CO_2 (alcoholic fermentation) is identical except for the manner in which pyruvic acid is metabolized. Organisms such as yeast, which carry out alcoholic fermentation, contain the enzyme **pyruvic decarboxylase**—this catalyzes the decarboxylation of pyruvate to acetaldehyde and CO_2 by an irreversible reaction. The enzyme, which is not present in animal tissues,

$$
\begin{array}{cc}
\underset{\text{Pyruvic acid}}{\overset{\displaystyle \begin{matrix} CO_2H \\ | \\ C=O \\ | \\ CH_3 \end{matrix}}{}} & \xrightarrow[\text{TPP}]{Mg^{2+}} \quad \underset{\text{Acetaldehyde}}{\overset{\displaystyle \begin{matrix} H \diagdown \diagup O \\ C \\ | \\ CH_3 \end{matrix}}{}} + CO_2
\end{array} \qquad (10.11)
$$

requires thiamin pyrophosphate (TPP or cocarboxylase) and Mg^{2+} as cofactors. The mechanism for this reaction was discussed in Section 5.2.3. The reaction is quite exergonic, which means that, in contrast to glycolysis, the end products of alcoholic fermentation, C_2H_5OH and CO_2, cannot be converted back to glucose.

10.4.8 ALCOHOL DEHYDROGENASE

In the final reaction of alcoholic fermentation, acetaldehyde is reduced to ethanol by NADH in the presence of **alcohol dehydrogenase** as follows:

$$
\underset{\text{Acetaldehyde}}{\overset{\displaystyle \begin{matrix} H \diagdown \diagup O \\ C \\ | \\ CH_3 \end{matrix}}{}} + NADH + H^+ \rightleftharpoons \underset{\text{Ethanol}}{\overset{\displaystyle \begin{matrix} CH_2OH \\ | \\ CH_3 \end{matrix}}{}} + NAD^+ \qquad (10.12)
$$

The K_{eq} for this reaction (Section 5.4.3) greatly favors the reduction of acetaldehyde at pH 7.0. Again, being a reaction that involves a nicotinamide nucleotide coenzyme, the process is highly dependent on pH, and one can quantitatively convert alcohol to acetaldehyde, that is, reverse Reaction 10.12, at pH 9.5 in the presence of an excess of NAD^+. Note that the reoxidation of NADH in Reaction 10.12 compensates for the production of NADH in Reaction 10.5 and means that NADH would not accumulate in a tissue carrying out alcoholic fermentation. Alcohol dehydrogenase is widely distributed, having been found in the liver, retina, and sera of animals, in the seeds and leaves of higher plants, and in many microorganisms, including yeast. Clearly, the enzyme is not restricted to tissues that produce large amounts of ethanol.

10.4.9 ENZYME NOMENCLATURE

In describing the individual reactions of the glycolytic sequence, we have used the *trivial* names of the enzymes rather than the *systematic* ones. The latter are the names assigned to these enzymes by the Committee on Enzyme Nomenclature of the International Union of Biochemistry. The systematic names of an enzyme clearly identify the protein being considered. However, as shown in Table 10.1, the systematic name is usually longer than the trivial name. In the interest of brevity, and also because the trivial name is frequently the one first used in describing the enzyme, we will usually use the trivial names in this text.

TABLE 10.1   The Trivial and Systematic Names of Enzymes of the Glycolytic Pathway

REACTION	TRIVIAL NAME	SYSTEMATIC NAME
10.1	Hexokinase	ATP: D-Hexose-6-phospho-transferase (EC 2.7.1.1)
	Glucokinase	ATP: D-Glucose-6-phospho-transferase (EC 2.7.1.2)
10.2	Phosphoglucoisomerase	D-Glucose-6-phosphate-ketol isomerase (EC 5.3.1.9)
10.3	Phosphofructokinase	ATP: D-Fructose-6-phosphate-1-phosphotransferase (EC 2.7.1.11)
10.4	Aldolase	Fructose-1,6-bisphosphate D-glyceraldehyde-3-phosphate lyase (EC 4.1.2.13)
10.4a	Triose phosphate isomerase	D-Glyceraldehyde-3-phosphate ketol isomerase (EC 5.3.1.1)
10.5	D-Glyceraldehyde 3-phosphate dehydrogenase (Triose phosphate dehydrogenase)	D-Glyceraldehyde-3-phosphate: NAD$^+$ oxidoreductase (phosphorylating) (EC 1.2.1.12)
10.6	Phosphoglyceryl kinase	ATP: 3-Phospho-D-glycerate-1-phosphotransferase (EC 2.7.2.3)
10.7	Bisphosphoglyceryl mutase	D-Phosphoglycerate-2,3-phospho-mutase (EC 5.4.2.1)
10.8	Enolase	2-Phospho-D-glycerate hydrolyase (EC 4.2.1.11)
10.9	Pyruvic kinase	ATP: Pyruvate 2-O-phospho-transferase (EC 2.7.1.40)
10.10	Lactic dehydrogenase	L-Lactate: NAD$^+$ oxidoreductase (EC 1.1.1.27)
10.11	Pyruvic decarboxylase	2-Oxoacid carboxy-lyase (EC 4.1.1.1)
10.12	Alcohol dehydrogenase	Alcohol: NAD$^+$ oxidoreductase (EC 1.1.1.1)

SOME ANCILLARY ENZYMES		
10.25	Fructose-1,6-bisphosphate phosphatase or fructose bis-phosphatase	D-Fructose-1,6-bisphosphate-1-phosphohydrolase (EC 3.1.3.11)
10.26	Glucose-6-phosphate phosphatase or glucose-6-phosphatase	D-Glucose-6-phosphate phospho-hydrolase (EC 3.1.3.9)
10.21	Phosphorylase a	1,4-α-D-Glucan: orthophos-phate α-D-glucosyl transferase (EC 2.4.1.1)
10.20	Phosphoglucomutase	D-Glucose-1,6-bisphosphate: α-D-Glucose-1-phosphate phos-photransferase (EC 2.7.5.1)

10.5 PRODUCTION OF ATP AND EFFICIENCY OF GLYCOLYSIS

Glycolysis (and alcoholic fermentation) results in a net production of ATP that is used for energy-consuming processes by the cells carrying out the anaerobic degradation of glucose. The net number of ATPs produced per molecule of glucose degraded is easily determined since only four reactions involve ATP.

Reaction 10.1

$$\text{Glucose} + \text{ATP} \longrightarrow \text{Glucose-6-phosphate} + \text{ADP}$$

Reaction 10.3

$$\text{Fructose-6-phosphate} + \text{ATP} \longrightarrow \text{Fructose-1,6-bisphosphate} + \text{ADP}$$

Reaction 10.6

$$2 \text{ 1,3-Bisphosphoglycerate} + 2 \text{ ADP} \longrightarrow 2 \text{ 3-Phosphoglycerate} + 2 \text{ ATP}$$

Reaction 10.9

$$2 \text{ Phosphoenolpyruvate} + 2 \text{ ADP} \longrightarrow 2 \text{ Pyruvate} + 2 \text{ ATP}$$

Note that Reactions 10.6 and 10.9 are doubled in order to accommodate the six carbon atoms in the molecule of glucose. Adding up these four reactions shows that two molecules of ATP are produced for each molecule of glucose degraded in the glycolytic sequence.

To understand the source of the energy used in producing these two molecules of ATP, consider the overall energy relationships as follows.

The $\Delta G'$ for the conversion of glucose to two moles of lactic acid as it might occur in a test tube can be calculated from various thermodynamic data.

$$\underset{\text{Glucose}}{C_6H_{12}O_6} \xrightarrow{\text{In a test tube}} 2 \underset{\text{Lactic acid}}{CH_3CHOHCOOH}$$
$$\Delta G' = -47,000 \text{ cal}$$

If we neglect entropy changes, this amount of energy would be liberated as heat (ΔH). In the biological organism performing glycolysis, however, the reaction must be corrected to show precisely what occurs—namely that, as glucose is converted to two moles of lactic acid, two moles of ATP are produced from ADP and inorganic phosphate. That is,

$$C_6H_{12}O_6 + 2 \text{ ADP} + 2 \text{ H}_3PO_4 \xrightarrow{\text{In the cell}} 2 \text{ CH}_3COHOCOOH + 2 \text{ ATP} + 2 \text{ H}_2O$$
$$\Delta G' = -32,400 \text{ cal}$$

Since the two ATP represent a conservation of 14,600 cal (2×7300), the $\Delta G'$ for the second equation is less by that amount [$-47,000 - (-14,600)$ or $-32,400$]. Moreover, one can speak of the *efficiency* with which the ATP has been produced in glycolysis. Since $-47,000$ cal are available and two ATPs are produced, the efficiency corresponds to $-14,600/-47,000$ or 31%.

At this point, note that ΔG for the hydrolysis of ATP under the conditions that exist in a cell may be more negative, by as much as 4000 cal, than the standard free-energy change ($\Delta G' = -7300$ cal). That is due, of course, to the fact that the concentrations of reactants in the formation of ATP are not at standard values (see Chapter 9). If the ΔG for formation of ATP is indeed

$+12,000$ cal, the efficiency of energy conservation will be $-24,000/-47,000$ or 51%.

The $\Delta G'$ values for the conversion of glucose to ethanol and CO_2 and the corresponding reaction as it occurs in the yeast cell may be written as

$$C_6H_{12}O_6 \xrightarrow{\text{In a test tube}} 2\ CH_3CH_2OH + 2\ CO_2$$

Glucose   Ethanol

$$\Delta G' = -40,000 \text{ cal}$$

$$C_6H_{12}O_6 + 2\ ADP + 2\ H_3PO_4 \xrightarrow{\text{In a yeast cell}} 2\ CH_3CH_2OH + 2\ CO_2 + 2\ ATP + 2\ H_2O$$

$$\Delta G' = -25,400 \text{ cal}$$

10.6 UTILIZATION OF OTHER CARBOHYDRATES

10.6.1 UTILIZATION OF FRUCTOSE AND MANNOSE

Sugars other than glucose are metabolized in the glycolytic sequence following their conversion by auxiliary enzymes to intermediates in that sequence. Thus, fructose and mannose can be phosphorylated by ATP in the presence of hexokinase and converted into fructose-6-phosphate and mannose-6-phosphate. The former is an intermediate in glycolysis; mannose-6-phosphate is converted to fructose-6-phosphate by the enzyme phospho-mannose isomerase in a reaction analogous to that catalyzed by phosphoglucoisomerase (Reaction 10.2).

Fructose may also be phosphorylated by a specific fructokinase, found in liver, that catalyzes the formation of fructose-1-phosphate.

Fructose + ATP $\longrightarrow$ Fructose-1-phosphate + ADP

This sugar phosphate is not convertible to either fructose-6-phosphate or fructose-1,6-bisphosphate, but can be cleaved by fructose-1-phosphate aldolase. The reaction is strictly analogous to Reaction 10.4.

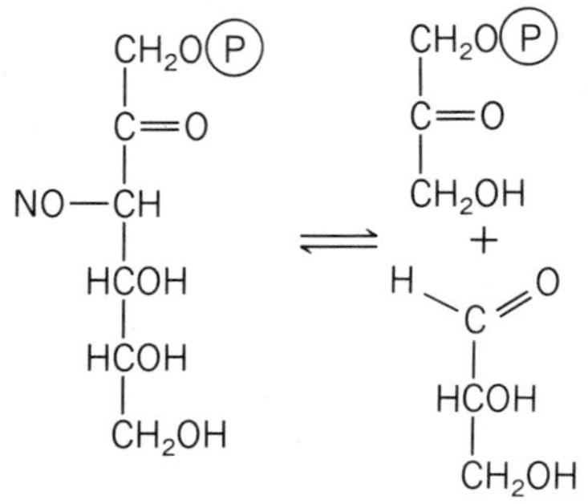

Fructose-1-phosphate

Dihydroxyacetone phosphate is an intermediate of glycolysis, but the glyceraldehyde must be additionally modified before it can be metabolized. This occurs by its oxidation to glyceric acid, followed by phosphorylation, to yield 2-phosphoglyceric acid.

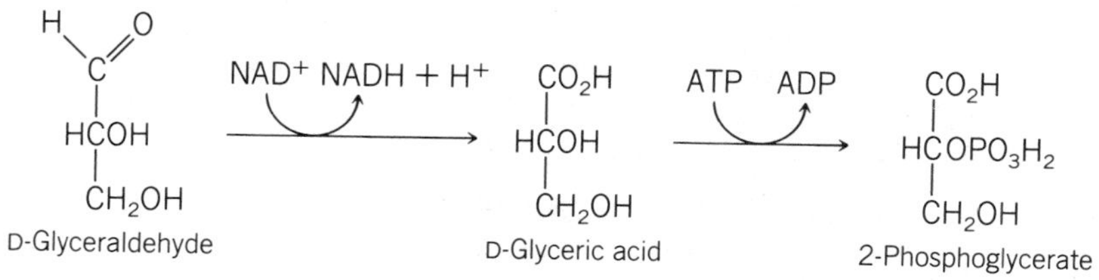

D-Glyceraldehyde   D-Glyceric acid   2-Phosphoglycerate

10.6.2 UTILIZATION OF GALACTOSE

Disaccharides such as lactose and sucrose are extremely common sources of carbohydrate in the diet of animals. The initial steps in their utilization involve hydrolysis to the component monosaccharides by specific glycosidases, lactase and sucrase (invertase), found in the animal's digestive tract. The subsequent metabolism of glucose and fructose obtained on hydrolysis of sucrose has been previously discussed. The metabolism of galactose formed (together with glucose) on the hydrolysis of lactose involves phosphorylation by ATP in the presence of a specific galactokinase that produces galactose-1-phosphate.

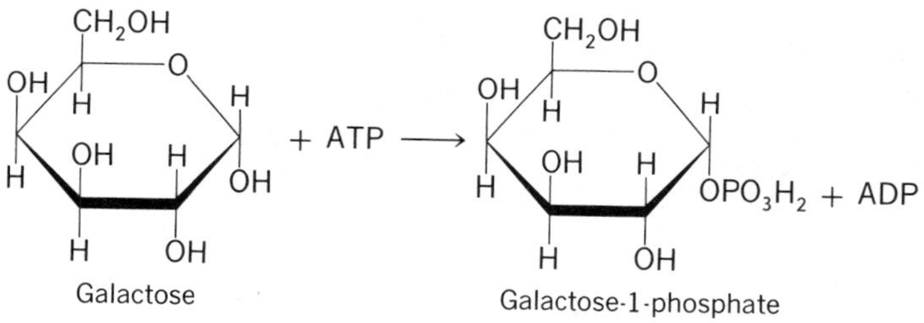

Galactose   Galactose-1-phosphate   (10.13)

Further metabolism of galactose-1-phosphate involves uridine triphosphate (UTP) and a uracil derivative of that sugar known as uridine diphosphate galactose (UDP-galactose).

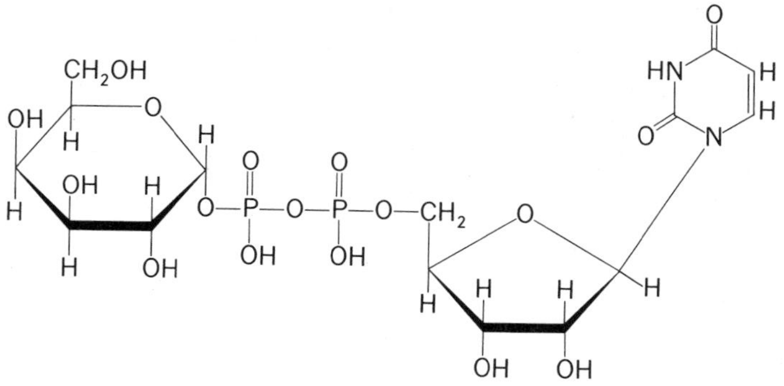

Uridine diphosphate galactose
(UDP-galactose)

The galactose-1-phosphate formed in Reaction 10.13 is converted to UDP-galactose by the enzyme **UDP-galactose pyrophosphorylase**, which is present in the liver of adult humans.

$$\underset{\text{Galactose-1-phosphate}}{\text{Gal—P}} \; + \; \underset{\text{UTP}}{\text{U—R—P—P—P}} \; \rightleftharpoons \; \underset{\text{UDP-galactose}}{\text{U—R—P—P—Gal}} \; + \; \underset{\text{Pyrophosphate}}{\text{P—P}} \qquad (10.14)$$

The various components of the UTP and sugar phosphate molecules have been identified (R, ribose; P, phosphate) to indicate the nature of the reaction. The reaction is readily reversible, as could be anticipated, since the one (interior) pyrophosphate bond is utilized to form the pyrophosphate in the sugar nucleotide; the number of energy-rich structures in the reactants and products is consequently the same. This reaction is a model one for forming these **nucleoside diphosphate sugars** or **sugar nucleotides.** As another example, ADP-glucose would be formed from ATP and glucose-1-phosphate in the presence of the specified pyrophosphorylase.

In the next step, the galactose moiety in UDP-galactose is isomerized to a glucose moiety, thereby forming UDP-glucose. The enzyme that catalyzes this reaction is known as **UDP-glucose 4-epimerase.**

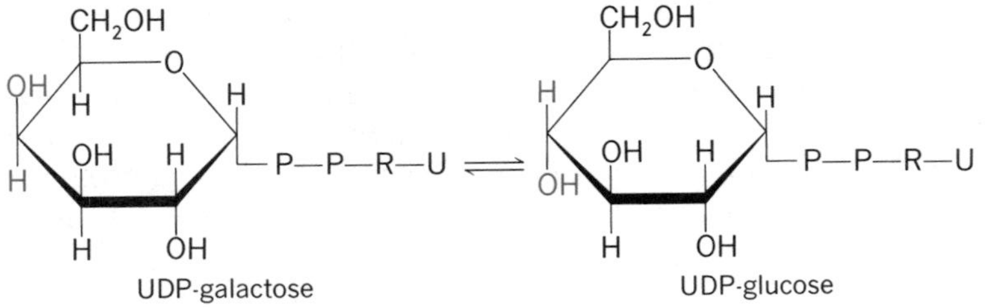

UDP-galactose ⇌ UDP-glucose   (10.15)

Finally, the action of a third enzyme **UDP-glucose pyrophosphorylase** liberates the glucose (formerly the galactose) moiety from UDP-glucose as glucose-1-phosphate.

$$U—R—P—P—Glu + P—P \rightleftharpoons U—R—P—P—P + Glu—P \qquad (10.16)$$

UDP-glucose   UTP   Glucose-1-phosphate

Note this reaction is the same as Reaction 10.14 except that glucose is the sugar involved. The sum of Reactions 10.14 through 10.16 is the conversion of galactose-1-phosphate into glucose-1-phosphate. The metabolism of the latter by glycolysis (see the following discussion) accounts for the metabolism of galactose in adult humans.

As noted above, the enzyme catalyzing the formation of UDP-galactose from galactose-1-phosphate is found only in the liver of adults. How, then, does an infant metabolize galactose? This is a pertinent question, because the main energy source that an infant has is the sugar lactose in the milk that it consumes.

Studies have shown that fetal and infant liver tissue contains the enzyme **phosphogalactose uridyl transferase.**

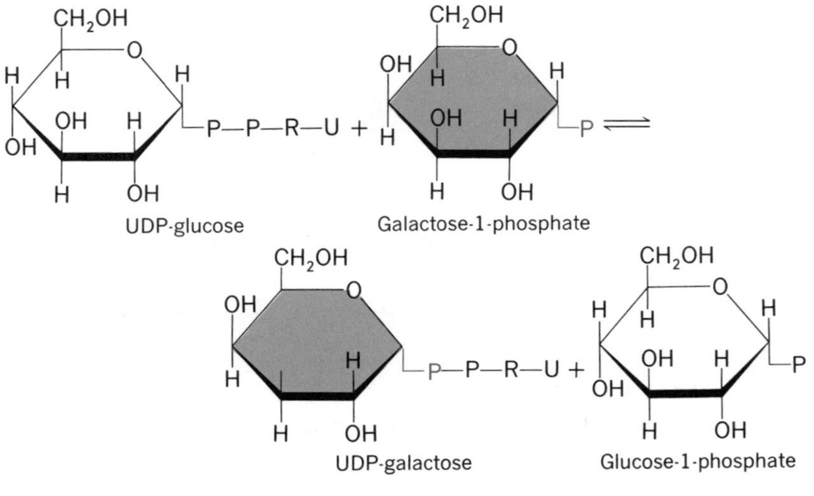

UDP-glucose + Galactose-1-phosphate ⇌ UDP-galactose + Glucose-1-phosphate   (10.17)

The coupling of this reaction with Reaction 10.15 accounts for the net conversion of galactose-1-phosphate into glucose-1-phosphate and is the normal route for galactose metabolism in infants. This series of reactions has attracted much attention because of a hereditary disorder known as **galactosemia.** Infants that have this defect cannot metabolize galactose and they exhibit a high level of galactose in the blood. The sugar is excreted in the urine and, if the condition is not attended to, the infant can develop cataracts and may become mentally retarded. The simple remedy, once the condition is identified, is to remove the source of galactose, usually the milk in the infant's diet, and supply a galactose-free diet.

Galactosemic individuals lack the uridyl transferase (Reaction 10.17), and this accounts for their failure to metabolize galactose. Only after the individual has reached puberty does an adequate amount of UDP-galactose pyrophosphorylase appear in the liver, thereby providing the adolescent with the capacity to metabolize galactose (Reaction 10.14).

It should be pointed out that the sugar nucleotides (e.g., UDP-glucose or UDP-galactose) are precursors of important cellular constituents such as glycogen, cell wall components, hyaluronic acids. Since the galactosemic infant needs a source of UDP-galactose to produce these cellular constituents, it will convert glucose-1-phosphate to UDP-galactose by reversing Reactions 10.16 and 10.15. The adult human, on the other hand, will have available the pyrophosphorylase (Reaction 10.14) for the direct synthesis of UDP-galactose from galactose.

10.6.3   UTILIZATION OF GLYCEROL

Another example of a common metabolite that is metabolized by means of the glycolytic sequence is the compound glycerol. Glycerol is produced during the breakdown of triacylglycerols (Chapter 13) and phosphorylated by ATP in the presence of glycerol kinase.

$$
\begin{array}{ccc}
\text{CH}_2\text{OH} & & \text{CH}_2\text{OH} \\
| & & | \\
\text{HOCH} \quad + \text{ATP} \xrightarrow{\text{Mg}^{2+}} & & \text{HOCH} \quad + \text{ADP} \qquad (10.18) \\
| & & | \\
\text{CH}_2\text{OH} & & \text{CH}_2\text{OPO}_3\text{H}_2 \\
\text{Glycerol} & & \textit{sn}\text{-Glycerol-3-phosphate}
\end{array}
$$

The phosphoglycerol produced in this reaction can then be oxidized to dihydroxy acetone phosphate by the enzyme **glycerol phosphate dehydrogenase.** This enzyme of the cytoplasm utilizes NAD^+ as the oxidant. The dihydroxy acetone phosphate can enter directly into the second stage of glycolysis.

$$
\begin{array}{ccc}
\text{CH}_2\text{OH} & & \text{CH}_2\text{OH} \\
| & & | \\
\text{HOCH} \quad + \text{NAD}^+ \longrightarrow & & \text{C}{=}\text{O} \quad + \quad \text{NADH} + \text{H}^+ \quad (10.19) \\
| & & | \\
\text{CH}_2\text{OPO}_3\text{H}_2 & & \text{CH}_2\text{OPO}_3\text{H}_2 \\
\textit{sn}\text{-Glycerol-3-phosphate} & & \text{Dihydroxy acetone phosphate}
\end{array}
$$

This reaction is readily reversible, thereby providing a means of making *sn*-glycerol-3-phosphate from glycolytic intermediates. *sn*-Glycerol-3-phosphate is a precursor of triacylglycerols in adipose tissue (Chapter 13).

Another glycerol phosphate dehydrogenase in mitochondria, a flavoprotein, utilizes FAD as the primary oxidant. These two enzymes play an important role in the transport of reducing equivalents from cytoplasmic NADH into the interior of the mitochondria (see Section 14.8).

10.6.4 UTILIZATION OF GLUCOSE-1-PHOSPHATE

An important intermediate in the utilization of polysaccharides, to be discussed in Section 10.6.5, is glucose-1-phosphate. This compound, produced by the action of phosphorylases on starch, glycogen, and other α-1,4-glucans, is converted to an intermediate in glycolysis by the action of the enzyme **phosphoglucomutase**. This enzyme catalyzes the interconversion of glucose-1-phosphate and glucose-6-phosphate.

$$ (10.20) $$

Glucose-1-phosphate   Glucose-6-phosphate

The K_{eq} of the reaction (left to right) is 19 at pH 7 and favors the formation of glucose-6-phosphate. Phosphoglucomutase requires as a cofactor glucose-1,6-bisphosphate that serves as a donor of phosphate groups to the enzyme. The mechanism of this reaction is identical with that described for phosphoglyceryl mutase.

10.6.5 UTILIZATION OF POLYSACCHARIDES

The polysaccharides starch and glycogen encountered in plants and animals, respectively, are important fuel molecules. These polymers of glucose can be broken down to glucose by two different processes in order that they be accommodated in the glycolytic sequence. In one process, the polysaccharide is hydrolyzed to produce, ultimately, D-glucose that can be phosphorylated and metabolized in the glycolytic sequence. This process occurs in the digestive tract of animals where food polysaccharides are broken down via dextrins to the disaccharides maltose, isomaltose, and glucose. These fragments then enter the mucosal cell where the disaccharides are split by maltase and isomaltase to glucose. The monosaccharide can be absorbed into the hepatic portal blood system and thus be carried to the cells. In the cell, the glucose is phosphorylated and can be catabolized in the glycolytic sequence. A similar degradative process occurs in germinating plant seeds where starch is broken down to simpler sugars in the malting process. Those sugars can then be hydrolyzed to monosaccharides that, in turn, can serve as substrates for alcoholic fermentation in the presence of yeasts.

The enzymes that catalyze the hydrolytic reactions are known as **amylases.** One of these, α-1,4-glucan-4-glucanhydrolase (α-amylase just described), is an endoamylase that hydrolyzes the interior bonds of the linear polysaccharide amylose in a random manner to yield a mixture of glucose and maltose. When the substrate is the branched polysaccharide amylopectin, the hydrolysis products consist of a mixture of branched and unbranched oligosaccharides in which α-1,6 bonds are found. The other hydrolytic enzyme, known as α-1,4-

glucan maltohydrolase (β-amylase), is an exoamylase that forms maltose units from the (linear) nonreducing end of the polysaccharide. Thus, β-amylase acts on amylose to yield maltose quantitatively. When the branched amylopectin (or glycogen) is the substrate, maltose and a highly branched dextrin are the products since the enzyme can act on α-1,4 linkages only up to two or three residues from the α-1,6 linkage. The amylases are found in animals, plants, and microorganisms; in animals, specifically, the amylases occur in digestive juices (saliva and pancreatic secretion).

   Storage polysaccharides can also be degraded by phosphorylases that cata-lyze the phosphorylytic cleavage of the α-1,4-glucosidic linkage at the nonre-ducing end of the starch of glycogen chain. The reaction, which is reversible, is represented as

CH$_2$OH   CH$_2$OH   CH$_2$OH   CH$_2$OH

$+$ H$_3$PO$_4$

Nonreducing end   Amylose   Reducing end

CH$_2$OH   CH$_2$OH   CH$_2$OH   CH$_2$OH

(10.21)

$+$

—OPO$_3$H$_2$

α-D-Glucose-1-phosphate

   As written from left to right, the reaction is a **phosphorolysis** (in contrast to hydrolysis) resulting in the formation of α-D-glucose-1-phosphate and the loss of one glucose unit from the nonreducing end of the polysaccharide chain. In the reverse reaction, inorganic phosphate is liberated from α-D-glucose-1-phosphate as the sugar residue is transferred to a preexisting oligo- or polysac-charide chain.

   Phosphorylase will catalyze the stepwise removal of glucose units from a

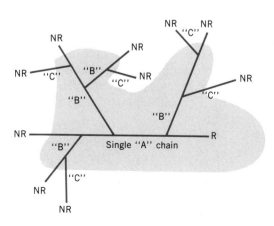

FIGURE 10.2 Action of phospho-rylase on the branched-chain poly-saccharide, amylopectin. Phospho-rylase degrades amylopectin until the vicinity of a branching point is reached. Limit dextrin on which phosphorylase cannot act is repre-sented by the shaded area. NR, nonreducing ends; R, reducing end; "A," the only chain that has both a reducing (R) and a nonreducing (NR) end; "B," chains that are branched and connected to the A chain by an $\alpha(1 \rightarrow 6)$ glucosidic link; "C," chains that are unbranched, connected only to B chains by an $\alpha(1 \rightarrow 6)$ glucosidic link, and also have a nonreducing (NR) end.

linear portion of a starch or glycogen molecule until it approaches within four units of an α-1-6 branch point. This branch point constitutes an area in which the enzyme is inactive. Highly branched polysaccharides such as amylopectin will therefore be degraded only about 55%, leaving a highly branched residue known as a **limit dextrin** (Figure 10.2). The highly branched dextrin is then modified by the action of an oligo (α-1,4 → α-1,4) glucan transferase that catalyzes the transfer of a trisaccharide attached to the α-1,6-linked glucose to the nonreducing end of a different chain.

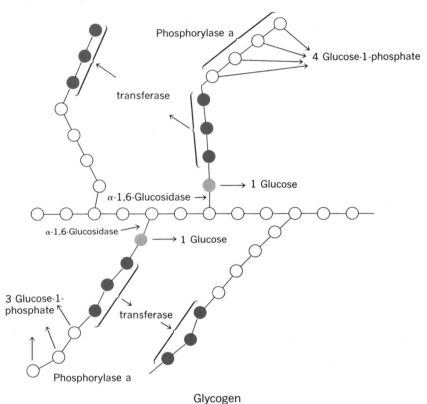

Glycogen

The elongated chain can then be acted on by phosphorylase, while the single glucose unit at the α-1,6 branch point is hydrolytically removed by a specific α-1,6-glucosidase.

The equilibrium of the phosphorylase-catalyzed reaction is independent of the polysaccharide concentration, provided a certain minimum concentration is exceeded. Thus, in the following expression for K_{eq}, the polysaccharide concentrations represent the number of nonreducing chain termini, a *number that does not change*. It then follows that at any pH, the extent of reaction is determined by the relative concentrations of glucose-1-phosphate and inorganic phosphate. At pH 7.0,

$$K_{eq} = \frac{[C_6H_{10}O_5]_{n-1}[\text{Glucose-1-phosphate}]}{[C_6H_{10}O_5]_n[H_3PO_4]}$$

$$= \frac{[\text{Glucose-1-phosphate}]}{[H_3PO_4]}$$

$$= 0.3$$

Muscle phosphorylase exists in two forms, *a* and *b* that are interconverted by two converter enzymes, synthase-phosphorylase kinase (SPK) and phosphoprotein phosphatase (PP–1). Rabbit muscle phosphorylase *a* (MW 400,000) is a dimer consisting of two identical polypeptide chains. Each chain contains a serine residue whose hydroxyl group is esterified with phosphate. Converter enzyme PP–1 catalyzes the hydrolysis of this phosphate group, thereby converting phosphorylase *a* into phosphorylase *b*.

$$\text{Phosphorylase } a \xrightarrow[\text{Phosphatase}]{\text{Phosphoprotein}} 2 \text{ Phosphorylase } b + 4 \text{ } H_3PO_4$$

Converter enzyme SPK, on the other hand, catalyzes the phosphorylation of phosphorylase *b*, thereby forming phosphorylase *a*.

$$2 \text{ Phosphorylase } b + 4 \text{ ATP} \xrightarrow[\text{Mg}^{2+}]{\substack{\text{Synthase phosphorylase}\\\text{Kinase,}}} \text{Phosphorylase } a + 4 \text{ ADP}$$

The two processes constitute a cycle by which phosphorylase can interchange between two forms that have different catalytic properties.

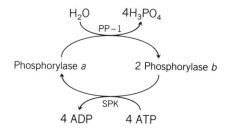

These two forms of phosphorylase are allosteric proteins whose catalytic activity is modified by a number of different effectors. In particular, however, phosphorylase *b* is inactive in the absence of AMP and is activated by glycogen. Phosphorylase *b* is inhibited by ATP, glucose-6-phosphate, and UDP-glucose. Phosphorylase *a* is inhibited by UDP-glucose and glucose. The interconversions of the *a* and *b* forms of the animal phosphorylases are of prime importance in regulating polysaccharide breakdown in intact tissues (see Section 10.8.1).

10.7 REVERSAL OF THE GLYCOLYTIC SEQUENCE

10.7.1 REVERSAL IN PHOTOSYNTHESIS

Organisms are capable of synthesizing carbohydrates from simpler precursors by, in effect, reversing the glycolytic sequence. To do so, however, they must bypass or otherwise circumvent the steps (Reactions 10.1, 10.3, and 10.9) that have been described as irreversible.

Perhaps the most important example of reversal is that part of photosynthesis in which CO_2 is utilized to make the storage polysaccharides of plants. We shall learn (Chapter 15) that the two reactions of the **reduction phase** and several reactions of the **regeneration phase** of the CO_2 reduction cycle are identical with some reactions of glycolysis.

10.7.2 GLUCONEOGENESIS

Another important example of reversal is the regeneration of glucose (and glycogen), termed **gluconeogenesis,** from lactic acid produced by higher animals during exercise. Skeletal muscle, with its poorer O_2 supply but enriched supply of glycolytic enzymes, utilizes glycolysis to meet its short-term needs for ATP. In doing so, the muscle produces relatively large amounts of lactic acid that is secreted into the blood and transported to the liver. In this organ primarily, about 80% of the lactic acid is resynthesized to glucose and returned to the muscle for glycogen synthesis. Oxidation of the remaining 20% via the tricarboxylic acid cycle (Chapter 12) provides the necessary energy for reversing the glycolytic sequence and converting the majority of the lactic acid back to glycogen.

Gluconeogenesis can also occur from compounds other than lactic acid. Pyruvic acid and oxalacetate, an intermediate of the tricarboxylic acid cycle, can be converted to glucose. Since several amino acids (alanine, cysteine, serine, aspartic acid) can give rise to pyruvate and oxalacetate, these compounds also can produce glucose. This occurs, however, in animals only during starvation, or under other wasting conditions, when an organism urgently requires glucose and has no other source of carbon than proteins available. Regardless of the precursor, all of these compounds must "bypass" Reactions 10.9, 10.3, and 10.1 in forming glucose.

The bypassing of Reaction 10.9 involves the participation of two key enzymes that catalyze reactions known as **CO_2-fixation** reactions, a group of reactions that have an important role in the functioning of the Krebs cycle, soon to be discussed (Chapter 12). The first of these enzymes, **pyruvic carboxylase,** which is found only in animal tissues, is located in the mitochondria; thus, pyruvic acid produced in the cytosol from lactate (or alanine or serine) must enter the mitochondria as a first step. The reaction catalyzed is

$$
\begin{array}{c}
CO_2H \\
| \\
C{=}O \\
| \\
CH_3
\end{array}
+ H_2O + CO_2 + ATP
\xrightarrow[\text{Mg}^{2+} \atop \text{Biotin}]{\text{Acetyl CoA}}
\begin{array}{c}
CO_2H \\
| \\
C{=}O \\
| \\
CH_2 \\
| \\
CO_2H
\end{array}
+ ADP + H_3PO_4 \quad (10.22)
$$

Pyruvic acid   Oxalacetic acid

$$\Delta G' = -500 \text{ cal (pH 7.0)}$$

The $\Delta G'$ is small; therefore, the reaction is readily reversible. Nevertheless, the enzyme in the cell appears to function only from left to right generating oxalacetate. Pyruvate carboxylases (MW 400,000 to 500,000) are tetramers with identical subunits. Each subunit contains an active site that binds Mg^{2+} and covalently bound biotin as well as an allosteric site for acetyl CoA. Biotin is covalently linked to the ε-amino group of lysyl moiety in the carboxylase subunit; it functions as the initial acceptor of CO_2, subsequently transferring it to pyruvic acid. The partial reactions are

$$\text{Enz-biotin} + CO_2 + ATP \rightleftharpoons \text{Enz-biotin-}CO_2 + ADP + Pi$$
$$\text{Enz-biotin-}CO_2 + \text{pyruvate} \rightleftharpoons \text{Enz-biotin} + \text{oxalacetate}$$

The structure of the carboxylated enzyme-biotin complex has been described in Section 5.7.3.

The second enzyme involved in reversing this part of the glycolytic sequence is known as phosphoenol pyruvic (PEP) carboxykinase.

$$
\begin{array}{c}
\text{CO}_2\text{H} \\
| \\
\text{C}=\text{O} \\
| \\
\text{CH}_2 \\
| \\
\text{CO}_2\text{H}
\end{array}
\quad + \text{ GTP} \xrightarrow{\text{Mg}^{2+}}
\begin{array}{c}
\text{CO}_2\text{H} \\
| \\
\text{C}-\text{OPO}_3\text{H}_2 \\
\| \\
\text{CH}_2
\end{array}
\quad + \text{ CO}_2 + \text{GDP} \quad (10.23)
$$

Oxalacetic acid   Phosphoenol pyruvic acid

$\Delta G' = +700$ cal (pH 7.0)

In this reaction, oxalacetate produced in Reaction 10.22 is converted to PEP by a reaction involving little change in free energy, but one in which CO_2 is produced, a "reverse CO_2 fixation." The cellular distribution of this enzyme varies greatly in different species. In those tissues (the liver of pig, guinea pig, and rabbit) where significant quantities occur in the mitochondria, the PEP produced subsequently diffuses out and then can be converted to fructose-1,6-bisphosphate, provided the ATP and NADH needed to reverse Reactions 10.6 and 10.5, respectively, are available.

The overall reaction for converting pyruvate to PEP can be obtained by adding Reactions 10.22 and 10.23.

$$
\begin{array}{c}
\text{CO}_2\text{H} \\
| \\
\text{C}=\text{O} \\
| \\
\text{CH}_3
\end{array}
+ \text{ ATP} + \text{GTP} \rightleftharpoons
\begin{array}{c}
\text{CO}_2\text{H} \\
| \\
\text{C}-\text{OPO}_3\text{H}_2 \\
\| \\
\text{CH}_2
\end{array}
+ \text{ ADP} + \text{GDP} + \text{H}_3\text{PO}_4 \quad (10.24)
$$

$\Delta G' = +200$ cal

Note that the large $\Delta G'$ of Reaction 10.9 has now been overcome but that two moles of nucleoside triphosphate have been expended in order for the overall reaction (Reaction 10.24) to have a negligible $\Delta G'$ of $+200$ cal.

In many species, this coupled sequence of reactions for reversing (actually bypassing) pyruvate kinase, Reaction 10.9, is complicated by the fact that PEP carboxykinase is a cytoplasmic enzyme. Since the inner mitochondrial membrane is impermeable to oxalacetate, this compound must be "modified" to permit the transfer of its four carbon atoms into the cytosol.

There are two series of coupled reactions that accomplish the net transfer of oxalacetate into the cytosol. One of these involves reduction of the oxalacetate to malate that then moves through the membrane and is reoxidized by cytoplasmic malic dehydrogenase.

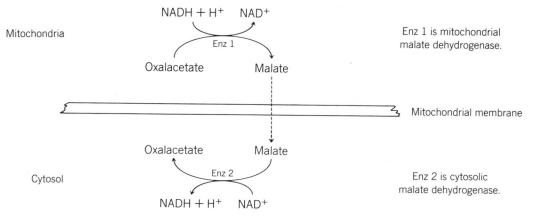

These reactions are part of the "malate shuttle" that allows the exchange of reducing equivalents between the cytosol and the mitochondrial matrix (Section 14.8).

The other way in which the carbon atoms in oxalacetate can be transferred to the cytosol involves the conversion of oxalacelate to aspartic acid and the passage of the amino acid out of the mitochondria.

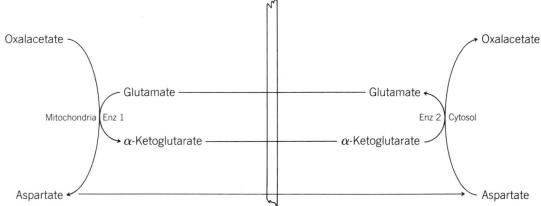

Enz 1 and Enz 2 are aspartate : glutamate
amino transferases located in the mitochondrial matrix
and the cytosol, respectively.

When aspartate traverses a membrane, it also carries with it an amino group. Note that the free passage of glutamate and α-keto-glutarate involved in the transamination process across the membrane prevents the net accumulation of amino groups in the cytosol.

In gluconeogenesis, Reactions 10.3 and 10.1 involving ATP and ADP must also be bypassed. In the former instance, this is accomplished by a **phosphatase** that catalyzes the **hydrolysis** of fructose-1,6-bisphosphate to form fructose-6-phosphate.

$$H_2O_3POCH_2 \quad CH_2OPO_3H_2 \quad + H_2O \longrightarrow \quad H_2O_3POCH_2 \quad CH_2OH \quad + H_3PO_4$$

Fructose-1,6-bisphosphate   Fructose-6-phosphate

(10.25)

$$\Delta G' = -4000 \text{ cal (pH 7.0)}$$

Fructose-1,6-bisphosphatase is an allosteric enzyme whose activity is controlled by several compounds. Together with phosphofructokinase (Reaction 10.3), these two enzymes effectively determine whether carbon atoms flow from glucose (and glycogen) to pyruvate, yielding energy to the cell, or whether the buildup of glucose (and glycogen) as a result of gluconeogenesis can occur. AMP is a strong negative effector of fructose-1,6-bisphosphatase and fructose-2,6-bisphosphate is a strong competitive inhibitor ($K_I = 0.5 \ \mu M$). The bisphosphate also enhances the action of AMP. The interplay between these compounds and the two key enzymes just mentioned is discussed in Section 10.8.

Glucose-6-phosphate must be hydrolyzed to complete the "reversal" of glycolysis.

$$\Delta G' = -3300 \text{ cal (pH 7.0)}$$

Glucose-6-phosphate + $H_2O \longrightarrow$ Glucose + H_3PO_4 (10.26)

In animals, the enzyme responsible, glucose-6-phosphatase, is found primarily in liver and kidney where it is responsible for maintaining the glucose level of the blood. In these organs and intestinal tissue, glucose-6-phosphatase is located in the endoplasmic reticulum (ER). In plants, the enzyme also occurs in the ER where its role is less clear since few plants accumulate free glucose.

Starting with two moles of lactate, then, and proceeding through Reactions 10.22 and 10.23, Reactions 10.8 through 10.4, and Reactions 10.25, 10.2, and 10.26, we can write the overall equation accounting for the reversal of glycolysis.

$$2 \text{ Lactate} + 4 \text{ ATP} + 2 \text{ GTP} + 6 \text{ } H_2O \longrightarrow \text{Glucose} + 4 \text{ ADP} + 2 \text{ GDP} + 6 \text{ } H_3PO_4$$

From this, it is apparent that a total of six energy-rich phosphates are required to convert the six carbon atoms in two moles of lactate to glucose. It is interesting to note that the energy made available by hydrolysis versus these energy-rich compounds (6×8000 to 10,000 cal) is in excess of that (47,000 cal) estimated to be released when glucose is converted to lactate (Section 10.4). In Section 13.12, we compare the paths for the synthesis of glucose from pyruvate and the synthesis of fatty acids from acetyl CoA. In both systems, the release of CO_2 is a key component of the reversal process.

10.8 THE REGULATION OF GLYCOLYSIS

Glycolysis is carefully regulated in all cells so that energy is released from carbohydrates only as it is needed by those cells. Evidence that this is so is the effect of O_2, first noted by Pasteur, on tissues that possess not only the capacity to convert glucose to lactate by glycolysis, but also can oxidize pyruvic acid completely to CO_2 and H_2O via the Krebs cycle (Chapter 12). Such tissues utilize glucose much more rapidly in the absence of O_2 than they do when O_2 is present. The functional significance of this inhibition of glucose consumption by oxygen—known as the Pasteur effect—is appreciated when we recognize that much more energy is made available as ATP when glucose is oxidized aerobically to CO_2 and H_2O than when it is anaerobically converted only to lactic acid or alcohol and CO_2. Since more ATP is formed under aerobic conditions, less glucose needs to be consumed to do the same amount of work in the cell.

Glycolysis is regulated at the reactions catalyzed by hexokinase, phosphofructokinase, and pyruvic kinase. In a reciprocal manner, gluconeogenesis, in which carbon flows back into carbohydrate from lactate, alanine, and other noncarbohydrate precursors or pyruvate, is regulated at the steps catalyzed by pyruvic carboxylase and fructose-1,6-bisphosphatase. Thus, one sees that regulation occurs at those reactions in glycolysis and gluconeogenesis that are unidirectional (i.e., operate in only one direction). The advantages of *control* of a step that is unidirectional are obvious since the passage of carbon through the unidirectional step can be turned on or off (also see Section 18.2).

Table 10.2 lists the activators and inhibitors of these regulatory enzymes. It can be seen that compounds (ATP, citrate, and acetyl CoA) that should be present in excess when glucose is being rapidly catabolized for production of energy serve to decrease the flow of carbon from glucose to pyruvate, lactate, and acetyl CoA and instead stimulate gluconeogenesis. Similarly, AMP (and ADP) and Pi that would be present in cells that have not been degrading glucose for energy will serve to increase glycolysis and inhibit gluconeogenesis.

The role of fructose-2,6-bisphosphate, an activator of phosphofructokinase and an inhibitor of fructose-1,6-bisphosphatase, deserves additional comment. This compound, which is effective at very low concentrations, is produced in animal tissues when

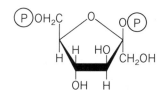

Fructose-2,6-bisphosphate

glucose is readily abundant and glycolysis needs to proceed. When the amount of glucose (in the blood) is low, glucagon is released by the pancreas as a signal to the liver to restore the level of blood glucose. This occurs by inactivation of the enzyme that synthesizes fructose-2,6-bisphosphate and a lowering of the concentration of this cofactor. In the absence of fructose-2,6-bisphosphate gluconeogenesis will proceed in response to a low level of ATP and relatively high levels of AMP and Pi.

The enzyme that *synthesizes* fructose-2,6-bisphosphate in animal tissues also catalyzes its hydrolysis (to fructose-6-phosphate). Whether the protein functions as a **kinase** to produce the bisphosphate or acts as a **phosphatase** to destroy the bisphosphate depends on whether the protein is phosphorylated or

TABLE 10.2   Regulatory Enzymes of Glycolysis and Gluconeogenesis

	ACTIVATOR	INHIBITOR
Hexokinase	—	Glucose-6-phosphate
Phosphofructokinase	Pi, AMP, ADP	ATP, citrate
	Fructose-6-phosphate	NADH
	Fructose-2,6-bisphosphate	
Fructose-1,6-bisphos-phatase	ATP	Fructose-2,6-bisphosphate
		AMP
Pyruvic kinase	K⁺, AMP	ATP, alanine
	Fructose-2,6-bisphosphate	Acetyl CoA
Pyruvate carboxylase	Acetyl CoA	

not (the phosphorylated form is the phosphatase form). The interconversion of the phosphorylated and dephosphorylated forms is catalyzed by a cyclic AMP-dependent protein kinase (C_2). (See Chapter 18.)

10.8.1 CONTROL OF GLYCOGEN METABOLISM

The control of glycogen metabolism by the enzymes phosphorylase and glycogen synthase is effected in animal tissues mainly through the interconversions of phosphorylated and nonphosphorylated forms of these two enzymes. However, the mechanism is more elaborate in that it operates on the **cascade** principle (Figure 10.3), which provides for amplification of the primary signal (Section 18.5). The interconversion of phosphorylase b and phosphorylase a catalyzed by synthase-phosphorylase kinase (SPK_a) and phosphoprotein phosphatase (PP–1) has already been described (Section 10.6.5). SPK, in turn, also exists in a more active (phosphorylated) form (SPK_a) and a less active (nonphosphorylated) form (SPK_b) that are interconvertible. The nonphosphorylated, low-activity form is converted to the former in the presence of ATP and another enzyme called **protein kinase A** that is activated by cyclic AMP. The high-activity form (SPK_a) is, in turn, converted back to the low-activity form (SPK_b) by the same converter enzyme PP–1 that converts phosphorylase a to phosphorylase b. This interconversion of the two forms of SPK is shown in Fig. 10.3 also.

Protein kinase A exists in an inactive form (R_2C_2) that, in turn, is activated by cyclic AMP. Protein kinase (skeletal muscle) has two regulatory units (R) and two catalytic units (C), and is represented as R_2C_2 when present as the tetramer. Cyclic AMP binds directly with the R subunits, releasing the C units, which now can catalyze the phosphorylation of SPK_b.

$$R_2C_2 + 2\ cAMP \xrightarrow{\ Mg^{2+}\ } R_2\ (cAMP)_2 + C_2\ \text{(active kinase)}$$

Mg^{2+} ions are also an essential part of this process.

A similar cascade phenomenon (Figure 10.3) exists for regulation of glycogen synthase in muscle in that SPK_a is the enzyme that phosphorylates glycogen synthase I, converting it to glycogen synthase D. These two forms of glycogen synthase are discussed in Section 11.3. Note that the single stimulus, the release of epinephrine, results not only in the increased glycogen breakdown by phosphorylase but equally important, decreased glycogen synthesis through the formation of glycogen synthase D.

Other hormones besides epinephrine are known to affect carbohydrate metabolism and the metabolism of glycogen. One example is glucagon, a polypeptide produced by the α-cells of the islets of Langehans in the pancreas. The action of glucagon is identical with that of epinephrine, its binding to hormone receptor sites initiating a similar cascade phenomenon through the formation of cAMP. Active in very low concentrations, it activates without the increase in blood pressure observed with epinephrine.

The hormone insulin, secreted by the β-cells of the islets of Langehans, also acts by influencing the level of cyclic AMP within cells. In this case, the cyclic AMP level is lowered and the immediate results are the opposite of epinephrine and glucagon. However, insulin also affects enzymes of lipid and amino acid metabolism, thereby giving rise to indirect effects on carbohydrate metabolism.

While glycogen synthesis and breakdown in mammalian muscle are clearly

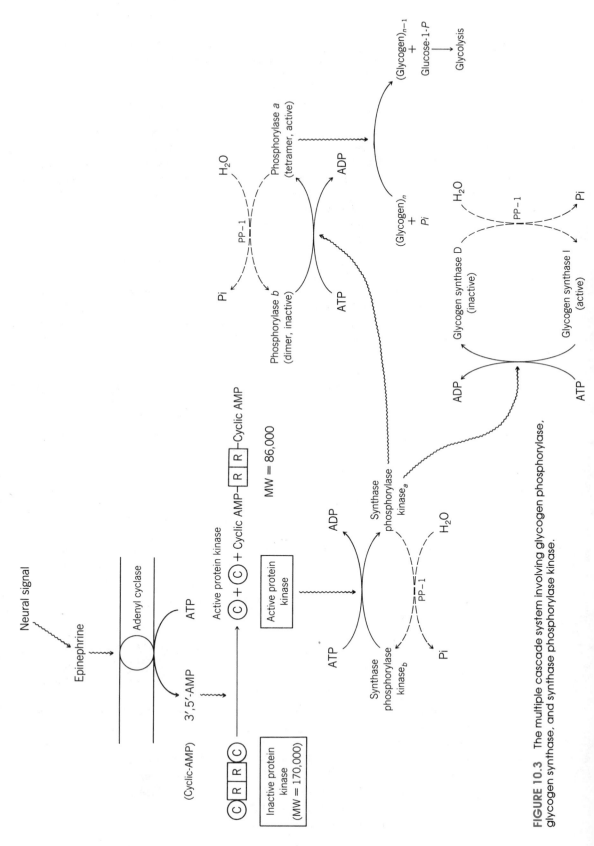

FIGURE 10.3 The multiple cascade system involving glycogen phosphorylase, glycogen synthase, and synthase phosphorylase kinase.

under the hormonal control just described, the same control is not possible in bacteria and plants. In these organisms, glycogen and starch synthesis is regulated at the reaction that produces the glycosyl donor, ADP-glucose, for polysaccharide synthesis. Moreover, the control is mediated through allosteric effectors rather than the active–inactive enzyme mechanism of muscle phosphorylase. Thus, by regulating the supply of ADP-glucose, the production of bacterial glycogen is, in turn, controlled. A similar mechanism for control of UDP-glucose production is not feasible in animal tissues since UDP-glucose is used not only for glycogen synthesis, but also for galactose and glucuronic acid synthesis. Instead, the control site is one step closer to glycogen production, namely the reaction in which UDP-glucose is used to make the polymer (Reaction 11.16).

10.9 ENERGY CHARGE

In glycolysis, we have encountered the interconversion of ATP and ADP by reactions that produce and consume these compounds. There are also numerous reactions in lipid and nucleic acid metabolism in which ATP is converted to AMP (instead of ADP) when it provides the driving force for a reaction. These three derivatives of adenosine are further inconvertible by an enzyme **adenylic kinase** that is widely distributed in nature.

$$2 \text{ ADP} \xrightleftharpoons{\text{Adenylic kinase}} \text{ATP} + \text{AMP} \tag{10.27}$$

This enzyme, operating from left to right, provides a mechanism for converting *half* of the ADP in a cell back to ATP that can then be used for further endergonic reactions. Obviously, in the cell, the relative amounts of ATP, ADP, and AMP will depend on the metabolic activities that predominate at any one time.

Atkinson has proposed the term **energy charge** to describe the energy state of the ATP–ADP–AMP system and has pointed out the analogy with the charge of an electromotive cell. In mathematical terms, he defines the energy charge as that fraction of the adenylic system (ATP + ADP + AMP) that is composed of ATP.

$$\text{Energy charge} = \frac{[\text{ATP}] + 0.5[\text{ADP}]}{[\text{AMP}] + [\text{ADP}] + [\text{ATP}]}$$

From this equation, we can see that the energy charge will be 1.0 when all of the AMP and ADP in the cell have been converted to ATP. This condition would be approached, for example, when a cell is carrying out oxidative phosphorylation at a rapid rate and few biosynthetic reactions are occurring. Under these conditions, the maximum number of energy-rich phosphate bonds would be available in the adenylic system. When all of the adenosine compounds are present as ADP, the energy charge will be 0.5 and half as many energy-rich bonds will be contained in the adenylic system. When all the ATP and ADP have been converted to AMP, the energy charge will be 0 and the adenylic system will be devoid of energy-rich structures.

The energy charge of the cell exerts its control of metabolism through allosteric regulation of specific enzymes by ATP, ADP, and AMP. The stimulation and inhibition of the regulatory enzymes in glycolysis and gluconeogenesis by these adenine nucleotides were described earlier in Section 10.8. These compounds also activate or inhibit enzymes that regulate lipid, amino acid, and protein metabolism.

NADH also contributes to the energy charge of a cell in that most of the NAD^+ in a cell will be present as NADH if glucose or lipids are being oxidized. Therefore, it is not surprising that NADH often acts on allosteric enzymes in the same way as ATP. When we consider the tricarboxylic acid (Krebs) cycle in Chapter 12, we shall see that citrate accumulates in cells with a high energy charge. It also can act as an allosteric modifier of proteins and exhibits the same qualitative effects as ATP.

While these multiple effects of ATP, AMP, citrate, and NADH may seem confusing, they can be beautifully integrated with controls in the degradation of fatty acids and reactions of the tricarboxylic acid cycle. Our discussion of these will be deferred until Chapter 12.

REFERENCES

1. *B. Axelrod, "Glycolysis." In* Metabolic Pathways, *3rd ed., Vol. 1, D. M. Greenberg, ed., New York: Academic Press, 1967.*

2. *S. Dagley and D. E. Nicholson,* Metabolic Pathways. *New York: Wiley, 1970.*

3. *H. A. Krebs and H. L. Kornberg,* Energy Transformations in Living Matter. *Berlin: Springer, 1957.*

4. *D. E. Atkinson,* Cellular Energy Metabolism and Its Regulation. *New York: Academic Press, 1977.*

5. *E. A. Newsholme and C. Start,* Regulation in Metabolism. *London: Wiley, 1973.*

6. *P. Hochachka,* Living Without Oxygen. *Cambridge, Mass.: Harvard University Press, 1980.*

7. *W. A. Wood, ed., "Carbohydrate Metabolism." In* Methods in Enzymology, *Vols. 89 and 90, New York: Academic Press, 1982.*

REVIEW PROBLEMS

1. Write balanced chemical equations for the following enzyme-catalyzed reactions (or reaction sequences). Use structures for all substrates and products except for complicated ones (nucleotides, coenzymes, etc.) in which case, use standard abbreviations. Name the enzymes involved.

    (a) The conversion of pyruvate to PEP during the reversal of glycolysis

    (b) The first two reactions in the utilization of glycerol as a carbon-energy source

    (c) The first reaction of glycolysis in which an energy-rich compound is made from energy-poor precursors

2. Compare the efficiency (i.e., net moles of ATP produced per mole of glucose utilized) of the glycolytic pathway with that of the Entner–Doudoroff pathway (Chapter 11). Assume that in both pathways, glucose is converted completely to lactate.

3. Describe the type of reaction catalyzed by
    (a) A mutase
    (b) An isomerase
    (c) An epimerase
    (d) A kinase
    (e) A phosphatase
    (f) A pyrophosphorylase
    (g) A lactonase

4. Write all the reactions involved in the conversion of lactic acid to glucose that are not simply reversals of the reactions involved in the conversion of glucose to lactic acid. Use structural formulas and name the enzymes involved.

CARBOHYDRATES II: BIOSYNTHETIC ROLE OF GLYCOLYSIS AND PENTOSE PHOSPHATE METABOLISM

In Chapter 10, we considered the process by which glucose is degraded anaerobically during glycolysis to produce energy. Another important role of glycolysis is to produce a number of specific compounds required by the cell for its growth and development. In doing so, glycolysis plays an important anabolic function as well as a catabolic role.

11.1 THE ANABOLIC ROLE OF GLYCOLYSIS

Several examples of the anabolic role of glycolysis have already been mentioned. Thus, *sn*-glycerol-3-phosphate, required for the synthesis of triacyl glycerides, can be obtained by reduction of dihydroxyacetone phosphate in the presence of glycerol phosphate dehydrogenase (Section 10.6.3). 3-Phosphoglyceric acid (Section 10.4.2) is converted by both plants and animals to the amino acid serine that, in turn, can be converted to glycine and cysteine. Similarly, pyruvic acid can be transaminated to form alanine. Thus, the carbon skeletons of these amino acids can be derived from carbohydrates and need not be furnished to animals in their diet.

Phosphoenolpyruvic acid when condensed with erythrose-4-phosphate obtained in the pentose phosphate pathway (Section 11.4) produces the first intermediate, a seven-carbon carboxylic acid, in the **shikimic acid pathway.** The latter is a biosynthetic pathway in plants and microorganisms that leads to the amino acids phenylalanine, tyrosine, and tryptophan. Animals are unable to carry out this synthesis of the seven-carbon intermediate, and for this reason, the amino acids mentioned are essential and must be supplied in the diet of animals.

A third example of a key intermediate obtained in glycolysis is pyruvic acid that can give rise to oxalacetate in the presence of pyruvic carboxylase (Section 10.7.2). This reaction is the key to enabling the tricarboxylic acid cycle (Chapter 12) to function in an anabolic manner and synthesize many important cellular components. This relationship will be described in detail in Chapter 12.

11.2   GLUCOSE CONVERSION TO OTHER MONOSACCHARIDES

Glucose serves as the precursor of the storage polysaccharides starch and glycogen found in plant and animal cells. These polymers are synthesized by the sequential addition of monosaccharide units to the nonreducing ends of a preformed chain. The sugar nucleotide UDP-glucose (Section 10.6.2) plays a key role in this process and such sugar derivatives are also involved in the conversion of glucose to other monosaccharides.

11.2.1   THE ROLE OF SUGAR NUCLEOTIDES

Two general reactions are important in understanding how sugar nucleotides function in carbohydrate metabolism. The first of these, catalyzed by a **pyrophosphorylase,** is

$$\underset{\substack{\text{(XTP)} \\ \text{Nucleoside} \\ \text{triphosphate}}}{\text{X—R—P—P—P}} + \underset{\substack{\text{Glycosyl-1} \\ \text{phosphate}}}{\text{P—Gly}} \rightleftharpoons \underset{\substack{\text{(XDP—Gly)} \\ \text{Nucleoside} \\ \text{diphosphate sugar} \\ \text{(sugar nucleotide)}}}{\text{X—R—P—P—Gly}} + \underset{\substack{\text{Pyrophosphate}}}{\text{P—P}} \qquad (11.1)$$

The K_{eq} for this reaction is 0.15. The reaction is freely reversible since the internal pyrophosphate bond in XTP is simply exchanged for the single pyrophosphate bond in the sugar nucleotide. In the cell, however, the reaction may be displaced to the right in the presence of **pyrophosphatases** that hydrolyze the inorganic pyrophosphate as it is formed.

The second general reaction accomplishes the transfer of the sugar moiety in the sugar nucleotide to an acceptor molecule. These enzymes are **glycosyl transferases.**

$$\underset{\substack{\text{Sugar nucleotide}}}{\text{X—R—P—P—Gly}} + \text{Acceptor} \longrightarrow \underset{\substack{\text{Nucleotide} \\ \text{diphosphate}}}{\text{X—R—P—P}} + \text{Gly-acceptor} \qquad (11.2)$$

The equilibrium for Reaction 11.2 is usually far to the right since the linkage between the C-1 atom of the monosaccharide and phosphate in the sugar nucleotide molecule is an energy-rich structure ($\Delta G'$ of hydrolysis of UDP-glucose to UDP and glucose is -8000 cal/mole).

A supply of glycosyl-1-phosphates and nucleoside triphosphates is obviously required for the synthesis of polysaccharides. We have already described the formation of glucose-1-phosphate by hexokinase or glucokinase coupled with phosphoglucomutase (Reactions 10.1 and 10.20).

$$\text{Glucose} + \text{ATP} \xrightarrow{\text{Kinases}} \text{Glucose-6-phosphate} \xrightarrow{\text{Phosphoglucomutase}} \text{Glucose-1-phosphate}$$

Glucose-1-phosphate can also be formed from glycogen or starch by the action of phosphorylase *a*.

$$\text{(Glucose)}_n + H_3PO_4 \rightleftharpoons \text{(Glucose)}_{n-1} + \text{Glucose-1-phosphate}$$

Galactose-1-phosphate is produced by the action of galactokinase (Reaction 10.13) and mannose-1-phosphate can be produced from mannose-6-phosphate by phosphomannomutase, an enzyme analogous to phosphoglucomutase (Reaction 10.20).

The nucleoside triphosphates are generated by a widespread enzyme called **nucleoside diphosphate kinase.**

$$\text{NDP} + \text{ATP} \rightleftharpoons \text{NTP} + \text{ADP}$$

11.2.2 BIOSYNTHESIS OF SUCROSE

The role of sugar nucleotides in sugar transformations can be illustrated by the synthesis of sucrose in plants. One of the early products of photosynthesis, dihydroxyacetone phosphate, is translocated from the chloroplast to the cytosol where it is rapidly converted to fructose-6-phosphate and ultimately to glucose-1-phosphate by the enzymes of glycolysis described in Chapter 10. Glucose-1-phosphate is then converted to UDP-glucose by glucose-1-phosphate-UTP pyrophosphorylase.

$$\text{Glucose-1-phosphate} + \text{UTP} \longrightarrow \text{UDP-Glucose} + P\text{—}P \qquad (11.3)$$

The glucose moiety in UDP-glucose is then transferred to fructose-6-phosphate, producing sucrose-6-phosphate and UDP.

$$\text{UDP-Glucose} + \text{Fructose-6-phosphate} \xrightleftharpoons[\text{Synthetase}]{\text{Sucrose-6-phosphate}}$$

$$\text{Glucose-fructose-6-phosphate} + \text{UDP} \quad (11.4)$$
$$\text{(Sucrose-6-phosphate)}$$

A specific sucrose phosphatase then catalyzes the hydrolysis of the phosphate and free sucrose is formed.

$$\text{Sucrose-6-phosphate} + H_2O \xrightarrow[\text{Phosphatase}]{\text{Sucrose-6-phosphate}} \text{Sucrose} + H_3PO_4 \quad (11.5)$$

The latter reaction is irreversible and drives the synthesis of sucrose by the coupling of Reaction 11.3 with 11.4. All these reactions occur in the cytosol of the leaf cell.

The $\Delta G'$ of hydrolysis of the bond between the two sugars in sucrose is remarkably high for a glycoside, about -7 kcal. (That for the glycosidic bond in starch and glycogen is -5.5 kcal and in other glycosides, for example, methyl glucoside, is only about -3 kcal.) This means that Reaction 11.4 proceeds with little decrease in free energy and the reaction is readily reversible, in contrast to most other reactions in which glycosides are produced. The energy in the glycosidic bond of sucrose can be utilized in another reaction catalyzed by the enzyme **sucrose synthetase** found in plants

$$\text{Sucrose} + \text{UDP} \rightleftharpoons \text{UDP-Glucose} + \text{Fructose} \qquad (11.6)$$
$$\text{Sucrose} + \text{ADP} \rightleftharpoons \text{ADP-Glucose} + \text{Fructose} \qquad (11.7)$$

These reactions are exceptions to the principle that glucose-1-phosphate is the source of sugars of sugar nucleotides. Plants make use of this reaction to synthesize the UDP-glucose required for cellulose formation and other transformations described next. The ADP-glucose produced in Reaction 11.7 is specifically utilized for starch synthesis.

11.2.3 BIOSYNTHESIS OF OTHER MONOSACCHARIDES

The role of the sugar nucleotide UDP-glucose in the synthesis of UDP-galactose (Section 10.6.2) has already been described. The galactose unit in UDP-galactose can then be transferred to acceptor molecules or released as galactose-1-phosphate in the presence of a pyrophosphorylase (Reaction 11.1).

In another important reaction involving UDP-glucose, the primary alcohol is oxidized to a glucuronic acid derivative by a dehydrogenase that catalyzes two consecutive oxidations.

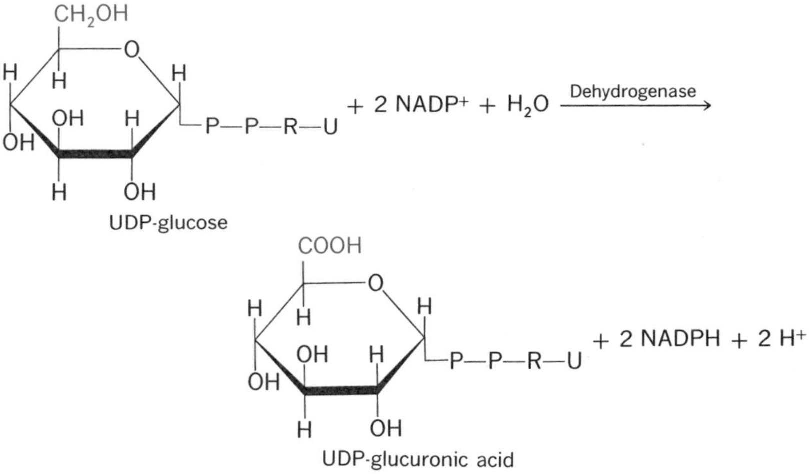

The UDP-glucuronic acid can be decarboxylated by a specific decarboxylase to form UDP-xylose required in the synthesis of oligosaccharides (i.e., xylans) containing that pentose.

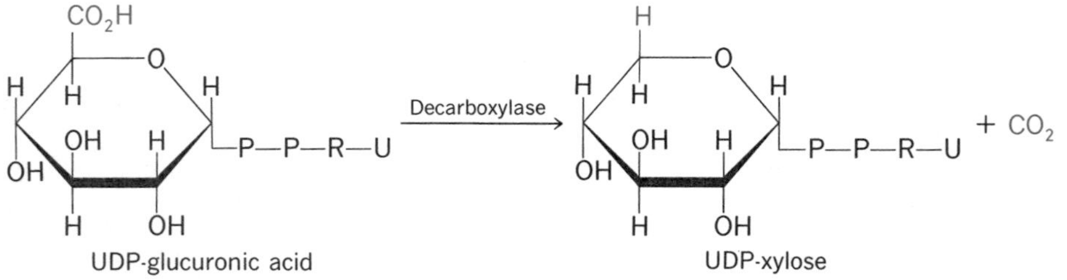

The glucuronic acid produced in Reaction 11.8 can be released by hydrolysis and participate in a series of reactions leading to ascorbic acid (Vitamin C). This **glucuronic acid oxidation pathway** is present in most mammals, but is lacking in man and the guinea pig.

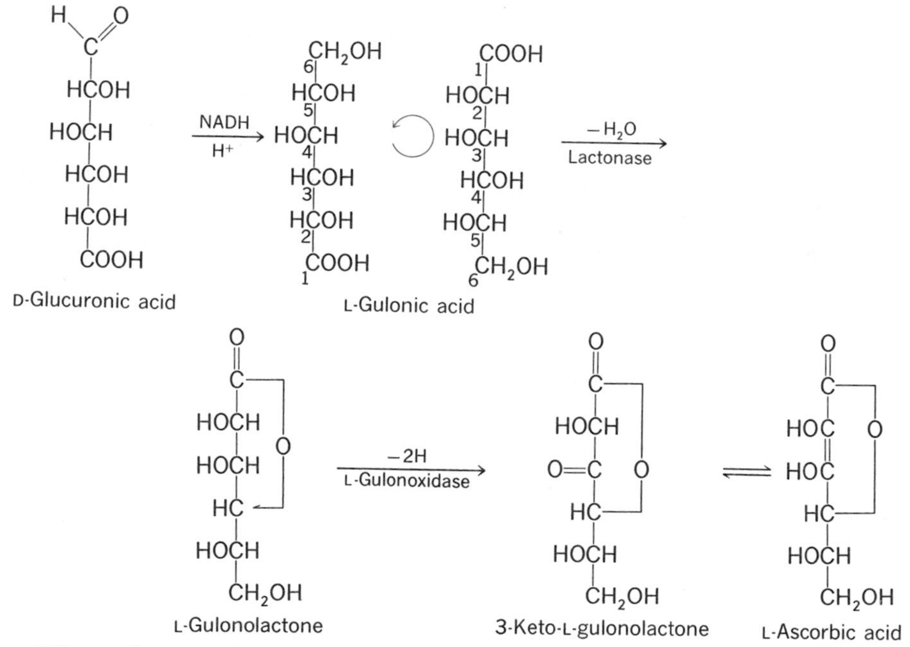

NADH
H+
−H2O
Lactonase
−2H
L-Gulonoxidase
D-Glucuronic acid
L-Gulonic acid
L-Gulonolactone
3-Keto-L-gulonolactone
L-Ascorbic acid

The amino sugars, found in structural polysaccharides, have their biosynthetic origin in fructose-6-phosphate. In the initial step, fructose-6-phosphate acquires an amino group from glutamine, forming glucosamine-6-phosphate in the presence of an **amidotransferase.**

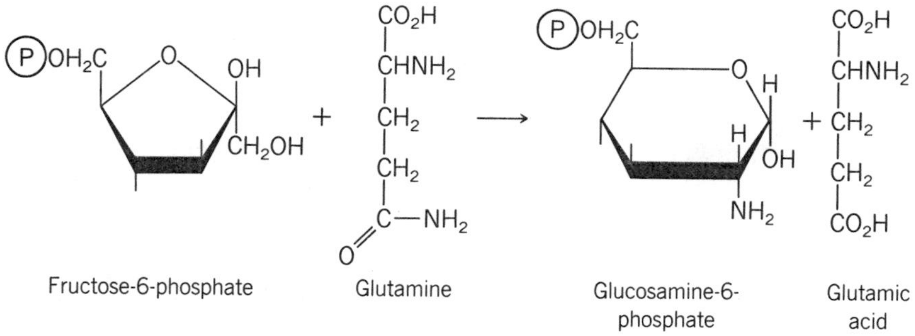

Fructose-6-phosphate
Glutamine
Glucosamine-6-phosphate
Glutamic acid

(11.10)

The amino group is then acetylated in the presence of acetyl CoA and an **acyltransferase.**

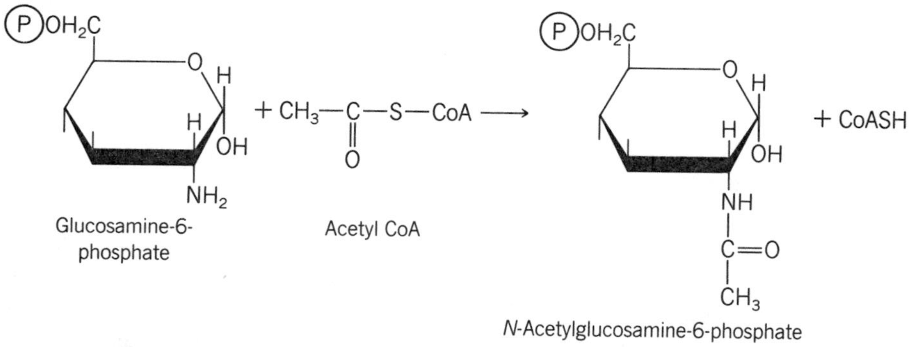

Glucosamine-6-phosphate
Acetyl CoA
N-Acetylglucosamine-6-phosphate

(11.11)

The *N*-acetylglucosamine-6-phosphate is converted to a sugar nucleotide after first forming the sugar-1-phosphate isomer.

$$N\text{-Acetylglycosamine-6-phosphate} \xrightleftharpoons{\text{Mutase}} N\text{-Acetylglucosamine-1-phosphate} \quad (11.12)$$

$$N\text{-Acetylglucosamine-1-phosphate} + \text{UTP} \xrightleftharpoons{\text{Pyrophosphorylase}} \text{UDP-}N\text{-Acetylglucosamine} + P \sim P \quad (11.13)$$

This sugar nucleotide of *N*-acetylglucosamine can be utilized as a precursor of that amino sugar unit in polysaccharides. The sugar nucleotide can also be acted on by the appropriate epimerases to form the corresponding sugar nucleotide of galactosamine or mannosamine.

$$\text{UDP-}N\text{-Acetylglucosamine} \underset{\searrow}{\overset{\nearrow}{}} \begin{array}{l} \text{UDP-}N\text{-Acetylgalactosamine} \\[2mm] \text{UDP-}N\text{-Acetylmannosamine} \end{array}$$

One final example, illustrative of the role of glycolytic intermediates in the synthesis of other carbohydrates, is the formation of GDP-L-fucose from D-fructose-6-phosphate through sugar nucleotide derivatives of mannose and galactose. GDP-L-Fucose is used in the synthesis of oligosaccharides containing fucose.

$$\text{Fructose-6-phosphate} \rightleftharpoons \text{Mannose-6-phosphate} \rightleftharpoons \text{Mannose-1-phosphate} \quad (11.14)$$

$$\text{Mannose-1-phosphate} + \text{GTP} \rightleftharpoons \text{GDP-mannose} + P \sim P \quad (11.15)$$

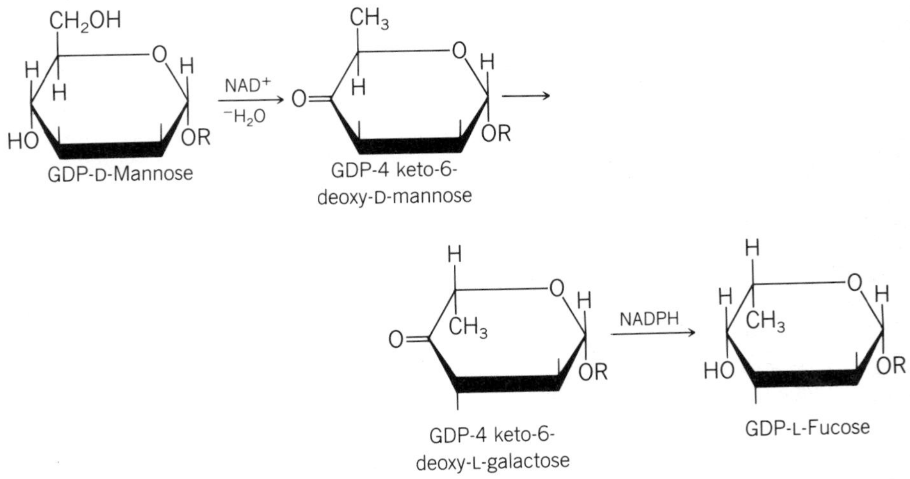

GDP-D-Mannose

GDP-4 keto-6-deoxy-D-mannose

GDP-4 keto-6-deoxy-L-galactose

GDP-L-Fucose

11.3   POLYSACCHARIDE BIOSYNTHESIS

Enzymes that catalyze the synthesis of branched polysaccharides are widespread in nature. The enzymes (glycogen synthase) in mammalian muscle and bacteria may be considered as models. These enzymes catalyze the transfer of glucose moieties from UDP-glucose (muscle) or ADP-glucose (bacteria) to the nonreducing end of an α-1,4-glucan to form a new α-1,4-glucosyl-glucan.

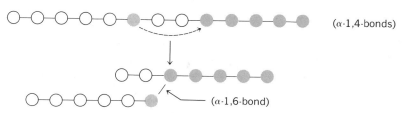

Amylose   UDP-Glucose

$$+ \quad P—P—R—U \quad (11.16)$$
UDP

$$\Delta G' = -2000 \text{ cal (pH 7.0)}$$

The $\Delta G'$ for this reaction ($\Delta G' = -2000$ cal) is somewhat smaller than that for the synthesis of most glycosides since the linkage between the glucose units in the glycogen molecule has an intermediate value for the $\Delta G'$ of hydrolysis ($\Delta G' = -5000$ cal). Glycogen synthase requires a primer molecule to accept the glucose units. A different enzyme is required to form the α-1-6 linkage found in glycogen. This enzyme, known as an **amylo transglycosylase** or **branching enzyme** (1,4 $\rightarrow$ 1,6), catalyzes the transfer of an oligosaccharide unit of six or seven residues in length to another point in the amylose chain to make the α-1-6 branch point.

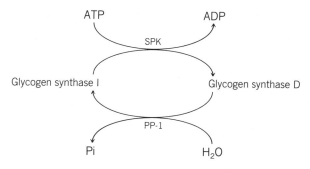

The glycogen synthase of animal tissues, like glycogen phosphorylase, also exists in phosphorylated and dephosphosphorylated forms that play a similar but *reciprocal* role in regulating the amount of glycogen in the cell. The dephosphorylated I form is active and *independent* of glucose-6-phosphate concentrations, while the phosphorylated, low-activity D form is allosterically rendered more active only at sufficient concentrations of glucose-6-phosphate. These two forms are interconverted by the same enzymes that interconvert phosphorylase a and phosphorylase b, synthase-phosphorylase kinase (SPK) and phosphoprotein phosphatase (PP-1).

ATP   ADP

SPK

Glycogen synthase I   Glycogen synthase D

PP-1

Pi H_2O

Note, however, that the phosphorylation (of serine hydroxyl groups) of glycogen synthase I converts a more active form of the enzyme to the less active D form, while phosphorylation of phosphorylase *b* converts a *less* active form of that enzyme to the *more* active form. Thus, the presence of the dephosphorylated I form of glycogen synthase occurs in resting muscle together with phosphorylase *b*, whereas contracting muscle contains more phosphorylase *a* and the low-activity D form of glycogen synthase. The importance of the phosphorylation-dephosphorylation cycles in the regulation of metabolism is discussed further in Sections 18.4 and 18.5.

11.3.1   COMPARISON OF THE ANIMAL AND PLANT SYSTEMS

There are important differences in the synthesis and regulation of glycogen in animal tissues and of starch in the plant systems. The sugar nucleotide donor is always UDP-glucose in animals and in plants ADP-glucose. Regulation in animal systems occurs at the UDP-glucose transfer reaction, but in plants at the level of the generation of ADP-glucose as indicated next.

$$\textit{Animals}\quad \text{UTP} + \text{Glucose-1-p} \xrightarrow[\quad\quad]{\text{PP}} \text{UDPG} \xrightarrow[\substack{\text{regulation}\\\text{site}}]{\text{primer}} \text{Glycogen} + \text{UDP}$$

$$\textit{Plants}\quad \text{ATP} + \text{Glucose-1-p} \xrightarrow[\substack{\text{regulation}\\\text{site}}]{\text{PP}} \text{ADPG} \xrightarrow[\text{primer}]{} \text{Starch} + \text{ADP}$$

Moreover, in plants, inorganic phosphate is a negative effector that inhibits ADP-glucose pyrophosphorylase; 3-phosphoglyceric acid is a positive effector.

11.4   PHOSPHOGLUCONATE PATHWAY AND PENTOSE PHOSPHATE METABOLISM

The glycolytic sequence described in Chapter 10 is a mechanism for partially degrading glucose and obtaining energy as ATP for the cell. This sequence undoubtedly was the first of several metabolic processes to evolve and meet the needs of evolving life forms. As organisms became more complex, there developed a need for biosynthetic capacities beyond those represented by intermediates of the glycolytic sequence and, specifically, the need for a source of reducing power in biosynthesis developed. Since the reducing agent NADH produced in one part of the glycolytic scheme is consumed in another part, reactions that were capable of producing a different reductant were presumably selected. The phosphogluconate pathway, to be described now, contains two reactions capable of producing the reductant NADPH. Furthermore, this pathway also produces a number of different sugar phosphates.

The existence of an alternate route for the metabolism of glucose is indicated by the fact that in some tissues the classical inhibitors of glycolysis, iodoacetate and fluoride, have little or no effect on the utilization of glucose. In addition, the discovery of $NADP^+$ by Warburg in 1935 and the oxidation of glucose-6-phosphate to 6-phosphogluconic acid led the glucose molecule into an unfamiliar area of metabolism. Moreover, with carbon-14, it can be shown in some instances that glucose labeled in the C-1 carbon atom is more readily oxidized to $^{14}CO_2$ than glucose labeled in the C-6 position. If the glycolytic sequence is the only means whereby glucose can be converted to [3-^{14}C]pyruvate and subsequently broken down to CO_2, then $^{14}CO_2$ should be produced at an equal rate from [1-^{14}C]glucose and [6-^{14}C]glucose. These observations stimulated work that resulted in the delineation of the phosphogluconate pathway. The pathway, shown inside the back cover of this text, consists of an oxidative part that is irreversible, and a nonoxidative, reversible part.

11.4.1 GLUCOSE-6-PHOSPHATE DEHYDROGENASE

The irreversible oxidative part of the pathway starts with the reaction catalyzed by the enzyme **glucose-6-phosphate dehydrogenase.**

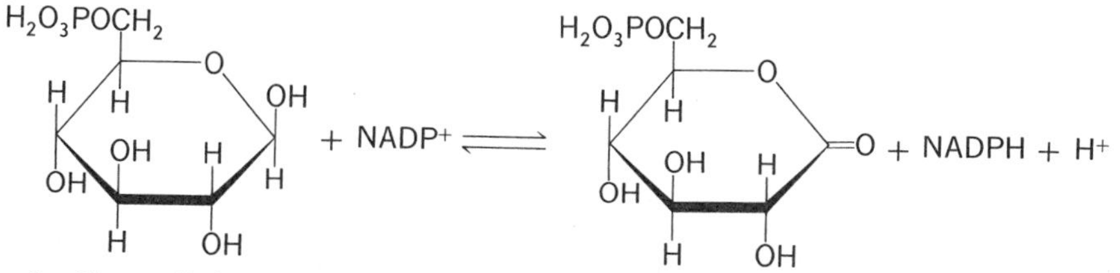

β-D-Glucose-6-phosphate   6-Phosphoglucono-δ-lactone

$$+ NADP^+ \rightleftharpoons \quad + NADPH + H^+ \quad (11.17)$$

The δ-lactone of phosphogluconic acid is formed initially. The reaction is reversible and the oxidation of NADPH will proceed in the presence of the enzyme and the lactone. It is easy to visualize that the oxidation of the pyranosyl form of the substrate involves the removal of two hydrogen atoms from the β anomer to form the lactone.

Not surprisingly, this reaction is subject to metabolic control; the NADP$^+$ required in the forward reaction is competitively inhibited by NADPH. The enzyme is also inhibited by fatty acids. Both types of inhibition are meaningful in terms of one of the functions of the phosphogluconate pathway.

11.4.2 6-PHOSPHOGLUCONOLACTONASE

Hydrolysis of the 6-phosphogluconic-δ-lactone produced in Reaction 11.17 occurs readily in the absence of any enzyme. However, a **lactonase** that ensures rapid hydrolysis of the lactone also exists.

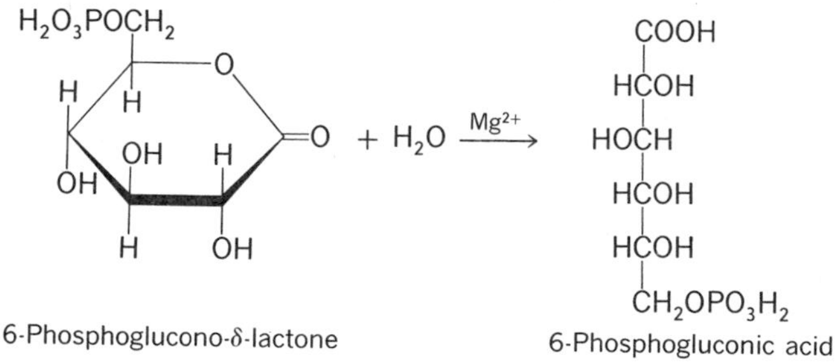

6-Phosphoglucono-δ-lactone   6-Phosphogluconic acid

$$+ H_2O \xrightarrow{Mg^{2+}} \quad (11.18)$$

The $\Delta G'$ for the hydrolysis of the lactone is large; therefore, the overall oxidation of glucose-6-phosphate to phosphogluconic acid is irreversible. Moreover, the next reaction also is irreversible, and together with Reactions 11.17 and 11.18 constitutes the *irreversible phase* of the phosphogluconate pathway.

11.4.3   6-PHOSPHOGLUCONIC ACID DEHYDROGENASE

The enzyme catalyzes the irreversible oxidative decarboxylation of 6-phos-phogluconate, forming D-ribulose-5-phosphate and CO_2. This enzyme also specifically utilizes $NADP^+$, and 3-keto-6-phosphogluconate is an enzyme-bound intermediate in the reaction.

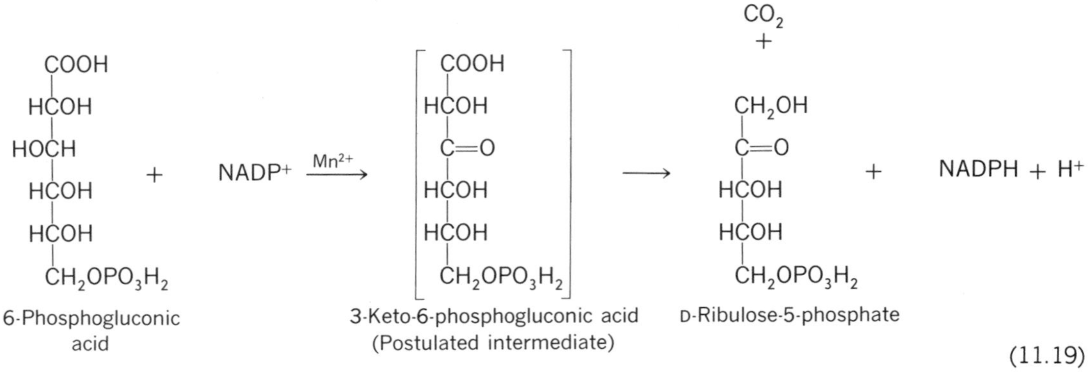

6-Phosphogluconic
acid

3-Keto-6-phosphogluconic acid
(Postulated intermediate)

D-Ribulose-5-phosphate

(11.19)

11.4.4   PHOSPHORIBOISOMERASE

Five-carbon atoms of glucose, initially contained in glucose-6-phosphate, are further metabolized as ribulose-5-phosphate in the *reversible* part of the phosphogluconate pathway. These reactions involve the interconversion of three-, four-, five-, six-, and seven-carbon sugars.

Initially, ribulose-5-phosphate undergoes two isomerization reactions to form products subsequently utilized in the pathway. **Phosphoriboisomerase** catalyzes the interconversion of the keto sugar and the aldopentose phosphate, ribose-5-phosphate. This reaction is analogous in its action to the phosphohexose isomerase (Section 10.3.2) encountered in glycolysis. The K_{eq} for the reaction from left to right is approximately 3.

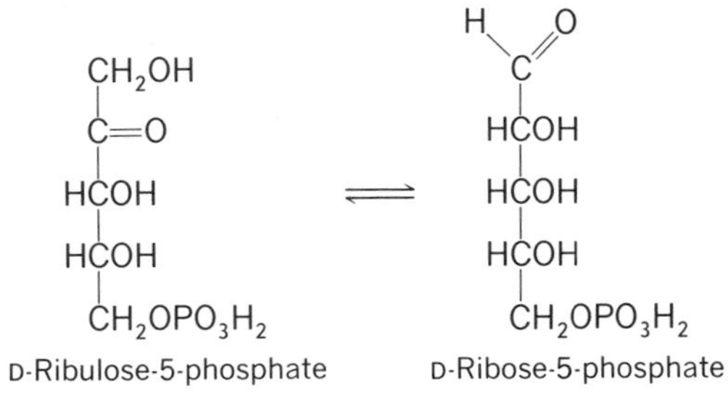

D-Ribulose-5-phosphate

D-Ribose-5-phosphate

(11.20)

11.4.5   PHOSPHOKETOPENTOSE EPIMERASE

The second isomerization involving ribulose-5-phosphate is catalyzed by the enzyme **phosphoketopentose epimerase**. The K_{eq} is 0.7.

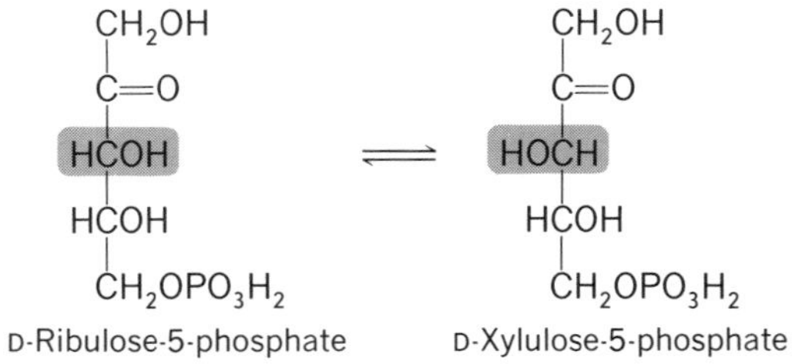

D-Ribulose-5-phosphate

D-Xylulose-5-phosphate

(11.21)

The mechanism involves a 2,3-enediol intermediate on the surface of the enzyme.

11.4.6 TRANSKETOLASE

The enzyme **transketolase** catalyzes the transfer of a "glycolaldehyde" or ketol unit from a donor molecule to an acceptor molecule. The generalized reaction may be written as

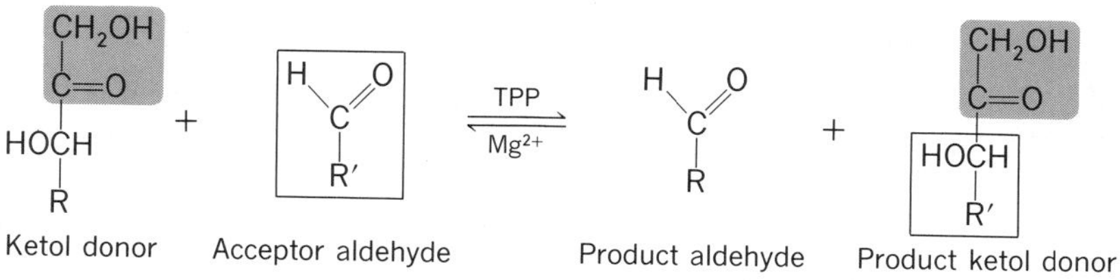

$$\text{(11.22)}$$

Ketol donor   Acceptor aldehyde   Product aldehyde   Product ketol donor

In the first reaction involving transketolase, the enzyme catalyzes the transfer of a ketol group from xylulose-5-phosphate to ribose-5-phosphate to form sedoheptulose-7-phosphate and glyceraldehyde-3-phosphate.

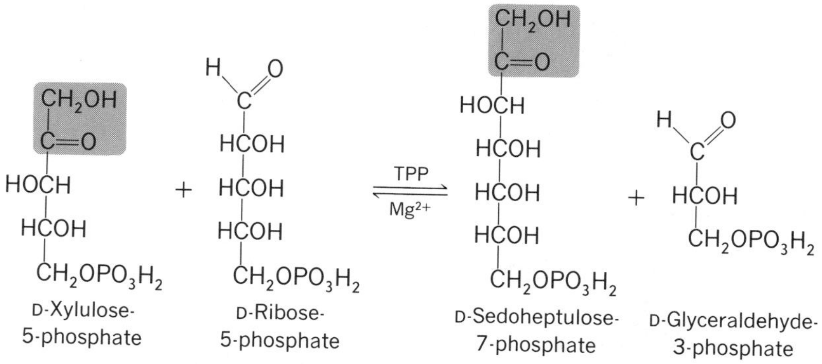

$$\text{(11.23)}$$

D-Xylulose-   D-Ribose-   D-Sedoheptulose-   D-Glyceraldehyde-
5-phosphate   5-phosphate   7-phosphate   3-phosphate

Transketolase (MW 70,000) consists of two identical subunits and utilizes both thiamin pyrophosphate (TPP) and Mg^{2+} as cofactors. The TPP functions as a carbanion by dissociation of a proton at the C-2 carbon atom of the thiazole ring (Section 5.2.3). The resultant carbanion can, in turn, react with the ketol donor to form an addition product (I), which by appropriate rearrangement of electrons, can dissociate in another manner to form the product aldehyde and leave the ketol group on the TPP, forming α,β-dihydroxyethyl thiamin pyrophosphate (II).

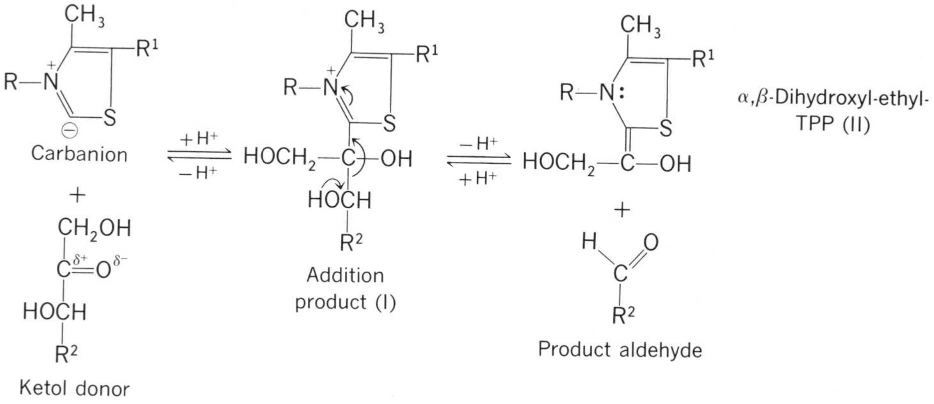

The ketol-TPP addition product (II) then reacts with an acceptor aldehyde to form the product ketol donor and regenerate the carbanion:

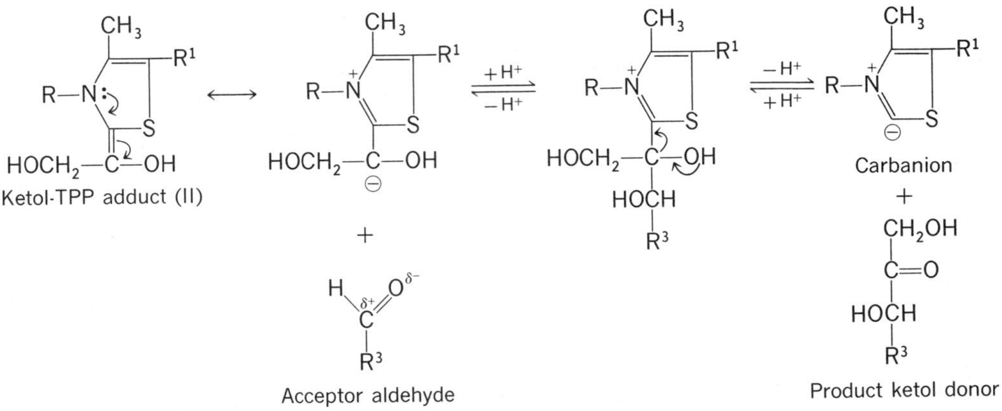

Transketolase also catalyzes the transfer of a ketol group from xylulose-5-phosphate to erythrose-4-phosphate to form fructose-6-phosphate and glyceraldehyde-3-phosphate (Reaction 11.25 and inside of cover). Since this reaction as well as Reaction 11.23 is readily reversible, we can list the following compounds that will serve as donor molecules and acceptor aldehydes for the enzyme.

KETOL DONORS (KETOSES)	ACCEPTOR ALDEHYDES (ALDOSES)
D-Xylulose-5-phosphate	D-Ribose-5-phosphate
D-Fructose-6-phosphate	D-Glyceraldehyde-3-phosphate
D-Sedoheptulose-7-phosphate	D-Erythrose-4-phosphate

Note that the ketose donor molecules all have the *trans* configuration on the two-carbon atoms adjacent to the carbonyl carbon atom.

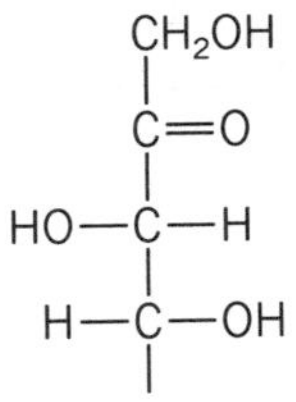

11.4.7 TRANSALDOLASE

This enzyme, like transketolase, functions as a transferring enzyme by catalyzing the transfer of the dihydroxy acetone moiety of fructose-6-phosphate or sedoheptulose-7-phosphate to a suitable aldose. As represented in the scheme for pentose phosphate metabolism, the acceptor aldose may be glyceraldehyde-3-phosphate or, in the reverse direction, erythrose-4-phosphate.

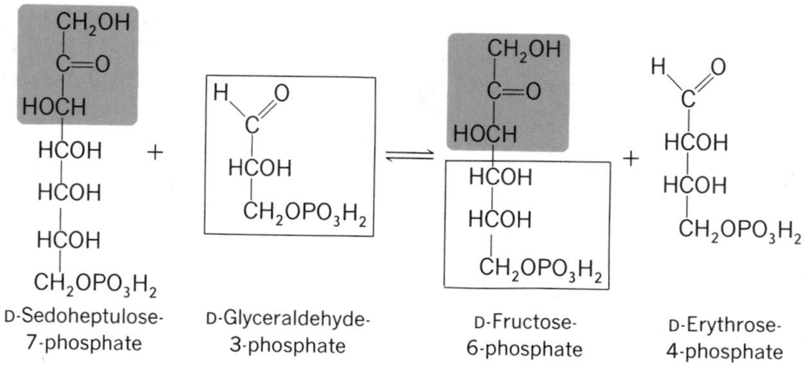

The chemical structures for reaction (11.24):

D-Sedoheptulose-7-phosphate + D-Glyceraldehyde-3-phosphate ⇌ D-Fructose-6-phosphate + D-Erythrose-4-phosphate   (11.24)

The enzymatic mechanism involves the formation of an intermediate Schiff base between the carbonyl of the transferred dihydroxy moiety and an ε-amino group in a lysine residue in the enzyme. This mechanism is similar to that of aldolase in glycolysis (Section 10.3.4). However, transaldolase cannot use free dihydroxyacetone or its phosphate as substrates.

Finally, to complete the pentose phosphate pathway, the erythrose-4-phosphate produced in Reaction 11.24 accepts a C_2 unit from xylulose-5-phosphate in a reaction also catalyzed by transketolase to form fructose-6-phosphate and glyceraldehyde-3-phosphate.

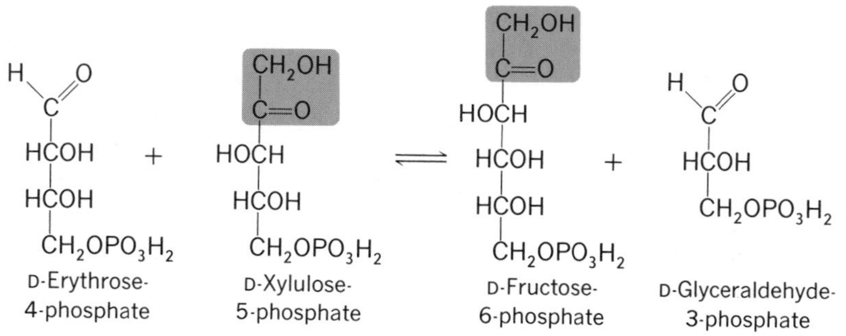

The chemical structures for reaction (11.25):

D-Erythrose-4-phosphate + D-Xylulose-5-phosphate ⇌ D-Fructose-6-phosphate + D-Glyceraldehyde-3-phosphate   (11.25)

11.5 SUMMARY OF THE NONOXIDATIVE PHASE

The interconversions of triose, tetrose, pentose, hexose, and heptose phosphate esters that occur in the nonoxidative part of the phosphogluconate pathway (Reactions 11.23, 11.24, and 11.25) can be confusing. Another way of representing the reactions involving these compounds is

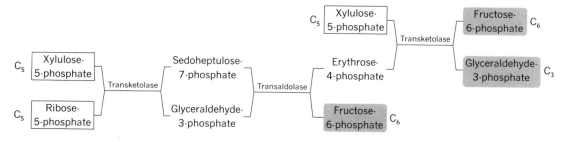

By drawing rectangles around three pentose molecules that can be considered as reactants in this scheme and by shading the three molecules thereby produced, we see that this scheme constitutes a readily reversible mechanism for making hexose and triose intermediates of glycolysis from pentoses derived from the oxidation of glucose-6-phosphate. That is, from left to right, 15 carbon atoms in three pentose molecules give rise to 15 carbon atoms in two hexose molecules and one triose. In the reverse direction, the scheme can account for the formation of pentose derivatives from intermediates of glycolysis.

11.6   FUNCTIONS OF THE PENTOSE PHOSPHATE PATHWAY

One of the most important roles of the phosphogluconate pathway is its generation of NADPH that, in contrast to NADH, plays an essential role as a reductant in many biosynthetic reactions. Whenever a biosynthetic step involves a reduction with a nicotinamide nucleotide, the coenzyme is NADPH with few exceptions.

As examples of this function, NADPH is specifically utilized in the biosynthesis of long-chain fatty acids and steroids. It is the reducing agent employed in the reduction of glucose to sorbitol, dihydrofolic acid to tetrahydrofolic acid, and glucuronic acid to L-gulonic acid. NADPH also plays a unique role in hydroxylation reactions involved in the formation of unsaturated fatty acids, the conversion of phenylalanine to tyrosine, and the formation of certain steroids. Evidence of the role of the phosphogluconate pathway in producing NADPH for biosynthetic purposes is found in the fact that the pathway enzymes are especially abundant in tissues such as adipose tissues, mammary gland, or adrenal cortex that carry out these biosyntheses.

Another important biosynthetic intermediate produced in this pathway is ribose-5-phosphate, required for the synthesis of both ribo- and deoxyribonucleotides. These compounds are precursors not only of RNA and DNA, but also numerous coenzymes and ATP itself.

Erythrose-4-phosphate, produced in Reactions 11.24 and 11.25, is required in the initial step of the shikimic acid pathway that in plants and microorganisms leads to the aromatic amino acids phenylalanine, tyrosine, and tryptophan. These amino acids, in turn, are precursors of many other cellular components, especially in plants.

While the ribose-5-phosphate and erythrose-4-phosphate can be produced from glucose-6-phosphate by the irreversible oxidative phase, they can also be formed from fructose-6-phosphate and glyceraldehyde-3-phosphate by the reversal of Reactions 11.25, 11.24, and 11.23; thus, the cell can use either an oxidative or nonoxidative process to form these important intermediates. We shall see that this latter process is also utilized by photosynthetic organisms to generate essential intermediates of the pentose phosphate reduction cycle of photosynthesis (Chapter 15).

It is conceivable that an organism might have a greater need for NADPH produced by the oxidative phase of the phosphogluconate pathway than for the pentoses that are simultaneously formed. The ready conversion of the pentoses so produced back to glycolytic intermediates clearly eliminates any difficulties due to excess pentose. By combining the reactions of the phosphogluconate pathway with three reactions of the glycolytic sequence, it is possible to repre-

TABLE 11.1 The Phosphogluconate Pathway as a Cyclic Process

REACTION NUMBER	REACTION
11.17, 11.18, and 11.19	6 Glucose-6-phosphate + 12 NADP$^+$ + 6 H$_2$O $\longrightarrow$ 6 Ribulose-5-phosphate + 6 CO$_2$ + 12 NADPH + 12 H$^+$
11.20, 11.21, 11.23, 11.24, and 11.25	6 Ribulose-5-phosphate $\rightleftharpoons$ 4 Fructose-6-phosphate + 2 Glyceraldehyde-3-phosphate
10.4a	Glyceraldehyde-3-phosphate $\rightleftharpoons$ Dihydroxyacetone-phosphate
10.4	Glyceraldehyde-3-phosphate + Dihydroxyacetone-phosphate $\rightleftharpoons$ Fructose-1,6-bisphosphate
10.25	Fructose-1,6-bisphosphate + H$_2$O $\longrightarrow$ Fructose-6-phosphate + H$_3$PO$_4$
10.2	5 Fructose-6-phosphate $\rightleftharpoons$ 5 Glucose-6-phosphate
Sum:	Glucose-6-phosphate + 12 NADP$^+$ + 7 H$_2$O $\longrightarrow$ 6 CO$_2$ + 12 NADPH + 12 H$^+$ + H$_3$PO$_4$

sent the oxidation of one mole of glucose-6-phosphate to six moles of CO$_2$ (Table 11.1). This cyclic oxidation, which results in the production of 12 moles of NADPH, can occur to a significant extent in mammary gland, adipose tissue, and the arenal cortex of animals. The reactions occur in the cytosolic portion of cells in these tissues that, in turn, have a great need for NADPH to support their specialized biosynthetic activities (e.g., lipid and steroid synthesis).

11.7 ENTNER–DOUDOROFF PATHWAY

Some bacteria (e.g., *Pseudomonads, Azotobacter sp.*) lack phosphofructokinase and therefore cannot degrade glucose by the glycolytic sequence. These organisms, instead, initiate glucose catabolism by producing 6-phosphogluconic acid by Reactions 11.17 and 11.18. The acid then undergoes a dehydration and rearrangement to form an α-ketodeoxy sugar phosphate that, in turn, is cleaved by an aldolase-type enzyme into pyruvate and glyceraldehyde-3-phosphate.

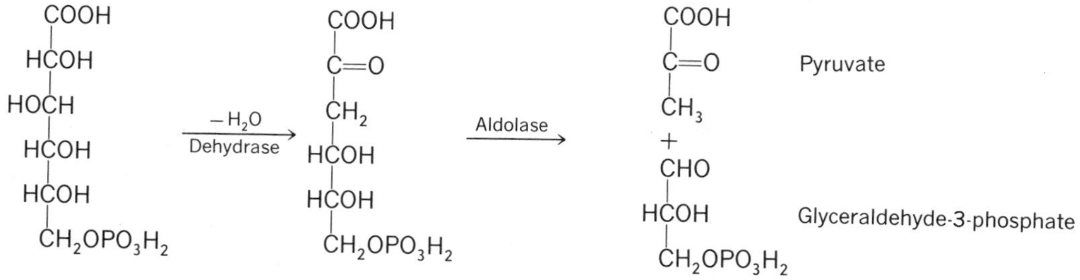

Modification of this scheme permits other sugars (galactose) and sugar acids (D-glucuronic acid and D-galacturonic acid) to be metabolized, but an essential feature is the production of a 2-keto-3-deoxy intermediate that can be cleaved after phosphorylation.

Other routes are known for the metabolism of glucose and other sugars, but they are beyond the purpose of this text.

REFERENCES

1. *B. Axelrod, "Other Pathways of Carbohydrate Metabolism." In* Metabolic Pathways, *3rd ed., D. M. Greenberg, ed. New York: Academic Press, 1967.*

2. *T. ap Rees, "Integration of Pathways of Synthesis and Degradation of Hexose Phosphate. In* Biochemistry of Plants, *Vol. 3, J. Preiss, ed., pp. 1–42. New York: Academic Press, 1981.*

3. *R. W. Hanson and M. A. Mehlman, eds.* Gluconeogenesis. *New York: Wiley-Interscience, 1974.*

4. *E. G. Krebs and J. A. Beavo, "Phosphorylation and Dephosphorylation of Enzyme."* Annual Review of Biochemistry, *Vol. 48 (1979), pp. 923–959.*

REVIEW PROBLEMS

1. What are the products of the reaction catalyzed by transketolase if fructose-6-phosphate and erythrose-4-phosphate are used as the reactants?

2. Outline experiments to determine whether a given tissue uses glycolysis, the pentose phosphate pathway, or a mixture of both in the degradation of glucose for energy. (Include at least four distinct types of experiments.)

CARBOHYDRATES III: THE TRICARBOXYLIC ACID CYCLE

The tricarboxylic acid or citric acid cycle is the process by which the acetyl moiety in acetyl CoA is oxidized completely to carbon dioxide and H_2O. The carbon atoms in the acetyl CoA can arise from carbohydrate via pyruvic acid produced in glycolysis (Chapter 10) or the catabolism of fatty acids (Chapter 13) and certain amino acids (Chapter 17). The thiol ester therefore represents an important, common intermediate in the process by which the oxidation to CO_2 and H_2O of carbohydrates, fatty acids, and certain amino acids is completed. The electrons removed from the substrates as they are oxidized are transferred eventually to molecular oxygen; thus, the process is an aerobic one. The evolution of this metabolic sequence had to wait until photosynthesis by green plants evolved and the oxygen content of the atmosphere increased sufficiently to support respiration. The processes that were selected (and now constitute the reactions of aerobic respiration) are highly effective in releasing the chemical energy of organic substrates because they are capable of oxidizing the carbon atom all the way to CO_2.

The enzymes catalyzing the tricarboxylic acid cycle, in contrast to those of glycolysis, are located in the mitochondria of the eucaryotic cell. Their location in this organelle permits the direct interaction of the electron pairs removed during oxidation of the cycle intermediates with the mitochondrial electron transport chain and the eventual reduction of oxygen. Moreover, as the electron pairs (reducing equivalents) are transported to oxygen, ADP is phosphorylated to form ATP. The process may be represented as

$$SH_2 + \tfrac{1}{2}O_2 + nADP + nH_3PO_4 \longrightarrow S + nH_2O + nATP$$

where S is an oxidizable substrate in the cycle and n varies (either 2 or 3) with the nature of the substrate.

12.1 ENERGETICS OF GLUCOSE OXIDATION

It is of interest to consider the relative amounts of energy liberated as glucose is degraded first through the anerobic process of glycolysis and then by aerobic oxidation of pyruvate. The free-energy change for the complete oxidation of one mole of glucose to CO_2 and H_2O has been given as -686 kcal.

$$C_6H_{12}O_6 + 6\ O_2 \longrightarrow 6\ CO_2 + 6\ H_2O$$
$$\Delta G' = -686,000 \text{ cal/mole (pH 7.0)}$$

In Section 10.5, we indicated that the $\Delta G'$ for the formation of lactic acid from glucose was about -47 kcal.

$$C_6H_{12}O_6 \longrightarrow 2\ CH_3CHOHCOOH$$
$$\Delta G' = -47,000 \text{ cal (pH 7.0)}$$

This means that only about 7% of the available energy of the glucose molecule has been released when lactic acid is formed in glycolysis and about $-639,000$ cal remain to be released when the two moles of lactate from the original glucose molecule are oxidized to completion. Thus, the $\Delta G'$ per mole of lactate oxidized can be estimated as $-319,000$ cal.

$$CH_3CHOHCOOH + 3\ O_2 \longrightarrow 3\ CO_2 + 3\ H_2O$$
$$\Delta G' = -319,000 \text{ cal/mole (pH 7.0)}$$

The student should also appreciate that the aerobic oxidation of glucose to CO_2 and H_2O by living organisms does not normally involve the formation of lactic acid as an intermediate step. Instead, the key compound produced in glycolysis, which can be either reduced to lactic acid or instead be oxidized completely to CO_2 and H_2O, is pyruvic acid. The individual reactions of the sequence that accomplishes this oxidation will now be considered.

BOX 12.A   THE KREBS' TRICARBOXYLIC ACID CYCLE

The tricarboxylic acid cycle was assigned its name by the individual whose research described the essential features of this key metabolic sequence, Sir Hans Krebs. The specific use of **tricarboxylic acid** is due to the fact that the well-known citric and isocitric acids were early identified as the products of oxidation of pyruvic acid in homogenates of animal tissues. Because of Krebs' many studies and publications during the 1940s and 1950s in this area of investigation, however, the tricarboxylic acid cycle is also commonly known as the Krebs' tricarboxylic acid cycle or, shortened, the Krebs cycle.

Krebs was also the discoverer, nearly a decade earlier, of another fundamental metabolic process known as the urea cycle (Section 17.8). This cycle accomplishes the disposal of excess ammonia in mammals and also accounts for the biosynthesis of the amino acid arginine. The cyclic nature of this process, sometimes denoted as the Krebs–Henseleit urea cycle, undoubtedly assisted Krebs in sorting out the cyclical nature of the tricarboxylic acid cycle. (To add to the nomenclatural confusion in this subject, the tricarboxylic acid is also sometimes referred to as the citric acid cycle.)

12.2 OXIDATION OF PYRUVATE TO ACETYL CoA

Pyruvic acid, produced in the cytosol, is oxidized to acetyl CoA after entering the mitochondria. This reaction, which is the oxidative decarboxylation of an α-keto acid, is catalyzed by a multienzyme complex known as the **pyruvic**

dehydrogenase complex. The reaction involves two dissociable coenzymes, coenzyme A and NAD^+, and three bound coenzymes plus Mg^{2+} ion.

$$CH_3-\underset{\underset{O}{\|}}{C}-CO_2H + CoA-SH + NAD^+ \xrightarrow[\substack{TPP \\ FAD}]{\substack{Lipoic\ acid \\ Mg^{2+}}}$$

$$CH_3-\underset{\underset{O}{\|}}{C}-S-CoA + NADH + H^+ + CO_2 \quad (12.1)$$

$$\Delta G' = -8000\ cal\ (pH\ 7.0)$$

The reaction proceeds with a large decrease in free energy ($\Delta G' = -8000$) and results in the formation of the energy-rich thioester, acetyl CoA.

The pyruvic dehydrogenase complex in all organisms consists of three separate proteins that catalyze a series of partial reactions that add up to the overall reaction given in Reaction 12.1. The three proteins have molecular weights ranging between 50,000 and 200,000, and there can be as many as 60 units of one of the three proteins and as few as 12 in the native complex. If the average number of each subunit in the complex is 24 and the average size is 10^5, the molecular weight of the complex can be very large, that is, $3 \times 24 \times 10^5$ or 7.2×10^6.

The protein that catalyzes the initial step in the oxidation of pyruvic acid is known as **pyruvic decarboxylase.** In mammalian tissue, this protein (MW 152,000) contains two peptide chains (β-chains), each of which binds a molecule of thiamin pyrophosphate (TPP). Two other chains (α-chains) are involved in transferring a hydroxyethyl (acetol) group to a lipoyl residue in the lipoyl transacetylase subunit of the complex. The pyruvic decarboxylase subunit also contains the serine residues that are phosphorylated by **pyruvic dehydrogenase kinase** (Section 12.5), thereby converting pyruvic dehydrogenase to an inactive form in a monocyclic cascade (Section 18.4.2).

The **lipoyl acetyl transferase** (MW 52,000) subunit contains a covalently bound lipoic acid residue that shuttles between the pyruvic dehydrogenase subunit and the third protein in the complex, **dihydrolipoyl dehydrogenase.** This third component contains a covalently bound FAD that accepts electrons from the dihydrolipoyl dehydrogenase and passes them on to NAD^+.

The partial reactions catalyzed by the three subunits of the pyruvic dehydrogenase complex can now be described. In the initial reaction, pyruvic acid is decarboxylated to form CO_2 and α-hydroxyethyl thiamin pyrophosphate that is tightly bound to the pyruvic decarboxylase.

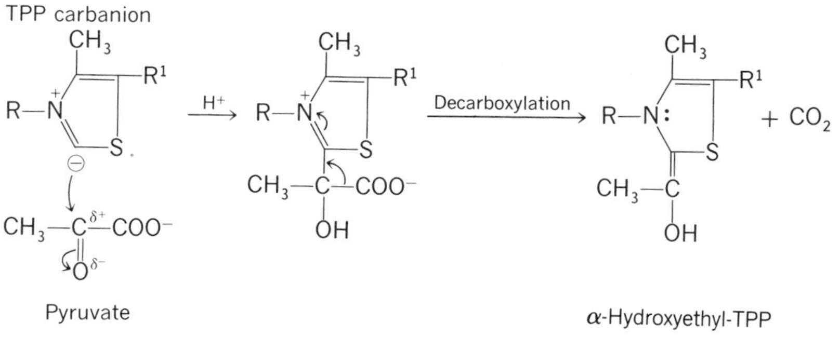

$$\text{Pyruvate} \xrightarrow{} \text{α-Hydroxyethyl-TPP} \quad (12.1a)$$

The two-carbon α-hydroxyethyl group is next transferred to an oxidized lipoic acid moiety that is covalently bound to the second enzyme of the complex, *dihydrolipoyl acetyl transferase* (Reaction 12.1*b*):

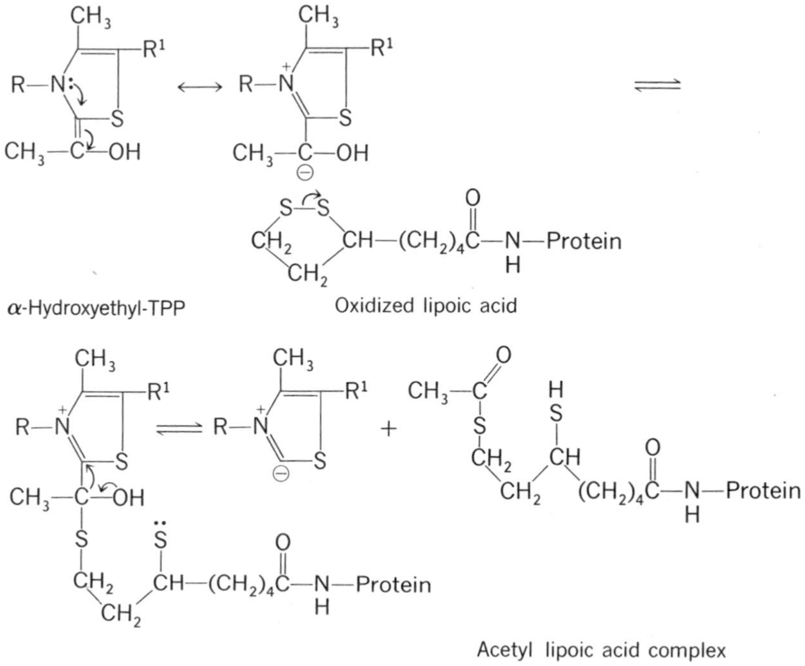

α-Hydroxyethyl-TPP   Oxidized lipoic acid

(12.1*b*)

Acetyl lipoic acid complex

SCHEME 12.1

Note that, as a result of this reaction, a high-energy thioester (of reduced lipoic acid) has been formed, and that the two-carbon unit is now at the oxidation level of acetic acid rather than acetaldehyde.

In a third reaction, the acetyl group is transferred to coenzyme A to form acetyl CoA, which dissociates from the enzyme in a free form, being one of the products of the overall reaction (Reaction 12.1).

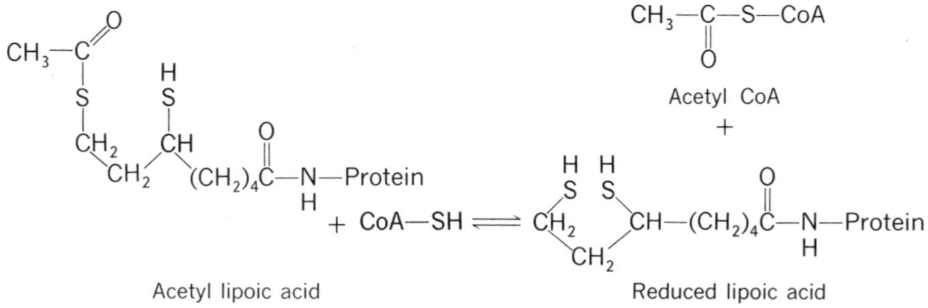

Acetyl CoA

(12.1*c*)

Acetyl lipoic acid   Reduced lipoic acid

The transfer of the acetyl group to coenzyme A (Reaction 12.1*c*) leaves the lipoic acid moiety of the dihydrolipoyl dehydrogenase in the reduced or dithiol form. This reduced form is then reoxidized to the cyclic lipoyl form by the third enzyme of the complex, dihydrolipoyl dehydrogenase, a flavoprotein that contains FAD.

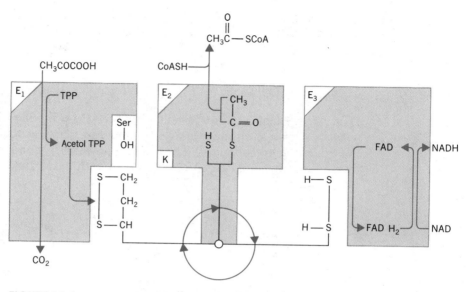

FIGURE 12.1 Mechanism of the pyruvate dehydrogenase complex. E_1, pyruvate decarboxylase; E_2, dihydrolipoyl transacetylase; E_3, dihydrolipoyl dehydrogenase. K, site of the pyruvate dehydrogenase kinase, a converter enzyme that phosphorylates a serine residue of E_1.

$$\text{Reduced lipoic acid} + \text{FAD} \rightleftharpoons \text{Oxidized lipoic acid} + \text{FADH}_2$$

Reduced lipoic acid   Oxidized lipoic acid   (12.1d)

Finally, the reduced flavin coenzyme is reoxidized by NAD^+, one of the reactants in the overall process (Reaction 12.1), and NADH is produced.

$$FADH_2 + NAD^+ \rightleftharpoons FAD + NADH + H^+$$

Note that all of the partial reactions catalyzed by the pyruvic dehydrogenase complex are reversible except the initial decarboxylation. The irreversible nature of this first step makes the overall reaction (Reaction 12.1) irreversible and the $\Delta G'$ has been estimated as approximately -8000 cal.

Figure 12.1 illustrates the role of the three subunits of the pyruvic dehydrogenase complex in catalyzing Reaction 12.1. In particular, the role of the lipoic acid residue of the acetyl transferase subunit in transferring the C_2 unit, becoming reduced, and then transferring the electrons first to FAD and ultimately to NAD^+ should be noted.

It should be pointed out that the pyruvic dehydrogenase complex, because of its key role in generating acetyl CoA, is a highly regulated system. (See Section 18.4.2).

12.3 THE CITRIC ACID CYCLE

As mentioned earlier, acetyl CoA is a metabolite shared in common by the metabolism of carbohydrates, lipids, and some amino acids. Its production by the pyruvic dehydrogenase complex, just described, is not a part of the citric acid cycle per se. Acetyl CoA, however, is the *fuel* for the cycle and, as will now be described, the two carbon atoms constituting the acetyl moiety of the thioester are, in effect, oxidized to CO_2 and H_2O by the sequential operation of the eight enzymes involved.

12.3.1   CITRATE SYNTHASE

The enzyme that catalyzes the entry of acetyl CoA into the tricarboxylic acid cycle is known as **citrate synthase** or condensing enzyme and is found in the matrix of the mitochondria. The reaction proceeds by condensation of a carbanion at the methyl group of acetyl CoA with oxalacetate in an aldol-type condensation.

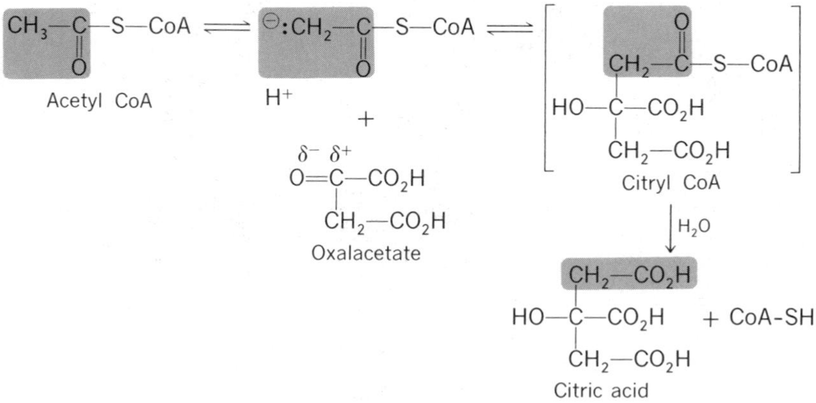

Carbon atoms that originate from the acetyl CoA are shaded in the reaction shown here and subsequent reactions. The equilibrium constant for the reaction is 3×10^5 and therefore the equilibrium is far in the direction of citrate synthesis. (Note in Reaction 12.2 that there is a formation of a carbon–carbon bond and free coenzyme A at the expense of the thioester.) Indirect evidence indicates that citryl CoA is formed as an intermediate on the enzyme, but does not dissociate as such until it is cleaved to free citrate and coenzyme A.

12.3.2   ACONITASE

Aconitase catalyzes the reversible equilibria between citric acid, *cis*-aconitic acid, and isocitric acid.

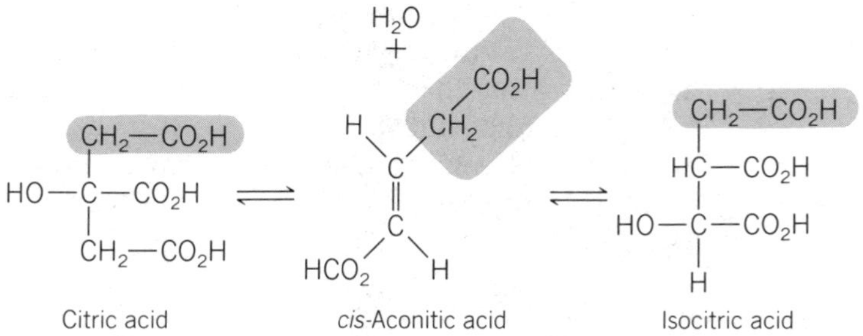

The enzyme catalyzes the reversible *trans*-addition of H_2O to the double bond of *cis*-aconitic acid that is an enzyme-bound intermediate in the interconversion of the two citric acid isomers. At equilibrium, the ratio of citric acid to isocitric acid is about 15.

Note that when isocitric acid is formed from citric acid (Reaction 12.3), the symmetric molecule citric acid is acted on in an asymmetric manner by the enzyme aconitase. That is, the hydroxyl group in isocitric acid is located on a carbon atom derived initially from oxalacetate rather than the methyl group of acetyl CoA. This important phenomenon is discussed in detail in Chapter 4.

12.3.3 ISOCITRIC DEHYDROGENASE

Isocitric dehydrogenase catalyzes the oxidative β-decarboxylation of isocitric acid to α-ketoglutaric acid and CO_2 in the presence of a divalent cation (Mg^{2+} or Mn^{2+}). A nicotinamide nucleotide serves as the oxidant and the β-keto acid oxalosuccinic acid is an enzyme-bound intermediate.

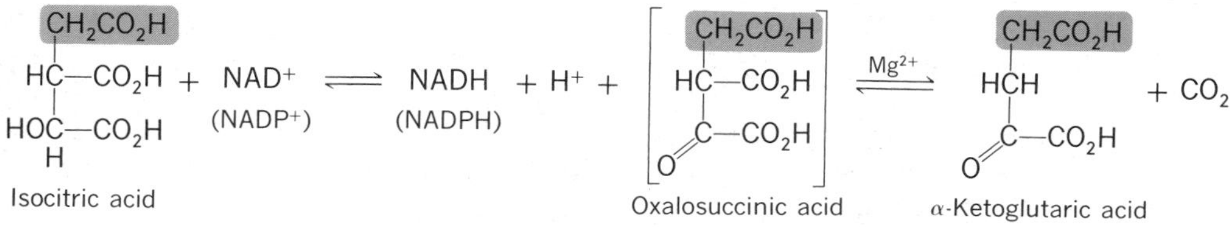

Although the equilibrium favors α-ketoglutarate formation, the reaction is demonstrably reversible and α-ketoglutarate can be reductively carboxylated in the presence of NADH or NADPH. The reaction is formally analogous to the oxidation of another α-hydroxy acid, malic acid, by **malic enzyme** (Section 12.6.1).

Most tissues contain two kinds of isocitric dehydrogenases. One of these requires NAD^+ and Mg^{2+} and is found only in the mitochondria. The other enzyme requires $NADP^+$ and occurs both in mitochondria and the cytoplasm. The NAD^+-specific enzyme is involved in the functioning of the tricarboxylic acid cycle; the $NADP^+$-requiring enzymes are associated with the generation of NADPH for use in biosynthetic reactions that require that coenzyme. The mitochondrial NAD^+-enzyme is activated by ADP that increases the affinity of the enzyme for isocitrate. The enzyme is inhibited by NADH (and NADPH).

12.3.4 α-KETOGLUTARIC ACID DEHYDROGENASE

The next step of the tricarboxylic acid cycle involves the formation of succinyl CoA by the oxidative α-decarboxylation of α-ketoglutaric acid. This reaction is catalyzed by the **α-ketoglutaric dehydrogenase** complex that contains three subunit proteins that require TPP, Mg^{2+}, NAD^+, FAD, lipoic acid, and coenzyme A as cofactors. The mechanism is analogous to that of the pyruvic acid dehydrogenase complex (Section 12.2). The overall process can be written as the sum of individual reactions in a manner entirely analogous to the partial reactions written for Reaction 12.1.

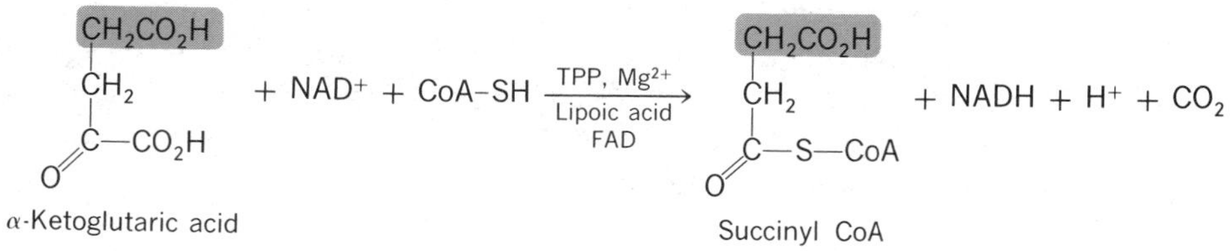

Because of the large negative ΔG, the reaction is not reversible. Succinyl CoA, produced in the reaction, is a competitive inhibitor and strongly inhibits its own formation from α-ketoglutarate unless it is removed. Unlike the pyruvic acid dehydrogenase complex, the α-ketoglutaric dehydrogenase system does not undergo a phosphorylation–dephosphorylation cycle and therefore is not regulated by the protein kinase-phosphatase system.

12.3.5 SUCCINIC THIOKINASE

In the preceding reaction, the high-energy bond of a thioester was formed as the result of an oxidative decarboxylation. The enzyme **succinic thiokinase** catalyzes the formation of a high-energy phosphate structure at the expense of that thioester (Reaction 12.6).

$$
\begin{array}{c}
\text{CH}_2\text{CO}_2\text{H} \\
| \\
\text{CH}_2 \\
| \\
\underset{\text{O}}{\text{C}}\text{—S—CoA}
\end{array}
\quad + \text{ GDP} + \text{H}_3\text{PO}_4 \rightleftharpoons
\overset{\text{Randomization of carbon atoms}}{
\begin{array}{c}
\text{CH}_2\text{CO}_2\text{H} \\
| \\
\text{CH}_2\text{CO}_2\text{H}
\end{array}}
\ + \text{ GTP} + \text{CoA–SH} \qquad (12.6)
$$

Succinyl CoA   Succinic acid

Because Reaction 12.6 involves the formation of a new high-energy phosphate structure and the utilization of a thioester, the total number of high-energy structures on each side of the reaction is equal. Therefore, the reaction is readily reversible; the K_{eq} is 3.7. The GTP formed in Reaction 12.6 can, in turn, react with ADP to form ATP and GDP in a reaction catalyzed by a **nucleoside diphosphokinase** located in the mitochondrial intermembrane space. Because the pyrophosphate linkages in GTP and ATP have approximately the same $\Delta G'$ of hydrolysis, the reaction is readily reversible, with an K_{eq} of about 1.

$$\text{GTP} + \text{ADP} \rightleftharpoons \text{GDP} + \text{ATP}$$

12.3.6 SUCCINIC DEHYDROGENASE

This enzyme catalyzes the removal of two hydrogen atoms from succinic acid to form fumaric acid.

$$
\begin{array}{c}
\text{CO}_2\text{H} \\
| \\
\text{HCH} \\
| \\
\text{HCH} \\
| \\
\text{CO}_2\text{H}
\end{array}
\ + \text{ FAD–Enz} \rightleftharpoons
\begin{array}{c}
\text{H}\quad\text{CO}_2\text{H} \\
\diagdown\ \diagup \\
\text{C} \\
\| \\
\text{C} \\
\diagup\ \diagdown \\
\text{HO}_2\text{C}\quad\text{H}
\end{array}
\ + \text{ FADH}_2\text{–Enz} \qquad (12.7)
$$

Succinic acid   Fumaric acid

The immediate acceptor (oxidizing agent) of the electrons is a flavin coenzyme (FAD) that is covalently bound through a histidine residue to the larger of two subunits of the enzyme. Succinic dehydrogenase of mammalian heart muscle is a heterodimer (MW of 70,000 and 27,000) found as an integral protein in the mitochondrial inner membrane.

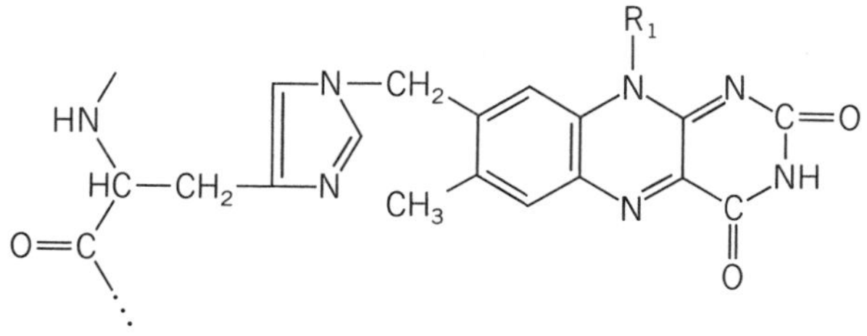

R_1 = Remainder of the FAD molecule

Both subunits contain nonheme iron atoms as (FeS) clusters, and the electrons are known to flow from the substrate through the flavin to the iron cluster in the larger protein subunit and then into iron of the smaller subunit before finally being transferred to the mitochondrial electron transport chain. Succinic dehydrogenase is competitively inhibited by malonic acid, a homologue of succinic acid. Extensive use was made of this fact by those who were concerned initially with working out the details of the tricarboxylic acid cycle.

12.3.7 FUMARASE

The next reaction is the addition of H_2O to fumaric acid to form L-malic acid.

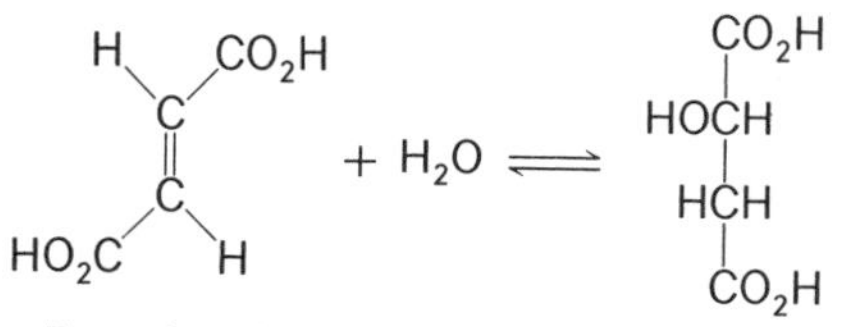

Fumaric acid   L-Malic acid   (12.8)

The equilibrium for this reaction is about 4.5. The enzyme that catalyzes the reaction, **fumarase,** has been crystallized (MW 200,000) from pig heart. It is a tetramer of four identical polypeptide chains. Its kinetics and mechanism of action have been extensively studied.

12.3.8 MALIC DEHYDROGENASE

The tricarboxylic cycle is completed with the oxidation of L-malic acid to oxalacetic acid by the enzyme **malic dehydrogenase.** The reaction is the fourth oxidation–reduction reaction to be encountered in the cycle; the oxidizing agent is NAD^+.

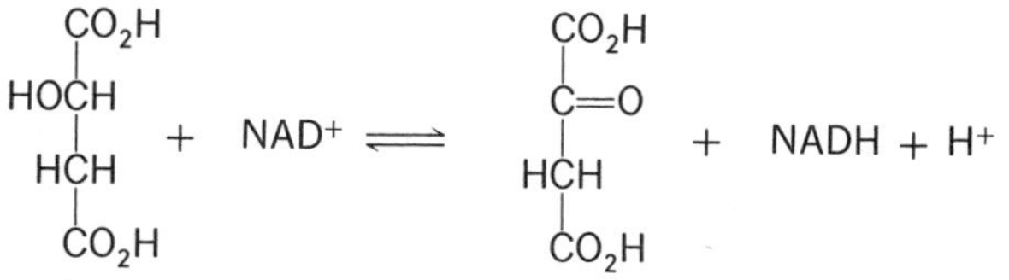

L-Malic acid   Oxalacetic acid   (12.9)

At pH 7.0, the equilibrium constant is 1.3×10^{-5}; thus, the equilibrium favors malate production. On the other hand, the further reaction of oxalacetate with acetyl CoA in the condensation reaction (Reaction 12.2) is strongly exergonic in the direction of citrate synthesis. This tends to drive the conversion of malate to oxalacetate by displacing the equilibrium through the continuous removal of oxalacetate.

The malic dehydrogenase of the mitochondrial matrix just described is distinct from its isoenzyme counterpart in the cytosol. The cytosolic malic dehydrogenase which also utilizes $NAD^+/NADH$, plays an important role in the reversal of glycolysis (Section 10.7.2) and the malatate–aspartate shuttle (Fig. 14.9a).

12.4 FEATURES OF THE TRICARBOXYLIC ACID CYCLE

12.4.1 STOICHIOMETRY

The balanced equation for the complete oxidation of acetyl CoA to CO_2 and H_2O may be written

$$CH_3-\underset{\underset{O}{\|}}{C}-S-CoA + 2\ O_2 \longrightarrow 2\ CO_2 + H_2O + CoA-SH \qquad (12.10)$$

Since this is accomplished in a stepwise manner by the reactions of the tricarboxylic acid cycle (Reactions 12.2 through 12.9), it is useful to examine the stoichiometry in detail.

1. There are four oxidation steps: Reactions 12.4, 12.5, 12.7, and 12.9. In each of these, a pair of hydrogen atoms is removed from the substrate and transferred to either a nicotinamide coenzyme or a flavin coenzyme. As we shall see in Chapter 14, the reoxidation of these four reduced coenzymes by means of the cytochrome electron-transport system results in the reduction of four atoms or two moles of oxygen.

The stoichiometry for pyruvate oxidation involves an additional electron pair removed from pyruvic acid in Reaction 12.1.

$$CH_3-\underset{\underset{O}{\|}}{C}-CO_2H + 2\tfrac{1}{2}\ O_2 \longrightarrow 3\ CO_2 + 2\ H_2O \qquad (12.11)$$

2. Each electron pair released during the oxidation of pyruvic acid ultimately is used to reduce O_2, and H_2O is formed.

$$\tfrac{1}{2}\ O_2 + 2\ H^+ + 2\ e^- \longrightarrow H_2O$$

By inspection, one can see that two moles of H_2O have been consumed directly in Reactions 12.2 and 12.8. To account for the net production of only two moles of H_2O in pyruvate oxidation (Reaction 12.11), a third mole of H_2O must be accounted for. This discrepancy involves the production of GTP from GDP and H_3PO_4 in Reaction 12.6 of the cycle, a reaction that involves no *net* consumption or production of H_2O. In order to write the overall reaction of 12.11 as corresponding to the sum of Reactions 12.1 through 12.9, the GTP produced in 12.6 must be balanced out (since it does not appear in Reaction 12.11) by consuming a third mole of H_2O to convert the GTP back to GDP and H_3PO_4.

3. Finally, two moles of CO_2 are produced in the tricarboxylic acid cycle proper and a third is formed during the oxidation of pyruvate to acetyl CoA

(Reaction 12.1). These are equivalent to the three carbon atoms in the pyruvic acid, but note that only the CO_2 produced in Reaction 12.1 arises *directly* from the pyruvic acid. The other two moles of CO_2 produced (Reactions 12.4 and 12.5) have their origin in the two carboxylic acid groups of oxalacetate (note shading).

All the reactions of the tricarboxylic acid cycle are reversible except the oxidative α-decarboxylation of α-ketoglutarate (Reaction 12.5). As pointed out earlier, this reaction is entirely analogous to the irreversible oxidative α-decarboxylation of pyruvic acid. This then means that the cycle cannot be made to proceed in a reverse direction, although individual sections are reversible (e.g., from oxalacetate to succinate or from α-ketoglutarate to citrate). Similarly, acetyl CoA and CO_2 cannot be converted to pyruvate by a reversal of Reaction 12.1.

12.4.2 ENERGY PRODUCTION

In Chapter 14, the process of oxidative phosphorylation resulting from the transport of electrons through the electron transport chain will be described. These processes occur in the mitochondrial inner membrane where electrons released in the oxidation of acetyl CoA (and pyruvic acid) are transferred to the transport chain. We shall see that 12 energy-rich bonds are formed (as ATP) when acetyl CoA is oxidized to CO_2 and H_2O by the Krebs cycle. The corresponding figure for pyruvic acid is 15 ATP. Because each bond of ATP produced requires approximately 8000 cal/mole, the amount of energy trapped as ATP is $\sim$ 120,000 cal/mole of pyruvic acid oxidized.

12.4.3 OXIDATION OF KREBS CYCLE INTERMEDIATES

Up to this point, we have stressed the catabolic nature of the Krebs cycle, that is, its ability to accomplish the complete oxidation of pyruvic acid, or more precisely, acetyl CoA derived from pyruvate (Reaction 12.1), to CO_2 and H_2O. It should be noted that the acetyl CoA can be derived from other sources, for example, from the breakdown of fatty acids (Chapter 13) or certain amino acids (Chapter 17).

The Krebs cycle obviously can serve as a mechanism for oxidizing the seven tri- and dicarboxylic intermediates of the cycle itself. As an example, consider the sequence of reactions by which succinate, produced in the breakdown of isoleucine, would be oxidized completely to CO_2 and H_2O. Initially, the succinate can be converted to oxalacetate by Reactions 12.7 through 12.9. At this point, the oxalacetate could then condense with a mole of acetyl CoA, but this, in effect, would simply represent accelerated oxidation of acetate by the cycle, because of the increased levels of oxalacetate produced from succinate. To accomplish the complete oxidation of oxalacetate, this compound would be converted back to phosphoenolpyruvate, as it is in gluconeogenesis (Section 10.7.2). That is, the oxalacetate would be reduced to malate by mitochondrial malate dehydrogenase (Reaction 12.9), and the malate would diffuse into the cytosol and be converted back to oxalacetate (Section 10.7.2) by the cytosolic malic dehydrogenase. The oxalacetate could then be converted to phosphoenol pyruvic acid by PEP-carboxykinase (Reaction 10.23 or 12.14). At this point, the phosphoenol pyruvate could be utilized to form glucose, or as just proposed, converted to pyruvate by pyruvic kinase (Section 10.4.5).

The latter compound would then reenter the mitochondria and be oxidized to CO_2 and H_2O by the cycle of reactions described. Note that this process accomplishes the production of CO_2 from one of the four carbon atoms of oxalacetate in Reaction 10.14, the other three being converted to CO_2 during the oxidation of pyruvate itself. The oxidation of oxalacetate by this combination of CO_2 fixation, glycolytic, and Krebs cycle enzymes obviously calls for coordination between the cytoplasm and the mitochondria.

12.5   REGULATION OF THE TRICARBOXYLIC ACID CYCLE

In this chapter and in Chapter 10, we have pointed out the role of glucose (i.e., carbohydrate) as a foodstuff providing energy for the needs of the cell. Since most of that energy is provided by the reactions of the tricarboxylic acid cycle, there has to be some mechanism for regulating the flow of carbon into the tricarboxylic acid cycle. Moreover, if carbohydrate supply is in excess and the energy level of the cell is saturated, there must be available some means for diverting the carbon into storage lipids.

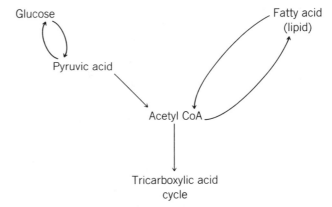

These processes are regulated by the ratios of ATP/ADP, NADH/NAD$^+$, acetyl CoA/CoA, and succinyl CoA/CoA. If the ratios of these different pairs of compounds is high (e.g., 10 : 1) then the cell is fully "charged" with energy-rich compounds (i.e., ATP, thioesters, and NADH) that can be used to carry out energy-consuming processes such as biosynthesis, cell-division, and nerve conduction. On the other hand, if those ratios are low (e.g., 1 or less), then the cell has been depleted of the compounds it needs to carry out energy-consuming processes and foodstuffs need to be metabolized to "recharge" the cell.

We may well ask how the levels of ADP and NAD$^+$ can regulate the flow of carbon. One way in which they regulate is as participants in the reaction under consideration. Clearly, a continuing supply of oxidized NAD$^+$ is required if Reactions 12.4, 12.5, and 12.9 are to function; only if NAD$^+$ is available for reduction can the tricarboxylic acid cycle continue to function. The enzymes of the electron transport sequence and oxidative phosphorylation (Chapter 14) couple the reoxidation of NADH produced in the reactions cited to the phosphorylation of ADP, producing ATP. If the concentration of ADP is low, then the reoxidation of NADH and operation of the cycle will be slowed.

In addition, however, ATP, ADP, NADH, acetyl CoA, and succinyl CoA can greatly influence the activity of key regulatory enzymes. Pyruvic dehydrogenase, situated at a central point between glycolysis, lipid metabolism, and the tricarboxylic acid, is regulated by two basic mechanisms. First, the dehydrogenase is subject to end-product inhibition by acetyl CoA and NADH. Second, NADH and acetyl CoA participate in a monocyclic cascade that controls the activity of pyruvic dehydrogenase. As pointed out in Section 18.4.2, the converter enzyme pyruvic dehydrogenase kinase can phosphorylate serine residues in the decarboxylase component of the pyruvic dehydrogenase complex and convert it into an inactive form. NADH and acetyl CoA stimulate the activity of the kinase, thereby facilitating the phosphorylation and inactivation of the complex. ATP also speeds up the rate of phosphorylation, while pyruvate inhibits it. Thus, if the energy level of the cell is high because of carbohydrate catabolism, the flow of carbon into the tricarboxylic acid cycle will be slowed, and acetyl CoA could be diverted into the synthesis of fatty acids (lipid). Similarly, if fatty acids are being degraded, the oxidation of pyruvate to acetyl CoA would be inhibited and pyruvate could be resynthesized into carbohydrate.

When the energy level of the cell falls, the levels of ATP, NADH, and acetyl CoA will decrease relative to those of ADP, NAD$^+$, and coenzyme A, and the kinase will become inactive. However, the converter enzyme, pyruvic decarboxylase phosphatase, will remain active and convert inactive carboxylase into its active form, thereby favoring pyruvate oxidation.

The rate-limiting step of the cycle is catalyzed by isocitric dehydrogenase, which is allosterically activated by ADP. ATP and NADH, on the other hand, competitively inhibit the enzyme. α-Ketoglutaric acid dehydrogenase is also inhibited by its products succinyl CoA and NADH. However, the dehydrogenase is not regulated by the phosphorylation–dephosphorylation system.

12.6 THE ANABOLIC NATURE OF THE TRICARBOXYLIC ACID CYCLE

The tricarboxylic acid cycle is a major source of important biosynthetic intermediates of the cell. An excellent example is α-ketoglutarate formed in Reaction 12.4. α-Ketoglutarate provides the carbon skeleton for the biosynthesis of glutamic acid, glutamine, ornithine (and therefore five-sixths of the carbon of citrulline and arginine), proline, and hydroxyproline. Another important biosynthetic intermediate is succinyl CoA, which is utilized in the synthesis of the porphyrins found in hemoglobin, myoglobin, and the cytochromes. Other more specialized examples can be cited: The citric acid that accumulates in the vacuoles of citrus species or the isocitric and malic acids that are found in high concentration in certain *Sedums* and apple fruit would have their biosynthetic origin in the cycle.

In order for α-ketoglutarate, or any of the other Krebs cycle intermediates just mentioned, to function in an anabolic role, one simple requirement must be met. Both the C_2 unit (acetyl CoA) and the C_4 unit (oxalacetate) that combine and give rise in the Krebs cycle to α-ketoglutarate (or other intermediate) must be provided in a *stoichiometric amount* equivalent to the α-ketoglutarate (or other intermediate) being removed for anabolic purposes. Thus, if a cell over a period of time needs to make 5.76 μmole of glutamic acid from a α-ketoglutarate, it must provide 5.76 μmole of acetyl CoA *and* 5.76 μmole of oxalacetate to "balance the books," so to speak.

This consideration immediately introduces the question of the "normal" sources of the C_2 and C_4 units, questions that we have already considered. As noted in this chapter, the acetyl CoA can be derived from pyruvate (Reaction 12.1) and therefore have its origin in carbohydrates that are converted to pyruvate during glycolysis. The C_2 unit can also be derived from fatty acids during β-oxidation (Chapter 13). The C_4 unit oxalacetate can be derived from a number of sources; we have considered its production from pyruvate through the action of pyruvic carboxylase (Reactions 10.22 or 12.12). It could also be produced by the action of PEP-carboxykinase (Reactions 10.23 or 12.14), although physiologically, this reaction operates to take carbon atoms out of the Krebs cycle rather than into it. A very important source of oxalacetate is the intermediates of the Krebs cycle itself. Thus, malate, produced in the glyoxylic acid cycle soon to be described, could be oxidized to oxalacetate. Finally, oxalacetate can be produced by the transamination of aspartic acid (Section 17.2.3).

12.6.1 ANAPLEROTIC REACTIONS

The preceding section on the anabolic nature of the Krebs cycle has raised the question of how the level of intermediates can be replenished when, for example, certain of those intermediates are removed for anabolic purposes. H. L. Kornberg has proposed the term *anaplerotic* for these replenishing or "filling-up" reactions. A relisting of those we have already considered and a listing of additional reactions not yet described are appropriate at this point.

1. Pyruvic carboxylase: The single most important anaplerotic reaction in animal tissues is catalyzed by pyruvic carboxylase, a mitochondrial enzyme.

$$CO_2 + \begin{matrix} CO_2H \\ | \\ C=O \\ | \\ CH_3 \end{matrix} + ATP + H_2O \underset{\text{Biotin}}{\overset{\underset{\text{Acetyl CoA}}{Mg^{2+}}}{\rightleftharpoons}} \begin{matrix} CO_2H \\ | \\ C=O \\ | \\ CH_2 \\ | \\ CO_2H \end{matrix} + ADP + H_3PO_4 \quad (12.12)$$

Pyruvic acid   Oxalacetic acid

The properties of this enzyme, which occurs in animals but not in plants, were described in detail in Section 10.7.2. The reaction it catalyzes links intermediates of the glycolytic sequence and the tricarboxylic acid cycle.

2. Phosphoenol pyruvic acid carboxylase (PEP-carboxylase) catalyzes Reaction 12.13.

$$CO_2 + \begin{matrix} CO_2H \\ | \\ C-OPO_3H_2 \\ || \\ CH_2 \end{matrix} + H_2O \longrightarrow \begin{matrix} CO_2H \\ | \\ C=O \\ | \\ CH_2 \\ | \\ CO_2H \end{matrix} + H_3PO_4 \quad (12.13)$$

Phosphoenol pyruvic   Oxalacetic acid
acid

The enzyme requires Mg^{2+} for activity; the reaction is irreversible. PEP-carboxylase occurs in higher plants, yeast, and bacteria (except pseudomonads), but not in animals. It presumably has the same function as pyruvic carboxylase, namely, to ensure that the Krebs cycle has an adequate supply of oxalacetate. The enzyme in some species is activated by fructose-1,6-bisphosphate; this is consistent with the function of seeing that the Krebs cycle can adequately oxidize the pyruvate being formed from glucose. The carboxylase is inhibited by aspartic acid; this effect is understandable when it is recognized that oxalacetate is the direct precursor of aspartic acid (by transamination). Thus, the biosynthetic sequence

$$\text{Phosphoenol pyruvate} \longrightarrow \text{Oxalacetate} \longrightarrow \text{Aspartic acid}$$

is a simple means for synthesizing aspartate from PEP, and aspartate can control its own production by inhibiting the first step in the sequence.

3. Phosphoenolpyruvic acid carboxykinase (described in Section 10.7.2) in theory could catalyze the replenishment of oxalacetate from phosphoenol pyruvate. However, the affinity of the enzyme for oxalacetate is very great ($K_m = 2 \times 10^{-6}$), while that for CO_2 is low. Thus, the enzyme strongly favors phosphoenol pyruvate formation.

$$CO_2 + \begin{array}{c} COOH \\ | \\ C-OPO_3H_2 \\ \| \\ CH_2 \end{array} + GDP \rightleftharpoons \begin{array}{c} COOH \\ | \\ C=O \\ | \\ CH_2 \\ | \\ CO_2H \end{array} + GTP \qquad (12.14)$$

$$\text{Phosphoenol pyruvic acid} \qquad\qquad \text{Oxalacetic acid}$$

The role of the carboxykinase in reversing glycolysis has been described in Section 10.7.2.

4. Malic enzyme catalyzes the readily reversible formation of L-malate from pyruvate and CO_2; the K_{eq} for the reaction at pH 7 is 1.6.

$$CO_2 + \begin{array}{c} CO_2H \\ | \\ C=O \\ | \\ CH_3 \end{array} + NADPH + H^+ \rightleftharpoons \begin{array}{c} CO_2H \\ | \\ HOCH \\ | \\ CH_2 \\ | \\ CO_2H \end{array} + NADP^+ \qquad (12.15)$$

$$\text{Pyruvic acid} \qquad\qquad\qquad \text{L-Malic acid}$$

Malic enzymes occur both in the mitochondria and the cytosol of plant and animal cells. The cytosolic enzyme is important because of its ability to produce NADPH required for biosynthetic purposes (e.g., fatty acid synthesis, Section 13.9). It, together with the two dehydrogenases of pentose phosphate metabolism (Chapter 11) and the NADP$^+$-specific isocitrate dehydrogenase (Section 12.3.3), are the main sources of NADPH in the cell.

12.6.2 CO_2-FIXATION REACTIONS

Reactions 12.12 through 12.15 are examples of "CO_2-fixation reactions." The initial observations that stimulated work on these reactions were made by Wood and Werkman in 1936. They observed that when propionic acid bacteria fermented glycerol to propionic and succinic acids, more carbon was found in the products than had been added as glycerol. Carbon dioxide, moreover, proved to be the source of the extra carbon atoms or the carbon that was "fixed." Today, the physiological significance of CO_2 fixation extends beyond the metabolism of propionic acid bacteria and includes not only the anaplerotic reactions listed above but also such enzymes as *acetyl CoA carboxylase* (Section 5.7.3), *propionyl CoA carboxylase* (Section 13.7), and *ribulose-1,5-bisphosphate carboxylase* (Section 15.10.2.1).

12.7 THE GLYOXYLIC ACID PATHWAY

Two major roles of the tricarboxylic acid cycle have now been described: the complete oxidation of acetyl CoA (and compounds convertible to acetyl CoA) and multiple anabolic activities, for example, the synthesis of glutamic acid, succinyl CoA, and aspartic acid. Since the reactions of the cycle (Reactions 12.2 to 12.9) can only degrade acetate, there remains the basic question of how some organisms (many bacteria, algae, and some higher plants at a certain stage in their life cycle) can utilize acetate as the only carbon source for all the carbon compounds of the cell. Or put another way, since acetate can only be oxidized to CO_2 and H_2O by the Krebs cycle, how can acetate, in some organisms, give rise to carbohydrates as well as amino acids derivable from the tricarboxylic acid cycle?

This challenging problem was most successfully pursued by H. L. Kornberg, who together with others, showed that acetate undergoes, in those organisms that convert acetate to carbohydrate, an **anabolic** sequence called the **glyoxylate pathway** (Figure 12.2). In effect, the glyoxylate pathway bypasses Reactions 12.4 through 12.8 of the tricarboxylic acid cycle, thereby omitting the two reactions in which CO_2 is produced (Reactions 12.4 and 12.5), and forms malic acid that can be used in anaplerotic processes.

The two enzymes that accomplish the net synthesis of malate (from two moles of acetyl CoA) are isocitrate lyase (isocitritase) and malate synthase. Isocitrate lyase catalyzes the aldol cleavage of isocitrate to form succinic acid and glyoxylic acid.

$$
\begin{array}{ccc}
& & \text{Succinate} \\
& & \text{CH}_2\text{—COOH} \\
\text{H}_2\text{C—COOH} & & \text{CH}_2\text{—COOH} \\
\text{HC—COOH} & \rightleftharpoons & + \\
\text{HOC—COOH} & & \\
\text{H} & & \text{O=C—COOH} \\
& & \text{H} \\
\text{Isocitrate} & & \text{Glyoxylate}
\end{array}
\tag{12.16}
$$

Malate synthase then catalyzes the condensation of glyoxylic acid with one mole of acetyl CoA to produce L-malic acid. The mechanism of the reaction is completely analogous to that of citrate synthase (Reaction 12.2) discussed earlier.

$$
\begin{array}{l}
\text{Acetyl–CoA} \\
\text{CH}_3\text{—C—S—CoA} \\
\qquad \| \\
\qquad \text{O} \\
\qquad + \qquad + \text{H}_2\text{O} \longrightarrow \text{HOC—COOH} \quad + \text{CoA-SH} \\
\text{O=C—COOH} \qquad\qquad\qquad \text{H} \\
\quad \text{H} \\
\text{Glyoxylate} \qquad\qquad\qquad\qquad \text{L-Malate}
\end{array}
\tag{12.17}
$$

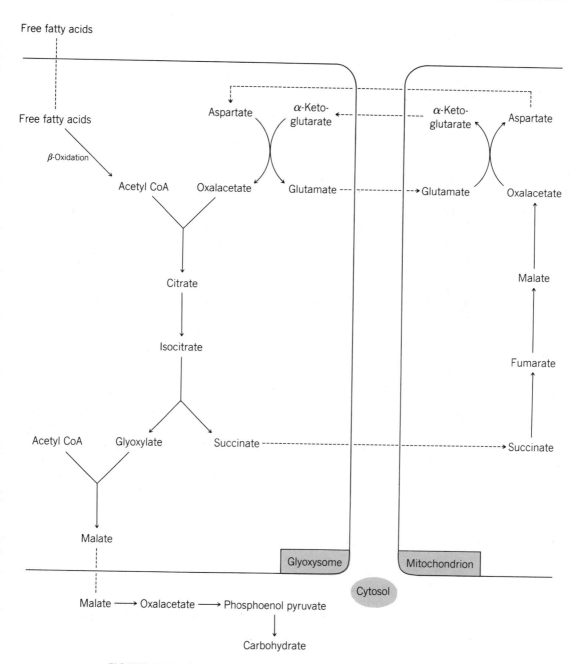

FIGURE 12.2 Compartmentalization of the glyoxylic acid cycle. Enzymes in both glyoxysomes and mitochondria are required to produce malic acid.

The glyoxylic acid pathway in plants functions as diagrammed in Figure 12.2 and involves a close cooperation between the organelle called the **glyoxysome** and mitochondria. The glyoxysms contain a complete set of β-oxidation enzymes that rapidly degrade fatty acids from storage lipids into acetyl CoA (Figure 13.5). Glyoxysomes appear in the cotyledons of lipid-rich seeds shortly after germination begins and at a time when lipids are being utilized as the major source of carbon for carbohydrate synthesis. Thus, the lipid-rich seed (e.g., peanut or castor bean) can convert lipid to carbohydrate, a synthesis that animals and most plants are incapable of performing since they lack the glyoxylate pathway. When the seedling is capable of photosynthesis and CO_2 can be used for the synthesis of carbohydrate, the glyoxysome disappears. Thus, only certain plant species (those having oil-rich seeds) carry out the glyoxylate pathway and do so only for a short, finite time during their life cycle.

As indicated (Figure 12.2), the glyoxysome contains citrate synthase (Reaction 12.2), aconitase (Reaction 12.3), isocitric lyase, and malate synthase. The succinate produced by the lyase is used to regenerate oxalacetate required by citrate synthase, but only after the succinate is transferred to the mitochondria. In the latter, the succinate is acted on by enzymes of the Krebs cycle and an aminotransferase to produce aspartate. The latter, in contrast to oxalacetate, can leave the mitochondria and enter the glyoxysome where oxalacetate can be produced by transamination.

The full significance of the glyoxalate pathway becomes apparent when the fate of the malate produced by malate synthase (Reaction 12.17) is considered. For example, the malate can be oxidized in the cytosol to oxalacetate and then converted to phosphoenol pyruvate and undergo gluconeogenesis (Section 10.7.2). In this way, two molecules of acetyl CoA converted to malate by the glyoxalate pathway are converted to carbohydrate, *a conversion that can not occur in animals.*

The malate can also enter the mitochondria, as part of an anaplerotic process, and be converted to oxalacetate, aspartate, and compounds (e.g., pyrimidines) derived from aspartate. The oxalacetate can instead combine in the mitochondria with another molecule of acetyl CoA (Reaction 12.2) and meet the requirements specified earlier (Section 12.6) for the Krebs cycle to function in an anabolic manner. Finally, the malate can be utilized in the formation of succinate and succinyl CoA by the reversal of Reactions 12.7, 12.6, and 12.5.

12.8   MITOCHONDRIAL COMPARTMENTATION

At this point, we have considered the enzymes that oxidize carbohydrates completely to CO_2 and H_2O. We have seen that the enzymes involved occur in two different parts of the eucaryotic cell, that is, the cytosol and the mitochondria, and that a major product of glycolysis, pyruvate, must enter the mitochondria for its further oxidation. Similarly, metabolites produced in the mitochondria may need to leave the mitochondria in order to perform a specific function (Sections 10.7.2 and 12.6). Finally, we have just considered a process —the glyoxylic acid pathway—that calls for an integrated interplay between two organelles, the glyoxysome and the mitochondrion, and can involve the cytosol in additional biosynthetic roles. These activities necessitate a discussion of the permeability of the mitochondria membranes.

The outer membrane is freely permeable to all molecules, even those as large as 8000 to 10,000 daltons. The inner membrane consists of a hydrophobic lipid bilayer that is impermeable to all large molecules as well as charged ions. Only water and O_2 enter freely as do a few anions (Cl^- and acetate), provided they are accompanied by cations such as K^+ or Na^+. Nevertheless, many metabolites do enter and leave the matrix, their movement across the inner membrane being assisted by carrier proteins functioning in an *antiport* manner. That is, the carrier proteins assist the movement of one compound *into* the mitochondrial matrix, but only if another molecule, of similar charge, moves out of the matrix. Thus, a phosphate carrier transports phosphate ion across the membrane in one direction in exchange for hydroxyl ions moving across in the opposite direction. The ADP–ATP **translocase** carrier moves ADP into the matrix where it can be phosphorylated, and in exchange, ATP comes out of the mitochondria. Malate can be moved across the membrane by the dicarboxylic acid carrier in exchange for aspartate, succinate, or phosphate.

The tricarboxylic acid carrier moves citrate and isocitrate in exchange for malate. This carrier plays an essential role in the transport of acetate units from the mitochondrial matrix into the cytosol where they are used in the biosynthesis of fatty acid. Acetyl CoA produced in the mitochondria by the oxidative decarboxylation of pyruvic acid is converted to citrate by Reaction 12.2. The citrate moves into the cytosol where it is cleaved by the ATP-dependent citrate cleavage enzyme (citrate-ATP lyase).

$$\underset{\text{Citrate}}{\begin{array}{c} CH_2-CO_2H \\ | \\ HO-C-CO_2H \\ | \\ CH_2-CO_2H \end{array}} + ATP + CoASH \longrightarrow \underset{\text{Acetyl CoA}}{CH_3-\underset{\underset{O}{\|}}{C}-S-CoA} + \underset{\text{Oxalacetate}}{\begin{array}{c} COOH \\ | \\ C=O \\ | \\ CH_2 \\ | \\ COOH \end{array}} + ADP + H_3PO_4 \quad (12.18)$$

The carbon atoms in the oxalacetate produced in this reaction can make their way back into the mitochondria after being reduced to malate by the NAD^+ malic dehydrogenase that occurs in the cytoplasm (Section 10.7.2). In the mitochondrial matrix, the malate will be oxidized to oxalacetate, which can then pick up another mole of acetyl CoA to form citrate and repeat the process.

REFERENCES

1. *T. W. Goodwin, ed.,* The Metabolic Roles of Citrate. *London: Academic Press, 1968.*

2. *P. A. Srere, "The Enzymology of the Formation and Breakdown of Citric Acid."* Advances in Enzymology, *Vol. 43 (1975), pp. 57–101.*

3. *H. A. Krebs, "The History of the Tricarboxylic Acid Cycle."* Perspectives in Biology and Medicine, *Vol. 14 (1970), pp. 154–170.*

4. *N. E. Tolbert,* Annual Review of Biochemistry, *Vol. 50 (1981), pp. 133–137.*

5. *R. G. Hansford, "Control of Mitochondrial Substrate Oxidation."* Current Topics in Bioenergetics, *Vol. 10 (1980), pp. 217–279.*

6. *L. J. Reed, "Multi-enzyme Complexes."* Account of Chemical Research, *Vol. 7 (1974), pp. 40–46.*

REVIEW PROBLEMS

1. If one considers only the fatty acid degradative sequence, the tricarboxylic acid cycle, and glycolysis, it is clear that a *net* conversion of fatty acids to carbohydrates (glucose) cannot occur. Therefore, explain why glycogen and blood glucose become

radioactive when radioactive acetate ($^{14}CH_3COOH$) is fed to a rat. Show which carbon atoms of the hexose unit are labeled.

2. An animal was injected with radioactive pyruvate labeled with ^{14}C in carbon 2.

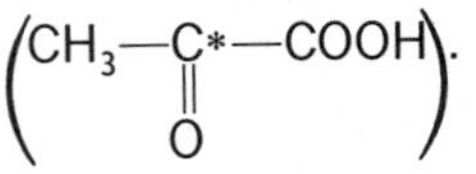

$$\left(CH_3-\underset{\underset{O}{\|}}{C}*-COOH \right).$$

After a few minutes, the carbon dioxide exhaled by the animal was trapped and found to be highly radioactive. Use equations to outline the series of enzyme-catalyzed reactions that would account for the appearance of ^{14}C in the exhaled CO_2.

3. Suggest a likely or possible enzyme-catalyzed reaction sequence by which α-ketoadipic acid (the α-keto, six-carbon dicarboxylic acid) may be synthesized from acetyl CoA and α-ketoglutaric acid. Use structures and show all cofactors required.

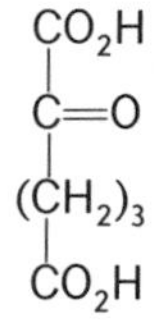

$$\begin{array}{c} CO_2H \\ | \\ C=O \\ | \\ (CH_2)_3 \\ | \\ CO_2H \end{array}$$

α-Ketoadipic acid

4. The *E. coli* bacterium obtains its glutamic acid (and other five-carbon amino acids) from α-ketoglutarate; the latter, in turn, can be synthesized from either pyruvic acid or acetic acid as the sole carbon source. Write the metabolic pathways by which the microorganism can achieve the net synthesis of α-ketoglutarate *from each of the two carbon sources.*

5. One reaction of the glyoxylate cycle is formally analogous to the reaction catalyzed by the condensing enzyme (citrate synthase) of the tricarboxylic acid cycle. Write the reaction of the glyoxylate cycle using structures for the organic acids and abbreviations for the factors (e.g., NAD^+).

LIPID
METABOLISM

In both plants and animals, lipids are stored in large amounts as neutral, highly insoluble triacyl glycerols; they can be rapidly mobilized and degraded to meet the cell's demands for energy. In the complete combustion of a typical fatty acid, palmitic acid, there is a large negative free-energy change.

$$C_{16}H_{32}O_2 + 23\ O_2 \longrightarrow 16\ CO_2 + 16\ H_2O$$
$$\Delta G' = -2340\ kcal/mole$$

This negative change is due to the oxidation of the highly reduced hydrocarbon chain attached to the carboxyl group of the fatty acid. Of all the common foodstuffs, only the long-chain fatty acids possess this important chemical feature. Thus, lipids have quantitatively the best caloric value of all foods; that is, 9.3 kcal/g for lipids in contrast to 4.1 kcal/g for carbohydrates and proteins.

Lipids also function as important insulators of delicate internal organs. In addition, nerve tissue, plasma membrane, and all membranes of subcellular particles such as mitochondria, endoplasmic reticulum, and nuclei have complex lipids as essential components. Moreover, the vital electron transport system in mitochondria and the intricate structures found in chloroplasts, the site of photosynthesis, contain complex lipids in their basic architecture.

As we have indicated, the chief storage form of available energy in the animal cell is the triacylglycerol molecule. When the caloric intake exceeds utilization, excess food is invariably stored as fat; the body cannot store any other form of food in such large amounts. Carbohydrates are converted to glycogen, for example, but the capacity of the body to store this polysaccharide as a potential source of energy is strictly limited. In a normal liver, the average amount of glycogen is 5 to 6% of the total weight, and in skeletal muscle, the glycogen content averages only 0.4 to 0.6%. Blood glucose is present at a level of 60 to 100 mg per 100 ml of whole blood. Only under pathological conditions are these values drastically altered. The normal animal therefore very carefully regulates, by hormonal and metabolic controls, the carbohydrate concentration in its various tissues, and this class of compounds can serve only to a limited extent as a storage form of energy.

Proteins, the third major class of foodstuffs, differ considerably from carbohydrates and fats in their biological function; they serve as a source of 20-odd amino acids required for *de novo* protein synthesis and for carbon skeletons essential for the synthesis of purines, pyrimidines, and other nitrogenous compounds. Moreover, in an adult organism in which active growth has ceased, nitrogen output is more or less geared to nitrogen intake, and the organism shows no tendency to store surplus proteins from the diet.

13.1   THE FATE OF DIETARY LIPIDS

Figure 13.1 presents, in a schematic form, the flow of lipids in the animal body. Three important compartments are the liver, blood, and adipose tissue. Both liver and adipose tissue are the principal sites of lipid metabolism while the blood serves as a transport system. Other compartments, which include cardiac and skeletal muscle, are important utilizers of fatty acids and ketone bodies.

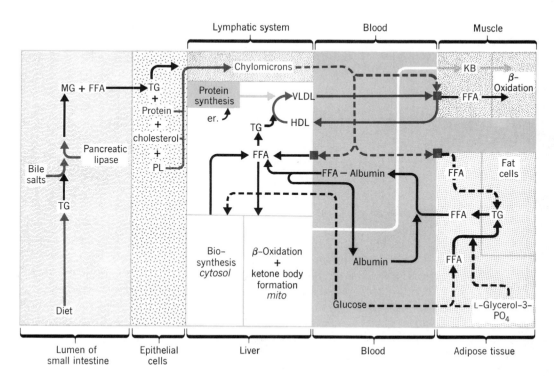

FIGURE 13.1   Scheme depicting role of compartments in the utilization of lipids in animals. TG, Triacylglycerol; MG, monoacylglycerol; FFA, free fatty acids; Chol., cholesterol; PL, phospholipids; KB, ketone bodies, VLDL, very light-density lipoprotein; ■, lipoprotein lipase; HDL, high-density lipoprotein.

13.2   LUMEN

In the lumen of the small intestines, triacylglycerols are degraded to free fatty acids (FFA) and monoacylglycerols in the presence of conjugated bile acids and pancreatic lipase.

Conjugated bile acids are detergents that consist of a lipid soluble (steroid) and a polar (taurine, glycine) part. Together, the bile acids, fatty acids, and monoacylglycerols form micelles. In the micelles, the nonpolar fraction is located centrally and the polar fraction is on the surface. Micelles are also the absorption vehicle for fat-soluble vitamins and cholesterol. However, the bile acids are not absorbed via the lymphatic route but through the portal blood vessel to the liver and are recycled back to the lumen via the gall bladder.

13.3   EPITHELIAL CELLS AND CHYLOMICRONS

The free fatty acids, monoacylglycerols, and the remaining triacylglycerols are absorbed as micelles into the epithelial cells of the small intestine where the following reactions are catalyzed by enzymes in the endoplasmic reticulum:

(a) *Acyl CoA synthetase*

$$RCH_2COOH + ATP + CoASH \xrightarrow{Mg^{2+}} RCH_2CO\text{—}SCoA + AMP + PPi$$

(b) *Monoacyl glycerol acyl transferase*

$$R^1CH_2CO\text{—}SCoA + \underset{\text{Monoacylglycerol}}{RCOCH} \longrightarrow \underset{\text{Diacylglycerol}}{RCOCH} + CoASH$$

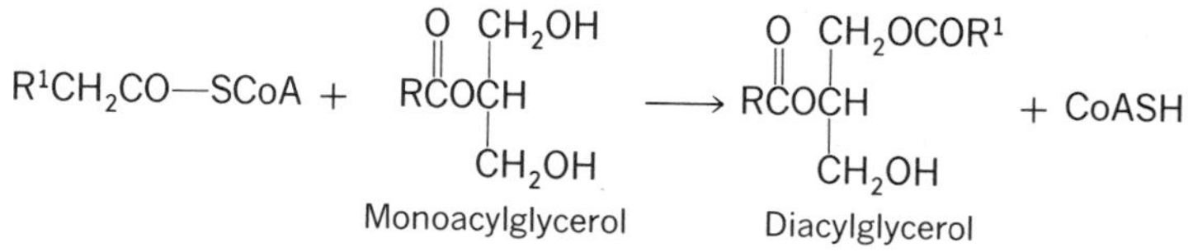

(c) *Diacylglycerol acyl transferase*

$$R^2CH_2CO\text{—}SCoA + RCOCH \longrightarrow \underset{\text{Triacylglycerol}}{RCOCH} + CoASH$$

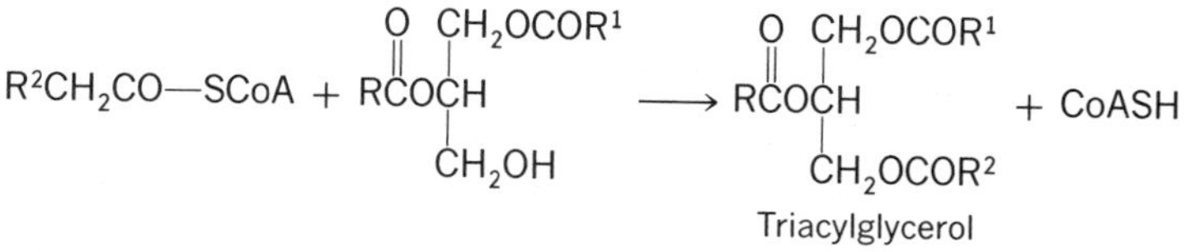

The newly synthesized triacylglycerol, dietary cholesterol, and newly synthesized phospholipids as well as specific newly synthesized proteins are combined in the endoplasmic reticulum of the epithelial cells and excreted into lacteals as chylomicrons. These particles are stable, about 200 nm in diameter, have about 0.2 to 0.5% protein, 6 to 10% phospholipid, 2 to 3% cholesterol + cholesterol esters, and about 80 to 90% triacylglycerols. These particles pass from the intestinal lacteals into the lymphatic system and then finally into the thoracic duct to be discharged into the blood system at the left sublavian vein as a milky suspension. The removal of chylomicrons from the blood is very rapid, the half-life of these particles being about 10 minutes.

Under isocaloric conditions, most of the chylomicrons are transported to adipose tissue for fat storage. However, under starvation conditions, when fat storage would be of considerable disadvantage to the animal, chylomicrons are utilized primarily by red skeletal muscle, cardiac muscle, and the liver for energy demands. Since chylomicrons must first have their triacylglycerol components degraded to free fatty acids before the target tissue can utilize them, an important regulatory factor, the lipoprotein lipase, participates directly in that degradative process.

$$\begin{array}{c}\text{Cholesterol}\\\widehat{\text{Protein}}\text{—Triacylglycerol}\\\text{Phospholipid}\end{array} \xrightarrow[\text{lipase}]{\text{Lipoprotein}} \begin{array}{c}\text{Cholesterol}\\\widehat{\text{Protein}} + \text{Glycerol} + \text{fatty acids}\\\text{Phospholipid}\end{array}$$

$$\downarrow$$

$$\begin{array}{c}\text{Absorption by}\\\text{target tissues}\end{array}$$

Under isocaloric conditions, this enzyme, which is localized on the walls of the capillary beds of the target tissues, has high levels of activity in adipose tissue. Thus, free fatty acids derived from chylomicrons will be transported into adipose tissues for storage. In sharp contrast, under starvation conditions, the activity falls markedly in the vascular system of adipose tissue but rises in muscle, liver, and cardiac tissues, all of which require fuel for their energy demands. Now, any chylomicrons in the blood will not be utilized by the adipose tissue for storage but rather will be diverted to important target tissues such as muscle for energy purposes.

13.4   ADIPOSE TISSUE

As free fatty acids enter the fat cells of adipose tissue from the action of lipoprotein lipase in the adjacent capillary walls, these acids are rapidly converted to triacylglycerols as depicted in Figure 13.2. The mature fat cell consists of a thin envelope of cytoplasm stretched over a large droplet of triacylglycerol that occupies up to 99% of the total cell volume. A full complement of all the eucaryotic organelles is found in fat cell cytoplasm. These fat cells in mammals and birds have the very important role of serving as an energy storehouse. The main fat depots in humans are the subcutaneous tissue, the muscle and mesenteric tissues. Sufficient lipids are stored as triacylglycerols in the fat cells to allow an individual to survive up to 40 days of starvation. In contrast, some fish, such as the cod, employ their liver for lipid storage.

Fat deposits are not stationary; that is, lipids are continuously being mobilized and deposited. Normally, the quantity of body lipids is kept constant over long periods of time by regulation of appetite by an unknown mechanism. When stress conditions develop in the animal such as starvation, prolonged exercise, or rapid fear responses in terms of violent exercise, adrenalin from the blood stream binds to a specific receptor in the fat cell surface and triggers a response as diagrammed in Figure 13.3. A hormone-sensitive lipase is activated, rapidly converting triacylglycerols to diacylglycerols and FFA. The FFA are ultimately transferred to the blood where they combine with serum albumin to form soluble, stable FFA-albumin complexes. Serum albumin makes up about 50% of the total plasma proteins. With a molecular weight of 69,000, this

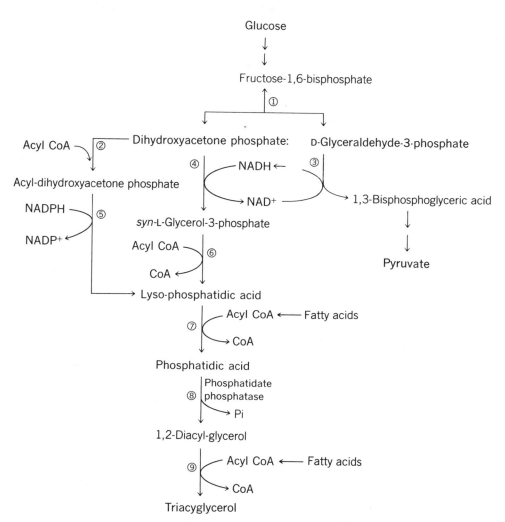

FIGURE 13.2 Enzymes involved in synthesis of triacylglycerols. 1, Aldolase; 2, dihydroxyacetone phosphate acyltransferase; 3, triose phosphate dehydrogenase; 4, glycerol 3-phosphate dehydrogenase; 5, acyl dihydroxyacetone phosphate reductase; 6, glycerol phosphate acyltransferase; 7, lyso-phosphatidate acyltransferase; 8, phosphatidate phosphatase; 9, diacylglycerol acyltransferase; 10, triose phosphate isomerase.

protein is principally concerned with osmotic regulation in blood. Because of its high solubility and unique binding sites for fatty acids (seven to eight sites per molecule of albumin), it plays in addition a very important transport role for fatty acids that would otherwise be highly insoluble and toxic (would lyse red cells). Once bound to serum albumin, the FFA-albumin complex is highly soluble, nontoxic, and is rapidly transported to the liver for further utilization. Although only about 2% of the total plasma lipid is associated with serum albumin as a FFA-albumin complex, the turnover of FFA in the blood is very high. Once the FFA-albumin complex enters the liver, a rapid transfer of the FFA into liver cells takes place with a simultaneous return of the fatty acid-free albumin into the blood stream. By this mechanism, the actual concentration of FFA as such is very low in blood plasma.

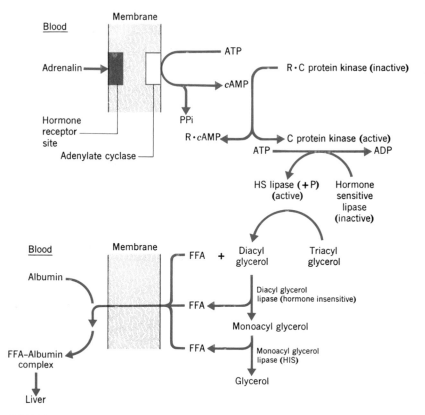

FIGURE 13.3   Events in fat cell of adipose tissue.

13.5   LIVER

In this organ, as in many other tissues and cells of plants and animals, several metabolic routes can now occur and they will be described in detail.

13.5.1   β-OXIDATION

All enzymes associated with the β-oxidation system are localized in the inner membranes and the matrix of liver and other tissue mitochondria. Because the inner membrane is also the site of the electron transport and oxidative phosphorylation systems, this arrangement is of fundamental importance to the efficient release and conservation of the potential energy stored in the long-chain fatty acid. When acetyl CoA is produced in the breakdown of fatty acids, it may be subsequently oxidized to CO_2 and H_2O by means of the tricarboxylic acid cycle enzymes that are localized as soluble enzymes in the matrix. An unusual property of liver and other tissues mitochondria is their inability to oxidize externally added fatty acids or fatty acyl CoA's unless (−)-carnitine (3-hydroxyl-4-trimethyl ammonium butyrate) is added in catalytic amounts. Evidently, free fatty acids or fatty acyl CoA's cannot penetrate the inner membranes of liver and other tissue mitochondria, whereas acylcarnitine readily passes through the membrane and is then converted to acyl CoA in the matrix. Figure 13.4 outlines the translocation of acyl CoA from outside the mitochon-

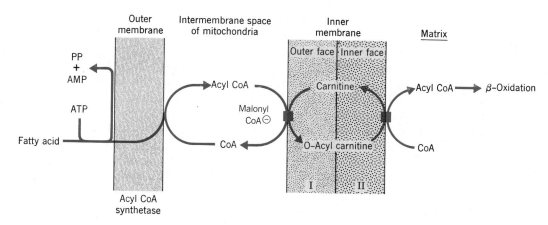

FIGURE 13.4 Transport mechanism for fatty acids from the cytosol to the β-oxidation site in the mitochondrion. ■, Carnitine: acyl CoA transferase I (outer face) and carnitine: acyl CoA transferase II (inner face), two distinct enzymes that catalyze the same reaction; (malonyl CoA)⊖ indicates inhibition of transferase I.

drion to the internal site of the β-oxidation system. The key enzymes are carnitine-acyl CoA transferase I and II, and their location is indicated in Figure 13.4. Transferase I is 50% inhibited by 1.2 μM levels of malonyl CoA. The implications of this effect will be considered when we discuss ketogenesis in this chapter.

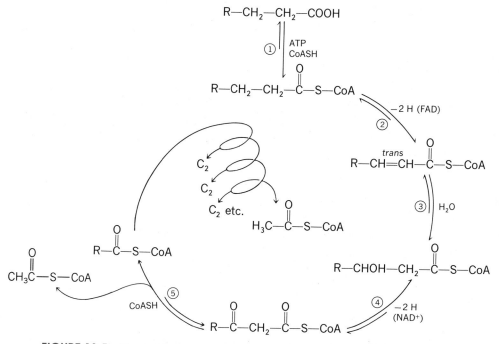

FIGURE 13.5 The β-oxidation helical scheme. (1) fatty acid CoA synthetase; (2) fatty acyl CoA dehydrogenases; (3) enoyl CoA hydrase; (4) β-hydroxyacyl CoA dehydrogenase; (5) β-ketoacyl CoA thiolase.

The overall β-oxidation scheme is presented in Figure 13.5. Note that only one molecule of ATP is required to activate a fatty acid for its complete degradation to acetyl CoA regardless of the number of carbon atoms in its hydrocarbon chain. In other words, whether we wish to oxidize either a C_4 or C_{16} acid, only one equivalent of ATP is needed for activation. This makes for great economy and efficiency in the oxidation of fatty acids.

Four enzymes catalyze the reactions depicted in Figure 13.5 while the fifth enzyme, acyl CoA synthetase, is involved in the initial activation step. These enzymes will now be described briefly. The student should review Section 5.10.3, which deals with the reactivity of acyl thioesters.

1. *Formation of acyl-S-CoA's by acyl CoA synthetase.* The overall type reaction is depicted as

$$\text{RCOOH} + \text{ATP} + \text{CoASH} \underset{\Delta G' = \sim 0 \text{ kcal/mole}}{\overset{\text{Mg}^{2+}}{\rightleftharpoons}} \text{RCO—SCoA} + \text{AMP} + \text{PPi}$$

There is good evidence that the reaction actually takes place in two steps.

(a) RCOOH + ATP + Enzyme $\rightleftharpoons$ Enzyme-acyladenylate + PPi
(b) Enzyme-acyladenylate + CoASH $\rightleftharpoons$ Enzyme + Acyl-S-CoA + AMP

Three different synthetases occur in the cell: One activates acetate and propionate to the corresponding thioesters, another activates medium-chain fatty acids from C_4 to C_{11}, and the third activates fatty acids from C_{10} to C_{20}. The two first mentioned are located in the outer membranes of mitochondria and the third synthetase is associated with endoplasmic reticuli membranes (microsomes).

2. *α-, β-Dehydrogenation of acyl CoA.*

$$\overset{O}{\overset{\|}{\text{RCH}_2\text{CH}_2\text{C}}}\text{—S—CoA} + \text{FAD} \xrightarrow[\Delta G' = -4.8 \text{ kcal/mole}]{} \underset{\beta}{\text{RCH}} = \underset{\alpha}{\overset{\text{trans}}{\text{CH}}}\text{—}\overset{O}{\overset{\|}{\text{C}}}\text{—SCoA} + \text{FADH}_2$$

Three acyl CoA dehydrogenases are found in the matrix of mitochondria. They all have FAD as a prosthetic group. The first has a specificity ranging from C_4 to C_6 acyl CoA's, the second from C_6 to C_{14}, and the third from C_6 to C_{18}. The $FADH_2$ is not directly oxidized by oxygen but follows the path.

3. *Hydration of α-, β-unsaturated acyl CoA's*

$$\overset{\text{trans}}{\text{RCH}} = \text{CH}\overset{O}{\overset{\|}{\text{C}}}\text{—SCoA} + \text{H}_2\text{O} \underset{\Delta G' = -0.75 \atop \text{kcal/mole}}{\overset{\pm\text{H}_2\text{O}}{\rightleftharpoons}} \underset{\overset{|}{\text{OH}}}{\text{RCH}}\text{—CH}_2\overset{O}{\overset{\|}{\text{C}}}\text{—SCoA}$$

L(+)-β-Hydroxy-acyl CoA

The enzyme, enoyl CoA hydrase, catalyzes this reaction; it possesses broad specificity. It should be noticed that hydration of the *trans-α,β*-acyl CoA results in the formation of the L(+)β-hydroxyacyl CoA. It will also hydrate α,β-*cis*-unsaturated-acyl CoA, but in this case, D(−)β-hydroxyacyl CoA is formed.

4.   *Oxidation of L-β-hydroxyacyl CoA.*

$$L(+)\ RCHOHCH_2\overset{\overset{\displaystyle O}{\parallel}}{C}\!-\!SCoA + NAD^+ \underset{\Delta G' = +3.75\ kcal/mole}{\rightleftarrows} RCCH_2\overset{\overset{\displaystyle O}{\parallel}}{C}\!-\!S\!-\!CoA + NADH + H^+$$

β-Ketoacyl CoA

A broadly specific L-β-hydroxyacyl CoA dehydrogenase catalyzes this reaction. It is specific for the L form.

5.   *Thiolysis.* The enzyme, thiolase, carries out a thiolytic cleavage of the β-ketoacyl CoA. It has broad specificity.

$$RC\overset{\overset{\displaystyle O}{\parallel}}{C}H_2\overset{\overset{\displaystyle O}{\parallel}}{C}\!-\!SCoA + CoASH \underset{\Delta G' = -6.65\ kcal/mole}{\rightleftarrows} R\overset{\overset{\displaystyle O}{\parallel}}{C}\ SCoA + CH_3\overset{\overset{\displaystyle O}{\parallel}}{C}SCoA$$

The enzyme protein has a reactive SH group on a cysteinyl residue that is involved in the following series of reactions:

$$RCOCH_2\overset{\overset{\displaystyle O}{\parallel}}{C}\!-\!SCoA + Enz\!-\!SH \rightleftarrows R\overset{\overset{\displaystyle O}{\parallel}}{C}\!-\!S\!-\!Enz + CH_3\overset{\overset{\displaystyle O}{\parallel}}{C}\!-\!SCoA$$

β-Ketoacyl CoA   Thiolase   Acyl—S—Enz   Acetyl CoA

$$R\overset{\overset{\displaystyle O}{\parallel}}{C}\!-\!S\!-\!Enz + CoA\!-\!SH \rightleftarrows R\overset{\overset{\displaystyle O}{\parallel}}{C}\!-\!SCoA + Enz\!-\!SH$$

Acyl CoA

The student should note that for the shortening of an acyl CoA by two carbon atoms, acetyl CoA, the net ΔG' is −8.45 kcal/mole. Therefore, thermodynamically, the cleavage of a C_2 unit (acetyl CoA) is highly favored.

6.   *Localization.* The β-oxidative system is found in all organisms. However, in bacteria grown in the absence of fatty acids, the β-oxidative system is practically absent but is readily induced by the presence of fatty acids in the growth medium. The bacterial β-oxidative system is completely soluble and hence is not membrane-bound. However, in *E. coli,* three enzymes, namely, enoyl CoA hydratase, L-3-hydroxyl-acyl CoA dehydrogenase, and 3-keto-acyl CoA thiolase, are all associated with a single protein with a molecular weight of 270,000.

Curiously, in germinating seeds possessing a high lipid content, the β-oxidation system is exclusively located in microbodies called glyoxysomes (see Sections 12.7 and 8.10), but in seeds and leaf cells with a low lipid content, the enzymes are associated with peroxisomes. Recently, it has been shown that a long-chain fatty acid β-oxidation system also exists in liver peroxisomes. Although the final product is acetyl CoA, and the reactions are identical to those described for mitochondrial β-oxidation, the enzymes responsible are quite different with the exception of thiolase that is the same as the mitochondrial enzyme (see Section 8.10). Moreover, since peroxisomes do not have the TCA electron transport-oxidative phosphorylation systems of the mitochondrion, the energy of β-oxidation is lost in the peroxisomal system and acetate is translocated out of the peroxisome. Thus, the peroxisome can be viewed as an organelle concerned primarily for the non-energy-coupled degradation of fatty acids. The important function of the peroxisomes in plants and animals is considered in more detail in Sections 8.10 and Chapter 15.

The universality of the β-oxidative system implies the prime importance of this sequence as a means of degrading fatty acids.

13.5.2   ENERGETICS OF β-OXIDATION

In the total combustion of palmitic acid, considerable energy is released.

$$C_{16}H_{32}O_2 + 23\ O_2 \longrightarrow 16\ CO_2 + 16\ H_2O$$
$$\Delta G' = -2340\ kcal/mole$$

Palmitic acid
$$C_{15}H_{31}COOH + 8\ CoASH + ATP + 7\ FAD + 7\ NAD^+ + 7\ H_2O \longrightarrow$$
$$8\ CH_3CO \sim SCoA + AMP + PPi + 7\ FADH_2 + 7\ NADH + H^+$$
Acetyl CoA

$$8\ CH_3CO \sim SCoA + 16\ O_2 \xrightarrow{\text{TCA cycle}} 16\ CO_2 + 16\ H_2O + 8\ CoASH$$

How much of this potential energy is actually made available to the cell? When one mole of palmitic acid is degraded enzymically, one ATP is required for the primary activation, and eight energy-rich acetyl CoA thioesters are formed. Each time the helical cycle (Figure 13.5) is traversed, one mole of FAD—H$_2$ and one mole of NADH are formed; they may be reoxidized by the electron-transport chain. Since, in the final turn of the helix, two moles of acetyl CoA are produced, the helical scheme must be traversed only seven times to degrade palmitic acid completely. In this process, seven moles each of reduced falvin and pyridine nucleotide are formed. The sequence can be divided into two steps:

Step 1:

$$\text{Palmitic acid} \longrightarrow 8\ \text{Acetyl–S–CoA} + 14\ \text{Electron pairs}$$

7 electron pairs via Flavin system at 2 ~ P/one electron pair = 14 ~ P
7 electron pairs via NAD$^+$ system at 3 ~ P/one electron pair = 21 ~ P
Total = 35 ~ P
Net = 35 ~ P − 1 ~ P
= 34 ~ P

Step 2:

$$8\ \text{Acetyl–CoA} + 16\ O_2 \xrightarrow{\text{TCA cycle}} 16\ CO_2 + 16\ H_2O + 8\ CoA–SH$$

If we assume that for each oxygen atom consumed, 3 ~ P are formed during oxidative phosphorylation (refer to Chapter 14 for this assumption), then

$$32 \times 3 = 96 \sim P$$

Thus, step 1 (34 ~ P) and step 2 (96 ~ P) = 130 ~ P; and

$$\frac{130 \times 7300 \times 100}{2,340,000} = 40\%$$

In the complete oxidation of palmitic acid to CO_2 and H_2O, 40% of the available energy can theoretically be conserved in a form (ATP) that is utilized by the cell for work. The remaining energy is lost, probably as heat. It hence becomes clear why, as a food, fat is an effective source of available energy. In this calculation, we neglect the combustion of glycerol, the other component of a triacylglycerol.

While the β-oxidative system undoubtedly is the primary mechanism for degrading fatty acids, the student should be aware of a number of other systems that attack the hydrocarbon chain oxidatively. A brief survey of these mechanisms and possible functions will now be given.

13.5.3 α-OXIDATION

This system, first observed in seed and leaf tissues of plants, is also found in brain and liver cells. The mechanism for this reaction in plants is depicted as

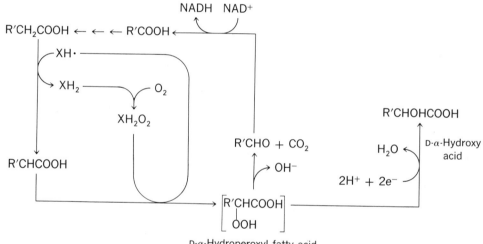

D-α-Hydroperoxyl fatty acid

Note that in this system, only free fatty acids serve as substrates and molecular oxygen is indirectly involved. The products may be either a D-α-hydroxyl fatty acid or a fatty acid containing one less carbon atom. This mechanism explains the occurrence of α-hydroxy fatty acids and odd-numbered fatty acids. The latter may, in nature, also be synthesized *de novo* from propionate. The α-oxidation system has been shown to play a key role in the capacity of mammalian tissues to oxidize phytanic acid, the oxidation product of phytol, to CO_2 and water. Normally, phytanic acid is rarely found in serum lipids because of the ability of normal tissue to degrade the acid very rapidly. It has now been observed that patients with Refsum's disease, a rare inheritable disease, have lost their α-oxidation system; hence, their normally functioning β-oxidation system cannot cope with the degradation of the phytanic acid. It is believed that the sequence shown in Figure 13.6 explains the disease on a molecular level. Here, α-oxidation makes possible the bypassing of a blocking group in a hydrocarbon chain that otherwise could prevent the participation of the β-oxidation system.

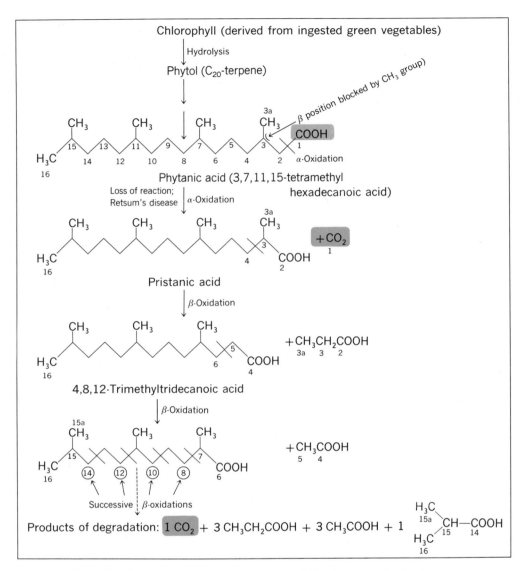

FIGURE 13.6   The metabolism of phytanic acid by the normal animal cell.

13.5.4   MIXED-FUNCTION OXYGENASES

Microsomes in hepatic cells rapidly catalyze the oxidation of hexanoic, octanoic, decanoic, and lauric acids to corresponding dicarboxylic acids via a cytochrome P-450 oxidation system. Cytochrome P-450 is a specialized cytochrome that has the unusual property of hydroxylating a wide variety of hydrocarbons, sterols, and drugs in the presence of molecular oxygen and a reductant, usually NADPH. It is classified as a mixed function oxygenase or a hydroxylase. Its participation in a hydroxylation is readily detected because in the reduced state, cytochrome P-450 binds very strongly to carbon monoxide and is thereby inhibited, as well as displaying a very strong spectral band at

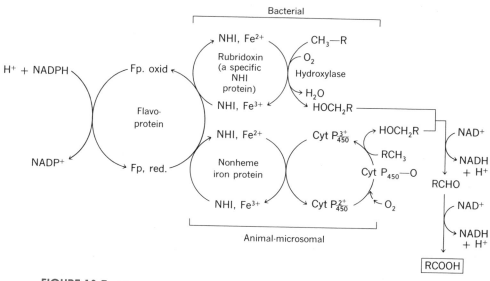

FIGURE 13.7 The oxidative system responsible for the oxidation of alkanes in bacteria and animal systems. In bacteria, rubridoxin is the intermediate electron carrier that feeds electrons to the hydroxylase system. In animals, the cytochrome P_{450} system is the hydroxylase responsible for alkane hydroxylation. The immediate product, RCH_2OH, is oxidized to an aldehyde by an alcohol dehydrogenase that, in turn, is oxidized to a carboxylic acid by an aldehyde dehydrogenage in both systems. NHI, nonheme iron protein.

450 nm—hence, the name cytochrome *P*-450. Cytochrome *P*-450 plays an important role in detoxifying drugs by the hydroxylating reaction and modifying sterols and vitamin D_3 in the liver.

A number of aerobic bacteria have been isolated from oil-soaked soil and these contain very active mixed function oxygenases; these rapidly degrade hydrocarbons or fatty acids to water-soluble products. The reactions involve an initial hydroxylation of the terminal methyl group to a primary alcohol and subsequent oxidation to a carboxylic acid (Figure 13.7). Thus, straight-chain hydrocarbons are oxidized to fatty acids and fatty acids, in turn, are β-oxidized to acetyl CoA. These series of reactions, which at first glance were of mild interest, now have assumed an extremely important scavenging role in the bacterial biodegradation of both detergents derived from fatty acids and even more important oil spilled over the ocean surface. It has been estimated that the rate of bacterial oxidation of floating oil under aerobic conditions may be as high as 0.5 g/day per square meter of oil surface.

13.6 OXIDATION OF UNSATURATED FATTY ACIDS

Although the β-oxidation system readily explains the degradation of saturated fatty acids, it offers no explanation for the oxidation of mono- or polyunsaturated fatty acids. Two additional mitochondrial enzymes, Δ^3-*cis/trans*, Δ^2-*trans*-enoyl CoA isomerase and 2,4-dienoyl CoA 4-reductase, make possible the β-oxidation of these acids (Figure 13.8).

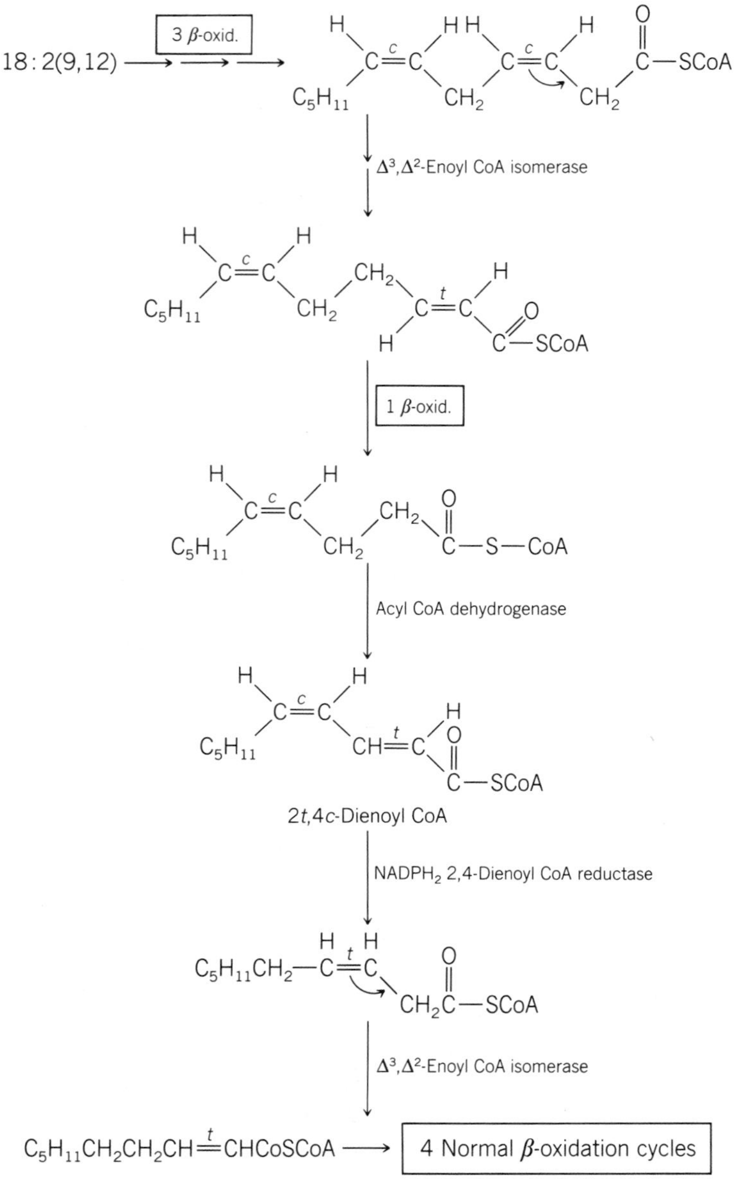

FIGURE 13.8 The mechanism for β-oxidizing unsaturated fatty acids.

With these enzymes incorporated in an extended β-oxidation scheme, the student can readily construct series of reactions for the β-oxidations of oleic, linoleic, and α-linolenic acids. For example, with linoleic acid as substrate, we would employ three normal β-oxidation enzymes for three cycles (3 C$_2$); then, use the Δ^3,Δ^2-enoyl CoA isomerase reaction; one β-oxidation cycle; then, the normal acyl CoA dehydrogenase, the 2,4-dienoyl CoA reductase, and again, the Δ^3,Δ^2-enoyl CoA isomerase to convert the *trans* double-bond barrier to a Δ^2 system; and finally, conclude with four normal β-oxidation cycles.

13.7 OXIDATION OF PROPIONIC ACID

The oxidation of propionic acid presents an interesting problem, since at first glance, the acid would appear to be a substrate unsuitable for β-oxidation. However, the substrate is handled by two strikingly dissimilar pathways. The first pathway is found only in animal tissues and some bacteria and involves biotin and vitamin B_{12}, while the second pathway, widespread in plants, is a modified β-oxidation pathway (Figure 13.9).

The plant pathway, which is ubiquitous in plants, nicely resolves the problem of how plants can cope with propionic acid, the product of oxidative degradation of valine and isoleucine, by a system not involving vitamin B_{12}. Since plants have no B_{12} functional enzymes, the animal system is absent; thus, the modified β-oxidation system of plant tissues bypasses the B_{12} barrier in an effective manner.

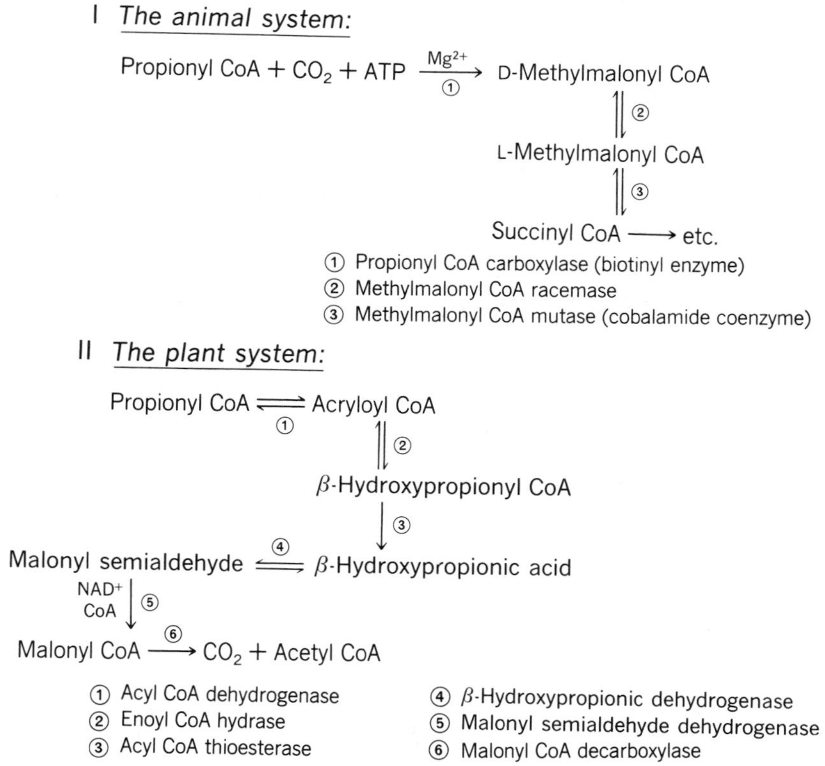

I *The animal system:*

$$\text{Propionyl CoA} + CO_2 + \text{ATP} \xrightarrow[\text{①}]{Mg^{2+}} \text{D-Methylmalonyl CoA}$$

$$\Big\Updownarrow \text{②}$$

$$\text{L-Methylmalonyl CoA}$$

$$\Big\Updownarrow \text{③}$$

$$\text{Succinyl CoA} \longrightarrow \text{etc.}$$

① Propionyl CoA carboxylase (biotinyl enzyme)
② Methylmalonyl CoA racemase
③ Methylmalonyl CoA mutase (cobalamide coenzyme)

II *The plant system:*

$$\text{Propionyl CoA} \underset{\text{①}}{\rightleftharpoons} \text{Acryloyl CoA}$$

$$\Big\Updownarrow \text{②}$$

$$\beta\text{-Hydroxypropionyl CoA}$$

$$\Big\downarrow \text{③}$$

$$\text{Malonyl semialdehyde} \underset{\text{④}}{\rightleftharpoons} \beta\text{-Hydroxypropionic acid}$$

$$\underset{\text{CoA}}{\overset{NAD^+}{\Big\downarrow}} \text{⑤}$$

$$\text{Malonyl CoA} \xrightarrow{\text{⑥}} CO_2 + \text{Acetyl CoA}$$

① Acyl CoA dehydrogenase ④ β-Hydroxypropionic dehydrogenase
② Enoyl CoA hydrase ⑤ Malonyl semialdehyde dehydrogenase
③ Acyl CoA thioesterase ⑥ Malonyl CoA decarboxylase

FIGURE 13.9 The animal and plant systems for degradation of propionic acid.

13.8 FORMATION OF KETONE BODIES

Having reviewed the several oxidative degradative pathways available to a cell, let us now continue to trace the route of a fatty acid in the animal. Free fatty acids enter liver cells from chylomicrons and FFA-albumin complexes originating from adipose tissue fat cells. Fatty acids formed *de novo* from glucose in the liver also are major contributors to this dynamic pool.

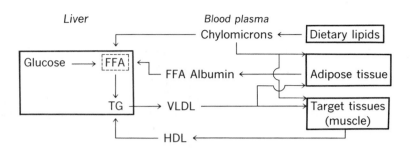

Under normal nutritional conditions, these fatty acids have several fates (Figure 13.1).

   1.   The acids are esterified to triacylglycerols. However, the liver has a limited capacity for triacylglycerol storage, and any excess combines with HDL (high-density lipoprotein), cholesterol esters, and phospholipids to form VLDL particles (very light-density lipoproteins) (see Section 7.11). These are now excreted into the blood system and transported via the vascular system to target tissues such as muscle and adipose tissue. Here, lipoprotein lipase removes and converts triacylglycerols to free fatty acids, which are then absorbed by the tissues and utilized. The residual VLDL particles in the meantime convert to HDL particles that presumably return to the liver via the blood system to pick up excess triacylglycerols and repeat the cycle.

   2.   The free fatty acids enter the mitochondria and/or peroxisomes to be β-oxidized and then converted by the TCA cycle to CO_2 and H_2O or acetate.

   3.   The product of β-oxidation, acetyl CoA is converted only in liver and kidney mitochondria to ketone bodies that are then transported to target tissues like red muscle, brain, and cardiac muscles to be burned to CO_2 and H_2O. Recent evidence strongly suggests that ketone bodies are major fuels for peripheral muscles and become important sources of energy in muscles involved in prolonged muscle exercises such as long distance running, and so on.

   Ketone bodies are D-α-hydroxy-butyric acid, acetoacetic acid, and acetone. They are formed by a series of unique reactions, summarized in Figure 13.10. The enzymes involved in the synthesis of ketone bodies are localized primarily in liver and kidney mitochondria. Ketone bodies cannot be utilized in the liver since the key utilizing enzyme, 3-oxoacid:CoA transferase, is absent in this tissue but is present in all those tissues metabolizing ketone bodies, namely, red muscle, cardiac muscle, brain, and kidney. Ketone bodies are alternative substrates to glucose for energy sources in muscle and brain. The precursors of ketone bodies, namely free fatty acids, are toxic in high concentrations, have very limited solubility, and readily saturate the carrying capacity of the plasma albumin. Ketone bodies, on the other hand, are very soluble, low in toxicity, tolerated at high concentrations, diffuse readily through membranes, and are rapidly metabolized to CO_2 and H_2O.

   In starvation, after the glucose level of the blood has been reduced to the physiologically allowable limit of 70% of the normal fasting level of about 90 mg/100 ml blood, a massive mobilization of storage fat occurs with a subsequent flooding of fatty acids into the liver and kidney. The mitochondria of

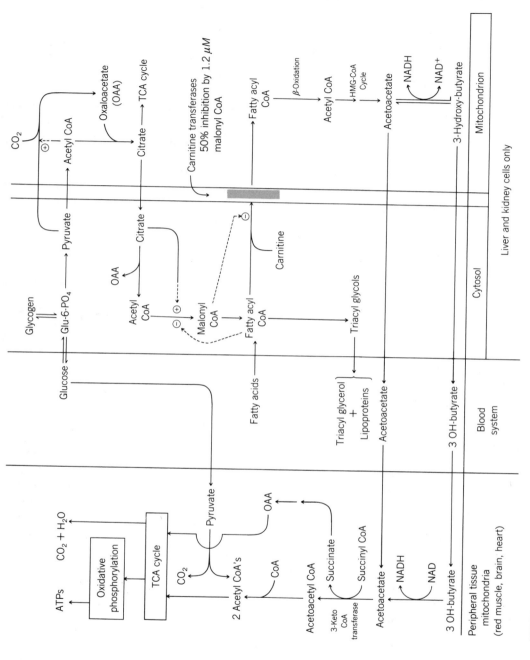

FIGURE 13.10 Generation of ketone bodies and their utilization.

these tissues will have limits as to the amount of fatty acids they can convert to CO_2 and H_2O because of decreased amounts of oxaloacetic acid. As a result, massive amounts of ketone bodies are produced. β-Hydroxy-butyrate normally is less than 3 mg/100 ml blood and the daily excretion is about 20 mg. As much as a 50 to 500-fold increase of ketone bodies will occur in the blood during starvation. Diabetic patients, who cannot utilize glucose for energy, depend on the catabolic utilization of fatty acids for energy. Once again, ketone bodies characteristically accumulate in the blood of diabetic people. Prolonged exercise by sedentary persons will also result in elevated ketone bodies in the blood. Of interest, athletes rarely show ketosis since they have elevated levels of the enzymes that utilize ketone bodies in their peripheral muscle tissues.

Let us examine in more detail the implications of Figure 13.10. If the animal is on a high carbohydrate diet, glucose will be converted to pyruvate in the liver that then enters the liver mitochondria to be converted to citrate. Part of the citrate enters the TCA cycle, the remainder is transported back to the cytosol, there to be converted to acetyl CoA by the ATP-dependent citrate lyase. Because of high levels of citrate, the cytosolic acetyl CoA carboxylase will be in part activated (see Section 5.2.3), acetyl CoA will be converted to malonyl CoA, and this substrate is then employed for fatty acid synthesis. The newly synthesized fatty acids are not β-oxidized because (1) the malonyl CoA formed under these conditions inhibits very strongly carnitine transferase I and, thus, fatty acids cannot be transported into the matrix of the mitochondria where the β-oxidation enzymes are localized and (2) the acyl CoA's so formed are rapidly trapped as triacylglycerols and are transported via the lipoprotein carrier systems into the vascular system for distribution to target tissues. No ketone bodies are formed.

If, on the other hand, the organism is fed on a high fat diet or is starving and therefore forced to use its storage triacylglycerols, high levels of acyl CoA's will be formed. Citrate will not be generated because of the restricted supply of glucose; acetyl CoA carboxylase will be inhibited by high levels of acyl CoA's. Under these conditions, fatty acid synthesis is essentially turned off. Simultaneously, because of the low levels of malonyl CoA in the cytosol, the inhibition of carnitine transferase I is relieved and now, acyl CoA's are rapidly transported into the mitochondrial matrix, there to be converted by β-oxidation to acetyl CoA. Acetyl CoA now enters the hydroxymethyl-glutaryl (HMG) CoA cycle (system not shown) that generates free acetoacetic acid. This unstable β-keto acid is rapidly converted to β-hydroxy-butyric acid by its specific dehydrogenase, β-hydroxy-butyric dehydrogenase.

The generation of these two ketone bodies acetoacetic acid and β-hydroxy-butyric acid occurs only in kidney and liver mitochondria because of the specific location of the HMG-CoA cycle enzymes in these organelles. These ketone bodies are now transported into the blood system to be excreted or carried to target tissues to be metabolized for fuel, as indicated in Figure 13.10. If the TCA cycle in the target tissues is overloaded by a lack of oxalocetate (as a consequence of a low carbohydrate diet), the target tissue will not metabolize the ketone bodies and they will be excreted.

These biochemical events then, in part, explain the observation that with a high carbohydrate diet, fatty acids are synthesized and stored rather than being β-oxidized, whereas in starvation (or a high fat diet), fatty acid synthesis is turned off and high levels of ketone bodies are generated.

13.9 BIOSYNTHESIS OF FATTY ACIDS

So far in this chapter, we have traced the fate of dietary lipids from the lumen of the small intestines to the several tissue masses in the body. Although much of the lipid requirements are of a dietary source, all of us are aware of the observation that carbohydrates are readily converted into fatty acids and then to adipose tissue for storage. We shall now trace the biochemical events involved in this sequence.

In the *de novo* synthesis of palmitic acid from acetyl CoA, the total overall reaction can be written as

$$8 \text{ Acetyl CoA} + 7 \text{ ATP} + 14 \text{ NADPH} + 14 \text{ H}^+ \longrightarrow$$
$$\text{Palmitate} + 7 \text{ ADP} + 7 \text{ Pi} + 14 \text{ NADPH} + 8 \text{ CoA} + 6 \text{ H}_2\text{O} \quad (13.1)$$

Let us first examine the sources of ATP, NADPH, and acetyl CoA before we consider the actual mechanism of biosynthesis of fatty acids.

Blood glucose readily enters the liver cells where it can undergo two important degradative pathways: (1) glycolysis and (2) the pentose phosphate pathway. Glycolysis is of major importance since it allows the formation of (1) 2 ATPs per 2 moles of pyruvate formed, (2) 2 NADH per glucose that was converted to 2 pyruvate, and (3) 2 moles of pyruvate per glucose utilized. The pentose phosphate pathway provides 2 NADPH per glucose utilized. As indicated in Equation 13.1, for the formation of palmitate, we need 8 acetyl CoA's, 7 ATPs, and 14 NADPH. Therefore, we have utilized 4 glucoses to generate 8

acetyl CoA's and 32 ATPs, more than sufficient to fulfill the energy requirements for fatty acid synthesis. However, 14 NADPHs are required. Experimental results tell us that up to 60% of the total NADPH requirements are fulfilled by the pentose phosphate system. The remaining reductive power is obtained by converting NADH (from glycolysis) indirectly to NADPH by cytosolic enzymes.

(a) $\text{NADH} + \text{H}^+ + \text{oxaloacetate} \xrightarrow[\text{dehydrogenase}]{\text{Malic}} \text{malic} + \text{NAD}^+$

(b) $\text{malic} + \text{NADP}^+ \xrightarrow[\text{Enzyme}]{\text{Malic}} \text{pyruvate} + \text{CO}_2 + \text{NADPH} + \text{H}^+$

Sum of (a + b): $\text{NADH} + \text{NADP}^+ + \text{Oxaloacetate} \longrightarrow$
$$\text{Pyruvate} + \text{CO}_2 + \text{NADPH} + \text{H}^+ + \text{NAD}^+$$

The final question remains. How does acetyl CoA, formed in the mitochondrion, become available in the cytosol where fatty acid synthase is located. Pyruvate moves by passive diffusion from the cytosol into the matrix of the mitochondrion where it is (1) oxidized by pyruvic dehydrogenase to acetyl CoA and (2) carboxylated by pyruvic carboxylase to oxaloacetate. Both acetyl CoA and oxaloacetate are condensed by the citrate synthase to citrate, which is then

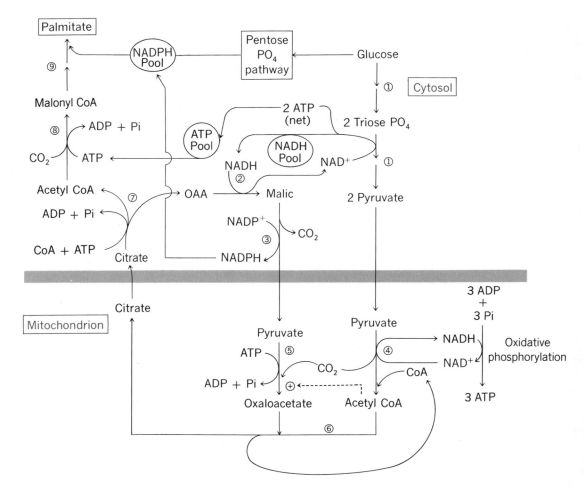

FIGURE 13.11   Origin of ATP, NADPH, and acetyl CoA for fatty acid biosynthesis in liver cell. 1, Glycolysis; 2, malic dehydrogenase (cytosolic); 3, malic enzyme; 4, pyruvic dehydrogenase; 5, pyruvic carboxylase (activated by acetyl CoA); 6, citrate synthase; 7, citrate lyase; 8, acetyl CoA carboxylase; 9, fatty acid synthetase.

transported out of the mitochondria to the cytosol. In the cytosol, citrate is cleaved.

$$\text{citrate} + \text{ATP} + \text{CoA} \xrightarrow[\text{Lyase}]{\text{Citrate}} \text{Acetyl CoA} + \text{oxaloacetate} + \text{ADP} + \text{Pi}$$
$$\Delta G^1 = -3400 \text{ cal/mole}$$

Acetyl CoA is now ready to serve as a substrate with the required amounts of ATP and NADPH to form palmitate. The entire sequence of events is described in Figure 13.11.

In the cytosol of the liver cell, acetyl CoA must be converted to malonyl CoA by acetyl CoA carboxylase. We have described the mechanism of carboxylation of acetyl CoA in Section 5.7.3

$$\text{Acetyl CoA} + \text{HCO}_3^- + \text{ATP} \longrightarrow \text{Malonyl CoA} + \text{ADP} + \text{Pi}$$

We can now refine Equation 13.1 by writing a more precise reaction.

Acetyl CoA + 7 Malonyl CoA + 14 H$^+$ ⟶

$$\text{Palmitate} + 8\ \text{CoA} + 7\ CO_2 + 14\ \text{NADP}^+ + 6\ H_2O \quad (13.2)$$

Because we now supply malonyl CoA to the equation, the ATP requirement in Equation 13.1 drops out. However, the cell must utilize at least 23 ATPs to form 1 palmitate. This can be accounted for as follows:

Mitochondria

8 ATPs to convert 8 pyruvate to 8 oxaloacetates ⟶ 8 citrates
8 ATPs to cleave 8 citrates to 8 acetyl CoA's + 8 oxaloacetates
<u>7 ATPs</u> to carboxylate 7 acetyl CoA's to 7 malonyl CoA's
Total   23 ATPs

The origin of carbon atoms in palmitic acid is then as follows:

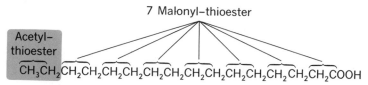

7 Malonyl–thioester

Acetyl–
thioester

$\overline{CH_3CH_2}\overline{CH_2}\overline{CH_2}\overline{CH_2}\overline{CH_2}\overline{CH_2}\overline{CH_2}\overline{CH_2}\overline{CH_2}\overline{CH_2}\overline{CH_2}\overline{CH_2}\overline{CH_2}\overline{CH_2}COOH$

The chemical events in the synthesis of long-chain fatty acids from acetyl CoA and malonyl CoA are identical in all organisms. However, in most bacteria and all higher plants, the fatty acid synthase (FAS) system consists of individual enzymes that can be readily separated, purified, and examined in detail. The sequence of reactions that occur with these soluble FAS systems in both bacteria and plants is as follows:

FAS Enzymes

1. $CH_3COSCoA + ACP \cdot SH \xrightleftharpoons{\text{Acetyl-transferase}} CH_3CO \cdot S \cdot ACP + CoASH$

2. $HO_2C \cdot CH_2COSCoA + ACP \cdot SH \xrightleftharpoons{\text{Malonyl-transferase}}$

$$HO_2C \cdot CH_2CO \cdot S \cdot ACP + CoASH$$

3. $CH_3COS \cdot ACP + HO_2C \cdot CH_2CO \cdot S \cdot ACP \xrightleftharpoons[\text{ACP synthetase I}]{\text{β-Keto-acyl-}}$

$$CH_3COCH_2CO \cdot S \cdot ACP + ACP \cdot SH + CO_2$$

4. $CH_3COCH_2CO \cdot S \cdot ACP + NADPH + H^+ \xrightleftharpoons[\text{ACP reductase}]{\text{β-Keto-acyl-}}$

$$D(-)CH_3CH(OH)CH_2CO \cdot S \cdot ACP + NADP^+$$

5. $CH_3CH(OH) \cdot CH_2CO \cdot S \cdot ACP \xrightleftharpoons{\text{β-Hydroxy-acyl-ACP dehydratase}}$

$$CH_3CH\overset{trans}{=\!=\!=}CHCO \cdot S \cdot ACP + H_2O$$

6. $CH_3CH=CHCO \cdot S \cdot ACP + NADPH + H^+ \xrightleftharpoons{\text{Enoyl-ACP reductase}}$

$$CH_3CH_2CH_2CO \cdot S \cdot ACP + NADP^+$$
$$\text{butyryl-ACP}$$

7.   Butyroyl·S·ACP now reacts with a second molecule of malonyl·S·ACP and proceeds through Reaction 3 to 6 to form hexanoyl·S·ACP, and so on until palmitoyl·S·ACP is formed.

8. $CH_3(CH_2)_{14}CO \cdot S \cdot ACP + HO_2CCH_2COSACP \xrightarrow[\substack{\text{β-Keto-acyl-ACP} \\ \text{synthetase II}}]{\text{Enzymes 4, 5, 6}}$

$$C_{17}H_{35}COSACP + CO_2$$

Characteristic of these FAS systems is the presence of the soluble, small molecular weight acyl carrier protein (see Section 5.10.3 for a description). These FAS systems are known as **nonassociated FAS systems.**

In sharp contrast, the vertebrate FAS system has an entirely different molecular structure. The system is a homodimer, consisting of two identical subunits, each with a molecular weight of 240,000, each with eight domains or activities associated with specific amino acid residues in the polypeptide chain, each subunit totally inactive in synthesizing a fatty acid but each containing the activities of each of the eight enzymes required for fatty acid synthesis. Only when the two identical subunits combine in an antiparallel orientation to form the homodimer is activity expressed. As noted in the homodimer structure, both a functional division and a subunit division exist. In the functional divi-

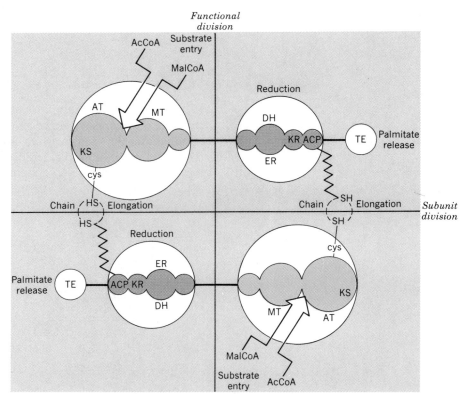

The homodimer system (after Wakil)

sion, three domains—namely, the acetyl- and malonyl-transferases and the keto-acyl synthetase present in the N-terminal subunit—line up adjacent to the remaining five domains—namely, ACP, keto-acyl reductase, enoyl-reductase, the dehydrase, and the acyl-thioesterase of the other subunit polypeptide located at the COOH terminal end. Now, with an input of acetyl CoA and malonyl CoA, the formation of free palmitic acid readily takes place. Unlike the bacterial and plant FASs where the product is palmitoyl ACP, in vertebrate FAS, the terminal product is free palmitic acid formed as follows:

$$\text{Palmitoyl-S}\boxed{\text{FAS}} + \text{H}_2\text{O} \xrightarrow{\text{TE}} \text{Palmitic} + \text{HS}\boxed{\text{FAS}}$$
$$\text{acid}$$

With yeast, the FAS system becomes even more complicated. The molecular structure is very large with a molecular weight of 2.4×10^6, a heterodimer consisting of six α subunits, each with a molecular weight of 212,000, containing three domains of keto-acyl synthetase, ACP, and β-keto-acyl reductase and six β subunits each with a molecular weight of 203,000, containing the remaining five domains—namely, hydroxyl-acyl-dehydrase, enoyl-reductase, acetyl-transacylase, malonyl-transacylase, and acyl-transferase. Unlike the other systems, the termination step involves the transfer of palmitoyl-S-FAS to free CoA to form palmitoyl CoA.

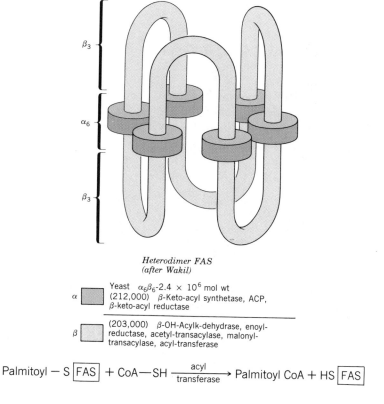

Heterodimer FAS
(after Wakil)

Yeast $\alpha_6\beta_6$-2.4×10^6 mol wt
α (212,000) β-Keto-acyl synthetase, ACP, β-keto-acyl reductase

β (203,000) β-OH-Acylk-dehydrase, enoyl-reductase, acetyl-transacylase, malonyl-transacylase, acyl-transferase

$$\text{Palmitoyl} - \text{S}\,\boxed{\text{FAS}} + \text{CoA—SH} \xrightarrow[\text{transferase}]{\text{acyl}} \text{Palmitoyl CoA} + \text{HS}\,\boxed{\text{FAS}}$$

13.10   ELONGATION OF PALMITIC ACID TO STEARIC ACID

In plants and animals, the most important fatty acids are the C_{16} fatty acid, palmitic acid, and the C_{18} fatty acids; namely stearic (18:0), oleic [18:1(9)], linoleic [18:2(9,12)], and α-linolenic acid [18:3(9,12,15)]. The systems described in Section 13.9 are called the *de novo* systems in that palmitic acid is constructed from acetyl CoA (ACP) and malonyl CoA (ACP). The C_{18} series are synthesized by elongation systems that differ from the *de novo* system.

In animals:

Endoplasmic reticulum membrane

$$\text{Palmitoyl CoA} \xrightarrow[\text{NADPH}]{\text{Malonyl CoA}} \text{Stearoyl CoA}$$

In plants:

Soluble system

$$\text{Palmitoyl ACP} \xrightarrow[\substack{\text{NADPH} \\ \text{ketopalmitoyl ACP} \\ \text{synthetase}}]{\text{Malonyl ACP}} \text{Stearoyl ACP}$$

The student should realize that the single arrow ($\rightarrow$) implies a full complement of enzymes, which catalyze the condensation of palmitoyl-thioester with the C_2 unit, the reduction, dehydration, and reduction to the final product, the stearoyl-thioester.

13.11   LOCALIZATION OF THE FAS SYSTEM IN THE CELL

In the bacterial cell, the nonassociated FAS system is situated in the cytosol, probably reasonably close to the inner face of the plasma membrane.

In higher plants, the location is more specific. In the leaf cell, the entire machinery for stearic acid synthesis is localized in the chloroplast; that is, the enzymes next listed

1. $\text{CH}_3\text{COOH} + \text{ATP} + \text{CoA} \xrightarrow[\text{Mg}^{2+}]{\text{Acetyl CoA synthetase}} \text{CH}_3\text{COCoA} + \text{AMP} + \text{PPi}$

2. $\text{CH}_3\text{COCoA} + \text{ATP} + \text{CO}_2 \xrightarrow[\text{Mg}^{2+}]{\text{Acetyl CoA carboxylase}} \text{COOHCH}_2\text{COCoA} + \text{ADP} + \text{Pi}$

3. $\text{Oleoyl-ACP} + \text{H}_2\text{O} \xrightarrow[\text{Oleoyl-ACP hydrolase}]{} \text{Oleic acid} + \text{ACP}$

as well as the FAS system enzymes and the stearoyl-ACP desaturase are all in the chloroplast compartment. In developing and germinating seeds, all these enzymes are in the proplastid compartment. In other words, in plants, the enzymes for the synthesis of palmitic, stearic, and oleic acids are localized in discrete organelles. In contrast, the vertebrate homodimeric FAS system is located in the cytosolic compartment as is the heterodimeric FAS system in yeast.

13.12   COMPARISON OF GLUCOSE AND FATTY ACID SYNTHESIS

It is worthwhile to compare the biosynthesis of glucose and a fatty acid. These compounds, although completely dissimilar in structure, share striking similarities for their syntheses.

For gluconeogenesis, we know that pyruvate cannot be converted to phosphoenolpyruvate because of an unfavorable $\Delta G'$ of the pyruvic kinase reaction, but by the use of the gluconeogenic enzymes pyruvic carboxylase and PEP carboxykinase, this barrier is resolved. A key element is the involvement of ATP and CO_2 for the conversion of pyruvate to phosphenol pyruvate (Chapter 10).

For fatty acid synthesis, acetyl CoA does not convert to acetoactyl CoA by the reversal of the thiolase reaction because of the unfavorable $\Delta G'$. However, once again, the involvement of ATP and CO_2 overcomes this barrier to form malonyl CoA.

Thus, pyruvate and acetyl CoA are the starting substrates for the synthesis of

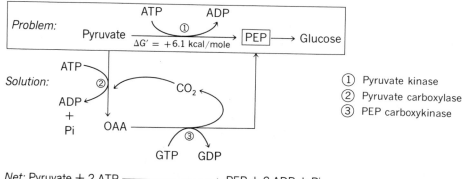

Glucose biosynthesis

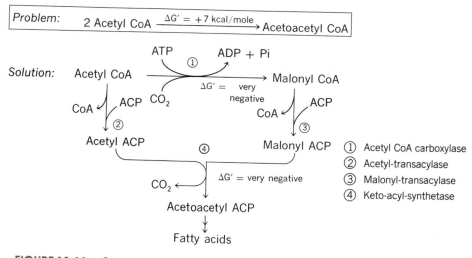

FIGURE 13.12   Comparison of initial stages of glucose and fatty acid synthesis. Note the involvement of ATP in both systems and the recycling of CO_2.

glucose and fatty acids, respectively; phosphoenolpyruvate and malonyl CoA are the primary intermediates; the driving force for the synthesis of glucose and fatty acids are ATP and CO_2; and CO_2 is not incorporated into the final product but is recycled over and over again.

These ideas are summarized in Figure 13.12.

13.13   BIOSYNTHESIS OF UNSATURATED FATTY ACIDS

13.13.1   INTRODUCTION OF THE FIRST DOUBLE BOND: AEROBIC PATHWAY

When stearoyl CoA has been synthesized in the liver cell, this substrate can be readily desaturated to oleoyl CoA in the presence of molecular oxygen, NADPH, and a membrane-bound enzyme system (associated with endoplasmic reticulum) according to the following sequence:

$$\text{Stearoyl CoA} + O_2 + \text{NADPH} \longrightarrow \text{Oleoyl CoA} + \text{NADP}^+ + \text{H}^+ + 2\,H_2O$$

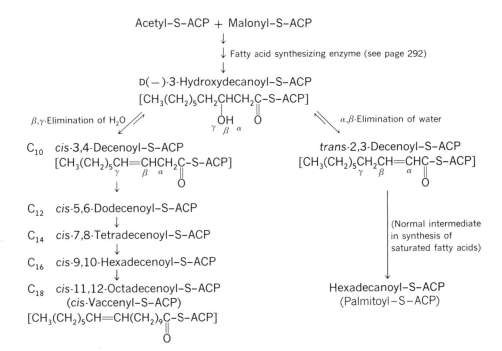

NADPH + H$^+$ → 2 Fe^{3+} ← → Oleoyl CoA

2 Cytochrome b_5

→ 2 H$_2$O   Desaturase

O$_2$

NADP$^+$ ← → 2 Fe^{2+} → Stearoyl CoA

This sequence is called the aerobic mechanism and is totally responsible for oleic acid synthesis in all animal cells. In plants, a very similar event occurs except the ferredoxin replaces cytochrome b_5, the desaturase is soluble, and the substrate is stearoyl ACP. While the precise chemical mechanism for desaturation is not known, experiments clearly support a *cis* elimination of the 2H$_D$ atoms of the C$_9$—C$_{10}$ carbon atoms of stearoyl-thioester by both the animal and plant systems.

13.13.2   ANAEROBIC PATHWAY

Because a large number of eubacteria can synthesize monoenoic acids under anaerobic conditions, the direct aerobic desaturation mechanism is obviously inoperative in these organisms. The work of Bloch and his group clearly showed that a single *cis* double bond is introduced at the C$_{10}$ level and finally positioned in the hydrocarbon chain by chain elongation. Specifically, in the synthesis of C$_{18}$ fatty acids, the branching point is at the D(−)-β-hydroxydecanoyl-S-ACP level. Although this thioester can serve as a substrate for several dehydrases, it

Acetyl–S–ACP + Malonyl–S–ACP

↓

↓ Fatty acid synthesizing enzyme (see page 292)

↓

D(−)-3-Hydroxydecanoyl–S–ACP

[CH$_3$(CH$_2$)$_5$CH$_2$CHCH$_2$C–S–ACP]

β,γ-Elimination of H$_2$O ⬋   OH O   ⬊ α,β-Elimination of water

γ β α

C$_{10}$ *cis*-3,4-Decenoyl–S–ACP   *trans*-2,3-Decenoyl–S–ACP
[CH$_3$(CH$_2$)$_5$CH=CHCH$_2$C–S–ACP]   [CH$_3$(CH$_2$)$_5$CH$_2$CH=CHC–S–ACP]
            γ   β α O   γ   β α O

↓

C$_{12}$ *cis*-5,6-Dodecenoyl–S–ACP

↓

C$_{14}$ *cis*-7,8-Tetradecenoyl–S–ACP   (Normal intermediate
↓   in synthesis of
C$_{16}$ *cis*-9,10-Hexadecenoyl–S–ACP   saturated fatty acids)
↓

C$_{18}$ *cis*-11,12-Octadecenoyl–S–ACP   Hexadecanoyl–S–ACP
          (*cis*-Vaccenyl–S–ACP)   (Palmitoyl–S–ACP)
[CH$_3$(CH$_2$)$_5$CH=CH(CH$_2$)$_9$C–S–ACP]
                              O

FIGURE 13.13   The anaerobic pathway in procaryotic organisms.

serves as a highly specific substrate for β-hydroxydecanoyl-S-ACP dehydrase, which introduces a single *cis-β,γ* double bond to form a *cis*-3,4-decenoyl-S-ACP; this intermediate is then extended as next indicated to form a monoenoic acid. Other dehydrases will form a *trans-α,β* double-bond system that is, however, readily reduced to form saturated fatty acids.

Figure 13.13 summarizes these observations. Each arrow (→) represents the sequential events of reduction, dehydration, reduction, and further condensation, as summarized in Section 13.9.

13.13.3 INTRODUCTION OF ADDITIONAL DOUBLE BONDS

One of the remarkable metabolic blocks in animal tissue is their inability to desaturate the monoenoic acid, oleic acid, toward the methyl end of the fatty acid, whereas in the plant kingdom, this reaction readily occurs. Thus, linoleic acid, 18:2(9,12), which is required by animals, must be obtained from plant dietary sources.

In the liver cell, a microsomal enzyme desaturates linoleyl CoA to γ-linolenyl CoA, which is then elongated to homolinolenyl CoA, and finally again desaturated to arachidonyl CoA. This acyl-thioester is then transferred to appropriate acceptors to form phospholipids, and so on. We can diagram these events as follows:

$$18:2(9,12) \xrightarrow{-2\,H} 18:3(6,9,12) \xrightarrow{+C_2} 20:3(8,11,14)$$

(Diet)	γ-Linolenic	Homo-γ-linolenic
Linoleic		

$$\Big\downarrow {-2\,H}$$

$$20:4(5,8,11,14)$$
$$\text{Arachidonic}$$

In all desaturation steps in the mammalian system, the desaturases are microsomal, the substrates are acyl CoA's, and NADPH or NADH and O_2 are essential components. Also, note that the desaturation steps are in the direction of the carboxyl group.

The role of polyunsaturated fatty acids is probably related to their occurrence in the lipoproteins and polar lipids of eucaryotic membranes. Another very important function in the animal kingdom is the role of a number of polyunsaturated fatty acids as precursors for the synthesis of prostaglandins. Thus,

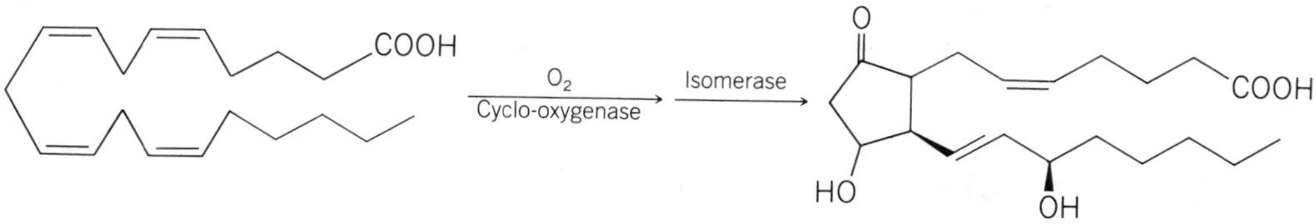

Arachidonic acid → (O_2, Cyclo-oxygenase) → (Isomerase) → Prostaglandin E_2

The family of prostaglandins, of which E_2 is only one of many derivatives, exerts profound physiological and pharmacological effects at very low concentrations.

In higher plants, the synthesis of linoleic acid employs as its substrate 2-

oleoyl-phosphatidyl choline and the reaction is catalyzed by a Δ^{12} membrane-bound desaturase.

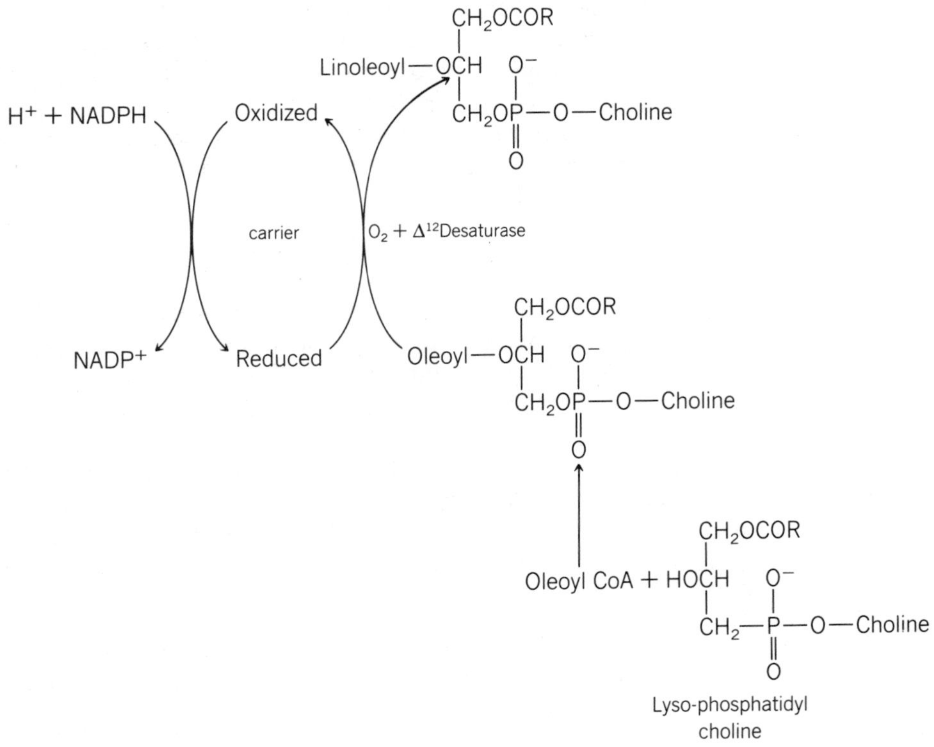

$$
\begin{array}{ccc}
 & & \begin{array}{l} CH_2OCOR \\ | \end{array} \\
 & \text{Linoleoyl}-OCH \quad O^- \\
H^+ + NADPH \qquad \text{Oxidized} & \quad | \\
 & \qquad CH_2OP-O-\text{Choline} \\
 & \qquad\quad \| \\
 & \qquad\quad O \\
\\
\text{carrier} & O_2 + \Delta^{12}\text{Desaturase} \\
\\
 & & \begin{array}{l} CH_2OCOR \\ | \end{array} \\
NADP^+ \qquad \text{Reduced} & \text{Oleoyl}-OCH \quad O^- \\
 & \quad | \\
 & \qquad CH_2OP-O-\text{Choline} \\
 & \qquad\quad \| \\
 & \qquad\quad O \\
\end{array}
$$

$$
\begin{array}{l}
CH_2OCOR \\
| \\
\text{Oleoyl CoA} + HOCH \qquad O^- \\
\qquad\qquad\qquad | \\
\qquad\qquad\quad CH_2-P-O-\text{Choline} \\
\qquad\qquad\qquad\quad \| \\
\qquad\qquad\qquad\quad O
\end{array}
$$

Lyso-phosphatidyl choline

A few comments should be made concerning the importance of unsaturated fatty acids. It is now recognized that unsaturated fatty acids play important roles in the structure of membrane systems in all living cells. All membrane systems contain in their complex lipids, fatty acids with different degrees of unsaturation. In procaryotic cells, for example, although polyunsaturated fatty acids are absent, monoenoic acids are important components of the membrane lipids. Procaryotic organisms are incapable of synthesizing polyunsaturated fatty acids. In eucaryotic cells, polyunsaturated fatty acids are important acyl components of phospholipids. Moreover, linoleic acid has been classified as an "essential fatty acid," since its omission in normal diets leads to pathological changes that can be reversed by the reintroduction of the acid to the diet. In photosynthetic tissues in which molecular oxygen is released by the photooxidation of water, highly unsaturated fatty acids are found, with very few exceptions, to be associated with the lamellar membrane lipids of the chloroplast.

13.14 PHOSPHOLIPID BIOSYNTHESIS

We have already discussed the synthesis of triacylglycerols (see Sections 13.3 and 13.4). As indicated in Chapter 7 and 8, phospholipids are the basic lipid components of all membranes. Therefore, a brief consideration of their biosynthesis will be presented here.

The biosynthetic enzymes are associated with the endoplasmic reticulum of all eucaryotic cells. In procaryotic cells, all the enzymes for the biosynthesis of phosphatidyl ethanolamine, the principal phospholipid in procaryotic plasma membranes, are associated with the plasma membrane.

If we examine the structure of phosphatidyl choline (lecithin), we note that its basic units are long-chain fatty acids ($-OCOR^1$ and $-OCOR^2$; the latter is almost always polyunsaturated), glycerol, phosphate, and choline. How are they assembled?

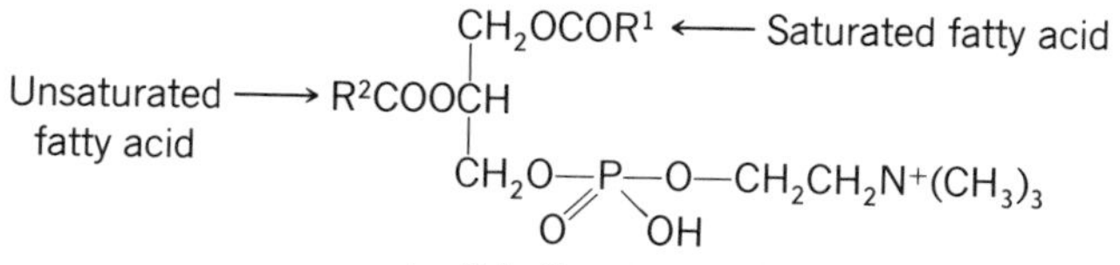

$$\begin{array}{c} CH_2OCOR^1 \longleftarrow \text{Saturated fatty acid} \\ \text{Unsaturated} \longrightarrow R^2COOCH \\ \text{fatty acid} \quad\quad\quad CH_2O-\underset{\underset{O}{\parallel}}{P}-O-CH_2CH_2N^+(CH_3)_3 \\ \quad\quad\quad\quad\quad\quad\quad OH \end{array}$$

Lecithin (3-*sn*-phosphatidyl choline)

In the eucaryotic cell, an orderly sequence of events, all catalyzed by specific enzymes, brings the separate units together. The first reaction is the phosphorylation of choline.

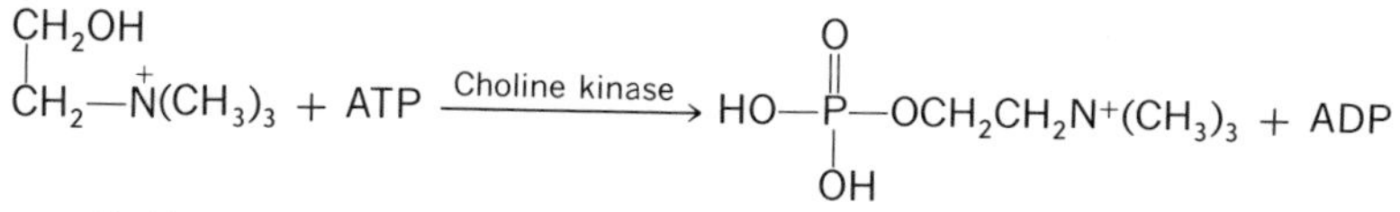

$$\begin{array}{c} CH_2OH \\ CH_2-\overset{+}{N}(CH_3)_3 + ATP \xrightarrow{\text{Choline kinase}} HO-\underset{\underset{OH}{\mid}}{\overset{\overset{O}{\parallel}}{P}}-OCH_2CH_2N^+(CH_3)_3 + ADP \end{array}$$

Choline   Phosphorylcholine

Phosphorycholine then reacts with CTP to form cytidine-diphosphate choline (CDP-choline).

$$\text{Phosphorylcholine cytidyltransferase}$$
$$CTP + \text{Phosphorylcholine} \longrightarrow \text{CDP-choline} + \text{Pyrophosphate}$$

The complete structure of CDP-choline is

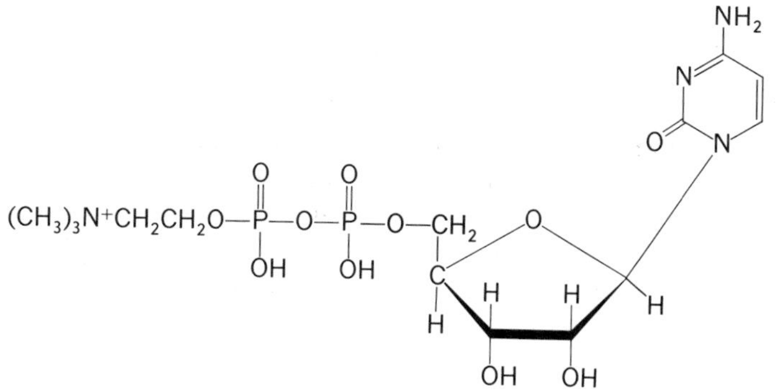

$$(CH_3)_3N^+CH_2CH_2O-\underset{\underset{OH}{\mid}}{\overset{\overset{O}{\parallel}}{P}}-O-\underset{\underset{OH}{\mid}}{\overset{\overset{O}{\parallel}}{P}}-O-CH_2 \cdots$$

The final assemblage is depicted in Figure 13.14 with the transfer of the phosphorycholine moiety to diacylglycerol to form phosphatidyl choline. Phosphatidyl ethanolamine and phosphatidyl serine are also formed by similar mechanisms. Phosphatidyl choline may also be produced by methylation of phosphatidyl ethanolamine by means of *S*-adenosyl-methionine.

Phosphatidyl ethanolamine + 3-*S*-Adenosyl-methionine $\longrightarrow$
Phosphatidyl choline + 3-*S*-Adenosyl-homocysteine

It should be noted, however, that while these mechanisms are important, alternative mechanisms for the synthesis of the phospholipids have been de-

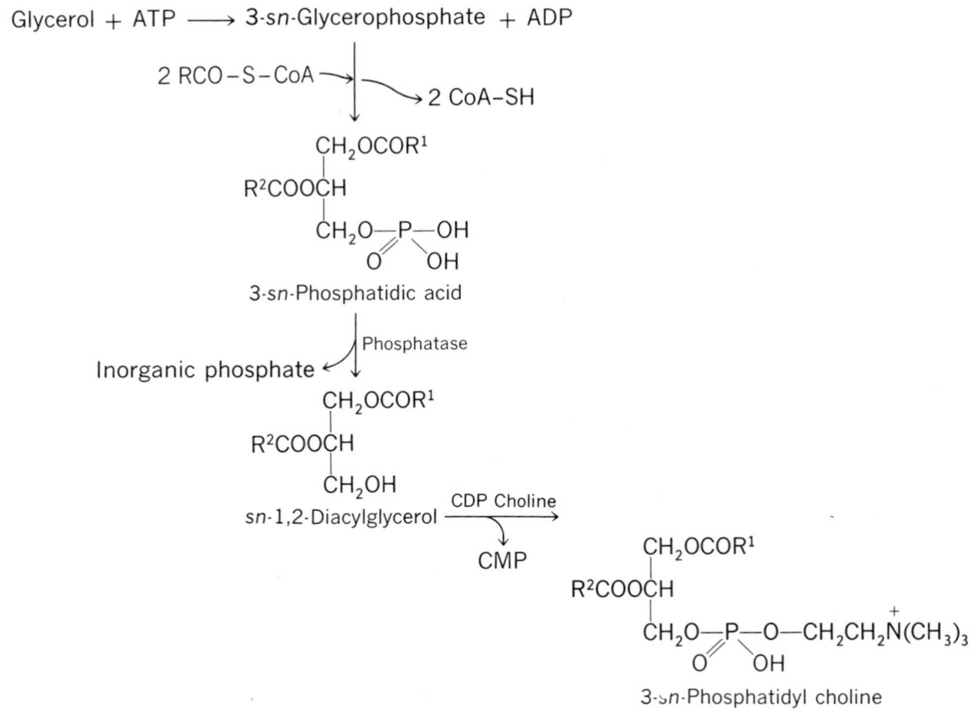

Glycerol + ATP $\longrightarrow$ 3-sn-Glycerophosphate + ADP

2 RCO−S−CoA $\longrightarrow$ 2 CoA−SH

$$\begin{array}{l} CH_2OCOR^1 \\ R^2COOCH \\ CH_2O-P-OH \\ \quad\quad O \quad OH \end{array}$$

3-sn-Phosphatidic acid

Phosphatase

Inorganic phosphate $\longleftarrow$

$$\begin{array}{l} CH_2OCOR^1 \\ R^2COOCH \\ CH_2OH \end{array}$$

sn-1,2-Diacylglycerol $\xrightarrow[\quad CMP \quad]{\text{CDP Choline}}$

$$\begin{array}{l} CH_2OCOR^1 \\ R^2COOCH \\ CH_2O-P-O-CH_2CH_2\overset{+}{N}(CH_3)_3 \\ \quad\quad O \quad OH \end{array}$$

3-sn-Phosphatidyl choline

FIGURE 13.14　Biosynthesis of phosphatidyl choline.

scribed and can be examined by referring to the references listed at the end of this chapter. Also, note the biosynthesis and important function of phosphatidyl-inositol triphosphate that is described in Section 18.6.2.

13.15 BIOSYNTHESIS OF CHOLESTEROL

Ever since the early 1930s, when the chemical structure of cholesterol was determined, the biogenesis of this complex ring system has been an intriguing puzzle. The solution was achieved almost 40 years later by the efforts of many investigators.

The ring structure of the molecule is planar and can be written as

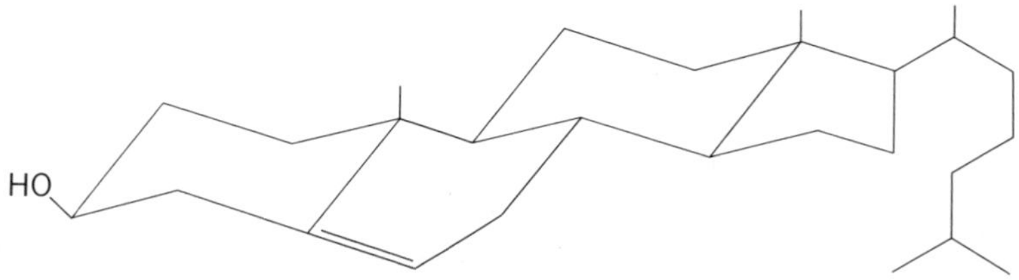

Cholesterol

All the carbon atoms of cholesterol derive directly from acetate. Reactions very different from those involved in straight-chain fatty acid synthesis were discovered in its biosynthesis. Its biosynthesis can be subdivided into three sets of reactions.

(1) Formation of mevalonic acid
(2) Conversion of mevalonic acid to squalene
(3) Conversion of squalene into lanosterol and then to cholesterol

(1) Acetate ⟶ mevalonate (microsomal enzymes):

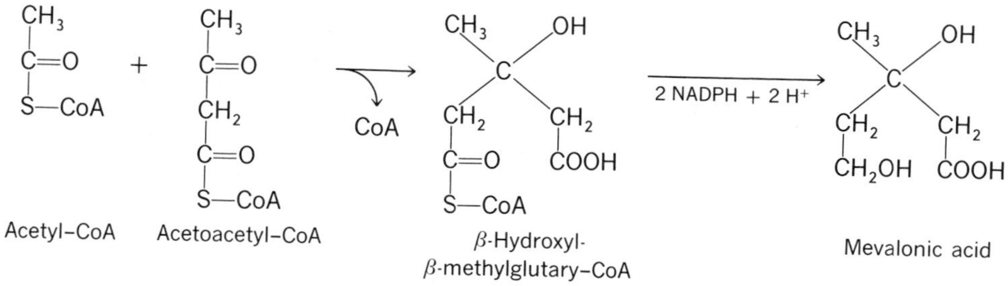

CoA
2 NADPH + 2 H+
Acetyl–CoA
Acetoacetyl–CoA
β-Hydroxyl-
β-methylglutary–CoA
Mevalonic acid

(2) Mevalonate to squalene (soluble enzymes):

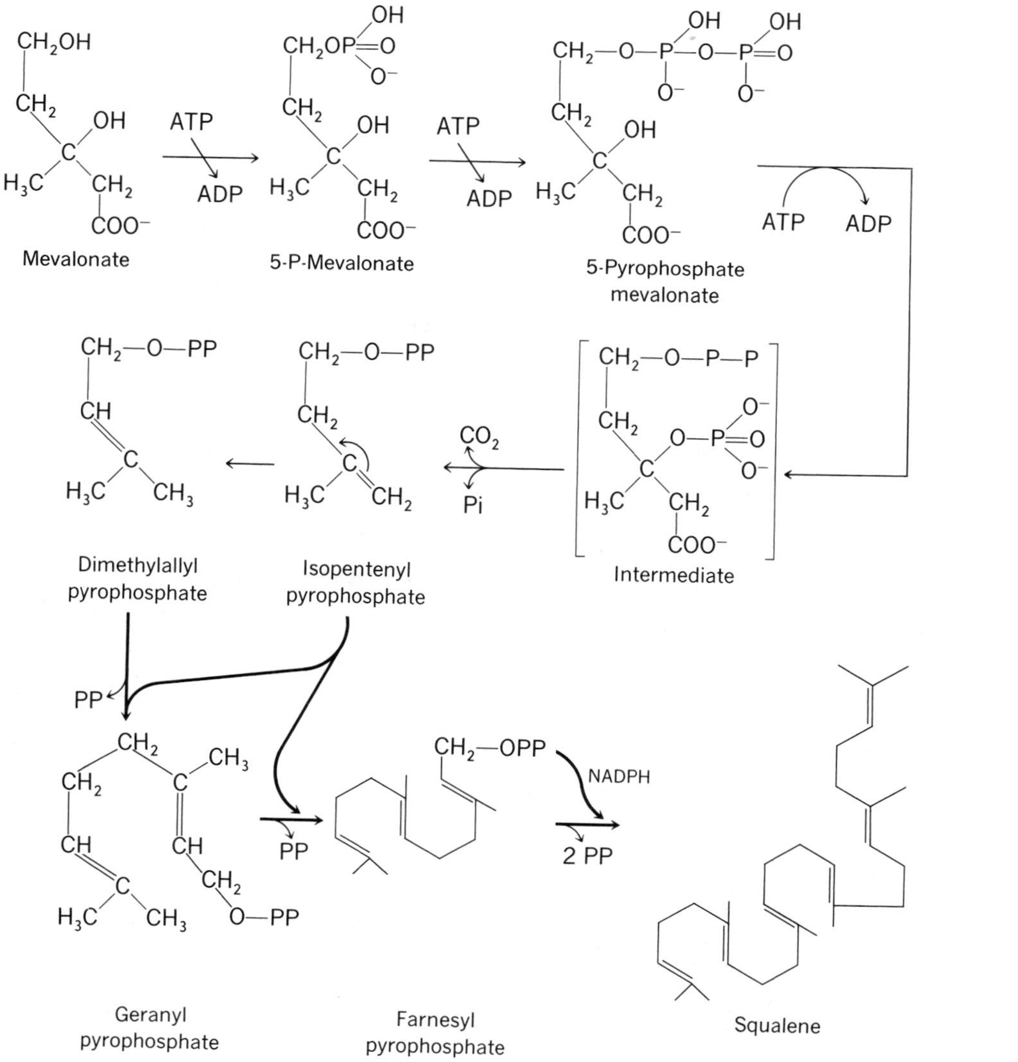

ATP
ADP
ATP
ADP
ATP
ADP
Mevalonate
5-P-Mevalonate
5-Pyrophosphate
mevalonate
CO2
Pi
Intermediate
Isopentenyl
pyrophosphate
Dimethylallyl
pyrophosphate
PP
PP
NADPH
2 PP
Geranyl
pyrophosphate
Farnesyl
pyrophosphate
Squalene

(3) Squalene ⟶ Lanosterol ⟶ Cholesterol (aerobic and microsomal):

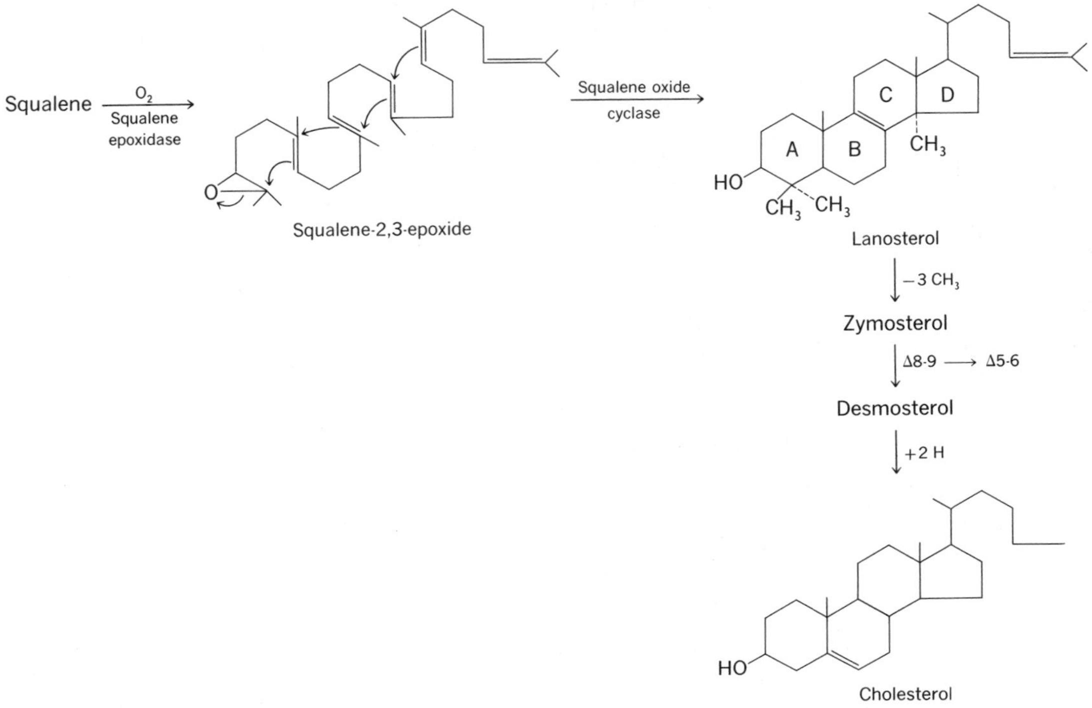

Squalene $\xrightarrow[\substack{\text{Squalene} \\ \text{epoxidase}}]{O_2}$

Squalene-2,3-epoxide

$\xrightarrow[\text{cyclase}]{\text{Squalene oxide}}$

Lanosterol

$\downarrow -3\ CH_3$

Zymosterol

$\downarrow \Delta 8\text{-}9 \longrightarrow \Delta 5\text{-}6$

Desmosterol

$\downarrow +2\ H$

Cholesterol

13.16 REGULATION OF CHOLESTEROL SYNTHESIS

Cholesterol is synthesized in all animal tissues. In addition, a variable quantity of dietary cholesterol contributes to the total concentration in humans. Its importance relates to its role in the stabilization of membrane structures because of its rigid planar structure. Approximately 140 g of cholesterol are present in humans, of which 85% are associated into cell membranes. It also serves as a precursor for the synthesis of steroid hormones.

In recent years, much research has centered on the regulation of the synthesis as well as the content of its levels in the body. There is good evidence that the rate-determining step in hepatic cholesterol biosynthesis involves the microsomal HMG (hydroxy-methyl glutaryl) reductase system. The catalysis is regulated by a reversible phosphorylation–dephosphorylation cycle (see Section 18.5), by which the reductase is converted from an active form (dephospho) to an inactive form (phospho) through the action of a specific protein kinase (cyclic AMP-independent) and reactivated by a broad specificity protein phosphatase.

The levels of cholesterol in the blood are controlled, in part, by a sophisticated LDL (light-density lipoprotein) pathway involving a receptor-mediated endocytosis (the student should review Section 8.9 on endocytosis). LDL is a large particle composed of cholesterol esters, proteins, phospholipid, and triacylglycerol (see Table 7.3). It binds to specific receptors in coated pits of the plasma membrane of cells. These pits when fully loaded invaginate to convert to coated vesicles that now move internally and attach to endosomes. Because of the low pH milieu of the endosome matrix, the LDL-receptor complex dissociates with the receptor recycling back to the plasma membrane, while the LDL component is routed to lysosomes, there to be degraded with the release of cholesterol from the cholesterol ester component of LDL. Free cholesterol may now participate (1) in the general metabolic requirement of the cell, (2) can also serve as a signal to inhibit HMG reductase activity by as yet a poorly understood mechanism, and (3) may also reduce the expression of LDL receptors in the plasma membrane.

The significance of the LDL pathway was dramatically demonstrated by M. S. Brown and J. L. Goldstein (1985 Nobel laureates). Patients who suffer from a genetic disease called familial hypercholesterolemia (FH) have an excessively high blood cholesterol content and exhibit premature atherosclerotic disease. It was shown that FH patients have abnormally low LDL receptor activity; as a result, these patients have elevated plasma LDL and suffer a dramatic loss in the utilization of cholesterol by their cells. There is good evidence that a mutation(s) had occurred in the structural gene for LDL receptors and hence a defective LDL pathway results.

Although the steroid ring is readily synthesized in plants and animals, procaryotic organisms cannot synthesize the ring system although they readily form polyisoprenoid pigments. Insects have lost the ability to synthesize sterols and hence employ exogenous sources for further conversion to important insect hormones such as ecdysome, an oxygenated derivative of cholesterol. In vertebrates, cholesterol is the substrate for a complex of modifications of the side chains and the ring system to form progesterone, androgens, estrogens, and corticosteroid, all extremely important mammalian hormones.

REFERENCES

1. *P. Boyer*, The Enzymes, *Vol. XVI, 3rd ed. New York: Academic Press, 1983.*
2. *S. Numa, ed.*, Fatty Acid Metabolism and Its Regulation. *Amsterdam: Elsevier, 1984.*

REVIEW PROBLEMS

1. When long-chain fatty acids are oxidized to CO_2 and H_2O by β-oxidation and the Krebs cycle, the following types of reactions are encountered. Give an example of each reaction using $R—CH_2—CH_2—COOH$ for the fatty acid structural formulas for other compounds and abbreviations for complicated cofactors such as NAD^+, CoASH, and so on. *Balance the equations and indicate all necessary cofactors. Names of enzymes are not required.*

   (a) A reaction that involves the breaking of a carbon-carbon bond
   (b) A reaction that requires FAD as a cofactor
   (c) Another reaction that requires FAD as a cofactor
   (d) A reaction in which a carbon-sulfur bond is formed
   (e) A reaction in which H_2O is consumed
   (f) Another reaction in which H_2O is consumed

(g) A reaction inhibited by malonic acid

(h) A reversible oxidative decarboxylation

(i) A reaction in which an isomerization occurs

2. Write the series of enzyme-catalyzed reactions by which pyruvic acid can be converted to caproic (hexanoic) acid. Use structural formulas, except for the coenzymes involved.

3. Explain *in detail* why an animal cannot effect the *net* conversion of lipid to carbohydrate while a plant or microorganism can. Illustrate your answer with the key enzyme-catalyzed reactions involved. (*Hint:* See Section 12.7)

4. Explain in six to eight short, concise, complete statements why ketone bodies may accumulate in an animal existing on a pure lipid diet, but not in a plant or microorganism utilizing lipids as their sole carbon-energy source.

5. **(a)** Glucose and caproic acid (hexanoic acid) are both six-carbon compounds. Which would you expect to yield the more ATP (per mole) on complete oxidation to CO_2 and H_2O by a living cell? *Why* (in very general chemical terms)?

   (b) Calculate the number of moles of ATP that could be produced from the complete oxidation of one mole of caproic acid by an aerobic organism. Show your work clearly.

   (c) Briefly list the major differences between the fatty acid oxidation pathway and the fatty acid biosynthesis pathway.

   (d) Define "ketosis."

   (e) Describe the biochemical defect that produces "ketosis."

6. **(a)** Describe in detail the biosynthesis of linoleic acid in higher plants, starting with free oleic acid.

   (b) Although animals cannot synthesize linoleic acid *de novo,* what mono unsaturated fatty acid, other than oleic acid, will be converted to linoleic acid in the animal?

7. Write out the reactions for the β-oxidation of elaedic acid [$18:1(9t)$] by liver mitochondria.

8. Although at first glance, only 1 ATP is required for the complete β-oxidation of a fatty acid, in actual fact, 2 ATPs are involved. (*Hint:* Note that pyrophosphate is one of the products in the conversion of a fatty acid to acyl CA.)

9. What role do coated vesicles play in the control of blood cholesterol levels?

ELECTRON TRANSPORT AND OXIDATIVE PHOSPHORYLATION

The oxidation of pyruvate and acetyl CoA as it occurs in mitochondria is frequently called the **aerobic** phase of carbohydrate metabolism. However, the title is misleading, since both pyruvate and acetyl CoA can also be obtained from noncarbohydrate sources. In addition, the term aerobic is not strictly precise when the reactions are described as on the inside of the front cover, because the immediate oxidizing agents are nicotinamide and flavin nucleotides rather than oxygen. As in the glycolytic sequence, the amounts of nicotinamide and flavin nucleotides in the cell are limited, and the reactions cease when the supply of oxidized nucleotides is exhausted. Hence, in order for the oxidation of organic substrates to continue, the reduced nicotinamide and flavin nucleotides must be reoxidized.

14.1 OXIDATION OF REDUCED COENZYMES

In procaryotic cells, the oxidation of $NADH$ and $FADH_2$ (and $FMNH_2$) is accomplished by enzymes located on the plasma membrane; in eucaryotes, the necessary catalysts are in the inner membrane of the mitochondrion adjacent to the matrix where the nucleotides are reduced (see Chapter 8). In all aerobic organisms, the ultimate oxidizing agent is molecular oxygen and, for the case of $NADH$, we may write the overall reaction as follows:

$$NADH + H^+ + \tfrac{1}{2}O_2 \longrightarrow NAD^+ + H_2O \qquad (14.1)$$
$$\Delta G' = -52{,}500 \text{ cal (pH 7.0)}$$

The enzymes accomplishing this oxidation constitute an **electron-transport chain** in which a series of electron carriers are alternately reduced and oxidized.

This reoxidation of $NADH$ by O_2 is accompanied by a large decrease in free energy (Section 14.8). The amount is sufficient to produce several moles of ATP per mole of $NADH$ oxidized. The enzymes that catalyze the production of ATP as the $NADH$ is oxidized are also localized in the mitochondrial inner membrane. Although the process is known as **oxidative phosphorylation,** it is perhaps better described as **respiratory chain phosphorylation.** This process will be described after the electron-transport chain is discussed.

14.2   COMPONENTS INVOLVED IN ELECTRON TRANSPORT

There are five different kinds of electron carriers that participate in the transport of electrons from substrates as they are oxidized in the mitochondria. In addition, Cu^{2+} is present and functions in the enzyme, cytochrome oxidase, that catalyzes the reduction of O_2. A brief description of each is in order before the electron-transport chain itself is described.

14.2.1   NICOTINAMIDE NUCLEOTIDES

The general properties of these coenzymes (cosubstrates) and their related dehydrogenases (apoenzymes) were described in some detail earlier (Section 5.4). Two of the oxidations in the tricarboxylic acid cycle involve the removal of the equivalent of two hydrogen atoms from the substrates, malate and isocitrate. In two others, those catalyzed by pyruvic dehydrogenase and α-keto-glutarate dehydrogenase, the electrons are transferred first to lipoic acid and then via a flavoprotein to NAD^+.

$$SH_2 + NAD^+ \rightleftharpoons S + NADH + H^+ \qquad (14.2)$$

14.2.2   FLAVOPROTEINS

The prosthetic groups of flavoproteins are the flavin coenzymes FAD and FMN (Section 5.3). These cofactors, in contrast to the nicotinamide nucleotide coenzymes, are much more firmly associated with the protein moiety, and in some instances (e.g., succinic dehydrogenase), are covalently bonded to that protein.

In their simplest form, the flavin cofactors accept two electrons and a proton from NADH or two electrons and two protons from an organic substrate such as succinic acid. The reaction with NADH or succinic acid may be represented as

$$NADH + H^+ + FMN \longrightarrow NAD^+ + FMNH_2 \qquad (14.3)$$
$$Succinate + FAD \longrightarrow Fumarate + FADH_2 \qquad (14.4)$$

The flavoproteins of the mitochondrial respiratory chain are complex in that they contain or are closely associated with iron–sulfur proteins, also called nonheme iron (NHI) proteins. Because the flavin cofactors can accept one electron at a time forming a semiquinone, the flavoproteins represent a point in the respiratory chain where electrons can be transferred one at a time rather than in pairs.

14.2.3   IRON–SULFUR PROTEINS

This type of protein was first encountered as ferredoxin, a reducing agent involved in nitrogen fixation and photosynthesis in plants, before it was recognized to function in mitochondrial electron transport in animals. Their most characteristic chemical feature is the release of H_2S on acidification (acid labile sulfur), a treatment that also removes the iron. The iron atoms are arranged in pairs in an iron–sulfur bridge that, in turn, is bonded to the sulfur atoms of cysteine residues in the protein. Some iron–sulfur proteins such as spinach ferredoxins contain only two iron atoms (Fe_2S_2) while others contain four (Fe_4S_4).

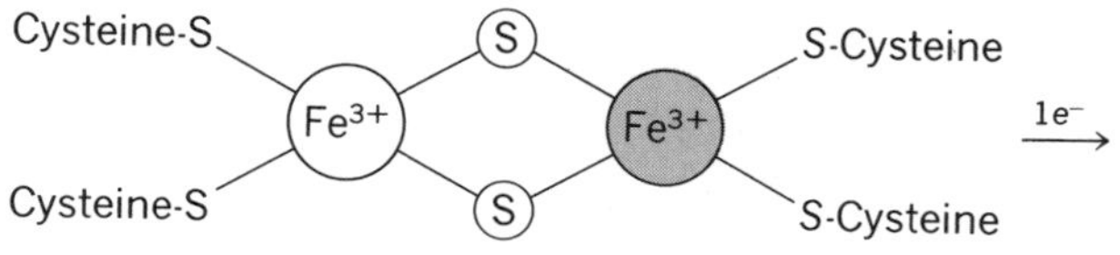

Oxidized form

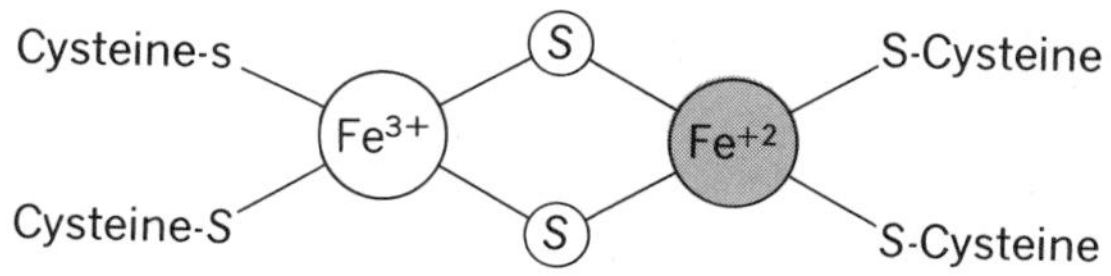

Reduced form

In the oxidized state, both iron atoms in the model are in the ferric state. When reduced, one iron becomes Fe^{2+} and can be detected by a characteristic EPR (electron paramagnetic resonance) signal. The standard reduction potentials (E_0') range from -0.420 V to as high as 0.300 V. Since each Fe^{3+} atom can become reduced, a center with four iron atoms (Fe_4S_4) can store up to four electrons.

14.2.4 QUINONES

Mitochondria contain a quinone called ubiquinone that has the general structure

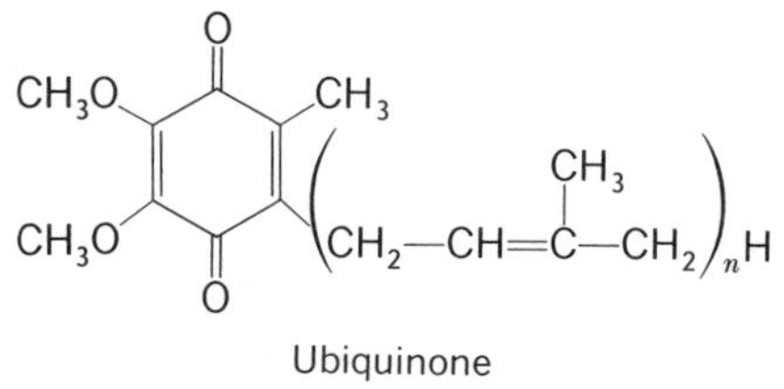

Ubiquinone

The length of the side chain (R) varies with the source of the mitochondria; in animal tissues, the quinone possesses 10 isoprenoid units in its side chain and is called coenzyme Q_{10} (CoQ_{10}). Because of its long aliphatic side chain, ubiquinone is lipid soluble and can be removed from the inner mitochondrial membrane with lipid solvents (e.g., butanol). This is in contrast to all other of the enzymes in the respiratory chain. When the quinone is extracted from the mitochondria, the transport of electrons from substrates to oxygen is prevented; the activity is restored when the quinone is added back.

The complete reduction of one molecule, of course, requires two electrons, but coenzyme Q also readily undergoes one electron reduction to form a free-radical intermediate

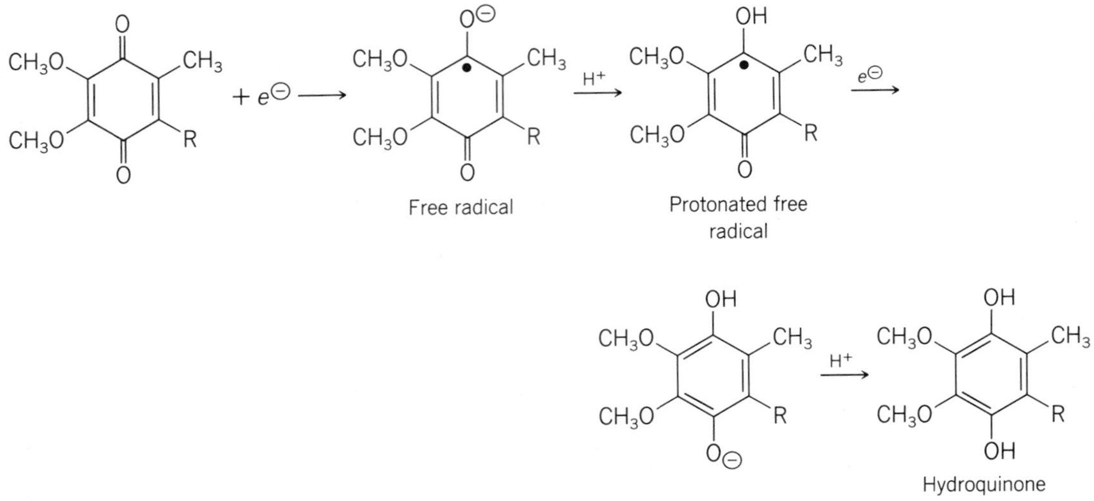

Free radical

Protonated free radical

Hydroquinone

Coenzyme Q_{10} occupies a pivotal position in the electron-transport chain where it accepts electrons not only from NADH dehydrogenase, but also from the flavin components of succinic dehydrogenase, glycerol phosphate dehydrogenase, and fatty acyl CoA dehydrogenase (see Figure 14.2).

14.2.5   THE CYTOCHROMES

These respiratory carriers were discovered in animal cells in 1886 by McMunn. He called the compounds myo- or histo-hematins and thought they were important for respiratory processes. His results were severely criticized and then forgotten until these components were rediscovered by Keilin. Keilin's classical experiments in England in 1926–1927 demonstrated that these cell pigments (cytochromes) were found in almost all living tissues and implied an essential role for these substances in cellular respiration.

Keilin's research showed that in every tissue there usually are three types of cytochrome to which he assigned the letters a, b, and c. The amount seemed to be proportional to the respiratory activity of the tissue, heart and other active muscles containing the largest amounts of these pigments. The research on the cytochromes was facilitated by the fact that they absorb light of different wavelengths in a characteristic manner. The absorption spectra of oxidized and reduced cytochrome c are shown in Figure 14.1; note the positions of the maxima of the α, β, and γ bands of the reduced carrier. The three types of cytochromes (a, b, and c) were distinguished by the positions of the absorption maxima of the α, β, and γ bands. As additional cytochromes were discovered and described, they were named c_1, b_1, b_2, and so on. More informative is a classification based on the absorption maximum of the α band; thus, c_1 may also be called c_{554} and b_5, b_{557}.

The absorption spectra together with other properties of the cytochromes indicate that these compounds are conjugated proteins having an iron por-

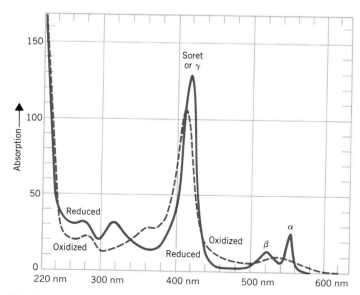

FIGURE 14.1 Absorption spectra of oxidized and reduced cytochrome c. (*Source:* Reproduced with permission from R. E. Dickerson and R. Timkovich, in *The Enzymes,* P. D. Boyer, ed., 3rd ed. Vol. 11, p. 399. New York: Academic Press, 1975.)

phyrin as a prosthetic group. The structure of the prosthetic group for cytochrome *c* is shown.

Cytochrome *c*

It is protoporphyrin IX linked to cysteine residues in the protein through thioether linkages with the vinyl groups of rings III and IV (Section 17.12). The prosthetic group of cytochrome *b* is protoporphyrin IX itself. The porphyrin of cytochrome *a* is porphyrin A, characterized chiefly by a long hydrophobic chain of hydrogenated isoprenoid units. In this regard, porphyrin A resembles the porphyrin of chlorophyll.

Porphyrin A

The cytochromes tend to form complexes with HCN, CO, and H_2S, and such complexes are readily detected by their characteristic absorption spectra. These reagents react by virtue of their ability to occupy one or both of the two coordination positions of the Fe atom that are not occupied by the nitrogen atoms of the pyrrole rings of the porphyrin. In cytochrome c, where those two positions are occupied by other structures, complexes with HCN, CO, and H_2S are not formed at neutral pH.

The studies on soluble cytochrome c confirmed what Keilin had observed in intact tissues. The cytochromes are capable of being alternately reduced and oxidized. The iron of the oxidized cytochromes is ferric iron; it is reduced to ferrous iron by the incorporation of one electron into the valence shell of the iron atom. It is this property that allows the cytochromes to function as carriers in the electron-transport process.

14.3 THE RESPIRATORY CHAIN

The electron-transport chain in the mitochondrial membrane has been separated into four components or complexes. When arranged as shown in Figure 14.2, the chain accomplishes the transfer of electrons from NADH or succinic acid to O_2. CoQ_{10} and cytochrome c are not included in any of the complexes because these two components can be removed with relative ease from the membrane. Cytochrome c is a small, peripheral protein (MW 12,000) that is readily extracted by treatment of the membrane with aqueous salt solutions. CoQ_{10} can be extracted with butanol. In either case, the flow of electrons is interrupted at the point in the chain from which these carriers have been removed. When they are added back, the activity is restored.

14.3.1 COMPLEX I: NADH-CoQ REDUCTASE

This complex catalyzes the transfer of electrons between NADH and CoQ.

$$NADH + H^+ + CoQ \longrightarrow NAD^+ + CoQ \cdot H_2$$

The complex, to which is coupled the synthesis of ATP at coupling site 1 (Section 14.4), contains four iron–sulfur centers and one FMN. The E_0' of the iron–sulfur center ranges from -0.330 to -0.020 V; therefore, the flow of electrons is from NADH to FMN to the iron–sulfur proteins where they flow from the iron–sulfur center with the lowest E_0' to that having the highest.

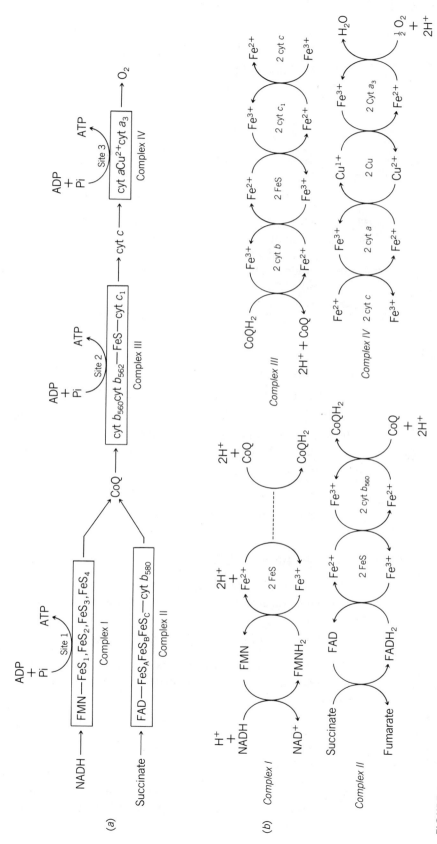

FIGURE 14.2 The cytochrome electron-transport chain. (a) Complexes I, II, III, and IV are shown together with the specific carrier components for each complex. (b) The sequential oxidation–reduction of each component in the four complexes.

$$NADH \longrightarrow FMN \longrightarrow (Fe-S_1) \longrightarrow (Fe-S_2) \longrightarrow (Fe-S_3) \longrightarrow (Fe-S_4) \longrightarrow CoQ$$

Complex I is inhibited by amytal and rotenone. It also contains phosphorylation site 1.

14.3.2   COMPLEX II: SUCCINATE-CoQ REDUCTASE

Complex II catalyzes the reduction of CoQ by electrons removed from succinate.

$$Succinate + CoQ \longrightarrow Fumarate + CoQ \cdot H_2$$

This complex, which contains FAD, is composed of four polypeptides with molecular weight of 70,000, 27,000, 15,000, and 13,000. The two larger peptides contain the catalytic site for the oxidation of succinate; the smaller (27,000) subunit of the two contains an Fe_4S_4 center. The two smallest peptides (15,000 and 13,000) contain a Fe_2S_2 center and a b-type cytochrome whose role is not clear.

14.3.3   COMPLEX III: CoQ-CYTOCHROME c REDUCTASE

The reaction catalyzed by Complex III is

$$CoQ \cdot H_2 + 2 \text{ cyt } c \text{ (Fe}^{3+}) \longrightarrow CoQ + 2 \text{ cyt } c \text{ (Fe}^{2+}) + 2 H^+$$

Complex III contains two b-type cytochromes, b_{560} and b_{562}; a c-type cytochrome, c_1; and an iron–sulfur center. It also contains phosphorylation site 2 that produces ATP as electrons flow through the complex. The order in which the electrons flow is

$$Co \cdot QH_2 \longrightarrow b_{560} \longrightarrow b_{562} \longrightarrow FeS \longrightarrow \text{cyt } c_1 \longrightarrow c$$

The complex is inhibited by the antibiotic antimycin.

14.3.4   COMPLEX IV: CYTOCHROME c OXIDASE

Cytochrome oxidase has long been recognized as the terminal oxidase of all aerobic cells. It catalyzes the oxidation of reduced cytochrome c by molecular O_2.

$$4 \text{ cyt } c \text{ (Fe}^{2+}) + 4 H^+ + O_2 \longrightarrow 4 \text{ cyt } c \text{ (Fe}^{3+}) + 2 H_2O$$

ATP is also generated as electrons flow through cytochrome oxidase. The complex contains cytochrome a, two Cu^{2+} ions, and cytochrome a_3; the flow of electrons is as follows:

$$\text{cyt } c \longrightarrow \text{cyt } a \longrightarrow Cu^{2+} \longrightarrow \text{cyt } a_3 \longrightarrow O_2$$

Complex IV is inhibited by cyanide anion, carbon monoxide, and sodium azide.

The clarification of the sequence of carriers in the electron-transport chain aided the study of the phosphorylation reactions that accompany the transport of electrons from substrate to O_2. This process of respiratory-chain phosphorylation can now be discussed.

14.4   OXIDATIVE PHOSPHORYLATION

A major objective of the degradation of carbon substrates by a living organism is the production of energy for the development and growth of that organism. In the anaerobic degradation of sugars to lactic acid, some of the energy available in the sugar molecule was conserved in the formation of energy-rich phosphate compounds that are made available to the organism. As pointed out in Chapter 12, over 90% of the energy available in glucose is released when pyruvate is oxidized to CO_2 and H_2O through the reactions of the tricarboxylic acid cycle. In that process however, there was only one energy-rich compound, namely succinyl CoA, synthesized by reactions involving the substrates of the cycle itself; in the presence of succinic thiokinase, this thioester was utilized to convert GDP to GTP.

When the production of energy-rich compounds in biological organisms was investigated in more detail, two different types of phosphorylation process were recognized. In one of the processes, phosphorylated or thioester derivatives of the substrate were produced initially and subsequently utilized to produce ATP. Examples of these are the reactions of glycolysis in which 1,3-bisphosphoglyceric acid and phosphoenol pyruvic acid are formed and react with ADP to form ATP (Sections 10.4.1, 10.4.2, 10.4.4, and 10.4.5), as well as the reaction catalyzed by succinic thiokinase in the Krebs cycle (Section 12.3.5). These phosphorylation processes have been referred to as **substrate-level phosphorylations** and are to be distinguished from the phosphorylations associated with electron transport that are usually referred to as **oxidative phosphorylation.**

In 1937, Belitzer in Russia and Kalckar in the United States observed that phosphorylation occurred during the oxidation of pyruvic acid by muscle homogenates. Although the subsequent fate of the pyruvate molecule was not clear at that time, oxygen was consumed by the homogenate, and inorganic phosphate was esterified as hexose phosphates. If the reaction were inhibited by cyanide or removal of O_2, both the phosphorylation and the oxidation ceased. Thus, the synthesis of a sugar phosphate bond was dependent on a biological oxidation in which molecular oxygen was consumed.

Several important advances occurred that simplified the study of this important process. First, in 1948, Kennedy and Lehninger showed that isolated rat liver mitochondria catalyzed oxidative phosphorylation coupled to the oxidation of Krebs cycle intermediates. Today, it is recognized that the inner membrane of mitochondria is the locus of this type of phosphorylation enzymes. In bacteria, smaller units in the cell membrane contain the phosphorylation assemblies (Chapter 8).

Second, it was found that the only phosphorylation reaction that could be identified in the mitochondrion was the incorporation of inorganic phosphate into ADP to form ATP.

$$ADP + H_3PO_4 \longrightarrow ATP + H_2O$$

This is clearly a reaction that requires energy; if all reactants are in the standard state, the ΔG ($\Delta G'$ by definition) would be $+7300$ cal/mole. Since the reactants are undoubtedly not present at concentrations of 1 M, the ΔG will be considerably larger, perhaps as much as $+12,000$ cal/mole.

Third, the composition of the electron-transport chain of mitochondria was investigated in some detail; and fourth, the oxidation of NADH itself by O_2 in the presence of mitochondria was shown to lead to the formation of ATP by the incorporation of inorganic phosphate into ADP. This extremely significant observation by Friedkin and Lehninger is described in Box 14.A.

BOX 14.A   OXIDATIVE PHOSPHORYLATION DURING OXIDATION OF NADH BY ISOLATED MITOCHONDRIA

Having shown that isolated liver mitochondria could synthesize ATP from ADP and Pi while oxidizing intermediates of the tricarboxylic acid cycle, Lehninger's laboratory simplified the system by showing that the oxidation of NADH could also support oxidative phosphorylation. When NADH was added to a reaction mixture containing ADP, inorganic phosphate, Mg^{2+}, and rat liver mitochondria, it was oxidized to NAD^+, O_2 was reduced to H_2O, and inorganic phosphate was esterified as ATP. This classical system has now been demonstrated in mitochondria isolated from many animal and plant tissues. The oxidation of NADH and the reduction of O_2 occur because, as described in Section 14.3, carefully isolated mitochondria contain the intact electron-transport chain. Simultaneously with this oxidation, inorganic phosphate reacts with ADP to form ATP. Under ideal conditions, between two and three moles of ATP will be formed per atom of O_2 consumed. Since the mitochondria contain ATPase and also can catalyze side reactions that utilize ATP, it is believed that three moles of ATP are formed per mole of NADH oxidized or atom of oxygen consumed. This may be represented schematically as

$$NADH + H^+ + \tfrac{1}{2} O_2 + 3\ ADP + 3\ H_3PO_4 \longrightarrow NAD^+ + 3\ ATP + 4\ H_2O$$

This reaction is also said to have a **P:O ratio** of 3.0, a term used to describe the ratio of the atoms of *phosphate esterified* to the atoms of *oxygen consumed* in the oxidation. Because the oxidation of malate and isocitrate exhibited P:O ratios of 3.0, the phosphorylations associated with these substrates were assumed to occur after the NAD^+ that serves as oxidant was reduced. A P:O ratio of 2.0 for succinate similarly indicated that one fewer phosphorylation step was involved when this compound is oxidized.

Much experimental evidence supports the conclusion that phosphorylation occurs as electrons make their way along the electron-transport chain pictured in Figure 14.2. For example, phosphorylation occurs when reduced cytochrome c is oxidized by molecular oxygen by the enzymes in Complex IV. Similarly, Complex I and III, but not Complex II, contain phosphorylation sites and, as noted earlier, ADP and H_3PO_4 are converted to ATP as electrons move along the carriers in those complexes. In each complex, there is at least one carrier whose E_0' depends on the concentration of ATP. These carriers [$(Fe-S)_4$ in Complex I; b_{562} in Complex III; a_3 in Complex IV] have therefore been proposed as the carrier specifically associated with the phosphorylation site in the complex.

14.5   THE ENERGY CONVERSION PROCESS

The phosphorylation mechanism(s) associated with electron transport are fundamentally different from the phosphorylation step in glycolysis in that energy-rich phosphorylated forms of substrates (e.g., 1,3-bisphosphoglyceric acid or phosphoenol pyruvate) or electron transport proteins have not been identified.

Despite the efforts made in several productive laboratories (those of Boyer, Chance, Green, Lardy, Lehninger, Mitchell, Racker, and Slater, to mention a few alphabetically!), there is still much to learn about the details of the actual chemical mechanism by which ADP and inorganic phosphate are joined to form ATP and H_2O.

The difficulty of this problem is associated with the fact that oxidative phosphorylation is a process intimately associated with the structure of the inner mitochondrial membrane. It is within this membrane that the transfer of electrons and the initial phosphorylation processes occur. The membrane also is involved in the transport of ions in and out of the mitochondrial matrix. As reducing equivalents move from NADH to O_2 in electron transport, protons are transported out of the matrix. Simultaneously, K^+ can be transported back into the matrix to maintain charge neutrality. Ca^{+2} and other divalent cations can also be accumulated in the matrix during respiration in processes that are sensitive to different inhibitors. However, it is the gradient of protons across the membrane that results in the phosphorylation of ADP to form ATP.

14.6 MECHANISMS OF OXIDATIVE PHOSPHORYLATION

The first theory seeking to explain oxidative phosphorylation proposed that, as in glycolysis, energy-rich intermediates were formed that, in turn, could be used to synthesize ATP from ADP. However, it was never possible to detect or identify any such energy-rich intermediates. An alternative hypothesis was then proposed by P. Mitchell that has experimental support and, in principle, is generally accepted. This is the **chemiosmotic hypothesis** that has, as a major experimental observation, the fact that a proton gradient is established as electrons move from NADH to O_2. Mitchell has proposed that the inner membrane of the mitochondria is impermeable to protons (H^+). However, as those steps in the electron-transport chain that produce or consume H^+ occur, the spatial arrangement of those enzymes in the inner membrane is such that the proton (H^+) consumed comes from the matrix. On the other hand, any protons (H^+) produced are released into the intermembrane space (Figure 14.3). In this way, a proton gradient is formed across the membrane as the oxidation of NADH proceeds.

The proton gradient, in turn, has two components: a pH gradient and a membrane potential. These two components constitute a **proton motive force** that is the source of the energy consumed in synthesizing ATP. Thus, the proton gradient is established as electrons flow along the respiratory chain and the disestablishment of this gradient results in the synthesis of ATP.

One of the most useful approaches in studying oxidative phosphorylation has been the fractionation of the inner membrane into two components that carry out oxidative phosphorylation when recombined. Racker and associates have separated the inner membrane into membranous vesicles and a separate protein fraction. The vesicles (inside-out submitochondrial particles) can carry out electron transport between NADH and O_2 but do not synthesize ATP. When the separate protein fraction is added back to the vesicles, ATP is produced during electron transport. Surprisingly, the separated protein, called the F_1 factor, when examined in the test tube, is an ATPase. The ability of the F_1

Inner membrane of
mitochondrion

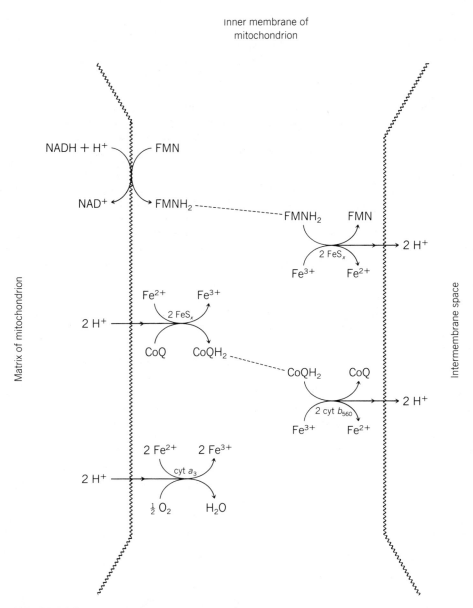

FIGURE 14.3   Components of the electron-transport chain that involve a gain or loss of protons as oxidation–reduction occurs.

factor to act as an ATPase is believed to be the *opposite* of its normal function, which would be ATP synthesis.

Figure 14.4 is a schematic drawing of the inside-out vesicle when recombined with the F_1-ATPase. A proton gradient is established as the result of electron transport. The role of F_1-ATPase together with its base F_0 protein in the membrane is to provide a channel through which protons can flow as the gradient is disestablished and ATP is synthesized.

Once synthesized, the ATP is transferred to the cytosol in exchange for ADP and Pi that enter prior to phosphorylation. In view of the selectively permeable nature of the inner membrane, the student may well ask how ADP, Pi, and ATP can move across this barrier when oxalacetate, NAD^+, NADH, coenzyme A, acetyl CoA, to name a few, cannot. The answer is that there are two carrier systems that affect the passage of ADP, ATP, and Pi in or out of the mitochondria.

The first of these, named the adenine nucleotide translocator, is an intrinsic membrane protein that extends across the membrane binding ADP on the outside and ATP at the surface of the inner membrane. This protein transports ADP inward and ATP outward; it will not function with any other nucleotide

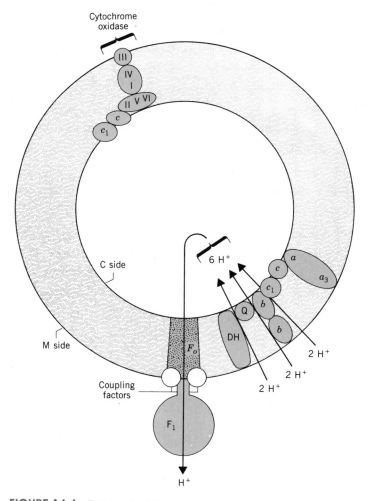

FIGURE 14.4 Transport of H^+ across the membrane of a submitochondrial inside–out vesicle as the result of oxidation–reduction. When the protons move back out of the vesicle through the F_0–F_1—ATPase complex, ATP is synthesized. Reproduced with permission from E. Racker in the Annual Review of Biochemistry, Vol. 46, p. 1008. © 1977 by Annual Reviews Inc.

such as GDP, GTP, or AMP. A second transporter protein, called phosphate carrier, transfers $H_2PO_4^-$ together with a proton (H^+) into the mitochondrial matrix. These carrier systems are separate and distinct from the carrier systems for pyruvate, the dicarboxylic acid carrier (for malate and succinate), the tricarboxylate transporters (for citrate and isocitrate), and the amino acids (for aspartate and glutamate).

14.7   UNCOUPLERS AND INHIBITORS OF OXIDATIVE PHOSPHORYLATION

Several compounds are known that **uncouple** oxidative phosphorylation from electron flow through the electron-transport chain. These compounds have been studied because they have provided information about the process of oxidative phosphorylation. In the presence of these compounds, the synthesis of ATP ceases but substrate oxidation and the reduction of O_2 continue or may even accelerate. One such uncoupler is 2,4-dinitrophenol, a lipophilic compound that is believed to act by transporting protons across the membrane in such a way that ATP formation is not coupled.

In mitochondria that are carefully prepared, the esterification of ADP and inorganic phosphate to form ATP is obligatorily linked to the transport of electrons through the electron-transport chain. Clearly, if there is no source of oxidizable substrates, electrons will not be introduced into the chain and ATP production will not occur. On the other hand, electrons will not be passed along the chain in intact mitochondria unless there is an adequate supply of ADP to be phosphorylated. In such mitochondria, the processes of electron transport and oxidative phosphorylation are said to be **tightly coupled.**

Other compounds such as oligomycin are true inhibitors of oxidative phosphorylation. Oligomycin inhibits the binding of F_1-ATPase to other proteins in the vesicle and thereby the translocation of protons that results in ATP generation. It should be stressed that the action of 2,4-dinitrophenol and oligomycin is different than that of inhibitors of electron transport such as cyanide, rotenone, or amytal that act directly on the carriers of the electron-transport chain.

14.8   ENERGETICS OF OXIDATIVE PHOSPHORYLATION

In Chapter 9, the $\Delta G'$ for the oxidation of one mole of NADH by molecular O_2 was calculated as approximately $-52,000$ cal from the reduction potentials of $NAD^+/NADH$ and O_2/H_2O. Because the oxidation of NADH by O_2 through the cytochrome electron-transport system leads to the formation of three high-energy phosphate bonds, the efficiency of the process of energy conservation may be calculated as $-21,900$ (i.e., 3×-7300) divided by $-52,000$ or 42%.

It is now possible to summarize the esterification of inorganic phosphate that accompanies the oxidation of acetyl CoA to CO_2 and H_2O by means of the tricarboxylic acid cycle. As shown in Table 14.1, three oxidation steps (Reactions 12.4, 12.5, and 12.9) in the cycle itself produce NADH while one (Reaction 12.7) produces $FADH_2$. When these are reoxidized by means of the elec-

TABLE 14.1 Formation of Energy-Rich Phosphate during the Oxidation of Acetyl CoA in the Tricarboxylic Acid Cycle

REACTION	COENZYME	ATP PRODUCED
Isocitrate $\longrightarrow \alpha$-Ketoglutarate + CO_2	NAD^+	3
α-Ketoglutarate $\longrightarrow$ Succinyl CoA + CO_2	NAD^+	3
Succinyl CoA $\longrightarrow$ Succinate	GDP	1
Succinate $\longrightarrow$ Fumarate	FAD	2
Malate $\longrightarrow$ Oxalacetate	NAD^+	3
	Total	12

tron-transport system of the mitochondria, 3 ATP will be produced for each NADH and 2 ATP for the $FADH_2$, or a total of 11 ATP. The GTP (equivalent to ATP) produced in Reaction 12.6 makes a total of 12 ATP for each mole of acetyl CoA oxidized.

The oxidation of pyruvate to acetyl CoA generates an additional NADH; its oxidation in the mitochondrial membrane will therefore produce an additional 3 ATP or a total of 15 ATP per mole of pyruvate. Since the oxidation of pyruvate to CO_2 and H_2O results in a free-energy change of $-273,000$ cal (Chapter 12), the efficiency of energy conservation in this process is at least $-109,000$ or (-7300×15) divided by $-273,000$ or 40%.

In line with this calculation, it is possible to estimate the total number of high-energy phosphate bonds that may be synthesized when glucose is oxidized to CO_2 and H_2O aerobically as illustrated in Figure 14.5. The conversion of one mole of glucose to two moles of pyruvic acid forms two high-energy phosphates as a result of substrate-level phosphorylation in the glycolytic sequence. The further oxidation of the two moles of pyruvic acid in the tricarboxylic acid cycle forms 30 high-energy phosphates. However, there are 4 to 6 more high-energy phosphates to be added to the 32 just listed. When glucose is converted to two molecules of pyruvate in glycolysis and the latter is not reduced to lactic acid, two molecules of NADH remain in the cytoplasm to be accounted for. While the NADH might be reoxidized by other cytoplasmic dehydrogenases, in a tissue that is actively oxidizing glucose completely to CO_2 and H_2O, these two molecules of NADH will contribute their electrons to the electron-transport chain of the organism to be passed to oxygen.

In procaryotic organisms, there is no permeability problem, as the NADH has ready access to the plasma membrane with its respiratory assemblies that contain the electron-transport chain and phosphorylation enzymes. A total of 38 ATP would therefore be formed in the complete oxidation of glucose to CO_2 and H_2O.

In eucaryotic organisms, the cytosolic NADH cannot pass through the inner membrane. Instead, the electrons in the NADH must be carried as a reducing equivalent by a reduced substrate in a **shuttle** process. Two such shuttles, known as the malate–aspartate shuttle and the glycerol phosphate shuttle, exist in plant and animal cells for the transport of reducing equivalents across the mitochondrial membrane.

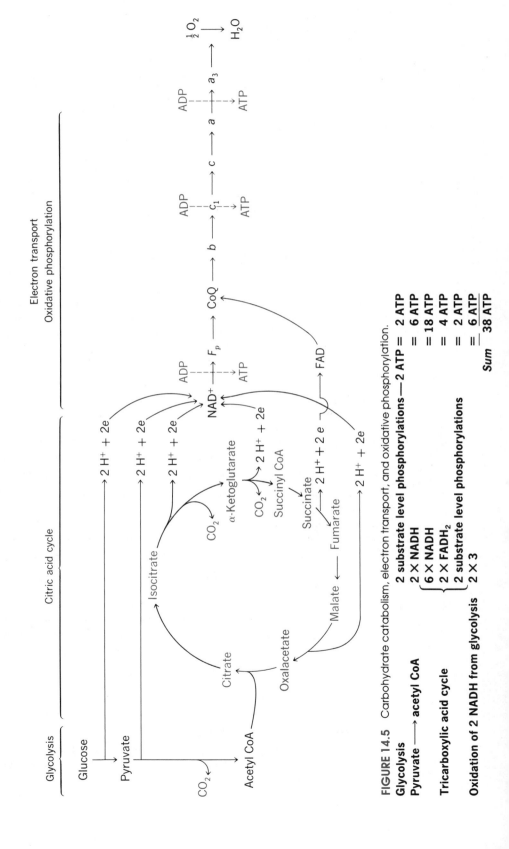

FIGURE 14.5 Carbohydrate catabolism, electron transport, and oxidative phosphorylation.

Glycolysis	2 substrate level phosphorylations—2 ATP =	2 ATP
Pyruvate ⟶ acetyl CoA	2 × NADH	= 6 ATP
	6 × NADH	= 18 ATP
Tricarboxylic acid cycle	2 × FADH$_2$	= 4 ATP
	2 substrate level phosphorylations	= 2 ATP
Oxidation of 2 NADH from glycolysis	2 × 3	= 6 ATP
	Sum	38 ATP

The malate–aspartate shuttle is a major mechanism for transporting reducing equivalents that can function in either direction. The following discussion of reactions involved accomplishes the transport of a reducing equivalent into the mitochondrial matrix from the cytosol. Initially, a reducing equivalent is used in converting oxalacetate to malate in the presence of cytosolic malate dehydrogenase.

$$
\begin{array}{ll}
\text{CO}_2\text{H} & \text{CO}_2\text{H} \\
| & | \\
\text{C}{=}\text{O} \quad + \text{NADH} + \text{H}^+ \rightleftharpoons \text{HOCH} \quad + \text{NAD}^+ \\
| & | \\
\text{CH}_2 & \text{CH}_2 \\
| & | \\
\text{CO}_2\text{H} & \text{CO}_2\text{H}
\end{array}
\qquad (14.5)
$$

Oxalacetate   Malate

The malate can then enter the matrix where it is oxidized by the mitochondrial malate dehydrogenase.

$$
\begin{array}{ll}
\text{CO}_2\text{H} & \text{CO}_2\text{H} \\
| & | \\
\text{HOCH} \quad + \text{NAD}^+ \rightleftharpoons \text{C}{=}\text{O} \quad + \text{NADH} + \text{H}^+ \\
| & | \\
\text{CH}_2 & \text{CH}_2 \\
| & | \\
\text{CO}_2\text{H} & \text{CO}_2\text{H}
\end{array}
\qquad (14.6)
$$

Malate   Oxalacetate

In order to keep the cycle functioning and avoid the net transfer of carbon into the mitochondria, the 4 carbon atoms in oxalacetate must be returned to the cytoplasm. Since oxalacetate cannot pass through the inner membrane, it is converted by transamination to aspartate that can move across. Glutamate provides the amino group for transamination and is exchanged for aspartate in order to prevent a net transfer of amino group from one compartment to the other.

$$
\begin{array}{llll}
\text{CO}_2\text{H} & \text{CO}_2\text{H} & \text{CO}_2\text{H} & \text{CO}_2\text{H} \\
| & | & | & | \\
\text{C}{=}\text{O} & \text{HCNH}_2 \rightleftharpoons \text{HCNH}_2 \;+ & \text{C}{=}\text{O} \\
| & | & | & | \\
\text{CH}_2 & \text{CH}_2 & \text{CH}_2 & \text{CH}_2 \\
| & | & | & | \\
\text{CO}_2\text{H} & \text{CH}_2 & \text{CO}_2\text{H} & \text{CH}_2 \\
& | & & | \\
& \text{CO}_2\text{H} & & \text{CO}_2\text{H}
\end{array}
\qquad (14.7)
$$

Oxalacetate   Glutamate   Aspartate α-Keto-glutarate

The malate–aspartate shuttle is diagramed in Figure 14.9a.

When a reducing equivalent is transported into the mitochondria by the malate–aspartate shuttle, the NADH will be introduced into Complex I of the electron-transport chain and 3 ATP will be produced. Thus, the complete oxidation of glucose to CO_2 and H_2O with the transfer of the reducing equivalent by the malate–aspartate shuttle can yield a total of 38 ATP in eucaryotic cells.

In the glycerol phosphate shuttle, NADH produced in glycolysis (or in any other cytoplasmic oxidation–reduction reaction) is first reoxidized by dihydroxy acetone phosphate in the presence of the cytoplasmic *sn*-glycerol-3-phosphate dehydrogenase.

$$\begin{array}{ccccc}
\text{CH}_2\text{OH} & & & \text{CH}_2\text{OH} & \\
| & & & | & \\
\text{C}{=}\text{O} & + & \text{NADH} + \text{H}^+ \rightleftharpoons & \text{HOCH} & + & \text{NAD}^+ \quad (14.8) \\
| & & & | & \\
\text{CH}_2\text{OPO}_3\text{H}_2 & & & \text{CH}_2\text{OPO}_3\text{H}_2 & \\
\text{Dihydroxy acetone} & & & \textit{sn}\text{-Glycerol-3-phosphate} & \\
\text{phosphate} & & & &
\end{array}$$

The *sn*-glycerol-3-phosphate formed is readily permeable to the mitochondrial membranes and passes through the inner membrane to the matrix where it is oxidized, this time by a dehydrogenase that utilizes FAD instead of NAD$^+$.

$$\begin{array}{ccccc}
\text{CH}_2\text{OH} & & & \text{CH}_2\text{OH} & \\
| & & & | & \\
\text{HOCH} & + & \text{FAD} \longrightarrow & \text{C}{=}\text{O} & + & \text{FADH}_2 \quad (14.9) \\
| & & & | & \\
\text{CH}_2\text{OPO}_3\text{H}_2 & & & \text{CH}_2\text{OPO}_3\text{H}_2 & \\
\textit{sn}\text{-Glycerol-3-phosphate} & & & \text{Dihydroxy acetone phosphate} &
\end{array}$$

The FADH$_2$ produced by this flavoprotein then contributes electrons to the electron-transport chain at CoQ (see Figure 14.2). Thus, as with succinate, two moles of ATP are formed when the α-glycerol phosphate is oxidized. To keep the shuttle operating, the dihydroxy acetone phosphate produced in Reaction 14.9 then passes back out of the mitochondria into the cytoplasm, where it can repeat the process (see Figure 14.9*b* in Section 14.12). In contrast to the malate–aspartate shuttle, this shuttle operates only in the manner described; namely, to transport reducing equivalents *into* the mitochondria because of the irreversible nature of the reaction catalysed by the mitochondrial *sn*-glycerol-3-phosphate dehydrogenase. The transfer of reducing equivalents from the two cytoplasmic NADHs produced when one mole of glucose is converted to pyruvate and the electrons transferred into animal mitochondria will therefore result in the formation of four high-energy phosphates or a total of 36 ATP for the complete oxidation of glucose to CO$_2$ and H$_2$O in a eucaryote.

As discussed previously, the $\Delta G'$ for the oxidation of glucose by O$_2$ to CO$_2$ and H$_2$O has been estimated from calorimetric data.

$$C_6H_{12}O_6 + 6\ O_2 \longrightarrow 6\ CO_2 + 6\ H_2O$$
$$\Delta G' = -686{,}000 \text{ cal (pH 7.0)} \qquad (14.10)$$

If there were no mechanism for trapping any of this energy, it would be released to the environment as heat, for the entropy term (see Chapter 9) is negligible. The cell can conserve a large portion of this energy, however, by coupling the energy released to the synthesis of the energy-rich ATP from ADP and H$_3$PO$_4$. If 38 moles of ATP are formed during the oxidation of glucose, this represents a total of 38×-7300 or $-277{,}000$ cal. The amount of energy that would be liberated as heat in Reaction 14.10 is thus reduced by this amount, and the overall oxidation and phosphorylation may now be written as

$$C_6H_{12}O_6 + 6\ O_2 + 38\ ADP + 38\ H_3PO_4 \longrightarrow 6\ CO_2 + 38\ ATP + 44\ H_2O$$
$$\Delta G' = -409{,}000 \text{ cal (pH 7.0)} \qquad (14.11)$$

The conservation of 277,000 cal as energy-rich phosphate represents an efficiency of conservation of 277,000 divided by $-686{,}000$ or 40%. The trapping of this amount of energy is a noteworthy achievement for the living cell.

14.9   OXYGENASES

Cellular oxidations can occur in three different ways; until now we have encountered only two of these: (1) removal of hydrogen as a hydride ion plus one proton (or two protons plus two electrons), and (2) removal of electrons (as in the cytochromes). The third important way is by the uptake of oxygen atoms. Oxygenases are a group of enzymes that catalyze the insertion of one or both of the atoms of the O_2 molecule into the substrate. The former are called **monooxygenases** (hydroxylases or mixed function oxidases) and the latter **dioxygenases.** The substrates that are oxidized tend to be the more reduced metabolites, such as steroids, fatty acids, carotenes, and amino acids, rather than the more oxidized carbohydrates and organic acids. The enzymes, in turn, contain iron or other metals; the iron may be present as inorganic iron, heme, or in a nonheme iron–sulfur protein. As typical examples, we may cite the dioxygenase that catalyzes the first step in the catabolism of tryptophan. In this reaction, the heterocyclic ring is opened.

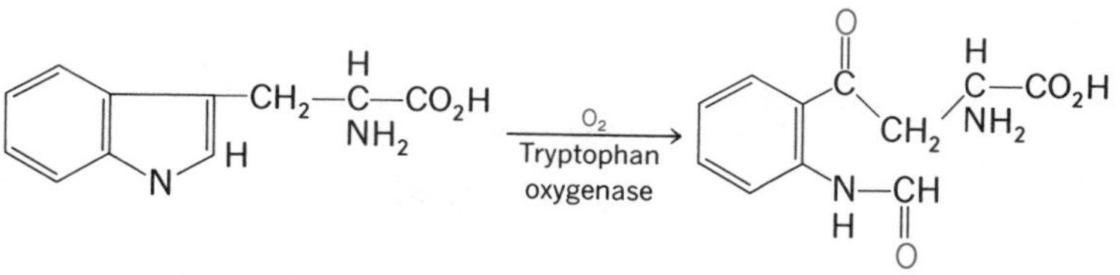

$$(14.12)$$

L-Tryptophan   *N*-Formyl kynurenine

Another important dioxygenase is cyclooxygenase that catalyzes the peroxidation of arachidonic acid to prostaglandin G_2 (PGG$_2$).

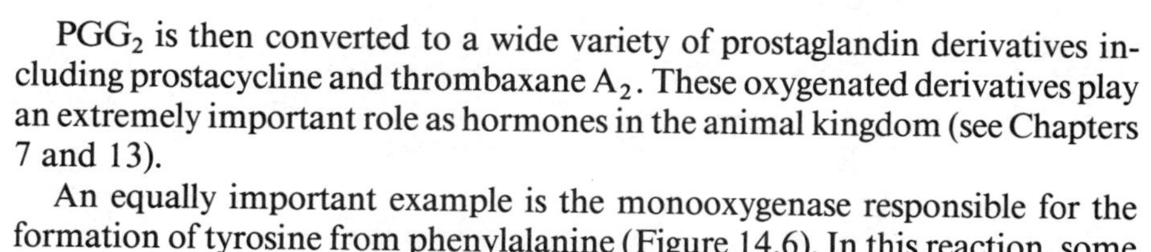

Arachidonic acid   PGG$_2$

PGG$_2$ is then converted to a wide variety of prostaglandin derivatives including prostacycline and thrombaxane A_2. These oxygenated derivatives play an extremely important role as hormones in the animal kingdom (see Chapters 7 and 13).

An equally important example is the monooxygenase responsible for the formation of tyrosine from phenylalanine (Figure 14.6). In this reaction, some reducing agent must contribute two electrons for the reduction of the oxygen atom that does not enter into the substrate. The reducing agent is tetrahydrobiopterin that, when oxidized to the dihydroform, is in turn reduced by NADPH.

As may be seen from the examples presented, dioxygenases incorporate both oxygen atoms (O) of O_2 into the oxidized compound. On the other hand, monooxygenases (hydroxylases) incorporate only one of the oxygen atoms into the oxidized compound while the other oxygen atom is reduced to water.

BOX 14.B BIOSYNTHESIS OF THE CATECHOL AMINES

The biosynthesis of the catechol amines—norepinephrine (noradrenaline), epinephrine (adrenaline), and dopamine—illustrates the role of amino acids as metabolic precursors of essential compounds in living matter. The catechol amines, which serve as neurotransmitters in the nervous system of animals, are produced in the adrenal medulla from the amino acid tyrosine.

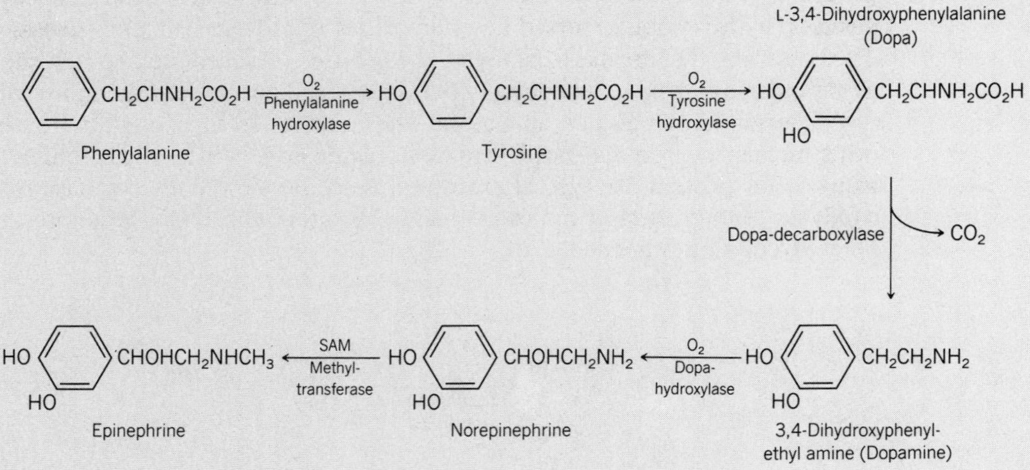

Since several monooxygenases, including the one that produces tyrosine from phenylalanine, are involved in this biosynthesis; the sequence illustrates the importance of this class of enzymes in nature. It is interesting to note that L-Dopa (L-3,4-dihydroxyphenylalanine), which is used in the treatment of Parkinson's disease, is especially abundant in the seeds of certain leguminous plants.

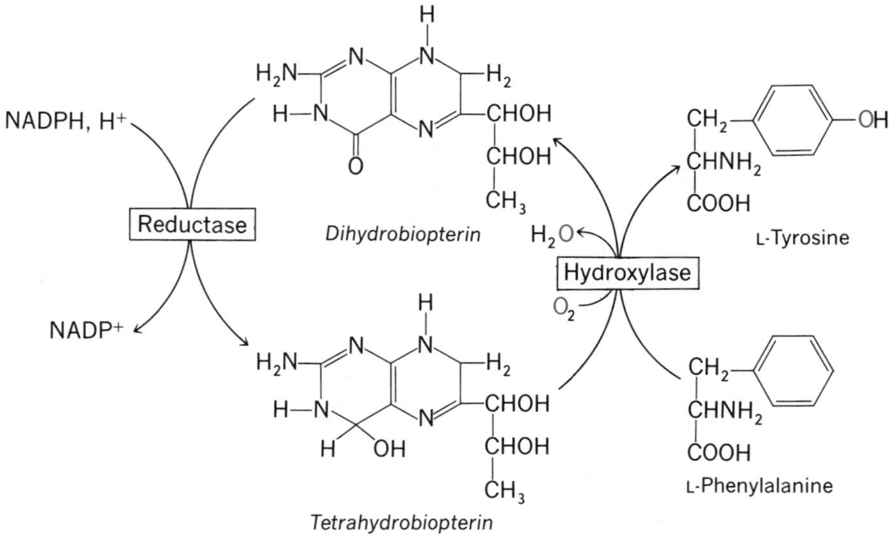

FIGURE 14.6 Reaction sequence for the hydroxylation of L-phenylalanine to L-tyrosine.

The oxidases are of special interest in that they activate the O_2 molecule and can reduce one or both atoms of oxygen to H_2O. Cytochrome oxidase, the terminal oxidase of all aerobic organisms, reduces one mole of O_2 to $2 H_2O$, a process requiring a total of four electrons. Other oxidases, especially those containing flavin cofactors, reduce O_2 to hydrogen peroxide, H_2O_2, a two-electron process. The nature of the intermediates and the mechanism involved in reducing O_2 remain unsettled but the step-wise reduction may be written.

$$O_2 \xrightarrow[\text{H}^+]{1\,e^-} \underset{\text{Superoxide}}{HO_2} \xrightarrow[\text{H}^+]{1\,e^-} \underset{\substack{\text{Hydrogen} \\ \text{peroxide}}}{H_2O_2} \xrightarrow[\text{H}^+\ \text{H}_2\text{O}]{1\,e^-} \underset{\text{radical}}{HO\cdot} \xrightarrow[\text{H}^+]{1\,e^-} H_2O \quad (14.13)$$

All three postulated intermediates are highly reactive chemically and it is interesting to note that living forms contain enzymes that tend to protect the cell against these intermediates. The enzyme, superoxide dismutase, which catalyzes the disproportionation of superoxide anion, is widespread in all aerobic organisms.

$$2H^+ + O_2^- + O_2^- \longrightarrow H_2O_2 + O_2 \quad (14.14)$$

It is identical with proteins previously isolated whose function was not known but which had been called hemocuprein and erythrocuprein. Catalase is present in almost all animal cells. It catalyzes the decomposition of two moles of H_2O_2 into $2 H_2O + O_2$ and thus protects the cells against noxious H_2O_2. It has been suggested that, as oxygen became available on the primitive earth and was used as an oxidant, the very reactive intermediates listed in Equation 14.13 proved toxic to developing life forms. These life forms therefore had to evolve some means of destroying these intermediates and acquired the capacity to form superoxide dismutase, catalases and peroxidases that utilize H_2O_2 as a substrate. Peroxidases are relatively rare in animal cells (exceptions are leucocytes, erythrocytes, liver, and kidney) but are common in all higher plants. Peroxidases oxidize dihydroxyphenols in the presence of H_2O_2, forming the quinone and H_2O.

14.10 INTEGRATION OF CARBOHYDRATE, LIPID, AND AMINO ACID METABOLISM

At this point, it will be useful to integrate some of the information on energy production from carbohydrates and lipids that has been discussed. In addition, we will anticipate some of the general features of amino acid metabolism, although this subject is not treated until Chapter 17.

Krebs and H. L. Kornberg have pointed out that many different compounds that may be classified roughly as carbohydrates, lipids, or proteins can serve as sources of energy for living organisms. These authors have also emphasized that

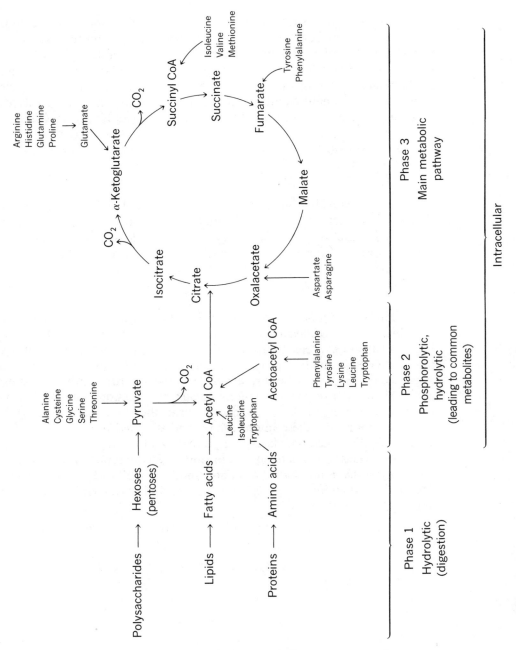

FIGURE 14.7   Main phases in the catabolism of foodstuff.

the number of reactions involved in obtaining energy from these compounds is astonishingly small, whether the organism involved is animal, higher plant, or microorganism. Thus, nature has practiced great economy in the processes developed for handling these compounds. These authors divide substrate degradation into three phases, as indicated in Figure 14.7.

In phase 1, polysaccharides, which serve as an energy source for many organisms, are hydrolyzed to monosaccharides, usually hexoses. Similarly, proteins can be hydrolyzed to their component amino acids, and triacylglycerols, which make up the major fraction of the lipid food sources, are hydrolyzed to glycerol and fatty acids. These processes are hydrolytic, and the energy released as the reactions occur is made available to the organism as heat.

In phase 2, the monosaccharides, glycerol, and fatty acids are further degraded to acetyl CoA by processes that may result in the formation of some energy-rich phosphate compounds. That is, in glycolysis, the hexoses are converted to pyruvate and then to acetyl CoA by reactions involving the formation of a limited number of high-energy phosphate bonds, as described in Chapter 10. Similarly, in phase 2, the long-chain fatty acids are oxidized to acetyl CoA (Chapter 13), while glycerol, obtained from hydrolysis of triacylglycerols, is converted to pyruvate and acetyl CoA by means of the glycolytic sequence.

For the amino acids, the situation is somewhat different. In phase 2, some amino acids (alanine, serine, and cysteine) are converted to pyruvate on degradation, and thus, acetyl CoA formation is predicted if these amino acids are utilized by an organism for energy production. Other amino acids (the prolines, histidine, and arginine) are converted to glutamic acid on degradation; this amino acid, in turn, undergoes transamination to yield α-ketoglutarate, a member of the tricarboxylic acid cycle. Aspartic acid is readily transaminated to form oxalacetate, another intermediate of the cycle. The branch-chain amino acids and lysine yield acetyl CoA or succinyl CoA on degradation, whereas phenylalanine and tyrosine, on oxidative degradation, produce both acetyl CoA and fumaric acid.

Thus, the carbon skeletons of the amino acids yield either an intermediate of the tricarboxylic acid cycle or acetyl CoA, the same product obtained from carbohydrate or lipid. During the oxidation of this compound in phase 3 by means of the cycle, energy-rich ATP is produced by oxidative phosphorylation. Specifically, 12 energy-rich bonds are produced for each mole of acetyl CoA oxidized. Hence, hundreds of organic compounds that biological organisms may consume as food are utilized by their conversion to acetyl CoA or an intermediate of the tricarboxylic acid cycle and their subsequent oxidation by the cycle.

In considering the actual steps involved in making energy available to the organisms, the reactions of oxidative phosphorylation that occur during electron transport through the cytochrome system are quantitatively the most significant. Even here, an economy in the number of reactions is involved. As discussed in Chapter 12, the oxidation of substrates in the tricarboxylic acid cycle is accompanied by the reduction of either a nicotinamide or a flavin nucleotide. It is the oxidation of the reduced nucleotide by molecular oxygen in the presence of mitochondria that results in the formation of the energy-rich ATP. As pointed out, three phosphorylations, all presumably of the same mechanism, occur during the transfer of a pair of electrons from NADH to O_2. We have discussed only three other reactions leading to the production of energy-rich compounds where none previously existed. These are (1) the formation of acylphosphate in the oxidation of triose phosphate (Reaction 10.5), (2) the formation of phosphoenol pyruvate (Reaction 10.8) and (3) the formation of thioesters (Reactions 12.1 and 12.5). It is indeed a beautiful design that permits the energy in the myriad foodstuffs to be trapped in only four different processes. Even here, a single compound, ATP, is the energy-rich substance ultimately formed.

14.11   INTERCONVERSION OF CARBOHYDRATE, LIPID, AND PROTEIN

The interconversions among the three major foodstuffs may be summarized with the help of Figure 14.8 as follows. In this figure, two reactions that are effectively irreversible are indicated by heavy unidirectional arrows. These irreversible processes in turn determine the interconversions of the foodstuffs as summarized here.

Carbohydrates are convertible to fats through the formation of acetyl CoA. Carbohydrates may also be converted to certain amino acids (alanine, aspartic, and glutamic acids), provided a supply of dicarboxylic acid is available for the formation of the keto acid analogs of those amino acids. Specifically, a supply of both oxalacetate (or other C_4-dicarboxylic acid) and acetyl CoA are required in an amount stoichiometrically equivalent to the amino acid being synthesized.

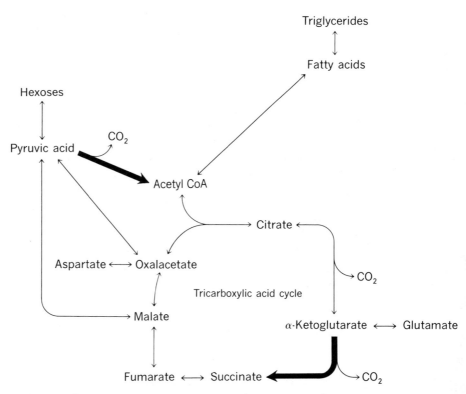

FIGURE 14.8   The possible interconversions between carbohydrate, lipids, and certain amino acids.

Several mechanisms exist for forming the C_4-dicarboxylic acids; the principal one is the formation of oxalacetic acid from pyruvic acid, a process catalyzed by pyruvic carboxylase in animals and PEP-carboxylase in plants. Another is the formation of malic acid from pyruvic acid, the reaction catalyzed by malic enzyme. These reactions have been described in detail in Chapters 10 and 12.

Fatty acids may be similarly converted to certain amino acids, provided a source of dicarboxylic acid is available. However, fatty acids cannot be converted to carbohydrate by the reactions shown in Figure 14.8. This inability is due to the fact that the equivalent of the two carbon atoms acquired in acetyl CoA will be lost as CO_2 prior to the production of the dicarboxylic acids. Note, however, that the glyoxylate cycle (discussed in Chapter 12) can enable an organism to form carbohydrate from fat, as it does, for instance, in some plants, bacteria, and fungi.

The naturally occurring amino acids are readily converted to carbohydrates and lipids. Each of the 20 protein amino acids may be classified as *glucogenic, ketogenic,* or both *glucogenic* and *ketogenic,* depending on the specific metabolism of the amino acid. As an example, aspartic acid is glucogenic through the formation of oxalacetic acid and its subsequent conversion to phosphoenol pyruvic acid. Similarly, glutamic acid is glucogenic by virtue of its eventual conversion to oxalacetic acid in the tricarboxylic acid cycle and the conversion of oxalacetate to phosphoenol pyruvic acid. The carbon skeleton of leucine is degraded to acetoacetyl CoA and acetyl CoA. Thus, it is a ketogenic amino acid. Examples of amino acids that are both gluco- and ketogenic are tyrosine, phenylalanine, isoleucine, and lysine.

14.12 THE PERMEABILITY OF MITOCHONDRIA

As pointed out in Chapter 8, the permeability of the inner and outer mitochondrial membranes is quite different. While the outer membrane is fully permeable to molecules up to 10,000 in molecular weight, the inner membrane exhibits selective permeability. Due to the inability of NAD^+ and NADH to pass through the inner membrane, shuttle mechanisms that accomplish the transfer of reducing equivalents across the membrane are available (Section 14.8). These are shown in Figure 14.9. Because of the facile reversibility of all of the enzymes involved in the malate–aspartate shuttle (Figure 14.9a) reducing equivalents can be transported in either direction across the membrane. The α-glycerol phosphate shuttle, (Figure 14.9b) by contrast, operates only to transfer reducing equivalents into the membrane due to the irreversible nature of Reaction 14.9.

Two other shuttle mechanisms are shown in Figure 14.9. One of these (Figure 14.9c) is the process whereby pyruvate (or lactate) is converted back to glucose during gluconeogenesis (Section 10.7.2). This shuttle is necessary due to the fact that pyruvic acid can freely cross the inner membrane for oxidation while oxalacetate is not permeable. As pointed out earlier, oxalacetate, produced by pyruvic carboxylase (Reaction 10.22), is first reduced to malate that passes out of the mitochondria. In the cytosol, the malate is then oxidized back to oxalacetate and converted to phosphoenol pyruvate by PEP-carboxykinase (Reaction 10.23). Note that in this shuttle, which accomplishes the first steps of gluconeogenesis, energy is required both in the form of ATP and NADH.

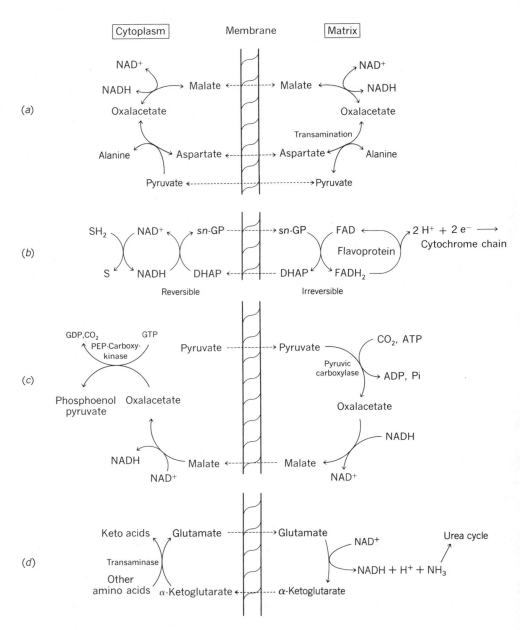

FIGURE 14.9 Four shuttle systems operating across the mitochondrial inner membrane. (a) The reversible malate–oxalacetate–aspartate shuttle for reducing equivalents. (b) The unidirectional sn-glycerol phosphate (sn-GP) and dihydroxyacetophosphate (DHAP) shuttle. (c) Shuttle for converting pyruvate (or lactate) to phosphoenol pyruvate in gluconeogenesis. (d) The glutamate shuttle for the transport of amino nitrogen.

The final shuttle represented in Figure 14.9d is encountered during the oxidative deamination of amino acids (Section 17.6.1). In the cytosol, amino acids transfer their amino group to α-ketoglutarate to form glutamate. When this amino acid enters the mitochondria, it can be oxidatively deaminated to release the NH_3 for urea synthesis (Section 17.8). The α-ketoglutarate can then return to the cytosol to continue the process.

REFERENCES

1. *H. A. Krebs and H. L. Kornberg,* Energy Transformation in Living Matter, A Survey. *Berlin: Springer, 1957.*

2. *E. Racker,* A New Look at Mechanisms in Bioenergetics. *New York: Academic Press, 1976.*

3. *P. D. Boyer, ed., "Oxidation–Reduction." The Enzymes, 3rd ed., Vols. XI to XIII, New York: Academic Press, 1975, 1976.*

4. *S. Fleisher and L. Packer, eds. "Biomembranes." Methods in Enzymology, Parts C, D, and F, Vols. 52, 53, and 55. New York: Academic Press, 1978, 1979.*

5. *P. D. Boyer, B. Chance, L. Ernster, P. Mitchell, E. Racker, and E. C. Slater, "Oxidative Phosphorylation and Photophosphorylation." Annual Review of Biochemistry, Vol. 46, (1977) pp. 955–1026.*

REVIEW PROBLEMS

1. The free energies of the following reactions are known to be approximately:

$$C_6H_{12}O_6 + 6\ O_2 \longrightarrow 6\ H_2O + 6\ CO_2 \qquad \Delta G' = -686{,}000 \text{ cal/mole glucose}$$
$$ATP + H_2O \longrightarrow ADP + H_3PO_4 \qquad \Delta G' = -8000 \text{ cal/mole ATP}$$

   Assume that 15 high-energy phosphate bonds are formed per mole of pyruvic acid ($CH_3COCOOH$) oxidized to CO_2 and H_2O via the Krebs cycle and that this occurs with an efficiency of 40%. Calculate the $\Delta G'$ for the two following reactions:

   (a) $C_6H_{12}O_6 + O_2 \longrightarrow 2\ CH_3COCOOH + 2\ H_2O$ $\qquad \Delta G' = ?$

   (b) $C_6H_{12}O_6 + 6\ O_2 + 38\ ADP + 38\ H_3PO_4 \longrightarrow$
   $$6\ CO_2 + 44\ H_2O + 38\ ATP \qquad \Delta G' = ?$$

2. The hydroxy fatty acid shown below is oxidized completely to CO_2 and H_2O by the β-oxidation sequence and the Krebs cycle. Calculate the *net* number of moles of ATP that will be produced.

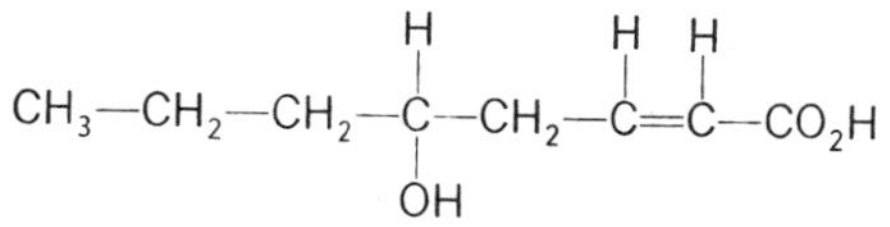

$$CH_3-CH_2-CH_2-\overset{\overset{\displaystyle H}{|}}{C}-CH_2-\overset{\displaystyle H}{C}=\overset{\displaystyle H}{C}-CO_2H$$
$$\underset{\displaystyle OH}{|}$$

3. The production of energy as ATP is under careful control in an animal that utilizes glucose as its chief source of energy and burns that compound to CO_2 and H_2O via glycolysis and the Krebs cycle.

   (a) Identify four enzymes (or enzyme reactions) that are affected when the ratio of ATP:AMP increases and *state how they are affected.*

   (b) Which enzyme (or enzyme reaction) is particularly affected when the ratio of $NADH:NAD^+$ is high?

   (c) What is the physiological significance of the activation of pyruvic carboxylase by acetyl CoA?

4. Calculate the *net* number of moles of energy-rich phosphate that should be produced when tricaproin is oxidized completely to CO_2 and H_2O.

$$CH_2OCCH_2CH_2CH_2CH_2CH_3$$
$$\underset{O}{\|}$$

$$CHOCCH_2CH_2CH_2CH_2CH_3$$
$$\underset{O}{\|}$$

$$CH_2OCCH_2CH_2CH_2CH_2CH_3$$
$$\underset{O}{\|}$$

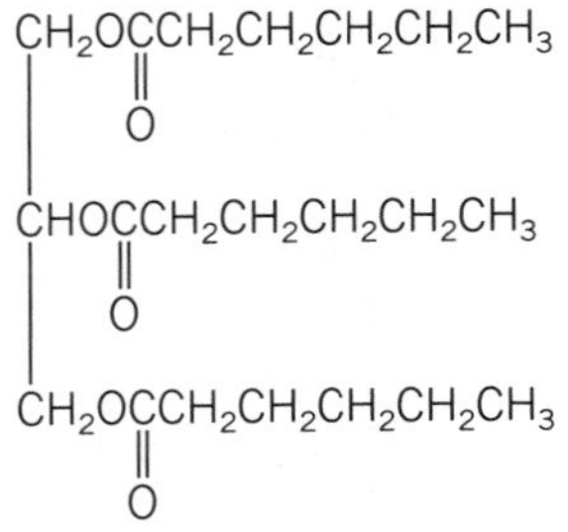

5. The following compounds are oxidized *completely* to CO_2 and H_2O by familiar reactions. Calculate the number of energy-rich phosphate bonds that should be produced when each compound is oxidized, accounting for the consumption of energy-rich phosphate if any such reactions occur.

(a) Lactic acid $CH_3CHOHCO_2H$

(b) Aspartic acid $HO_2CCH_2CHNH_2CO_2H$

CHAPTER 15

PHOTOSYNTHESIS

All life on the planet earth is dependent on **photosynthesis,** the process by which CO_2 is converted into the organic compounds found not only in photosynthetic organisms but in all living cells. The conversion of CO_2 and H_2O to glucose, for example, may be presented as the reverse of the process in which glucose is oxidized to CO_2 and water (Section 12.1). Therefore, the conversion will require the input, as a minimum, of the same amount of energy that is released when glucose is oxidized to CO_2 and H_2O.

$$6 CO_2 + 6 H_2O \longrightarrow C_6H_{12}O_6 + 6 O_2$$
$$\Delta G' = +686,000 \text{ cal} \tag{15.1}$$

As the term photosynthesis implies, the energy for this process is provided by light.

The evolution of photosynthesis followed after the evolution of glycolysis and pentose phosphate metabolism. It could not occur, however, until pigments such as the chlorophylls could be formed. These pigments possessed the ability to absorb solar radiation, transferring some of that energy into chemical forms (ATP). Thus, the process of photophosphorylation evolved.

When significant quantities of CO_2 had accumulated, it became the substrate for photosynthesis as reactions of photophosphorylation and pentose phosphate metabolism combined to yield a light-dependent reduction of CO_2. The electrons for reduction of the CO_2 were furnished by H_2S and H_2, components present in the primordial atmosphere; certain photosynthetic bacteria remain today as evidence of those early forms of photosynthesis. As photosynthetic organisms continued to evolve, they acquired the capacity to use H_2O as a source of electrons. When this occurred, O_2 was produced and evolving forms were then provided with a new oxidant for a form of respiration, aerobic respiration, previously unknown.

15.1   THE MAGNITUDE OF PHOTOSYNTHESIS

It is important to consider the magnitude of photosynthesis for it exceeds any other "manufacturing process" on earth. It is estimated that the amount of carbon assimilated in a year throughout the world is 200 billion tons (200×10^9 tons) of C. If we assume that 114,000 cal are required per mole of CO_2 fixed (see Reaction 15.1), the energy required for the production of this amount of biomass and stored for use of nonphotosynthetic organisms is 2×10^{21} cal. This amount of energy corresponds to only about 0.1% of the total radiant energy that the sun transfers to the earth in a year. Of course, not all of the light actually falling on plants is used in photosynthesis; estimates suggest that only about one-tenth is absorbed by the chloroplasts and thereby available for biosynthesis. Thus, as much as 1% of the sun's energy may be consumed in producing the biomass cited. Since much of this biomass is in the form of two polymers, cellulose and lignin, which cannot be directly used by humans as food, it is clear why man and microorganisms have evolved other mechanisms for recycling this biomass, ranging from burning it for heat energy to highly evolved microorganisms (the bacteria in the gut of termites) that can utilize cellulose.

15.2   EARLY STUDIES ON PHOTOSYNTHESIS

15.2.1   LIGHT AND DARK REACTIONS

In studies initiated in 1905, Blackman showed that photosynthesis consists of two processes, a **light-dependent** phase that is limited in its rate by light-independent or **dark reactions.** The light-dependent processes exhibit the expected independence of temperature characteristic of photochemical reactions, whereas the dark reactions were sensitive to different temperatures. Today, the light-dependent processes are recognized as those in which light energy is converted into chemical energy in the form of ATP and NADPH. The dark reactions, on the other hand, are those enzymatic reactions that incorporate CO_2 into reduced carbon compounds previously encountered in carbohydrate metabolism.

Evidence was first provided by Emerson in the 1930s to show that the light-dependent phase of photosynthesis consists of two light reactions. When Emerson measured the amount of photosynthesis carried out by the green alga, *Scenedesmus,* as a function of the wavelength of light, he observed that photosynthesis did not proceed at wavelengths greater than 700 nm. This was surprising, since light of this far-red wavelength was still being absorbed by the algal cells. Emerson subsequently showed that this decrease in the far-red region — the so-called "far-red drop" — could be reversed in varying amounts if he supplemented the light at 700 nm with a second source of light having a wavelength of 650 nm. This *enhancement* of the amount of photosynthesis caused Emerson to postulate that, in the case of *Scenedesmus,* the assimilation of CO_2 during photosynthesis required light of two different wavelengths. Today, it is known that these algae and most other photosynthetic organisms have two photosystems (PS I and PS II) that are activated by light of far-red wavelength (680–700 nm) and shorter wavelength (650 nm), respectively.

15.2.2 BACTERIAL PHOTOSYNTHESIS

Studies on photosynthetic bacteria provided much useful information and were the basis for a major hypothesis that stimulated research in photosynthesis for many years. The two classes of purple bacteria, sulfur and nonsulfur, have been extensively used. To compare the process of photosynthesis in these organisms, consider writing the overall reaction of photosynthesis as carried out in green plants on the basis of one mole of CO_2. This may be done by dividing Reaction 15.1 by six to give

$$CO_2 + H_2O \xrightarrow{hv} C(H_2O) + O_2 \tag{15.2}$$

Oxidized (top bracket over CO_2 and O_2)
Reduced (bottom bracket under CO_2 and $C(H_2O)$)

$$\Delta G' = +118{,}000 \text{ cal}$$

We observe that this is an oxidation–reduction reaction in which the oxidizing agent CO_2 is reduced to the level of carbohydrate represented by $C(H_2O)$. The reducing agent in this reaction is H_2O that, in turn, is oxidized to O_2. Since the reaction is highly endergonic, it will only proceed when the necessary energy is supplied by light (hv). While this reaction is balanced with regard to the carbon, hydrogen, and oxygen atoms involved, we shall see that it is not balanced with regard to the number of electrons used in reducing CO_2.

The purple sulfur bacteria, for example, *Chromatium*, utilize H_2S instead of H_2O as a reducing agent in photosynthesis. Elemental sulfur, S, is produced, but no oxygen is formed.

$$CO_2 + 2 H_2S \xrightarrow{hv} C(H_2O) + 2 S + H_2O \tag{15.3}$$

Note that two moles of H_2S are required to balance the equation, the S^{2-} ions in the H_2S furnishing the total of four electrons required to reduce CO_2 to $C(H_2O)$. Thiosulfate can also serve as the reductant for photosynthesis by purple sulfur bacteria.

$$2 CO_2 + Na_2S_2O_3 + 3 H_2O \xrightarrow{hv} 2 C(H_2O) + 2 NaHSO_4$$

In this reaction, the sulfur atoms are oxidized from S^{2+} in $Na_2S_2O_3$ to S^{6+} in $NaHSO_4$. Thus, each mole of $Na_2S_2O_3$ contributes 2×4 or 8 electrons, sufficient for the reduction of two moles of CO_2 to $C(H_2O)$. This reaction also demonstrates that the reducing agent need not contain hydrogen itself but simply be capable of furnishing electrons.

The nonsulfur purple bacteria (e.g., *Rhodospirillum rubrum*) utilize organic reductants such as ethanol, isopropanol, or succinate as electron donors. The balanced equation with ethanol, for example, may be written as

$$CO_2 + 2 CH_3CH_2OH \xrightarrow{hv} C(H_2O) + 2 CH_3CHO + H_2O \tag{15.4}$$

with the four electrons required for reduction of CO_2 being furnished by the oxidation of two moles of ethanol to acetaldehyde.

C. B. van Niel, a pioneer in the study of photosynthesis, pointed out the similarity of these reactions to the one that occurs in green plants and he suggested that a general reaction for photosynthesis may be represented as

$$CO_2 + 2\ H_2A \xrightarrow{h\nu} C(H_2O) + 2\ A + H_2O \qquad (15.5)$$

where H_2A is a general expression for a reducing agent that, as we have seen, may be a variety of compounds.

Since H_2S is a much stronger reducing agent than $Na_2S_2O_3$ or H_2O, it was expected that less light energy would be required for photosynthesis with H_2S as the reducing agent than with $Na_2S_2O_3$ or H_2O. Experimentally, however, it was observed that the same amount of light energy was required, regardless of the nature of the external reducing agent. This caused van Niel to postulate that the primary reaction is the same in all organisms and that it consists of the *splitting* of a molecule of H_2O to yield both a reducing agent [H] and an oxidizing agent [OH].

$$H_2O \xrightarrow{h\nu} [H] + [OH] \qquad (15.6)$$

This productive hypothesis stimulated much experimental work that led to greater understanding of the process of photosynthesis. The realization that four electrons are required to reduce CO_2 to $C(H_2O)$ meant that Equation 15.2 had to be rewritten to involve two moles of H_2O as reductant, each atom of oxygen in H_2O providing two electrons.

$$CO_2 + 2\ H_2{}^{18}O \xrightarrow{h\nu} C(H_2O) + H_2O + {}^{18}O_2 \qquad (15.7)$$

Further, this revised equation would indicate that the two oxygen atoms produced in green plant photosynthesis should come only from H_2O. This was confirmed experimentally by Ruben and Kamen in a classical experiment in which H_2O labeled with the isotope ^{18}O was utilized in photosynthesis by algae. The oxygen produced under these conditions contained the same concentration of ^{18}O as the $H_2{}^{18}O$. Later developments concerning the role of H_2O in photosynthesis made it necessary to abandon van Niel's hypothesis of the photolytic cleavage of H_2O. However, his proposal that the initial photosynthetic act involves the production of an oxidant and a reductant is retained in the current descriptions of the energy-conversion process.

15.2.3   THE HILL REACTION

In 1937, R. Hill of Cambridge University initiated cell free studies on photosynthesis by working with isolated chloroplasts rather than intact plants. He reasoned that more information might be obtained if grana or chloroplasts, which contain the pigments that absorb the solar energy, were studied separately from the cell. It would have been ideal if the chloroplasts could have carried out both the oxidation of H_2O and the reduction of CO_2 to organic carbon compounds. This was not accomplished at that time. Nevertheless, chloroplasts were able to produce O_2 photochemically in the presence of another oxidizing agent, potassium ferric oxalate. In this reaction, the ferric ion substitutes for CO_2 as an oxidant during the photooxidation of H_2O.

$$4\ Fe^{3+} + 2\ H_2O \xrightarrow[\text{Chloroplasts}]{h\nu} 4\ Fe^{2+} + 4\ H^+ + O_2$$

Molecular oxygen is evolved in an amount stoichiometrically equivalent to the oxidizing agent added. This observation was of fundamental importance, for it permitted the study of the role of H_2O as a reducing agent in photosynthesis. The reaction is known as the **Hill reaction,** and potassium ferric oxalate is known as a **Hill reagent.** Other compounds such as benzoquinone were subsequently shown to serve as Hill reagents in studies on isolated chloroplasts.

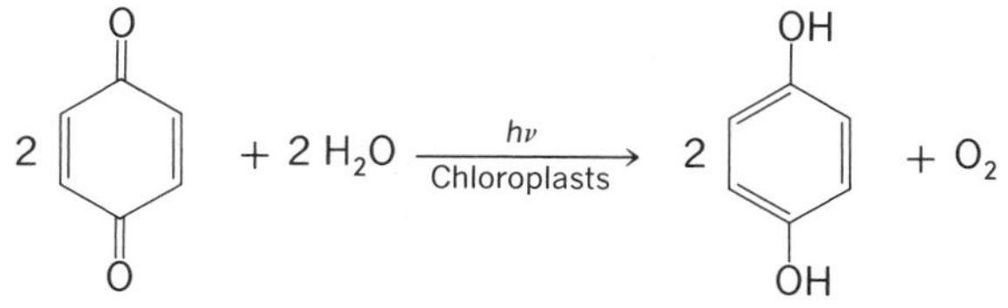

Benzoquinone   Hydroquinone

Oxidized dyes were later shown to function as Hill reagents by being reduced. Although this approach was criticized because the substances that could serve as Hill reagents were not physiologically important compounds, the properties of these reactions were studied extensively.

In 1952, three U.S. laboratories reported that $NADP^+$ (and NAD^+ under certain conditions) could serve as Hill reagents in the presence of spinach grana and light. Thus, for the first time, a physiologically significant compound could function as a Hill reagent. This observation was of prime importance; it constituted a mechanism whereby a reduced nicotinamide nucleotide was produced as the result of a light-dependent reaction.

$$2\ NADP^+ + 2\ H_2O \xrightarrow[\text{Chloroplasts}]{h\nu} 2\ NADPH + 2\ H^+ + O_2 \qquad (15.8)$$

Numerous examples have been given earlier in this text of the ability of NADPH and NADH to reduce various substrates in the presence of the proper enzyme. Thus, reducing power produced in Reaction 15.8 can be coupled to the biosynthesis of these reduced substrates.

15.2.4 PHOTOPHOSPHORYLATION

Work on the path of carbon (Section 15.10) during this time had shown that both NADPH and ATP were required to convert CO_2 to carbohydrates in photosynthesis. With the NADPH obtained via a Hill reaction, it was thought that reoxidation of the reduced nicotinamide nucleotide by oxygen through the cytochrome electron-transport system of plant mitochondria would produce ATP. In the intact plant cell containing chloroplasts and mitochondria, both of these organelles would be required to produce the two coenzymes NADPH and ATP needed to drive the photosynthetic carbon reduction cycle (Figure 15.4). In 1954, Arnon and his associates questioned whether ATP is so produced when they discovered that chloroplasts alone, when isolated by special techniques, could convert CO_2 to carbohydrates in the light. Further studies in Arnon's laboratory showed that chloroplasts, in the *absence* of mitochondria, could synthesize ATP in two types of light-dependent phosphorylation reactions. The first type, **cyclic photophosphorylation,** yields ATP only and produces no net change in any external electron donor or acceptor.

$$ADP + H_3PO_4 \xrightarrow{h\nu} ATP + H_2O \tag{15.9}$$

The second type, **noncyclic photophosphorylation,** involves a process in which ATP formation is coupled with a light-driven transfer of electrons from water to a terminal electron acceptor such as $NADP^+$ with the resultant evolution of oxygen.

$$2\,NADP^+ + 2\,H_2O + 2\,ADP + 2\,H_3PO_4 \xrightarrow{h\nu}$$
$$2\,NADPH + 2\,H^+ + O_2 + 2\,ATP + 2\,H_2O \tag{15.10}$$

This reaction deserves further comment for two reasons: First, note that the movement of electrons would appear to be the opposite of that encountered in the electron-transport system of mitochondria. In the latter, electrons flow from NADH ($E_0' = -0.32$) to O_2 ($E_0' = 0.82$) along a potential gradient that releases energy, some of which is trapped in the form of ATP (Chapter 14). According to Reaction 15.10, electrons arising in the oxygen atom of H_2O somehow *appear* to make their way to $NADP^+$ and reduce it to NADPH. This movement of electrons against the potential gradient clearly requires energy; this is the function of light in photosynthesis. Second, Reaction 15.10 is even more remarkable in that, as electrons are apparently made to flow from H_2O to $NADP^+$, energy is *also* made available as ATP. These observations will be explained after the photosynthetic apparatus and the photochemistry of photosynthesis are described.

15.3   THE PHOTOSYNTHETIC APPARATUS

Photosynthesis is carried out by both procaryotic and eucaryotic cells. The procaryotes include the cyanobacteria (blue-green algae) and the purple and green bacteria; in these organisms, the light-trapping process takes place in small structures called **chromatophores.** In eucaryotic organisms that photosynthesize (higher green plants; the multicellular red, green, and brown algae; dinoflagellates; and diatoms), the **chloroplast** is the site of the photosynthetic process.

The chloroplasts, whose structure and composition were described in Section 8.6, contain the photosynthetic pigments; these are chlorophylls *a* and *b* in the higher green plants together with certain carotenoids, one of which is β-carotene (Section 5.13).

Chlorophyll is a magnesium-containing porphyrin that has an aliphatic alcohol **phytol** esterified to a propionic acid residue on ring IV of a tetrapyrrole. The structures for both chlorophyll *a* and *b* are given here. The

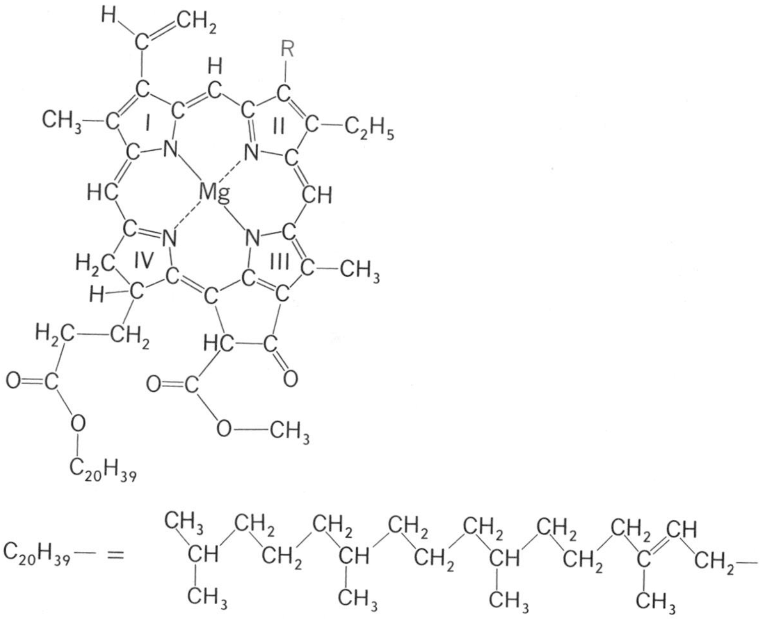

Chlorophyll a, R = CH₃
Chlorophyll b, R = CHO

Chlorophyll *a*, R = CH_3
Chlorophyll *b*, R = CHO

red and blue-green algae contain, in addition to chlorophyll *a*, blue or red pigments known as phycobilins, tetrapyrroles related to the chlorophylls but lacking Mg^{2+} and the *cyclic* structure of those compounds.

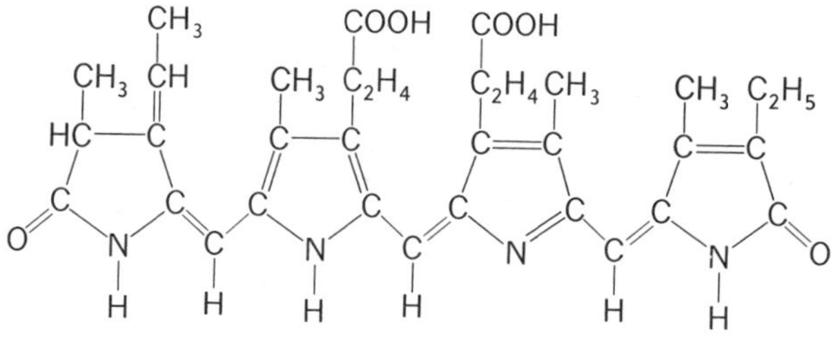

Phycobilin

While photosynthetic organisms contain a variety of photosynthetic pigments, a distinction can be made between those that play a *primary* role in the photosynthetic act and those that perform a *secondary* function. Only chlorophyll *a* (or the bacterial chlorophyll of the *a* type that is found in the photosynthetic bacteria) undergoes excitation and the charge separation process in which electron transport is initiated and energy-rich chemical compounds are generated. The other pigments do not participate directly in this energy-conversion process but instead collect light (of shorter wavelength and higher energy content) and pass it along to chlorophyll *a* by a process called *resonance transfer*.

The photosynthetic pigments, necessary enzymes, and structural components in green plants are organized into two photosynthetic reaction centers called photosystem I and II. These centers accomplish the overall photosynthetic process composed of the following events: light absorption, charge separation, electron transport, and phosphorylation of ADP. Photosystem I (PS I), for example, contains approximately 200 molecules of antennae chlorophyll *a* molecules together with one molecule of chlorophyll (P_{700}) that undergoes the charge separation process. PS I also contains cytochrome f (cyt c_{552}) and cytochrome b_6 (cyt b_{563}), ferredoxin (bound and soluble), and ferredoxin-NADP reductase. The roles of these different components will become apparent when the details of the overall photosynthetic process are described.

15.4   PROPERTIES OF LIGHT

The study of radiant energy has disclosed that light may be treated as a wave of particles known as **photons**. The energy of these photons may be calculated from the equation

$$E = Nh\nu = \frac{Nhc}{\lambda}$$

where E is the energy (in calories) of one mole or Einstein of photons; N is Avogadro's number (6.023×10^{23}); h is Planck's constant (1.58×10^{-34} cal-sec); c is the velocity of light (3×10^{10} cm/sec); and λ is the wavelength (in nanometers). As the equation indicates, the energy of photons is inversely proportional to the wavelength of the light under consideration. Thus, the energy of blue light of short wavelength is greater than that of a corresponding amount of red light of longer wavelength. Table 15.1 lists the energy contents of Einsteins (6.023×10^{23} of photons) of different types of light.

One of the important properties of matter is its ability to absorb light. Briefly, the ability of a substance to absorb light is dependent on its atomic structure. In a stable atom, the number of electrons surrounding the nucleus is equal to the positive charges (the atomic number) in the nucleus. These electrons are arranged in different orbitals around the nucleus and those in the outer orbitals are less strongly attracted to the nucleus. Additional orbitals further out from the nucleus can also be occupied by these electrons, but energy is required to place an electron into these outer, unoccupied orbitals, because the placement involves moving a negative charge further away from the positively charged nucleus.

One way in which the electron can acquire this energy and be moved into an outer or higher orbital is to absorb a photon of light. When this occurs, the atom is said to be in an excited state. There are numerous possible excited states that a given molecule can attain; the particular one reached depends on the wavelength (and therefore the energy) of the quantum of light absorbed.

TABLE 15.1   Energy Content of Light of Different Wavelengths

COLOR OF LIGHT	WAVELENGTH (nm)	ENERGY (cal/EINSTEIN)
Far-red	750	38,000
Red	650	43,000
Yellow	590	48,000
Blue	490	58,000
Ultraviolet	395	72,000

An atom in an excited state is not stable; the tendency is for the electron to return from the outer orbital to the lower energy level. Its return is done in stages and is accompanied by the release of some of the energy acquired during excitation. The initial act is to return to a slightly lower energy level (transitional level) from the excited state, a process accompanied by the production of heat. When the electron returns to its original or ground state, the remainder of the excitation energy is released in a form of light known as fluorescence and phosphorescence. [Fluorescence is the immediate ($\approx 10^{-8}$ sec) emission while phosphorescence is the delayed release of absorbed radiation.]

15.5 ABSORPTION OF LIGHT BY CHLOROPHYLL

Chlorophyll *a* absorbs light in both the blue and the red region of the visible spectrum. Because the energy content of blue light (Table 15.1) is about 50% greater than red light, one might expect that the former would be more effective in photosynthesis. That this is not so is explained by the energy levels reached during excitation of the chlorophyll *a* molecule. Blue light is sufficiently energetic to give rise to the second excited singlet state (Figure 15.1). However, the chlorophyll molecule always decays to its lowest (first) excited singlet state before charge separation occurs. Because this lower energy state can also be reached directly by the less energetic red light, blue light is no more effective in driving photosynthesis.

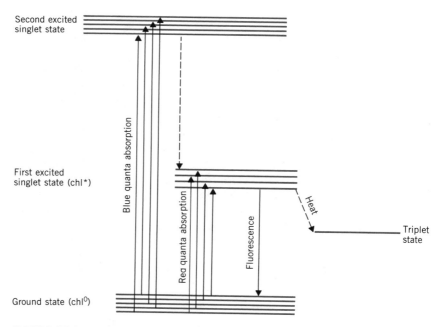

FIGURE 15.1 Diagram of energy states in the chlorophyll molecule. The ground, first, and second excited singlet states of the molecule possess a series of energy sublevels. Heat loss occurs when the excited molecule reverts to the lowest sublevel energy of the excited state. The fluorescence emission is shifted to the red end of the spectrum relative to the excitation spectrum because the light emitted in fluorescence is less than that absorbed during excitation (Stokes shift). (*Source:* Reproduced with permission from C. H. Foyer, *Photosynthesis*, p. 7. New York: Wiley, 1984.)

15.6   THE ENERGY CONVERSION PROCESS

The energy of the excited states of chlorophyll a described in Section 15.5 is rarely released as fluorescence. Instead, it is transferred through the antenna network, by a process called **inductive resonance** until it encounters an energy trap in the form of the specialized chlorophylls P_{700} or P_{680}, also called reaction center chlorophylls. Similarly, energy transfer between unlike molecules such as carotenoids or chlorophyll b and chlorophyll a occurs by the process of **resonance transfer**. Energy transfer into P_{700} or P_{680} causes these molecules to undergo **change separation** that results in the formation of a reducing agent and an oxidant.

The charge separation process involves an adjacent acceptor molecule (A) and an adjacent donor molecule (D). When P_{700} receives energy by inductive resonance from antenna chlorophyll a, it is elevated to its excited stage P_{700}^*. This excited state is capable of acting as both an oxidant and a reductant. As a reductant ($P_{700}^{\bullet}$), it can donate an electron; as an oxidant (P_{700}^+), it will accept electrons. An electron in a high-energy orbital can be captured by an acceptor molecule that then becomes reduced (A$^-$). The electron deficient P_{700}^+ then accepts an electron from a donor molecule D to fill the vacancy in the lower energy orbital and return to its initial state. The overall process may be represented as

$$D \cdot P_{700} \cdot A \xrightarrow{\ hv\ } D \cdot P_{700}^* \cdot A$$
$$D \cdot P_{700}^* \cdot A \longrightarrow D \cdot P_{700}^+ \cdot A^-$$
$$D \cdot P_{700}^+ \cdot A^- \longrightarrow D^+ \cdot P_{700} \cdot A^-$$

The result of the above reactions is that a reductant A$^-$ has been produced as well as an oxidant D$^+$. The chemical nature of these donor and acceptor molecules (D and A) varies with the two photosystems.

At this point, it is useful to introduce the energy-conversion scheme or "Z scheme" (Figure 15.2) first proposed by Hill and Bendall and subsequently modified by other investigators. This scheme describes the role of photosystems I and II found in green plants (and any other organisms that utilize H_2O as the reducing agent) that are activated by far-red (680–700 nm) and red light (650 nm), respectively. In photosystem I (PS I), light energy absorbed by the 200 molecules of chlorophyll a and accessory pigments in the light harvesting antenna system of PS I is transferred to the specialized reaction center chlorophyll P_{700}. As described in the preceding section, P_{700} is excited and transfers an electron to an early acceptor molecule, possibly a chlorophyll or quinone (designated as A_0, in Figure 15.2). The electron-donating form of P_{700} is a strong reductant and is placed at $E'_0 = \sim -1.2$ V. The reduced acceptor then passes its electron on to one or more carriers including bound ferredoxins ($Fd_{b_{1,2,\ldots}}$) imbedded in the thylakoid membrane. The electron-deficient P_{700} (P_{700}^+, $E'_0 = +0.5$ V) is restored to its reduced state by accepting an electron from the electron-transport chain linking PS I and PS II. A similar trapping process occurs in PS II where another reaction center chlorophyll a (Chl a_{680} or P_{680}) is excited. On excitation, an electron is transferred from $P_{680}^{\bullet}$ ($E'_0 = -0.8$ V) to an acceptor that is a pheophytin (Ph), known primarily from its ability to quench fluorescence produced by illuminating PS II. This acceptor then donates its

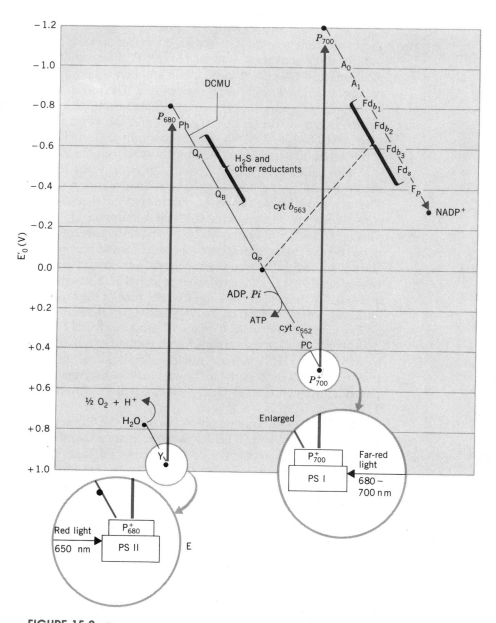

FIGURE 15.2 The energy conversion process of photosynthesis. (*Source:* Based on a scheme originally proposed by Hill and Bendall.)

electron to a series of quinones (Q_A, Q_B, . . .) including one called plasto-quinone (Q_p, $E_0' = 0.0$ V) that is similar in structure to ubiquinone (Section 14.2.4). The electron-deficient P_{680}^+ ($E_0' = +1.0$ V) accepts an electron from Y, to form a strong oxidant Y^+ that, in turn, is capable of oxidizing H_2O to O_2 ($E_0' = 0.82$ V).

An essential feature of the Hill–Bendall scheme is that electrons always flow from reductants having a lower potential to oxidants that have a more positive

E_0'. The various carriers have been positioned in the Z scheme on the basis of their reduction potentials, if known, as well as spectral changes undergone in red (for PS II activation) and far-red (for PS I activation) light. Moreover, as electrons flow from PS II to PS I, Hill and Bendall postulated that energy-rich phosphate in the form of ATP is generated. The role of light is to produce the reductants (P_{680}^* and P_{700}^*) required to donate electrons to such strong reducing agents as ferredoxin ($E_0' = -0.420$ V) and the oxidants (P_{680}^+ and P_{700}^+) to oxidize H_2O ($E_0' = 0.82$ V) and plastocyanin ($E_0' = 0.37$ V).

Information is available regarding the carriers in the Z scheme. Cytochrome c_{552}, also called cytochrome f (from the Latin *frons,* leaf) is a c-type of cytochrome; its E_0' is $+0.36$ V and it has an absorption maximum at 552 nm. Plastocyanin is a blue, copper-containing protein that also undergoes one-electron reductions; its E_0' is about 0.37. Plastoquinone (Q_p) has an E_0' of approximately 0.00 V when bound in the thylakoid membrane. A pool of other quinone (Q pool) molecules also present receives and passes on electrons from PS II. In Figure 15.2, the arrows are placed on the diagram in such a way that the tips designate the E_0' of the **reducing** potential of the excited P_{700} and P_{680} (-1.2 and -0.8 V, respectively). Similarly, the base of the arrows are located at the **oxidizing** potential of the excited P_{700} and P_{680} ($+0.5$ and $+1.0$ V, respectively).

The nature of the strong oxidant Y^+ produced by PS II is not known, and the details of the mechanism whereby that oxidant oxidizes H_2O [more likely the hydroxyl ion (OH^-)] remain unclear. Mn^{2+} ion is, however, known to be essential to this process. Note, however, that electrons flow from a more reduced compound (i.e., H_2O; $E_0' = 0.82$ V) to a stronger oxidant (P_{680}^+; $E_0' = 1.0$ V).

More is known about the process in which the PS I accomplishes the reduction of $NADP^+$. When PS I is activated, acceptors A_0 and A_1 and then a series of membrane-bound molecules of ferredoxin (Fd_b) are reduced. These compounds ultimately transfer electrons to a molecule of soluble ferredoxin (Fd_s, $E_0' = -0.42$ V). When reduced, this protein can, in turn, in the presence of the flavin-containing enzyme *ferredoxin-NADP$^+$ reductase* (F_p), reduce $NADP^+$ ($E_0' = -0.320$ V). Again, the flow of electrons is from compounds of lower potential (Fd_b) to those of higher potential ($NADP^+$) along the potential gradient.

There is much evidence in support of this Z scheme and the electron-transport chain linking the two photosystems. It is possible, for example, to obtain fractions enriched in their capacity to carry out either the oxidation of H_2O (PS II) or the reduction of $NADP^+$ (PS I) when properly supplemented with suitable electron acceptors or donors. Then, also, when chloroplasts are illuminated by light of longer wavelength (the kind that activates PS I), the oxidant produced accepts electrons from cytochrome-552 and plastocyanin, and the oxidized forms of those pigments predominate in the plastid. When light of shorter wavelength activates PS II, the reduced pheophytin produced feeds electrons into the chain leading to PS I, and all the carriers including cytochrome-552 become reduced.

The herbicide dichlorophenyldimethyl urea (DCMU) exerts its action as a weed killer by blocking the flow of electrons along the electron chain at the point indicated in Figure 15.2. In the presence of DCMU, chloroplasts will oxidize the carriers in that chain in the presence of far-red light (PS I), but when shorter wavelength light is used (PS II), the carriers are not reduced.

BOX 15.A FERREDOXIN, A NONHEME IRON PROTEIN

The nonheme iron protein ferredoxin (Section 14.2.3) has been isolated from a large number of photosynthetic organisms—bacteria, algae, higher plants. It was initially discovered, however, by Carnahan and Mortenson, who were studying nitrogen-fixation in *Clostridium pasteurianum*. Ferredoxin from this organism contains seven atoms of iron per molecule of protein and seven sulfide groups per mole of protein. It has a molecular weight of 6000 and the remarkably low redox potential at pH 7.55 of -0.42 V. When isolated from spinach chloroplasts, it contains two iron atoms linked to two specific sulfur atoms that are released as H_2S on acidification. In spinach, its molecular weight is 11,600 and its redox potential is -0.43 V. Oxidized ferredoxin has characteristic absorption bands at 420 and 463 nm. When acidified, these bands disappear, and the protein loses its biochemical activity.

15.7 NONCYCLIC PHOSPHORYLATION

The flow of electrons in noncyclic phosphorylation can now be outlined in terms of the energy-conversion process first described. According to Reaction 15.10, both ATP and NADPH are produced as H_2O is oxidized to O_2. Starting then with PS II, an electron from OH^- can enter P_{680}^+ and restore it to its ground state as another electron passes along the chain to the oxidant at PS I. To complete the process, electrons flow from bound ferredoxin (Fd_b) to $NADP^+$, accomplishing the reduction of that compound.

As electrons flow along the electron-transport chain linking PS II to PS I, protons are moved across the thylakoid membrane *into* the inner membrane space, and a pH gradient is established. As protons flow back across the membrane, the synthesis of ATP from ADP and Pi occurs. Lamellar fragments enriched in ATPase and coupling factors have been isolated from chloroplasts, and some of the strongest evidence for Mitchell's chemiosmotic theory (Section

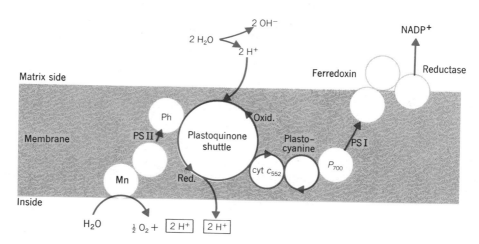

FIGURE 15.3 Photosynthetic electron flow from water to $NADP^+$ in a zigzag across a membrane. The plastoquinone shuttle represents alternate reduction and oxidation of plastoquinone. (*Source:* Based on Figure 5 in A. Trebst, *Annual Review of Plant Physiology.* Vol. 25, 1974. p. 423.)

14.6) of oxidative phosphorylation is furnished by work on photosynthetic phosphorylation. In this connection, it is possible to arrange the components of the Z scheme in a membrane so that one can see how protons are moved across the membrane during electron transport (Figure 15.3).

15.8   CYCLIC PHOSPHORYLATION

The essential feature of cyclic phosphorylation is that ATP is produced in this process without any **net transfer** of electrons to $NADP^+$. This process, studied extensively by Arnon and his associates, is driven by light of longer wavelengths that activates only PS I. Because neither NADPH nor any other reduced compound accumulates in this process, electrons made available by the action of light on PS I make their way back to the electron-transport chain linking PS II to PS I through still another cytochrome (cyt b_{563}). The point of entry into the chain is not clear, but in Figure 15.2 it is placed at plastoquinone (Q_p). As electrons flow through this cyclic path, driven by the action of light on PS I, ATP is produced from ADP and H_3PO_4, presumably by mechanisms also analogous to those occurring in oxidative phosphorylation.

15.9   ELECTRON FLOW IN BACTERIAL PHOTOSYNTHESIS

Those photosynthetic bacteria that utilize inorganic (H_2S, $Na_2S_2O_3$) and organic reducing agents (succinate, acetate) instead of H_2O do not possess PS II that is present in organisms using H_2O. The electron path utilized by these organisms is represented in Figure 15.2, where electrons derived from the primary chemical reductants feed into the electron-transport chain at one of the early quinones in the chain. Thus, these organisms require light (for PS I) only in order to reduce ferredoxin and $NADP^+$. The oxidant (P_{700}^+) produced by light will accept electrons originally contributed by the primary reductant.

These processes, then, describe how NADPH and ATP are generated by light during photosynthesis. The next section discusses the dark reactions responsible for the assimilation of CO_2 into the organic compounds of photosynthetic organisms.

15.10   PATH OF CARBON

15.10.1   METHODOLOGY

The series of reactions whereby CO_2 is assimilated by plants and converted to carbohydrates and other organic compounds was the result of work in several laboratories. The problem was not extensively pursued, however, until the first product into which CO_2 is incorporated in photosynthesis was identified by Calvin and his associates. This research is an outstanding example of the application of new techniques to the solution of an extremely complicated problem.

The basic experimental approach was as follows: In a plant that is carrying out photosynthesis at a steady rate, CO_2 is being converted to carbohydrates and other plant compounds through a series of intermediates.

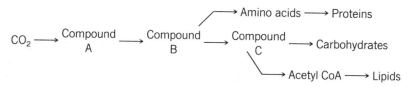

If, at time zero, radioactive $CO_2(^{14}CO_2)$ is introduced into the system, some of the labeled carbon atoms will be converted to these different products, and during the time it takes for this to occur all the intermediates will become labeled. If, after a relatively shorter period of time, the photosynthesizing plant is plunged into hot alcohol to inactivate its enzymes and stop all reactions, the labeled carbon atom will have had time to make its way through only the first few intermediates. If the time interval is short enough, the labeled carbon atoms will have made their way only into the first stable intermediate, compound A, and only the **first product** of CO_2 fixation will be labeled.

In 1946, carbon-14 was made available in appreciable amounts from the Atomic Energy Commission. Moreover, the technique of paper chromatography was in full development and provided a means for separating the large number of cellular constituents that occur in a plant. With these tools, Calvin's group was able to identify the early stable intermediates in the path of carbon from CO_2 to plant metabolites. They used suspensions of algae, *Scenedesmus* or *Chlorella*, that were grown at a constant rate in the presence of light and CO_2. Radioactive CO_2 was introduced into the reaction mixture at zero time, and a period of time was allowed to elapse. The cells were then extracted with boiling alcohol, and the soluble constituents of the alcohol solution were analyzed by paper chromatography. When the algae were exposed to $^{14}CO_2$ for 30 seconds, hexose phosphates, triose phosphates, and phosphoglyceric acid were labeled. With longer periods, these compounds as well as amino acids, organic acids, lipids, and proteins were labeled. With 5-second exposure, however, most of the radioactive carbon was located in 3-phosphoglyceric acid and, within this compound, the carboxyl group contained the majority of the radioactivity.

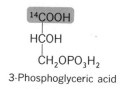

3-Phosphoglyceric acid

This result suggested that 3-phosphoglyceric acid was formed by the carboxylation of some unknown compound containing two carbon atoms. Attempts to demonstrate any such acceptor molecule failed, however. More careful examination of the early products of photosynthesis disclosed that sedoheptulose-7-phosphate and several phosphorylated pentoses were also present as labeled compunds, and this, in turn, suggested that these sugars might be involved in producing the acceptor molecule for CO_2.

15.10.2   THE REDUCTIVE PENTOSE PHOSPHATE (RPP) OR CALVIN CYCLE

During this period, the reactions of the pentose phosphate pathway (Chapter 11) were being clarified in the laboratories of Horecker, Racker, and others and the relationships between trioses, tetroses, pentoses, hexoses, and heptoses were being established. More careful study of the ^{14}C labeling of the sugars produced during short periods of photosynthesis permitted Calvin's laboratory to postulate the operation of a reductive pentose phosphate (RPP) cycle (Figure 15.4) during photosynthesis. This cycle involves essentially only one new reaction, the carboxylation of ribulose-1,5-bisphosphate, discussed next; the remainder of the reactions are identical or similar to reactions encountered previously in glycolysis and pentose phosphate metabolism. All of the enzymes required to catalyze the reactions postulated in the RPP cycle are known to occur in the stromal phase of chloroplasts.

15.10.2.1 Carboxylation Phase. This rather bewildering scheme can be grouped into three phases. The first of these, the **carboxylation phase,** involves the single reaction catalyzed by **ribulose-1,5-bisphosphate carboxylase-oxygenase** (also called *rubisco*). This key reaction involves the carboxylation, not of a *two-carbon compound,* but a *five-carbon compound,* ribulose-1,5-bisphosphate, to yield two moles of 3-phosphoglyceric acid.

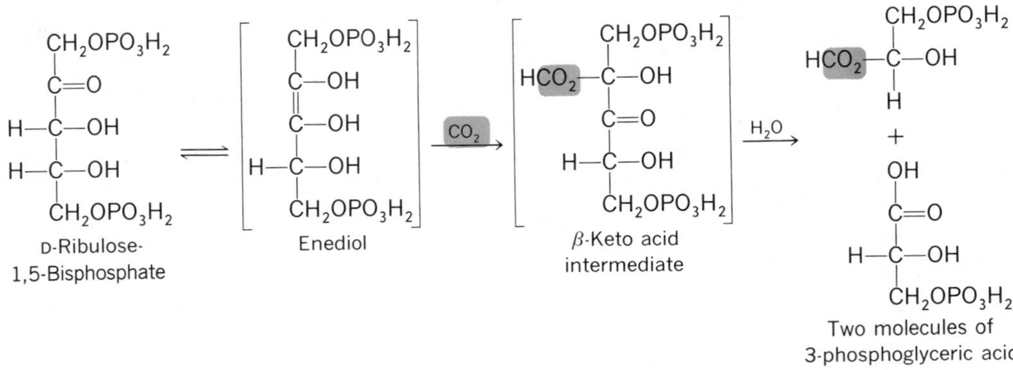

(15.11)

In the presence of the enzyme rubisco, CO_2 adds to the enediol form of ribulose-1,5-bisphosphate to form an unstable β-keto acid that undergoes hydrolytic cleavage to form two molecules of 3-phosphoglyceric acid. The equilibrium of the reaction is far to the right. Rubisco is the most abundant protein on earth; in spinach leaves, it constitutes 5 to 10% of the soluble protein. It has a molecular weight of 550,000 and is an oligomer composed of eight small monomers (MW 12–16,000) and eight large units (MW 54–60,000). The larger subunits contain the catalytic site, while the smaller subunit carries the regulatory site. As will be pointed out in Section 15.15, rubisco also catalyzes an oxygenation reaction.

15.10.2.2 Reduction Phase. A second phase of the RPP cycle, termed the **reduction phase,** consists of two reactions previously encountered in glycolysis. In these reactions, ATP and a reduced nicotinamide nucleotide are consumed. The first involves the phosphorylation of 3-phosphoglycerate by ATP to form 1,3-bisphosphoglycerate.

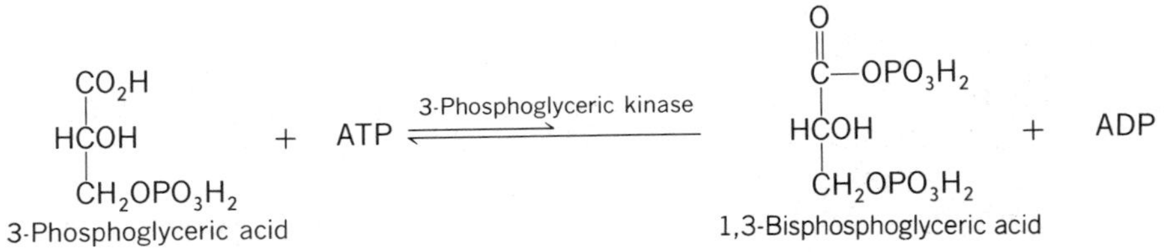

(15.12)

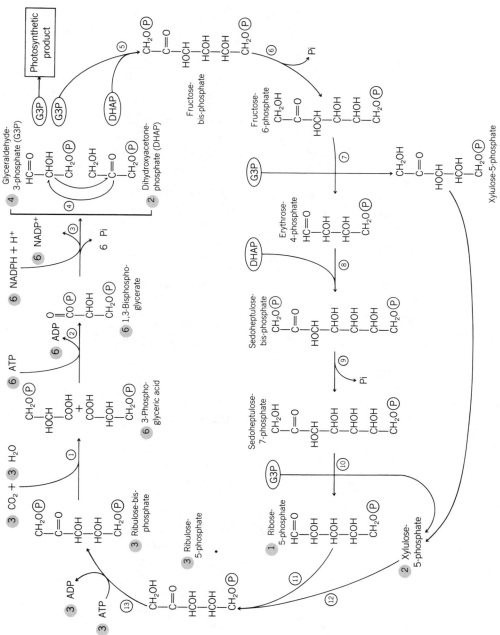

FIGURE 15.4   The reductive pentose phosphate cycle. The enzymes involved are (1) ribulose biphosphate carboxylase, (2) phosphoglycerate kinase, (3) NADP-G3P dehydrogenase, (4) triose-phosphate isomerase, (5) aldolase, (6) fructose bisphosphatase, (7) transketolase, (8) aldolase, (9) sedoheptulose-1,7-bisphosphatase, (10) transketolase, (11) ribose-5-phosphate isomerase, (12) ribulose-5-phosphate epimerase, and (13) phosphoribulokinase. (*Source:* Reproduced with permission from C. H. Foyer, *Photosynthesis*, p. 8. New York: Wiley, 1984.)

The second reaction involves reduction of the 1,3-bisphosphoglyceric acid by NADPH in the presence of a NADP-specific glyceraldehyde-3-phosphate dehydrogenase.

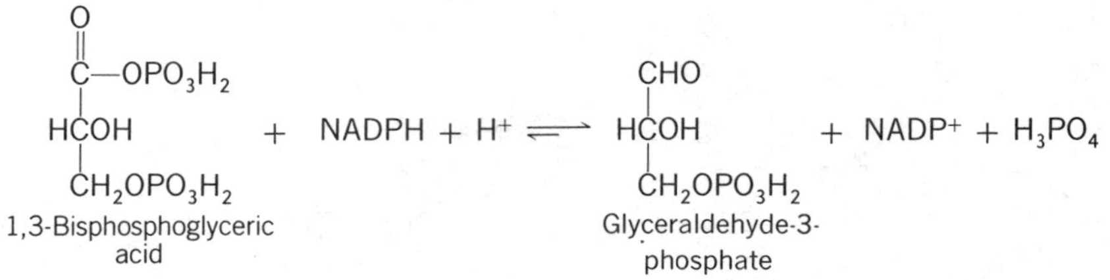

$$\begin{array}{c} O \\ \| \\ C-OPO_3H_2 \\ HCOH \\ CH_2OPO_3H_2 \\ \text{1,3-Bisphosphoglyceric} \\ \text{acid} \end{array} + NADPH + H^+ \rightleftharpoons \begin{array}{c} CHO \\ | \\ HCOH \\ CH_2OPO_3H_2 \\ \text{Glyceraldehyde-3-} \\ \text{phosphate} \end{array} + NADP^+ + H_3PO_4 \quad (15.13)$$

The chloroplast enzyme is activated by ATP and NADPH. It is in these two reactions that all of the NADPH and two-thirds of the ATP required to drive the RPP cycle are utilized.

TABLE 15.2   Stoichiometry of the Reductive Pentose Phosphate Cycle

Reaction[a]

Carboxylation phase

1. **3** Ribulose-1,5-bisphosphate + **3** CO_2 + **3** $H_2O \longrightarrow$ **6** 3-Phosphoglycerate

Reduction phase

2. **6** 3-Phosphoglycerate + **6** ATP $\longrightarrow$ **6** 1,3-Bisphosphoglycerate + **6** ADP
3. **6** 1,3-Bisphosphoglycerate + **6** NADPH + **6** $H^+ \longrightarrow$
    6 Glyceraldehyde-3-phosphate + **6** $NADP^+$ + **6** H_3PO_4

Regeneration phase

4. **2** Glyceraldehyde-3-phosphate $\longrightarrow$ **2** Dihydroxy-acetone phosphate
5. Glyceraldehyde-3-phosphate + dihydroxy-acetone phosphate $\longrightarrow$
    Fructose-1,6-bisphosphate
6. Fructose-1,6-bisphosphate + $H_2O \longrightarrow$ Fructose-6-phosphate + H_3PO_4
7. Fructose-6-phosphate + glyceraldehyde-3-phosphate $\longrightarrow$
    Xylulose-5-phosphate + erythrose-4-phosphate
8. Erythrose-4-phosphate + dihydroxy-acetone phosphate $\longrightarrow$
    Sedoheptulose-1,7-bisphosphate
9. Sedoheptulose-1,7-bisphosphate + $H_2O \longrightarrow$
    Sedoheptulose-7-phosphate + H_3PO_4
10. Sedoheptulose-7-phosphate + Glyceraldehyde-3-phosphate $\longrightarrow$
    Ribose-5-phosphate + xylulose-5-phosphate
11. Ribose-5-phosphate $\longrightarrow$ Ribulose-5-phosphate
12. **2** Xylulose-5-phosphate $\longrightarrow$ **2** Ribulose-5-phosphate
13. **3** Ribulose-5-phosphate + **3** ATP $\longrightarrow$ **3** Ribulose-1,5-bisphosphate + **3** ADP

Sum

3 CO_2 + **9** ATP + **6** NADPH + **6** H^+ + **5** $H_2O \longrightarrow$
    Glyceraldehyde-3-phosphate + **9** ADP + **6** $NADP^+$ + **8** H_3PO_4

[a] *The numbers of the reactions correspond to those shown in the legend for Figure 15.4.*

15.10.2.3 Regeneration Phase.

The remainder of the reactions in the RPP cycle compose the third, or **regeneration phase,** that accomplishes the regeneration of ribulose-1,5-bisphosphate necessary to keep the cycle operating. In Table 15.2 are listed the reactions of the three phases with the stoichiometry illustrated with three moles of CO_2 being assimilated. This stoichiometry is also specified in Figure 15.4, and at the end of the reduction phase, a total of 18 carbon atoms will be present in six molecules of glyceraldehyde-3-phosphate.

One of these can be considered as the photosynthetic product produced when 3 CO_2 are put through the RPP cycle. The 15 carbon atoms in the other five molecules of glyceraldehyde-3-phosphate are, in turn, converted by the reactions of the regeneration phase into three moles of ribulose-5-phosphate. The phosphorylation of these pentose phosphates into ribulose-1,5-bisphosphate completes the RPP cycle and produces the substrates for assimilation of another three moles of CO_2. This last step also is the other reaction in which ATP is consumed in the RPP cycle. Examination of steps 2, 3, and 13 show that a total of 2 NADPH and 3 ATP are required for each CO_2 assimilated by this process.

15.11 QUANTUM REQUIREMENT OF PHOTOSYNTHESIS

Returning to the energy-conversion process in Figure 15.2, we can now place a lower limit on the number of light quanta required to make the two molecules of NADPH needed to assimilate one molecule of CO_2. In green plants that utilize H_2O, both PS I and PS II will have to be activated four times each to produce the four electrons required to reduce 2 $NADP^+$. Therefore, a total of eight quanta of light would appear to be required as a minimum, since it is generally assumed that at least one quantum is required for each photoactivation process that makes an electron available at A_0 and Ph in the energy-conversion scheme. It should also be apparent that noncyclic photophosphorylation can only produce two-thirds of the ATP required if only one phosphorylation site exists in the electron-transport chain linking PS II and PS I. Under those conditions, cyclic phosphorylation presumably could furnish the additional ATP, provided further light is supplied. If, however, two phosphorylation sites exist in the chain linking PS I and PS II, sufficient ATP would then be available.

One final aspect of the energy requirements of photosynthesis deserves comment. In Table 15.1, light of 650 nm was shown to have an energy content of 44,000 cal/Einstein. It is generally assumed that about 75% of this energy (approximately 30,000 cal/Einstein) is available for photosynthesis, the remaining being dissipated as heat as electrons pass from the first excited state to transitional levels. If, however, eight quanta per two molecules of NADPH (eight Einsteins for two moles of NADPH) are the minimum required to drive the energy-conversion process, we see that 8 × 30,000 or 240,000 cal are available to convert one mole of CO_2 into carbohydrate. From Equation 15.2, we have seen that 118,000 cal are required as a minimum to accomplish this conversion. The overall efficiency of photosynthesis with these calculations would therefore be 118,000/240,000 or 49%.

15.12 THE C₄ PATHWAY

Many plants, including important crop plants of tropical origin — the tropical grasses sugar cane, corn, and sorghum — possess an interesting variation for CO_2 assimilation. Early studies on these plants indicated that the CO_2 was incorporated initially into the dicarboxylic acids, malic acid, and aspartic acid, rather than phosphoglyceric acid. M. D. Hatch and C. R. Slack initiated a series of studies on sugar cane in 1966 that stimulated work in many laboratories and resulted in our knowledge of the C_4 (or Hatch–Slack) pathway for CO_2 assimilation.

Plants that utilize the C_4 pathway also possess a common feature of leaf anatomy in which the vascular elements (phloem and xylem) are surrounded by a row of bundle-sheath cells and then, in turn, by one or more layers of mesophyll cells (Figure 15.5). This characteristic anatomy (Kranz-type) has long been cited as a mechanism whereby plants living in dry or hot regions could minimize their loss of tissue water by transpiration, since the conducting elements were separated from the stomata on the surface of the leaf by one or more layers of mesophyll cells as well as the bundle-sheath layer. This structural feature obviously also restricts the amount of CO_2 available for photosynthesis, and it is argued that the C_4 pathway represents the adaptation of tropical and desert plants to this stress. The C_4 pathway will be described in terms of the reactions occurring in the mesophyll and the bundle-sheath cells.

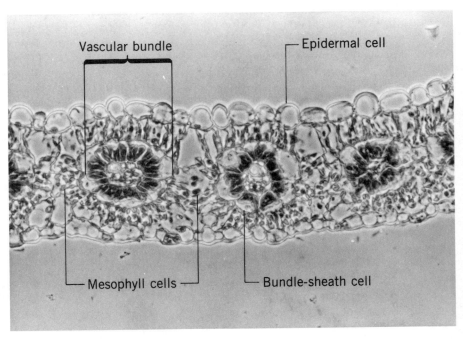

FIGURE 15.5 Microphotograph of the vascular bundle, bundle-sheath cells, and mesophyll layer of *Amaranthus edulis*. (*Source:* Courtesy of W. M. Laetsch.)

15.12.1 MESOPHYLL CELLS

CO_2 entering the leaf of a C_4 plant during stomatal opening will diffuse to the mesophyll where it serves as a substrate for **phosphoenol pyruvic acid carboxylase.**

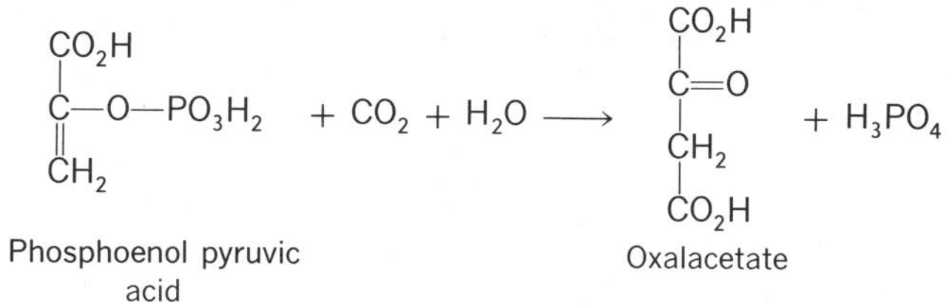

$$
\begin{array}{c}
CO_2H \\
| \\
C-O-PO_3H_2 \\
| \\
CH_2
\end{array}
+ CO_2 + H_2O \longrightarrow
\begin{array}{c}
CO_2H \\
| \\
C=O \\
| \\
CH_2 \\
| \\
CO_2H
\end{array}
+ H_3PO_4 \qquad (15.14)
$$

Phosphoenol pyruvic acid   Oxalacetate

This enzyme, which has a much higher affinity for CO_2 than ribulose-1,5-bisphosphate carboxylase-oxygenase (rubisco), is localized in chloroplasts of the mesophyll cells. It therefore serves as a more efficient trap for the low levels of CO_2 that are encountered, and oxalacetate is produced.

The C_4 plants can be divided into those that have a high concentration of malic dehydrogenase in the mesophyll cells and those that have an active alanine-aspartic transaminase. In the former (Reaction 15.15), oxalacetate is reduced to malate and in the latter (Reaction 15.16), aspartic acid is formed. These two dicarboxylic acids then serve as carriers of CO_2 into the bundle-sheath cells.

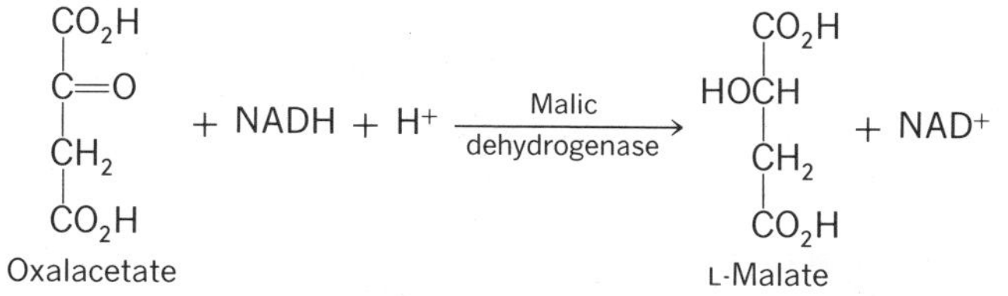

$$
\begin{array}{c}
CO_2H \\
| \\
C=O \\
| \\
CH_2 \\
| \\
CO_2H
\end{array}
+ NADH + H^+ \xrightarrow[\text{dehydrogenase}]{\text{Malic}}
\begin{array}{c}
CO_2H \\
| \\
HOCH \\
| \\
CH_2 \\
| \\
CO_2H
\end{array}
+ NAD^+ \qquad (15.15)
$$

Oxalacetate   L-Malate

$$
\begin{array}{c}
CO_2H \\
| \\
C=O \\
| \\
CH_2 \\
| \\
CO_2H
\end{array}
+
\begin{array}{c}
CO_2H \\
| \\
NH_2CH \\
| \\
CH_3
\end{array}
\underset{\text{aminase}}{\overset{\text{Trans-}}{\rightleftharpoons}}
\begin{array}{c}
CO_2H \\
| \\
NH_2CH \\
| \\
CH_2 \\
| \\
CO_2H
\end{array}
+
\begin{array}{c}
CO_2H \\
| \\
C=O \\
| \\
CH_3
\end{array}
\qquad (15.16)
$$

Oxalacetate   Alanine   Aspartic acid   Pyruvate

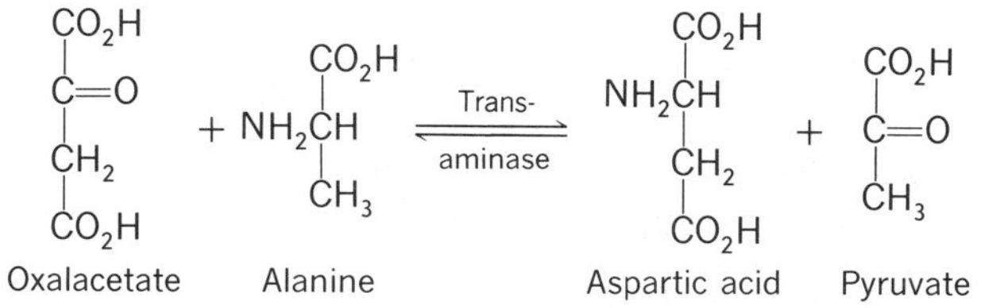

The other unique reaction of the mesophyll is the one in which the CO_2 trapping agent, phosphoenol pyruvic acid, is formed from pyruvate (that returns from the bundle-sheath cells eventually). The enzyme that catalyzes this reaction is **pyruvate, phosphate dikinase** (Reaction 15.17).

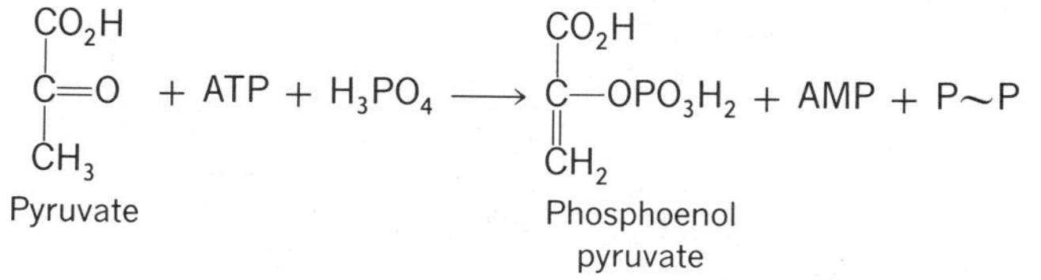

$$
\begin{array}{c}
CO_2H \\
| \\
C=O \\
| \\
CH_3
\end{array}
+ ATP + H_3PO_4 \longrightarrow
\begin{array}{c}
CO_2H \\
| \\
C-OPO_3H_2 \\
\| \\
CH_2
\end{array}
+ AMP + P\sim P \qquad (15.17)
$$

Pyruvate   Phosphoenol pyruvate

This enzyme is also found only in the mesophyll cells.

15.12.2   BUNDLE-SHEATH CELLS

The plants that utilize malate as a carrier of CO_2 have a high level of a NADP-specific malic enzyme (Section 12.6.1) in the bundle-sheath chloroplasts. This enzyme catalyzes the formation (i.e., the release) of CO_2 from malate, which is then incorporated by means of the RPP cycle. The enzymes of the RPP cycle are found only in the bundle-sheath chloroplasts, together with the NADP-malic enzyme. Pyruvate formed in this reaction returns to the mesophyll.

$$
\begin{array}{c}
CO_2H \\
| \\
HOCH \\
| \\
CH_2 \\
| \\
CO_2H \\
\text{L-Malate}
\end{array}
\;+\; NADP^+ \;\longrightarrow\;
\begin{array}{c}
CO_2H \\
| \\
C{=}O \\
| \\
CH_3 \\
\text{Pyruvate}
\end{array}
\;+\; NADPH + H^+ + CO_2 \qquad (15.18)
$$

The plants that utilize aspartic acid as a CO_2 carrier contain a transaminase in the bundle-sheath cells that convert the aspartic acid back to oxalacetic acid.

$$\text{Aspartic acid} \xrightarrow{\text{Transaminase}} \text{Oxalacetic acid}$$

The fate of the oxalacetate depends again on the particular plant. One major group of aspartic formers contain a NAD^+-specific malic dehydrogenase (Section 12.3.8) and a NAD^+-specific malic enzyme (Section 12.6.1). These two enzymes thereby convert the oxalacetic acid first to malic acid and then to CO_2 and pyruvic acid.

$$\text{Oxalacetate} + NADH + H^+ \longrightarrow \text{L-Malate} + NAD^+ \qquad (15.19)$$
$$\text{L-Malate} + NAD^+ \longrightarrow \text{Pyruvate} + NADH + H + CO_2 \qquad (15.20)$$

Another group of aspartic-carrier plants convert the oxalacetate to PEP and CO_2 due to their content of **PEP carboxykinase** (Section 10.7.2).

$$\text{Oxalacetate} + ATP \longrightarrow \text{Phosphoenol pyruvate} + CO_2 + ADP$$

In both the malate- and aspartic-carrier plants, the CO_2 is released in the bundle-sheath cells where it serves as the substrate for ribulose-1,5-bisphosphate carboxylase-oxygenase of the RPP cycle. In the aspartic-carrier plants, an additional step is called for since, as the amino acid moves from mesophyll to bundle sheath, it carries not only the CO_2 but also an amino ($-NH_2$) group. In order to avoid a net transfer of amino nitrogen into the bundle-sheath cells, the pyruvate formed in Reaction 15.20 undergoes transamination to alanine that then moves out to the mesophyll, thereby balancing the $-NH_2$ groups. In the mesophyll, the alanine can transaminate with the oxalacetate formed initially. These relationships are shown in Figure 15.6.

In summary, the fixation of CO_2 into the C_4 dicarboxylic acids is viewed as an efficient mechanism for trapping CO_2 and concentrating it in the bundle sheath for assimilation by means of the RPP cycle. The C_4 pathway presumably has evolved in response to ecological conditions characterized by the combination of high radiation, higher temperatures, and a limited supply of H_2O. The capacity of C_4 species to survive and grow under these conditions is due to the ability of PEP-carboxylase to operate with very low concentrations of CO_2.

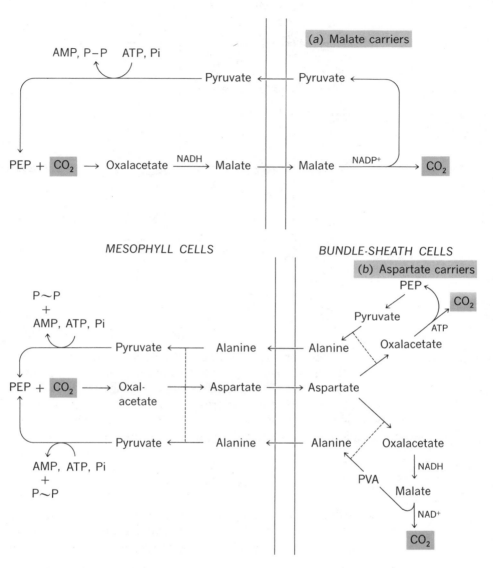

FIGURE 15.6 The role of malate and aspartate as CO_2 carriers in C_4 photosynthesis. The dotted lines represent a transfer of amino groups.

15.13 TRANSFER OF ATP AND ELECTRONS FROM CHLOROPLAST TO CYTOSOL

During photosynthesis, NADPH and ATP are formed in the chloroplast but are not directly available for reactions outside the chloroplast since neither nucleotide can pass through the inner membrane of the chloroplast. A shuttle system has been proposed that involves dihydroxyacetone phosphate (DHAP) and 3-phosphoglyceric acid (3-PGA), both of which are freely permeable. Thus, 3-PGA synthesized in the chloroplast stroma phase by the carboxylation of ribulose bisphosphate is phosphorylated in the presence of ATP (Reaction 15.12) and is reduced to glyceraldehyde-3-phosphate by NADPH (Reaction

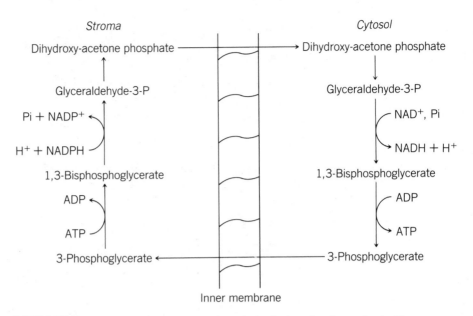

FIGURE 15.7   The dihydroxyacetone phosphate-3-phosphoglycerate shuttle.

15.13). After isomerization to dihydroxyacetone phosphate, the latter moves out of the chloroplast on the phosphate transporter into the cytosol.

In the cytosol, 3-phosphoglycerate is regenerated by a reversal of the sequence of reactions just described, producing NADH and ATP. These compounds are then available in the cytosol for biosynthesis while the 3-PGA enters the chloroplast and completes the shuttle (Figure 15.7).

15.14   REGULATION OF THE RPP CYCLE

The assimilation of CO_2 by the enzymes of the RPP cycle is regulated primarily by the activity of the carboxylation enzyme, rubisco. This enzyme exists in an active form that is a complex of CO_2 and Mg^{2+} formed sequentially as follows.

$$\text{Enzyme} + CO_2 \underset{}{\overset{\text{Slow}}{\rightleftharpoons}} \text{Enz-}CO_2 \overset{Mg^{2+}}{\underset{Mg^{2+}}{\rightleftharpoons}} \text{Enz-}CO_2\text{-}Mg^{2+}$$

Inactive   Inactive   Active

The preliminary binding of CO_2 is a relatively slow process, but the reaction with Mg^{2+} is more rapid and is accelerated at higher pH (8.5). Light strongly influences the process just described since the stromal compartment becomes more alkaline in the light as protons move into the thylakoids. The acidification of the intermembrane space of the thylakoid is accompanied by movement of Mg^{2+} ions out into the stroma where they can complex to form the active form of rubisco. Since formation of the active form of rubisco is reversible, the enzyme is converted to its inactive form in the dark when protons move out of the thylakoids and Mg^{2+} leaves the stroma.

The flow of carbon through the RPP cycle is also regulated at the steps catalyzed by fructose-1,6-bisphosphate phosphatase, sedoheptulose-1,7-bisphosphate phosphatase, and phosphoribulose kinase. The regulation involves the conversion of inactive forms of these enzymes to active enzymes by a small molecular weight protein, **thioredoxin**. This protein, discovered in plants by Buchanan, contains a cysteine residue that can be oxidized and reduced.

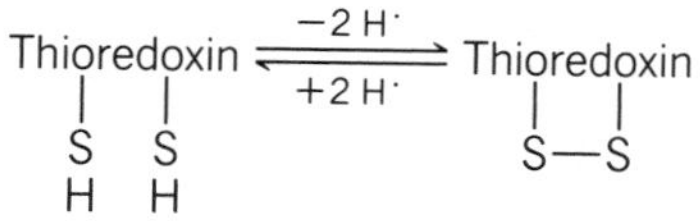

$$\text{Thioredoxin} \underset{+2\,H^{\cdot}}{\overset{-2\,H^{\cdot}}{\rightleftharpoons}} \text{Thioredoxin}$$

The reduced form of thioredoxin can activate the three enzymes listed above, presumably by reducing essential thiol groups in these proteins. Thioredoxin in turn is reduced in the light in the presence of thioredoxin reductase, an enzyme that catalyzes the reduction of oxidized thioredoxin by reduced ferredoxin produced by photosystem I. The overall process whereby these three enzymes are activated in the light may be represented as

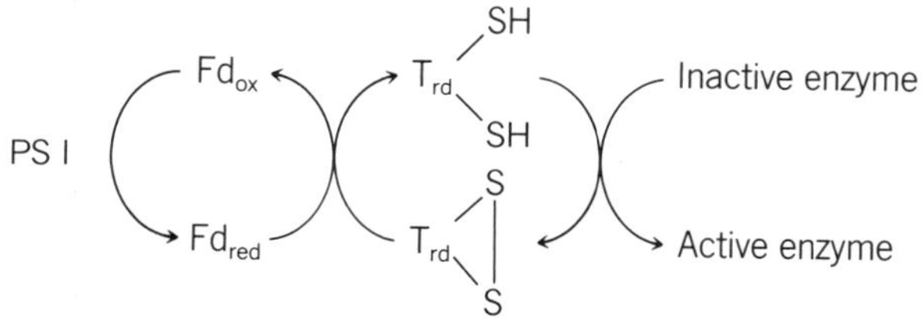

The operation of the RPP cycle, of course, requires NADPH and ATP, both of which are produced in the light. Thus, the concentration of these cosubstrates required in the reduction phase of the cycle increases in the light and is available for the reactions described (Reactions 15.12 and 15.13).

15.15 PHOTORESPIRATION

Plants, of course, carry out the same general respiratory processes as animals and microorganisms in that they degrade carbohydrates by means of glycolysis and the Krebs cycle. They also catalyze the general reactions of protein and amino acid metabolism. These reactions occur in the same organelles of the plant cell (mitochondria, cytoplasm, microsomes, etc.) as in animal cells. To give some indication of the amount of this process of dark respiration, the rate of O_2 consumption in the dark is frequently only about 1% of the rate of O_2 evolution in light resulting from photosynthesis.

All plants can exhibit an additional metabolic activity termed **photorespiration** that occurs only when the plants are illuminated. Because photorespiration, like dark respiration, results in CO_2 evolution and O_2 consumption, it is the opposite of photosynthesis and has the net effect of decreasing plant growth and crop yield. For this reason, the phenomenon has been extensively studied.

Photorespiration results from the ability of rubisco to catalyze the **oxygenation** of ribulose-1,5-bisphosphate as well as its carboxylation. This occurs because O_2 and CO_2 can occupy the same catalytic site and compete in reacting with the sugar phosphate. The production of one mole of phosphoglycolic acid and *only* one mole of phosphoglyceric acid in Reaction 15.21 prevents the functioning of the RPP cycle. The reactions of the photorespiratory cycle (Fig-

$$
\begin{array}{c}
CH_2OPO_3H_2 \\
| \\
C=O \\
| \\
HCOH \\
| \\
HCOH \\
| \\
CH_2OPO_3H_2
\end{array}
\longrightarrow
\left[
\begin{array}{c}
CH_2OPO_3H_2 \\
| \\
C-OH \\
\| \\
C-OH \\
| \\
HCOH \\
| \\
CH_2OPO_3H_2
\end{array}
\right]
\xrightarrow{\ O_2\ }
\tag{15.21}
$$

Ribulose-1,5-bisphosphate

$$
\left[
\begin{array}{c}
CH_2OPO_3H_2 \\
| \\
H-O-O-C-OH \\
| \\
C=O \\
| \\
HCOH \\
| \\
CH_2OPO_3H_2
\end{array}
\right]
\xrightarrow[OH^-]{}
\begin{array}{c}
CH_2OPO_3H_2 \\
| \\
O=C-OH \\
\text{Phospho-} \\
\text{glycolic} \\
\text{acid} \\
+ OH^- \\
HO-C=O \\
| \\
HCOH \\
| \\
CH_2OPO_3H_2 \\
\text{3-Phospho-} \\
\text{glyceric} \\
\text{acid}
\end{array}
$$

ure 15.8) accomplish the recycling of 75% of the carbon in phosphoglycolate (25% is lost as CO_2) back to phosphoglyceric acid to permit the continued operation of the RPP cycle. The enzymes that accomplish this salvage operation are located in three separate organelles. It is interesting to ponder the evolutionary pressure that resulted in the compartmentalization of these enzymes in three separate cellular compartments.

Photorespiration is readily observed in crops that are an important part of the world's food supply such as wheat and rice, many legumes, and sugar beets. In some of these, it has been estimated that the net assimilation of CO_2 by photosynthesis may be reduced by as much as 50% by photorespiration. For this reason, it has been proposed that it might be possible to increase the crop yield of such plants by finding a means to inhibit the photorespiration they exhibit. Other equally important food crops — corn, sorghum, sugar cane — exhibit much less photorespiration. This difference can be explained, in part, by the fact that these species are C_4 plants carrying out C_4 photosynthesis (Section 15.12). In this process, the high concentration of CO_2 in the bundle-sheath chloroplasts will selectively favor *carboxylation* (Reaction 15.11) rather than oxygenation (Reaction 15.21), and little phosphoglycolate will be produced.

This brief discussion of the primary life process of photosynthesis can only outline the knowledge of the subject. Undoubtedly, further research will clarify the details of the energy-conversion process and provide basic information that will allow humans to regulate photosynthesis for the optimum production of food.

15.16   THE MULTIFUNCTIONAL CHLOROPLAST

It is valuable to list the many metabolic activities of the intact chloroplast of the leaf cell. Figure 15.9 summarizes the activities of this multifunctional organelle that is the sole site for the photoreduction of CO_2 to phosphoglyceric acid, the

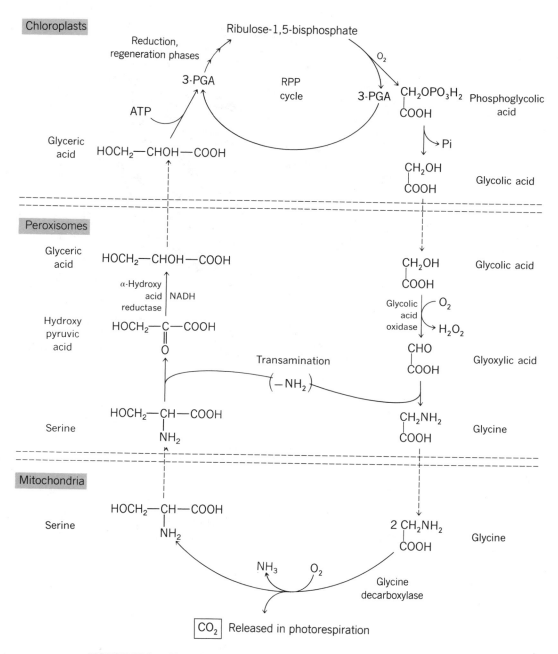

FIGURE 15.8 Glycolic acid metabolism and photorespiration, a process involving three organelles.

photooxidation of water, and photophosphorylation. The leaf chloroplast is also the sole site for fatty acid synthesis, nitrite reduction to ammonia, reduction of sulfate to sulfhydryl compounds, limited protein synthesis, amino acid biosynthesis, chlorophyll biosynthesis, and carotenoid biosynthesis. In addition, the synthesis of complex lipids occurs on the outer envelope of the chloroplast. Thus, in addition to fixing CO_2, the chloroplast carries out many other

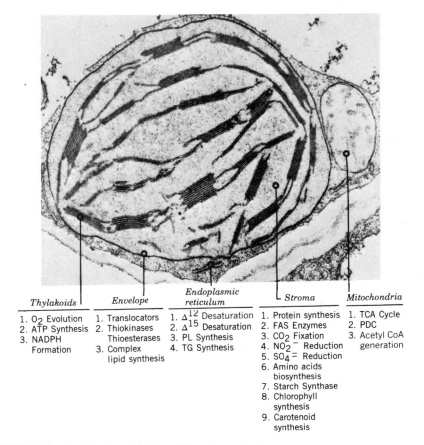

Thylakoids	Envelope	Endoplasmic reticulum	Stroma	Mitochondria
1. O_2 Evolution	1. Translocators	1. Δ^{12} Desaturation	1. Protein synthesis	1. TCA Cycle
2. ATP Synthesis	2. Thiokinases	2. Δ^{15} Desaturation	2. FAS Enzymes	2. PDC
3. NADPH Formation	Thioesterases	3. PL Synthesis	3. CO_2 Fixation	3. Acetyl CoA generation
	3. Complex lipid synthesis	4. TG Synthesis	4. NO_2^- Reduction	
			5. $SO_4^=$ Reduction	
			6. Amino acids biosynthesis	
			7. Starch Synthase	
			8. Chlorophyll synthesis	
			9. Carotenoid synthesis	

FIGURE 15.9   Metabolic activities of the chloroplast.

important functions without which the plant could not survive. The student should review Section 8.6 in which the structure of the chloroplast is described.

REFERENCES

1. *R. Hill, "The Biochemists' Green Mansions: the Photosynthetic Electron Transport Chain in Plants." In* Essays in Biochemistry, *Vol. I. London: Academic Press, 1965.*

2. *J. A. Bassham and M. Calvin,* The Path of Carbon in Photosynthesis. *Englewood Cliffs, N.J.: Prentice-Hall, 1957.*

3. *C. H. Foyer,* Photosynthesis. *New York: Wiley, 1984.*

4. *M. D. Hatch and N. K. Boardman, eds.,* Biochemistry of Plants, *Vol. 8. New York: Academic Press, 1981.*

5. *N. E. Tolbert, "Photorespiration." In* Biochemistry of Plants, *Vol. 2, New York: Academic Press, 1981. pp. 488–523.*

REVIEW PROBLEMS

1. A wheat plant was placed in an atmosphere containing radioactive carbon dioxide ($^{14}CO_2$). After 30 seconds of photosynthesis, the plant was killed and the monosaccharide glucose was isolated. Upon degrading the glucose, radioactive carbon (^{14}C) was found predominantly in two of its six carbon atoms. Indicate which carbon atoms of glucose were labeled and the most probable sequence of reactions that can account for the labeling.

2. In the same experiment described in Problem 1, the amino acid alanine was isolated. Upon degrading the alanine, radioactive carbon (^{14}C) was found predominantly in one of its three carbon atoms. Indicate which carbon atom of alanine was labeled and the most probable sequence of reactions that can account for the labeling.

3. The RPP cycle for the path of carbon photosynthesis is composed of three phases. Identify the three phases and write a balanced equation for one enzyme reaction that occurs in each. Name the enzyme involved.

4. Explain why the Hill–Bendall or Z scheme of energy transfer in photosynthesis predicts a quantum requirement of eight quanta per mole of CO_2 consumed.

5. Write the equations for one reversible and one irreversible enzyme-catalyzed CO_2 fixation reaction. Name the enzymes and cofactors involved.

6. In a short period of time, oxalacetate becomes labeled in *Chlorella* cells that are photosynthesizing radioactive $^{14}CO_2$ by the RPP pathway. In sugar cane (which utilizes the Hatch–Slack pathway), oxalacetate also becomes labeled in a short period of time when the plant is administered $^{14}CO_2$. Show how the two samples (from *Chlorella* and sugar cane) of oxalacetate will be labeled.

THE NITROGEN AND SULFUR CYCLES

The third fundamental process in nature that is carried out by living cells in addition to photosynthesis and respiration is **nitrogen fixation.** This process, in turn, is part of the cycle of reactions known as the **nitrogen cycle.** Many constituents of the living cell contain nitrogen; they include proteins, amino acids, nucleic acids, purines, pyrimidines, porphyrins, alkaloids, and vitamins. The nitrogen atoms of these compounds eventually travel the nitrogen cycle, in which the nitrogen of the atmosphere serves as a reservoir. Nitrogen is removed from the reservoir by the process of fixation; it is then returned by the process of denitrification (see Figure 16.4).

16.1 COMPONENTS OF THE NITROGEN CYCLE

Several inorganic nitrogen compounds, as well as a myriad of organic nitrogen compounds, can be considered components of the nitrogen cycle. The former include N_2 gas, NH_3, nitrate ion (NO_3^-), nitrite ion (NO_2^-), and hydroxylamine (NH_2OH). At a glance, it is apparent that the nitrogen atom can possess a variety of oxidation numbers. Some of these are the following:

	NITRATE ION NO_3^-	NITRITE ION NO_2^-	HYPONITRITE ION $N_2O_2^{2-}$	NITROGEN GAS N_2	HYDROXYL-AMINE NH_2OH	AMMONIA NH_3
Oxidation number	$+5$	$+3$	$+1$	0	-1	-3

Thus, in nature, nitrogen may exist in either a highly oxidized form (NO_3^-) or a highly reduced state (NH_3).

For plant growth, nitrogen must be present in an available form, namely either as ammonia or the nitrate anion. Ammonia is provided from two principal sources, namely biological nitrogen fixation and nitrogenous fertilizers. The former contributes over 175×10^6 metric tons of fixed nitrogen per year worldwide; approximately 40×10^6 metric tons of nitrogenous fertilizer are produced per year worldwide. Still another 40×10^6 metric tons of fixed nitrogen are obtained from nitrate mining as well as from nitrogen fixed through electrical discharges that occur during lightning storms. During the discharge, oxides of nitrogen are formed that are subsequently hydrated by water vapor and carried to earth as nitrites and nitrates.

$$N_2 + O_2 \longrightarrow 2 \text{ NO} \xrightarrow{O_2} 2 \text{ NO}_2$$

16.2   NONBIOLOGICAL NITROGEN FIXATION

The term fixation is defined as the conversion of molecular N_2 into one of the inorganic forms listed above. The distinguishing feature of this process is the separation of the two atoms of N_2 that are triply bonded ($N\equiv N$); N_2 is an extremely stable molecule. An indication of the difficult nature of this reaction is seen in the conditions for the fixation of nitrogen in the Haber process, developed in Germany during World War I. The English naval blockade of Germany prevented German access to the Chilean nitrate fields, and it was necessary to develop another source of nitrate for their industry. The Haber process involves the reaction of N_2 and H_2 at extreme temperatures and pressures to form NH_3. The Haber process is used today for the fixation of N_2 by the chemical industry in the production of chemical fertilizer.

$$N_2(g) + 3 \text{ H}_2(g) \xrightarrow[200 \text{ atm}]{450°C} 2 \text{ NH}_3(g)$$
$$\Delta G° = -8 \text{ kcal/mole } N_2$$

Although thermodynamically favorable, the reaction does not proceed spontaneously but requires high temperatures and pressures to proceed.

16.3   BIOLOGICAL NITROGEN FIXATION

In sharp contrast to the chemical fixation of nitrogen is biological fixation, which occurs at one atmosphere of pressure and the temperature of living cells in the presence of appropriate enzymes.

$$N_2 + 3 \text{ H}_2 \xrightarrow[1 \text{ atm}]{25°C} 2 \text{ NH}_3$$

Biological fixation of nitrogen is accomplished either by nonsymbiotic microorganisms that can live independently or certain bacteria living in **symbiosis** with higher plants. The former group includes aerobic organisms of the soil (e.g., *Azotobacter*), soil anaerobes (e.g., *Clostridium* sp.), photosynthetic bacteria (e.g., *Rhodospirillum rubrum*), and cyanobacteria (e.g., *Anabaena* sp.). The symbiotic system consists of bacteria (*Rhizobia*) living in symbiosis with members of the *Leguminoseae* such as clover, alfalfa, and soy beans. Legumes are not the only higher plants that can fix nitrogen symbiotically; about 190 species of shrubs and trees, including the Sierra Sweet Bay, ceanothus, and alder, are nitrogen fixers. Indeed, the fertility of high-altitude mountain lakes may be determined by the number of alder trees growing near their inlets.

An essential feature of symbiotic fixation is the development of nodular tissue that forms on the roots of legumes after infection by a strain of *Rhizobia* specific for the given legume. The legume alone is unable to fix nitrogen; free living *Rhizobia* bacteria can fix N_2 only when grown with a limiting supply of organic nitrogen and oxygen. In symbiosis, however, the *Rhizobia* organisms and the legume interact in a remarkable relationship to attain the formation of organic nitrogen from nitrogen gas.

16.3.1 FREE LIVING ORGANISMS

Until 1960, many workers were unsuccessful in obtaining cell-free preparations that could fix nitrogen. In that year, J. E. Carnahan and his group at du Pont announced the first successful reduction of nitrogen gas *in vitro* to ammonia by a water-soluble extract of *Clostridium pasteurianum*. They discovered that in order for fixation to occur, large amounts of pyruvic acid had to be added to the extracts, whereupon the keto acid underwent phosphorolytic degradation to acetyl phosphate, CO_2, and H_2. It was soon discovered that the extract could be fractionated into two systems (Figure 16.1). One of these, the **hydrogen-donat-**

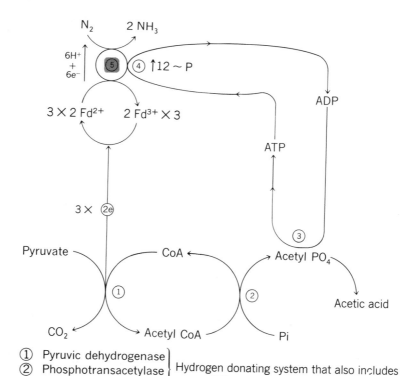

1. Pyruvic dehydrogenase ⎫
2. Phosphotransacetylase ⎬ Hydrogen donating system that also includes
3. Acetic kinase ⎭   an ATP-generating system for 12 ~ P
4. Reductant-dependent ATPase ⎫
5. Nitrogenase   ⎬ Nitrogenase system

FIGURE 16.1 Reduction of dinitrogen to ammonia by enzyme systems of *Clostridium pasteurianum.* Enzymes 1 to 3 comprise the hydrogen-donating system, and 4, 5, the nitrogenase system.

ing component, is responsible for the flow of electrons from the dissimilation of pyruvic acid via ferredoxin to the second component, called the **nitrogenase** system, which participates in the conversion of nitrogen to ammonia.

Thus, pyruvate did not participate directly in nitrogen fixation but served as a source of electrons and ATP during its metabolism. A second important observation was that the nitrogenase system was absent in extracts of *Clostridia* grown in the presence of NH_3 as the sole source of nitrogen although the hydrogen-donating system was still present in normal amounts. The third was the discovery of ferredoxin, the first nonheme-iron protein isolated and characterized (see Section 14.2.3).

The research by Carnahan and his colleagues stimulated a new effort by a number of investigators into a detailed analysis of this important series of reactions. As a result, it is now known that, regardless of whether extracts are prepared from anaerobes, aerobes, facultative anaerobes, cyanobacteria, or legume nodules, the essential reaction components are (1) an electron donor, (2) an electron acceptor (i.e., nitrogen gas), (3) ATP together with a divalent cation, Mg^{2+}, and (4) two protein components, the first being a molybdenum nonheme iron protein of about 220,000 MW (MoFe protein) and a nonheme iron protein component of 55,000 MW (Fe protein). (See Figure 16.2.) Each component alone is ineffective in catalyzing nitrogen fixation, but combined they form the nitrogenase complex. There is a specific requirement for ATP. In the absence of nitrogen as an electron acceptor, ATP can readily undergo hydrolysis to ADP and inorganic phosphate (reductive ATPase activity), with the evolution of hydrogen gas, by the reaction

$$4 \text{ ATP} + 4 \text{ H}_2\text{O} + 2 \text{ } e^- + 2 \text{ H}^+ \longrightarrow 4 \text{ ADP} + 4 \text{ Pi} + \text{H}_2$$

There is now good evidence that ATP binds specifically to the Fe protein; the redox potential of the Fe protein alone is -280 mV but shifts to -490 mV on

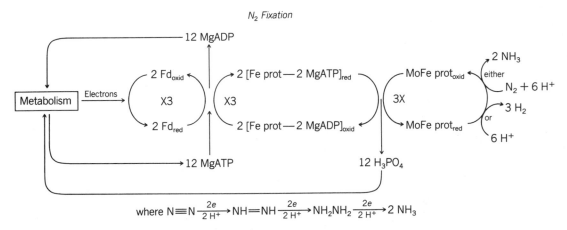

where $N{\equiv}N \xrightarrow[2\,H^+]{2e} NH{=}NH \xrightarrow[2\,H^+]{2e} NH_2NH_2 \xrightarrow[2\,H^+]{2e} 2\,NH_3$

FIGURE 16.2 Current scheme for nitrogen fixation by the nitrogenase complex (MoFe protein + Fe protein). Electrons derived from metabolic processes reduce ferredoxin (Fd). Reduced Fd in turn reduces an iron-protein-(Mg ADP)$_2$ complex to a very negative redox potential. In this process, MgADP is converted to MgATP as indicated. The reduced Feprot-(Mg ATP)$_2$ complex transfers electrons to MoFe protein with concomitant conversion to ADP and Pi. Reduced MoFe protein now reduces dinitrogen to NH_3 in the presence of protons. In the absence of N_2, protons are reduced to H_2.

complexing with 2(Mg ATP)s. As a result, the Fe-protein-(Mg ATP)$_2$ complex becomes a powerful reductant. As indicated in Figure 16.2, electrons are supplied by metabolic processes to a suitable acceptor that, in the case of *Clostridia,* is ferredoxin, and, with *Azobacter,* a flavoprotein called flavodoxin. The source of electrons in the nodular nitrogenase system is not clear. The metabolic processes also serve to generate ATP from ADP + Pi. The reduced Fe-protein-(Mg ATP)$_2$ complex with its very negative redox potential transfers its electrons to the MoFe protein that, in turn, reduces nitrogen to ammonia as well as protons to hydrogen gas. The overall reaction can be written as

$$N_2 + 6\ e^- + 12\ ATP + 12\ H_2O \longrightarrow 2\ NH_4^+ + 12\ ADP + 12\ Pi + 4\ H^+$$
$$\Delta G' = -136\ \text{Kcal/mole}\ N_2\ \text{reduced}$$

The nitrogenase system is rather nonspecific as next indicated. Protons are readily reduced to hydrogen as already indicated. Indeed, in all nitrogenase systems, about 30% of the electrons are diverted from nitrogen to convert protons to hydrogen, a wasteful process.

Reaction	Relative rate
$N_2 \xrightarrow{\ 6\ e^-\ } 2\ NH_3$	1.0
$\underset{\text{Acetylene}}{C_2H_2} \xrightarrow{\ 2\ e^-\ } \underset{\text{Ethylene}}{C_2H_4}$	3.4
$HCN \xrightarrow{\ 6\ e^-\ } CH_4 + NH_3$	0.6
$N_2O \xrightarrow{\ 2\ e^-\ } N_2 + H_2O$	3.0
$2\ H^+ \xrightarrow{\ 2\ e^-\ } H_2$	—

Acetylene is also an excellent substrate for the nitrogenase complex. This observation has led to the development of an ingenious microassay for nitrogen fixation. By measuring the rate of reduction of acetylene to ethylene by a soil or water sample under standard conditions, a field analysis can be rapidly conducted to reveal the capacity of that sample to fix nitrogen. This information can provide a basis for evaluating the effect of different environmental (bacterial and plant) factors on nitrogen fixation. This information, in turn, can be of great value in agricultural practices.

16.3.2 SYMBIOTIC N$_2$ FIXATION

The concepts of nitrogen fixation have been obtained, for the most part, from research with free-living organisms such as *Clostridium pasteurianum,* an anaerobe, or *Azotobacter vinlandii,* an aerobe. However, symbiotic nitrogen fixation, involving both legumes and *Rhizobia* bacteria, is unique and ecologically the most important contributor to biological nitrogen fixation. The root system of legumes such as clover, peas, and beans are infected by specific strains of free living, gram-negative bacteria, the *Rhizobia.* Once the *Rhizobia* enters the root hairs of the legume root system, a series of events occur, leading to the formation in the roots of specialized tumorlike tissues called nodules in which are found swollen, nonmotile, nonviable derivative cells of the original infecting *Rhizobia,* called **bacteroids.** Bacteroids contain a complete nitrogenase system that is very similar in its biochemical properties to those systems described in Section 16.3.1.

The plant cells in which the bacteroids reside contain a pigment, called leghemoglobin, that reversibly combines with oxygen in a manner similar to the interaction of oxygen with hemoglobin. The oxygenated leghemoglobin then transports the oxygen under low free oxygen tension to the sites of oxidative phosphorylation in the bacteroid where it is used during the generation of ATP. Leghemoglobin therefore serves as a buffer to control the oxygen levels. The biochemical events believed to occur in the bacteroid and the plant cell are outlined in Figure 16.3.

16.3.3   UNIQUE PROPERTIES OF THE NITROGENASE SYSTEM

This very important system is so unusual that a few additional comments should be made. Why were investigators prior to Carnahan's discovery unsuccessful in isolating a soluble nitrogenase system? The reasons are now apparent.

1.   The Fe protein is irreversibly inactivated by O_2 ($t_{1/2} = 30$ seconds). The MoFe protein is slightly more stable ($t_{1/2} = 4$ minutes).

2.   The Fe protein is extremely labile at $0°C$ but not at $20°C$.

3.   There must be a **large** input of ATP and a simultaneous maintenance of very low levels of ADP. ADP is a strong competitive inhibitor of the nitrogenase system. Note in Figures 16.1, 16.2, and 16.3 how the different systems maintain high levels of ATP and low levels of ADP.

4.   The requirement for electrons at a very negative redox potential must be met. Of further interest is the fact that the MoFe protein and the Fe protein that make up the nitrogenase complex in different species are almost identical regardless of their origin.

Much has been written about the efforts to convert cereal plants into nitrogen fixating plants. When molecular biologists attempt to transfer the nitrogenase (Nif) genes from procaryotic organisms to eucaryotic cells, they must transfer to the plant cell the 17 Nif genes and all the enzymes that enable the systems to generate ATP and electrons that help protect the nitrogenase from oxygen inhibition, as well as the systems needed to construct the nodules and transport the molybdenum and iron. All these factors must be provided to the plant so that it can fix nitrogen gas efficiently. In other words, the molecular biologist will have to convert a non-legume into a legume.

16.3.4   CONTROL OF NITROGENASE ACTIVITY

Control is exerted at two levels. The first or coarse control involves the repression of the synthesis of nitrogenase by ammonia. It is well known that in the presence of ammonia, nitrogen fixation ceases rather quickly. Ammonia as such has no direct inhibitory effect on nitrogenase *in vitro*. Thus, once an excess of ammonia accumulates in the organism, the synthesis of more nitrogenase is repressed; as the ammonia is utilized by the growing cell and falls to a low level, derepression will result and renewed synthesis of the enzyme begins again.

The second or fine control involves ADP as a competitive inhibitor of nitrogenase. At an ADP/ATP ratio of 0.2, 53% inhibition will occur under *in*

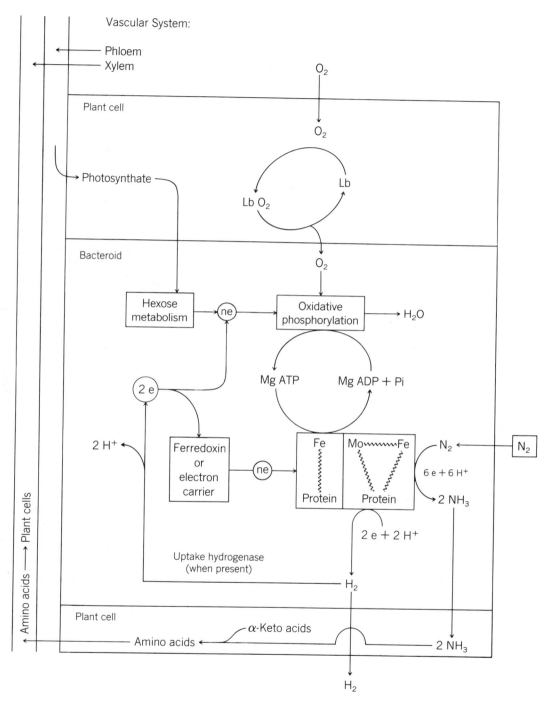

FIGURE 16.3 Fixation by a nodule cell composed of the original plant cell and the bacterial cell.

vitro conditions; at a ratio of 2.0, complete inhibition of nitrogenase is observed. Thus, as ATP levels fall and ADP levels rise in a cell, the cell signals for cessation of nitrogenase activity, which requires large amounts of ATP. In this way, the limited supply of ATP is redirected for other important cell functions.

16.3.5   ECOLOGY OF BIOLOGICAL N_2 FIXATION

The organisms that carry out nitrogen fixation utilize the NH_3 to produce the nitrogenous components (proteins, nucleic acids, pigments) of their tissues. The excess nitrogen fixed may be excreted into the soil or other media in which the nitrogen fixer is growing. For example, there is evidence that legumes and alders growing in sand culture excrete NH_3 and some amino acids into the sand surrounding their roots. The cyanobacteria also excrete NH_3 as well as amino acids and peptides. If NH_3 is excreted into the soil, it may be subjected to the process of nitrification next described, or it may be utilized by other living forms (soil bacteria or higher plants) incapable of nitrogen fixation. If the fixing organism is a higher plant, the excess fixed nitrogen may be synthesized into asparagine or glutamine and stored as these compounds.

When the nitrogen-fixing organism perishes, the proteins of its cells will be hydrolyzed to amino acids and these subsequently deaminated by decay bacteria through the action of amino acid oxidases, or transaminases and glutamic dehydrogenase. The reactions of interest, which result in the release of NH_3, are described in Chapter 17. Obviously, the nitrogenous components of the non-fixing organisms encounter the same fate on the death of the organism. Thus, the fertility of the soil is built up by the acquisition of NH_3 directly from nitrogen-fixing systems and indirectly after the nitrogen atom has made a cycle into the amino acids and proteins of the nitrogen fixers.

16.4   AMMONIA ASSIMILATION

There are three reactions that can catalyze the incorporation of the nitrogen atom as NH_3 into organic compounds. These reactions are those catalyzed by glutamic dehydrogenase (Equation 16.1), glutamine synthetase (Equation 16.2), and carbamyl phosphate synthetase (Equation 16.3). The details of these reactions are discussed in Section 17.2 in which the enzymes are described in the context of the metabolism of amino acids.

Glutamic dehydrogenase

$$
\begin{array}{ccc}
\text{COOH} & & \text{COOH} \\
| & & | \\
\text{C}{=}\text{O} \quad + \text{NH}_3 + \text{NADH} + \text{H}^+ \rightleftharpoons & \text{H}_2\text{NCH} \quad + \text{NAD}^+ + \text{H}_2\text{O} \quad (16.1)\\
| & & | \\
\text{CH}_2 & & \text{CH}_2 \\
| & & | \\
\text{CH}_2 & & \text{CH}_2 \\
| & & | \\
\text{COOH} & & \text{COOH} \\
\alpha\text{-Ketoglutaric acid} & & \text{L-Glutamic acid}
\end{array}
$$

Glutamine synthetase

$$
\underset{\text{Glutamic acid}}{\begin{array}{c}\text{COOH}\\ |\\ \text{H}_2\text{NCH}\\ |\\ \text{CH}_2\\ |\\ \text{CH}_2\\ |\\ \text{COOH}\end{array}} + \text{NH}_3 + \text{ATP} \xrightarrow{\text{Mg}^{2+}} \underset{\text{Glutamine}}{\begin{array}{c}\text{COOH}\\ |\\ \text{H}_2\text{NCH}\\ |\\ \text{CH}_2\\ |\\ \text{CH}_2\\ |\\ \text{CONH}_2\end{array}} + \text{ADP} + \text{H}_3\text{PO}_4 \quad (16.2)
$$

Carbamyl phosphate synthetase

$$
\text{NH}_3 + \text{CO}_2 + 2\text{ ATP} \xrightarrow{\text{Mg}^{2+}} \underset{\text{Carbamyl phosphate}}{\text{H}_2\text{N}\!-\!\overset{\displaystyle\text{O}}{\underset{\displaystyle\|}{\text{C}}}\!-\!\text{OPO}_3\text{H}_2} + 2\text{ ADP} + \text{H}_3\text{PO}_4 \quad (16.3)
$$

The reaction catalyzed by glutamic dehydrogenase was at one time considered to be the principal mechanism for conversion of NH_3 into the α-amino group of glutamic acid and, in turn, into other amino acids (see transamination, Section 17.2.3). However, the K_m for NH_4^+ for a number of glutamic dehydrogenases from different sources varies from 5 to 40 mM, that is, concentrations that approach toxic levels in the cell.

In higher plants and microorganisms, the enzyme glutamate synthase (Equation 16.4) performs this extremely important function of accomplishing the synthesis of glutamic acid required for transamination and many other biosynthetic roles.

$$
\underset{\text{Glutamine}}{\begin{array}{c}\text{CONH}_2\\ |\\ \text{CH}_2\\ |\\ \text{CH}_2\\ |\\ \text{CHNH}_2\\ |\\ \text{CO}_2\text{H}\end{array}} + \underset{\alpha\text{-Ketoglutarate}}{\begin{array}{c}\text{CO}_2\text{H}\\ |\\ \text{CH}_2\\ |\\ \text{CH}_2\\ |\\ \text{C}\!=\!\text{O}\\ |\\ \text{CO}_2\text{H}\end{array}} + \begin{array}{c}\text{NADPH}\\ +\\ \text{H}^+\end{array} \longrightarrow \underset{\text{Glutamate}}{\begin{array}{c}\text{CO}_2\text{H}\\ |\\ \text{CH}_2\\ |\\ \text{CH}_2\\ |\\ \text{CHNH}_2\\ |\\ \text{CO}_2\text{H}\end{array}} + \underset{\text{Glutamate}}{\begin{array}{c}\text{CO}_2\text{H}\\ |\\ \text{CH}_2\\ |\\ \text{CH}_2\\ |\\ \text{CH NH}_2\\ |\\ \text{CO}_2\text{H}\end{array}} + \text{NADP}^+ \quad (16.4)
$$

In this reaction, which is a reductive amination, glutamine serves as a source of the amino nitrogen atom. By coupling Equation 16.2 with Equation 16.4, we note that the organism must spend one ATP and one NADPH in the formation of one glutamate and that the system is essentially irreversible. Moreover, since the K_m value of glutamine synthetase for NH_4^+ is less than 0.5 mM and that for glutamine

$$
\text{NH}_3 + \text{Glutamate} + \text{ATP} \xrightarrow{\text{Mg}^{2+}} \text{Glutamine} + \text{ADP} + \text{Pi} \quad (16.2)
$$

Glutamine + α-Ketoglutarate + NADPH + H$^+$ $\longrightarrow$

2 Glutamate + NADP$^+$ + H$_2$O $\quad$ (16.4)

Sum: $\quad$ NH$_3$ + ATP + α-Ketoglutarate + NADPH + H$^+$ $\longrightarrow$

Glutamate + ADP + Pi + NADP$^+$ + H$_2$O

about 0.3 mM, it is now understood how plants and microorganisms can utilize lower, nontoxic levels of ammonia for the synthesis of organic nitrogen compounds. The occurrence of glutamate synthase in animal tissue has not yet been demonstrated.

16.5   NITRIFICATION

Despite the fact that NH_3 is the form in which nitrogen is normally added to the soil, little NH_3 is found there. Studies have shown that it is rapidly oxidized to nitrate ion; the latter represents the chief source of nitrogen for nonfixing plants. The oxidation of NH_3 is carried out by two groups of bacteria called the nitrifying bacteria. One group, *Nitrosomonas,* converts NH_3 to nitrite ion with O_2 as the oxidizing agent.

$$NH_3 + \tfrac{3}{2} O_2 \longrightarrow NO_2^- + H_2O + H^+$$
$$\Delta G' = -66,500 \text{ cal}$$

The other group, *Nitrobacter,* oxidizes nitrite to nitrate.

$$NO_2^- + \tfrac{1}{2} O_2 \longrightarrow NO_3^-$$
$$\Delta G' = -17,500 \text{ cal}$$

Both reactions are exergonic; the first involves the oxidation of nitrogen from -3 to $+3$; the second is a two-electron oxidation from $+3$ to $+5$. Both groups of organisms are **autotrophs;** that is, they make all their cellular carbon compounds (protein, lipids, carbohydrates) from CO_2. As was indicated in Chapter 15, the conversion of CO_2 to carbohydrate requires energy. In photosynthesis, that energy is supplied by light; in the cases of *Nitrosomonas* and *Nitrobacter,* the energy for the reduction of CO_2 to carbohydrate and other carbon compounds is furnished by the oxidation of NH_3 and NO_2^- ion, respectively. Since the organisms obtain their energy for growth by the oxidation of simple inorganic compounds, they are termed **chemoautotrophs.**

Little is known about the intermediates in the oxidation of NH_3 to NO_2^- by *Nitrosomonas,* nor is there much information on the intermediary metabolism of the carbon compounds found in these bacteria. The lack of knowledge is due chiefly to the difficulty encountered in growing adequate amounts of the bacteria for experimentation. From the standpoint of comparative biochemistry, it may be predicted that the carbon compounds will undergo reactions resembling those described for animals, plants, and other microorganisms. The unique reactions, if any, may be expected to involve NH_3 and NO_2^-, the compounds that supply the energy for the growth of these bacteria.

16.6   UTILIZATION OF NITRATE ION

With NO_3^- as the most abundant form of nitrogen in the soil, plants and soil organisms have developed an ability to utilize the anion as the nitrogen source required for their growth and development. In Chapter 17, however, it will be pointed out that the major route for incorporating inorganic nitrogen into organic nitrogen compounds involves the reactions catalyzed by glutamine synthetase and glutamate synthase. Therefore, higher plants and microorganisms after taking up NO_3^- must reduce it to NH_3 so that it can be assimilated. This reduction occurs in two steps.

The first step is catalyzed by the enzyme **nitrate reductase** that is found in the cytosol of plant cells and catalyzes the following reaction.

$$NO_3^- + NADPH + H^+ \longrightarrow NO_2^- + NADP^+ + H_2O$$

Nitrate reductases have been purified from bacteria, higher plants (soya beans), and the bread mold, *Neurospora*. In each case, one of the reduced nicotinamide nucleotides (NADPH or NADH) serves as a source of electrons for the reduction. The enzymes are flavoproteins that contain FAD and require the metal molybdenum as a cofactor. Both cofactors undergo oxidation – reduction during the reaction.

The second step in the reduction of NO_3^- to NH_3 is catalyzed by the enzyme **nitrite reductase**

$$NO_2^- + 3NADPH + 3\,H^+ \xrightarrow{\text{Nitrite reductase}} NH_3 + 3NADP^+ + H_2O + OH^-$$

The enzyme accomplishes the six-electron reduction of NO_2^- to NH_3. It is found in the chloroplasts of higher plants where it is associated with the thylakoid membrane; in this location it would have easy access to the reducing agent NADPH produced by light. Nitrite reductase contains an iron-heme protein *siroheme* that is presumably involved in the reduction. Hyponitrite ion ($N_2O_2^{2-}$) and hydroxylamine (NH_2OH) may be enzyme-bound intermediates.

This utilization of nitrogen, in which aerobic microorganisms and higher plants reduce a nitrate ion to NH_3 in order to incorporate it into cell protein, is referred to as **nitrate assimilation.** It is perhaps difficult to understand why in nature NH_3 is readily oxidized to NO_3^- that, in turn, must be again reduced to NH_3 before incorporation into amino acids. One possibility is that ammonia is fairly active metabolically and cannot be stored as such in plant tissues. Nitrate, on the other hand, can be absorbed and stored in the vacuole to be reduced to NH_3, as needed, which is then assimilated.

Many microorganisms, including *E. coli* and *B. subtilis,* reduce NO_3^- to NH_3 for another purpose; they utilize NO_3^- as a terminal electron acceptor instead of O_2. The NO_3^-, with its high oxidation reduction potential of 0.96 V at pH 7.0, can accept electrons released during the oxidation of organic substrates. The intermediates are probably NO_2^-, $N_2O_2^{2-}$, and NH_2OH, as in nitrate assimilation, but this process is known as **nitrate respiration.** Moreover, the enzymes involved are firmly associated with the insoluble matter of the cell, principally the plasma membrane. In the case of *Achromobacter fischeri,* the reduction of NO_3^- has been coupled with the oxidation of reduced cytochrome *c*; the presence of a cytochrome electron-transport chain that can react with NO_3^- rather than O_2 as the terminal oxidase is therefore indicated.

Some bacteria (e.g., *Pseudomonas denitrificans* and *Denitrobacillus*) that utilize nitrate respiration produce N_2 instead of NH_3. In this process, the return of the nitrogen atom as nitrogen gas to the atmosphere is accomplished. This sequence is referred to as **denitrification.** There is little detailed information on the enzyme systems involved.

The different processes that constitute the nitrogen cycle are diagrammed in Figure 16.4.

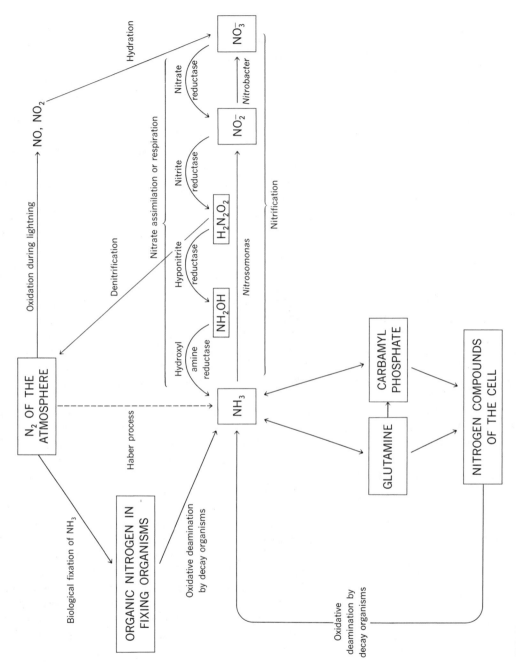

FIGURE 16.4   The nitrogen cycle.

16.7 THE SULFUR CYCLE

There are many similarities between the biochemistry of the sulfur atom and that of the nitrogen atom. Because of the occurrence of sulfur in many essential biochemical compounds, the metabolism of this element will be discussed briefly.

The sulfur atom exists in inorganic forms such as sulfate (SO_4^{2-}), sulfite (SO_3^{2-}), thiosulfate ($S_2O_3^{2-}$), elemental sulfur (S), and sulfide (S^{2-}). The oxidation state for these compounds ranges from $+6$ for sulfate to -2 for sulfide. Before the sulfur atom can enter into organic combination, it must be reduced to the level of sulfide (H_2S). When the sulfur atom is released from organic combination, it can be oxidized by soil organisms to its highest oxidation state (SO_4^{2-}). Thus, one may speak of the **sulfur cycle** in nature and identify certain aspects of that cycle.

16.7.1 SULFATE ACTIVATION

The sulfate anion can be utilized by both plants and animals for the formation of the numerous sulfate esters found in nature — the steroid and phenol sulfates, choline sulfate, the polysaccharide sulfates of animals (chondroitin sulfate and heparin) and algae (agar). Before it can be utilized, the sulfate anion must be activated, in analogy with the activation of phosphate. This occurs in two steps, the first being the formation of adenosine-5′-phosphosulfate (APS) followed by the formation of adenosine-3′-phosphate-5′-phosphosulfate (PAPS).

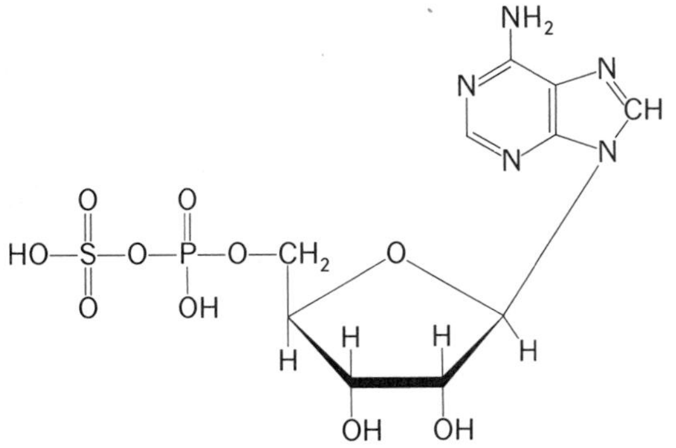

Adenosine-5′-phosphosulfate
(APS)

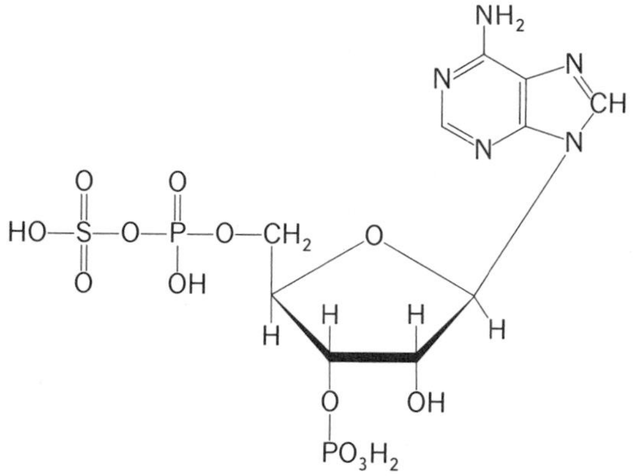

Adenosine-3′-Phosphate-5′-phosphosulfate
(PAPS)

The initial reaction catalyzed by **ATP sulfurylase** is analogous to the activation of the carboxyl group of fatty acids or amino acids in that the adenylic acid moiety of ATP is transferred to the sulfate and inorganic pyrophosphate is formed.

$$SO_4^{2-} + A\!-\!R\!-\!P\!-\!P\!-\!P \rightleftharpoons A\!-\!R\!-\!P\!-\!S + P\!-\!P \qquad (16.5)$$

$$\underset{\text{ATP}}{} \qquad \underset{\text{APS}}{}$$

The equilibrium for this reaction is highly unfavorable for APS formation because the $\Delta G'$ of hydrolysis for the phosphate–sulfate anhydride is much more negative (estimated at $-19,000$ cal/mole) than that of the phosphate anhydride bond in ATP. The phosphate–sulfate linkage in APS is stabilized by a further activation reaction involving ATP and the formation of PAPS.

$$APS + ATP \rightleftharpoons PAPS + ADP \qquad (16.6)$$

The kinase involved here, APS kinase, has a high affinity for APS and this, together with the presence of the ubiquitous pyrophosphatases, tends to pull Reaction 16.5 to the right and favor PAPS formation in Reaction 16.6.

PAPS is the sulfate donor in the formation of the sulfate esters cited above. The general reaction may be represented as

$$ROH + PAPS \longrightarrow R-OS + PAP$$

Little detail is available, however, on the enzymes involved in these sulfurylation reactions.

16.7.2   SULFATE REDUCTION

The reduction of sulfate to the oxidation level of sulfide ion must occur before the sulfur atom can be utilized to form cysteine. This assimilation process, which is performed only by higher plants and microorganisms, is analogous to the assimilation of nitrate to form NH_3. As in the nitrogen cycle, some anaerobic bacteria (e.g., *Desulfovibrio* sp.) utilize sulfate as a terminal oxidizing agent in analogy with those microorganisms that similarly use nitrate respiration. This process is called **dissimilatory sulfate reduction** and H_2S is the reduced form of sulfur produced.

Higher plants perform the essential role of assimilating sulfur into organic compounds that can then be made available to animals. This process involves the reduction of the sulfur atom from the oxidation state of $+6$ to -2 and its incorporation into cysteine. The reduction takes place in the chloroplast where synthesis of cysteine (and many other amino acids) occurs. Cysteine then serves as the source of organic sulfur for the rest of the biosphere.

The reduction of sulfur occurs in the presence of a membrane-bound carrier (CAR) that contains an active sulfhydryl group. Initially, the carrier is charged with a sulfate group from APS to form a thiosulfonate.

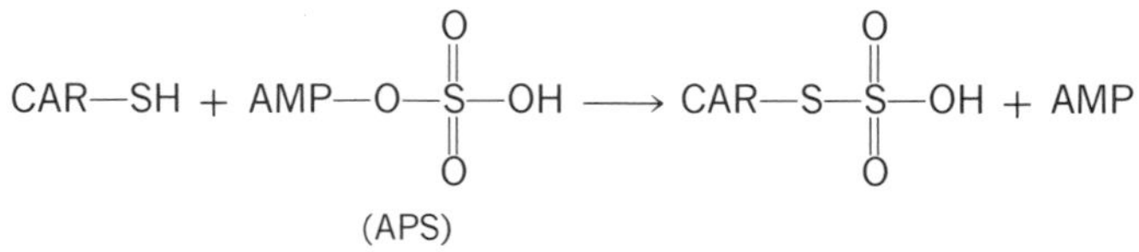

(APS)

While bound to the carrier, the thiosulfonate is reduced by thiosulfonate reductase to form a dithiol form of the carrier. Reduced ferredoxin provides the six electrons required for this process.

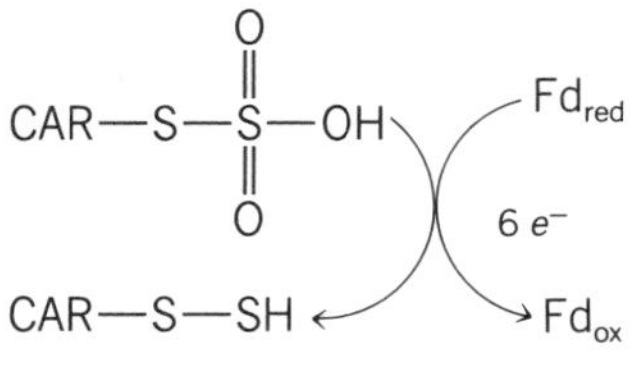

16.7.3 INCORPORATION OF SULFUR INTO ORGANIC COMPOUNDS

The dithio form of the carrier then reacts with O-acetylserine to produce cysteine, acetate, and the thiol group in the carrier. The two electrons required for this process are also provided by ferredoxin.

$$
\text{CAR—S—SH} + \text{AcO—CH}_2 \quad\text{HS—CH}_2 \qquad\qquad (16.7)
$$

$$
\underset{\text{O-Acetylserine}}{\overset{\displaystyle\text{AcO—CH}_2}{\underset{\displaystyle\text{CO}_2\text{H}}{\overset{\displaystyle\mid}{\underset{\displaystyle\mid}{\text{CHNH}_2}}}}} \xrightarrow{2\,e^-} \underset{\text{Cysteine}}{\overset{\displaystyle\text{HS—CH}_2}{\underset{\displaystyle\text{CO}_2\text{H}}{\overset{\displaystyle\mid}{\underset{\displaystyle\mid}{\text{CHNH}_2}}}}} + \begin{array}{c}\text{CAR—SH}\\ + \\ \text{Acetate}\end{array}
$$

If plants and microorganisms are provided with reduced sulfur in the form of H_2S, this compound can also react with O-acetylserine to produce cysteine. The enzyme that catalyzes this reaction, **cysteine synthase**, is widely distributed in plants, but is unknown in animals.

$$
\text{H}_2\text{S} + \underset{\text{O-Acetyl serine}}{\overset{\displaystyle\text{AcO—CH}_2}{\underset{\displaystyle\text{CO}_2\text{H}}{\overset{\displaystyle\mid}{\underset{\displaystyle\mid}{\text{CHNH}_2}}}}} \longrightarrow \text{Acetate} + \underset{\text{Cysteine}}{\overset{\displaystyle\text{HS—CH}_2}{\underset{\displaystyle\text{CO}_2\text{H}}{\overset{\displaystyle\mid}{\underset{\displaystyle\mid}{\text{CHNH}_2}}}}} \qquad (16.8)
$$

It is not clear whether Reactions 16.7 and 16.8 are catalyzed by the same enzyme.

An analogous reaction involving homoserine (as O-succinyl homoserine) and H_2S to form homocysteine is catalyzed by a separate enzyme found in higher plants and some bacteria.

$$
\text{H}_2\text{S} + \underset{\text{O-Succinyl homoserine}}{\overset{\displaystyle\text{Suc—O—CH}_2}{\underset{\displaystyle\text{CO}_2\text{H}}{\overset{\displaystyle\mid}{\underset{\displaystyle\mid}{\underset{\displaystyle\text{CHNH}_2}{\overset{\displaystyle\mid}{\text{CH}_2}}}}}}} \longrightarrow \text{Succinate} + \underset{\text{Homocysteine}}{\overset{\displaystyle\text{HS—CH}_2}{\underset{\displaystyle\text{CO}_2\text{H}}{\overset{\displaystyle\mid}{\underset{\displaystyle\mid}{\underset{\displaystyle\text{CHNH}_2}{\overset{\displaystyle\mid}{\text{CH}_2}}}}}}}
$$

Free H_2S can also be produced by the reduction of free sulfite; the enzyme responsible, sulfite reductase, is widely distributed in plants where it is found in the chloroplasts. It utilizes reduced ferredoxin as the reductant.

$$
\text{SO}_3^{2-} + 6\text{Fd}_{\text{red}} \longrightarrow \text{S}^{2-} + 6\text{Fd}_{\text{ox}}
$$

16.7.4 CYSTEINE AS A PRIMARY SULFUR SOURCE

Cysteine serves as the primary source of sulfur for the formation of methionine in plants and microorganisms. Intermediates in this process described in Section 17.5.1 are cystathionine and homocysteine; two enzymes, **cystathionine synthase I** and **cystathionase**, are required. The synthesis of cysteine from methionine in animals is also discussed in Section 17.5.1.

16.8   THE RELEASE OF SULFUR FROM ORGANIC COMPOUNDS

In animal tissues, the major route for cysteine catabolism is catalyzed by cysteine oxidase. In this reaction, the sulfur atom is oxidized while remaining in organic combination and cysteine sulfinic acid is the product.

$$HS-CH_2-\underset{\underset{NH_2}{|}}{\overset{\overset{H}{|}}{C}}-CO_2H + O_2 \longrightarrow HO-\overset{\overset{O}{\|}}{S}-CH_2-\underset{\underset{NH_2}{|}}{\overset{\overset{H}{|}}{C}}-CO_2H$$

Cysteine   Cysteine sulfinic
                            acid

This reaction, which has no analogy in the metabolism of the other amino acids, can be followed by transamination to yield pyruvic acid, SO_2, and the product amino acid.

$$HO-\overset{\overset{O}{\|}}{S}-CH_2-\underset{\underset{NH_2}{|}}{\overset{\overset{H}{|}}{C}}-CO_2H + R-\underset{\underset{O}{\|}}{C}-CO_2H \longrightarrow CH_3-\underset{\underset{O}{\|}}{C}-CO_2H + SO_2 + R-\underset{\underset{NH_2}{|}}{\overset{\overset{H}{|}}{C}}-CO_2H$$

Cysteine sulfinic   Keto acid   Pyruvic   Amino
     acid   acid   acid

The transamination of cysteine also readily occurs, forming β-mercaptopyruvate. This keto acid, in turn, serves as donor of the sulfur atom to nucleophiles such as CN^-, RSH, or SO_3^{2-} to form SCN^-, $R-S-SH$, and $S_2O_3^{2-}$. The enzyme is **β-mercaptopyruvate transsulfurase.** The $R-S-SH$ in the presence of excess thiol (RSH) is reduced to H_2S.

$$
\begin{array}{l}
SH \\
| \\
CH_2 \\
| \\
CHNH_2 \\
| \\
CO_2H
\end{array}
\xrightarrow{\text{Transamination}}
\begin{array}{l}
SH \\
| \\
CH_2 \\
| \\
C{=}O \\
| \\
CO_2H
\end{array}
\begin{array}{c}
\xrightarrow[R-SH]{CN^-} SCN^- \\
\xrightarrow{SO_3^{2-}} R-S-SH + \\
\searrow S_2O_3^{2-}
\end{array}
\begin{array}{l}
CH_3 \\
| \\
C{=}O \\
| \\
CO_2H
\end{array}
$$

Cysteine β-Mercapto   Pyruvate
                            pyruvate

$$R-S-SH + R-SH \longrightarrow R-S-S-R + H_2S$$

The main route of cysteine catabolism in plants involves its oxidation, followed by cleavage in the presence of **S-alkylcysteine lyase** to form thiocysteine, pyruvate, and NH_3.

$$
2\begin{array}{l}
SH \\
| \\
CH_2 \\
| \\
CHNH_2 \\
| \\
CO_2H
\end{array}
\xrightarrow{-2 H}
\begin{array}{l}
CO_2H \\
| \\
CHNH_2 \\
| \\
CH_2-S-S-CH_2
\end{array}
\begin{array}{l}
CO_2H \\
| \\
CHNH_2
\end{array}
\xrightarrow{H_2O}
\begin{array}{l}
CO_2H \\
| \\
CHNH_2 \\
| \\
CH_2-S-SH
\end{array}
+ \begin{array}{l}
CO_2H \\
| \\
C{=}O \\
| \\
CH_3
\end{array} + NH_3
$$

Cysteine   Cystine   Thiocysteine

The thiocysteine can react with a mole of cysteine to produce cystine and H_2S.

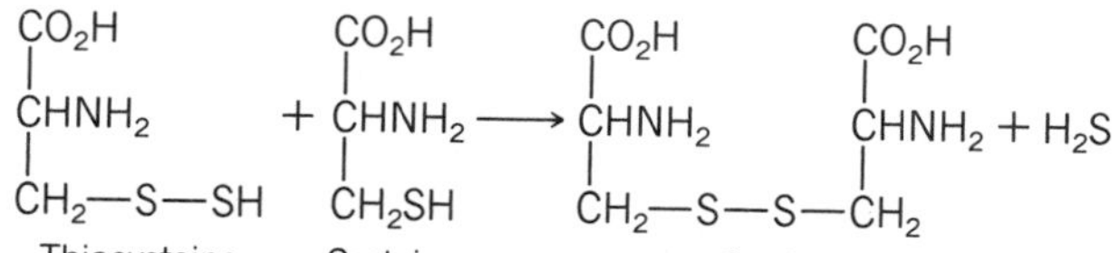

$$\underset{\text{Thiocysteine}}{\overset{\displaystyle \begin{array}{c}CO_2H \\ | \\ CHNH_2 \\ | \\ CH_2-S-SH \end{array}}{}} \quad + \quad \underset{\text{Cysteine}}{\overset{\displaystyle \begin{array}{c}CO_2H \\ | \\ CHNH_2 \\ | \\ CH_2SH \end{array}}{}} \quad \longrightarrow \quad \underset{\text{Cystine}}{\overset{\displaystyle \begin{array}{c}CO_2H \\ | \\ CHNH_2 \\ | \\ CH_2-S-S-CH_2 \end{array}}{}} \quad \overset{\displaystyle \begin{array}{c}CO_2H \\ | \\ CHNH_2 \\ | \\ \end{array}}{} + H_2S$$

Since sulfur is included among the six most abundant elements found in the biosphere, it is not surprising that mechanisms have evolved for conserving this element.

REFERENCES

1. *B. J. Miflin, ed., Chapters 1–5. In* Biochemistry of Plants, *Vol. V, New York: Academic Press, 1980.*

2. *C. Veeger and W. Newton, eds.* Advances in Nitrogen Fixation Research *Amsterdam: Nejhoff/Junk, 1984.*

REVIEW PROBLEMS

1. Explain the role of pyruvate in biological nitrogen fixation.

2. Define the following terms:
   (a) Nitrification   (c) Nitrogen fixation
   (b) Denitrification

3. Distinguish between *nitrate assimilation* and *nitrate respiration*.

4. Compare nitrate assimilation and sulfate assimilation with regard to the requirements for ATP, reducing power, and the participation of low molecular weight proteins that act as "giant coenzymes."

5. What are two physiological functions of nitrate reduction in microorganisms?

6. Compare and contrast the interconversion of cysteine and methionine in animals and plants.

7. What function does the acetylation of serine play in the biosynthesis of cysteine?

8. Write balanced, enzyme-catalyzed reactions for the biosynthesis of cysteine via a pathway utilizing H_2S as the sulfur source (as it occurs in a plant or microorganisms).

9. What role does the acetylation of serine play in the biosynthetic pathway for cysteine? Give two reasons or functions.

THE METABOLISM OF AMMONIA AND NITROGEN-CONTAINING COMPOUNDS

The role of higher plants in assimilating CO_2 during photosynthesis and the dependence of all animal life on this process was discussed in Chapter 15. Higher plants play a similar essential role in assimilating nitrogen as it is recycled through NH_3 (and nitrate) in the biological nitrogen cycle. Finally, plants also play a unique role in the assimilation of inorganic sulfur in the form of H_2S into the sulfur-containing amino acids (Chapter 16). Thus, animals are totally dependent on the plant kingdom for providing combined forms of carbon, nitrogen, and sulfur.

17.1 SOURCES OF NITROGEN

Whereas the higher animal can utilize NH_3 for synthesis of glutamine and glutamic acid, only the former is of consequence because the animal's principal source of nitrogen is the plant or animal protein that it consumes in its diet. The protein is hydrolyzed to amino acids by enzymes in the gastrointestinal tract and these are absorbed into the blood and transported to the liver. This organ will remove a portion of the amino acids for specific biosynthetic tasks, while the remainder pass on to extrahepatic tissues where they can be synthesized into proteins. The liver is the site of synthesis of several blood proteins (plasma albumin, the globulins, fibrinogen, and prothrombin). It also metabolizes any amino acids in excess of hepatic needs for protein synthesis, converting the nitrogen atoms into urea and the carbon skeleton into intermediates previously encountered in the metabolism of carbohydrates and lipids.

Plants, on the other hand, utilize only inorganic nitrogen (as nitrate or ammonia) in the synthesis of organic nitrogen compounds. Plants therefore have the ability to synthesize glutamine from NH_3, and glutamate in turn from glutamine and, as discussed in Section 17.2, these two molecules play key roles in the formation of all the other amino acids and a large number of other organic nitrogen compounds.

17.2 ASSIMILATION OF NH$_3$

17.2.1 GLUTAMINE SYNTHETASE

One of the two key reactions in the assimilation of NH$_3$ is that catalyzed by **glutamine synthetase,** an enzyme that is ubiquitous in nature.

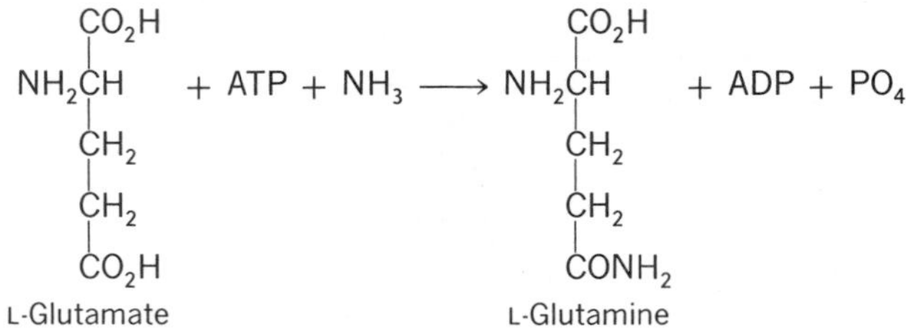

$$
\begin{array}{l}
\text{CO}_2\text{H} \\
|\\
\text{NH}_2\text{CH} \\
|\\
\text{CH}_2 \\
|\\
\text{CH}_2 \\
|\\
\text{CO}_2\text{H}
\end{array}
\;+\; \text{ATP} + \text{NH}_3 \longrightarrow
\begin{array}{l}
\text{CO}_2\text{H} \\
|\\
\text{NH}_2\text{CH} \\
|\\
\text{CH}_2 \\
|\\
\text{CH}_2 \\
|\\
\text{CONH}_2
\end{array}
\;+\; \text{ADP} + \text{PO}_4 \qquad (17.1)
$$

L-Glutamate   L-Glutamine

The first step in the reaction is the formation of a γ-glutamyl-phosphate-enzyme complex. In the second step, ammonia, a good nucleophile, attacks the complex and displaces the phosphate group to form glutamine and inorganic phosphate. Note that only the γ-amide is formed. Isoglutamine, the compound with the α-carboxyl group amidated, is never formed. The enzyme is also highly specific, since aspartic acid cannot replace glutamic acid as a substrate.

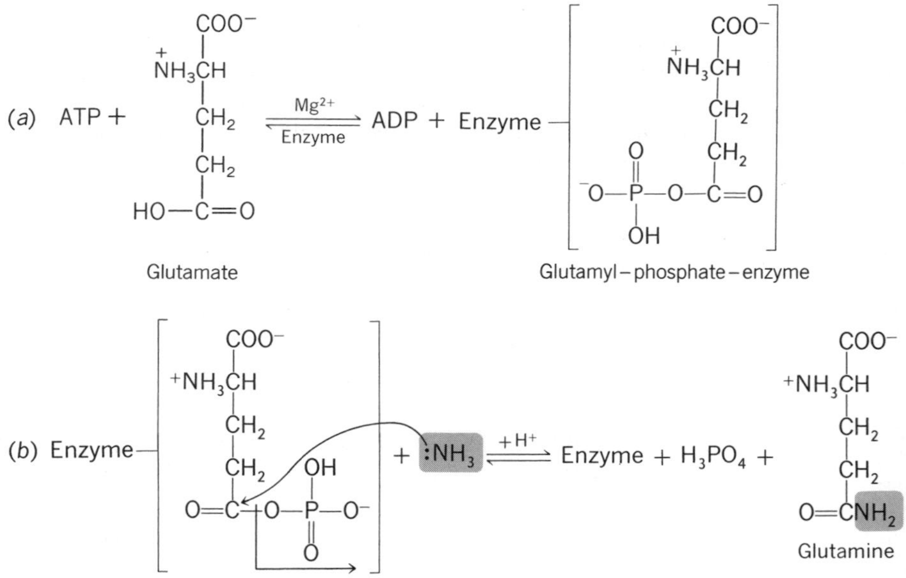

(a) ATP + [Glutamate structure] $\xrightleftharpoons[\text{Enzyme}]{\text{Mg}^{2+}}$ ADP + Enzyme—[Glutamyl–phosphate–enzyme structure]

Glutamate   Glutamyl–phosphate–enzyme

(b) Enzyme—[intermediate structure] + :NH$_3$ $\xrightleftharpoons{+\,\text{H}^+}$ Enzyme + H$_3$PO$_4$ + [Glutamine structure]

Glutamine

Overall reaction:

$$\text{Glutamate} + \text{ATP} + \text{NH}_3 \xrightarrow[\text{synthetase}]{\text{Glutamine}} \text{Glutamine} + \text{ADP} + \text{H}_3\text{PO}_4$$

The full significance of glutamine in nitrogen metabolism becomes apparent when it is recognized that the amide nitrogen atom serves as the ultimate source of the nitrogen atoms in *all* other nitrogen compounds, at least in plants. The only other way in which NH$_3$ can be *directly* incorporated into organic nitrogen compounds involves enzymes that catalyze the synthesis of glutamate (Reaction 17.8) or carbamyl phosphate (Reaction 17.5), and these are limited to a few specific animal tissues or organs.

Because glutamine is a multifunctional precursor, it is not surprising that the enzyme catalyzing its synthesis in bacteria is an allosteric enzyme subject to control by a variety of different compounds. No fewer than eight products of glutamine metabolism — tryptophan, histidine, glycine, alanine, glucosamine-6-phosphate, carbamyl phosphate, AMP, and CTP — have been shown to serve as independent negative feedback inhibitors of the enzyme in *E. coli*. Similar regulatory properties have not yet been demonstrated for glutamine synthetases of animal or plant origin.

The biosynthesis of asparagine and its metabolism are briefly described in Box 17.A.

BOX 17.A   BIOSYNTHESIS OF ASPARAGINE

The metabolism of the amino acid amide, asparagine, contrasts in several ways with that of the other amido amino acid glutamine. In the first place, the biosynthesis of asparagine in plants is not completely sorted out although asparagine was the first amino acid discovered (in 1806 in asparagus). In contrast to glutamine, asparagine is not known to donate its amide nitrogen atom directly as described for glutamine in Steps 1 and 4, Figure 17.5 and Reaction 17.2. Nevertheless, asparagine plays an important role as a transporter of nitrogen in many plants. In particular, when the seeds of some legumes germinate, the nitrogen in the protein-rich seed is mobilized in the form of asparagine and made available to the growing seedling as a source of nitrogen needed for amino acid synthesis. Asparaginases are known that catalyze the release of NH$_3$ from asparagine (Reaction 17.12).

The biosynthesis of asparagine from aspartate involves glutamine as a source of the amide nitrogen. The enzyme asparagine synthetase found in plant tissues catalyzes the ATP-dependent synthesis shown.

$$
\begin{array}{c}
\text{CO}_2\text{H} \\
|\\
\text{CH}_2 \\
|\\
\text{CHNH}_2 \\
|\\
\text{CO}_2\text{H}
\end{array}
+ \text{ATP} +
\begin{array}{c}
\text{CONH}_2 \\
|\\
\text{CH}_2 \\
|\\
\text{CH}_2 \\
|\\
\text{CHNH}_2 \\
|\\
\text{CO}_2\text{H}
\end{array}
\longrightarrow
\begin{array}{c}
\text{CONH}_2 \\
|\\
\text{CH}_2 \\
|\\
\text{CHNH}_2 \\
|\\
\text{CO}_2\text{H}
\end{array}
+
\begin{array}{c}
\text{CO}_2\text{H} \\
|\\
\text{CH}_2 \\
|\\
\text{CH}_2 \\
|\\
\text{CHNH}_2 \\
|\\
\text{CO}_2\text{H}
\end{array}
+
\begin{array}{c}
\text{AMP} \\
+ \\
\text{PPi}
\end{array}
$$

Aspartate   Glutamine   Asparagine   Glutamate

This enzyme, first discovered in plants, contrasts greatly with the synthesis of glutamine (Reaction 17.1) that utilizes inorganic NH$_3$ as a source of nitrogen. This fact again stresses the importance and unique nature of the reaction catalyzed by glutamine synthetase.

17.2.2 GLUTAMATE SYNTHASE

The synthesis of amino acids, in plant or animals, requires that the nitrogen atom in NH$_3$ ultimately make its way into the α-amino groups of these compounds. In plants and microorganisms, this is accomplished by the enzyme glutamate synthase that catalyzes the **reductive amination** of the keto group of α-ketoglutarate with the amide group of glutamine serving as the source of the nitrogen atom.

$$\begin{array}{c} \text{CO}_2\text{H} \\ | \\ \text{C}{=}\text{O} \\ | \\ \text{CH}_2 \\ | \\ \text{CH}_2 \\ | \\ \text{CO}_2\text{H} \end{array} + \text{NADPH} + \text{H}^+ + \begin{array}{c} \text{CO}_2\text{H} \\ | \\ \text{NH}_2\text{CH} \\ | \\ \text{CH}_2 \\ | \\ \text{CH}_2 \\ | \\ \text{NH}_2{-}\text{C}{=}\text{O} \end{array} \longrightarrow \begin{array}{c} \text{CO}_2\text{H} \\ | \\ \text{NH}_2\text{CH} \\ | \\ \text{CH}_2 \\ | \\ \text{CH}_2 \\ | \\ \text{CO}_2\text{H} \end{array} + \text{NADP}^+ + \begin{array}{c} \text{CO}_2\text{H} \\ | \\ \text{NH}_2\text{CH} \\ | \\ \text{CH}_2 \\ | \\ \text{CH}_2 \\ | \\ \text{CO}_2\text{H} \end{array} \qquad (17.2)$$

| α-Ketoglutarate | L-Glutamine | L-Glutamate | L-Glutamate |

In bacteria, where it was first discovered, the enzyme utilizes NADPH as the reducing agent. In plants, two glutamate synthases are known; the one located in root tissue also utilizes NADPH. The importance of this enzyme in the assimilation of NH_3 produced in legumes by N_2 fixation has been described (Section 16.4). An equally important glutamate synthase is the enzyme in chloroplasts where most amino acids in plants are synthesized. This chloroplast enzyme utilizes reduced ferredoxin as the reductant; the reduced ferredoxin, in turn, is readily produced by the action of light on the thylakoid membranes.

It should be stressed that glutamate synthase is highly specific for L-glutamine and α-ketoglutarate; it does not utilize any other keto acid for acceptor nor does it use asparagine as a donor of amide nitrogen. Glutamate synthase has never been described in animal tissues.

17.2.3   TRANSAMINATION

The synthesis of all other amino acids can be accomplished by the process of **transamination;** the reaction of transamination involves the transfer of the amino group from an amino acid to a keto acid (the carbon skeleton) to form the analogous amino acid and produce the keto acid (the carbon skeleton) of the original amino donor.

$$\text{R}^1{-}\underset{\underset{\text{NH}_2}{|}}{\overset{\overset{\text{H}}{|}}{\text{C}}}{-}\text{CO}_2\text{H} + \text{R}^2{-}\underset{\underset{\text{O}}{\|}}{\text{C}}{-}\text{CO}_2\text{H} \rightleftharpoons \text{R}^1{-}\underset{\underset{\text{O}}{\|}}{\text{C}}{-}\text{CO}_2\text{H} + \text{R}^2{-}\underset{\underset{\text{NH}_2}{|}}{\overset{\overset{\text{H}}{|}}{\text{C}}}{-}\text{CO}_2\text{H}$$

| Amino acid$_1$ | Keto acid$_2$ | Keto acid$_1$ | Amino acid$_2$ |
| (donor) | (acceptor) | | |

Amino transferases capable of reacting with nearly all of the amino acids have been reported, but especially important are **glutamic amino transferase** and **alanine amino transferase.** Glutamic amino transferase is specific for glutamic and α-ketoglutaric acid as one of its two substrate pairs, but will react, at different rates, with nearly all of the other proteinaceous amino acids.

$$\text{R}^1{-}\underset{\underset{\text{O}}{\|}}{\text{C}}{-}\text{CO}_2\text{H} + \text{H}_2\text{N}{-}\underset{\begin{array}{c}|\\\text{CH}_2\\|\\\text{CH}_2\\|\\\text{CO}_2\text{H}\end{array}}{\overset{\overset{\text{CO}_2\text{H}}{|}}{\text{C}}}{-}\text{H} \underset{\underset{\text{transferase}}{\overset{\text{Amino}}{\rightleftharpoons}}}{\overset{\text{Glutamate}}{\rightleftharpoons}} \text{R}^1{-}\underset{\underset{\text{NH}_2}{|}}{\overset{\overset{\text{H}}{|}}{\text{C}}}{-}\text{CO}_2\text{H} + \underset{\begin{array}{c}|\\\text{CH}_2\\|\\\text{CH}_2\\|\\\text{CO}_2\text{H}\end{array}}{\overset{\overset{\text{CO}_2\text{H}}{|}}{\text{C}}}{=}\text{O} \qquad (17.3)$$

| Acceptor keto acid | Glutamic acid | Amino acid | α-Ketoglutaric acid |

Similarly, **alanine amino transferase** is specific for alanine and pyruvic acid as one of its substrate pairs but reacts with almost any other amino acid. The reactions catalyzed by the transaminases have, as would be expected, an equilibrium constant of approximately 1.0; therefore, the reactions are readily reversible.

By combining the three reactions just described, it is possible to incorporate NH_3 into the α-amino position of any amino acid; that is, any of the protein amino acids can be synthesized by combining Reactions 17.1, 17.2 and 17.3, provided that the keto acid analog of the desired amino acid is available.

L-Glutamic acid + ATP + NH_3 $\longrightarrow$ L-Glutamine + ADP + H_3PO_4 (17.1)

α-Ketoglutarate + L-Glutamine + NADPH + H^+ $\longrightarrow$

$$2 \text{ L-Glutamic acid} + NADP^+ + H_2O \quad (17.2)$$

L-Glutamic acid + RCOCOOH $\longrightarrow$ $RCHNH_2COOH$ + α-Ketoglutarate   (17.3)

RCOCOOH + ATP + NH_3 + NADPH + H^+ $\longrightarrow$

$$RCHNH_2COOH + ADP + H_3PO_4 + NADP^+ + H_2O$$

All that is required is the necessary enzymes, the keto acid, ATP, NADPH (or other reductant), and NH_3. The overall synthesis will be driven in the direction of amino acid synthesis by the hydrolysis of ATP in Reaction 17.1. Because higher plants are capable of synthesizing *all* of their carbon compounds, including the corresponding keto acid, from CO_2, the synthesis of all plant amino acids can be accomplished by these three reactions. Similarly, since animals lack glutamate synthase (Reaction 17.2), one might conclude that animals cannot synthesize any amino acids and are therefore dependent ultimately on plants for these compounds. Such simplified relationships are not to be found and the synthesis of amino acids in animals is discussed in Section 17.3. It should also be stressed that not all amino acids are synthesized by transamination of their keto acids (see Reaction 17.7 for the synthesis of cysteine).

17.2.4   CARBAMYL PHOSPHATE SYNTHESIS

The other key compound for the incorporation of NH_3 into organic nitrogen compounds is **carbamyl phosphate**. This compound was first implicated in nitrogen metabolism when Lipmann and Jones described the presence of an enzyme in *Streptococcus faecalis* that catalyzed its formation from the ammonium salt of carbamic acid. The chemistry of carbamic acid is complex; the reaction is endergonic ($\Delta G' = +2000$ cal/mole).

$$[NH_4]^+ \left[O-\overset{\overset{\displaystyle O}{\|}}{C}-NH_2 \right]^- + ATP \xrightarrow[\text{Carbamyl kinase}]{Mg^{2+}} H_2O_3P-\overset{\overset{\displaystyle O}{\|}}{C}-NH_2 + ADP + NH_3 \quad (17.4)$$

    Ammonium carbamate   Carbamyl phosphate

Mammalian liver contains a mitochondrial synthetase I, which catalyzes the same reaction except that 2 ATP are required.

$$NH_3 + CO_2 + 2 ATP \xrightarrow[\text{phosphate synthetase I}]{Carbamyl} H_2O_3PO-\overset{\overset{\displaystyle O}{\|}}{C}-NH_2 + 2 ADP + H_3PO_4 \quad (17.5)$$

This reaction is not readily reversible because there is a decrease of one energy-rich bond as the reaction proceeds from left to right. Evidence indicates that the first ATP is used to activate the CO_2 as enzyme-bound carbonyl phosphate. The phosphate can then be displaced by NH_3 to form carbamate. The second mole of ATP would then contribute the phosphate to form carbamyl phosphate.

All other animal tissues and plant tissues contain carbamyl phosphate synthetase II that utilizes the amide nitrogen of glutamine as a source of nitrogen in place of NH_3. Again, 2 ATP are required and the partial reactions may be represented as follows:

$$ATP + Enz + CO_2 \longrightarrow Enz\text{---}\left[HOC\text{---}OPO_3H_2\right] + ADP$$

$$\text{Glutamine}$$

$$NH_2\text{---}C\text{---}OPO_3H_2 + Enz \longleftarrow \overset{\frown}{\underset{ADP \quad ATP}{}} \longleftarrow Enz\text{---}\left[HO\text{---}C\text{---}NH_2\right] + \text{Glutamic acid} + Pi$$

Sum: L-Glutamine $+ CO_2 + 2$ ATP $\longrightarrow$

$$NH_2\text{---}C\text{---}OPO_3H_2 + 2\,ADP + H_3PO_4 + \text{L-Glutamate} \qquad (17.6)$$

The significance of carbamyl phosphate is that it is a donor of the carbamyl $(NH_2\text{---}C\text{---})$ group in two ubiquitous, fundamental biosynthetic pathways.

These are the synthesis of urea (Figure 17.2) and the synthesis of pyrimidines (Reaction 1, Figure 17.6). In this process, the nitrogen atom that is transferred is accompanied by a carbon atom.

Aspartic acid is the third amino acid that plays a unique role in the synthesis of nitrogenous compounds. Its unique ability to function as a source of nitrogen atoms not involving transamination is seen in the synthesis of argininosuccinic acid in the urea cycle (Reaction 17.17) and in the synthesis of purines (Figure 17.5) and pyrimidines (Figure 17.6).

17.3   AMINO ACID BIOSYNTHESIS IN ANIMALS

Although animals can and must assimilate NH_3 by converting it to glutamine, they acquire their nitrogen in the form of dietary proteins (and to a limited extent as dietary amino acids). Moreover, the protein must contain those amino acids that early nutritional studies showed animals cannot synthesize. As they evolved, animals have lost the ability to synthesize about half of the amino acids found in their protein. Thus, the adult human (and albino rats) must obtain arginine, histidine, isoleucine, leucine, lysine, methionine, phenylalanine, threonine, tryptophan, and valine in specific amounts on a daily basis in their diet. These *essential* amino acids stand in sharp contrast to the other 10 nonessential amino acids that the animal can synthesize.

In view of the ubiquitous occurrence of amino transferases (Reaction 17.3), one may ask if animals lack the ability to synthesize the essential amino acid per se or whether they are simply unable to synthesize the carbon skeleton of the amino acid. Research has shown that the keto acids corresponding to the essential amino acids will substitute for the latter, provided extra nitrogen is furnished for synthesis (by transamination). More specifically, we can identify structural features of the carbon skeletons of the essential amino acids that animals cannot construct. In the case of phenylalanine, tyrosine, and tryptophan, it is the benzenoid ring. Plants and microorganisms have a fascinating biosynthetic pathway—the shikimic acid pathway—which accomplishes the synthesis of the benzene ring in a series of discrete, logical reactions. [Although animals lack the shikimate pathway, they can make tyrosine, from phenylalanine, by hydroxylation of the aromatic ring (Figure 14.6).] Similarly, the branch-chain amino acids, valine, leucine, and isoleucine are essential because animals lack the capacity to produce the branched carbon structure in those amino acids.

One of the major nutritional problems facing the world is the provision of an adequate (in the sense of containing the essential amino acids) protein diet for its human population. Carbohydrate, in the form of plant starch, is readily available from the major food crops—rice, corn, wheat, and cassava—but an adequate protein supply is much more difficult to obtain. In countries where animal protein is consumed, the problem is insignificant because it contains the essential amino acids. However, most of the world's population subsists on plant food, and if the amount of protein is low or inadequate in quality (as in corn and rice that are low in lysine), the diet is inferior. On the other hand, the seeds of legumes (e.g., beans, peas, and soybeans) are an excellent source of the essential amino acids.

It is a tragic fact that in Third World countries, growing children are particularly susceptible to protein deficiency. As long as a baby is nursed by its mother, its protein intake will be adequate in amount and quality. When, however, the older child is displaced by a new baby and fed a diet rich in starch, the effects of an inadequate plant diet are noticed and the disease *Kwashiorkor* develops, characterized by bloated stomachs, discolored hair and skin, and a generalized malaise. Such children are weakened in their resistance to the normal complement of diseases and infections in their environment and usually die in the first few years of life.

17.4 ANABOLIC ASPECTS OF AMINO ACID METABOLISM

In the preceding sections, we have seen how NH_3 can be assimilated into two primary metabolites, glutamine and carbamyl phosphate, and the nitrogen atom then passed on to other nitrogenous compounds. The importance of glutamate in transamination was discussed and we shall now consider, in general terms only, the origin of the carbon skeletons of the amino acids. The description of the detailed biosynthetic sequence for each of the proteinaceous amino acids is beyond the purpose of this text, but certain obvious sources of those carbon skeletons can be mentioned.

The keto acids, pyruvic, oxalacetic, and α-ketoglutaric, are intermediates of glycolysis (Chapter 10) or the tricarboxylic acid cycle (Chapter 12). By transamination, these compounds are converted to alanine, aspartic, and glutamic acids, respectively. Since their keto acids can be produced from carbohydrate precursors (see Chapter 12 for specific conditions for α-ketoglutarate and oxalacetate formation), it is not surprising that alanine, aspartic, and glutamic acids are nonessential amino acids. Since aspartic and glutamic acids can be con-

TABLE 17.1 Families of Amino Acids Related by Biosynthesis

GLUTAMATE	ASPARTATE	PYRUVATE	PHOSPHOENOL PYRUVATE	3-PHOSPHO-GLYCERATE
Glutamate	Aspartate	Alanine	Phenylalanine	Serine
Glutamine	Asparagine	Leucine	Tyrosine	Glycine
Proline	Lysine	Valine	Tryptophan	Cysteine
Arginine	Methionine	(Isoleucine)		
Ornithine	Threonine			
	Isoleucine			

verted to their amides (Reaction 17.1 and Box 17-A), the amides are also dispensible. In addition, glutamic acid can be converted to proline and ornithine [and indirectly therefore to hydroxyproline, citrulline, and arginine (see Section 17.8)], and these amino acids are classified as nonessential. This ability of the carbon skeleton of glutamic acid to give rise to these amino acids has led to the concept of a "glutamate family" of amino acids (Table 17.1).

Four other families are recognized from biosynthetic studies, mainly with microorganisms that can make all of the protein amino acids from glucose or other simple precursors such as acetic acid. Presumably, these same family relationships exist in higher plants, which make all of these compounds ultimately from CO_2. Note that pyruvate, phosphoenol pyruvate, and 3-phosphoglycerate, intermediates in glycolysis, are the carbon precursors of some of the amino acids. In these cases, the parent compound lacks the amino nitrogen atom and it usually is supplied by transamination.

The enzymes that are missing in animals, which cannot make the indispensible amino acids, are well known. The student is encouraged to explore this classical area of intermediary metabolism on his own. The student is also referred to Section 18.3.2 in which there are described a number of feedback mechanisms involved in the regulation of the synthesis of these amino acids.

17.5 METABOLISM OF THE SULFUR-CONTAINING AMINO ACIDS

17.5.1 BIOSYNTHESIS

The metabolism of the sulfur-containing amino acids will be considered briefly because of its somewhat unusual nature.

The relationships between cysteine, homocysteine, and methionine can best be appreciated if we recall that the primary reaction for incorporation of sulfur into organic compounds is the formation of cysteine by cysteine synthase (Section 16.7.3), a reaction that occurs in bacteria and higher plants but not in animals.

$$
\begin{array}{l}
\text{CH}_2\text{—O-Acetyl} \\
| \\
\text{CHNH}_2 \\
| \\
\text{CO}_2\text{H}
\end{array}
\quad + \text{H}_2\text{S} \longrightarrow
\begin{array}{l}
\text{CH}_2\text{SH} \\
| \\
\text{CHNH}_2 \\
| \\
\text{CO}_2\text{H}
\end{array}
+ \text{CH}_3\text{COOH}
\qquad (17.7)
$$

O-Acetyl serine   Cysteine   Acetic acid

Bacteria and higher plants can utilize cysteine as a sulfur source to form methionine; the process known as **transsulfurylation** has its analogy in the manner that the nitrogen atom of aspartic acid makes its way into the guanidinium group in arginine (Reactions 17.17 and 17.18) and into inosinic acid (Reactions 17.7 and 17.8, in Figure 17.5). In the presence of the enzyme **cystathionine synthase I,** a sulfur-containing addition product known as **cystathionine** is formed.

$$
\begin{array}{llll}
\underset{\text{Cysteine}}{\begin{array}{l}\text{CH}_2\text{—SH}\\ \text{CHNH}_2\\ \text{CO}_2\text{H}\end{array}}
+
\underset{\substack{\textit{O}\text{-Succinyl}\\ \text{homoserine}}}{\begin{array}{l}\text{Succinyl}\\ \text{O—CH}_2\\ \text{CH}_2\\ \text{CHNH}_2\\ \text{CO}_2\text{H}\end{array}}
\xrightarrow[\text{(bacteria; plants)}]{\text{Cystathionine synthase I}}
\underset{\text{Cystathionine}}{\begin{array}{ll}\text{CH}_2\text{—S—CH}_2\\ \text{CHNH}_2\quad \text{CH}_2\\ \text{CO}_2\text{H}\quad \text{CHNH}_2\\ \qquad\quad \text{CO}_2\text{H}\end{array}}
+ \text{Succinate}
\end{array}
$$

In this reaction, the three-carbon unit is contributed by cysteine and the four-carbon unit by *O*-succinyl-homoserine (derived from aspartic acid). Note that the hydroxy group of homoserine has been activated by acylation, in analogy with Reaction 17.7. In the presence of cystathionase, the cystathionine is hydrolytically cleaved on the *opposite* side of the sulfur atom to produce **homocysteine,** pyruvic acid, and NH_3.

$$
\underset{\text{Cystathionine}}{\begin{array}{ll}\text{CH}_2\text{—S—CH}_2\\ \text{CHNH}_2\quad \text{CH}_2\\ \text{CO}_2\text{H}\quad \text{CHNH}_2\\ \qquad\quad \text{CO}_2\text{H}\end{array}}
+ \text{H}_2\text{O}
\xrightarrow[\substack{\text{(bacteria;}\\ \text{plants)}}]{\text{Cystathionase}}
\underset{\text{Pyruvate}}{\begin{array}{l}\text{CH}_3\\ \text{C}=\text{O}\\ \text{CO}_2\text{H}\end{array}}
+ \text{NH}_3 +
\underset{\text{Homocysteine}}{\begin{array}{l}\text{HS—CH}_2\\ \text{CH}_2\\ \text{CHNH}_2\\ \text{CO}_2\text{H}\end{array}}
$$

The homocysteine is subsequently methylated (by tetrahydrofolic acid) to form methionine (see Section 5.8.3.3).

The reactions and relationships just described for bacteria and plants are almost reversed in animals, which cannot make cysteine (or homocysteine) from H_2S or SO_4^{2-}. Instead, animals synthesize their cysteine from methionine, which is an indispensible amino acid. Indeed, the "essentiality" of methionine is due to this inability. Since the sulfur of methionine can be utilized to make cysteine, however, the latter is not an essential amino acid. Cystathionine just encountered is again an intermediate in this process. In the presence of mammalian **cystathionine synthase II** homocysteine (derived from the demethylation of methionine) reacts with serine to produce cystathionine.

$$
\underset{\text{Homocysteine}}{\begin{array}{l}\text{CH}_2\text{—SH}\\ \text{CH}_2\\ \text{CHNH}_2\\ \text{CO}_2\text{H}\end{array}}
+
\underset{\text{Serine}}{\begin{array}{l}\text{HO—CH}_2\\ \text{CHNH}_2\\ \text{CO}_2\text{H}\end{array}}
\xrightarrow[\text{(mammalian)}]{\substack{\text{Cystathionine}\\ \text{synthase II}}}
\underset{\text{Cystathionine}}{\begin{array}{ll}\text{CH}_2\text{—S—CH}_2\\ \text{CH}_2\quad\ \text{CHNH}_2\\ \text{CHNH}_2\ \ \text{CO}_2\text{H}\\ \text{CO}_2\text{H}\end{array}}
+ \text{H}_2\text{O}
$$

This compound is then hydrolyzed to yield cysteine, α-ketobutyric acid, and NH_3.

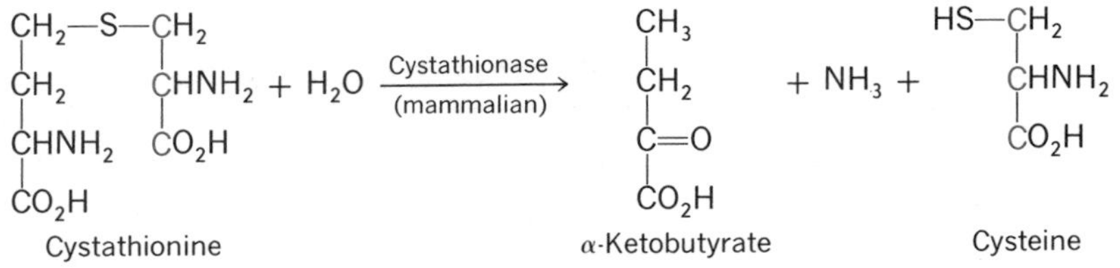

$$
\begin{array}{ccc}
\underset{\substack{| \\ \text{CH}_2 \\ | \\ \text{CHNH}_2 \\ | \\ \text{CO}_2\text{H}}}{\text{CH}_2\text{—S—CH}_2} \quad + \text{H}_2\text{O} \xrightarrow[\text{(mammalian)}]{\text{Cystathionase}} & \underset{\substack{| \\ \text{CH}_2 \\ | \\ \text{C=O} \\ | \\ \text{CO}_2\text{H}}}{\text{CH}_3} + \text{NH}_3 + & \underset{\substack{| \\ \text{CHNH}_2 \\ | \\ \text{CO}_2\text{H}}}{\text{HS—CH}_2} \\
\text{Cystathionine} & \alpha\text{-Ketobutyrate} & \text{Cysteine}
\end{array}
$$

Note that this time the three carbon atoms of cysteine originate in serine, while the sulfur atom comes from homocysteine and therefore indirectly from methionine. The four enzymes just described that are involved in the formation and hydrolysis of cystathionine utilize pyridoxal phosphate as a cofactor.

17.5.2 ACTIVE METHIONINE

Two additional reactions will round out the relationships between the sulfur amino acids just described and, in addition, illustrate a different way in which ATP can serve to activate a substrate. There are numerous examples in which the methyl group of methionine is transferred to acceptor molecules to form methylated derivatives. The methyl donor is known as a *S*-adenosyl methionine; it is formed by the activating enzyme, **methionine adenosyl transferase,** in the presence of ATP and methionine.

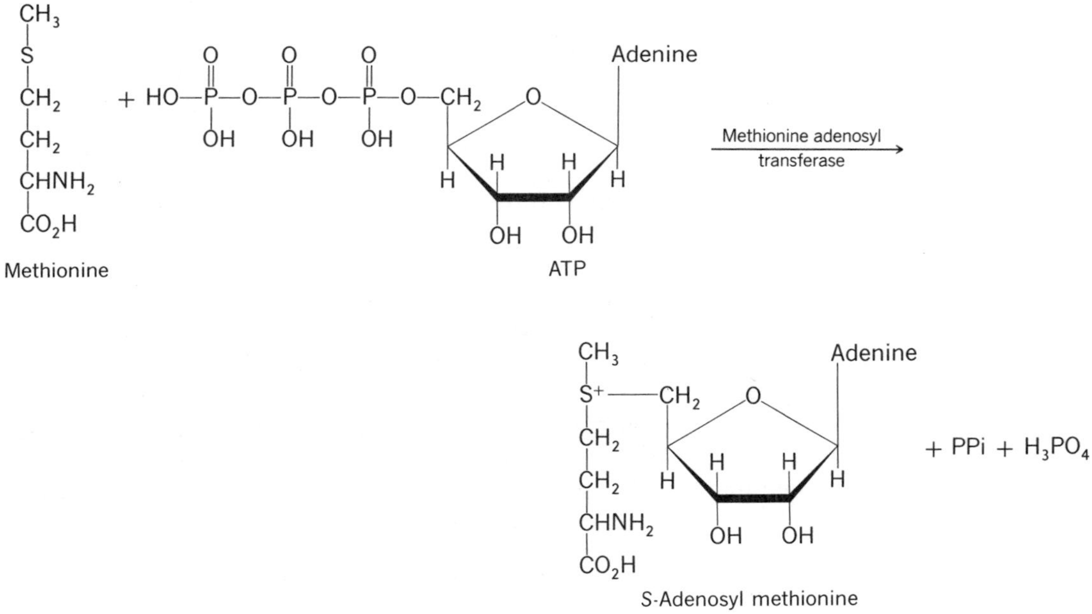

Methionine + ATP $\xrightarrow{\text{Methionine adenosyl transferase}}$ *S*-Adenosyl methionine + PPi + H_3PO_4

In this unusual reaction, the three phosphate groups of ATP are removed, as inorganic phosphate and pyrophosphate, and the adenosine residue is attached to the sulfur atom to form a sulfonium derivative. This compound is a high-energy compound that readily transfers its methyl group to acceptor molecules (e.g., guanidoacetic acid); in the process, *S*-adenosyl homocysteine is formed.

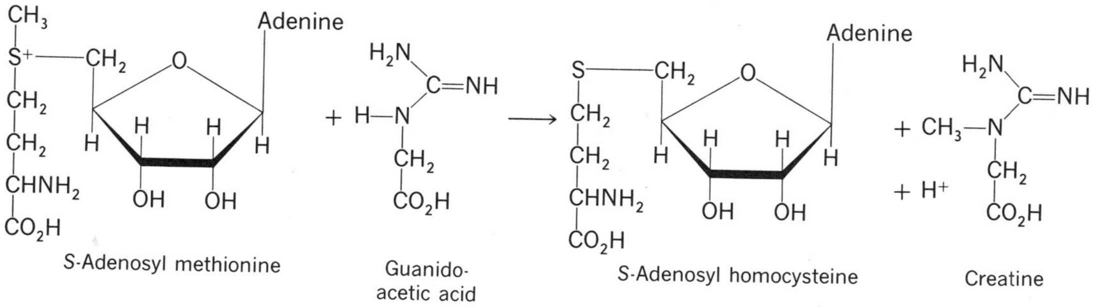

S-Adenosyl methionine + Guanido-acetic acid → S-Adenosyl homocysteine + Creatine

S-Adenosyl methionine   Guanido-acetic acid   S-Adenosyl homocysteine   Creatine

The *S*-adenosyl homocysteine can be hydrolyzed to form adenosine and homocysteine, and again methylated (by tetrahydrofolic acid in plants and betaine in animals) to reform methionine.

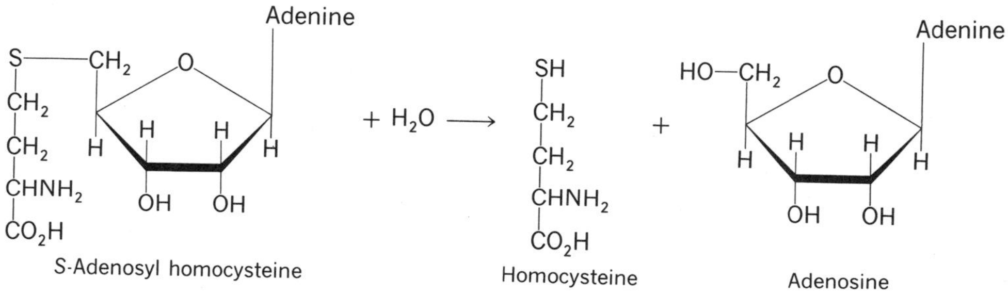

S-Adenosyl homocysteine + H_2O → Homocysteine + Adenosine

S-Adenosyl homocysteine   Homocysteine   Adenosine

BOX 17.B   BIOSYNTHESIS OF ETHYLENE IN PLANTS

S-Adenosyl methionine (SAM) undergoes an important reaction in plants in which the bond between the sulfur atom and the four-carbon chain contributed by methionine is broken instead of the bond between the sulfur atom and the methyl group. The reaction, catalyzed by the plant enzyme ACC synthase, involves cleavage to form 1-aminocyclopropane-1-carboxylic acid (ACC) and 5′-methylthioadenosine.

SAM $\xrightarrow{\text{ACC Synthase}}$ 1-Aminocyclopropane-1-carboxylic acid (ACC) + 5′-Methylthioadenosine

SAM   1-Aminocyclopropane-1-carboxylic acid (ACC)   5′-Methylthioadenosine

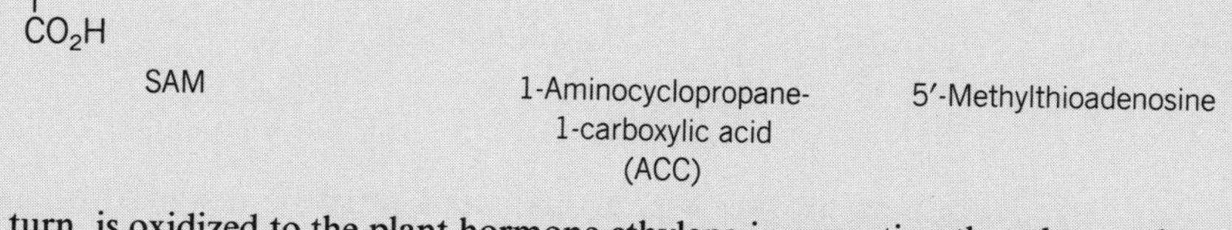

ACC, in turn, is oxidized to the plant hormone ethylene in a reaction that also produces HCN.

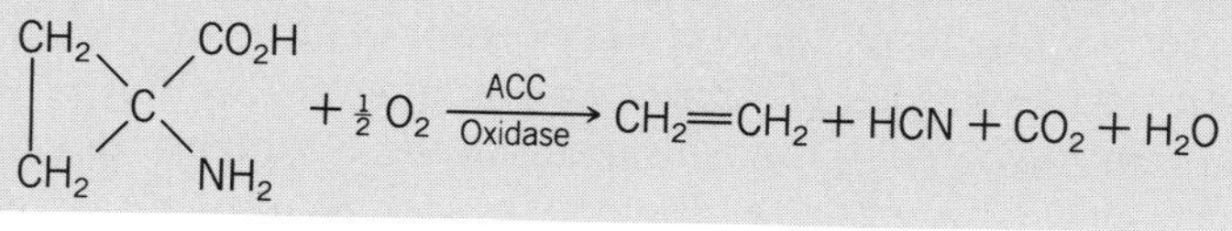

$$\text{ACC} + \tfrac{1}{2}O_2 \xrightarrow{\text{ACC Oxidase}} CH_2=CH_2 + HCN + CO_2 + H_2O$$

These two reactions are presumably ubiquitous in the plant kingdom since all plants are capable of making ethylene. The 5'-methylthioadenosine is also recycled in plants and the sulfur atom utilized again in methionine synthesis.

17.6 CATABOLISM OF AMINO ACIDS

17.6.1 DEAMINATION

17.6.1.1 By Glutamic Dehydrogenase.
L-Glutamic acid plays a key role in the metabolism of amino acids because of the widespread occurrence of the enzyme **glutamic dehydrogenase.** This enzyme catalyzes the reversible **oxidative deamination** of L-glutamate to form α-ketoglutaric acid and NH_3.

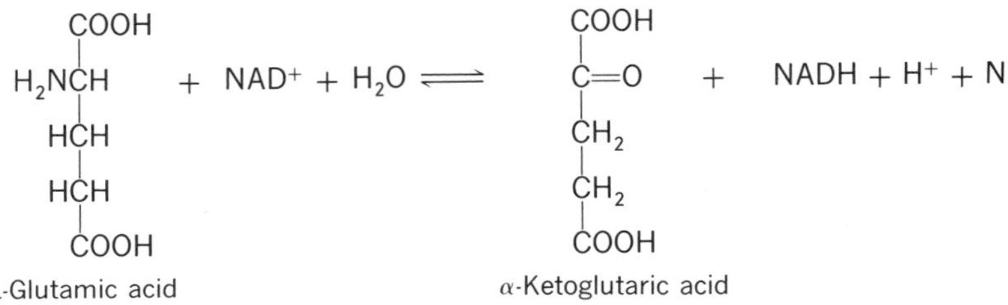

$$
\begin{array}{c}
\text{COOH} \\
| \\
H_2N\text{CH} \\
| \\
H\text{CH} \\
| \\
H\text{CH} \\
| \\
\text{COOH}
\end{array}
\;+\; NAD^+ + H_2O \rightleftharpoons
\begin{array}{c}
\text{COOH} \\
| \\
\text{C}=\text{O} \\
| \\
\text{CH}_2 \\
| \\
\text{CH}_2 \\
| \\
\text{COOH}
\end{array}
\;+\; NADH + H^+ + NH_3 \quad (17.8)
$$

L-Glutamic acid α-Ketoglutaric acid

The liver enzyme functions with either NAD^+ or $NADP^+$ as oxidant and is present in the mitochondria.

Reaction 17.8 is readily reversible and, at equal concentrations of NAD^+ and NADH, the equilibrium actually favors synthesis of glutamate. However, the relatively high K_m for NH_3 ($1-5$ mM) required for reversal, together with the rapid reoxidation of NADH by means of the mitochondrial electron-transport system, suggest that glutamic dehydrogenase will function almost exclusively in the direction of glutamate oxidation and NH_3 production.

The coupling of Reaction 17.8 with transamination of glutamic acid (Reaction 17.3) creates a mechanism for deaminating all of the other amino acids except proline.

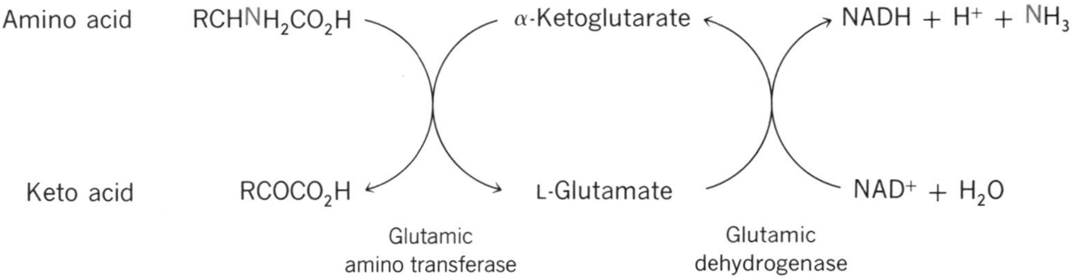

Amino acid $RCHNH_2CO_2H$ ⟶ α-Ketoglutarate ⟵ $NADH + H^+ + NH_3$

Keto acid $RCOCO_2H$ ⟵ L-Glutamate ⟶ $NAD^+ + H_2O$

Glutamic   Glutamic
amino transferase   dehydrogenase

NH_3 produced in this way is toxic and must be disposed of. In animals, elaborate mechanisms for detoxification exist (see the urea cycle, Section 17.8). In plants, which lack the excretory organs of animals, the NH_3 is converted to the nontoxic amides, glutamine and asparagine (Section 17.2.1 and Box 17.A) and remarkably high concentrations of these compounds can accumulate. On germination, protein-rich lupine seeds may accumulate asparagine in the seedling to the level of 20% of its dry weight.

While the production of NH_3 from glutamate in Reaction 17.8, or in other reactions described next, requires that it be disposed of, it is also a major means for producing the NH_3 needed for the synthesis of glutamine. In contrast to plants that can and do utilize NH_3 present in the soils, animals must make NH_3 as an essential metabolite for the production of glutamine and the numerous nitrogenous compounds derived only from glutamine.

Because of its importance in the metabolism of other amino acids, as well as the fact that glutamic acid serves as a precursor of proline and ornithine [and indirectly therefore of hydroxyproline, citruline, and arginine (see Section 17.8)], it is not surprising to learn that glutamic dehydrogenase is an allosteric enzyme. The beef liver enzyme, as an example, is inhibited by ATP and NADH and is stimulated by ADP and AMP.

17.6.1.2 By Ammonia Lyases.

In contrast to the reactions of oxidative deamination is the process of **nonoxidative deamination.** One type of nonoxidative deamination is the reaction catalyzed by the amino acid ammonia-lyases (trivial name, α-deaminases). Aspartic ammonia lyase (aspartase), which belongs to this group of enzymes, catalyzes the following reaction.

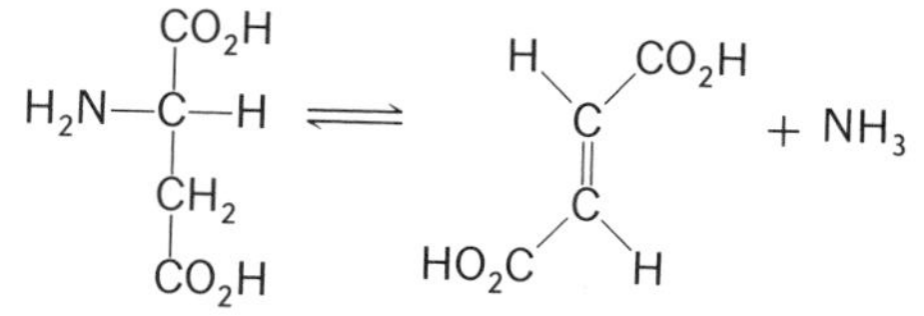

L-Aspartic acid   Fumaric acid   (17.9)

The enzyme, which is specific for L-aspartic acid and fumaric acid, has been found in *E. coli* and a few other microorganisms. Because the reaction catalyzed is readily reversible, this reaction constitutes a mechanism for incorporating inorganic nitrogen in the form of NH_3 into the α-amino position of an amino acid in microorganisms. Other ammonia lyases catalyze the deamination of histidine, phenylalanine, and tyrosine in animal and plant tissues. However, these reactions, in contrast to Reaction 17.9, are not readily reversible and therefore are of little significance in the biosynthesis of the amino acids whose deamination they catalyze.

17.6.1.3 By Specific Deaminases.

A somewhat different type of deamination is catalyzed by an enzyme in liver termed serine dehydratase

BOX 17.C AMINO ACID OXIDASES

Amino acids can also be oxidatively deaminated by enzymes known as **amino acid oxidases** that require FMN or FAD as cofactors. Liver and kidney contain the L-amino acid oxidase that utilizes FMN; some microorganisms and the venom of certain snakes are also excellent sources of the enzyme.

The overall reaction is

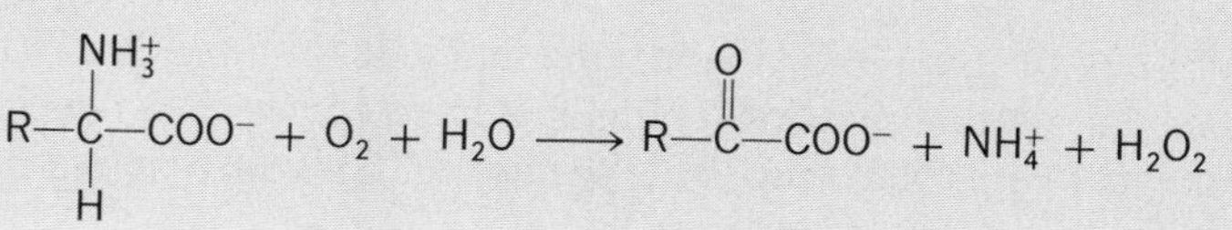

$$R-\underset{\underset{H}{|}}{\overset{\overset{NH_3^+}{|}}{C}}-COO^- + O_2 + H_2O \longrightarrow R-\overset{\overset{O}{\|}}{C}-COO^- + NH_4^+ + H_2O_2 \qquad (17.10)$$

Most of the proteinaceous amino acids serve as substrates and the reaction is not readily reversible. The overall reaction may be broken down into individual steps for which there is experimental evidence. In the first step, oxidation of the amino acid by a flavin enzyme leads to the corresponding imino acid.

$$\underset{\underset{H}{|}}{\overset{\overset{NH_3^+}{|}}{R-C-COO^-}} + FMN \rightleftharpoons \underset{}{\overset{\overset{NH}{\parallel}}{R-C-COO^-}} + FMNH_2 + H^+ \qquad (17.10a)$$

The imino acid, in turn, is spontaneously hydrolyzed in the presence of H_2O.

$$\overset{\overset{NH}{\parallel}}{R-C-COO^-} + H_2O + H^+ \longrightarrow \overset{\overset{O}{\parallel}}{R-C-COO^-} + NH_4^+ \qquad (17.10b)$$

The reduced flavin formed will, in turn, be reoxidized by molecular oxygen to form H_2O_2.

$$FMNH_2 + O_2 \longrightarrow FMN + H_2O_2 \qquad (17.10c)$$

Neither Reaction 17.10b or 17.10c is reversible. Therefore, the overall reaction (Reaction 17.10) that is the sum of Reactions 17.10a, 17.10b, and 17.10c is not. In the presence of a highly purified enzyme that contains no impurities to destroy H_2O_2, the reaction proceeds further. In this case, the H_2O_2 formed in Reaction 17.10c reacts nonenzymatically with the keto acid.

$$\underset{\underset{O}{\parallel}}{R-C-COO^-} + H_2O_2 \longrightarrow R-COO^- + CO_2 + H_2O$$

Liver and kidney also contain a highly active D-amino acid oxidase that oxidizes the D-enantiomers of numerous amino acids and requires FAD. Since D-amino acids are largely unknown in plant and animal metabolism (except for their limited occurrence in bacterial cell walls), the role of this enzyme is enigmatic. It is located in the peroxisomes of liver, kidney, and brain, together with catalase that decomposes the H_2O_2 produced in Reaction 17.10c. Since glycine is a good substrate for D-amino acid oxidase, its metabolic role may be associated with the oxidation of glycine to glyoxylic acid.

$$NH_2-CH_2-CO_2H + O_2 + H_2O \longrightarrow \overset{\overset{H}{|}}{O=C-CO_2H} + NH_3 + H_2O_2$$

$$\underset{\text{Glycine}}{} \qquad\qquad\qquad\qquad \underset{\substack{\text{Glyoxylic}\\\text{acid}}}{}$$

[systematic name, L-serine hydrolyase (deaminating)]. The reaction, which is specific for L-serine, involves the loss of NH_3 and rearrangement of the remaining atoms to yield pyruvate.

$$\underset{\underset{CH_2OH}{|}}{\overset{\overset{CO_2H}{|}}{H_2N-C-H}} \longrightarrow \underset{\underset{CH_3}{|}}{\overset{\overset{CO_2H}{|}}{C=O}} + NH_3$$

$$\underset{\text{L-Serine}}{} \qquad\qquad \underset{\text{Pyruvic acid}}{}$$

This enzyme and an analogous one that acts on threonine, forming α-ketobutyric acid, require pyridoxal phosphate as a coenzyme. The Schiff's base of the coenzyme with the amino acid (Section 5.5.3) is the reactive species that undergoes β-elimination, forming NH_3.

The deamination of cysteine is catalyzed by an enzyme found in animals, plants, and microorganisms.

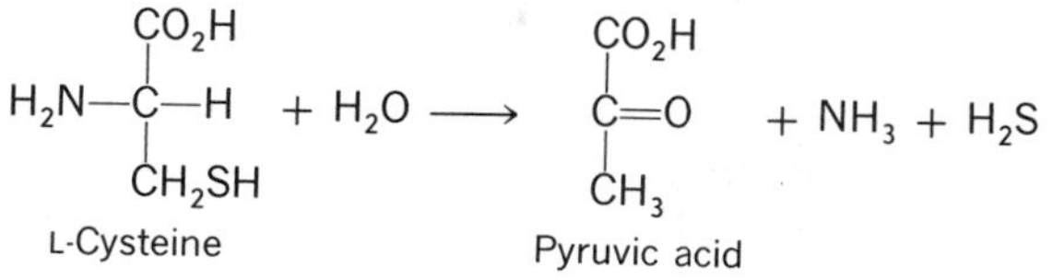

$$
\underset{\text{L-Cysteine}}{\underset{\overset{|}{\underset{\text{CH}_2\text{SH}}{}}}{\overset{\overset{\text{CO}_2\text{H}}{|}}{\text{H}_2\text{N}-\text{C}-\text{H}}}} + \text{H}_2\text{O} \longrightarrow \underset{\text{Pyruvic acid}}{\underset{\overset{|}{\underset{\text{CH}_3}{}}}{\overset{\overset{\text{CO}_2\text{H}}{|}}{\text{C}=\text{O}}}} + \text{NH}_3 + \text{H}_2\text{S}
$$

This enzyme, **cysteine desulfhydrase,** also requires pyridoxal phosphate as a coenzyme and presumably operates by a mechanism similar to serine dehydratase.

17.6.1.5 By Deamidases.

In addition to the above reactions, in which the α-amino groups of amino acids are released as NH_3, mention should be made of the reactions in which the amide nitrogen of glutamine and asparagine are liberated as ammonia. Specific hydrolytic enzymes catalyze the hydrolysis of these two amides and produce NH_3.

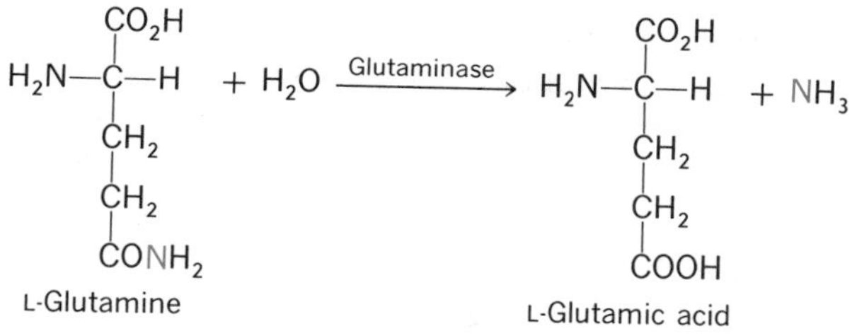

$$
\underset{\text{L-Glutamine}}{\overset{\text{CO}_2\text{H}}{\text{H}_2\text{N}-\text{C}-\text{H}}\atop{\text{CH}_2 \atop \text{CH}_2 \atop \text{CONH}_2}} + \text{H}_2\text{O} \xrightarrow{\text{Glutaminase}} \underset{\text{L-Glutamic acid}}{\overset{\text{CO}_2\text{H}}{\text{H}_2\text{N}-\text{C}-\text{H}}\atop{\text{CH}_2 \atop \text{CH}_2 \atop \text{COOH}}} + \text{NH}_3 \qquad (17.11)
$$

$$
\underset{\text{L-Asparagine}}{\overset{\text{CO}_2\text{H}}{\text{H}_2\text{N}-\text{C}-\text{H}}\atop{\text{CH}_2 \atop \text{CONH}_2}} + \text{H}_2\text{O} \xrightarrow{\text{Asparaginase}} \underset{\text{L-Aspartic acid}}{\overset{\text{CO}_2\text{H}}{\text{H}_2\text{N}-\text{C}-\text{H}}\atop{\text{CH}_2 \atop \text{COOH}}} + \text{NH}_3 \qquad (17.12)
$$

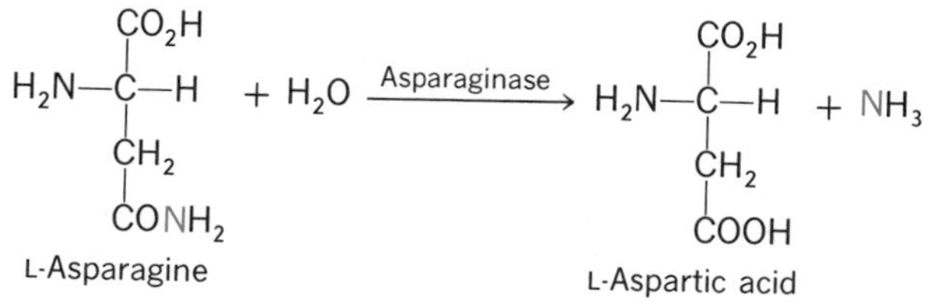

The fundamental role of glutamine in the assimilation of NH_3 and as a precursor of numerous nitrogenous compounds has been described (Section 17.2.1). Glutamine is also used to transport and store NH_3 in a nontoxic form before it is utilized. Thus, organisms have the means to degrade as well as synthesize this compound. Asparagine also serves as a transporter of NH_3, but in sharp contrast to glutamine, does not directly contribute its amide nitrogen in the synthesis of other nitrogen compounds. The amide nitrogen can be released as NH_3 (Reaction 17.12), but this compound can only reenter into organic carbon combination by undergoing Reactions 17.1 and 17.2.

17.6.2 DECARBOXYLATION

Amino acids generally undergo decarboxylation in the presence of **amino acid decarboxylases.**

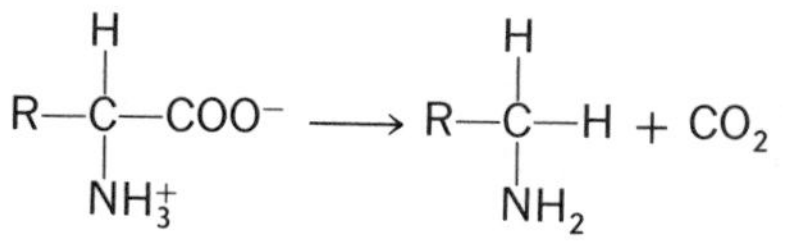

$$
\underset{\text{NH}_3^+}{\overset{\text{H}}{R-\text{C}-\text{COO}^-}} \longrightarrow \underset{\text{NH}_2}{\overset{\text{H}}{R-\text{C}-\text{H}}} + \text{CO}_2 \qquad (17.13)
$$

The enzymes that catalyze these reactions are widely distributed in nature and the amines produced frequently have important physiological effects. In a sense, therefore, these reactions are not catabolic in nature but rather allow the amino acid to serve as a source of important, indeed essential, compounds. Several examples will clarify this point.

γ-Aminobutyric acid, which is a neurotransmitter in brain, is produced from L-glutamate by the L-glutamate-α-decarboxylase of that organ.

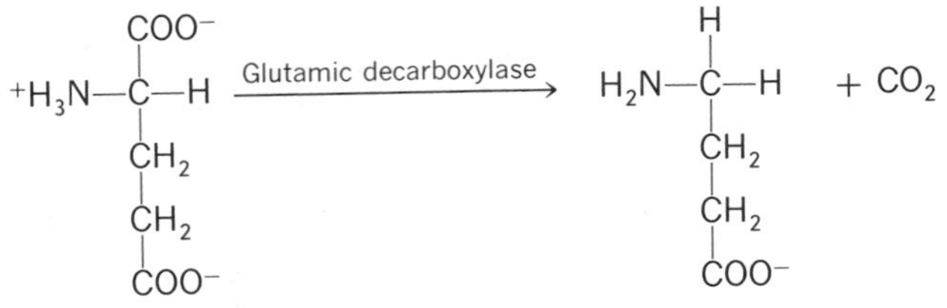

$$^+H_3N-\overset{\displaystyle COO^-}{\underset{\displaystyle CH_2}{\overset{\displaystyle |}{\underset{\displaystyle |}{C}}}}-H \xrightarrow{\text{Glutamic decarboxylase}} H_2N-\overset{\displaystyle H}{\underset{\displaystyle CH_2}{\overset{\displaystyle |}{\underset{\displaystyle |}{C}}}}-H \;+\; CO_2 \qquad (17.14)$$

L-Glutamate $\qquad\qquad\qquad\qquad\qquad$ γ-Amino butyrate

The decarboxylation of histidine produces histamine, a very powerful vaso-dilator. The decarboxylation of 3,4-dihydroxyphenylalanine (DOPA) produces dopamine, an intermediate in the formation of epinephrine. Serotonin (5-hydroxytryptamine), a neurohumoral compound in humans, is produced by the decarboxylation of 5-hydroxytryptophan.

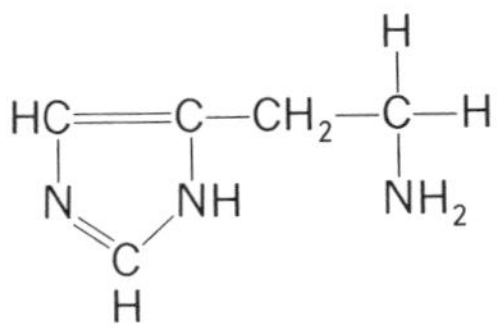

Histamine 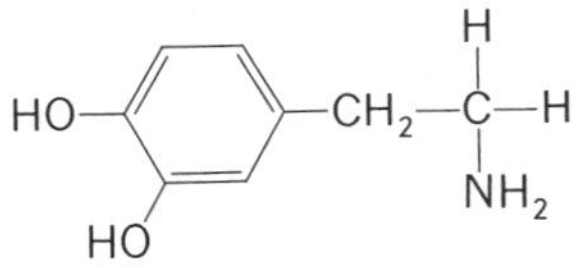 $\qquad\qquad$ 3,4-Dihydroxyphenylethylamine $\qquad$ 5-Hydroxytryptamine
$\qquad\qquad\qquad\qquad$ (dopamine) $\qquad\qquad\qquad\qquad\qquad$ (serotonin)

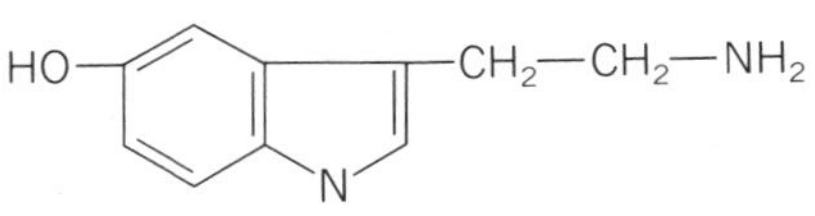

These physiologically active amines are produced by an **aromatic L-amino acid decarboxylase** acting on their respective amino acid precursor. The enzyme is found in several mammalian tissues. Decarboxylases that act specifically on only a single amino acid are known in both plants and animals.

Bacteria are a good source of amino acid decarboxylases. The enzymes, which are inducible, are formed when the bacteria are grown with amino acids in the culture medium. The amino acid decarboxylases require pyridoxal phosphate as a cofactor. A Schiff's base is again an intermediate, and it is possible to write a detailed mechanism, resulting in decarboxylation. (See Section 5.5.3 for the mechanism.)

BOX 17.D   BIOSYNTHESIS OF POLYAMINES

The polyamines spermine and spermidine are basic compounds that bind tightly to DNA and thereby exert regulatory roles in DNA replication. In so doing, they can influence protein synthesis and cell division. These amines are produced from the two amino acids ornithine and methionine, the latter compound being first converted to *S*-adenosyl methionine.

The biosynthesis proceeds from the diamine putrescine formed by the action of a specific decarboxylase acting on the amino acid ornithine.

$$NH_2—CH_2—CH_2—CH_2—CHNH_2—CO_2H \xrightarrow{\substack{\text{Ornithine} \\ \text{decarboxylase}}}$$

Ornithine

(17.15)

$$NH_2—CH_2—CH_2—CH_2—CH_2—NH_2 + CO_2$$

Putrescine

Propylamino groups, derived from "decarboxy" *S*-adenosyl methionine, are then transferred to both amino groups of putrescine to form spermidine first and then spermine. The decarboxylation of *S*-adenosyl methionine is also catalyzed by a specific decarboxylase.

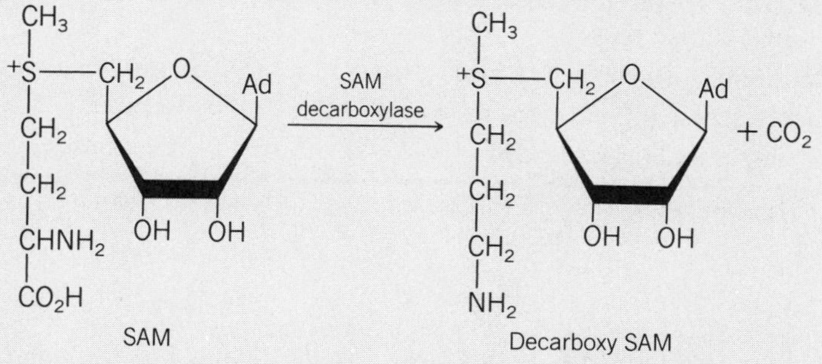

The transfer of the propylamino group to putrescine (and spermidine) is analogous to the reaction discussed in Box 17.B, in which the bond between the sulfur atom and the carbon chain contributed by methionine is cleaved, permitting three of the four carbons in methionine to serve as precursors of an important group of compounds.

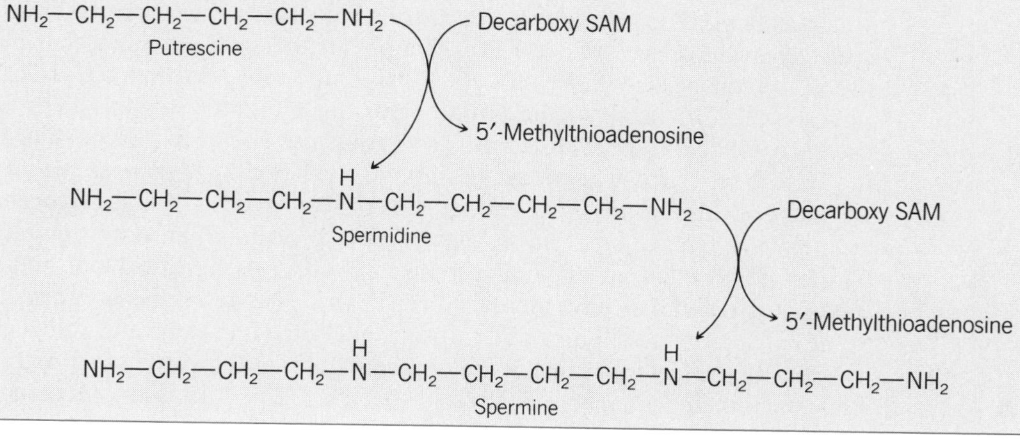

17.7 THE METABOLIC FATE OF THE AMINO ACIDS

As noted earlier, proteins (and amino acids) are not usually degraded for energy production if carbohydrates or lipids are available to the organism. Instead, the amino acids are used (1) in the synthesis of peptides and proteins; (2) as a source of nitrogen atoms (by transamination) for the synthesis of other amino acids; and (3) in the synthesis of other nitrogenous and nonnitrogenous compounds (see Sections 17.6.1 and 17.6.2). Any amino acids in excess of the amounts required for these three activities will be degraded by deamination and the carbon skeleton formed will be metabolized. The NH_3 produced, if in excess, will be eliminated as a nitrogenous waste product.

TABLE 17.2   End Products of Amino Acid Metabolism

AMINO ACIDS[a]	END PRODUCT
Alanine, serine, cysteine (cystine), glycine, and threonine (2)	Pyruvic acid
Leucine (2)	Acetyl CoA
Phenylalanine (4), tyrosine (4), leucine (4), lysine (4), and tryptophan (4)	Acetoacetic acid (or its CoA-ester)
Arginine (5), proline, histidine (5), glutamine, and glutamic acid	α-Ketoglutaric acid
Methionine, isoleucine (4), and valine (4)	Succinyl CoA
Phenylalanine (4) and tyrosine (4)	Fumarate
Asparagine and aspartic acid	Oxaloacetic acid

[a] The figures in parentheses specify the number of carbon atoms in the amino acid that are actually converted to the end product listed.

The metabolic fate of the carbon skeleton of the amino acid has been extensively studied and it is possible to write detailed catabolic sequences for each proteinaceous amino acid. However, the description of such sequences falls outside the purpose of this text. Instead, Table 17.2 lists the catabolic end product of the 20 protein amino acids. It may be seen that nearly all of the amino acids yield on breakdown either an intermediate of the tricarboxylic acid cycle, pyruvate, or acetyl CoA. The exceptions are the five amino acids that give rise to acetoacetic acid. Since, however, this compound also forms acetyl CoA, carbon skeletons of all of the amino acids are ultimately oxidized via the tricarboxylic acid cycle. Those amino acids that give rise to an intermediate of the cycle (or to pyruvic acid) can, in turn, be converted to glucose (see Section 10.6.2). For this reason, those amino acids have been described as **glucogenic** amino acids. On the other hand, those which on degradation produce acetyl CoA or acetoacetic acid will, under some conditions, give rise to ketone bodies in the animal and they have therefore been described as **ketogenic** amino acids. Some, such as phenylalanine and tyrosine, are both glucogenic and ketogenic by virtue of the fact that some of their carbon atoms are converted to fumarate while the remainder are converted to acetoacetate. The utilization of acetoacetate in animals is described in Section 13.8.

17.8   THE UREA CYCLE

Ammonia in excess of that required for synthesis of organic nitrogen compounds is disposed of by animals in different ways (see Section 17.9). Mammals convert the nitrogen atoms of NH_3 into urea that is secreted in the urine. The cycle of reactions that accomplishes urea synthesis is also, except for one reaction, the biosynthetic pathway for arginine formation. Presumably, the sequence of reactions leading to arginine evolved first in the earliest forms of life and then was modified for use by animals as a means for disposal of NH_3.

H. Krebs, then of Germany, and K. Henseleit were among the first to study the formation of urea in animal tissues. They observed that rat liver slices could

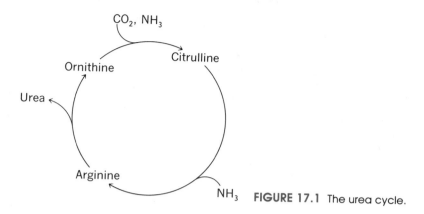

FIGURE 17.1 The urea cycle.

convert CO_2 and NH_3 (two moles of NH_3 per mole of CO_2) to urea, provided some energy source was available. The requirement for some oxidizable substance such as lactic acid or glucose was understandable, since the formation of urea from NH_3 and CO_2 required energy.

The amino acid arginine was also implicated in this process, since the enzyme arginase, which catalyzes Reaction 17.19, was known to form urea and ornithine on the hydrolysis of arginine. The exact relationship was clarified, however, when Krebs showed that catalytic quantities of arginine and ornithine, or citrulline as well, stimulated the formation of appreciable amounts of urea from ammonia. In 1932, Krebs proposed a cycle of reactions that accounted for the production of urea from NH_3 and CO_2 and explained the catalytic action of arginine, ornithine, and citrulline. That cycle, known as the urea or ornithine cycle, is shown in outline form in Figure 17.1.

The role of carbamylphosphate, as a carrier of the nitrogen atoms of NH_3 *and* the carbon atom of CO_2, is seen in the reaction catalyzed by **ornithine carbamoyl transferase.**

$$
\begin{array}{c}
NH_2 \\
| \\
CH_2 \\
| \\
CH_2 \\
| \\
CH_2 \\
| \\
HCNH_2 \\
| \\
COOH \\
\text{L-Ornithine}
\end{array}
+ \;
\underset{\text{Carbamyl phosphate}}{H_2N-\overset{\displaystyle O}{\overset{\|}{C}}-OPO_3H_2}
\longrightarrow
\begin{array}{c}
\overset{\displaystyle H_2N}{}\!\!\diagdown \\
C{=}O \\
HN\diagup \\
| \\
CH_2 \\
| \\
CH_2 \\
| \\
CH_2 \\
| \\
HCNH_2 \\
| \\
COOH \\
\text{L-Citrulline}
\end{array}
+ \; H_3PO_4
\qquad (17.16)
$$

The enzyme has no cofactors and exhibits extreme substrate specificity. The equilibrium strongly favors citrulline synthesis. Ornithine transcarbamylase occurs in mammalian liver mitochondria where it is closely associated with carbamylphosphate synthetase I as part of a multienzyme complex. The close association of these two proteins tends to provide for the efficient production and utilization of the labile carbamylphosphate.

The next step in the cycle, the formation of arginine from citrulline, was largely worked out by S. Ratner, who first showed that two enzymes were involved. The first of these, **argininosuccinic synthetase,** catalyzes the formation of argininosuccinic acid from citrulline and aspartic acid. This may be conveniently represented by picturing the enolic form of citrulline as reacting with the aspartic acid to form a new compound, argininosuccinic acid. Reaction 17.17 requires ATP and Mg^{2+}. The K_{eq} for this reaction is approximately 9 at pH 7.5; therefore, the reaction is readily reversible.

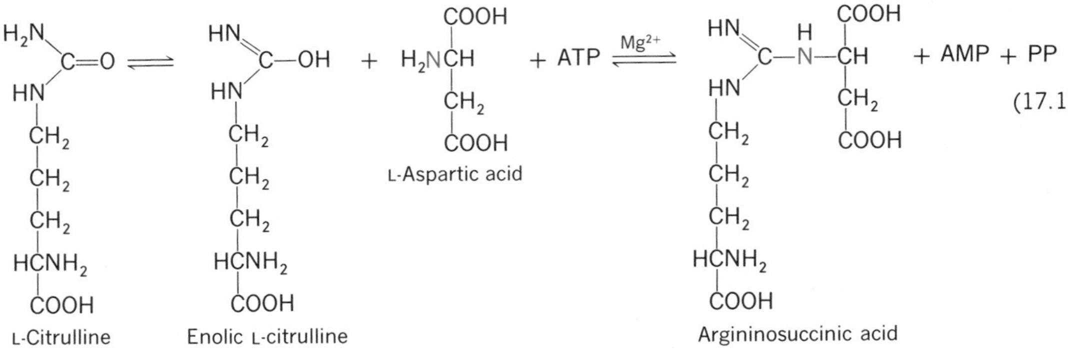

$$(17.17)$$

Note that the nitrogen atom that eventually becomes one of the two such atoms in urea is contributed by aspartic acid in this reaction and not by NH_3. Other examples of reactions in which aspartic acid contributes its nitrogen atoms in the biosynthesis of a new nitrogenous compound are shown in Figures 17.5 and 17.6.

The subsequent cleavage of argininosuccinic acid is catalyzed by the **argininosuccinic cleavage enzyme,** which has been purified from ox liver.

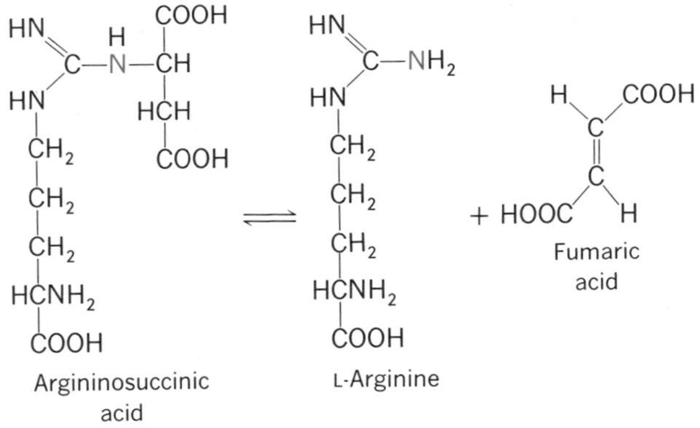

$$(17.18)$$

It has also been observed in plant tissues and microorganisms. Reaction 17.18 is formally analogous to the aspartic ammonia lyase reaction (Reaction 17.9), in which NH_3 or a substituted amine is eliminated to form fumaric acid. The K_{eq} for the reaction is 11.4×10^{-3} at pH 7.5. Since the reaction as written from left to right results in the formation of two products from a single reactant, this value of K_{eq} determines that argininosuccinic acid will predominate in concentrated solutions, whereas arginine and fumaric acid will predominate in dilute solution.

Arginase, which catalyzes the irreversible hydrolysis of L-arginine to ornithine and urea, is the enzyme that converts the unidirectional sequence for the biosynthesis of arginine into a cyclic process for making urea.

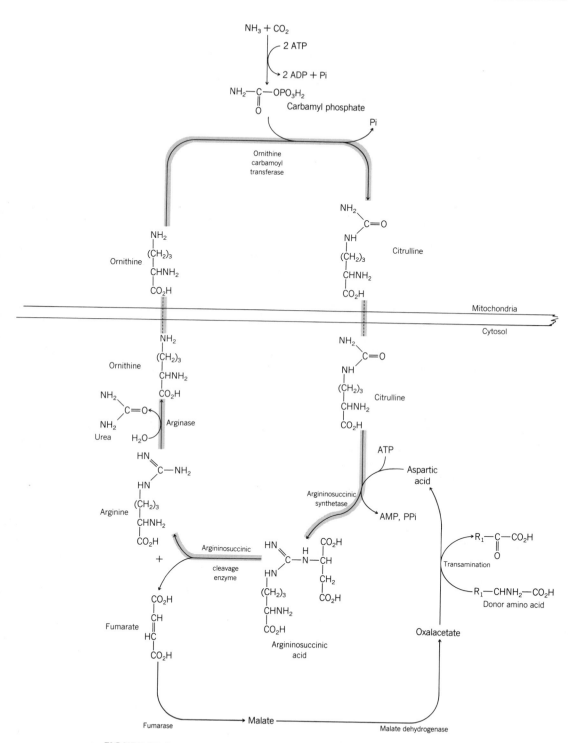

FIGURE 17.2 The urea cycle in detail.

$$\begin{array}{c} \text{HN} \\ \diagdown \\ \text{C}-\text{NH}_2 \\ \diagup \\ \text{HN} \\ | \\ \text{CH}_2 \\ | \\ \text{CH}_2 \\ | \\ \text{CH}_2 \\ | \\ \text{HCNH}_2 \\ | \\ \text{COOH} \end{array} + \text{H}_2\text{O} \longrightarrow \begin{array}{c} \text{H}_2\text{N} \\ \diagdown \\ \text{C}-\text{NH}_2 \\ \parallel \\ \text{O} \\ \text{Urea} \end{array} + \begin{array}{c} \text{NH}_2 \\ | \\ \text{CH}_2 \\ | \\ \text{CH}_2 \\ | \\ \text{CH}_2 \\ | \\ \text{HCNH}_2 \\ | \\ \text{COOH} \end{array} \qquad (17.19)$$

L-Arginine   L-Ornithine

Thus, Reactions 17.16 through 17.18 accomplish the formation of arginine, a widely occurring amino acid, from ornithine, NH_3, and CO_2. The three enzymes catalyzing these reactions presumably occur in a wide number of tissues in animals, plants, and microorganisms. However, the rate of synthesis is too low in most mammalian tissues, except in mammalian liver, to describe arginine as a nonessential amino acid. On the other hand, in liver where it is more rapidly synthesized, it is also more rapidly hydrolyzed by arginase and prevents its being used for protein synthesis. Arginase, whose activity makes urea formation possible, is found only in the liver of animals known to excrete urea. Liver is the major site of urea formation in mammals, although some urea synthesis apparently does occur in brain and kidney.

The sequence of reactions just discussed is shown in Figure 17.2 where the subcellular localization of the different enzymes is indicated. It will be noted that the synthesis of carbamylphosphate and citrulline occur in the mitochondria, while the remaining enzymes of the urea cycle are cytosolic. Note also the participation of the mitochondrial enzymes fumarase, malate dehydrogenase, and aspartic amino transferase in converting fumarate produced in Reaction 17.18 back into aspartate with the uptake of an amino group from donor amino acids.

17.9   COMPARATIVE BIOCHEMISTRY OF NITROGEN EXCRETION

If we survey the animal kingdom, we find that three nitrogen excretory products are common: NH_3, urea, and uric acid. An organism's preference for one of these forms depends, in part, on certain properties of the compounds: NH_3 is very toxic but it is also extremely soluble in H_2O; urea is far less toxic and is appreciably soluble in H_2O; uric acid is quite insoluble and, as such, is fairly nontoxic. There is abundant evidence to suggest that the form in which nitrogen is excreted by an organism is determined largely by the accessibility of H_2O to that organism.

Marine animals, living in H_2O, have large amounts of H_2O into which their waste products can be excreted. Although NH_3 is fairly toxic, it can be excreted and will be diluted out instantly in the H_2O of the environment. As a result, many marine forms excrete NH_3 as their major nitrogenous end product, although there are important exceptions to this among the bony fishes.

Land-dwelling animals no longer have an unlimited supply of H_2O in intimate contact with their tissues. Since NH_3 is toxic, it cannot be conveniently accumulated. As a result, most terrestrial animals have developed procedures for converting NH_3 into either urea or uric acid.

According to Needham, the choice between urea and uric acid is determined by the conditions under which the embryo develops. The mammalian embryo develops in close contact with the circulatory system of the mother. Thus, urea, which is quite soluble, can be removed from the embryo and excreted. On the other hand, the embryos of birds and reptiles develop in a hard-shelled egg in an external environment. The eggs are laid with enough water to see them through the hatching period. Production of NH_3 or even urea in such a closed system would be fatal because they are toxic. Instead, uric acid is produced by these embryos and precipitates out as a solid in a small sac (the **amnion**) on the interior surface of the shell. These characteristics, which are so necessary for the development of the embryo, are then carried over to the adult organism.

There are interesting examples in support of the principles we have cited. The tadpole, which is aquatic, excretes chiefly NH_3. When it undergoes metamorphosis into the frog, however, it becomes a true amphibian and spends much of its time away from water. During the metamorphosis, the animal begins excreting urea instead of NH_3, and by the time the change is complete, urea is the predominant nitrogen-excretory product.

Lungfish are another interesting example; while in water, they excrete chiefly NH_3, but as the river or lake runs dry, the lungfish settles down in the mud, begins to aestivate, and accumulates urea as the nitrogen end product. When the rains return, the lungfish excretes a massive amount of urea and sets about excreting NH_3 again.

Within one group of animals, the chelonia (tortoises and turtles), there are totally aquatic species, semiterrestrial species, and a third group (the tortoises) that is wholly terrestrial. The aquatic forms secrete a mixture of urea and ammonia; the semiterrestrial species, on the other hand, excrete urea; and the tortoises excrete almost all their nitrogen as uric acid.

The topic of nitrogen excretion is one of the best examples of comparative biochemistry that has been developed.

17.10 FORMATION OF URIC ACID

Uric acid, referred to in the preceding section, is the form in which birds and terrestrial reptiles excrete the NH_3 produced in protein metabolism. It is also the chief end product of metabolism of purines in man and other primates, the Dalmatian coach hound, birds, and some reptiles. Thus, the birds and reptiles that have uric acid as their chief nitrogen waste product first must convert NH_3 into purines by reactions to be considered shortly.

The free purine bases are converted to uric acid as shown in Figure 17.3. Xanthine oxidase, which catalyzes the formation of uric acid, is found in the peroxisomes of the kidney along with other oxidases (see Box 17.C). Mammals other than the primates and most reptiles produce allantoin as their end product of purine metabolism. Such organisms contain the enzyme uricase that converts uric acid to allantoin. The teleost fish convert allantoin on to allantoic acid, while most fish and the amphybia degrade the allantoic acid further to urea and glyoxylic acid. The pyrimidine bases are broken down into NH_3, CO_2,

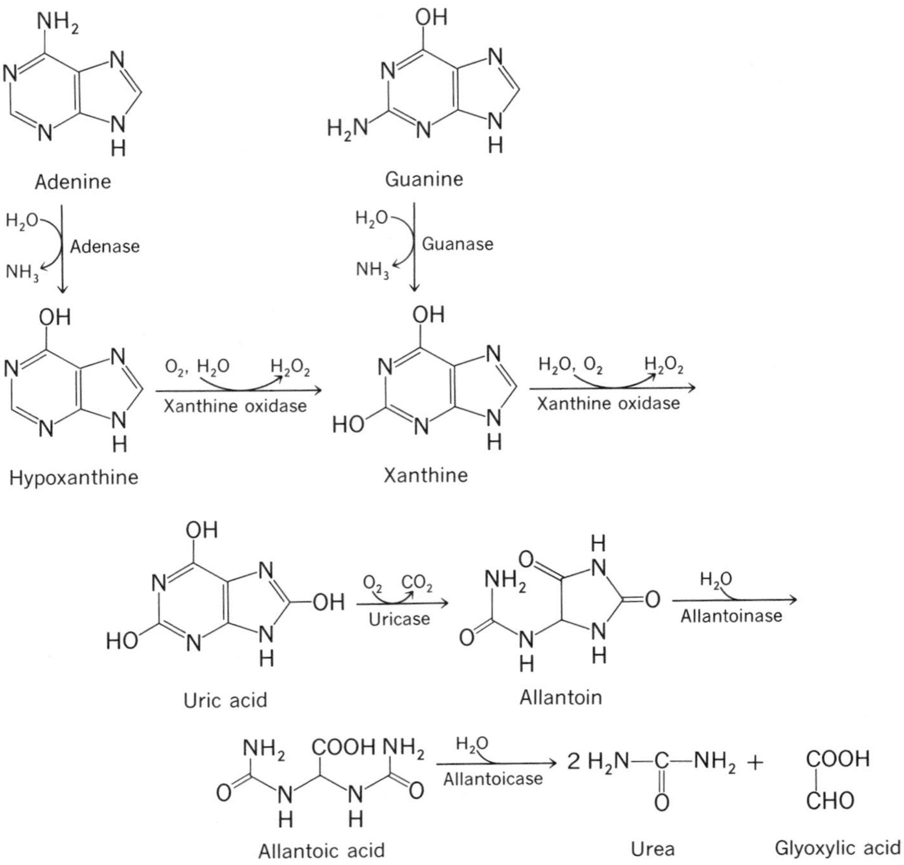

FIGURE 17.3   The metabolic degradation of adenine and guanine.

and propionic and succinic acids by reactions that will not be discussed in this text.

17.11   AMINO ACIDS AS PRECURSORS OF OTHER COMPOUNDS

It was noted earlier that amino acids function as precursors of important non-protein components. We have already referred to the synthesis of the physiologically active amines by decarboxylation (Section 17.6.2 and Box 17.D).

Amino acids serve as the primary precursors of a large number of natural products in plants. Thus, the plant alkaloids are derived primarily from aspartic acid, glutamic acid, lysine, tryptophan, phenylalanine, and tyrosine. The nitrogen-containing cyanogenic glycosides and mustard oil glucosides (glucosinolates) are derived from amino acids. Lignin, the second most abundant compound in nature (the first is cellulose), is produced from *trans*-cinnamic acid, as is a variety of flavonoids, phenolic acids, and coumarins. *trans*-Cinnamic acid, in turn, is produced by the action of **phenylalanine ammonia lyase** (PAL) on

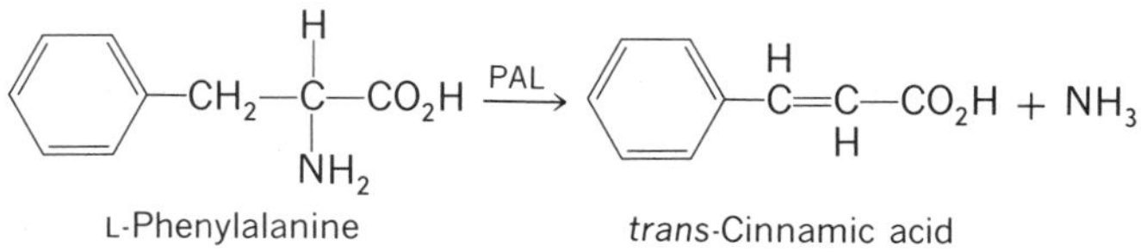

$$\text{L-Phenylalanine} \xrightarrow{\text{PAL}} \text{trans-Cinnamic acid} + NH_3 \qquad (17.20)$$

L-phenylalanine. The enzyme, which catalyzes this first step in the conversion of phenylalanine into a wide variety of natural products, is ubiquitous in plants and is found in many bacteria, yeasts, and fungi but never in animals.

BOX 17.E   BIOSYNTHESIS OF PHENYLPROPANOIDS

Many compounds bearing the phenylpropanoid carbon skeleton (C_6H_5–C–C–C) are found in plants. One of these, lignin, is second only to cellulose as the most abundant carbon compound in nature (most plants are composed of 20–40% lignin). Other phenylpropanoids such as flavonoids, encountered as pigments in flowers or as flavoring agents (tannin) found in wines, or phenolic acids (chlorogenic, caffeic) present in most plant foods, are familiar examples of this class of compounds.

The primary metabolic precursor of all of these phenylpropanoids is the amino acid phenylalanine.

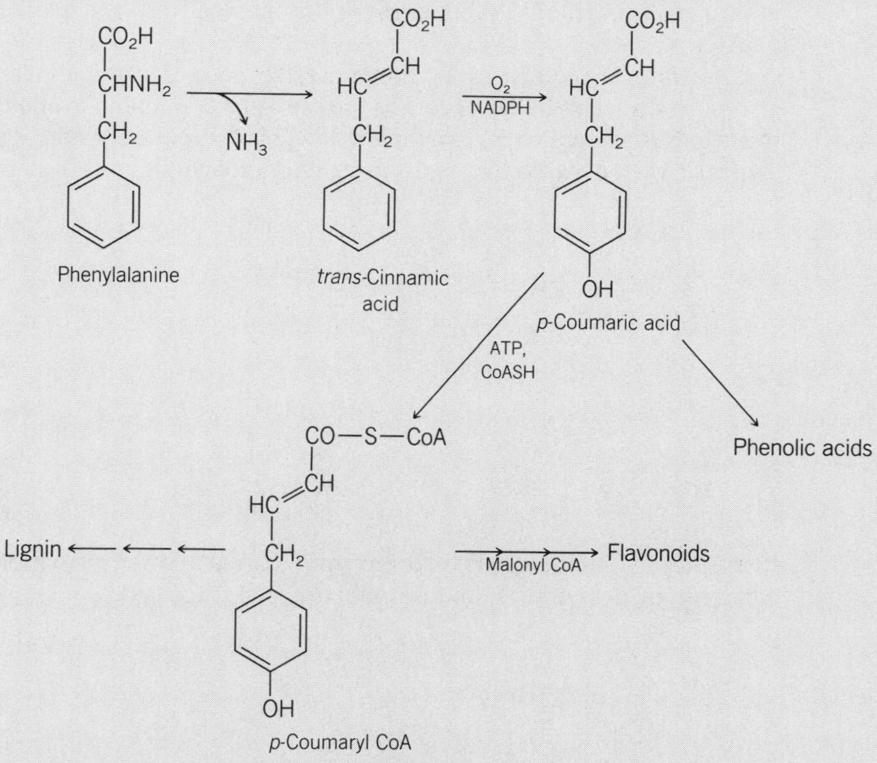

The initial steps of all phenylpropanoid biosyntheses involves the deamination of phenylalanine to form *trans*-cinnamic acid (Reaction 17.20). This deaminated acid is then hydroxylated by a monooxygenase, cinnamic acid-4-hydroxylase, to produce *p*-coumaric acid. The third common step then is the activation of the carboxyl group as a thiol ester to form 4-coumaryl CoA. This compound then is a branch point where the biosynthetic pathway diverges to form the compounds just mentioned and others. The magnitude of carbon flow from CO_2 through this pathway is second only to that encountered in the photosynthetic cycles described in Chapter 15. The biosynthetic process is an important example of the role of amino acids as precursors of other cell constituents and again emphasizes the role of monooxygenases as anabolic enzymes.

Other examples of amino acids serving as precursors of important, nonprotein molecules include glycine that is a precursor of porphyrins (Section 17.12) and of purines (Section 17.13), serine as a precursor of ethanolamine (found in phospholipids), and aspartic acid as a precursor of purines and pyrimidines (Section 17.14).

17.12   PORPHYRIN BIOSYNTHESIS

The heme proteins (catalase, cytochrome, hemoglobin, leghemoglobin, myoglobin, peroxidase) and chlorophyll have in common a cyclic tetrapyrrole structure called a porphyrin. The porphyrins, in turn, are synthesized from a monopyrrole called porphobilinogen. This compound is produced from δ-aminolevulinic acid (ALA) that, in turn, is derived from glycine and succinyl CoA.

Research carried out by D. Shemin established that the condensation of glycine and succinyl CoA occurs in the presence of **δ-aminolevulinic acid synthase** to form α-amino-β-ketoadipic acid. This compound, while bound to the synthase, then decarboxylates, forming δ-aminolevulinic acid.

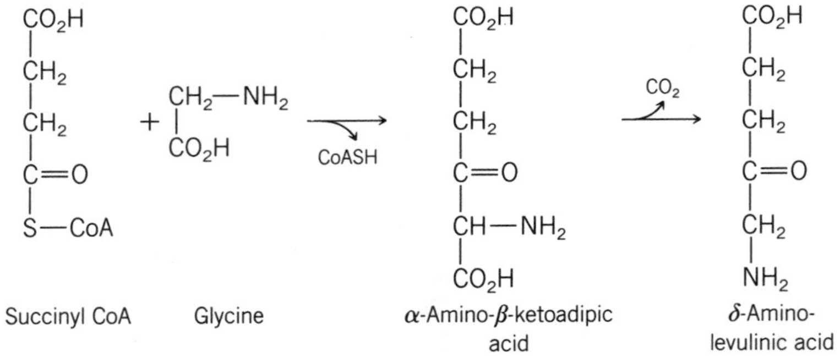

Succinyl CoA   Glycine α-Amino-β-ketoadipic acid δ-Amino-levulinic acid

In the presence of the appropriate enzyme (ALA dehydrase), two moles of ALA condense by dehydration and porphobilinogen is formed.

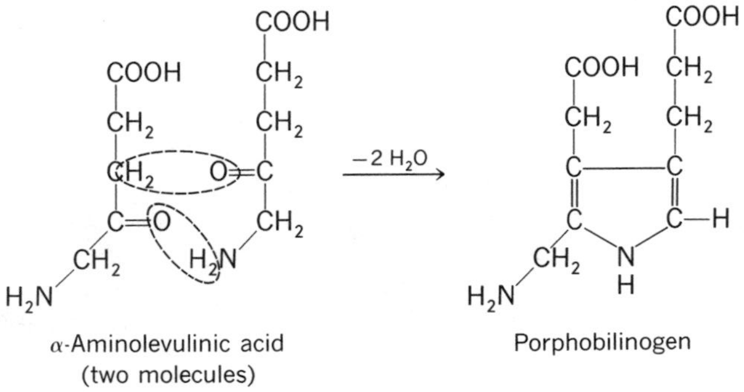

α-Aminolevulinic acid (two molecules)   Porphobilinogen

Note that the pyrrole ring is formed by this reaction and that it bears one propionic acid and one acetic acid residue.

Several additional enzymes acting sequentially combine four molecules of porphobilinogen into the cyclic tetrapyrrole structure and modify the residues attached to the pyrrole nuclei. Thus, decarboxylation of the acetate unit produces a methyl group on the pyrrole ring. Dehydrogenation and decarboxylation of the propionic residue yield the vinyl group, two of which are found in the structure of protoporphyrin IX.

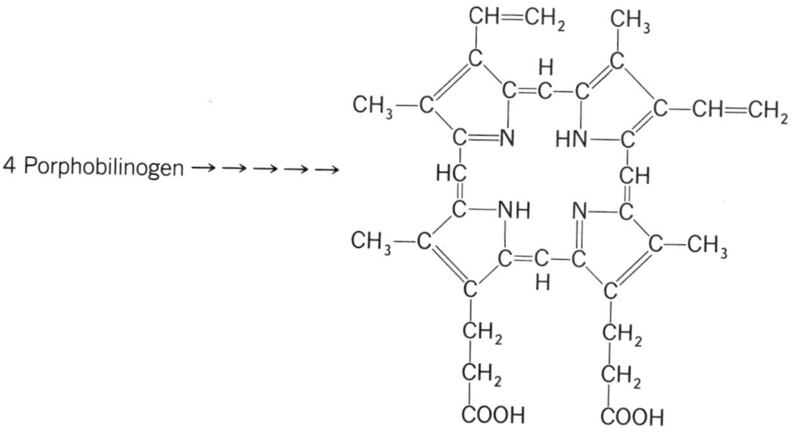

4 Porphobilinogen → → → → →

Protoporphyrin IX

Finally, a specific ferrochelatase in mitochondria inserts the ferrous ion into the tetrapyrrole ring to form heme. The heme moiety in cytochrome c is bonded to its specific protein by thioether bonds to a cysteine residue, and by methionyl and histidyl residues as described in Section 14.2.5.

17.13 PURINE BIOSYNTHESIS

Our discussion of the metabolism of nitrogenous monomers will be concluded with the purines and pyrimidines. In contrast to the essential or indispensible amino acids, the purines and pyrimidines can be formed from simple precursors by both plants and animals. Experiments with radioisotopes and nitrogen-15 have shown that the nine atoms of the purine nucleus are derived from five different precursors, each precursor contributing the atoms indicated in Figure 17.4. Work by a number of different researchers with mammals, birds, insects, and bacteria has shown that essentially the same biosynthetic pathway is followed in these diverse living forms. As will be seen, the pathway consists of a stepwise addition of individual atoms to the carbon-1 of ribose-5-phosphate to produce the key intermediate, **inosinic acid.** It is possible to describe most of the reactions in detail, and these will be given in order to demonstrate that certain principles of biochemical reactions, already encountered in the metabolism of carbohydrates, lipids, and the amino acids, can be applied to the synthesis of the nucleic acids and their derivatives.

17.13.1 FORMATION OF PRPP

Before we discuss the formation of purines in detail, the synthesis of 5'-phospho-α-D-ribose-1-pyrophosphate (PRPP) will be described. PRPP, which

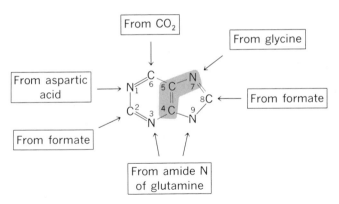

FIGURE 17.4   The metabolic origin of the purine nucleus.

serves as the source of the ribose-5-phosphate moiety of all purine and pyrimidine nucleotides, is derived from ATP and ribose-5-phosphate.

Ribose-5-phosphate

5'-Phospho-α-D-ribose-1-pyrophosphate
(PRPP)

Ribose-5-phosphate, in turn, is produced by either the irreversible part of the pentosephosphate pathway (Section 11.4) or the reversible sequence of reactions involving transketolase and transaldolase.

17.13.2   BIOSYNTHESIS OF INOSINE MONOPHOSPHATE (IMP)

The initial or committed step in purine biosynthesis is the reaction of PRPP with glutamine to form 5-phospho-β-D-ribosyl amine (Step 1, Figure 17.5). This reaction is another example in which glutamine donates its amide nitrogen for the synthesis of a new nitrogenous compound. This step is subject to feedback inhibition by purine nucleotides produced later in the pathway. Note that the configuration at the hemiacetal carbon of ribose is inverted during the reaction; the linkage of the pyrophosphate group in PRPP is α, while the amino group has the β-configuration. This, of course, is the configuration of the N-ribosyl bond in the purine nucleotides eventually formed by the pathway.

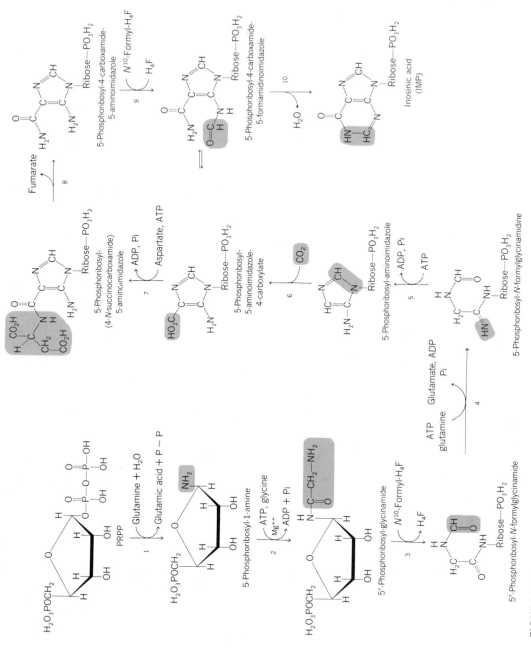

FIGURE 17.5 The biosynthesis of purines.

Step 1, Figure 17.5 is inhibited by the antibiotic **azaserine,**

$$\overset{-}{N}{=}\overset{+}{N}{-}CH_2{-}\underset{\underset{O}{\|}}{C}{-}O{-}CH_2{-}CHNH_2{-}CO_2H$$

Azaserine

a structural analog of the glutamine required by this reaction. Subsequent reactions that involve glutamine are also inhibited by azaserine.

In the next step, the amino acid glycine is linked to the ribosyl-amine in an amide linkage. It is not surprising, therefore, that the reaction should require a source of energy that is supplied by ATP. The 5′-phosphoribosyl glycinamide then reacts in the presence of a **transformylase** (Step 3) that catalyzes the transfer of a **formyl** group from the formyl transfer coenzyme, N^{10}-formyl tetrahydrofolic acid, H_4F (Section 5.8) to produce 5-phosphoribosyl-N-formylglycinamide. This reaction and a later one in the pathway (Step 9) that require folic acid coenzymes are inhibited by aminopterin and other antagonists of that vitamin. At this point, all the atoms of the imidazole ring of the purine nucleus have been attached to the phosphoribose moiety, and the latter will be represented in subsequent reactions by ribose-PO_3H_2.

While it would be reasonable to have ring closure at this point, the next reaction involves the addition of the nitrogen atom located at position 3 of the purine structure. As might be predicted, the nitrogen is provided by the amide group of glutamine in the presence of an energy source, ATP. The mechanism of Step 4, and several others in which a nitrogen atom is transferred to the purine or pyrimidine skeleton, presumably involves a phosphorylated intermediate. The ATP also required in these reactions, in effect, acts as a dehydrating agent.

The 5′-phosphoribosyl-N-formylglycinamidine then undergoes ring closure by a dehydration reaction that requires ATP. In this step, the imidazole ring of the purine nucleus is formed, and ATP is hydrolyzed to ADP and H_3PO_4. As the three remaining atoms of the purine skeleton have yet to be acquired, the carbon atom at position 6 is formed next by carboxylation of the imidazole nucleus with CO_2 (Step 6). This reaction might be expected to require a biotin coenzyme system and ATP, but neither appear to be involved.

The next step in the pathway is one of several reactions in intermediary metabolism in which a nitrogen atom is contributed by aspartic acid. The process is quite analogous to the synthesis of argininosuccinic acid in the urea cycle (Reaction 17.17), where ATP is required as an energy source. In the present instance, however, ATP is cleaved to ADP and H_3PO_4. Subsequently, the succinocarboxamide derivative is cleaved in an **aspartase** type of reaction (Step 8, Figure 17.5) to form fumaric acid in a manner quite analogous to the cleavage of argininosuccinic acid (Reaction 17.18).

One final carbon atom must now be acquired before the six-membered ring of the purine can be formed by ring closure. This atom is provided through the one-carbon metabolism of the folic acid system in the form of N^{10}-formyl tetrahydrofolic acid (Step 9). The reaction is also inhibited by sulfonamide antibiotics.

Ring closure (Step 10, Figure 17.5) then occurs in the presence of an enzyme that catalyzes the removal of H_2O in a reversible reaction. The product is

inosinic acid that does occur free in biological materials, but, of course, is not a component of RNA or DNA.

The nine steps described in Figure 17.5 can be summarized in the overall reaction

2 NH$_3$ + 2 Formic acid + CO$_2$ + Glycine + Aspartic acid + Ribose-5-phosphate $\longrightarrow$

Inosinic acid + Fumaric acid + 9 H$_2$O

The energy required to accomplish this process is, of course, provided by nine ATP molecules, all but one of which are cleaved to produce ADP and H$_3$PO$_4$. Thus, we have another example of the means by which the energy-richness of ATP can be utilized in discrete reactions to accomplish a biosynthetic sequence requiring energy.

17.13.3 PURINE NUCLEOTIDE INTERCONVERSIONS

The two purine nucleotides AMP and GMP are subsequently formed from inosinic acid, the initial product of the purine pathway. In the case of AMP, the nitrogen atom at position 6 is contributed by aspartic acid in a reaction involving the formation of a substituted succinic acid. While this reaction is strictly analogous to reaction Step 7 (Figure 17.5), note that a different nucleoside triphosphate (GTP) is required. The substituted succinic acid is then cleaved by an aspartase type of reaction to yield AMP and fumaric acid.

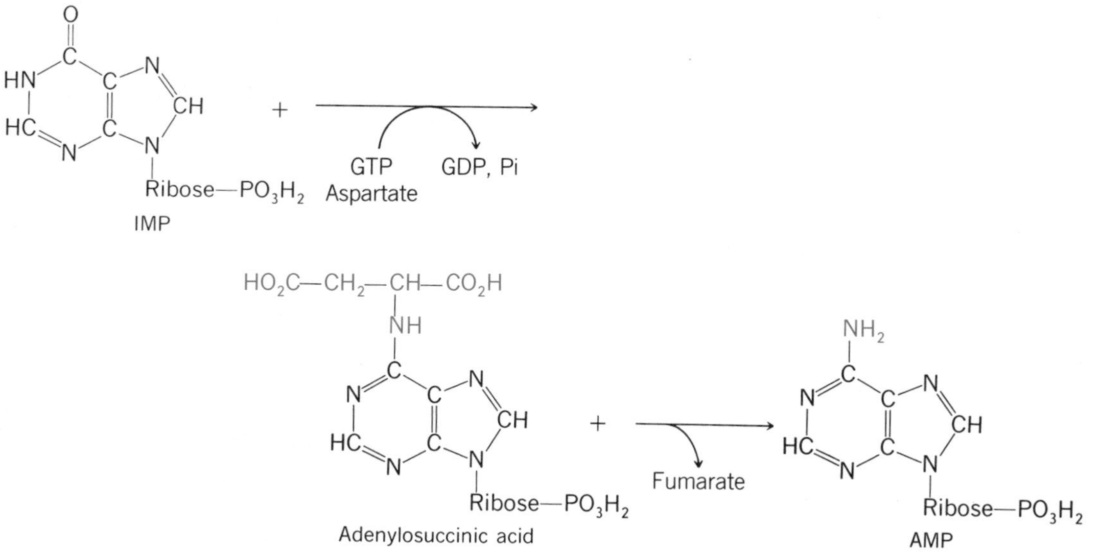

This reaction is similar to Step 8, Figure 17.5, and the enzymes catalyzing the two reactions are probably identical.

In the case of guanylic acid (GMP) synthesis, a nitrogen atom must be introduced in position 2 of the purine nucleus. To do this, the carbon atom at that position in inosinic acid must first be oxidized to the next higher oxidation state in order to acquire the amino group that is already at that oxidation level. The oxidation is accomplished by a dehydrogenase that requires the nicotinamide nucleotide NAD$^+$.

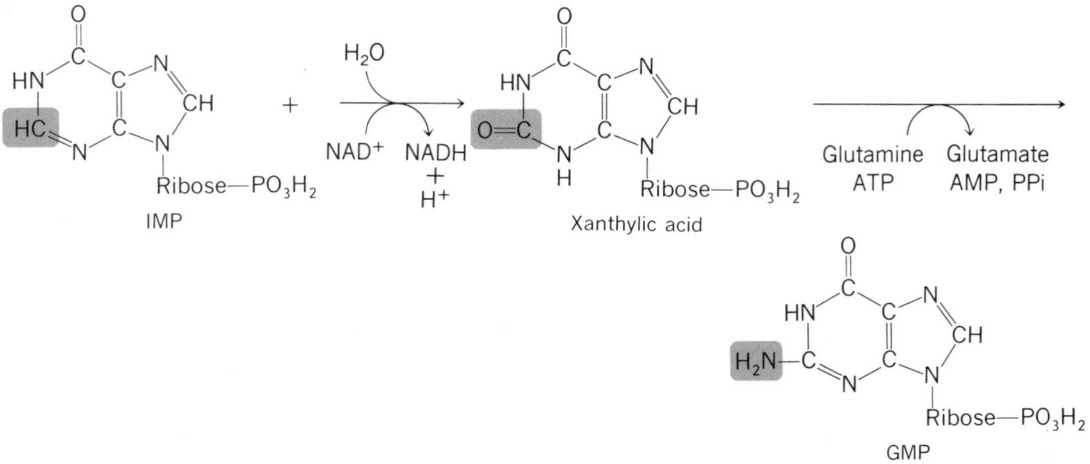

Once xanthylic acid is obtained, the nitrogen atom is acquired from glutamine in a reaction utilizing ATP; the latter is cleaved to AMP and PPi.

17.13.4   REGULATION OF PURINE NUCLEOTIDE BIOSYNTHESIS

The regulation of purine mononucleotide biosynthesis occurs at two different levels in the biosynthetic sequence. The first is at the synthesis of 5'-phosphoribosylamine (Step 1, Figure 17.5), which can be considered the initial step in the biosynthetic pathway. The enzyme catalyzing this reaction is inhibited at one regulatory site by AMP, ADP, and ATP and at another control site by GMP, GDP, and GTP.

The other control is at the branching compound, inosinic acid. It may be seen that the conversion of IMP to AMP requires GTP, while the formation of GMP from IMP requires ATP. Thus, when ATP is in excess, the higher concentration available simply leads to more GMP (and eventually GTP) being produced. Conversely, an excess of GTP leads to a higher production of AMP and therefore ATP.

17.14   PYRIMIDINE BIOSYNTHESIS

The atoms of the pyrimidine skeleton are derived from aspartic acid and carbamyl phosphate. The first step of the biosynthetic pathway (Figure 17.6) shows the metabolic origin of the skeleton from carbamyl phosphate and aspartic acid.

17.14.1   THE ASSEMBLING OF THE PYRIMIDINE NUCLEUS

The biosynthetic pathway commences with the transfer of a carbamyl group from carbamyl phosphate to aspartic acid to form N-carbamyl aspartic acid (ureidosuccinic acid). The energy for synthesis of the amide bond in the carbamyl aspartate is provided by the acylphosphate bond of carbamyl phosphate. The enzyme that catalyzes Step 1 (Figure 17.6) is known as **aspartic carbamoyl transferase;** it is the site of feedback inhibition by CTP that will be shown to be a product of the pathway under consideration. A detailed discussion of this enzyme as it relates to feedback inhibition is presented elsewhere (Section 18.3.2).

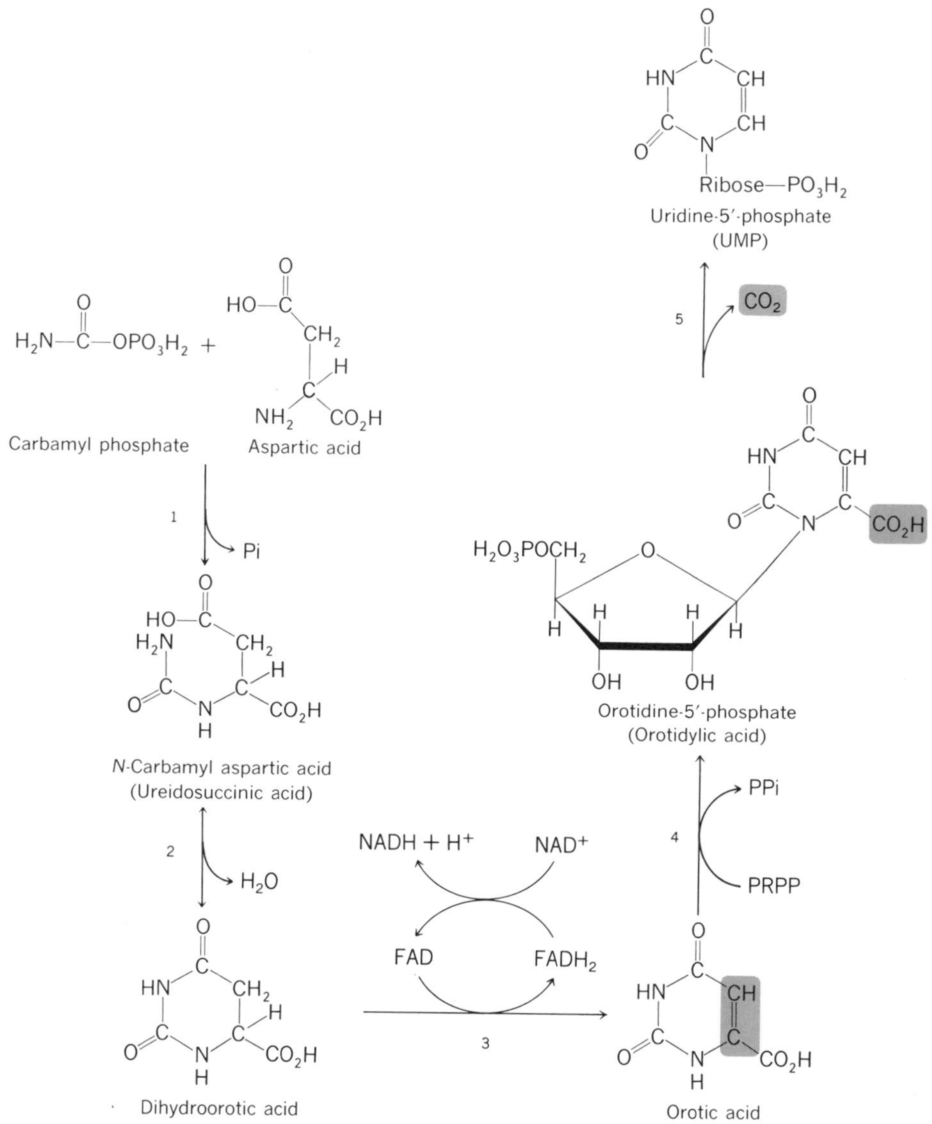

FIGURE 17.6 The biosynthesis of pyrimidines.

Ring closure of *N*-carbamyl aspartic acid catalyzed by the enzyme **dihydroorotase** leads to the formation of dihydroorotic acid (Step 2, Figure 17.6). This dehydration is freely reversible, with the open-chain compound predominating in the ratio of 2:1 at equilibrium. In the next step, a flavin enzyme, **dihydroorotic acid dehydrogenase,** catalyzes the formation of orotic acid by removing two hydrogen atoms from adjacent carbon atoms to form a carbon–carbon double bond (Step 3, Figure 17.6). The reduced flavin, in turn, is reoxidized by a NAD-dependent dehydrogenase.

Orotic acid reacts with phosphoribosyl pyrophosphate (PRPP) to acquire the 5′-phosphoribosyl moiety (-ribose-PO_3H_2) and become the nucleotide orotidine-5′-phosphate (Step 4, Figure 17.6). In this process, the ring nitrogen of orotic acid reacts as a nucleophile to displace the pyrophosphate group of PRPP and form the β-N-glycosyl bond. Finally, orotidylic acid is decarboxylated (Step 5, Figure 17.6) in the presence of a specific decarboxylase to yield uridine-5′-phosphate (UMP), the starting point for synthesis of the cytidine and thymidine nucleotides.

We might logically expect that some mechanism would exist for the amination of UMP to yield CMP, these monophosphates subsequently being phosphorylated to produce the di- and triphosphates. Instead, the formation of the cytidine derivative requires the triphosphate form of uridine. In animal tissues the source of the nitrogen is the amide group of glutamine; ATP provides the energy for transfer of the amide nitrogen.

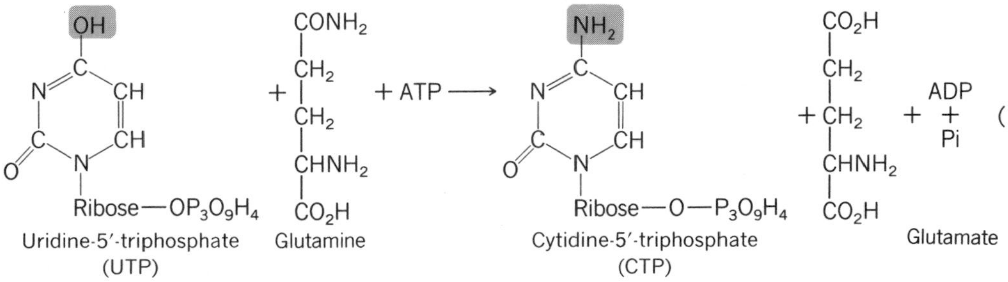

Uridine-5′-triphosphate   Glutamine   Cytidine-5′-triphosphate   Glutamate
(UTP)   (CTP)

As already noted, the biosynthesis of pyrimidine nucleotides is primarily regulated through the action of CTP on the enzyme aspartic carbamoyl transferase that catalyzes the first reaction in the biosynthetic sequence (Step 1, Fig. 17.6).

17.15   SYNTHESIS OF DIPHOSPHATES AND TRIPHOSPHATES

Once the synthesis of the monophosphates of purine and pyrimidine nucleosides is achieved, formation of the di- and triphosphates is readily accomplished. There are specific kinases that will catalyze the transfer of phosphate from ATP to the specific nucleoside monophosphates (NMP).

$$\text{NMP} + \text{ATP} \xrightarrow{\text{Mg}^{2+}} \text{NDP} + \text{ADP} \qquad (17.22)$$

These kinases are highly specific for the individual bases but utilize either the riboside or deoxyriboside. The synthesis of the nucleoside diphosphate is favored because of the removal of ADP by its subsequent phosphorylation to ATP during oxidative phosphorylation.

The triphosphates can, in turn, be formed by the phosphorylation of the nucleoside diphosphate in the presence of **nucleoside diphosphate kinase.**

$$\text{NDP} + \text{XTP} \rightleftharpoons \text{NTP} + \text{XDP}$$
$$\text{dNDP} + \text{XTP} \rightleftharpoons \text{dNTP} + \text{XDP}$$

The enzyme is ubiquitous in nature and quite nonspecific, showing no preference either for any particular base or ribose instead of deoxyribose. Again, the donor (XTP) is usually ATP and the synthesis of the other triphosphate (NTP) is driven by the ability of the cell to reform ATP from ADP by oxidative phosphorylation.

17.16 FORMATION OF DEOXYRIBOTIDES

The synthesis of DNA requires the availability of the four deoxyribonucleoside triphosphates (dATP, dGTP, dCTP, dTTP) that serve as precursors of that biopolymer. The unity (and simplicity) of biochemistry is readily apparent in this process, in which only one additional reaction is required to produce these compounds instead of a group of biosynthetic reactions analogous to those in the preceding two sections. The details of the reduction were first worked out in *Escherichia coli* where a single enzyme **ribonucleotide diphosphate reductase** catalyzes the replacement of the 2′-hydroxyl group in the ribose with H. This replacement therefore catalyzes the *reduction* of ribose to deoxyribose and a reducing agent is required. The ultimate source of the reducing equivalent is NADPH that reduces **thioredoxin** in the presence of a flavoprotein called thioredoxin reductase that also contains nonheme iron.

Thioredoxin is the small (MW 12,000) protein that contains two cysteine residues that can be alternately oxidized to a disulfide and reduced. Reduced thioredoxin, in turn, can reduce disulfide bonds in the ribonucleotide diphosphate reductase and the reduced form of the enzyme, in turn, accomplishes the reduction of the ribose unit.

NADPH → FAD → 2 Fe²⁺ → T—S₂ (Thioredoxin) → E—(SH)₂ (Ribonucleotide reductase) → [ribonucleoside diphosphate with base]

NADP⁺ ← FADH₂ ← 2 NHI / Fe³⁺ ← T—(SH)₂ ← E—S₂ → [2′-deoxyribonucleoside diphosphate with base]

Ribonucleoside diphosphate reductase of *E. coli* is a heteropolymer containing nonheme iron. It reduces the four natural ribonucleotides: ADP, GDP, CDP, and UDP.

The reductases found in animal tissues, tumor cells, and higher plants resemble that of *E. coli* in that a thioredoxin is the reducing agent and the substrates are the diphosphates. Another reductase in a variety of procaryotes differs in that the substrates for reduction are the nucleoside triphosphates rather than the diphosphates. Moreover, the reducing agent is a coenzyme form of vitamin B_{12}, 5,6-dimethylbenzimidazole cobamide coenzyme. This coenzyme accomplishes reductions by catalyzing hydrogen shifts involving the 5′-methylene group of the adenosyl moiety.

The conversion of a riboside to a deoxyriboside is, in effect, a **deoxygenation** that occurs only infrequently in biochemistry. While such reactions might occur, for example, by dehydration followed by hydrogenation, or by phosphorylation and displacement (with a hydride ion), neither of these mechanisms is involved in the deoxyriboside formation. The -OH group appears to be reduced directly without any intermediates, and the configuration on the carbon atom is retained. That is, D-ribose is reduced to 2-D-deoxyribose.

17.17   THYMINE BIOSYNTHESIS

Deoxythymidylic acid (dTMP) is produced from deoxyuridylic acid (dUMP) by the enzyme **thymidylate synthetase** and tetrahydrofolic acid (H_4F) previously described (Section 5.8.3.2). H_4F serves as a carbon and hydrogen donor. Furthermore, cobamide coenzyme is also implicated in this reaction. The folic acid requiring process is inhibited by aminopterin and amethopterin.

17.18   SALVAGE PATHWAY FOR PURINE AND PYRIMIDINE NUCLEOTIDES

The synthesis of purine and pyrimidine nucleotides described above requires significant quantities of energy because synthesis proceeds stepwise from the simple monomers NH_3 (or glutamine), CO_2 (or carbamylphosphate), formic acid, glycine, and aspartic acid. It is not surprising therefore that organisms have evolved a means for rescuing or *salvaging* these complex nitrogen bases if they are formed during breakdown of DNA and RNA. There are several reactions that serve to rescue these bases from further breakdown. One is the reaction catalyzed by enzymes called **phosphoribosyl transferases.**

$$Adenine + PRPP \rightleftharpoons AMP + PPi$$
$$Guanine + PRPP \rightleftharpoons GMP + PPi$$
$$Hypoxanthine + PRPP \rightleftharpoons IMP + PPi$$

In this reaction, PRPP again serves as a source of the ribose phosphate moiety as in the first reaction in purine biosynthesis (Step 1, Figure 17.5). The reaction is readily reversible but, in practice, operates from left to right because of the action of ubiquitous pyrophosphatases that hydrolyze the pyrophosphate formed.

A second type of salvage reaction is catalyzed by enzymes called **nucleoside phosphorylase.**

$$Base + ribose-1-phosphate \rightleftharpoons Nucleoside + Pi$$

This enzyme will work either with ribose-1-PO_4 or 2-deoxyribose-1-phosphate. Phosphorylases have been described that work either with purines (hypoxanthine and guanine) or pyrimidines (uracil and thymine).

A third salvage reaction is catalyzed by **nucleoside kinases** that are relatively specific for adenosine or thymidine and ATP. In this process, which is formally analogous to hexokinase (Reaction 10.1), an energy-rich phosphate is used to produce a low-energy phosphate ester.

$$Adenosine + ATP \longrightarrow AMP + ADP$$
$$Thymidine + ATP \longrightarrow TMP + ADP$$

The conversion of the monophosphates (NMP) to diphosphates (NDP) by specific kinases and their subsequent further phosphorylation to triphosphates (NTP) has already been described.

While other reactions of this sort can be listed, the aforementioned give some idea of the means that organisms have evolved to conserve the nitrogen-base monomers of the nucleic acids. In those instances in which a cell is temporarily unable to carry out the *de novo* formation of purines and pyrimidines, these

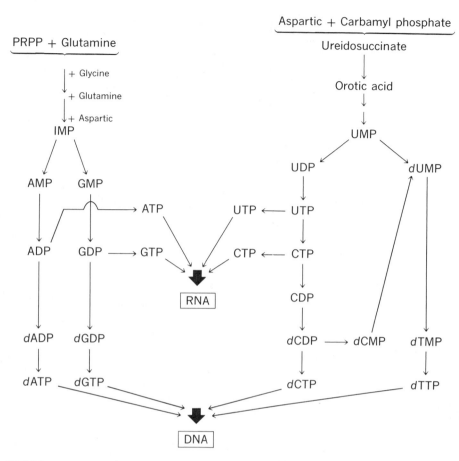

FIGURE 17.7 Relationship of synthesis of nucleotides and nucleic acids. The prefix *d* signifies deoxy-; thus, AMP contains the ribosyl moiety, and *d*AMP contains a deoxyribosyl moiety. Although in some organisms, the ribonucleoside triphosphates are directly reduced to their deoxy derivatives, for simplicity, these reactions are not included here.

salvage reactions are essential for maintaining a supply of the monomers for DNA and RNA synthesis.

Figure 17.7 summarizes the numerous reactions for the synthesis of nucleoside triphosphates required in the synthesis of RNA and DNA.

REFERENCES

1. *A. Meister,* Biochemistry of the Amino Acids, *Vols. I and II, 2nd ed. New York: Academic Press, 1965.*

2. *D. M. Greenberg,* Metabolic Pathways, *Vol. 3. New York: Academic Press, 1969.*

3. *J. O. Stanbury, J. B. Wyngaarden, and D. S. Fredrickson, eds.,* The Metabolic Basis of Inherited Disease, *4th ed. New York: McGraw-Hill, 1978.*

4. *D. Wellner and A. Meister, "A survey of inborn errors of amino acid metabolism and transport in man."* Annual Review of Biochemistry, *Vol. 50 (1981), pp. 911–968.*

5. *H. E. Umbarger, "Amino acid biosynthesis and its regulation."* Annual Review of Biochemistry, *Vol. 47 (1978), pp. 533–606.*

6. *H. A. Barker, "Amino acid degradation by anaerobic bacteria."* Annual Review of Biochemistry, *Vol. 50 (1981), pp. 23–40.*

7. *B. J. Miflin, ed., "Amino acids and derivatives."* In Biochemistry of Plants, *Vol. 5. New York: Academic Press, 1981.*

8. *G. A. Rosenthal,* Plant Nonprotein Amino and Imino Acids. *New York: Academic Press, 1982.*

9. *J. Mora and R. Palacios, eds.,* Glutamine: Metabolism, Enzymology and Regulation. *New York: Academic Press, 1979.*

10. *A. Kornberg,* DNA Synthesis. *San Francisco: W. H. Freeman, 1974.*

11. *M. E. Jones, "Pyrimidine nucleotide biosynthesis in animals: Genes, enzymes, and regulation of UMP-biosynthesis."* Annual Review of Biochemistry, *Vol. 49 (1980), pp. 253–279.*

REVIEW PROBLEMS

1. When alanine labeled with ^{15}N was administered to a rat, the animal excreted urea containing ^{15}N in both of its nitrogen atoms. By means of known enzymatic reactions, explain how this conversion can take place.

2. Describe two different enzyme-catalyzed reactions by which the amino group of aspartic acid may be lost (i.e., by which aspartic acid may be "deaminated").

3. What compound is the most likely immediate precursor of the ethanolamine portion of phosphatidylethanolamine (cephalin)? Write the enzyme-catalyzed reaction by which ethanolamine is produced. Name the enzyme.

4. Explain why plants that have no need to produce urea contain nearly all the enzymes of the urea cycle.

5. Write four distinctly different enzyme-catalyzed reactions by which inorganic nitrogen (as ammonia) can be incorporated into an organic molecule.

6. α-Aminoadipic acid is a six-carbon, α-amino dicarboxylic acid. Write a likely series of enzyme-catalyzed reactions by which α-aminoadipic acid might be synthesized from the usual intermediates found in cells (e.g., glycolytic intermediates, TCA cycle intermediates, β-oxidation intermediates, and common amino acids).

ENZYME-MEDIATED REGULATION OF METABOLISM

The growth and maintenance of a cell require a highly integrated coordination of anabolic and catabolic processes. Since the functioning unit of metabolic machinery is the enzyme-catalyzed reaction, the control of this unit becomes the essential feature in metabolic regulation.

Mechanisms of metabolic regulation have been intensively investigated in the past several years in both procaryotic and eucaryotic organisms. Unifying principles are beginning to emerge. Being a multifaceted term, metabolic regulation involves (1) compartmentation of enzymes; (2) alternate or separate pathways for catabolism and anabolism of a key substrate; (3) kinetic factors involving the interactions of substrates, cofactors, pH, and enzymes; and (4) the control of enzyme activity and enzyme concentration. The rates of enzyme synthesis and degradation, in turn, control the actual functioning concentration of the enzyme in the cell.

18.1 ENZYME COMPARTMENTATION

18.1.1 THE PROCARYOTIC CELL

The procaryotic cell has for over 40 years been employed by biochemists as a model cell to explore all aspects of cell metabolism. In terms of structure, it is a rather simple cell with a plasma membrane onto which an important number of key enzymes are associated (Chapter 8) and a cytoplasmic region in which the principal pathways of metabolism are carried out in an astonishingly orderly manner. The plasma membrane serves as a substitute for organelle enzymes in that the enzymes frequently associated with eucaryotic organelle membranes, such as those that participate in the respiratory chain, oxidative phosphorylation, and phospholipid biosynthesis, are found in the procaryotic plasma membranes. At first glance, the cytoplasm of the procaryote may exhibit little structure, but there is increasing evidence that even in this region, enzymes may assume a loose, fragile organizational structure.

18.1.2   THE EUCARYOTIC CELL

In the eucaryotic cell, however, an entirely different situation exists. In these cells, compartmentation of metabolic machineries occurs for very specific purposes. As in procaryotic organisms, the plasma membrane of eucaryotic organisms is involved in the selective transport of important cations, anions, and neutral compounds, as receptor sites for a whole host of hormones, as well as serving as a barrier from the external milieu. The nucleus is the site for storage of genetic information and the transcription of this information, that is, the biosynthesis of mRNA, the synthesis of tRNA in the nucleoplasm and of rRNA in the nucleolus, and the subsequent modification and transport of these molecules to the cytoplasm for translation of mRNA into catalytic units, namely, enzymes. The mitochondrion is characterized by its complex of enzymes involved in maintaining the energetics of the entire cell. The endoplasmic reticulum serves as a site of important membrane enzymes and protein-synthesizing systems. Lysosomes are specific compartments for a host of hydrolytic enzymes. The lysosomes function as cell "scavengers" and are active in the autolytic reactions post mortem. The Golgi apparatus in eucaryotic cells is involved in the formation of secretory bodies and also participates in the sorting out and modification of proteins for transfer to other organelles. In plants, the chloroplast is the prime organelle for the generation of oxygen, ATP, and the fixation of CO_2 for the plant cell. The student should consult Chapter 8 for further information.

Another compartmental consideration is the spatial separation of multienzyme systems from each other, Thus, in the degradation of glucose to carbon dioxide and water, at least three pathways are involved: glycolysis, the pentose phosphate cycle, and the tricarboxylic acid cycle. The glycolytic enzymes and the enzymes of the pentose phosphate cycle are found in the cytoplasm, whereas enzymes of the tricarboxylic acid cycle are located inside the mitochondria as are the tightly bound particulate enzymes of electron transport and oxidative phosphorylation. A close coordination must exist between the three metabolic sequences, and any interference in that coordination will result in a breakdown or modification of glucose metabolism. Furthermore, any change in the concentration of phosphate and magnesium ions, the ratio of ADP to ATP, $NADP^+$ to NADPH, NAD^+ to NADH, or the concentration of oxygen and carbon dioxide would also affect this partnership.

Still another factor in metabolic control and regulation is the ability of mitochondria, for example, to concentrate coenzymes, substrates, and enzymes far above the concentration found outside these particles. By this mechanism, the kinetic responses of enzyme-catalyzed reactions in mitochondria are greatly changed.

18.1.3   SUBSTRATE CYCLES

Whereas many metabolic reactions are reversible and catalyzed by a single enzyme, an important number of biochemical reactions appear to be reversible because of the involvement of two separate enzymes, one catalyzing the forward reaction and the other the backward reaction. These are called **substrate cycles** or **opposing unidirectional reactions** and may result in *futile* cycles.

$$A \underset{b}{\overset{a}{\rightleftharpoons}} B$$

Typical examples of varying complexity can be cited:

(1)
(a) Glucose + ATP $\xrightarrow{\text{Hexokinase}}$ Glucose-6-phosphate + ADP

(b) Glucose-6-phosphate + H_2O $\xrightarrow{\text{Glucose-6-phosphatase}}$ Glucose + Pi

---

(2)
(a) Fructose-6-phosphate + ATP $\xrightarrow{\text{Phosphofructokinase}}$ Fructose-1,6-bisphosphate + ADP

(b) Fructose-1,6-bisphosphate $\xrightarrow{\text{Fructo-1,6-bisphosphatase}}$ Fructose-6-phosphate + Pi

---

(3)
(a) Acetate + ATP + CoA $\xrightarrow{\text{Thiokinase}}$ Acetyl CoA + AMP + PPi

(b) Acetyl CoA + H_2O $\xrightarrow{\text{Thioesterase}}$ Acetate + CoA

---

(4)
(a) Acetyl CoA + CO_2 + ATP $\xrightarrow{\text{Acetyl CoA carboxylase}}$ Malonyl CoA + ADP + Pi

(b) Malonyl CoA $\xrightarrow{\text{Malonyl CoA decarboxylase}}$ Acetyl CoA + CO_2

---

(5)
(a) Phosphoenol pyruvate + ADP $\xrightarrow{\text{Pyruvic kinase}}$ Pyruvate + ATP

(b) Pyruvate + CO_2 $\xrightarrow{\text{ATP}}$ Oxalacetic acid $\xrightarrow{\text{GTP}}$ Phosphoenol pyruvate + CO_2

---

(6)
(a) Glucose-1-phosphate + UTP $\longrightarrow$ UDPG $\longrightarrow$ Glycogen

(b) Glycogen + Pi $\xrightarrow{\text{Phosphorylase}}$ Glucose-1-phosphate

In all cases, the forward reaction (a) is catalyzed by a specific enzyme, while the back reaction (b) is catalyzed by a completely different enzyme, which is usually hydrolytic and thus essentially irreversible. The cell utilizes these reactions involving two completely different sets of enzymes to allow fine regulation of Reactions a or b since it would be very difficult to control Reactions a and b by employing a single enzyme. However, controls must be imposed on these systems, since otherwise these opposing reactions would couple and lead to futile cycle activities. Thus, Reactions 1 through 6, if not coupled to other systems, could lead to a net hydrolysis of nucleoside triphosphates to nucleoside diphosphates and inorganic phosphate.

18.2 CATABOLIC AND ANABOLIC ROUTES

In a sequential series of reversible reactions $A \rightleftharpoons B \rightleftharpoons \ldots \rightleftharpoons X$, by which substrate A is converted eventually to product X with a concommitant release of energy and by which X may be converted back to A by the same reversible route, such routes may have thermodynamic difficulties. An input of energy would have to be inserted to return X to A, as well as imposing difficulties in the regulation of this type of route.

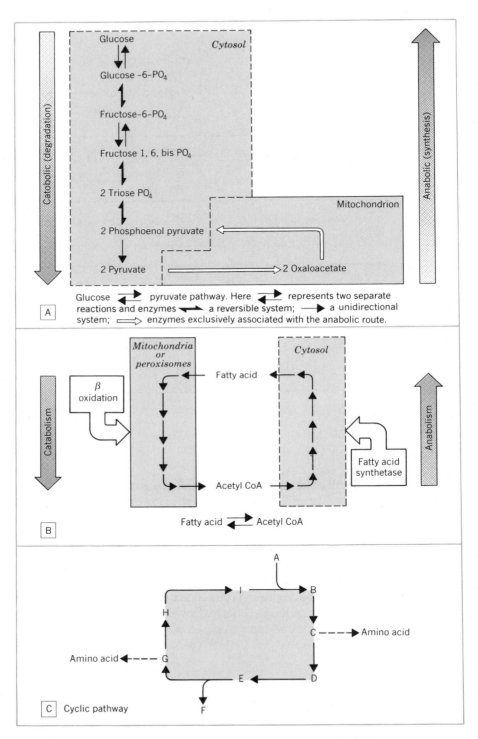

FIGURE 18.1   Three examples of different routes that can be taken for catabolic and anabolic processes.

Typical routes that have been already discussed in previous chapters are the breakdown of glucose to pyruvate (catabolic route) and the formation of glucose from pyruvate (gluconeogenesis in the anabolic route) (Chapter 10). Another example is the β-oxidation of fatty acids to acetyl CoA and the synthesis of fatty acids from acetyl CoA by a completely separate route (Chapter 13).

As summarized in Figure 18.1, catabolic and anabolic routes are organized by a variety of mechanisms. With the glucose $\rightleftharpoons$ pyruvate routes (Figure 18.1A), most of the enzymes function in both the catabolic and anabolic pathways with the exception of the conversion of pyruvate to phosphoenolpyruvate —not by pyruvate kinase (catabolic route) but by pyruvate carboxylase and phosphoenolpyruvate carboxylkinase to resolve the thermodynamic barrier of pyruvate $\rightarrow$ phosphoenolpyruvate (Chapter 10).

With the metabolism of fatty acids, a more complicated picture occurs. Here, the catabolic route is catalyzed by a set of enzymes that do not serve at all in the anabolic route. Furthermore, these enzymes are all localized in the mitochondrial or the peroxisome matrix of all animal cells. In contrast, the synthesis of fatty acids from acetyl CoA is catalyzed by a fatty acid synthetase and its ancilliary enzymes, and these are localized exclusively in the cytosol in animal cells. In plants, fatty acid synthesis occurs, however, in the chloroplast of the leaf cell or the proplastid of the seed cell. These schemes are depicted in Figure 18.1B and are also described in Chapter 13.

In addition, a number of catabolic routes employ cyclic pathways (Figure 18.1C) in which a substrate A is converted by a series of reactions to F and G, with F being a product; G then forms I that combines with A to form B, and so on. Two examples of this type of pathway are the TCA cycle and the glyoxylate bypass described in Chapter 12. Although one could regard a cyclic pathway as purely catabolic, in actual fact, intermediate products can be employed as carbon skeletons for important new biocompounds. Thus, in the TCA cycle, α-ketoglutaric acid is the precursor of glutamic acid, succinyl CoA is the precursor for porphyrin biosynthesis, and oxaloacetic acid can serve as the precursor for aspartic acid.

There are many other possible pathways that are utilized but those just discussed give the student a general idea of how the cell resolves the problem of catabolic and anabolic routes.

18.3 KINETIC FACTORS

We have already discussed throughout this book the effects of a number of compounds (effectors and modulators) on the activity of a group of enzymes called regulatory enzymes. The activity of regulatory enzymes are modulated by activation by compounds, called positive effectors, or inhibited by negative effectors. Inhibition may be manifested by any of three types of inhibitions, namely, competitive, noncompetitive, or uncompetitive or by a combination of any of these three types (see Chapter 4 for a discussion of these types of inhibition as well as a discussion on allostery). We shall now examine in more detail several important modulations of enzyme activity by metabolites (or effectors).

18.3.1   PRODUCT INHIBITION

A rather simple inhibition of a reaction is called product inhibition, in which the product of the reaction, by mass action effect, inhibits its own formation. Thus, in the conversion of glucose to glucose-6-phosphate by the enzyme hexokinase, as glucose-6-phosphate begins to accumulate, the reaction slows down. It is for this reason that enzyme assays should be carried out at the initial period of the reaction to avoid inhibition by the accumulating product.

18.3.2   FEEDBACK (END-PRODUCT) INHIBITION

An even more subtle type of control of enzyme action is designated as feedback inhibition. This is demonstrated most easily by considering the following sequence:

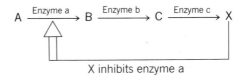

$$A \xrightarrow{\text{Enzyme a}} B \xrightarrow{\text{Enzyme b}} C \xrightarrow{\text{Enzyme c}} X$$

X inhibits enzyme a

Here, X, the ultimate product of the sequence, serves to prevent the formation of one of its own precursors by inhibiting the action of enzyme *a*. The first enzyme of the sequence, which is also called a monovalent **regulatory or allosteric enzyme,** namely enzyme *a*, can also be called the **pacemaker** since the entire sequence is effectively regulated by inhibiting it. A specific example is the formation in *E. coli* of cytidine triphosphate, CTP, from aspartic acid and carbamyl phosphate (Figure 18.2). As a critical concentration of CTP is built up, the triphosphate slows down its own formation by inhibiting the enzyme, aspartate carbamyltransferase (ATCase), which catalyzes the pacemaker step for the synthesis of carbamyl aspartate. When the concentration of the triphosphate is sufficiently lowered by metabolic utilization, inhibition is released, and its synthesis is renewed (Figure 18.2).

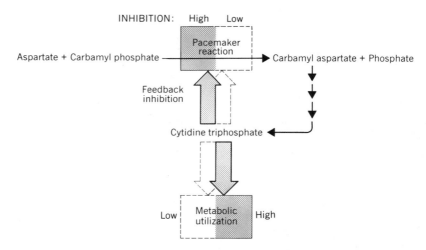

FIGURE 18.2   When metabolic utilization is low and CTP concentration is high, feedback inhibition operates. When metabolic utilization is high and CTP concentration is low, feedback inhibition is inoperative.

In all feedback inhibitions, the inhibitor (effector or modulator) usually has no structural similarity to the substrate of the enzyme it is regulating. Thus, CTP in no way resembles aspartic acid, the substrate for aspartic carbamyltransferase. Furthermore, all allosteric enzymes so far examined are **oligomeric** enzymes; that is, they have two or more distinct subunits. For example, aspartic carbamyltransferase is readily dissociated into two large subunits, one of which carries the catalytic site and the other the regulatory site. The first subunit, once separated from the second or regulatory subunit, has normal Michaelis–Menten kinetics rather than sigmoidal kinetics and now is no longer affected by CTP. The second subunit has no catalytic activity but binds CTP strongly.

Let us now consider variations of feedback inhibition of metabolic sequences. The regulation of the linear sequence referred to earlier is a straightforward end-product inhibition of the first enzyme in the sequence, a monovalent allosteric enzyme. However, regulation of a branched biosynthetic pathway by X and Y would lead to a situation in which an excess of one end product would lead not only to a decrease in the synthesis of X, but also the other end product, not a very good control system. However, a number of mechanisms exist that resolve this dilemma.

Monovalent feedback
(by X alone; Y inactive)
Divalent feedback
(by X and Y)

18.3.2.1 Isofunctional Enzymes.

In this mechanism, the first common step is catalyzed by two different or **isofunctional enzymes** that convert the same substrate to the same product. However, enzyme a is under the specific feedback control of X, while enzyme a' is insensitive to X but is under the specific control of Y.

Isofunctional enzymes a and a'

Since enzyme a in the latter instance would still be involved in the synthesis of B, C, and D, a secondary feedback control must be exerted by the two end products, namely, X on enzyme d and Y on enzyme d'. Thus, if an excess of X is formed, it will not only inhibit enzymes a and d, but also will not interfere with the synthesis of Y. An excellent example of this control mechanism has been described in the biosynthesis of lysine, methionine, threonine, and isoleucine from aspartic acid, as depicted in the simplified pathway shown in Figure 18.3.

18.3.2.2 Sequential Feedback Control.

In this mechanism, enzyme a is not regulated by either of the end products of the branched pathway. However, X will inhibit the enzyme that converts the last common substrate D to

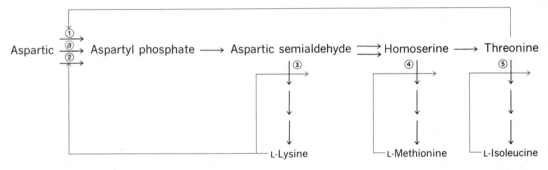

FIGURE 18.3   Feedback control of amino acid synthesis. ①; ②; ⓐ, isofunctional aspartic kinases; ①, monovalent feedback inhibited by threonine; ②, monovalent feedback inhibited by lysine; ⓐ, not a regulatory enzyme; ③; ④; ⑤, enzymes under feedback control by lysine, emthionine, and isoleucine, respectively.

precursors of X, and Y will inhibit

$$\text{Sequential feedback} \qquad A \xrightarrow{a} B \longrightarrow C \longrightarrow D \begin{smallmatrix} \xrightarrow{d} X \\ \xrightarrow{d'} Y \end{smallmatrix}$$

the enzyme that will convert D to precursors of Y. Thereby, D will accumulate and inhibit enzyme a, which shuts off the entire pathway. An example of this pathway is observed in the biosynthesis of aromatic acids in a number of bacteria and the regulation of threonine and isoleucine biosynthesis in *Rhodopseudomonas spheroides.*

18.3.2.3   Concerted Feedback Inhibition.

In this system, enzyme a is insensitive to X or Y alone, but when both are present, they act in concert to inhibit enzyme a. Again, both X and Y exert secondary controls by having X inhibit enzyme d and Y enzyme d'.

$$\text{Concerted feedback} \qquad A \xrightarrow{a} B \longrightarrow C \longrightarrow D \begin{smallmatrix} \xrightarrow{d} X \\ \xrightarrow{d'} Y \end{smallmatrix}$$

Thus, if there is an excess of X synthesized, it will only inhibit its own synthesis by controlling the activity of enzyme d, allowing Y to be synthesized. As Y accumulates, both X and Y can now, in concert, inhibit enzyme a, which is sensitive to inhibition only in the presence of both X and Y. A good example is the inhibition of aspartyl kinase from *Rhodopseudomonas capsulatus* by the combination of both threonine and lysine. Alone, these amino acids are ineffective inhibitors.

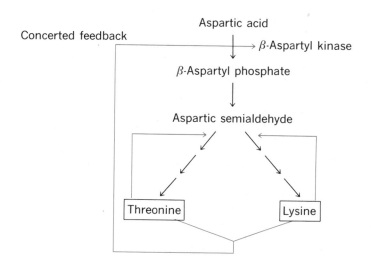

18.3.2.4   Cumulative Feedback Inhibition.   In this mechanism, X and Y, in saturating concentrations, only cause partial inhibition of enzyme a, but when they are both present simultaneously, a cumulative effect is observed. Thus, if X at saturating

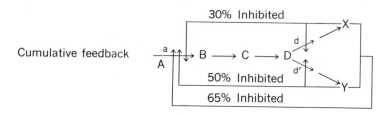

concentration inhibits a so that its residual activity is, for example, 70% and Y alone inhibits a by 50%, then when X and Y are both present in saturating concentrations, the residual activity will be 0.7×0.5 or 35% of the total activity. Inhibition will be 65%.

18.4   COVALENT MODIFICATION OF REGULATORY ENZYMES

We have seen that both linear as well as branched biosynthetic pathways can be selectively controlled by different types of feedback control. A number of these regulatory enzymes are controlled by having the effectors induce conformational changes in the protein structures that physically modify the catalytic site of the enzyme. An additional mechanism involves a covalent or **chemical modification of the regulatory enzyme** by the covalent attachment of specific groups to the regulated enzyme, leading to changes in primary and, thus, the tertiary structure of the enzyme. The modifying proteins are specific enzymes that are involved in *inserting* and *removing* the specific groups that include phosphoryl, adenyl, and other components. Covalent modifications now include the following.

1. *Phosphorylation of a specific serine, threonine, or tyrosine residue on an enzyme protein.*

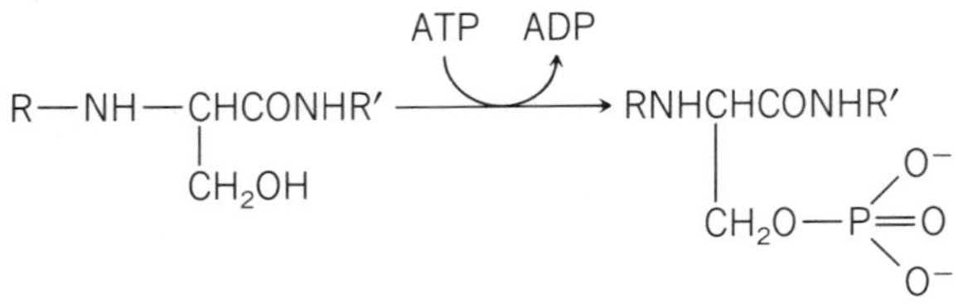

Example: Pyruvate dehydrogenase complex

2. *Nucleotidylation of a specific tyrosine residue.*

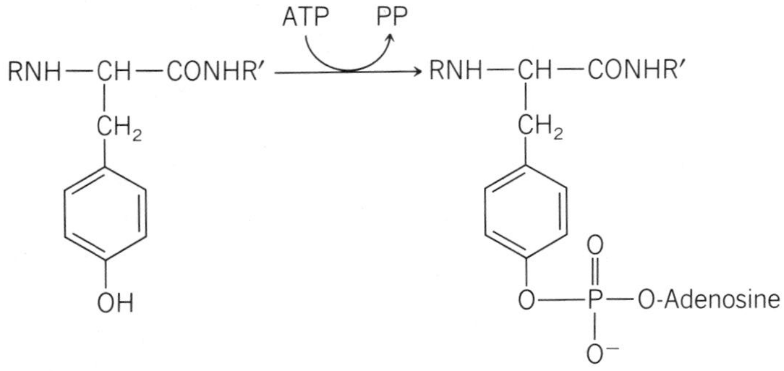

Example: Glutamine synthetase

3. *ADP ribosylation of specific arginine residue.*

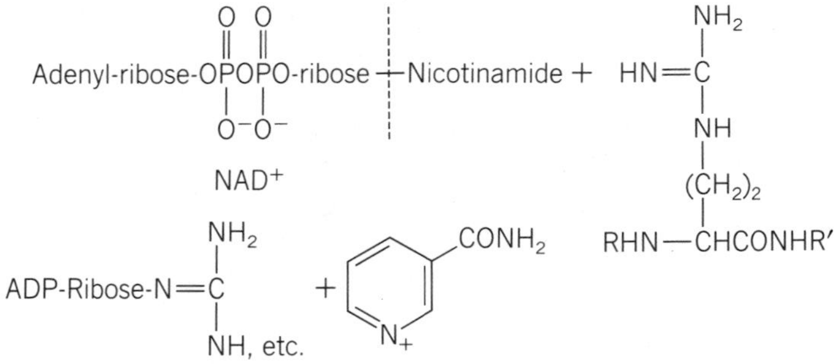

Example: Erthrocyte adenylate cyclase

4. A number of other covalent modifications that are not as well defined as those just listed can also be cited.

   (a) Methylation of glutamyl or aspartyl residues of membrane proteins.

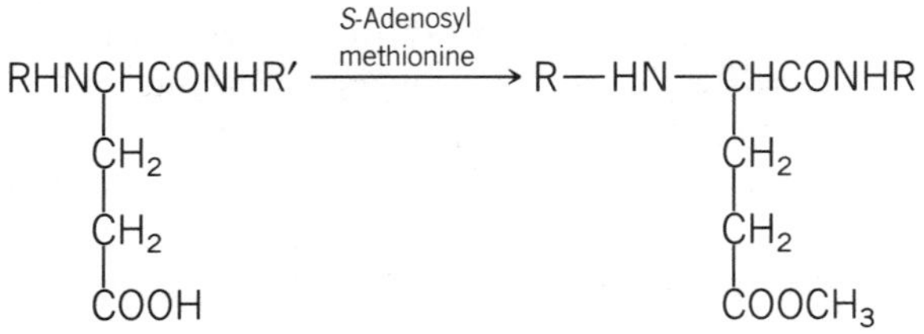

   (b) Acetylation of ε-amino-lysyl residues of histones.

(c) Acetylation of NH_2 terminal serine residue of cyclo-oxygenase. This reaction is of unusual interest because it involves the irreversible inactivation of the key enzyme that catalyzes the following reaction:

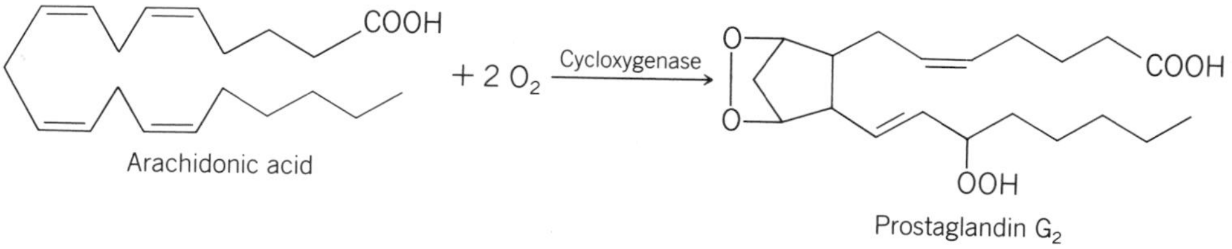

Arachidonic acid $+ 2 O_2$ Cycloxygenase →   Prostaglandin G_2

It has been known for several years that aspirin markedly inhibits the formation of blood clots in blood vessels. The reason for this effect can now be explained. Aspirin (acetyl-salicylate) irreversibly acetylates the amino group of the NH_2 terminal serine residue of cyclooxygenase to form the stable *N*-acetyl-serine derivative.

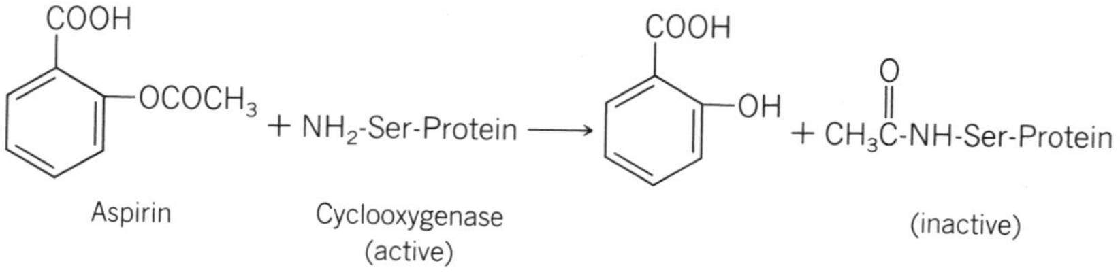

Aspirin   Cyclooxygenase (active)   →   (inactive)

The loss of activity of this enzyme closely parallels the extent of acetylation of its protein by aspirin. The enzyme is localized in blood platelets that aggregate to form the primary blood clot when a blood vessel is damaged. Cyclooxygenase converts arachidonic acid to prostaglandin G_2 that is further modified to form the extremely active thromboxane A_2, and it is this extremely unstable compound ($t_{1/2} = 30$ seconds) at 10^{-9} M level that causes the aggregation of blood platelets to form the primary blood clot. It is for this reason, that patients who are to have surgery are required not to take aspirin at least two weeks prior to surgery.

18.4.1 MONOCYCLIC CASCADES

How is reversible covalent modification employed by the cell to regulate an important enzyme? First of all, each modification reaction is catalyzed by an enzyme called a **converter enzyme;** another converter enzyme is involved in the demodification reaction. The enzyme that is undergoing the modification and demodification is called an **interconvertible enzyme.** The process is dynamic in that the fraction of modified enzymes can be varied over a wide range by changing the concentration of effectors that, in turn, regulate the converter enzymes. What we then have is a continuous cyclic process consisting of two tightly coupled cascades that operate in opposing directions. The simplest example is a monocyclic cascade system. The nomenclature describing these systems is as follows: (1) the most active form of an enzyme is identified by the subscript a and the least active by b; (2) the prefix o is the original form of the enzyme and m, the modified form. Figure 18.4 summarizes these ideas.

What is the importance of these cyclic systems on metabolism? By the interaction of the various effectors on the converter enzymes and the subsequent modification and demodification of the interconvertible enzymes, an

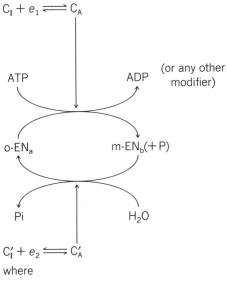

$$C_I + e_1 \rightleftharpoons C_A$$

ATP   ADP   (or any other modifier)

o-EN$_a$ m-EN$_b$(+P)

Pi H_2O

$$C_I' + e_2 \rightleftharpoons C_A'$$

where

C_I and C_I' = the inactive converter enzymes
e_1 and e_2 = the two effectors
C_A and C_A' = the active converter enzymes
o-EN$_a$ = the original active interconvertible enzyme
m-EN$_b$ = the modified inactive enzyme

FIGURE 18.4 A schematic presentation of a monocyclic cascade.

exquisite and sensitive control system becomes possible so that a wide range of activity can be expressed, both by fractional changes in the $C_1 \rightleftharpoons C_A$ system and the o-EN$_a \rightleftharpoons$ m-En$_b$ system. Thus, several properties of a cyclic cascade become evident.

1. A high capacity for **signal amplification** because a small activation of the converter enzyme can result in a large change in the o-EN $\rightarrow$ m-EN system.

2. These systems modulate the **amplitude** of the maximal response of the regulated enzyme with saturating concentrations of the effectors.

3. They modulate the **sensitivity** of the modification of the interconvertible enzymes by dynamic changes in the concentration of effectors e_1 and e_2.

18.4.2 PYRUVATE DEHYDROGENASE COMPLEX REGULATION

Let us now give an example of a monocyclic cascade system, namely, the important mammalian pyruvate dehydrogenase complex (PDC). The molecular structure of this enzyme has already been discussed in Section 12.2.

It is well known that oxidation of glucose is inhibited by starvation and diabetes and that oxidation of fatty acids (by β-oxidation) lowers glucose oxidation in the liver. It turns out that the regulation of the PDC system directly relates to these observations. Recent investigations have shown that PDC is carefully controlled by two mechanisms, namely, by end-product inhibition and covalent modification of the pyruvate dehydrogenase protein of PDC.

18.4.2.1 End-Product Inhibition. The reaction we are considering is

$$\text{Pyruvate} + \text{CoA} + \text{NAD}^+ \xrightarrow{\text{PDC}} \text{Acetyl CoA} + \text{NADH} + \text{H}^+ + \text{CO}_2$$

Because both acetyl CoA and NADH function as competitive inhibitors of the reaction, the ratios acetyl CoA/CoA and NADH/NAP$^+$ become very important in regulating PDC activity. High ratios will reflect the high inhibition of pyruvate breakdown, whereas low ratios will favor the forward reaction.

18.4.2.2 Covalent Modification.

The second mechanism of control involves a phosphorylation–dephosphorylation cycle in which the pyruvate dehydrogenase protein is phosphorylated (and thus inactivated) and dephosphorylated (and thus reactivated). Phosphorylation is catalyzed by the converter enzyme, a Mg^{2+}-dependent protein kinase that is associated with the pyruvate dehydrogenase protein (as a domain?); dephosphorylation is catalyzed by the second converter enzyme, a Ca^{2+}-dependent phosphatase that is loosely associated with the PDC system.

There are three serine residues in the dehydrogenase protein: sites 1, 2, and 3. Site 1 is rapidly phosphorylated by ATP, as a result of which up to 75% of the activity is lost when this site is fully phosphorylated. Slow phosphorylation of sites 2 and 3 further inactivate the residual activity of 25%.

The PDC kinase is stimulated by acetyl CoA and NADH and inhibited by pyruvate and ADP (other inhibitors are listed in Figure 18.5); the phosphatase is inhibited by NADH and activated by NAD; it further requires Ca^{2+} or Mg^{2+} for activation. One can now readily appreciate the multiple possibilities available to the cell to regulate PDC over a wide range of activity. Figure 18.5 summarizes the monocyclic cascade of PDC.

We can now write the following scenario. Remember the complete oxidation of pyruvate in the mitochondrion generates 15 ATPs (Chapter 14). High ATP levels will tend to turn off pyruvate oxidation, as would be predicted from Figure 18.5. Pyruvate would then be rerouted to other pathways such as gluconeogenesis. Low levels of ATP and high levels of ADP will stimulate pyruvate oxidation because ADP turns off kinase activity. Likewise, high levels of acetyl CoA and NADH activate the kinase, as well as being end-product inhibitors of PDC. However, high levels of CoA and NAD$^+$ inhibit the kinase and permit PDC to remain active.

We can now see that both the converter enzymes, the kinase and the phosphatase, through their effectors, sense fluctuations of pyruvate concentration and the flux of ATP/ADP, acetyl CoA/CoA, NADH/NAD and integrate these changes into a modulation of PDC activity.

Thus, in a starving animal or an animal on a high fat diet, low levels of pyruvate will be present; in addition, high levels of acetyl CoA, ATP, and NADH (obtained from β-oxidation of fatty acids) will turn off pyruvate oxidation. Pyruvate formed from lactate or alanine and other amino acids will then be rerouted by the acetyl CoA-dependent pyruvate carboxylase in the direction of gluconeogenesis (see Chapter 10 for an explanation of gluconeogenesis).

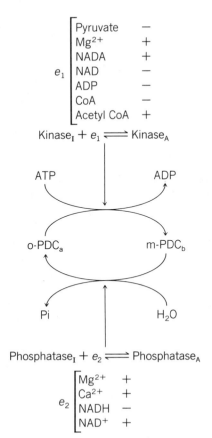

FIGURE 18.5 The mammalian PDC monocyclic cascade system. +, Activation; −, inactivation of converter enzymes.

18.5 MULTIPLE CASCADE SYSTEMS

A more complicated system is the multiple cascade system by which a monocyclic cascade locks into another cascade that, in turn, couples to still another (see Section 10.8.1 for a discussion on the regulation of glycogen synthesis). More important is the interaction of a multiple cascade system with a signal system applied externally to the cell and transmitted internally to turn on or off multiple cascade systems. The external signal is called the **first messenger** and includes physiological agents such as adrenalin, glucogon, seratonin, prostacyclin, acetylcholine, growth factors, and so on. These agents bind to a specific receptor protein buried in the external face of the cell's plasma membrane. The binding sets off a series of transduction reactions by which the signal is transmitted to the inner face of the plasma membrane by a series of intrinsic proteins called G proteins that, by undergoing conformational changes, activate an **amplifier enzyme** located in the inner face of the plasma membrane. This enzyme, in turn, converts a phosphorylated precursor to a product called a **second messenger** that activates a protein kinase. A number of second messengers have now been identified and the best studied are cyclic adenylate (cAMP) and both diacylglycerol (DG) and inositol triphosphate (IP_3), both derived from phosphotidyl-inositol 4,5-bisphosphate (PIP_2), a membrane lipid. Figure 18.6 describes the general sequences of this system.

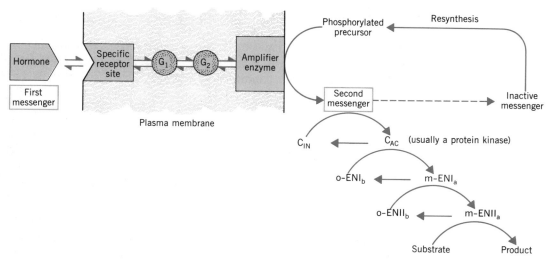

FIGURE 18.6 A general scheme that integrates the second messenger pathway with a multiple cascade system. See the legend of Figure 18.4 for a definition of terms. Note that in this pathway, the phosphorylated form of o-En, namely, m-En$_a$ is the active form of the enzyme. Conversion of m-En to the inactive o-En$_b$ results from phosphatase action on the phosphorylated m-En$_a$.

18.5.1 CYCLIC AMP

We will now describe the synthesis and function of cyclic AMP as it relates to the second messenger or signal pathway.

Cyclic AMP (cAMP) is synthesized by the intrinsic amplifier enzyme, adenylate cyclase, located in the inner face of all animal cell plasma membranes. The reaction is

cAMP is rapidly removed by the enzyme cAMP phosphodiesterase that is strongly inhibited by caffeine.

$$cAMP + H_2O \longrightarrow AMP$$

With both the synthesizing and the hydrolyzing enzyme present in the cell, the basal concentration of cAMP is usually quite low, approximately 10^{-6} M or less. The presence of caffeine blocks the hydrolysis reaction and permits the buildup of cAMP and suggests explanations for the physiological effect of caffeine in the body.

Cyclic AMP as the second messenger regulates the activity of a protein kinase referred to as protein kinase A. The enzyme consists of regulatory (R) and the catalytic (C) subunits in the proportions of R_2C_2. When cAMP binds to the R_2 subunit, the affinity of R to C is markedly lowered with the result that R_2C_2 dissociates.

$$C_2R_2 + 2 \text{ cAMP} \rightleftharpoons 2\text{-R}\text{—cAMP} + C_2$$

Inactive   Active

kinase   kinase

As already described in Section 10.8.1, the active kinase will now phosphorylate a number of important enzymes. For example, in skeletal muscle cells adrenalin, the first messenger, activates the formation of cAMP that, in turn, reacts as just described to form the cAMP-dependent protein kinase A. This protein now phosphorylates a phosphorylase protein kinase (+ ATP and Mg^{2+}) to convert it from the inactive (o-En) to the active phosphorylated (m-En) form. The active phosphorylase protein kinase, in turn, converts the inactive phosphorylase *b* to the active phospho-phosphorylase *a* that now catalyzes the degradation of glycogen to glucose-1-phosphate. The first messenger was adrenalin, the amplifier enzyme was adenylate cyclase, the second messenger was cAMP, the converter enzyme was protein kinase A, and the interconvertible enzymes were now turned on to allow glycogen breakdown.

The second messenger, cAMP, participates in a large number of cellular responses, including the activation of triacyl-glycerol lipase (Section 13.5), β-hydroxymethylglutaryl CoA reductase (Section 13.17) glycogen synthesis, and many physiological effects such as fluid secretion, thyroxine secretion, and so on. Also, note that the cell balances the synthesis of cAMP with its degradation to AMP. Transient increases in the concentration of cAMP will impinge on the appropriate converter enzymes, but will be rapidly brought back to basal levels.

18.5.2 PHOSPHOTIDYL INOSITOL DIPHOSPHATE

In recent years, a somewhat more sophisticated signal pathway has been described that involves the unique generation of two second messengers from one precursor. It has been observed for some time that a minor plasma membrane lipid, phosphatidylinositol 4,5-bisphosphate (PIP_2), turned over much more rapidly than other membrane phospholipids when the external signal such as acetylcholine was given to secretory cells of the pancreas. Ca^{2+} was also markedly mobilized.

The mechanism of this phenomenon has now been elucidated. It turns out that the first messenger, be it acetylcholine, seratonin, or antigens, activates an amplifier enzyme, namely, phospholipase C localized in the inner face of the cell plasma membrane. Transduction apparently is the same as for the cAMP system. Phospholipase C catalyzes the reaction.

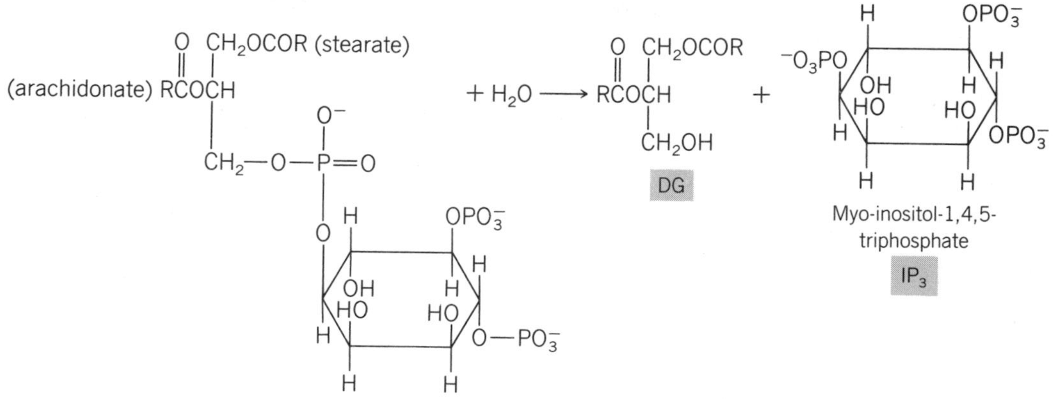
O CH2OCOR (stearate)
(arachidonate) RCOCH
CH2—O—P=O
+ H2O
RCOCH
CH2OH
DG
Myo-inositol-1,4,5-
triphosphate
IP3

Both IP_3 and DG serve as second messengers. IP_3 diffuses into the cytosol and mobilizes the release of Ca^{2+} from the endoplasmic reticulum. Ca^{2+} ions participate in the activation of a number of important cellular proteins such as troponin C involved in muscle contraction and by binding to the calcium-binding protein, calmodulin, activate what is called a Ca^{2+}/calmodulin protein kinase. DG remains in the membrane and activates a membrane protein kinase C that is cAMP independent but requires Ca^{2+} and membrane phosphatidyl serine for full activity. Kinase C is present in high concentrations as a membrane enzyme in brain tissue and although its function has not yet been defined, it probably will play an important role in nerve tissue functions. In nonnerve tissues, kinase C is a soluble protein and it phosphorylates a wide variety of proteins. Again, its function remains to be defined.

To complete the discussion, we should mention briefly the inositol-lipid cycle that replenishes PIP_2. The sequence is as follows.

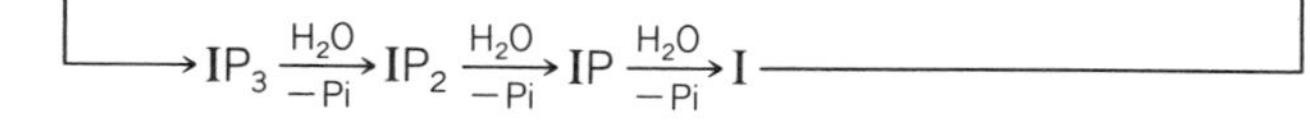
ATP ADP CTP PPi
DG
PA
CMPP-DG
ADP ATP ADP ATP
PIP2
P-IP2
PIP
PI + CMP
IP3 IP2 IP I
H2O
−Pi

DG = Diacylglycerol
PA = Phosphatidic acid
CTP = Cytidine triphosphate
CMPP-DG = Cytidine-diphosphodiacylglycerol
CMP = Cytidine monophosphate
PIP = Phosphatidylinositol-4-phosphate
PIP_2 = Phosphatidylinositol 4,5-bisphosphate
IP_3 = Inositol-1,4,5-triphosphate
IP_2 = Inositol-1,4-diphosphate
IP = Inositol-1-phosphate
I = Inositol

All the necessary enzymes for this cycle are present in all animal cells. Thus, a rapid resynthesis of PIP_2 and its insertion into the inner face of plasma membranes is assured. Also, note once more how the cell balances the synthesis and breakdowns of these second messengers. Any break in these balancing reactions would alter the concentrations of the second messenger.

In this chapter, we have outlined some of the mechanisms by which the cell can regulate metabolism. In addition to these systems that relate primarily to the modulation of activity of already existing enzymes, the equally sophisticated transcription – translational – post-translation processes that control the synthesis and degradation of enzyme proteins will be discussed in Chapters 20 and 21.

REFERENCES

1. *E. R. Stadtman,* The Enzymes. *P. D. Boyer, ed. New York: Academic Press, Vol. I, 3rd ed. 1970, p. 397; Vol. X, 1974, p. 755.*

2. *E. E. Snell, ed.,* Annual Review of Biochemistry. *Palo Alto, Calif.: Annual Reviews, Inc.*

3. *E. A. Newsholme and A. R. Leech,* Biochemistry for the Medical Sciences. *New York: Wiley, 1983.*

4. *M. J. Berridge. "The Molecular Basis of Communication within the Cell."* Scientific American, *Vol. 253, No. 4 (1985), pp. 142–152.*

REVIEW PROBLEMS

1. Define the differences between (a) cumulative, (b) concerted, and (c) cooperative feedback inhibition. Give examples.

2. In the multibranched pathway shown here, A is the precursor of the end products L, G, and J. Describe plausible mechanisms by which a living cell could independently regulate the rates of biosynthesis of the three end-products.

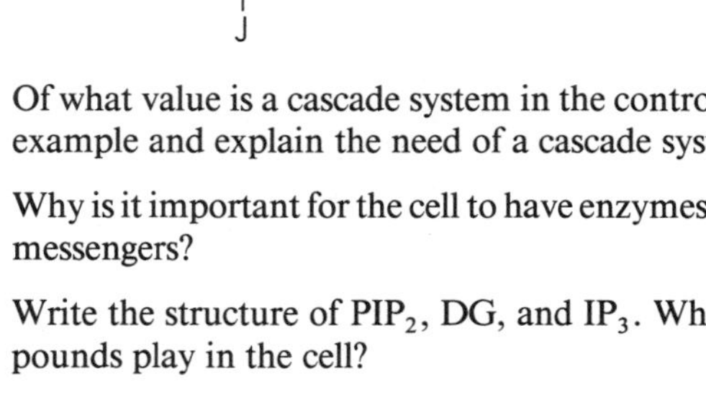

3. Of what value is a cascade system in the control of a metabolic sequence? Give an example and explain the need of a cascade system in this process.

4. Why is it important for the cell to have enzymes that synthesize and degrade second messengers?

5. Write the structure of PIP_2, DG, and IP_3. What functions do each of these compounds play in the cell?

PART 3

GENES, GENE EXPRESSION, AND THE METABOLISM OF INFORMATIONAL MACROMOLECULES

BIOSYNTHESIS OF NUCLEIC ACIDS

We begin this first chapter of Part 3 with an introduction to all of Part 3. Part 3 is concerned with the biosynthesis of nucleic acids (Chapter 19), the biosynthesis of proteins (Chapter 20), the regulation of nucleic acid and protein biosynthesis (Chapter 21), and finally, the study and manipulation of nucleic acids as they occur outside of cells (Chapter 22), primarily as plasmids, as the nucleic acids of viruses and as molecules formed by the *in vitro* recombination of DNA sequences. We urge the student to review Chapter 6 because it presents much of the background needed for Part 3.

The properties of living systems cannot be explained by reference only to fixed structures and metabolic pathways. Many of the dynamic characteristics of organisms, including reproduction, development, and adaptation, as well as the production of proteins and other cellular structures, require mechanisms for the transmission of genetic information, both within the organism and from organism to organism. Genetic information must be transferred from one cell to another during cell division and also must be used to specify the catalysts of the cell, the enzymes.

Remarkable progress in understanding the informational macromolecules of living systems has occurred within the last 40 years. This progress is based on several conceptual and technical advances, including (1) the demonstration that DNA is the genetic material, (2) the proposal of the Watson–Crick model of DNA structure and the resulting formulation of the central dogma of molecular biology next described, and (3) the development of an array of genetic and biochemical techniques to manipulate and create recombinant DNA molecules *in vitro* and to facilitate the expression of the newly created genes in homologous and heterologous biological systems.

In 1944, Avery and his colleagues demonstrated that pneumococcal bacterial strains with the rough, or R-type, of colony morphology could be transformed to the smooth, or S-type, of morphology by incubation of cells of the R-type with purified DNA from S-type cells. This **transformation** experiment indicated that genetic information for conferring the smooth (S) phenotype resided in the DNA preparation. Further proof of the genetic role of DNA was generated by the 1952 experiment of Hershey and Chase, who differentially radio-labeled the DNA and proteins of a bacterial virus, that is, a bacteriophage. They found that, during the process of initiating an infection, only the bacteriophage DNA, and not the bacteriophage protein, was injected into the bacterial host. Therefore, all the genetic information for bacteriophage replication must reside in the DNA of the bacteriophage.

19.1   PATHWAYS FOR THE TRANSFER OF GENETIC INFORMATION

Once DNA had been demonstrated to be the genetic material, the major problems in biological information transfer were to determine how the genetic material was replicated to assure that each daughter cell received the full complement of genetic information and how information in DNA was utilized to synthesize proteins. That is, how can a 4-character alphabet of deoxyribonucleotide residues in DNA be converted to a 20-character alphabet, that of the amino acids in proteins? Many of the most important early studies of molecular biology were concerned with the "central dogma," the concept of how biological information in cells proceeds from DNA to RNA to proteins. The central dogma for cells is depicted in the upper line of the following diagram.

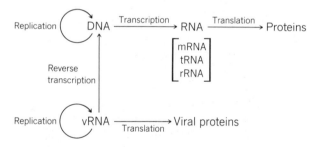

Replication is the process by which each strand of the parental DNA duplex is copied precisely by base pairing with complementary nucleotides. If the replication is errorless, the product is two duplexes identical in nucleotide sequence to the parental duplex.

Transcription is the process by which the information contained in DNA is copied, by base pairing, to form a complementary sequence of ribonucleotides, an RNA chain.

Translation is a complex process by which the information that has been transcribed from DNA into messenger RNA, abbreviated mRNA, directs the ordered polymerization of specific amino acids for the synthesis of proteins.

In the central and lower part of the diagram of the central dogma is presented arrangements for information transfer that were observed first for viruses. An important group of viruses, called the retroviruses, have RNA as the genetic material of virus particles. However, on infection of cells, the retrovirus RNA is transcribed into DNA by the action of the enzyme **reverse transcriptase.** The enzyme was named for its ability to transcribe, and therefore transfer, genetic information in a direction, from RNA to DNA, that is contrary to the usual flow of information in uninfected cells. New retrovirus RNA is transcribed from the retrovirus DNA after it has been integrated into the genome of the host cell. Several retroviruses are cancer inducing.

Other viruses replicate without the involvement of DNA, which is indicated by the bottom line of the diagram. Replication of such viruses is accomplished by the replication of RNA to RNA, always copied from one RNA strand to a complementary RNA strand. Both single-stranded and double-stranded RNA viruses are known. Some viruses have single-stranded DNA forming the genome of the virus in the virus particles. Such a single-stranded DNA is converted to double-stranded DNA, which serves as the template for the synthesis of single-stranded DNA and the mRNAs needed to express virus genes in the infected cell. Finally, a few viruses have DNA as the genetic material in virus particles but transcribe this DNA into RNA. The RNA then serves as the template for DNA synthesis through the action of a reverse transcriptase.

When Watson and Crick in 1953 proposed the double-stranded helical DNA model, based on hydrogen bonding between complementary bases present in two single DNA strands of opposite polarity, it became evident that DNA could be replicated by a semiconservative mode to ensure the passage of a complete genome to each daughter cell. In **semiconservative** replication, each of the two strands of DNA in the template becomes one of the two strands in the two daughter DNA molecules. Furthermore, genetic information for protein synthesis is transcribed to single-stranded RNA molecules, mRNAs, by making a complementary copy of one of the two strands of DNA. Thus, the information in the master plan (DNA) could be copied to a working blueprint (mRNA) for protein synthesis.

The next step in information transfer requires a mechanism to transfer the information present in mRNA to form the linear sequence of amino acids present in proteins. For this purpose, the cell has used an intermediary molecule called transfer RNA (tRNA) that acts as an adaptor. tRNA has the capability of covalently binding amino acids through an ester bond to form amino acyl-tRNAs. The amino acyl-tRNAs, in turn, form complementary base pairs with the transcribed genetic information in messenger RNA (mRNA). Thus, these two functions of tRNA allow the 4-letter messages in mRNA to be translated into the 20-letter message in protein. The translation process required the use of a mechanism that allows the proper arrangement of amino acyl-tRNAs on the mRNA and the formation of peptide bonds between adjacent amino acid residues. This complex machinery requires ribonucleoprotein complexes called ribosomes, already described in Section 6.10.2, and a large number (over 100) of enzymes and other proteins. Thus, the linear information present in DNA is transcribed and translated into a linear array of amino acids in a protein. The elucidation of this mechanism involved solving the genetic code, which is essentially a table of three nucleotide sets and the amino acids to which they correspond, understanding the transcription apparatus, and determining the complex processes of protein synthesis. Although the overall features of gene expression are understood, there are many aspects that are still being investigated.

A final aspect of information transfer requires an understanding of the regulation of gene expression, since not all genes are expressed equally nor at the same time, for example, differentiated cells may produce one protein in much greater abundance than all the other proteins normally found in the cell. Many genes, although present in cells, may not be expressed at all in particular cells. The regulation of gene expression is being investigated actively in order to understand problems such as cancer and normal development, cellular differentiation, and hormonal control of gene expression.

19.2   INTRODUCTION TO GENETIC ENGINEERING

Since 1973, the information accumulated from a variety of studies on informational macromolecules has been directed toward the development of genetic engineering. In this new approach, *in vitro* techniques allow researchers to create recombinant DNA molecules that do not occur in nature. One of the keys to developing genetic engineering was the discovery of the restriction–modification (RM) phenomenon in the interaction of bacteria with bacteriophages. Most bacterial strains have a strain-specific RM system consisting of both a site-specific restriction endonuclease and a methylase.

As is described in Box 6.D, the RM system protects the host from invasion by foreign DNA by cleaving the incoming DNA at unmethylated restriction sequences. The host DNA is protected from its restriction enzyme by having its restriction sites fully methylated. The use of restriction enzymes has enabled the scientist to cleave DNA at precise sites and obtain discrete DNA fragments. Thus, by having precise DNA fragments from two different genomes, one can recombine these fragments in a test tube, introduce them into a suitable host, and obtain the expression of foreign genes, for example, the expression of the human insulin gene in the bacterium *E. coli*. The critical enzymes and other tools for joining DNA fragments are described in Chapter 22.

Being able to cut and join DNA fragments allows the rearrangement of genetic information within one organism and the transfer of new arrangements of genetic information from one organism to another. This makes possible the investigation of basic genetic questions that in the past were unapproachable. The capability of chemically synthesizing DNA molecules of arbitrary sequence is described in Box 6.G. Combining synthetic and naturally occurring DNA sequences in new DNA molecules provides an even more powerful set of tools for the genetic engineer. Genetic engineering will have wide practical application in medicine, agriculture, and industry, for example, in the production of scarce human hormones or the rapid development of disease-resistant plants. The techniques in this new field have developed so rapidly within the past 10 years that virtually any DNA can be synthesized.

The new biotechnologies, based on *in vitro* recombination of DNA, were developed from basic knowledge derived from studies on bacteria, animals, plants, plasmids, bacteriophages, and other viruses, and from the subdisciplines of microbial genetics, enzymology, protein and nucleic acid chemistry, and analytical and preparative biochemistry. Rapid progress in this field has depended heavily on the concerted use of genetic and biochemical concepts and techniques. The most urgent current need is the development of new knowledge and ideas on how to utilize the techniques that have been developed to such a high state. Thus, our understanding of basic biological phenomena in bacteria, plants, and animals must be advanced through fundamental research before the new biotechnologies can be fully exploited.

In Part 3, we describe in outline our current understanding of the biochemi-

cal basis of molecular biology and its application to the development of biotechnology. In this chapter, we present the current knowledge concerning replication of DNA, the important mechanisms for the repair of DNA damaged by
physical processes (uv and x radiation), and a discussion of the transcription
apparatus and its requirements and specificity during RNA biosynthesis.

19.3   TEMPLATES, PRIMERS, AND SEMICONSERVATIVE REPLICATION

The concepts of template and primer are important for understanding the
mechanism of DNA replication. **Template** refers to the DNA chain that provides precise information for the synthesis of a complementary strand of nucleic acid. For DNA replication, a template is required in addition to a **primer.**
A primer, in biochemistry, refers to the initial portion of a linear molecule, onto
which additional units are added to produce the final polymeric chain. In DNA
replication, oligoribonucleotides are first formed with DNA as the template,
and they then serve as a primer for the addition of deoxyribonucleotides for the
synthesis of daughter DNA strands. These ideas are illustrated here.

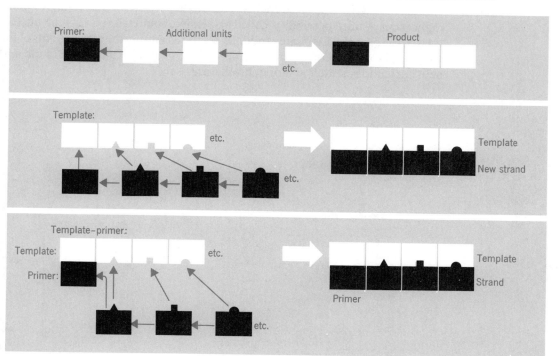

In replication, strands of DNA will separate and new complementary strands
of DNA will be assembled from the four available deoxyribonucleoside triphosphates on each of the two separate parental strands. If we assume the base
pairing is precise, the two new DNA molecules should be identical to the parent
molecule. This type of replication has been called **semiconservative** and is
illustrated in Figures 19.1 and 19.2.

An alternative possibility is that the final duplication product consists of a
double helix of the original two strands and a second double helix consisting

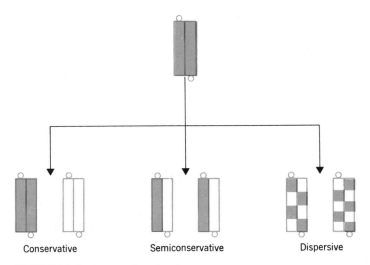

FIGURE 19.1   Types of replication.

entirely of newly synthesized chains. This process is called a **conservative** type of replication. A third possibility, called **dispersive,** could take place if the nucleotides of the parental DNA are randomly scattered among the components of the daughter DNA material so that the new DNA consists of a mixture of old and new nucleotides scattered along the chains.

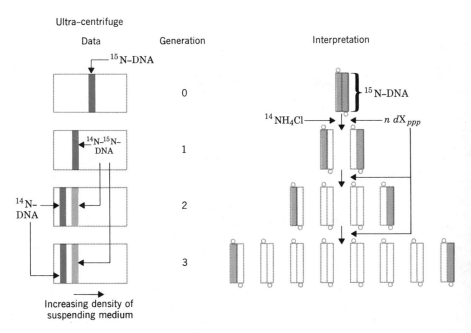

FIGURE 19.2   Meselson–Stahl experiment to demonstrate semiconservative replication of DNA. Small circles at each end of diagrammatic DNA indicate the antiparallel nature of DNA.

To test these possible mechanisms, Meselson and Stahl in 1958 grew *E. coli* in a medium in which the sole source of nitrogen was $^{15}NH_4Cl$. After several generations of growth, excess $^{14}NH_4Cl$ was added, and at short intervals, cells were removed; the DNA was carefully extracted and analyzed for relative ^{14}N and ^{15}N content by equilibrium density gradient centrifugation. The results depicted in Figure 19.2 definitely eliminated the dispersive mechanism. With other evidence, these results gave strong support to the semiconservative mode of replication.

19.4 REPLICATION CAN BE UNIDIRECTIONAL OR BIDIRECTIONAL

Replication of a DNA molecule can either be unidirectional or bidirectional. In the unidirectional mode, replication will be initiated at a site in the DNA and the replication will proceed in one direction from this origin. In the bidirectional mode, replication will proceed via two replication forks in opposite directions from the initiation site. The mode of bidirectional replication is illustrated in Figure 19.3. In most procaryotic and eucaryotic **replicons**—a replication unit—bidirectional replication is observed, although several cases of unidirectional replication have also been reported.

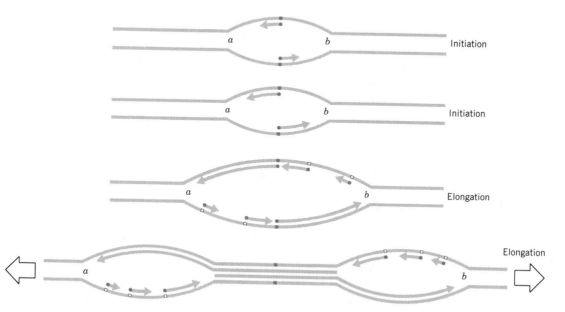

FIGURE 19.3 A schematic representation of bidirectional replication at the origin of replication (*ori*) and during elongation. The presence of two replication forks (*a* and *b*), discontinuous DNA synthesis, and Okazaki fragments (Section 19.7) are illustrated. The closed square represents initiation priming sites at *ori;* the closed circles represent primers; the open squares represent elongation priming sites. The small arrows represent the 5' to 3' direction of DNA synthesis. The heavy arrows represent bidirectional replication.

19.5   ORIGIN OF REPLICATION

Replication does not start randomly on a DNA molecule. In small procaryotic chromosomes, there is a unique site called the **origin of replication** or *ori* at which replication is initiated. In larger eucaryotic chromosomes, there are multiple sites of replication initiation. The *ori* in procaryotes consists of approximatel: 245 base pairs (bp) and contains information for the regulation of initiation of replication, sites for binding the initiation machinery, and sequences involved in the segregation of the daughter DNA molecules during cell division. Replication occurs usually in a bidirectional mode from the *ori* site in circular chromosomes and terminates at a locus about 180° from the *ori* on the chromosome circle. Although a substantial amount of information is now available concerning DNA replication, relatively little is known about the coordinated regulation of DNA synthesis and cell division.

19.6   INITIATION OF DNA REPLICATION

Replication is a complex process requiring more than 20 enzymatic functions. There are three major stages in the replication of DNA: initiation, elongation, and termination.

For initiation of DNA replication in *Escherichia coli*, the two strands of the double helix are separated at the *ori* site by a DNA **helicase** enzyme that unwinds the DNA molecule and expends one ATP for every base pair that is denatured. During unwinding of the DNA by DNA helicase, positive supercoils (see Figure 6.5) are inserted into the molecule that are compensated by the action of the enzyme DNA gyrase that places negative supercoils into the DNA in front of the site of DNA helicase activity. DNA gyrase requires ATP for its activity. Thus, to maintain the proper topology of the DNA, a large amount of energy is required. After strand separation, low molecular weight (approx. 35,000) DNA-binding proteins bind to the separated single-stranded DNA and prevent renaturation of the DNA. These DNA binding proteins are also important during the elongation process. A complex of six or seven initiation proteins called a **primosome** forms two priming sites on each of the separated complementary strands in the *ori*. This is an energy-requiring process and ATP is cleaved during primosome interaction with the DNA.

RNA polymerase, an enzyme that recognizes precise binding sites on DNA called promoter sites and uses DNA as a template to form RNA, binds to two promoters in the initiation priming sites and synthesizes two short RNA primer molecules up to 30 bases long and complementary to each of the two DNA strands. Rifampicin, an inhibitor of RNA polymerase, inhibits initiation of DNA replication *in vivo* and *in vitro*. The RNA primers are used for the bidirectional replication of DNA from the *ori* site that results in the formation of two replicating forks going in opposite directions. Why are primers necessary for DNA replication? It turns out that all DNA polymerases are incapable of initiating DNA synthesis and require a 3'-hydroxyl residue for the formation of the first phosphodiester bond. The DNA polymerases catalyze the attack by the 3'—OH group of the primer on the α-phosphate of a deoxynucleoside triphosphate carried to the primer to form a phosphodiester bond. The overall initiation process is depicted in Figure 19.3. Some of the features of this model are tentative.

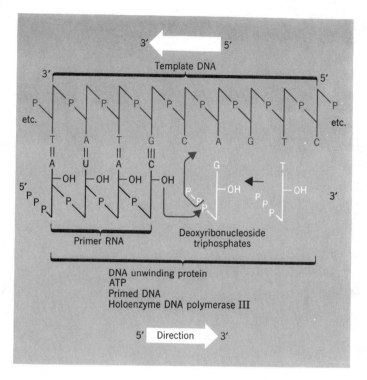

19.7 DISCONTINUOUS DNA SYNTHESIS DURING THE ELONGATION PHASE

After primer formation has been completed at the *ori* site, the elongation process is initiated by DNA polymerase III, a complex protein with a molecular weight of approximately 450,000 and consisting of seven subunits. This enzyme and DNA polymerase I both synthesize DNA *only* in the 5' to 3' direction.

The absence of an enzyme that can synthesize DNA in the 3' to 5' direction has forced the cell to develop a process called **discontinuous DNA synthesis.** This can be visualized in Figure 19.4. Although the leading strand of the replicating fork can be synthesized in a continuous fashion, the lagging strand can only be replicated in a discontinuous fashion. As at *ori* during the initiation phase, the DNA during elongation is unwound by helicase, and DNA-binding proteins serve to keep the complementary strands separated at the replication fork. The discontinuous mode of replication is initiated by the formation of priming sites by the primosome on the lagging strand. Primers are synthesized at the priming site during the elongation phase by another RNA synthesizing enzyme called the **DNA primase.** This enzyme differs from the RNA polymerase used during the initiation phase, since DNA primase is capable of using multiple nonspecific initiating sites and synthesizing mixed oligomers containing both deoxyribo- and ribonucleotide residues and is not inhibited by rifampicin.

After a primer ranging in size from 3 to 30 nucleotides is synthesized, DNA

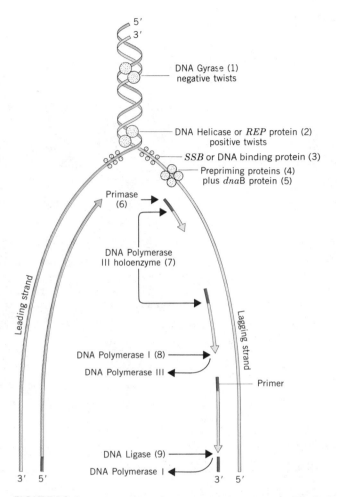

FIGURE 19.4 Scheme for enzymes operating at one of the forks in the bidirectional replication of an *E. coli* chromosome or of a plasmid (*oriC*) employing the unique origin of the *E. coli* chromosome. Replication is continuous for one strand (leading) and discontinuous for the other (lagging). Reprinted with permission from *DNA Replication* by Arthur Kornberg. W. H. Freeman and Company. Copyright © 1980.

polymerase III adds deoxynucleotide residues to extend the chain to about 1000. These short chains have been called "Okazaki pieces" after the Japanese discoverer of discontinuous DNA synthesis. Several Okazaki pieces are formed simultaneously on the lagging strand. Since RNA is not ordinarily linked to DNA in the cell and DNA is not found in small pieces, several reactions must follow. The RNA primers must be removed and replaced quickly by deoxyribonucleotide residues, and the short DNA pieces must be linked to form a continuous DNA chain complementary to the lagging strand. Processivity (lack of dissociation) of the DNA polymerase III holoenzyme in the continuous chain elongation and of the primosome in short RNA primer formation, for the discontinuous syntheses, assures rapid replication.

DNA polymerase I, a multifunctional enzyme (Figure 19.5), replaces DNA polymerase III at this point, since it has the capacity to remove the RNA

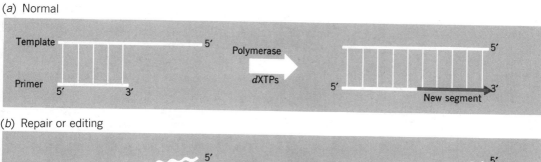

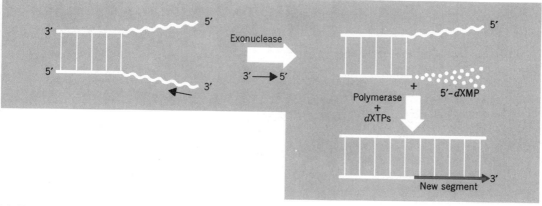

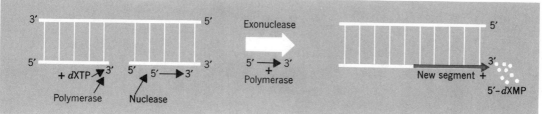

FIGURE 19.5   The trifunctional activity of *E. coli* DNA polymerase. (*a*) The polymerase action. (*b*) The 3′ → 5′ nuclease activity as well as the polymerase activity occurs in a repair mechanism in which an unpaired 3′ segment is digested back to the first paired base by the 3′ → 5′ nuclease activity of the polymerase, with a subsequent resynthesis by the polymerase. (*c*) Nick translation involves a very interesting synthesis and degradation at equivalent rates with a nicked DNA involving both the polymerase and the 5′ → 3′ nuclease and resulting in no net synthesis but a repair of the defective region.

primers by its 5′ to 3′ exonuclease activity and replace the primers with deoxy-ribonucleotides. The short pieces of DNA resulting from DNA polymerase I activity are finally linked together by the action of DNA ligase (Figure 19.6), a key enzyme in the replication of DNA and a useful enzyme for genetic engineering (Chapter 22). The half-life of an Okazaki piece is usually very short (a few seconds) at normal growth temperatures (37°C), indicating that the removal of primers, the replacement of primers with deoxyribonucleotides, and the ligation steps occur very rapidly. Thus, the elongation phase is carried out very efficiently by the cell.

The DNA ligase activity that forms phosphodiester bonds between two DNA strands requires NAD⁺ in *E. coli* or ATP in bacteriophage T4 and animal cells. The enzyme is involved in several types of functions including the linking of Okazaki fragments during DNA replication, the formation of closed circular

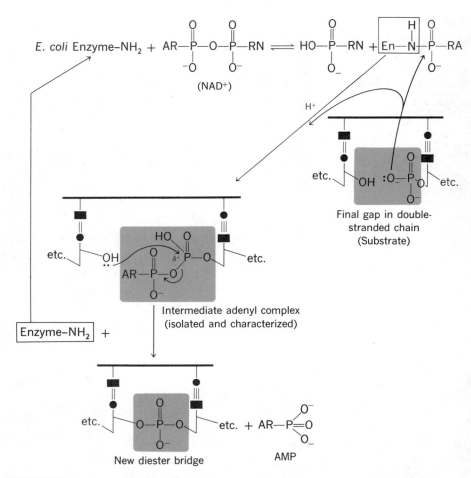

FIGURE 19.6   Mechanism of DNA ligase closing the final gap in DNA replication.

DNA molecules from linear molecules, the recombination of DNA *in vivo* and *in vitro,* and the closing of DNA gaps during DNA repair.

19.8   TERMINATION OF DNA REPLICATION

The termination event occurs during bidirectional replication when the two replicating forks in a circular chromosome approach each other at about 180° from the *ori* site. The last steps include the removal of the RNA primers and the filling in of the gap created during primer removal by DNA polymerase I, and fusion of the resulting DNA fragments by DNA ligase.

The multifunctional DNA polymerase I of *E. coli* has a molecular weight of 109,000 and is composed of a single polypeptide chain. This protein contains one binding site for all four deoxyribonucleotides, a binding site for template DNA, a site for the growing primer, a site for the 3'—OH group of the terminal nucleotide residue, a site for the 3' to 5', and a site for the 5' to 3' exonucleolytic activities. Furthermore, the polymerase binds strongly to nicks on a DNA duplex, at which point it can catalyze a nick translation sequence (see Figure 19.5).

The 3′ to 5′ exonuclease activity of the enzyme is very specific, in that only a 3′—OH deoxyribose configuration is recognized. The products are only 5′-mononucleotides. This function can remove incorrectly incorporated deoxyribonucleotide residues by an "editing" function. The 5′ to 3′ exonuclease activity is far less specific in that the 5′-hydroxyl, mono-, di-, and triphosphate termini of DNA or RNA can be recognized. Products are predominately 5′-mononucleotides although up to 20% of oligodeoxyribonucleotides can also accumulate.

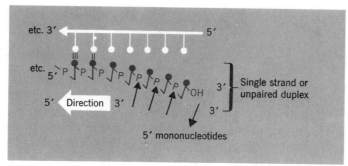

(a) 3′ ⟶ 5′-Exonuclease activity

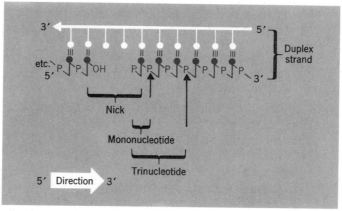

(b) 5′ ⟶ 3′-Exonuclease activity

19.9 MODIFICATION OF DNA

Glucosylation and methylation of DNA have been observed in various DNA preparations. 5-Methylcytosine is found in place of cytosine in animal and higher plant DNAs. Glycosyl hydroxymethyl cytosine residues were found in the DNA of *E. coli* T-even bacteriophages. Low amounts of methylated bases are found in almost all bacterial DNAs. In all these cases, the modification of the bases occurs after the DNA has been synthesized. Specific enzymes are involved that carry out these modification procedures. The DNA methylase in the bacterial restriction–modification system (Chapter 22 and Box 6.D) recognizes specific sequences in the DNA and plays an important role in protecting the host DNA from destruction from restriction endonucleases that attack foreign DNA that may gain entry into the cell.

19.10   REVERSE TRANSCRIPTION

Retroviruses (RNA tumor viruses) contain an enzyme called reverse transcriptase that is capable of using single-stranded viral RNA as a template to form a complementary single-stranded DNA or cDNA.

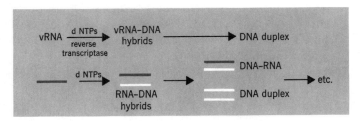

The enzyme also contains a ribonuclease (RNase H) activity that is capable of hydrolyzing RNA, which is hybridized to DNA. The RNase H activity removes the viral RNA, and the cDNA is copied into a double-stranded DNA by the reverse transcriptase. Thus, the single-stranded viral RNA can be converted into a double-stranded DNA containing the information originally present in the viral RNA. The incorporation of the double-stranded DNA copy of the original viral RNA into the chromosome of the infected cell can transform it to a cancerous cell.

For genetic engineering purposes, the reverse transcriptase has been useful for making double-stranded DNA copies of various messenger RNAs.

19.11   REPAIR MECHANISMS

All cells possess machinery by which damage to DNA can be eliminated and the original form of the DNA double helix restored. DNA damage occurs through misincorporation of deoxynucleotides during DNA replication, by spontaneous deamination of bases during normal genetic functions, from x radiation that cause nicks in the DNA, from ultraviolet (uv) radiation that causes thymine dimer formation, and from various chemicals that interact with DNA. In the case of the latter two types of damaging agents, conformational distortions in the secondary structure of DNA may occur. The rapid repair of DNA damages are necessary since they may be lethal to the cell or cause mutations that may result in abnormal cell growth. Basically, all repair mechanisms consist of endonucleases that can make an incision at the damaged or incorrect base, exonucleases that can remove the damaged or incorrectly incorporated base, DNA polymerase I activity that fills the gap that had been created during removal of the damaged bases, and DNA ligase activity that makes the final phosphodiester bond after DNA synthesis by DNA polymerase I.

Damage caused by the misincorporation of deoxynucleotides during DNA replication is repaired by the editing or proofreading 3' to 5' exonuclease function of DNA polymerase I that removes the incorrectly base-paired deoxynucleotide and replaces it with the correct deoxynucleotide by its polymerase activity. The small gap left after the polymerization activity is closed by the action of DNA ligase (see Figure 19.6).

The spontaneous deamination of cytosine or adenine, which occurs naturally in the cell, leaves either uracil or hypoxanthine in the DNA. A group of enzymes called DNA glycosylases catalyzes the hydrolysis of the N-glycosidic bonds, linking inappropriate or damaged bases to the deoxyribose-phosphate backbone with release of the free base from DNA. This action leaves apurinic or apyrimidinic (AP) sites on the DNA that are cleaved by 5'-AP-endonuclease (Figure 19.7), resulting in a gap with 3'-hydroxyl and 5'-phosphate groups. This gap is attacked by either a 5'- to 3'-exonuclease that releases an oligonucleotide excision product or a 3'-AP-endonuclease that releases a 5'-deoxyribosephosphate residue. The resultant gap is filled by the repair activities of DNA polymerase I and DNA ligase.

Ultraviolet radiation of DNA results in the formation of thymine dimers.

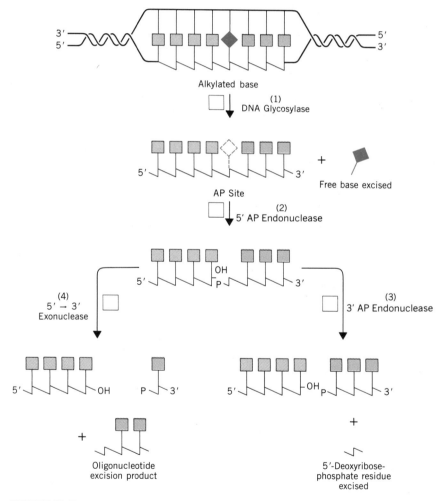

FIGURE 19.7 Excision repair of DNA. Inappropriate (e.g., uracil or hypoxanthine) or damaged (e.g., methylated) bases in DNA are recognized by specific DNA glycosylases that excise the base (1), leaving apurinic or apyrimidinic (AP) sites in the DNA. These sites are attacked by 5'-acting-AP-endonucleases (2), and the resulting 5'-terminal deoxyribose-phosphate moieties are excised by either a 5'- to 3'-exonuclease (3) or a 3'-AP-endonuclease (4). This leaves a gap in the affected strand. Reprinted with permission from *DNA Repair* by Errol C. Friedberg. W. H. Freeman and Company. Copyright © 1985.

The bonding between the 5 and 6 carbons of adjacent thymines in the same chain causes conformational distortions in the DNA structure. Damage-specific enzymes are mobilized by the cell to remove the altered deoxyribonucleotide residues. In *E. coli,* the damaged DNA can be removed and repaired by a two-step reaction involving the sequential action of incision on the 5′ side of the thymine dimer by *uvr*ABC endonuclease, followed by removal of the dimer by the 5′- to 3′-exonuclease activity of DNA polymerase I or other damage-induced 5′- to 3′-exonucleases (Figure 19.8). With DNA polymerase I, the damaged DNA is removed and replaced simultaneously by the action of the enzyme, and then the ends finally are fused with DNA ligase. A second mechanism involves the excision activity of the *uvr*ABC endonuclease that can cleave the DNA on both the 5′ and 3′ flanks of the thymine dimer and remove about 12 nucleotide residues. This gap is then repaired by DNA polymerase I and DNA ligase. Since *E. coli* mutants defective in DNA polymerase I had enhanced sensitivity to uv radiation, the correct conclusion was drawn that this polymerase was directly involved in repair mechanisms.

DNA repair in humans undergoes a very similar process of endonuclease incision, exonuclease removal of the damaged strand, polymerase closing of the gap, and fusion by ligase action. These conclusions derive from a study of the rare hereditary disease *Xeroderma pigmentosum.* Patients suffering from this disease are unusually sensitive to sunlight, resulting in severe skin reactions as

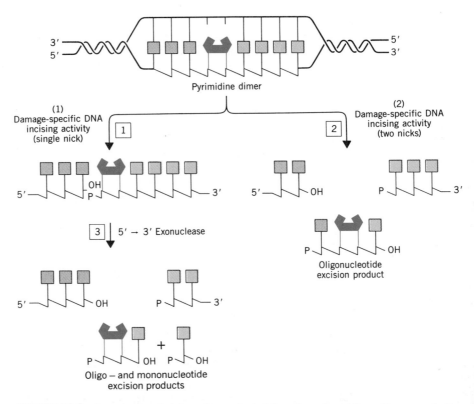

FIGURE 19.8   Repair of pyrimidine dimers. Pyrimidine dimer formation after uv radiation can be removed in two ways, which are discussed in the text. Reprinted with permission from *DNA Repair* by Errol C. Friedberg. W. H. Freeman and Company. Copyright © 1985.

well as skin tumors. When skin fibroblasts (tissue culture cells) were first irradiated with x rays and then with uv light to form the dimers, DNA repair occurred rapidly. Evidently, x radiation causes chain breaks, and then endogenous exonuclease, polymerase, and ligase activity repaired the damage. These experiments therefore demonstrated (1) that the molecular lesion for this rare disease is the absence of an endonuclease for incision, which made impossible the followup repair mechanism and (2) that DNA repair demonstrated in bacteria has essentially an identical mechanism in man.

19.12   RNA BIOSYNTHESIS

With the exception of the biosynthesis of RNAs in such organelles as mitochondria and chloroplasts, the site of DNA-dependent RNA biosynthesis in eucaryotic cells is the nucleus. While the nucleolus contains the enzymes and genes for ribosomal RNA (rRNA) synthesis, the enzymes responsible for the synthesis of messenger (mRNA) and transfer RNAs (tRNA) are localized in the nucleoplasm. In procaryotic organisms, that lack a nucleus, RNA synthesis occurs in the cytoplasm.

All RNAs in the cell except for those in RNA viruses are synthesized from a DNA template. This process is called **transcription.** The enzyme that carries out this process is called DNA-dependent RNA polymerase or RNA polymerase. In contrast to DNA polymerase, which requires a template and a primer, RNA polymerase requires a duplex DNA template, but not a primer to initiate RNA synthesis and can catalyze the linkage of the two initial ribonucleotides (Figure 19.9). Most RNA biosynthesis occurs in an **asymmetric** fashion in which only one of the two complementary strands of DNA is copied into an RNA. The RNA product or **transcript** is synthesized in the 5′ to 3′ direction, resulting in the presence of a 5′-triphosphate end. Thus, all known nucleic acid polymerases synthesize their products in the 5′ to 3′ direction.

The initiation of RNA synthesis on the DNA template does not occur at random locations, but at very precise sites on the DNA called **promoters.**

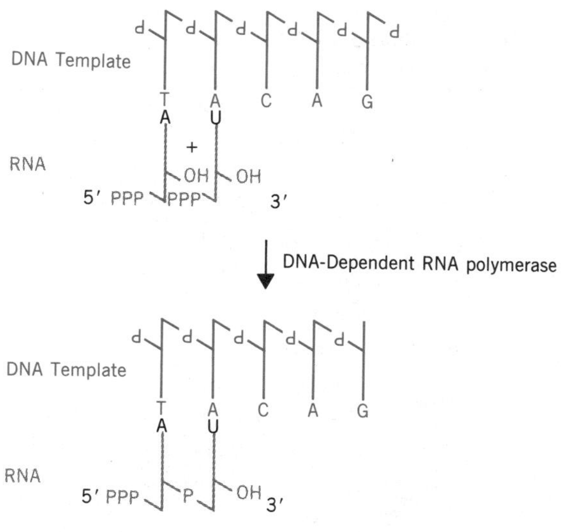

FIGURE 19.9 Transcription by RNA polymerase does not require a primer. Two nucleotides can be joined together to initiate RNA synthesis. Note the triphosphate at the 5′ end of the transcript. The growth of the RNA chain is therefore in the 5′ to 3′ direction.

Promoters are short segments of DNA that contain the information for binding RNA polymerase in a specific manner and generally precede the initiation point of transcription for each gene or **operon.** An operon is a genetic unit in which the expression of several genes is controlled by a single promoter. The initial base of the transcript is designated as the "+1" position and the procaryotic promoter covers approximately 60 base pairs of the DNA from positions -55 to $+5$. For most eucaryotic genes, the promoters also precede the $+1$ position; however, in some special cases, the information for binding RNA polymerase may occur far into the structural part of the transcript. In addition to the promoter site, there are several other types of sequence signals that either overlap or are near to promoters and that affect RNA polymerase activity. These signals are used to regulate transcription and will be discussed in more detail in Chapter 21.

19.13   PROCARYOTIC TRANSCRIPTION APPARATUS

The RNA polymerases of *E. coli* and *B. subtilis* have been studied extensively. The complex enzyme consists of two parts called the core enzyme (E) and the σ factor. The σ factor binds to the core enzyme to form the holoenzyme. The core enzyme contains four subunits with the composition $\alpha_2\beta\beta'$ and has a molecular weight of approximately 400,000 (Table 19.1). The β' subunit is capable of binding DNA. The β-subunit contains the catalytic site and also binds rifampicin, an inhibitor of RNA synthesis. The core enzyme alone can initiate transcription *in vitro* from nicks and single-stranded regions of DNA and synthesizes RNA from both complementary strands of DNA in a **symmetric** fashion. *In vivo* RNA synthesis is carried out essentially in an **asymmetric** fashion (see Section 19.12).

The σ factor determines the specificity of interaction of the RNA polymerase to promoter sites on the DNA and is absolutely essential for initiating RNA synthesis at the proper site and carrying out asymmetric RNA synthesis. In *B. subtilis,* a family of five σ factors have been identified with molecular weights

TABLE 19.1   Subunit Composition and Molecular Weights of Procaryotic RNA Polymerase

SUBUNITS	P ubtilis	E. coli
β'	130,000	160,000
β	140,000	150,000
α	45,000	42,000
σ (major)	43,000	70,000
σ^{37}	37,000	
σ^{32}	32,000	
σ^{29}	29,000	
σ^{28} (heat shock)	28,000	
σ^{32} (heat shock)		32,000
σ^{60} (nitrogen metabolism)		60,000

TABLE 19.2 Conserved Promoter Sequences Recognized by Procaryotic RNA
Polymerase Holoenzymes

ORGANISM	ENZYME FORM	-35	-10
B. subtilis	$E\sigma^{43}$	TTGACA	TATAAT
	$E\sigma^{37}$	AGGXTT	GGXATTGXT
	$E\sigma^{28}$	CTAAA	CCGATAT
	$E\sigma^{29}$	TTAATAAAT	CATATT
E. coli	$E\sigma^{70}$	TTGACA	TATAAT
	$E\sigma^{32}$	CTTGAA	CCCCATAT

E = core enzyme; $E\sigma$ = holoenzyme

ranging from 43,000 to 28,000 (Table 19.2). Each of the five holoenzyme forms recognizes a different promoter sequence, and thus, each transcribes a different family of genes. The major RNA polymerase form that comprises about 90% of the total enzyme contains a σ factor with a molecular weight of 43,000 (σ-43). The other σ factors are smaller in molecular weight and are present in much smaller amounts. The minor holoenzymes transcribe certain genes in response to starvation and heat shock.

In *E. coli,* three σ factors with molecular weights of 70,000 (σ-70), 32,000 (σ-32), and 60,000 (σ-60) have been observed. While the major σ-70 holoenzyme transcribes the vast majority of the *E. coli* genes, the σ-32 holoenzyme has a special function of transcribing heat shock protein genes (about 20 in number) in *E. coli* and the σ-60 holoenzyme controls the expression of the glutamine synthetase gene and other nitrogen metabolism genes. Heat shock protein genes that occur in limited number in virtually all organisms are activated by a sudden shift in temperature of about 10°C and are preferentially expressed during a heat shock period. The activity of these heat shock genes confers added heat resistance to the organism.

The minor σ factors of procaryotes appear to control the expression of stress-related genes and allow the organism to respond to nutrient depletion, heat shock, and possibly other stress conditions.

19.14 PROCARYOTIC PROMOTERS

In order to study the properties of promoters and their interactions with RNA polymerase holoenzyme, the sequence of more than 150 promoters from various procaryotic forms has been determined. The major procaryotic promoters (see Section 19.12) are characterized by having two highly conserved regions at the -10 and -35 regions with the sequence TATAAT and TTGACA, respectively (Figure 19.10). Usually, there are 17 base pairs (bp) between the two conserved regions and 4 to 7 bp between the -10 region and $+1$ position where transcription is initiated. The region upstream of the -35 region (-35 to -55) may be rich in AT base pairs. Generally, the promoter region is rich in AT base pairs. The promoters recognized by minor forms of RNA polymerase (see Table 19.2) have unique -10 and -35 regions. Chemical cross-linking studies

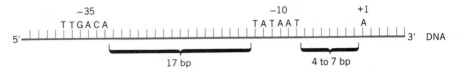

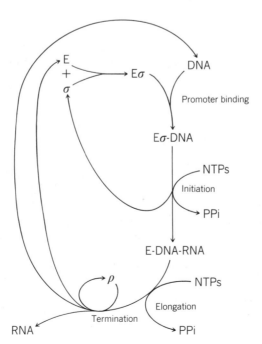

FIGURE 19.10 A typical procaryotic promoter recognized by the major RNA polymerase has two regions of conserved homology called the −10 and −35 regions with the consensus sequences indicated. The base incorporated at the 5′ end of the transcript is usually adenine or guanine. The DNA strand illustrated is complementary to the strand that is used as the template for RNA polymerase. It is customary to write DNA and RNA sequences that are complementary to the strand that is acting as the template.

and base protection studies have shown that the RNA polymerase subunits interact physically with the promoter region prior to initiation of RNA synthesis.

19.15   INITIATION, ELONGATION, AND TERMINATION OF TRANSCRIPTION

During initiation of transcription, the holoenzyme binds to the promoter to form an **open complex** in which the bound DNA is locally denatured between positions −8 and +3. After incorporation of the first few nucleotides complementary to one of the two strands of DNA (asymmetric RNA synthesis), the σ-factor is released. The released σ factor combines with another E and is ready to initiate transcription. The antibiotic rifampicin inhibits RNA synthesis at the stage in which the first phosphodiester bond has been formed and causes the formation of dinucleotides. The σ-cycle is illustrated in Figure 19.11.

FIGURE 19.11 The σ-cycle and ρ termination in RNA synthesis catalyzed by RNA polymerase of *E. coli.*

```
        G
      T = A
      T = A
      C ≡ G
      T = A
      G   T
      G ≡ C
      G ≡ C
      C ≡ G
      A = T
      A = T
      G ≡ C
  —GATG   TTCTTTTTTTAAAA—
```

FIGURE 19.12 A typical ρ-independent transcription termination signal contains a GC-rich hairpin structure followed by a string of T residues.

After release of the σ-factor, the core enzyme continues the **elongation** process. The antibiotic streptolydigin, which binds to the β-subunit of *E. coli* RNA polymerase, is an inhibitor of the elongation phase of RNA synthesis. Another antibiotic actinomycin D inhibits the RNA synthesis reaction by intercalating between the bases of the DNA template. Near the end of the transcriptional unit, there are ρ-dependent and ρ-independent signals in the DNA that cause the **termination** of RNA synthesis. ρ is a termination factor that causes precise termination of RNA synthesis by facilitating the release of the RNA product and the RNA polymerase from the template. The ρ-independent signal at the 3′ end of the gene is characterized by a GC-rich stem-and-loop structure preceding several thymine residues in the noncoding strand (Figure 19.12). The termination of RNA synthesis occurs at these locations in the absence of a ρ factor *in vitro*. The ρ-dependent termination signal is also characterized by a GC-rich region of dyad symmetry, but is not followed by several thymine residues. This type of termination signal is recognized by the RNA polymerase only in the presence of a ρ factor. The tetrameric ρ factor from *E. coli* has a molecular weight of 200,000 and exhibits an RNA-dependent ATPase activity that is necessary for its termination function.

19.16 EUCARYOTIC RNA POLYMERASES

The eucaryotic RNA polymerases occur in the nucleus in three forms called RNA polymerases I, II, and III. Each of the forms transcribes particular types of genes. RNA polymerase I transcribes the ribosomal RNA genes in the nucleolus and produces 18S, 5.8S and 26 to 28S rRNAs; RNA polymerase II transcribes unique structural genes in the nucleoplasm and produces heterogeneous nuclear RNAs (hnRNA) that are precursors to mature mRNAs and undergo a series of cleavage and splicing modifications. (There is some variation among eucaryotic organisms; e.g., for yeast, much less processing occurs with its mRNA.) RNA polymerase III transcribes tRNA and 5S rRNA in the nucleoplasm. The eucaryotic RNA polymerases are more structurally complex than the procaryotic RNA polymerases described previously and contain from 6 to 10 subunits whose molecular weights vary from 16,500 to 240,000. Their molecular weights vary from about 470,000 to 700,000.

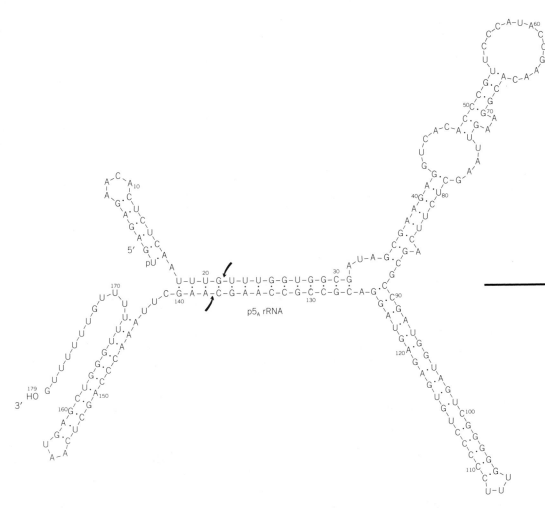

FIGURE 19.13   Processing of precursor 5S rRNA to the mature 5S rRNA. The two cleavage sites of RNAse M5 are indicated by the arrows. Reprinted with permission from *RNA Processing* by Norman R. Pace. CRC Press, Inc. Copyright © 1984.

A species of mushrooms called *Amanita phalloides* produces α-amanitin, a deadly poison, that causes more than a hundred deaths each year. This cyclic octapeptide inhibits RNA polymerase II by binding very tightly to the enzyme during the elongation phase and therefore inhibits the expression of structural genes. Since the toxin inhibits RNA polymerase III only at much higher concentrations (100 times) and RNA polymerase I is insensitive to the inhibitor, α-amanitin has been used to distinguish between these three nuclear enzymes.

19.17   EUCARYOTIC PROMOTERS AND REGULATORY SEQUENCES

Several eucaryotic promoters have been identified and sequenced. The features that are conserved among the various promoters include a TATA box at −25 to −30 and an upstream sequence usually −40 to −100. Certain sequences of

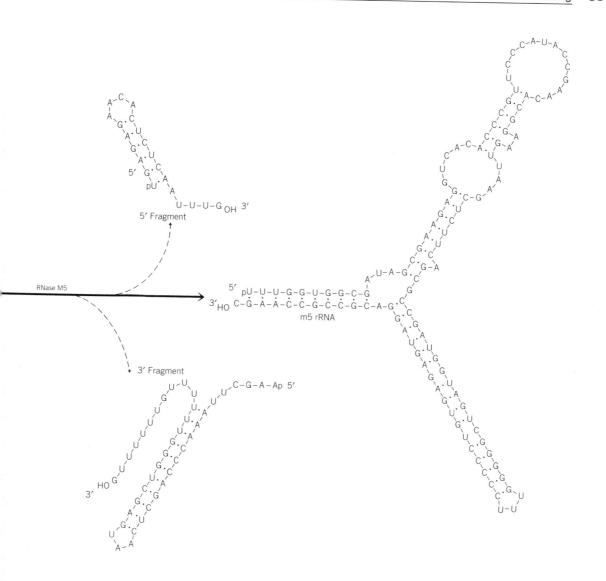

variable lengths called **enhancers** are usually located upstream from their corresponding promoter regions (up to several kilobases); enhancers facilitate the initiation of transcription from eucaryotic promoters.

19.18   TRANSCRIPTION PRODUCTS AND RNA PROCESSING

Most RNA products undergo two general types of post-transcriptional modification: covalent modification of the bases and cleavage of the phosphodiester bond with removal of bases from a precursor molecule, leading to formation of the mature RNA molecule.

In procaryotic cells, tRNA and rRNA are usually synthesized as a larger precursor form and are cleaved by endonucleases and exonucleases to produce the mature form of the molecule. The 16*S*, 23*S*, and 5*S* rRNAs of *E. coli* are produced sequentially from an rRNA operon. A large precursor containing all the rRNAs is not found under usual conditions because of the rapidity of the processing reactions. As the 16S rRNA is synthesized, it is cleaved to release a precursor 16S rRNA that is converted to mature 16S rRNA. The RNA polymerase continues to transcribe the 23S rRNA, and it is also cleaved from the growing chain in a precursor form that is modified to its mature form. Finally, the precursor 5S rRNA is synthesized and modified to its mature form. The tRNA of procaryotes may also be synthesized as a precursor that is cleaved by an endonuclease at the 5′ end and by an exonuclease at the 3′ end to form a mature tRNA molecule. Ribonuclease P, which cleaves tRNA precursors, is described in Section 6.11. Most mRNA in procaryotes are not made initially in a precursor form and therefore do not need modification. The linear information of the DNA is transferred to mature mRNA that then is translated directly into protein as described in Chapter 20.

As is indicated in Section 6.8, functional RNAs are not linear molecules or random coils of the polynucleotide chain. Rather, each seems to be folded into a specific but, at this time, poorly understood conformation apparently controlled by the sequence of nucleotide residues. Some insight into such structures is given by experimentally determined and theoretically calculated secondary structures: the pattern of hydrogen bonding between nucleotide bases. This is illustrated for tRNA (Figure 6.8), 16S rRNA (Figure 6.10) and 5S rRNA (Figure 19.13). All primary RNA transcripts produced in the eucaryotic nucleus are processed to mature form, and short- and long-distance interactions between nucleotide residues undoubtedly influence processing as well as the final conformation of the mature RNA molecule.

The 18S, 5.8S, and 27S rRNAs of a typical eucaryotic cell are synthesized by the action of RNA polymerase I as part of a large 45S rRNA precursor in the nucleolus. This precursor is cleaved by specific endonucleases that produce the final three rRNAs. The usual cleavage sites involve stem-and-loop structures of the rRNA precursor molecule with cleavage usually occurring at the base of the stems.

RNA polymerase II produces mRNA and heterogeneous nuclear RNA (hnRNA) from unique structural genes. The hnRNAs are precursors to mRNAs that are processed by cleavage and splicing activities to form mature mRNA. Many, but not all, eucaryotic genes contain intervening sequences or **introns,** which are noncoding sequences within the gene, and coding sequences called **exons.** The introns are removed from the hnRNA by a nucleoprotein complex that cleaves hnRNA and removes introns and splices together the exons into the mature mRNA (Figure 19.14).

In one form of RNA processing, that of the Group II introns, the intron forms an unusual *lariat* structure. A model has been proposed for the removal of an intron and the splicing of adjacent exons, with the formation of a lariat from the intron. In this two step process a ribonucleoprotein enzyme complex initially selects both the 5′ and 3′ splice sites before cleaving at the 5′ splice site in step 1 of Figure 19.15. The cleavage of the 5′ splice site is a phosphotransfer reaction resulting from the attack of a 2′-hydroxyl group in the intron on the phosphodiester bond that links L1 to the intron.

The result of step I is the formation of a circle from the 5′-most portion of the intron polynucleotide chain. This RNA circle with, its "tail" of the 3′ end of the intron and the entire 3′ exon (L2), constitutes a lariat structure. Thus the

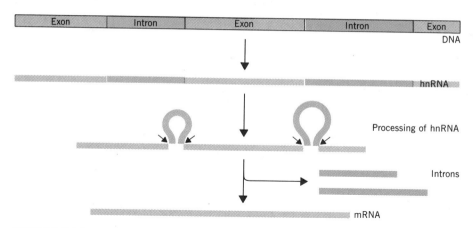

FIGURE 19.14 A schematic representation of a eucaryotic gene. The primary transcript (hnRNA) containing *exons* and *introns* and is processed by specific endonucleases that remove the introns and splice the exons to produce the mature mRNA.

metastable intermediate produced by reaction 1 of Figure 19.15 contains the free 5′ exon (L1) and the lariat structure. In step 2 the 3′-hydroxyl group of L1 attacks the phosphodiester bond of the 3′ splice site. The resulting phospho-transfer reaction joins L1 to L2 to produce the nucleotide sequence of the mature RNA and release the lariat form of the intervening sequence (IVS), or intron. Note that no change occurs in the number of phosphodiester bonds in either step 1 or step 2.

Another type of intron is that of Group I. Group I introns participate in another type of splicing reaction that, at least *in vitro,* occurs without the action of any enzyme. This reaction already has been mentioned in Section 6.11, and it is characteristic of the precursors of some nuclear rRNAs, mitochondrial rRNAs and mRNAs and chloroplast tRNAs. The reaction is diagrammed in Figure 19.16. In step 1 guanosine attacks the polyribonucleotide chain at the junction between the 5′ exon, L1, and the intron. The guanosine becomes incorporated as the 5′-terminal residue of the released intron-3′-exon polyri-bonucleotide chain.

In step 2 the newly exposed 3′-hydroxyl group of the 5′ exon attacks the junction between the intron and the 3′ exon, L2, producing the spliced RNA molecule and the guanosine-terminated intron, completing step 2. In step 3 the intron circularizes, releasing a short, G-terminated oligoribonucleotide of 1 to 15 residues, depending upon the intron. The function of step 3 may be to prevent the reversal of step 2. Note that as in the reactions of the Group II introns, there is no net change in the number of phosphodiester bonds in the Group I intron reaction, so there is no hydrolysis reaction. How this complex series of reactions can be accomplished by the RNA molecule alone is only dimly understood at this time. Nor do we have much information on the contributions, if any, of proteins to the *in vivo* reaction.

Eucaryotic mRNA precursors vary widely in the number of introns. Some have none. However, the genes that code for chicken ovalbumin and conalbu-min have been found to contain 7 and 17 introns, respectively. The mature mRNA is usually considerably smaller than the hnRNA. The introns can be

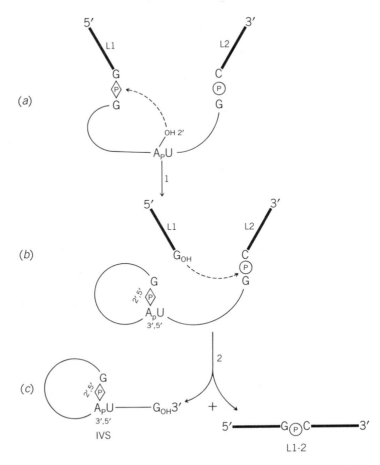

(a)

(b)

(c)

FIGURE 19.15 Model for lariat formation during excision of a Group II intron and the splicing of two exons (L1 and L2). See text for details. Reprinted with permission from P. A. Sharp, Cell 42:397–400 (1985). Copyright © 1985 is held by MIT.

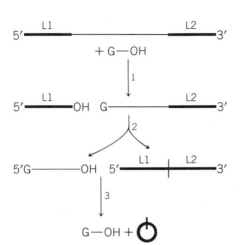

FIGURE 19.16 Excision of a Group I intron initiated by the reaction of the precursor RNA with guanosine, represented as G—OH. Each step in the reaction series produces a new 3′-terminal hydroxyl group, including step 3, and the circularization of the excised intron with release of the 5′-terminal portion of the polyribonucleotide chain.

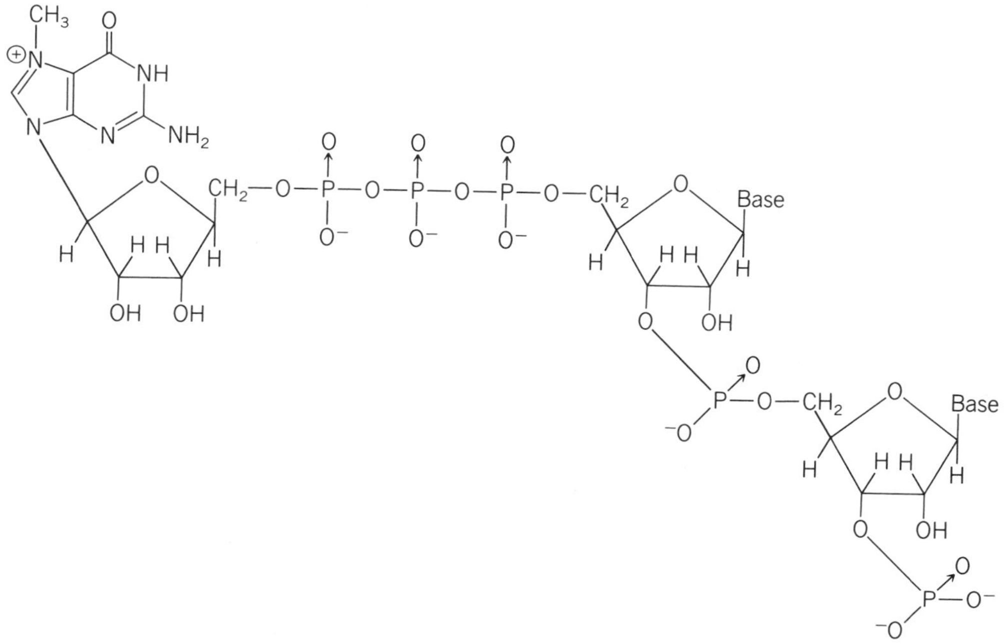

FIGURE 19.17   The presence of a 7-methyl guanylate residue at the 5′ end of a eucaryotic mRNA.

several thousand base pairs in length. The exact functions of the introns are still uncertain.

In addition to the cleavage and splicing steps, both ends of many eucaryotic mRNA are processed. A **capping** process adds a 7-methylguanylate residue attached by an unusual 5′-phosphoryl linkage to the 5′ end of the mRNA (Figure 19.17). The formation of the cap requires three steps. A phosphatase removes the terminal phosphate from the triphosphate present at the 5′ end of the mRNA, leaving a nucleoside diphosphate. The guanylate residue is added by transfer reaction from GTP. The 7-methyl group is added to the terminal guanylate group with S-adenosylmethionine as the methyl donor.

A "tailing" phenomenon occurs at the 3′ end of the mRNA. Up to 200 adenylate residues are added by a special adenylate transferase that uses ATP to form a long poly A tail. Thus, both the 5′ and 3′ ends are processed for most eucaryotic mRNAs.

The processed ends of the mRNA lend stability to the mRNA molecule against cellular nucleases and may designate the RNA as an mRNA for transport from the nucleus to the cytoplasm. Not all RNA made in the nucleus are processed into mRNA and most of the nuclear RNA is not transported into the cytoplasm. These RNA are rapidly degraded within the nucleus. The 5′ cap may also facilitate the translation of the mRNA during protein synthesis in the cytoplasm.

In addition to the modifications just described, many bases in mature tRNAs are modified by the covalent addition of small groups to their bases (see Chapter 6). About 50 different modifications of tRNA bases have been noted. The major covalent modification of large rRNAs consists of methylation of a few bases by highly specific methylating enzymes.

REFERENCES

1. *A. Kornberg*, DNA Replication. *San Francisco, Calif.: W. H. Freeman, 1980; 1982 Supplement.*

2. *E. C. Friedberg*, DNA Repair. *San Francisco: W. H. Freeman, 1985.*

3. *W. R. McClure, Mechanism and control of transcription initiation in prokaryotes.* Annual Review of Biochemistry, *Vol. 54 (1985), pp. 171–204.*

4. *O. Ribaud and M. Schwartz, Positive control of transcription initiation in bacteria.* Annual Review of Genetics, *Vol. 18 (1984), pp. 173–206.*

REVIEW PROBLEMS

1. Compare the substrate, primer, and enzymatic requirements and the initiation mechanisms for DNA and RNA synthesis in procaryotes.

2. If the antisense strand of DNA has the following sequence 3′—AGTACCGT-CAAAGTA—5′, what is the sequence of the sense strand? What is the sequence of the RNA transcribed from this region of the DNA?

3. What proteins and DNA sequences determine the specificity of gene transcription in procaryotes?

4. Define the following terms: primer, *ori*, DNA primase, "Okazaki piece," discontinuous DNA synthesis, σ-factor, promoters, posttranscriptional processing, and transcription terminator.

BIOSYNTHESIS OF PROTEINS

The elucidation of the mechanism of protein synthesis in both procaryotic and eucaryotic cells has occurred in the past 30 years and ranks as one of the major triumphs in modern biochemistry. The synthesis of the peptide bond is the principal chemical event in protein synthesis and is coupled to a sophisticated machinery for the precise translation of specific sequences programmed in the informational nucleic acids. However, a number of compounds are synthesized in the cell that have a "peptidelike" bond:

$$
\begin{array}{cc}
\overset{\displaystyle O}{\underset{\displaystyle \parallel}{R-C-NH-R}} & \overset{\displaystyle O}{\underset{\displaystyle \parallel}{R-C-NH_2}} \\
\text{Peptide bond} & \text{Amide} \\
\text{(an } N\text{-substituted amide)} & \text{(an unsubstituted amide)}
\end{array}
$$

and a discussion of their syntheses will be instructive. Amides and several small peptides are formed by enzyme systems that do not involve a complex mRNA–ribosome translation system. The synthesis of glutamine, hippuric acid, and glutathione will be examined.

Historically, biochemists selected these compounds as simple models to study protein synthesis; it soon became evident that these models did not reflect the complexity of protein synthesis and were not employed further for such studies.

20.1 GLUTAMINE

Glutamine is synthesized by glutamine synthetase, an enzyme found in plants, animals, and microorganisms. The first step is believed to be the formation of a γ-glutamyl acylphosphate–enzyme complex. In the second step, ammonia, a good nucleophile, attacks the complex and displaces the acylphosphate group to form glutamine and inorganic phosphate; this mechanism is outlined in Section 17.2. Only the γ-amide is formed. Isoglutamine, the compound with the α-carboxyl group amidated, is never found. The enzyme is not only highly specific, since aspartic acid cannot replace glutamic acid as a substrate, but is an extremely complex protein that is regulated by an interesting number of mechanisms (Section 17.2).

20.2   HIPPURIC ACID

The enzymatic synthesis of hippuric acid, a common urinary product in mammalian animals, involves the formation of a peptide bond between the carboxyl group of benzoic acid and the amino group of glycine by the sequence shown here.

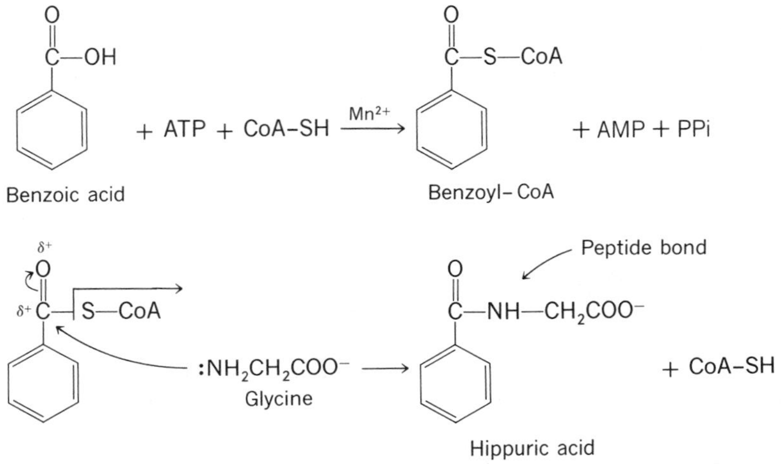

Benzoic acid   Benzoyl–CoA

Glycine

Hippuric acid

20.3   GLUTATHIONE

Glutathione a tripeptide occurs in yeast, plants, and animal tissues and requires two discrete enzyme systems to form its two peptide bonds. The first enzyme, γ-glutamyl cysteine synthetase (a) catalyzes the condensation of glutamic acid and cysteine with the formation of the first peptide bond. Then, a second enzyme, glutathione synthetase (b), adds glycine to the previously synthesized dipeptide to form the second peptide bond. In each step, the carboxyl group is presumably activated by ATP as already outlined for glutamine synthesis. Cysteine does not directly attack the γ-carboxyl group of glutamic acid since the $-O^-$ of the carboxyl group is a poor leaving group; if a phosphate group is placed on the carboxyl carbon at the expense of ATP, then, as with glutamine synthesis, we have an excellent leaving group.

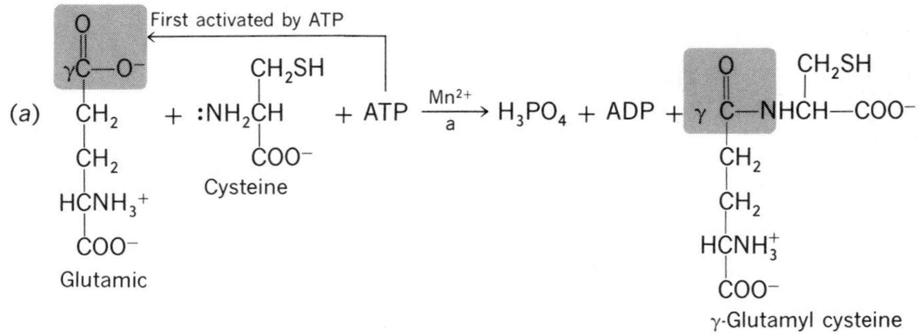

γ-Glutamyl cysteine

(b)

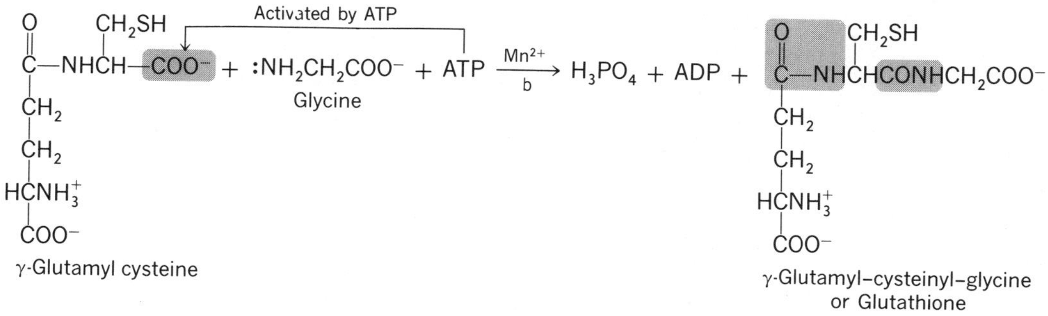

$$\text{C—NHCH—COO}^- + :\text{NH}_2\text{CH}_2\text{COO}^- + \text{ATP} \xrightarrow[b]{\text{Mn}^{2+}} \text{H}_3\text{PO}_4 + \text{ADP} + \text{C—NHCHCONHCH}_2\text{COO}^-$$

Activated by ATP

Glycine

γ-Glutamyl cysteine

γ-Glutamyl–cysteinyl–glycine or Glutathione

Note that CoASH is required in the synthesis of hippuric acid and that the reaction products include AMP and pyrophosphate; in the synthesis of glutamine and glutathione, ADP and inorganic phosphate are the reaction products, and CoASH is not required. This strongly indicates that the peptide bond in hippuric acid is synthesized by a mechanism unlike that found in the latter two examples. Note, moreover, that the sequence of these simple steps is controlled by the specificity of the enzymes involved. That is, in glutathione synthesis, the reverse peptide glycyl-glutamyl-cysteine is not produced because the specificities of the two enzyme systems control the order of addition. Thus, enzyme *a* catalyzes only Reaction *a* and enzyme *b* catalyzes only Reaction *b*. Hence, the order of reaction is glutamic acid with cysteine and not glutamic acid with glycine. Also, one ATP and one enzyme is required for each peptide bond formed.

20.4 CYCLIC POLYPEPTIDES

The biosynthesis of antibiotic cyclic polypeptides, like the biosynthesis of glutathione, occurs in the complete absence of polynucleotides, which suggests sequence determination by enzyme specificity.

As an example, let us examine the biosynthesis of gramicidin S, a cyclic decapeptide with the structure given in Section 3.5.

Extracts of *Bacillus brevis* strains that synthesize gramicidin S contain two protein fractions that, on the addition of ATP, Mg^{2+}, and the appropriate amino acids, readily catalyze the synthesis of the cyclic polypeptide. Protein I has a molecular weight of 280,000 and protein II, 100,000. Protein II activates and racemizes D- or L-phenylalanine. It also becomes charged with D-phenylalanine via a thioester linkage. Protein I activates the other four amino acids, namely, proline, ornithine, valine, and leucine, via the sequence

$$\text{Amino acid} + \text{ATP} \rightleftharpoons \text{Amino acyladenylate} + \text{PPi} \qquad (20.1)$$
$$\text{Amino acyladenylate} + \text{HS–protein} \rightleftharpoons \text{Amino acyl–S–protein} + \text{AMP} \qquad (20.2)$$

There is now evidence that Protein I is actually a polyenzyme consisting of at least four separate specific amino acid activating enzymes, each with a molecular weight of about 65,000 to 70,000. Each enzyme activates its specific amino acid according to the Reactions 20.1 and 20.2. Each amino acid is bound covalently as a thioester to the protein. A fifth protein is the additional component of the polyenzyme and it contains one 4′-phosphopantetheine functional group per protein (MW 17,000).

The following picture now emerges: L-Phenylalanine is first racemized to

D-phenylalanine by Protein II and then is activated to the D-phenylalanylthio-ester–Protein II complex. This complex now reacts with L-prolyl-S-activating enzyme component of Protein I to form a D-phenylalanyl–L-prolyl-S complex.

$$R^1\!-\!\underset{\underset{H}{|}}{\overset{\overset{NH_2}{|}}{C}}\!-\!\overset{\overset{O}{\|}}{C}\!-\!S\!-\!Prot\ II + R^2\!-\!\underset{\underset{H}{|}}{\overset{\overset{NH_2}{|}}{C}}\!-\!\overset{\overset{O}{\|}}{C}\!-\!S\!-\!Prot\ I \xrightarrow{\text{Transpeptidation}}$$

$$R^1\!-\!\underset{\underset{H}{|}}{\overset{\overset{NH_2}{|}}{C}}\!-\!\overset{\overset{O}{\|}}{C}\!-\!NH\!-\!\underset{\underset{H}{|}}{\overset{\overset{R^2}{|}}{C}}\!-\!\overset{\overset{O}{\|}}{C}\!-\!S\!-\!Prot\ I + \underset{\underset{SH}{|}}{Prot\ II} \quad (20.3)$$

The dipeptidyl–S complex is now transferred to the free SH group of 4′-phosphopantetheine–Protein I.

$$R^1\!-\!\underset{\underset{H}{|}}{\overset{\overset{NH_2}{|}}{C}}\!-\!\overset{\overset{O}{\|}}{C}\!-\!NH\!-\!\underset{\underset{H}{|}}{\overset{\overset{R^2}{|}}{C}}\!-\!\overset{\overset{O}{\|}}{C}\!-\!S\!-\!Prot\ I + HS\!-\!4'\text{-Phosphopantetheyl-Prot I} \xrightarrow[\text{thiolation}]{\text{Trans-}}$$

$$R^1\!-\!\underset{\underset{H}{|}}{\overset{\overset{NH_2}{|}}{C}}\!-\!\overset{\overset{O}{\|}}{C}\!-\!NH\!-\!\underset{\underset{H}{|}}{\overset{\overset{R_2}{|}}{C}}\!-\!\overset{\overset{O}{\|}}{C}\!-\!S\!-\!4'\text{-Phosphopantetheyl-Prot I} + \underset{\underset{SH}{|}}{Prot\ I} \quad (20.4)$$

The charged 4′-phosphopantetheine group is now translocated to the ornithyl–S–Protein I site, where a second transpeptidation occurs to form the tripeptide–S complex, returned to the 4′-phosphopantetheyl arm that translocates the tripeptide to the leucyl-S site, and so on until the final pentapeptide is constructed. Simultaneously, the second pentapeptidyl complex is synthesized. As soon as the two pentapeptidyl thioenzyme complexes are completed, they interact to form the complete cyclic decapeptide.

$$NH_2\!-\!D\text{-Phe} \longrightarrow Pro \longrightarrow Val \longrightarrow Orn \longrightarrow Leu\!-\!\overset{\overset{O^{\delta-}}{\|}}{\underset{\delta+}{C}}\!-\!S\!-\!Prot\ I$$

$$Prot\ I\!-\!S\!-\!\underset{\underset{O^{\delta-}}{\|}}{\overset{\delta+}{C}}\!-\!Leu \longleftarrow Orn \longleftarrow Val \longleftarrow Pro \longleftarrow D\text{-Phe}\!-\!\overset{..}{N}H_2$$

$$\Downarrow$$

$$D\text{-Phe} \longrightarrow Pro \longrightarrow Val \longrightarrow Orn \longrightarrow Leu$$
$$\uparrow \qquad\qquad\qquad\qquad\qquad\qquad\qquad \downarrow$$
$$Leu \longleftarrow Orn \longleftarrow Val \longleftarrow Pro \longleftarrow D\text{-Phe}$$

Gramicidin—S

Although more primitive than the nucleic acid directed ribosomal system for protein synthesis, this system does restrict the choice to only the correct stereo-isomeric amino acids, namely, D-phenylalanine, L-proline, L-valine, L-ornithine, and L-leucine are utilized; all other amino acids or the incorrect optical isomers are rejected. The reactive amino acyl thioester–protein complex is employed for activation of the carboxyl groups necessary for peptide bond formation. The student should note the startling similarity in the synthesis of this cyclic peptide with the synthesis of fatty acids (Section 13.10).

20.5 COMPONENTS OF PROTEIN SYNTHESIS APPARATUS

In examining the problem of protein synthesis, we are immediately faced with the gigantic task of synthesizing a very complex molecule containing hundreds of L-amino acid residues in exactly the same sequence each time the molecule is produced. The mechanisms of synthesis must have a precise coding system that automatically programs the insertion of only one specific amino acid residue in a specific position in the protein chain. The coding system, in determining the primary structure precisely, in turn establishes the secondary and tertiary structures of a given protein. This problem obviously does not exist in the area of synthesis of simple peptide or amides.

Of primary interest is the mechanism by which information from DNA, the genetic carrier of the coding system, is used in a precise manner for the biosynthesis of proteins. We have already discussed DNA and RNA synthesis (Chapter 19) and will summarize the orderly processes by which proteins are formed.

Genetic information from DNA is programmed to RNA for the synthesis of new proteins.

$$DNA \xrightarrow{\text{Transcription}} RNA \xrightarrow{\text{Translation}} Protein$$

We have already discussed the mechanism of transcription whereby the genetic information in DNA is employed to order a complementary sequence of bases in a new RNA chain (Chapter 19). In this process, three key RNAs are synthesized: (1) messenger RNA (mRNA), which carries the genetic message from DNA for the orderly and specific sequence of amino acids for the new protein and acts as a template for protein synthesis; (2) ribosomal RNAs (rRNA), which serve as structural components of ribosomes (ribonucleoprotein particles); and (3) transfer RNAs (tRNA), which carry activated amino acids to specific recognition sites on the mRNA template.

The machinery for protein synthesis thus consists of a large number of components including mRNA; ribosomes, which are the actual sites for protein synthesis; aminoacyl-tRNAs; and a number of enzymes, cofactors, and energy-rich compounds. The whole machinery operates in the cytoplasm of procaryotic and eucaryotic organisms. In eucaryotes, it can be tightly associated with the endoplasmic reticulum. Mitochondria and chloroplasts also have the complete machinery for protein synthesis, except for a more limited set of mRNAs and proteins.

20.5.1 ACTIVATION OF AMINO ACID

One of the major steps in protein synthesis, in which sequence specificity is determined, occurs during a two-step reaction involving activation of an amino acid and the transfer of the activated amino acid to a specific transfer RNA (tRNA). Each of the 20 amino acids commonly found in proteins must undergo an initial activation step by highly specific aminoacyl-tRNA synthetases. Thus,

D isomers and certain amino acids such as ornithine, citrulline, β-alanine, and diaminopimelic acid, which are used for other purposes in the cell, are rejected at this stage. The activation part of this reaction involves the following.

$$\text{ATP} + \text{amino acid} \underset{\text{Mn}^{2+}}{\overset{\text{Enzyme}}{\rightleftharpoons}} \quad \text{aa-AMP–Enz} \quad + \quad \text{PPi} \qquad (20.5)$$

$$\text{(aa)} \qquad\qquad \underset{\substack{\text{Amino acid – adenylate –}\\ \text{enzyme complex}}}{} \quad \underset{\text{Pyrophosphate}}{}$$

Adenine–ribose—O—P(=O)(OH)—O—P(=O)(OH)—O—P(=O)(OH)—OH + $^-$:O—C(=O)—CHR(NH$_3^+$) $\overset{\text{Enzyme}}{\rightleftharpoons}$

$$\left[\text{Adenine–ribose—O—}\overset{\delta-}{P}^{\delta+}\!\!\sim\!\! \text{O—}\overset{\delta+}{C}^{\delta-}\text{—CHR} \right] \text{Enzyme} + \text{PPi}$$

Very labile bond
(an acid anhydride)

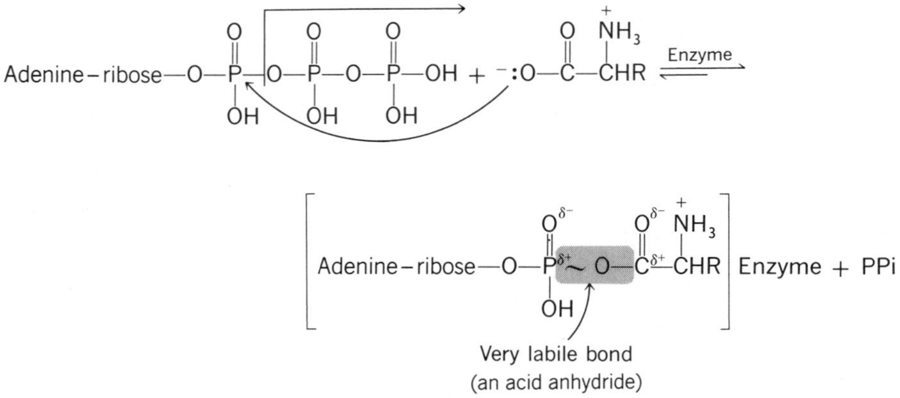

Aminoacyl adenylates are extremely reactive but are stabilized by remaining tightly associated with the parent enzyme. The great lability is associated with the large positive charge on the amino group adjacent to the positive phosphorus atom, resulting in a strong electrostatic repulsion and a subsequent labilization of the P—O—C bond. Acylation of the amino group of the amino acid will, thus, stabilize the P—O—C bond. A naturally occurring acylated aminoacyl-tRNA is N-formyl-methionyl-tRNA (f-Met-tRNA$_f^{\text{Met}}$) (see Section 20.6). In this compound, one of the protons of the methionine amino group is replaced by the HCO group of formic acid. (See Sections 20.9.1 and 20.9.3 for f-Met-tRNA$_f^{\text{Met}}$ functions.)

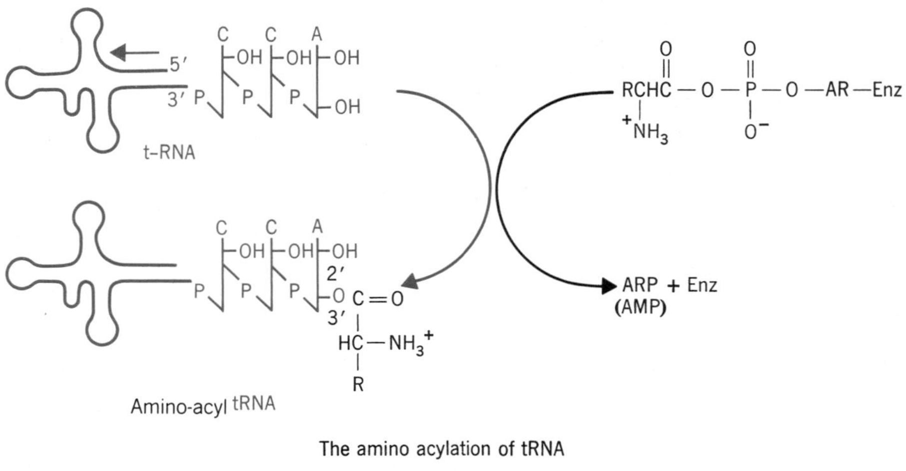

The amino acylation of tRNA

FIGURE 20.1   The aminoacylation of tRNA.

A large number of aminoacyl-tRNA synthetases have been extracted from a variety of tissues. Their molecular weights vary from 50,000 to 200,000, and they may be monomeric or oligomeric proteins.

The second step in the amino acid activation-transfer reaction is the transfer of the activated amino acyl-adenylate group to its specific tRNA (Figure 20.1).

20.5.2 TRANSFER RNA

The acceptors for the activated amino acids are the specific tRNAs. Each bacterial cell contains about 60 different tRNA species, whereas eucaryotic cells may have as many as 100 to 120 different tRNAs. Many of the tRNAs have been completely sequenced. Several of them have been crystallized. Chain lengths of known tRNA species vary between 73 and 88 nucleotide residues, with about 10 to 20% of the bases being modified (see Section 6.1). The molecular weights vary between 25,000 and 30,000. The nomenclature $tRNA_{yeast}^{ala}$ signifies a transfer RNA specific for alanine and obtained from yeast. A discussion of the structure of tRNA will be found in Sections 6.7 and 6.8.

Over 85% of the tRNAs have guanine at their 5′ terminus, while the remainder have cytosine as the terminal base. All tRNAs require the presence of the sequence cytidylate–cytidylate–adenylate (—CCA) at their 3′ terminus before they can be charged with an amino acid. Some tRNAs are synthesized with the —CCA attached to their 3′ ends; others require the addition of the —CCA to the 3′ end of the tRNA by the activity of a specific nucleotidyl transferase. Present evidence indicates that the aminoacyl group is linked to the 3′-OH ribosyl moiety of the 3′-terminal adenosyl group by a highly reactive ester linkage with a free energy of hydrolysis of about −8000 cal.

Apparently, the presence of the vicinal *cis*-2-hydroxyl group (see Figure 20.1) together with the protonated α-amino group renders the aminoacyl ester linkage very reactive. Acylation of the α-amino group to a *N*-acyl derivative greatly decreases the reactivity of the ester linkage.

Transfer RNAs have three specific functions:

1. Recognition by a specific aminoacyl-tRNA synthetase in order to accept the correct activated amino acid.
2. Interaction with the correct codon in the mRNA sequence with its own specific anticodon to thereby ensure that the correct amino acid is placed in the proper sequence in the growing polypeptide chain (see Figure 20.6).
3. Binding the growing peptide chain to the ribosome participating in the translation process.

As we have described in Section 19.18, in both procaryotes and eucaryotes, tRNAs are transcribed initially in larger precursor forms (120–130 bases) that are processed by specific endonucleases and exonucleases to form mature tRNA molecules (75–87 bases). In eucaryotes, the precursor forms are synthesized in the nucleus and then transported to the cytoplasm where they are processed by nucleases and modified by specific enzymes. More than 50 different modified bases have been observed in various tRNAs. Some of these structures are depicted in Section 6.1. An interesting correlation seems to exist between the extent of modified nucleosides contained in a tRNA of particular organism and the evolutionary development of that organism. Thus, tRNA of

Mycoplasma, the smallest free living organism known, contains only a low amount of modified nucleosides. However, *E. coli,* yeast, wheat germ, rat liver, and tumor cells contain an increasing number of modified bases in their tRNA species. Mammalian tRNAs have as much as 20% of their bases modified. Results from studies, in which some of the modified bases are replaced by other bases, imply that the modified bases induce subtle changes in tRNA structure that are necessary for the exacting, multiple functions of these molecules.

An interesting modified nucleoside contains isopentenyl adenosine (iA).

$$HN—CH_2—CH=\overset{\overset{\textstyle CH_3}{|}}{C}—CH_3$$

Ribose–etc.

Isopentenyl adenosine itself belongs to a class of cytokinins that are potent plant growth factors that promote cell division, growth, and organ formation in plants. Although the relationship of iA to these activities is not clear, iA always occurs adjacent to the base adenine in the anticodon region when adenine is the third base of the anticodon triplet sequence. Isopentenyl adenosine is readily synthesized by a cytoplasmic enzyme, isopentenyl pyrophosphate : tRNA isopentenyl transferase.

Isopentenyl pyrophosphate + Adenine-tRNA ⟶ PPi + Isopentenyl-adenine-tRNA

Mention was made earlier of about 60 different bacterial tRNA species that can be separated and purified. The existence of several tRNAs for the same amino acid is called multiplicity. For example, one type of multiplicity is attributed to the degeneracy of the genetic code; that is, three tRNA^{Ser} species are required for the six serine codons (see the genetic code, Table 20.1). A second type of multiplicity is due to separate cytoplasmic and organellar tRNAs, the mitochondrial tRNA species in animal and plant cells and the chloroplast tRNAs in chloroplasts of plants.

A third type is a specific tRNA species employed by the cell for a highly specialized function. For example, gram-positive procaryotic organisms synthesize large amounts of a cell wall component called peptidoglycan (see Sections 2.8 and 8.1.1). The interpeptide bridge that joins the separate strands of peptidoglycans in some organisms is a pentaglycyl peptide. *Staphylococcus epidermidis* is a typical organism that possesses a unique tRNA^{Gly} at relatively high concentrations. This tRNA^{Gly} has about 85 residues, but contains only one modified nucleoside; namely, 4-thiouridine. Although three additional isoaccepting tRNA^{Gly} species were isolated from this organism, which together with the specific tRNA^{Gly} were charged with glycine and could participate in the formation of the interpeptide bridge, the specific tRNA^{Gly} did not participate in protein synthesis and possess a glycine anticodon, whereas the three other tRNA^{Gly} species readily participated in protein synthesis. One can thus speculate that the organism, by synthesizing a highly specific tRNA, could depend on this tRNA species for the all-important task of forming the interpeptide bridges so essential for the synthesis of a complete peptidoglycan without competition from the protein-synthesizing machinery of the cell.

In summary, tRNAs serve as adaptors in directing the proper placement of amino acids according to the nucleotide sequence of the mRNA. Transfer RNA must have several recognition sites in its structure, namely, (1) the anticodon site, that is, the three-base site (Figures 6.8 and 6.9) responsible for the recognition of the complementary triplet codons in mRNA; (2) the synthetase site by which the specific aminoacyl-tRNA synthetase recognizes and charges the tRNA with a specific amino acid; (3) the amino acid attachment site, which in all tRNAs is the 3′-terminal-CCA nucleotide sequence; and (4) the ribosome recognition site that allows the aminoacyl-tRNA to attach to a mRNA–ribosome complex. These recognition sites allow the tRNA to convert a 4-letter code found in nucleic acids into a 20-letter code found in proteins.

20.6 THE GENETIC CODE

We have been discussing codons, anticodons, and codes in a number of Sections. It is now appropriate to define these terms more precisely.

Until about 25 years ago, one of the most intriguing puzzles in modern biology was how to code for 20 amino acids in an unambiguous manner with only four nucleotide residues. Obviously, a definite nucleotide sequence could serve as a code. How long should this sequence be? If each sequence is two residues long, only 4^2 or 16 possible different binary combinations would be available, which is less than the 20 amino acids that must be coded. If each sequence is three residues long, a total of 4^3 or 64 different combinations would be available that would be more than adequate. The solution to this puzzle is one of the most interesting chapters in modern biochemistry.

As we shall see, the sequence of bases on the mRNA directs the precise synthesis of the amino acid sequence of a protein. The **codon**, the unit that codes for a given amino acid, consists of a group of three adjacent nucleotide residues in mRNA; the next three nucleotide residues on the mRNA code for the next amino acid, and so on. In 1961, M. Nirenberg performed a classical experiment. He employed a system from *E. coli* that consisted of a subcellular, supernatant fraction and ribosomes supplemented with tRNA. To this system, he added a mixture of radioactive amino acids and polyuridylic acid [poly(U)] that had been prepared by the action of polynucleotide phosphorylase on UDP. From this complex mixture of amino acids, the only amino acid incorporated into an acid insoluble fraction consisting of newly synthesized protein was phenylalanine. The product proved to be polyphenylalanine. Nirenberg correctly concluded that the synthetic poly(U) was, in effect, serving as mRNA and providing the information that specified that only phenylalanyl-tRNA units should become associated with the ribosomal poly(U) complex.

Randomly mixed polynucleotides were then prepared by the addition of varying amounts of CDP, ADP, or GDP together with UDP to polynucleotide phosphorylase. When the synthetic RNA polymers of different base composition were added to the test system we have described, different amino acids were incorporated into the proteins. Knowing the concentration of each of the ribonucleoside diphosphates used in the reaction and assuming random incorporation of bases into the RNA synthesized, one can calculate the frequency of occurrence of particular trinucleotides in the RNA. One could therefore correlate the levels of amino acids incorporated with the frequency of codons in the mRNA and determine which codons were responsible for the incorporation of specific amino acids. A minimum coding ratio of three nucleotide bases for each amino acid was determined, although the precise order of the bases in each triplet code was not yet known.

In 1964, Nirenberg and Leder devised a simple method whereby trinucleotides of known sequence were employed to decipher the code. A trinucleotide of known sequence was mixed with a labeled aminoacyl-tRNA and ribosomes. After a given period of incubation, the suspension was filtered through a nitrocellulose filter. Binding of a particular aminoacyl-tRNA to ribosomes depended on the presence of a specific trinucleotide. If no binding occurred, the aminoacyl-tRNA would pass through the filter; however, if binding did occur, the aminoacyl-tRNA–ribosome complex would remain on the filter and could be easily counted for radioactivity. In effect, the synthetic trinucleotides served as model codons. Employing this trinucleotide binding assay, Nirenberg examined 64 trinucleotides with 20 aminoacyl-tRNAs.

During this same period, H. G. Khorana, employing organic synthetic techniques as well as enzymatic techniques, prepared synthetic polyribonucleotides with completely defined repeating sequences. Thus, the repeating sequence of CUC UCU CUC . . . , when added to the protein-synthesizing system and radioactive amino acids, yielded a polypeptide that contained only alternating residues of leucine and serine.

From these and other experiments, biochemists were finally able to assign specific amino acids to 61 out of 64 possible codons, with the remaining three

TABLE 20.1 The Genetic Code

FIRST POSITION (5′ END)	SECOND POSITION U	SECOND POSITION C	SECOND POSITION A	SECOND POSITION G	THIRD POSITION (3′ END)
U	Phe	Ser	Tyr	Cys	U
	Phe	Ser	Tyr	Cys	C
	Leu	Ser	Term[a]	Term	A
	Leu	Ser	Term	Trp	G
C	Leu	Pro	His	Arg	U
	Leu	Pro	His	Arg	C
	Leu	Pro	GlN	Arg	A
	Leu	Pro	GlN	Arg	G
A	Ile	Thr	AsN	Ser	U
	Ile	Thr	AsN	Ser	C
	Ile	Thr	Lys	Arg	A
	Met (initiation)	Thr	Lys	Arg	G
G	Val	Ala	Asp	Gly	U
	Val	Ala	Asp	Gly	C
	Val	Ala	Glu	Gly	A
	Val	Ala	Glu	Gly	G

[a] *Chain-terminating.*

being designated as nonsense or terminator codons. Table 20.1 summarizes these results, which define the genetic code.

We shall now list some brief generalizations concerning the code.

1. The code is universal; that is, all procaryotic and eucaryotic organisms use the same codons to specify each amino acid. Rare exceptions to this rule have been observed with codon use in yeast mitochondria and *Mycoplasma.*

2. The code is degenerate; that is, more than one arrangement of nucleotide triplets specify the same amino acid. Thus, UUA, UUG, CUU, CUC, CUA, and CUG all are codons for leucine. We immediately notice that for a given amino acid the first two bases are limited to one or two combinations, but the third base can have as many as four. This suggests that a change in the third base by mutation may still allow the correct translation of a given amino acid into protein. This degeneracy often involves only the third base in the codon. For example, the codons for phenylalanine are UUU and UUC. The third base need simply be a pyrimidine. Two of the codons for leucine also begin with UU that in this case must have its third base occupied by a purine. The general pattern of codon assignments suggests that the nucleotide at the 3' end of the codon may often be occupied by either of two bases, a purine or pyrimidine. This flexibility in the third position is termed **wobble**. Therefore, nearly all codons can be represented as xy_G^A or xy_C^U. Beside the usual four bases A, U, G, and C, a fifth base I (inosine) is frequently found to form part of the anticodon; I never occurs in codons. It, however, occurs as a base in the triplet anticodon of tRNA and invariably complements the third or flexible base of the codon. Thus, the codon CUU can be read by the anticodon, AAG, or IAG (read from the 5' to 3' direction). The reason for the flexibility of I is that I can hydrogen-bond with U, A, or C (Figure 20.2).

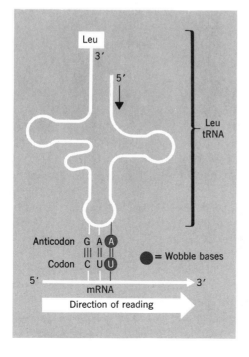

FIGURE 20.2 Association of aminoacyl tRNAs to mRNA.

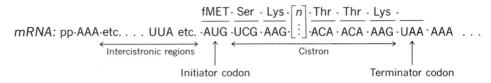

FIGURE 20.3 The AUG signal is usually not at the 5′ terminus nor does it directly follow an internal terminator signal (UAG, UAA, or UGA). There are regions along a polycistronic mRNA that are not translated. The function of such intercistronic regions is not clear.

3. The code is nonoverlapping; that is, adjacent codons do not overlap.

4. The code is commaless; that is, there are no special signals or commas between codons.

5. Of the 64 possible triplet codons, 61 are employed for encoding amino acids. Three UAA, UAG, and UGA had been originally called **nonsense** codons, but now are recognized as specific translation or chain termination codons for the COOH end of the peptide chain.

6. The codon AUG is of considerable interest, since it is the only codon for methionine regardless of whether f-Met-tRNA$_f^{Met}$ or Met-tRNA$_m^{Met}$ is employed as the methionyl carrier (i.e., whether or not the methionine α-amino group is formylated: Section 20.5.1). The tRNA$_f^{Met}$ is a special tRNA species that is involved only in the initiation of protein synthesis. The AUG serves as the extremely important **initiator** codon as well as for internal methionine codons. Presumably, the initiation factors IF1, IF2, and IF3 or the secondary structure of the mRNA discriminate as to the correct use of AUG as an initiator or internal codon. The roles of the initiator and terminator codons are depicted in Figure 20.3.

7. In general, amino acids with hydrocarbon residue side chains have U or C as the second base; those with branched methyl groups have U as the second base. Basic and acidic amino acids have A or G as the second base.

20.7 MESSENGER RNA

As has been mentioned on several occasions, a key component of the translation process is mRNA, which comprises only a few percent of the total RNA of a cell. It carries the genetic message from DNA to the site of protein synthesis, the ribosome, and hence is called messenger RNA. Messenger RNA is metabolically unstable with a high turnover rate in procaryotic cells; in eucaryotic cells, it has a slower turnover rate. It is synthesized by RNA polymerase (Section 19.11).

Messenger RNA varies greatly in chain length and thus in molecular weight. This great variation is related to the heterogenous size of proteins and the presence of operons in procaryotic cells. Operons consist of two or more adjacent genes (cistrons) whose expression is regulated by a single promoter. The relatively large mRNA from an operon contains information for several adjacent genes and is called a polycistronic mRNA. Since few proteins contain less than 100 amino acids, the mRNA coding for these proteins must have at least 100×3 or 300 nucleotide residues. In *E. coli,* the average size of mRNA is 900 to 1500 nucleotides, and it codes for more than one polypeptide chain. The metabolic instability of mRNA is characteristic of bacterial mRNAs that have half-lives from a few seconds to about two minutes. In mammalian systems, however, mRNA molecules are considerably more stable with a half-life ranging from a few hours to more than 24 hours. This has been interpreted to mean that bacteria, which must have greater flexibility in adjusting to their ever-changing environment, must be able to synthesize different enzymes to cope with their surroundings and hence, degrade their mRNAs much more rapidly than eucaryotic cells.

In eucaryotic cells, a precursor heterogeneous nuclear RNA (hnRNA) is synthesized in the nucleoplasm by RNA polymerase II. The hnRNA is processed by endonucleolytic cleavage and splicing enzymes, capping enzymes, and tailing enzyme to form the mature mRNA. This mature mRNA is translocated to the cytoplasm where it becomes associated with the ribosome system for translation. An important distinction between the mRNAs of procaryotes and eucaryotes is that almost all eucaryotic mRNAs are monocistronic, that is, code for only one polypeptide.

Procaryotic and eucaryotic mRNAs contain other information besides the coding sequence for the polypeptide. Some mRNAs contain relatively long leader sequences at the 5' end that control the translation rate of the message. The procaryotic mRNAs contain a short sequence (Shine–Dalgarno sequence) in the 5' region that is involved in binding the $30S$ ribosome during initiation of protein synthesis. In polycistronic mRNAs, there are intercistronic regions. The transcription termination signal is also transcribed into the 3' ends of mRNAs (Section 19.14).

20.8 RIBOSOMES

In the early 1950s, evidence began to accumulate suggesting that ribosomes (Sections 6.10.2 and 8.4.1) were the site for protein synthesis. For example, Zamecnik injected radioactive amino acids into a rat and then, within a short interval of time, homogenized the liver and fractionated the homogenate by differential centrifugation into nuclei, mitochondria, microsomes, and supernatant protein. The microsomal fraction had the highest specific activity. When these microsomes were treated with detergent to free the ribosomes from the vesicular matrix and again assayed for radioactivity, the ribosomes contained up to seven times more radioactivity per milligram of protein than the remainder of the microsomes. Clearly, the ribosomes served in some capacity as the locus for protein synthesis. At first, biochemists concluded that because of the high RNA content of ribosomes, rRNA could serve admirably as the template RNA. With the discovery of mRNA in 1957–1958, further modifications of the earlier views were necessary. What role do the ribosomes play? Let us first examine in some detail the chemistry of these particles.

Ribosomes are large ribonucleoprotein particles on which the actual process of translation occurs. In procaryotic cells, they occur in the free form as mono-ribosomes or are associated with mRNA as polyribosomes. An average bacterial cell contains about 10^4 ribosomes. In eucaryotic cells, they occur in forms similar to those found in procaryotic cells and also are associated with membranes of the rough endoplasmic reticulum (Section 8.4). About 10^6 to 10^7

ribosomes occur in these cells. Mitochondria and chloroplasts also possess ribosomes that differ slightly from those found in the cytoplasm of eucaryotic cells. Tables 6.3 and 20.2 summarize the physical properties of procaryotic ribosomes and plant and animal cytoplasmic ribosomes. As is indicated in Section 6.10.2, ribosomes may be purified by prolonged centrifugation of tissue extracts at high centrifugal speeds ($100,000 \times g$ for several hours). All ribosomes consist of a larger subunit and a smaller subunit. No proteins are common to both the large and small subunits. Metabolically, ribosomes are stable particles and do not turn over as rapidly as mRNAs, which have a relatively short half-life. Thus, rRNAs and tRNAs are considered metabolically stable forms of RNA in the cell.

Specific ribosome-associated proteins are directly involved in binding mRNA and tRNA. A major role of rRNAs is to serve as structural polymers holding the multiprotein particle in a compact configuration. The rRNAs also function in protein synthesis, but in ways not yet well defined. For example, the procaryotic 16S rRNA is involved in the interaction of 30S ribosomes with mRNA.

The two ribosomal subunits have different binding properties. Thus, the *E. coli* 30S subunit binds mRNA in the absence of the 50S subunit and the 30S—mRNA complex binds specific tRNAs. The 50S subunit does not associate with mRNA in the absence of the 30S subunit, but will nonspecifically bind tRNA. Each 70S ribosome contains two different binding sites for tRNA molecules. Site A (aminoacyl site) is involved in the positioning of the specific incoming aminoacyl-tRNA with its matching codon on the mRNA. Site P

TABLE 20.2   Sedimentation Coefficient Values for Ribosomes

RIBOSOMES	SUBUNITS	rRNA (MW)	NUMBER OF DISTINCT PROTEINS
Procaryotic			
Bacteria, actinomycetes, blue-green algae, mitochondria from eucaryotes			
70S	30S	16S (550,000)	21
	50S	5S (40,000)	
		23S (1,100,000)	35
Eucaryotic			
Plant kingdom[a]			
~80S	40S	16–18S (~700,000)	34
	60S	5S (40,000)	
		25S (~1,300,000)	ca. 50
Animal kingdom[a]			
~80S	40S	18S (~700,000)	34
	60S	5S (40,000)	ca. 50
		28–29S (1,400,000–1,800,000)	

[a] In general, organelle ribosomes (mitochondrial or chloroplastic) are in the 70S category.

(peptidyl site) binds the growing polypeptidyl-tRNA. The **peptidyl transferase** that is responsible for the formation of the peptide bond is located on the 50S ribosomal particle, presumably near the P site. The 50S subunit also has a site that hydrolyzes GTP to GDP during the **translocation** process.

A few comments should be made concerning the biosynthesis of ribosomes. In eucaryotes, the site of synthesis of the ribosomal RNA is the nucleolus (Section 19.18). In the nucleolus, the RNA polymerase I transcribes a large precursor rRNA from the rRNA cistron region (rDNA) of the nucleolar DNA. Precursor rRNA has a sedimentation value of 45S (MW 4.1×10^6) and is rapidly cleaved to three smaller RNAs, namely, a 32S, a 20S, and a 5.8S component. The 32S component is further modified to yield a mature 28S rRNA, while the 20S component is cleaved to the mature 18S rRNA. In the meantime, specific ribosomal proteins synthesized in the cytoplasm are translocated to the nucleolus and become associated with their respective rRNAs to form the completed 40S (18S rRNA) and the 60S ribosomal (28S and 5.8S rRNAs) subunits. At this point, the 40S unit is transferred to the cytoplasm, where it becomes associated with mRNA; shortly thereafter, the 60S subunit is translocated to the cytoplasm to become associated with the 40S–mRNA complex to form the complete 80S–mRNA complex. The 5S rRNA that is transcribed by RNA polymerase III may be involved in binding the two ribosomal subunits together. The residual fragments from the precursor rRNA are converted back to nucleotides and recycled into the nucleolar transcription machinery.

The 16S, 23S, and 5S rRNAs in procaryotic cells are processed from a large rRNA precursor to smaller precursors that are trimmed to achieve the correct mature structure. Nomura has demonstrated the sequential addition of ribosomal proteins to the 16S rRNA for the formation of mature 30S ribosomes. The biosynthesis of ribosomes, which involves the coordinate synthesis of 3 different rRNAs and 56 different ribosomal proteins, is a remarkable example of regulated gene expression.

20.9 PROTEIN SYNTHESIS

The translation of coded information from DNA via mRNA into the amino acid sequences of proteins involves the orderly interactions of over 100 different macromolecules. We shall now describe the current status of this highly complex sequence of reactions in which tRNA, mRNA, ribosomes, and many ancillary enzymes and proteins are involved.

An overall picture of the process is illustrated in Figure 20.4. The gene is transcribed into mRNA by RNA polymerase, and the mRNA is translated into a protein by the complex translational machinery that will be discussed in detail. Since the mechanism of protein synthesis has been investigated in great depth in extracts of *E. coli,* we shall use this organism as the model for our description but will introduce comments concerning eucaryotic systems at the appropriate points.

There are four major steps in the synthesis of a protein: (1) activation and transfer of amino acids to tRNAs, (2) initiation of the synthesis of the polypeptide chain, (3) its elongation, and (4) termination. Table 20.3 summarizes the important components of these reactions. We have already described step 1 in

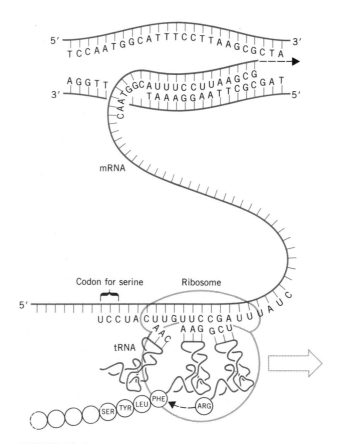

FIGURE 20.4 An overall view of genetic information flow. Reprinted with permission from Klaus Heckman, Alexander von Humboldt Stiftung Mitteilungen 46:21–28 (1985) and the Alexander von Humboldt Foundation.

some detail. Steps 2, 3, and 4 are illustrated in Figures 20.5 to 20.7. The student should refer to Table 20.3 and these figures as we describe the process of protein synthesis.

20.9.1   INITIATION

The first reaction in the *E. coli* system is the binding of mRNA to the 30*S* subunit in the presence of initiation factor 3 (IF3) to yield a mRNA–30*S*–IF3 complex with a 1:1:1 ratio of components. The interaction between the mRNA and 30*S* ribosome is facilitated by the presence of a Shine–Dalgarno (SD) sequence at the 3′ end of the 16*S* rRNA of the 30*S* ribosome that is partially complementary to sequences near the 5′ ends of mRNAs. The SD sequence is about 11 bases long and the efficiency of translation is partially related to the complementary fit between the SD sequence in the 16*S* rRNA and the sequence at the 5′ end of mRNAs.

Initiation factors 1 (IF1) and 2 (IF2) now participate in the binding of fMet-tRNA and GTP to the 30*S*—mRNA–IF3 complex to form the initiation

TABLE 20.3   Important Components of the *E. coli* Protein-Synthesizing System

SECTION NUMBER	STEP	COMPONENTS
(20.5.1)	Activation	tRNAs
		ATP
		Mg^{2+}
		L-Amino acids
		Amino acyl-tRNA synthetases
(20.9.1)	Initiation	30S Ribosomal unit
		50S Ribosomal unit
		mRNA with initiator codon AUG
		IF1 (MW 9400), IF2 (MW 80,000), IF3 (MW 21,000)
		Initiator tRNA = fMet–tRNA$_f$
		Formyl-tetrahydrofolic acid
		Met–tRNA$_f$ formylase
		GTP, Mg^{2+}
(20.9.2)	Elongation	EFTs factor (19,000)
		EFTu factor (40,000)
		EFG factor (80,000)
		Amino acyl–tRNAs
(20.9.3)	Termination	Terminator codons UAA, UAG, UGA
		R1 (44,000) factor
		R2 (47,000) factor
		S factor (40,000)
		TR factor
(20.9.3)	Deformylmethionylation	Deformylase
		Amino peptidase

complex of 30S–mRNA–fMet-tRNA–GTP with the release of IF3. The fMet-tRNA is located on the initiation codon AUG in the mRNA. Now, the 50S ribosome associates with this complex; GTP is hydrolyzed to GDP + Pi and both IF1 and IF2 are released. The final product is a 70S complex containing fMet-tRNA–mRNA with fMet-tRNA occupying the peptidyl site (P site) of the 70S ribosome. These events are summarized in Figure 20.5.

In eucaryotic translation systems, an analogous SD sequence has not been observed in 18S rRNA and many eucaryotic mRNAs have the initiating AUG codon so close to the 5′ end of the molecule that no sequence complementary to a putative eucaryotic SD sequence can be present. The simplest model for initiation in eucaryotes has the 40S ribosome binding at or near the 5′ end of the mRNA and migrating down to the first AUG codon at which initiation would occur. However, since initiation does not always occur with the first AUG encountered, the current model has the 40S ribosome binding near the 5′ end of the mRNA and migrating to the AUG with the following optimal

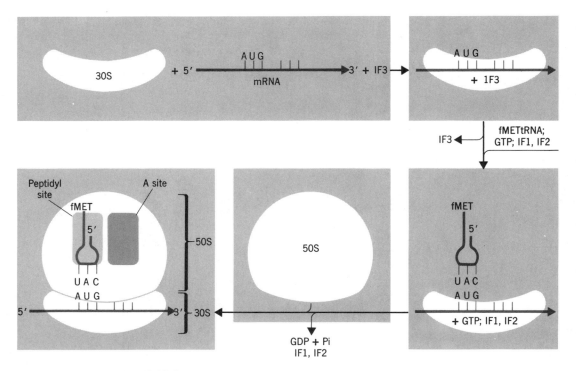

FIGURE 20.5   Steps by which the 70S initiation complex is formed in *E. coli*.

flanking sequence: ^A_GNN<u>AUG</u>G. Eucaryotic ribosomes would have the highest probability of initiating protein synthesis at this AUG.

A few comments should now be made about the fMet-tRNA. Some years ago it was observed that the major NH_2-terminal amino acid of total *E. coli* proteins was methionine. Later it was noted that the addition of methionine and in particular *N*-formylmethionine greatly stimulated protein synthesis in crude *E. coli* extracts. It was finally discovered that the starting amino acid in the synthesis of all proteins in procaryotic organisms is *N*-formylmethione, which in turn is associated with a specific $tRNA_f^{Met}$. *N*-formylmethionine of fMet-$tRNA_f^{Met}$ is formed by the reaction

$$\text{Formyl-tetrahydrofolate} + NH_2\text{-Met-tRNA}_f \xrightarrow{\text{Formylase}} N\text{-FormylMet-tRNA}_f$$

This is an extremely important reaction in which the α-NH_2 group of methionine is formylated.

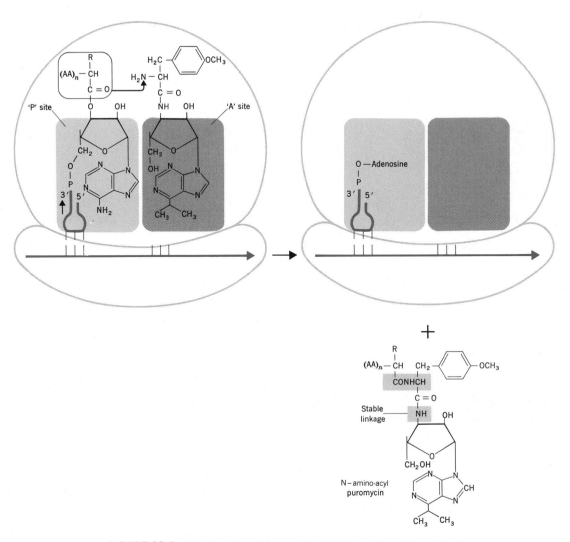

FIGURE 20.6 Mechanism of puromycin inhibition of protein synthesis.

Not all Met-tRNAs are formylated. The second type with a slightly different base sequence, Met-tRNA$_m^{Met}$, is inactive in the initiation step and is the specific methionine carrier for internal methionyl residues in the growing polypeptide chain. It is of considerable interest that in eucaryotic organisms, an unblocked Met-tRNA^{Met} serves as the specific initiator tRNA.

The P site on the 70S–mRNA complex is now occupied with fMet-tRNA$_f^{Met}$, and the A site is ready to receive the first aminoacyl-tRNA that is specified by the codon adjacent to the initiator codon AUG.

20.9.2 ELONGATION STEP

This step involves three stages: (1) the GTP-dependent codon-directed binding of a new aminoacyl-tRNA to site A of the 70S ribosome; (2) the transfer of the

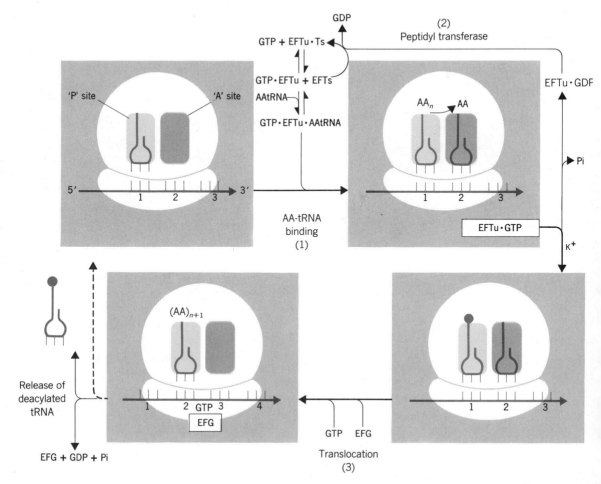

FIGURE 20.7   The elongation process. The symbol $(AA)_{n+1}$ denotes a formyl-methionyl polypeptide. See discussion in text.

peptidyl residue of the tRNA bound to the P site, to the newly bound amino-acyl-tRNA on site A, thereby forming a new peptide with one additional amino acid residue; and (3) translocation of the newly formed peptidyl$_{(n+1)}$-tRNA from site A to site P on the 70S ribosome by a movement of the 70S ribosome in a $5' \rightarrow 3'$ direction on the mRNA with the charged tRNA still being bound to its specific codon on mRNA.

20.9.2.1   Stage 1.   A new and specific aminoacyl-tRNA is bound to the A site of the 70S ribosome as determined by the codon on the mRNA at the A site. This is the second site at which amino acid specificity is determined, that is, at the codon–anticodon interaction. GTP and two elongation factors, EFTu and EFTs, are involved. EFTu in a complex with GTP, namely, EFTu–GTP, interacts with an aminoacyl-tRNA to form a ternary complex. The following events then occur.

EFTu—GTP + Amino-acyl-tRNA $\longrightarrow$ Amino-acyl-tRNA·EFTu·GTP

Amino-acyl-tRNA·EFTu·GTP + mRNA·Ribosome·fMet-tRNA $\longrightarrow$

Amino-acyl-tRNA·mRNA·fMet-tRNA + EFTu·GDP + Pi

(A site)   (P site)

EFTu·GDP + EFTs $\rightleftharpoons$ EFTu·EFTS + GDP

EFTu·EFTs + GTP $\rightleftharpoons$ EFTu·GTP + EFTs

Of some importance, all aminoacyl-tRNAs must react with EFTu–GTP in order to bind at the A site of the 70S ribosome. The significant exception is fMet-tRNA. Since this initiator aminoacyl-tRNA does not react with EFTu–GTP, its insertion into an internal position in the elongating polypeptide is avoided. The binding of an aminoacyl-tRNA to the A site is an energy requiring step, since GTP is hydrolyzed to GDP and Pi.

20.9.2.2 Stage 2. The formation of the new peptide bond is catalyzed by **peptidyl transferase** that is associated with the 50S ribosome subunit. The peptidyl moiety associated with its tRNA on the P site is transferred to the amino group of the aminoacyl-tRNA on the A site to form a new peptide bond, leaving a deacylated tRNA on the P site. The overall direction of protein synthesis is in the NH_2 to COOH direction. This peptide bond formation step does not require ATP, since the energy from the high-energy bond formed during amino acid activation is used during formation of the peptide bond. Relatively high concentrations of K^+ cations are required for this reaction. The student should recall the discussion of the NaK ATPase pump as it relates to protein synthesis (Section 8.12.4.2). Of interest, the antibiotic puromycin inhibits protein synthesis at this step. The peptidyl transferase can transfer the peptidyl residue from the charged tRNA, bound to the P site to puromycin to form a peptidyl puromycin that is released from the ribosome and is, of course, inactive (Figure 20.6). Note that the structure of puromycin resembles an amino acid attached to the 3'-adenosine residue of tRNA. Other antibiotics such as chloramphenicol, streptomycin, kanamycin, and neomycin interact with various components of the procaryotic ribosome and exert an inhibitory action on protein synthesis. These antibiotics do not interact with eucaryotic ribosome components and are therefore effective drugs against bacterial infections.

20.9.2.3 Stage 3. The **translocation process** involves the shift of the new peptidyl-tRNA from the A site to the P site with deacylated tRNA on the P site being released from the ribosome. Note carefully that in this shift, the peptidyl-tRNA remains bound to its codon-mRNA, but the ribosome moves relative to the peptidyl-tRNA in a 5' to 3' direction, thereby positioning its A site over the next codon on the mRNA. For this translocation to occur, a new elongation factor, EFG, is required as well as GTP, which is hydrolyzed to GDP + Pi.

The role of GTP is not clear at present. GTP is hydrolyzed to GDP +Pi in its involvement with IF2 (binding of initiator tRNA), EFTu (binding of amino-acyl-tRNAs), and EFG (translocation). At first, it was believed that the energy of hydrolysis of GTP was directly or indirectly utilized for peptide bond formation. The evidence now suggests that GTP hydrolysis is in someway involved in the dissociation of the three factors (just listed) from the ribosome for the purpose of recycling these factors for further protein synthesis.

In eucaryotic cells, a similar translocation factor has been found. Of considerable interest, the EF2 factor of eucaryotic cells, which is identical in function with the EFG translocation factor of procaryotic cells, is rapidly inactivated by diphtheria toxin. Apparently in the presence of NAD^+, the following reaction occurs with free EF2.

$$EF2 + ARPPR\text{—}nicotinamide \xrightarrow{\text{Diphtheria toxin}} ARPPR\text{—}EF2 + Nicotinamide$$
$$\text{(active)} \quad\quad NAD^+ \quad\quad\quad\quad\quad\quad \text{(inactive)}$$

In the intact eucaryotic cell, the toxin is bound to the cell membrane. However, EF2, bound to the ribosomal system, when released into the cytoplasm rapidly diffuses to the periphery of the cell and becomes inactivated by the mechanism described above. Thus, the effects of diphtheria, a dreadful disease, can be explained on a molecular level. Incidentally, all eucaryotic cells including yeast are sensitive to the diphtheria toxin in the presence of NAD^+, but procaryotic cells are completely insensitive.

Figure 20.7 summarizes the steps involved in the elongation process.

20.9.3   TERMINATION

The **termination** reaction during translation consists of two events: (1) the recognition of a termination codon in the mRNA and (2) the hydrolysis of the final peptidyl-tRNA ester linkage to release the nascent protein. The termination codons are UAA, UAG, and UGA. In several mRNAs, tandem termination codons have been observed. Three protein factors are required: R1, R2, and a S factor. The R1 factor is required for the recognition of the codons UAA and UAG, and R2 is required for the recognition of UAA and UGA. The third protein, S, has no release activity, but appears to aid in terminator codon recognition. The picture that is emerging suggests that the termination step can be divided into a terminator codon-dependent R1 or R2 factor binding reaction and a hydrolytic reaction in which either R1 or R2 converts the peptidyl transferase activity at site P into a hydrolytic reaction with the transfer of the peptidyl-tRNA to water rather than to another aminoacyl-tRNA. A final factor, TR, may be involved in discharging the residual tRNA from the P site. Once the tRNA is removed, the 70S ribosome dissociates from the mRNA into a 30S and a 50S subunit and is ready to reenter the ribosomal cycle for the synthesis of another protein molecule. IF3 combines with the 30S subunit, thereby preventing a reassociation of the 50S and 30S subunits and also prepares the 30S unit for recycling.

The nascent protein presumably has a formyl-methionyl-amino terminus that must be removed before the protein completes its folding sequence. Two enzymes may participate at this final stage.

1.   A specific deformylase

$$\text{Formyl-methionyl-peptide} \longrightarrow \text{Formic acid} + \text{Methionyl peptide}$$

2. A specific amino peptidase

Methionyl peptide ⟶ Methionine + Peptide

Figure 20.8 summarizes the various steps in the termination process.

20.9.4 ENERGY REQUIREMENTS TO FORM A PEPTIDE BOND

As we discussed in Section 20.3, the formation of a peptide bond in a small peptide requires the utilization of one ATP. In the case of ribosome-dependent peptide bond synthesis, the equivalent of four ATPs are required per peptide bond formed.

1. Activation of an amino acid = 2 ATPs (since AMP has to be converted back to ATP).

2. Binding of an aminoacyl-tRNA to the A site = 1 GTP.

3. Translocation of the peptidyl-tRNA from the P site to the A site = 1 GTP.

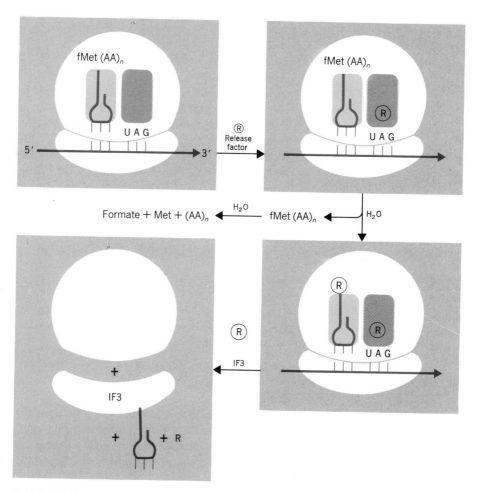

FIGURE 20.8 The termination step. R (release factor) denotes the combined action of R1 (or R2), S, and TR factors mentioned in Section 20.9.3.

This greater energy requirement for the synthesis of the peptide bonds of large proteins probably reflects the need for greater sequence specificity in these macromolecules.

20.10   *In vitro* SYNTHESIS OF COMPLETE PROTEINS

We have outlined the complex array of steps necessary for the complete synthesis of a protein. Until a few years ago, protein synthesis was observed by counting the incorporation of ^{14}C-labeled amino acids into a poorly defined trichloroacetic acid precipitate of denatured proteins. However, with the elucidation of the detailed steps involved in polypeptide biosynthesis, it is now possible to synthesize specific enzymes employing the appropriate template DNA, RNA polymerase, and a protein-synthesizing system enriched for all the necessary components in coupled transcription–translation systems. The addition of RNA polymerase assures the presence of high levels of mRNA in the synthesis system. For example, the gene for β-glucosyl transferase comprises 0.3 to 1.0% of the T4 phage DNA. Thus, the investigator can set up the following scheme.

$$\text{T4 phage DNA} \xrightarrow[\text{polymerase}]{\text{RNA}} \text{mRNA} \xrightarrow[\substack{\text{Complete protein-synthesizing} \\ \text{components (Table 20.3)}}]{\text{fMet-tRNA}_f^{Met}}$$
$$\longrightarrow \beta\text{-Glucosyl-transferase} + \text{Other proteins}$$

A number of eucaryotic mRNAs have been isolated and employed with suitable protein-synthesizing systems capable of translating these mRNAs into identifiable proteins. These systems include subcellular extracts and also whole cells into which mRNA preparations have been injected for translation and processing to form mature proteins. The translated mRNAs have included globin RNA, ovalbumin RNA, immunoglobin RNA, histone RNA, lens crystalline RNA, myosin RNA, silk RNA, avidin RNA, and protamine RNA. These experiments confirm the concepts developed to describe the mechanism of protein synthesis and provide powerful tools for the study of coupled transcription–translation phenomena.

20.11   CHEMICAL SYNTHESIS OF PROTEINS

In recent years, the chemical synthesis of polypeptides and proteins with molecular weights of up to 9000 have been successfully developed. The student should refer to Box 3.D for a detailed account of these syntheses.

20.12   BIOSYNTHESIS OF INSULIN

It is appropriate to outline the general biosynthetic aspects of the important hormone, insulin, to illustrate the intriguing complexities of eucaryotic protein synthesis. D. F. Steiner has described in a series of elegant experiments the biosynthesis of insulin by the β-cells of the islets of Langerhans of the pancreas in a number of species.

The classic work of F. Sanger of England on the precise amino acid sequence of insulin made possible a detailed molecular picture of insulin. Until 1965, insulin was believed to be synthesized as two separate polypeptides that in some manner were oriented to allow the specific formation of disulfide linkages between the two chains to yield insulin.

In 1967, Steiner demonstrated that a protein molecule larger than insulin was formed in pancreatic β-cells that exhibited all the properties of a precursor of insulin (see Section 3.15). Called proinsulin, its molecular weight was 9000 (insulin, MW 6500) and it had 81 amino acid residues (insulin, 51). It could be rapidly converted to a fully physiologically active hormone by the proteolytic action of trypsin.

The presence of an even more complex preproinsulin form of the hormone has now been established. The translation of insulin mRNA *in vitro* resulted in the formation of preproinsulin with a molecular weight of 11,500 containing a 24-amino acid residue **signal peptide** at the NH_2 terminus. The signal peptide (prepeptide) consists of a short N-terminus region usually containing charged amino acids, followed by a sequence of hydrophobic amino acid residues and containing a signal peptidase cleavage site at the C-terminus region between the prepeptide and propeptide.

$$\overset{+}{MetAlaLeuTrpMetArg}\underline{PheLeuProLeuLeuAlaLeuLeuValLeuTrp}\overset{-}{Glu}$$

$$\overset{+}{ProLysProAlaGlnAla}/Phe\cdots\cdots Propeptide\cdots\cdots\cdots$$

Signal peptidase cleavage site

The signal peptides from both procaryotic and eucaryotic cells are quite similar in amino acid composition and size. See Section 8.13 for another discussion of protein transfer across membranes.

The model for the biosynthesis of insulin that is now emerging is, in its general features, illustrated in Figure 20.9.

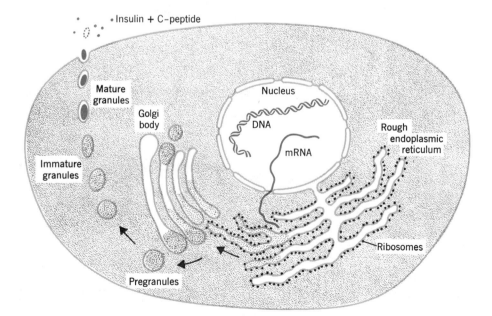

FIGURE 20.9 Schematic representation of the biosynthesis of insulin in a β cell of the islets of Langerhans in pancreatic tissue. (Modified from a diagram by permission of D. F. Steiner.)

In the presence of the protein-synthesizing enzymes and factors and the insulin mRNA, the ribosomes clustered around the rough endoplasmic reticulum (RER) synthesize preproinsulin. The single polypeptide rapidly folds and the disulfide bridges form as the signal peptide of the preproinsulin translocates the proinsulin into the cisternal (interior) spaces of the RER. A membrane-bound **signal peptidase** removes the signal peptide and releases the free proin-

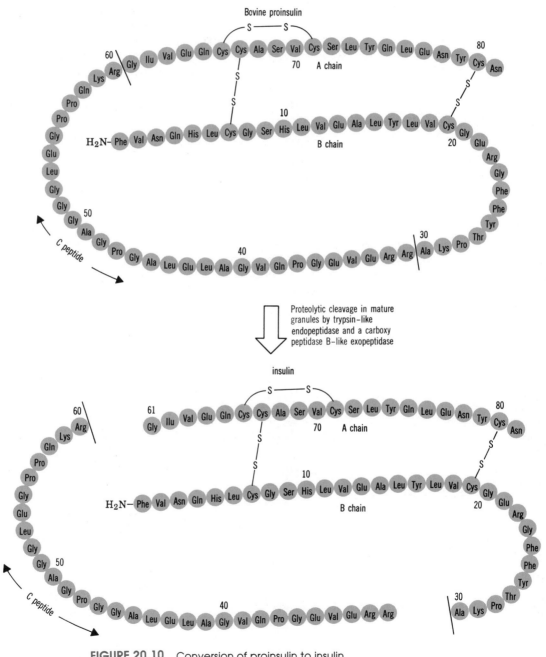

FIGURE 20.10   Conversion of proinsulin to insulin.

sulin that is transported via the vesicular tubules of RER to the contiguous Golgi apparatus. The time interval for these events is approximately 10 minutes. An hour later, immature secretory granules are formed by vesiculation from the periphery of the Golgi apparatus. Having a single limiting membrane, they contain proinsulin, proteolytic enzymes, and zinc ions. A rapid conversion to mature granules takes place in the periphery of the β-cells, with complete transformation of proinsulin to zinc insulin and the C peptide (see Figure 20.10). At the appropriate signal, the mature granules are secreted by reverse pinocytosis into the bloodstream, where the insulin is released. Not only insulin but other secreted proteins such as α-amylase, ribonuclease, and the like are synthesized in the pancreatic exocrine cells by this mechanism.

The question arises as to why the cell first forms preproinsulin. The prepeptide or signal peptide functions to facilitate passage of the proinsulin through the membrane. The signal peptide is recognized by a signal recognition protein (SRP) that takes the prepropeptide to a "docking protein" in the inner surface of the membrane. The hydrophobic region of the signal peptide then pulls the proinsulin through the membrane, and a signal peptidase at the outer surface of the membrane cleaves the signal peptide off, leaving the proinsulin. Most secreted proteins have been found to be synthesized initially with a signal peptide that is removed to produce either a proprotein or a mature protein. In the case of insulin, a proinsulin form is observed.

Why does the cell form proinsulin? As is described in Section 3.15, the amino acid sequence of many proteins is decisive in directing the folding of polypeptide chains into the native conformation. Insulin, with its two polypeptide chains, does not readily recombine to form the typical native structure. Thus, one can conclude that a major function of the C polypeptide chain of proinsulin in insulin biosynthesis is the facilitation of the proper folding of the insulin A and B chains under conditions that are thermodynamically favored.

20.13 PROTEIN DEGRADATION

Protein turnover—that is, synthesis and degradation—occurs constantly in eucaryotic cells, but it is a highly selective process with different rates of turnover for various proteins. Turnover of proteins can control the level of certain enzymes, furnish amino acids in times of need, and degrade faulty or damaged proteins that are produced during synthesis or arise from deleterious metabolic events in the cell.

In eucaryotes, degradation of proteins occurs in two steps. First, the protein undergoes an oxidative modification that leads to loss of enzymatic activity. Then, the enzyme is attacked by proteases in the second step. It appears that the modification of the enzyme marks it as a substrate for the protease. Stadtman has shown that inactivation of glutamine synthetase and a number of other key enzymes in metabolism is dependent on the presence of O_2 and NADH, is stimulated by Fe(III), and is inhibited by catalase and Mn^{2+}, typical conditions for mixed-function oxidation (MFO) systems (see Section 13.5.4 and Section 14.9).

Some of the MFO systems that can carry out the oxidative modification of proteins include (1) microbial NADH oxidase, (2) rabbit liver microsomal cytochrome *P*-450 reductase together with cytochrome *P*-450 isozyme 2, and (3) xanthine oxidase together with ferredoxin or putidaredoxin.

It has been proposed that the inactivation involves two steps: (1) the MFO-catalyzed synthesis of H_2O_2 and reduction of Fe(III) to Fe(II) followed by (2) the oxidation of enzyme-bound Fe(II) by H_2O_2 to generate oxygen radicals that attack histidine or other oxidizable amino acids at the metal binding site of the enzymes. Catalase blocks the modification procedure at the second step. Evidence has also been found that the rate of inactivation of glutamine synthetase in *E. coli* cell suspensions varied inversely with the intracellular level of catalase.

The oxidatively modified proteins are readily degraded *in vitro* by cell extracts, whereas native forms of the proteins are resistant to cellular proteases. This strongly supports the notion that oxidative inactivation of proteins is a prerequisite for degradation and that MFO systems are implicated in the process of intracellular protein turnover.

In *E. coli*, an ATP-dependent protease has been characterized as the product of the **lon** gene. This protease normally does not attack native proteins, but does degrade protein fragments produced from genes that have "nonsense" mutations that cause premature translation termination during protein synthesis, missense proteins resulting from mutations that cause a change in an amino acid in the protein, and proteins containing amino acid analogues. Thus, the lon gene product appears to be involved in protein turnover in *E. coli*.

REFERENCES

1. *K. Moldave, "Eukaryotic Protein Synthesis."* Annual Reviews of Biochemistry, *Vol. 54 (1985), pp. 1109–1149.*

2. *U. Maitra, E. A. Stringer and A. Chaudhuri, "Initiation Factors in Protein Biosynthesis."* Annual Reviews of Biochemistry, *Vol. 51 (1982), pp. 869–900.*

3. *M. Kozak, "Initiation of Protein Synthesis."* Microbiological Reviews, *Vol. 47 (1983), pp. 1–45.*

REVIEW PROBLEMS

1. Compare the energy (ATP) requirements for the synthesis of a peptide bond in glutathione versus that in a large protein made by the mRNA–ribosome system. Why is there a difference?

2. What reactions require ATP (or ATP equivalents) during peptide bond formation in the mRNA–ribosome system.

3. In the mRNA–ribosome system, what mechanisms or reactions determine the specificity of amino acid incorporation?

4. Draw a sequence of bases in the mRNA that codes for the following peptide: Met–Ala–His–Asp–Trp. What anticodons would you expect to find in the tRNAs that translate this mRNA?

REGULATION OF GENE EXPRESSION

During the last 30 years, considerable knowledge has accumulated concerning the metabolic pathways and the synthesis of biologically important macromolecules. However, in order to fully comprehend cellular functions, it is necessary to understand how cellular activities, particularly gene expression, are regulated. Although expression from many genes is relatively continuous especially during the active growth of the cell, many genes are expressed only under certain nutritional conditions, during differentiation and development, after certain physiological stimulations (nervous, hormonal, etc.), or during stressful situations (heat shock, uv radiation, chemical treatment, etc.). Under these same conditions, many other genes are turned off. Thus, elaborate mechanisms have evolved to allow the cell to express or turn off certain genes under specific cellular and environmental conditions.

21.1 SITES OF REGULATION OF GENE EXPRESSION

Regulation of gene expression ordinarily occurs at the **transcriptional, post-transcriptional, translational,** or **post-translational** levels. **Transcriptional regulation** includes all mechanisms that control the information transfer from DNA to RNA by RNA polymerase. **Post-transcriptional regulation,** which was discussed in Chapter 19, involves all modifications of the primary RNA transcript before it is translated into proteins. **Translational regulation** involves those factors that determine the rate of translation of mature mRNA molecules. **Post-translational regulation** involves mechanisms that control the processing of the primary translation product into the mature protein product and was discussed in Chapter 20.

The environmental and metabolic state of the cell has a direct and significant effect on the control of gene expression. Usually, small extracellular or intracellular metabolites trigger the complex mechanisms that result either in stimulation or inhibition of gene expression.

637

21.2   TRANSCRIPTIONAL REGULATION IN PROCARYOTES

The major locus for controlling gene expression in procaryotes is at the level of transcription, and this is the initial level at which expression can be regulated. The high rate of mRNA turnover in procaryotes also serves to change the metabolic machinery very quickly in response to variations in the microenvironment. Furthermore, RNA synthesis requires a high expenditure of energy (ATP equivalents); thus, if the cell transcribed RNA from genes whose expression was not required and regulation was at the post-transcriptional level, a large, but unnecessary expenditure of energy would have occurred. Several types of mechanisms that regulate gene expression at the transcriptional level in procaryotes have been identified and examples will be discussed next.

21.2.1   EFFICIENCY OF PROMOTER-RNA POLYMERASE INTERACTION

The simplest regulatory mechanism for transcription involves the inherent productive interaction between the RNA polymerase holoenzyme with a promoter. Depending on the promoter sequence, the RNA polymerase would have either a relatively strong or weak interaction. Since no two promoters with identical sequences have been found to date, the variation in promoter sequences results in a range of very strong to very weak promoter-RNA polymerase interactions and a relative degree of expression of genes. Most constitutive genes are expressed at rates that are directly correlated to their promoter strengths. A good example is the *lac* repressor gene in *E. coli*. The promoter for the *lac* repressor gene interacts poorly with *E. coli* RNA polymerase and this gene is transcribed only a few times per cell generation. This limited level of promoter utilization results in the production of about 10 to 20 *lac* repressor molecules per cell. This low level of constitutive gene expression is sufficient, however, to cause repression of the *lac* operon. Genes whose expression is controlled by this simple mechanism generally have low and constant rates of expression.

21.2.2   NEGATIVE AND POSITIVE CONTROL ELEMENTS FOR TRANSCRIPTION

More complex transcriptional control systems involve mechanisms that determine whether RNA polymerase can have a productive interaction with promoters. These mechanisms are reversible and can cause **repression** or **derepression** of genes. If RNA polymerase is prevented from interacting with the promoter and transcription is inhibited, this results in repression of the gene. When the conditions allow RNA polymerase to interact with the promoter, then transcription occurs, resulting in derepression of the gene. There are generally two types of mechanisms for repression and derepression, one classified as a negative and the other as a positive type control of gene expression (Figure 21.1).

For **negative control** of gene expression, a mechanism exists that prevents the RNA polymerase from having a productive interaction with the promoter. The usual case involves an **operator** sequence that overlaps or is very close to a promoter sequence. The operator is a binding site for a **repressor** molecule. A repressor is a protein that binds to a specific operator site and sterically inhibits the binding of the RNA polymerase to the promoter. In the case of the *lac* and

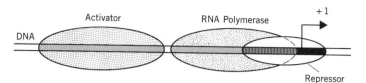

FIGURE 21.1   A schematic representation of a complex promoter region containing a promoter (color), operator (black), and an activator site (gray). The +1 indicates the transcription initiation site.

trp operons of *E. coli,* the operator overlaps the promoter. The binding of the repressor to the operator depends on the presence or absence of small metabolites that can interact with the repressor.

Repressors are allosteric proteins that have an active site capable of binding to specific DNA sequences (operators) and an allosteric (Section 4.12) site that recognizes small regulatory metabolites. Depending on the gene or operon, a small metabolite interacting with the repressor may facilitate either its binding to or release from the operator.

In positive control mechanisms, the action of an activator protein is required for the productive interaction of the RNA polymerase with the promoter. An activator protein binds to a region of DNA (an activator site) adjacent to and usually on the 5′ side of the promoter site. The physical attachment of the activator to the DNA facilitates the binding of RNA polymerase to the promoter site. In the absence of the activator, the RNA polymerase may not bind at all to the promoter. Several mechanisms for the role of the activator protein have been proposed: (1) the binding of the activator molecule to DNA changes the configuration of the DNA in the promoter region from an inaccessible to a more accessible form that can be recognized and bound by the RNA polymerase; (2) the DNA-bound activator touches the RNA polymerase, changes its conformation, and allows the modified RNA polymerase to bind to the promoter and initiate transcription; and (3) a combination of the two mechanisms just described. There are specific promoters that are not utilized unless an activator protein molecule is present. Usually, these promoters have fairly typical −10 regions, but do not have typical or consensuslike −35 regions (see Section 19.13).

The binding of the activator protein to the activator site is also dependent on the presence or absence of small regulatory metabolites. Thus, both negative and positive control mechanisms ultimately respond to small molecules of the cellular milieu.

21.2.3   THE *E. coli lac* OPERON

A classic example, which demonstrates both the negative and positive types of control mechanisms, is illustrated with the *E. coli lac* operon. This operon codes for a complex promoter region and three structural genes whose protein products are involved in the metabolism of lactose (Figure 21.2). The *z* gene codes for *β*-galactosidase, the *y* gene for a lactose permease that actively transports lactose into the cell, and the *x* gene for a transacetylase. In addition, a regulatory gene coding for the *lac* repressor is located in a separate locus from the *lac* operon.

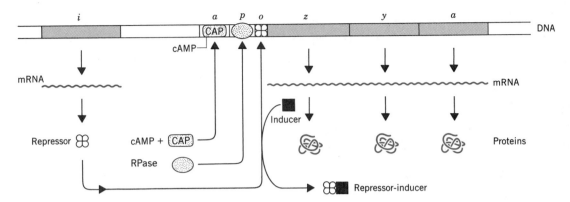

FIGURE 21.2 A schematic representation of the *E. coli lac* operon. The *i* gene codes for the *lac* repressor, *z* for *β*-galactosidase, *y* for *lac* permease, and *a* for transacetylase. The *a* represents the cAMP-CAP binding site, *p* the promoter, and *o* the operator. RNA polymerase (RPase) can bind to the promoter if a *lac* inducer interacts with the repressor and releases the repressor from the operator and cAMP-CAP binds to the activator site.

When *E. coli* is grown in a medium containing glucose and lactose, the *lac* genes are repressed, since there is preferential utilization of glucose over lactose and other carbohydrates. The *lac* genes are derepressed only after glucose has been depleted from the medium. The analysis of this phenomenon by Monod and Jacob and their co-workers laid the basis for understanding many of the genetic control mechanisms as they are known today. Investigators have found that the expression of the *lac* operon was controlled by both a negative repression system and a positive activator system. We will first discuss the negative repression system.

Extensive investigation of this phenomenon demonstrated that the *lac* operon was negatively controlled by a **repressor** that was able to bind to a specific locus on the DNA called an **operator** site. The *lac* operator site is a 35-bp long region of dyad symmetry (the sequence has a 180° rotational symmetry) that overlaps the promoter for the *lac* operon (Figure 21.3). The repressor when attached to the operator sterically interferes with the binding of the RNA polymerase to the promoter and therefore prevents transcription of the operon. The *lac* repressor is a tetramer composed of four identical subunits (MW 38,600) has a molecular weight of 154,400, and is an allosteric protein. The protein has an active site that is capable of binding the DNA operator site and also four allosteric sites that are capable of binding four molecules of *lac* inducer. In the absence of the inducer, the repressor can bind to the operator and

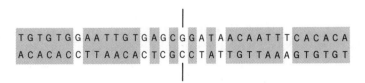

FIGURE 21.3 The *lac* operator sequence. Note the 180° rotational symmetry around the GC pair in the center. This sequence overlaps the *lac* promoter.

prevent transcription. When an inducer binds to the allosteric site, the protein conformation changes and the repressor is no longer capable of binding tightly to the operator.

Although the presence of an inducer is necessary for transcription of the *lac* operon, its presence is not sufficient, since the *lac* operon also requires a positive control element whose presence is essential for transcription. Thus, this operon is controlled by both positive and negative control elements.

The positive control element is composed of two parts: cAMP and an allosteric protein called CAP (catabolite gene-activating protein) that is capable of binding cAMP to form a cAMP-CAP complex. When CAP is not associated with cAMP, it cannot bind to a CAP-binding site on DNA located upstream from and adjacent to the promoter. When CAP is associated with cAMP, it can bind to the CAP-binding site. The presence of the cAMP-CAP complex on the CAP site facilitates the binding of RNA polymerase to the *lac* promoter and results in the transcription of the *lac* operon (Figure 21.2). The full expression from the *lac* promoter therefore requires the presence of a *lac* inducer that causes the repressor to be released from the operator and cAMP that allows the CAP protein to bind to the CAP-binding site.

The formation of cAMP that is the key to this complex mechanism of positive regulation occurs when the glucose level in the medium is depleted and an **adenylate cyclase** is activated. This enzyme catalyzes the formation of cAMP from ATP and the presence of high levels of cAMP results in the formation of the cAMP-CAP complex. The repression of the *lac* and other sugar utilization operons by the presence of glucose is called **catabolite repression,** and this phenomenon allows the cell to use glucose preferentially over many carbon sources, such as lactose, arabinose, and galactose.

21.2.4 THE *E. coli* ARABINOSE OPERON

The regulation of the arabinose, or *ara,* operon is also controlled by a positive and negative regulatory mechanism (Figure 21.4). The promoter for this operon (*araBAD*) is complex in that it contains a binding site for RNA polymerase, cAMP-CAP (see Section 21.2.3), and a *ara*-specific positive regulatory protein. The interesting differences between this system and the *lac* operon are that a single regulatory protein (the product of *araC*) can act either as a repressor or an activator, and two positive regulatory proteins (cAMP-CAP and P2) are required for the expression of the *ara* operon. The regulatory protein for this operon can exist in two states. In the absence of arabinose, it acts as a repressor (P1) and binds to the operator site of its own gene (*araC*) in an autoregulatory fashion. In the presence of arabinose, the regulatory protein acts as a positive effector (P2) and binds to an activator site located between the CAP site and the *araBAD* promoter and facilitates binding of RNA polymerase to the *araBAD* promoter. Thus, two positive regulatory proteins, P2 and cAMP-CAP, are required to express the *araBAD* operon. The CAP site that lies between the divergent promoters of the *araBAD* operon and the *araC* gene appears to control the expression of both the operon and the gene.

21.2.5 THE *E. coli* TRYPTOPHAN OPERON

Repressors also regulate many amino acid biosynthetic pathways including the tryptophan operon. When excess tryptophan is present in the growth medium,

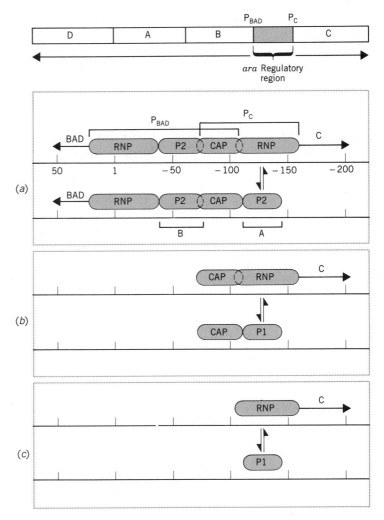

FIGURE 21.4   Schematic representation of the control region of the *ara* operon. The *araC* codes for the repressor (P1) that controls *araC* expression autogenously (*c*); CAP (cAMP-CAP) is required to express *araC* (*b*) and *araBAD* (*a*). In the presence of arabinose, PI is converted to P2 (a positive activator) that is required for the expression of *araBAD* (*a*). Thus, both P2 and CAP are required for the expression of *araBAD* (*a*). P_{BAD} and P_C represent the promoters for the operon BAD and C gene, respectively. Reprinted with permission from Nancy L. Lee, W. O. Gielow, and R. G. Wallace, *Proceedings of the National Academy of Science* 78:752–756 (1981). Copyright © 1981 by the National Academy of Sciences.

the *trp* operon is repressed. In contrast to the situation in the *lac* operon, the *trp* repressor binds to its operator when it is complexed to the small effector molecule—that is, tryptophan. When the intracellular concentration of tryptophan falls, the tryptophan dissociates from the repressor and the free repressor is released from the operator, resulting in the derepression of the *trp* operon. In the *trp* operon, the operator overlaps the *trp* promoter and the *trp* repressor blocks RNA polymerase binding by steric hindrance.

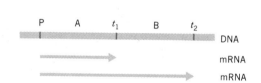

FIGURE 21.5 Schematic representation of an antitermination mechanism. Transcription is initiated at promoter (P) and transcribes gene A and terminates at t_1. However, if antitermination factors are present, the RNA polymerase is able to transcribe both genes A and B and terminates at t_2.

21.2.6 ANTITERMINATION

A novel mode of transcriptional control is demonstrated by the antitermination mechanism (Figure 21.5). Some operons contain a transcription termination site between two adjacent genes. When RNA polymerase arrives at this site, transcription is terminated by the termination factors including ρ, and *nusA* and *nusB* proteins (see Section 19.15). Thus, the gene distal to this site cannot be transcribed. However, if antitermination factors are present, then the RNA polymerase is able to bypass the intercistronic termination site and continue transcription through the distal gene. These antitermination factors include *nusA* and *nusB* proteins that appear to have roles in both termination and antitermination.

21.2.7 MULTIPLE RNA POLYMERASE σ-FACTORS

Another mode of transcriptional control involves the presence of multiple RNA polymerase holoenzymes. As discussed in Section 19.13, in procaryotes the different RNA polymerase holoenzymes are distinguished by the presence of different σ-factors that associate with the core enzyme. These σ-factors can be considered positive regulatory elements whose presence is required for the transcription of different families of genes. Current evidence suggests that the various σ-factors control the expression of stress-related genes such as those for heat shock, nutritional deprivation, irradiation damage, dehydration, and sporulation.

21.3 TRANSCRIPTIONAL CONTROL IN EUCARYOTES

Transcriptional control in eucaryotes is less well defined when compared to the information available for procaryotic control systems. Generally speaking, regulatory mechanisms involve multiple RNA polymerases, nuclear regulatory factors that control RNA polymerase specificity, and hormonal regulation of gene expression. The specificity of the multiple forms of eucaryotic RNA polymerases has been described in Chapter 19. The specificity to transcribe thousands of specific genes by RNA polymerase II is controlled by nuclear proteins that are either DNA-binding proteins or proteins that interact with and/or modify RNA polymerase II. These nuclear factors are somewhat analogous to the multiple σ-factors found in procaryotic cells. These nuclear factors, in turn, may be controlled by hormonal signals to the cell.

21.4   COUPLED TRANSCRIPTION–TRANSLATION CONTROL: ATTENUATION

A mechanism called **attenuation** involving coupled transcription–translation regulation has been described for the *E. coli* tryptophan operon by Yanofsky (Figure 21.6). The *trp* operon has a repressor system that measures the amount of free tryptophan available to the cell, but in addition, has an attenuation system that is sensitive to the level of *trp*-tRNA present in the cell. In the presence of excess tryptophan in the medium, the *trp*-repressor complex binds to the *trp* operator site and blocks the transcription by RNA polymerase to about one-seventieth of the derepressed level. However, some RNA polymerase molecules escape repression and continue transcribing the operon and come to an **attenuation site** that determines whether those molecules will be terminated or allowed to continue transcription of the operon.

This attenuation site is located in a **leader region** that lies between the promoter and the sequence of bases for the first structural *trp* gene, *trpE*. If termination occurs, a short **leader RNA** is synthesized; if termination does not occur, then a long *trp* mRNA is formed. The passage of the RNA polymerase through the attenuation region is determined by alternate secondary structures that are formed in the leader transcript and the level of *trp*-tRNA present in the cell (Figure 21.7).

The leader RNA is able to assume two alternative forms by intramolecular hydrogen bonding between complementary bases that result in the formation of stem and loop regions. The first form has two stem and loop regions consisting of hydrogen bonding between sequences 1 and 2 (stem 1 : 2) and sequences 3 and 4 (stem 3 : 4). The second form consists of a single stem and loop between sequences 2 and 3 (stem 2 : 3) with sequences 1 and 4 not participating in any stem formation. When stem 3 : 4 can form, the RNA polymerase is prevented from continuing transcription and the expression of the operon is reduced by a further sevenfold. If stem 2 : 3 can form, then transcription is allowed to proceed. While these stem structures are forming in the leader region of mRNA, the RNA polymerase pauses at a site slightly distal to these sequences.

What determines which complementary structures are formed? In the leader RNA, upstream of the sequences involved in stem formation, there occurs a ribosomal binding site followed by a small **open reading frame** (a sequence of bases that can be translated into an amino acid sequence and that does not contain a termination codon) containing the initiation codon, AUG, followed by 13 other codons including two tandem *trp* codons, UUGUUG, and a translation termination codon, UGA. Thus, information exists in the leader RNA for the synthesis of a small **leader peptide**. The formation of the alternative stem structures depends on whether *trp*-tRNA is available to translate the tandem *trp* codons in the leader mRNA. The ribosome will translate the newly transcribed leader RNA and synthesize the complete leader peptide. During this synthesis the ribosome will mask stems 1 and 2 of the RNA and prevent the formation of stem and loop 1–2 or 2–3. Stem and loop 3–4 will form and signal the RNA polymerase to terminate transcription. If there is insufficient *trp*-tRNA, then the ribosome complex translating the leader RNA pauses at the *trp* codons, prevents the formation of the 1 : 2 and 3 : 4 stems, but allows the formation of a 2 : 3 stem that is the signal for continuation of transcription. However, if suffi-

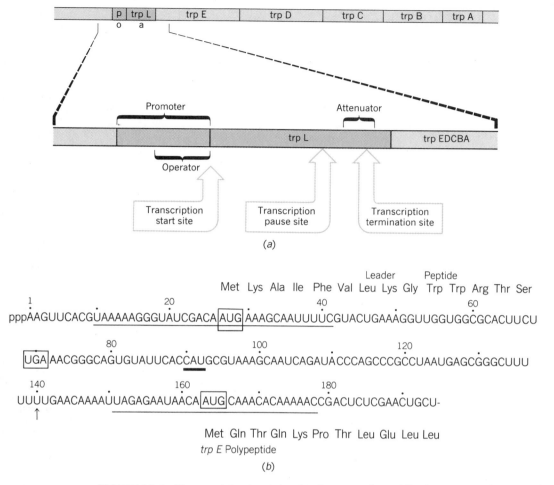

FIGURE 21.6 The regulatory and structural gene regions of the *trp* operon of *E. coli*. (*a*) Transcription initiation is controlled at a promoter–operator. Transcription termination is regulated at an attenuator, in the transcribed 162-base pair leader region, *trpL*. All RNA polymerase molecules transcribing the operon pause at the transcription pause site before proceeding further. (*b*) The nucleotide sequence of the 5′ end of *trp* messenger RNA. The nonterminated transcript is present. When transcription is terminated at the attenuator, a 140-nucleotide transcript is produced. Its 3′ terminus is marked by an arrow. The 3′ terminus of the pause transcript, at nucleotide 90, is underlined by a bar. The two AUG-centered ribosome binding sites in this transcript segment are underlined. The boxed AUGs are where translation starts and the boxed UGA where it stops. The predicted amino acid sequence of the *trp* leader peptide is shown. Reprinted with permission from *Nature* 289:751–758 (1981). Copyright © 1981 Macmillan Journals Limited.

cient *trp*-tRNA is available to translate the tandem *trp* codons, then the ribosome complex does not pause at the tandem *trp* codons and allows the formation of the alternate 1:2 and 3:4 stems, and the alternate 3:4 stems signals the paused RNA polymerase to terminate transcription at the **attenuator** site. The attenuator site resembles the terminator site and may involve the termination factors mentioned previously in Sections 19.15 and 21.2.6. Thus, attenuation regulates the expression of the *trp* operon at a finer level than repression.

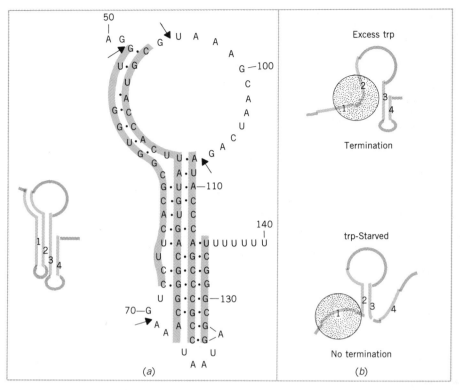

FIGURE 21.7   The proposed secondary structure of *E. coli* terminated *trp* leader RNA and its function in attenuation. (*a*) Four regions can base pair to form three stem and loop structures. The diagram shows the four stem-forming segments of the RNA chain, numbered 1 through 4. (*b*) The model for attenuation in the *E. coli trp* operon showing the alternating secondary structure of the leader RNA during binding and movement of a ribosome (shaded circle). Reprinted with permission from Nature 289:751–758 (1981). Copyright © 1981 Macmillan Journals Limited.

21.5   TRANSLATIONAL REGULATION

Several mechanisms have been observed for translational regulation at the mRNA level. These mechanisms determine the rate of initiation and elongation of protein synthesis. In a polycistronic mRNA, the rates of synthesis from the different cistrons can also be regulated, such that much more protein is made from one cistron than from another.

21.5.1   BINDING OF mRNA

A general type of control at this level includes the relative efficiency of interaction between the 30*S* ribosome and the mRNA (Section 20.9.1). This interaction involves the 5′ sequences in mRNA that are complementary to and neighbors to the SD sequence found at the 3′ end of 16*S* rRNA. Certain sequences favor the strong interaction of 30*S* ribosomes to mRNA and others induce weak interactions. Thus, an integral 5′ region of the mRNA determines how efficiently it will be translated. The complementary SD sequences play a role in this interaction, but they are not sufficient to explain completely the relative rates of initiation of protein synthesis. Thus, the secondary structure of the 5′ end of the mRNA and the initiation factors for protein synthesis also play roles in the efficient recognition of mRNA by the 30S ribosome.

21.5.2 ABUNDANT AND SCARCE CODONS

Another inherent property of mRNAs that regulates the efficiency of translation is the frequency of use of codons for which there are low concentrations of tRNA. Studies on mRNA populations have shown that certain codons are used more frequently than others and that the tRNAs for these codons are more plentiful. On the other hand, rarely used codons are also reflected in the small amount of tRNAs available to translate these codons. Thus, if rarely used codons are present frequently in a given mRNA, the rate of translation will be affected by the limiting amounts of tRNA available for the translation of these rare codons. This type of mechanism has been used to synthesize much more protein from one cistron (*rpoD*) than from another cistron (*dnaG*) in the polycistronic mRNA of the σ-70 operon of *E. coli*. The *dnaG* cistron that codes for DNA primase contains six rarely used codons that reduce its rate of translation significantly compared to that for the *rpoD* cistron that contains only frequently used codons. Thus, the cell contains much more σ-70 than DNA primase.

21.5.3 TRANSLATIONAL COUPLING

In polycistronic mRNAs, a phenomenon called **translational coupling** can regulate the expression of some distal cistrons. This occurs when the translation termination codon of the proximal cistron overlaps the translation initiator region of the distal cistron. In some polycistronic mRNA, the distal cistron may have its translation initiation region masked by a stem-and-loop structure that prevents access of the 30S ribosome to the SD sequence of the mRNA and/or the initiation codon AUG (Figure 21.8). Thus, the translation of this distal

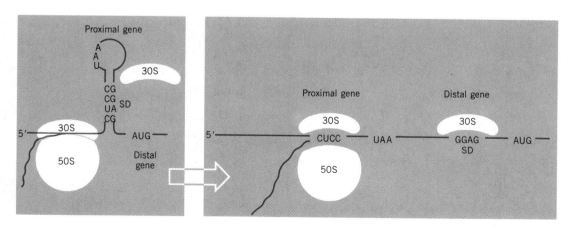

FIGURE 21.8 A schematic representation of translational coupling. The Shine–Delgarno (SD) sequence of the distal gene is masked in a stem region and is inaccessible to the 30S ribosome until a 70S ribosome translating the proximal gene "melts" the stem and makes the SD sequence available for interaction with the 30S ribosome for initiation of translation of the distal gene. The UAA is the translational stop signal for the proximal gene. AUG is the initiation codon for the distal gene.

cistron is dependent on passage of the ribosome complex from the adjacent upstream cistron into this stem-and-loop area. The passage of the ribosome "opens up" the stem region that is masking the initiation signals and allows translation to be initiated. Translational coupling also allows efficient translation of the distal cistron, since the ribosomes that have just finished translating the proximal cistron can immediately reinitiate translation at the ribosome binding site of the distal cistron.

21.5.4   COMPLEMENTARY OR ANTI-SENSE RNA

Translation can also be controlled by the synthesis of small RNA molecules that are complementary to the initiating region of a specific mRNA. These small micRNAs (mRNA inhibiting complementary RNA) hybridize to the Shine–Dalgarno region of the mRNA and prevent binding of 30S ribosomes to this region (Figure 21.9). This effectively blocks the translation process. The relative effectiveness of this type of regulation will depend on the degree of complementarity of the small RNA to the Shine–Dalgarno sequence and the number of small RNA molecules present in the cell relative to the number of mRNA.

21.5.5   TRANSLATIONAL CONTROL BY TRANSLATION PRODUCT

Translational regulation can also occur by an autogenous control mechanism that is best illustrated by the ribosomal protein operons. The synthesis of the 56 ribosomal proteins of *E. coli* has to be coordinated in order to form the 30S and 50S subunits. As the 21 small ribosome proteins are made, they interact with the 16S rRNA to form the 30S ribosome. The 50S ribosome is assembled from 23S and 5S rRNA and 35 large ribosome proteins. The cell, in part, coordinates ribosomal protein synthesis at the level of translation by an ingenious protein-mRNA interaction mechanism that can effectively block translation of the ribosomal protein mRNAs. The protein that binds near the 5′ end of the mRNA to block translation is a ribosomal protein encoded in the mRNA itself.

The ribosomal protein genes are encoded in a number of operons. Certain of the ribosomal proteins are able to interact with the naked 16S or 23S rRNAs to form a core rRNA-protein complex during ribosome synthesis. These proteins are usually coded near the 5′ end of the ribosomal protein operons. During ribosome synthesis, the newly made ribosomal proteins will interact with either the 16S or 23S rRNAs to initiate ribosome synthesis. However, if there is insufficient 16S or 23S rRNA available, the translation of the ribosomal protein mRNA is turned off by the interaction of the newly synthesized ribosomal protein to a 5′ region of the polycistronic mRNA. This 5′ region of the mRNA has a sequence and secondary structure very similar to that found in the 16S

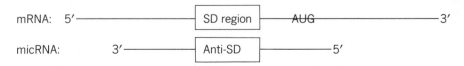

FIGURE 21.9   Regulation of mRNA translation by micRNA. The micRNA is complementary to the SD region of the mRNA and forms a double-stranded RNA hybrid in this region. This prevents the interaction of the SD region and the 30S ribosome and blocks the initiation of translation. AUG is the initiation codon for the gene.

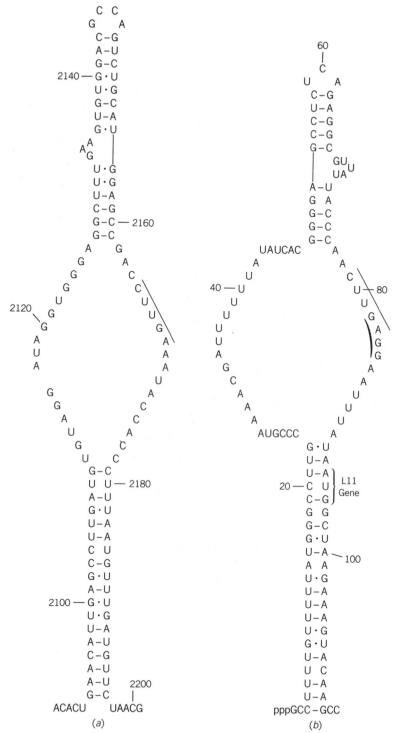

FIGURE 21.10 Regulation at the translational level. Models of secondary structures of L1 ribosomal protein binding sites on 23S rRNA (*a*) and on mRNA coding for L11 (*b*). The binding site on the mRNA is near the 5′ end. When L1 binds to this site, it blocks further translation of the mRNA and the synthesis of L11 and L1. The presumptive SD sequences GAGG and the initiation codon AUG are indicated as well as the common CUUGA sequence in both loops. Reprinted with permission from G. Baughman and M. Nomura, *Proceedings of the National Academy of Science USA* 81:5389–5393 (1984). Copyright © 1984 by National Academy of Sciences.

and 23S rRNAs (Figure 21.10). Thus, the newly made small and large ribosomal proteins will bind to a 16S rRNA-like region and a 23S rRNA-like region in their respective mRNAs if the supply of rRNAs is insufficient. If both rRNAs and mRNAs are present, the ribosomal proteins will preferentially bind to the rRNAs to form ribosomes. This autogenous control systems prevents the overproduction of ribosomal proteins when the level of rRNA synthesis has been reduced.

Thus, several transcriptional and translational mechanisms have evolved to regulate gene expression in procaryotes. It is likely that a number of additional mechanisms will be uncovered in procaryotes and particularly in eucaryotes.

REFERENCES

1. *O. Ribaud and M. Schwartz, Positive control of transcription initiation in bacteria.* Annual Review of Genetics, *Vol. 18 (1985), p. 173.*

2. *S. Gottesman, Bacterial regulation: global regulatory networks.* Annual Review of Genetics, *Vol. 18 (1985), p. 415.*

REVIEW PROBLEMS

1. What are the differences between
   (a) Positive and negative control mechanisms in bacteria
   (b) Transcriptional and translational regulation
   (c) Repression in the *lac* and *trp* operons of *E. coli*
   (d) Repression and attenuation

2. How are the following phenomena used in regulating gene expression:
   (a) Alternate secondary structures
   (b) Translational coupling
   (c) Stem and loop structures
   (d) Autogenous regulation
   (e) Catabolite repression

GENE REARRANGEMENTS AND NONCELLULAR ORGANISMS

As we have emphasized in Chapter 6 and in Part III, stability is a virtue of double-stranded DNA. As the repository of genetic information, DNA must maintain its sequence of nucleotide residues with great fidelity over time and as it is replicated. The chemical stability of DNA and its double-stranded structure are critical to DNA stability, as we indicated in Chapter 19, especially in Section 19.11. There are special situations, however, both natural and experimental, in which DNA is made unusually variable. The actions of certain enzymes allow the cell, or the genetic engineer, to rearrange DNA sequences. Natural rearrangements have important consequences for the survival of the organism; experimental rearrangements have revealed a great deal of what we know about gene expression, and they form the basis of many aspects of genetic engineering.

We take, as one very important example of naturally occurring nucleotide sequence rearrangement, the production of immunoglobulins in higher organisms. We then discuss the replication of bacterial and animal viruses and, finally, recombinant DNA methodologies.

22.1 BIOSYNTHESIS OF IMMUNOGLOBULINS

The ability to recognize, on the molecular level, foreign substances and to synthesize proteins that can bind to foreign molecules is an important adaptation of vertebrates and invertebrates. The consequences of this ability include acquired resistance to pathogens, graft rejection, and autoimmune diseases. The phenomena require complex interactions between molecules and between cells, most of which are beyond the scope of this text. We focus on just one aspect of the production of one type of circulating antibody molecule, the immunoglobulin G molecule already introduced in Section 3.14.4. Antibody molecules have received wide application as reagents for the qualitative and quantitative analysis of both high molecular weight and low molecular weight substances. We begin with some definitions.

651

Immunogen   A molecule that elicits antibody production by the immune system. The immunogen is usually a protein or glycoprotein or polysaccharide, but some other macromolecules are also immunogenic.

Antibody   Any of several kinds of protein (immunoglobulin) molecules produced by the immune system in response to the immunogen that have the capability of recognizing some chemical characteristic of the immunogen molecule

Antigen   A molecule with which an antibody molecule will react specifically. The immunogen that elicits a given antibody and the corresponding antigen may have the same structure; they are at least structurally related.

Hapten   A small molecule that corresponds to part of the structure of an antigen and that can react specifically with the corresponding antibody molecule — also, an isolated antigenic determinant.

Serum   Liquid portion of blood that remains after clotting has occurred.

Several types of antibody (immunoglobulin) molecules are produced after inoculation of an immunogen. Only two of these, IgM and IgG, are present in high concentrations in serum. IgM is seen in the early stages of the immune reaction, whereas IgG predominates later and is the principal immunoglobulin used as a reagent. The production of IgG that will be useful as a reagent for a small molecule, such as a peptide hormone or a nucleotide derivative, requires treating the small molecule as a hapten. In order to obtain antibodies against small molecules, which themselves are usually not immunogenic, the hapten is coupled to a protein to form the immunogen.

The antibodies formed upon inoculation of an experimental animal, such as a rabbit, with either a natural or synthesized immunogen, are polyclonal. That is, the antibodies are a mixture of several molecules that recognize different antigenic determinants or haptens. The fact that different IgG molecules in a polyclonal population have different specificities was discovered by showing differential reactivities and binding constants of IgG molecules to different haptens. These and other experiments also revealed that a single organism can produce more than a million different kinds of IgG molecules, each specific for a different hapten.

As is indicated in Figure 3.19, the IgG molecule is composed of four polypeptide chains linked by noncovalent associations and by disulfide bonds. The four chains are two identical *light* chains of about 215 amino acid residues and two identical *heavy* chains of about 250 amino acid residues. It is by the pairwise association of amino-terminal domains of light and heavy chains that the antigen-binding site is formed. This observation reveals a part of the secret of how a single organism can produce so many different IgG molecules. Because the antigen-combining site is derived from portions of a heavy chain and a light chain, the possible combinations of 1000 different heavy chains and 1000 different light chains will give a million different structures of IgG molecules.

The antibody population is not only diverse, it is also limited, for the most part, to reacting only with foreign antigens and not with the many antigens of the organism that is producing the antibodies. How can this amazing pattern of specificity be explained? Is the ability to make a particular antibody preexisting in the animal, or is it developed in response to the immunogen? The cells that produce IgG molecules form a class of white blood cells designated as the **B lymphocytes,** because of their origin in the bone marrow. The study of these cells, as well as the genetics and other aspects of the immune system, has revealed that the ability to produce a particular antibody molecule with a particular specificity — and the gene for that antibody molecule — exists before the organism is exposed to the corresponding immunogen. To encode all of these different types of antibody molecules each in its entirety would require a substantial fraction of the genome. It is through nucleotide sequence rearrangements that the cell is able to generate its IgG molecules.

Each B lymphocyte is devoted to the synthesis of an antibody molecule with a single antigenic specificity. This means that each B lymphocyte in a quiescent population of B lymphocytes is capable of producing one kind of light chain and one kind of heavy chain. The recognition and response to a given immunogen by the organism is a very complex affair. The result is proliferation of those formerly quiescent B lymphocytes that are capable of producing antibodies that correspond to the immunogen. The increased population of B lymphocytes synthesizes the required antibodies in large amounts. Thus, the basis of diversity of the antibody population resides in the population of B lymphocytes.

A very important modern technology to come out of the study of B lymphocytes is that of **monoclonal antibody** production. Although B lymphocytes are generally unable to proliferate outside of the organism, the fusion of a B lymphocyte with certain other cell types produces a new cell that can divide in culture and produce a single type of antibody in very large quantities. This technology has been exploited extensively for the production of analytical reagents — that is, monospecific antibodies — and for the study of the immune system.

A given B lymphocyte develops the ability to produce a given IgG without a large investment of genetic information in the chromosomal DNA of the organism. Each B lymphocyte has an arrangement of antibody-coding DNA sequences that is not present in the germ cells or other somatic cells of the organism. These sequences have undergone **somatic recombination** to generate the gene for the IgG polypeptide chains from segments of DNA that were separated, but on the same chromosome, in the germ line cells. Splicing of a precursor mRNA generates the mature mRNA. We consider these processes here only for one particular light chain, the κ light chain. Other light chains of mice and humans are designated as members of the λ family of light chains.

The nucleotide sequence segments that encode the 215 amino acid residues of a κ light chain fall into three groups, that are designated V, an abbreviation for variable, J, for joining, and C, for constant. This order of segments on the chromosome corresponds to the order of encoded amino acid residues from the amino terminus to the carboxyl terminus of the polypeptide chain. The assembly of the final sequence into the light chain mRNA begins with somatic recombination that joins any one of the approximately 300 V segments (as represented in the top line of Figure 22.1) to one of a few, much shorter J segments. A J segment encodes only about 13 amino acid residues. The V and J segments together encode what becomes the amino terminal region (Figure 3.19) of the light chain. The J segments are adjacent to the C segment in the chromosomal DNA.

A transcript of the somatically recombined DNA begins just upstream of the V segment in a leader region that has been omitted from Figure 22.1 for

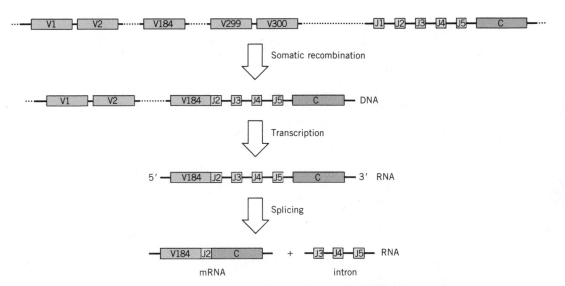

FIGURE 22.1   Formation of the messenger RNA for a κ type IgG light chain. The nucleotide sequences of the mRNA are derived from sequences in DNA segments that are widely
separated in the chromosomal DNA of germ line cells, as illustrated in the first line of the
diagram.

simplicity. It extends through the J segments and terminates after the C segment, producing a precursor mRNA. RNA splicing (Section 19.18) removes an
intron that was encoded by the remaining J segments, giving the mature mRNA
with a V-J-C nucleotide sequence. In the example given in Figure 22.1, the
specific segments of the mature mRNA are from V segment 184, J segment 2
and the single C segment of the κ light chain family. Clearly the approximately
300 V segments and 4 to 5 active J segments of a κ family can give rise to more
than 1000 different amino terminal regions of the light chain. Without somatic
recombination and RNA splicing, such diversity could have been achieved only
with a considerably larger investment of nucleotide sequences in the germ line
and other somatic cells.

The genes for the kappa and lambda light chains and for the heavy chains
reside on different chromosomes. Four kinds of DNA segments contribute to
the structure of heavy-chain mRNA allowing, after somatic recombination and
RNA splicing, a very large population of distinct heavy chains in the B lymphocytes. The separately synthesized heavy and light chains combine — one combination in each B lymphocyte — to give the mature IgG molecule. This scheme
allows for the production of more than a million kinds of IgG molecules in a
mouse or human.

However, the diversity is even greater than is indicated above because of two
other processes: variable V–J recombination and somatic mutagenesis. The
site at which a V segment and a J segment may combine is not fixed but can vary
over a few residues in the sequence. This can alter the particular codons specified at the V–J nucleotide sequence junction. In addition, the register (Section
20.6) of the codons can be changed. Some of these register changes, combined
with proper J–C joining will not introduce early terminator codons in the final
mRNA but will allow significant amino acid sequence variation in the J segment region of the final IgG polypeptide. Somatic mutations in the V region of
B lymphocytes or B lymphocyte progenitors also contribute to antibody diversity.

22.2 NONCELLULAR ORGANISMS: VIRUSES

Viruses are relatively simple nucleoprotein particles that contain their genetic information in either RNA or DNA and can replicate only by infecting living cells. Viruses can infect plants, animals, and bacteria and each virus has either a specific host or a limited host range. Those viruses that infect bacteria are called **bacteriophage** or **phage**. The genetic information found in the viral chromosome may code for as little as three proteins or as many as 100 proteins. Therefore, viruses can be considered very simple or reasonably complex structures depending on their genetic content. However, even the most complex viruses are very simple compared to cellular forms and are completely dependent on the metabolism and synthetic machineries of the host cell for viral nucleic acid replication and virus-specific protein synthesis.

Depending on the virus, the infection can be highly virulent resulting in the rapid replication of the virus and lysis and death of the host cell. In other cases, the infecting virus may invade the host cell and only replicate periodically and kill just a few of the infected cells. Some viruses can integrate into the chromosome of the host cell and alternate between a lytic or lysogenic cycle. Other viruses can infect the host and remain dormant for a long period (20 years or more) without any manifestations of a viral infection. Thus, the regulation of viral gene expression and replication can be extremely complex.

The relatively simple genetic composition of viruses has made them a favorite research organism for studying DNA and RNA replication, regulation of gene expression, and development. More recently, they have played a significant part as tools for recombinant DNA research, particularly as useful cloning vectors. Furthermore, since they cause diseases of plants and animals, they play a significant and direct role in human affairs. The properties of several viruses will be discussed and their significance examined in terms of basic concepts that have been derived from their study. Also, the usefulness of several viruses in genetic engineering and recombinant DNA studies will be illustrated.

22.2.1 BACTERIOPHAGES THAT INFECT *E. coli*

We will discuss three types of *E. coli* phages. Although these three phages illustrate a diversity among coliphages, they demonstrate only partially the total diversity of phages that infect this single bacterial species. *E. coli* phages include small single-stranded RNA and DNA phages, various sized double-stranded DNA phages, and temperate and virulent phages. A **virulent** phage infects a host cell, replicates and destroys the host cell in less than a generation time of the host, and may produce hundreds of phage progeny from a single host cell. A **temperate** phage can allow the joint multiplication of the phage and the host with only an occasional replication cycle resulting in phage progeny. An example of a temperate coliphage is λ, one of the most highly studied phages during the past 40 years. A typical virulent middle-sized phage is represented by T7 phage. λ and T7 contain double-stranded DNA genomes. M13 phage is a single-stranded DNA phage that has become very useful for the cloning of DNA.

22.2.2 BACTERIOPHAGE T7

Bacteriophage T7 is an example of a virulent phage. It is an icosahedral particle (a polyhedron having 20 faces) with a short cone-shaped tail and contains a

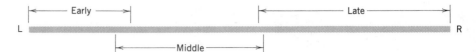

FIGURE 22.2   Schematic representation of the location of the T7 genes that are transcribed during early, middle, and late stages of phage infection.

linear double-stranded DNA genome of 40 kb and a molecular weight of 25.2×10^6. This is sufficient genetic information for about 30 proteins (Figure 22.2). The genes for T7 are arranged according to function with early genes on the left end involved in establishing the infection and regulating later gene expression, middle genes involved in DNA replication, and late genes involved in the synthesis of head and tail proteins of the phage particle.

T7 phage infects *E. coli* and the phage replication cycle starts immediately. The host-specific transcription and replication functions are impaired early in the infection and the phage functions begin to take over. Early phage gene transcription occurs within the first 4 to 8 minutes. The synthesis of phage DNA replication functions occurs from 6 to 16 minutes, and the synthesis of phage head and tail proteins occur from 7 to 30 minutes. Lysis occurs at about 25 to 30 minutes after infection was initiated.

The early genes are transcribed by the host RNA polymerase into a polycistronic mRNA that is processed by host RNase III into five monocistronic mRNAs. This is a relatively rare event in *E. coli,* since most polycistronic procarotic messages are not processed. One of the early phage genes codes for a T7 RNA polymerase that is used to transcribe the middle and late genes for phage replication. The T7 RNA polymerase is a monomeric protein with a molecular weight of 110,000, is capable of carrying out all the transcription functions and is much simpler in structure than the *E. coli* RNA polymerase (see Section 19.13). The promoters for the middle and late genes are different from the promoters of the early genes and can be recognized only by the T7-specified RNA polymerase. Thus, T7 RNA polymerase is a positive regulator of gene expression. Furthermore, transcription occurs from only one of the two complementary strands of the T7 DNA.

The replication of the linear T7 DNA occurs by bidirectional replication from a point about 17% from the left end of the T7 genome. Since the rate of replication from the origin of replication (*ori*) is the same in both directions, the replication of the left end is completed first, followed by the completion of the right side of the molecule.

The analysis of this virulent phage illustrated the presence of an ordered sequence of events that led to the development of mature phage particles. Other novel information obtained from the study of T7 includes the occurrence of a phage-specific RNA polymerase and phage-specific promoters, the processing of phage polycistronic mRNA to monocistronic mRNA by RNase III, and an example of asymmetric bidirectional DNA replication.

22.2.3   *E. coli* λ-PHAGE

The λ phage of *E. coli* has been studied more extensively than any other virus. This phage contains a linear double-stranded DNA genome with a molecular weight of 30.8×10^6 corresponding to 46.5 kb. The linear phage DNA has two cohesive ends that allow the phage DNA to circularize after it is injected into an *E. coli* cell. The cohesive ends consist of 12 bases that are complementary to each other.

GGGCGGCGACCT-----------------------
-----------------------CCCGCCGCTGGA

λ-Phage is the prime example of a temperate phage. A temperate phage after infection of the host follows two possible paths: (1) it can enter the virulent productive cycle, replicate, lyse the host cell, and release new phage progeny or (2) it can enter the latent prophage cycle by integration of the phage genome into a specific site on the host chromosome and replicate passively as part of the host chromosome.

In the prophage cycle, the linearly injected *λ*-DNA is circularized by its cohesive ends, and the circular molecule can be inserted or excised from the *E. coli* chromosome by the mechanism diagrammed in Figure 22.3.

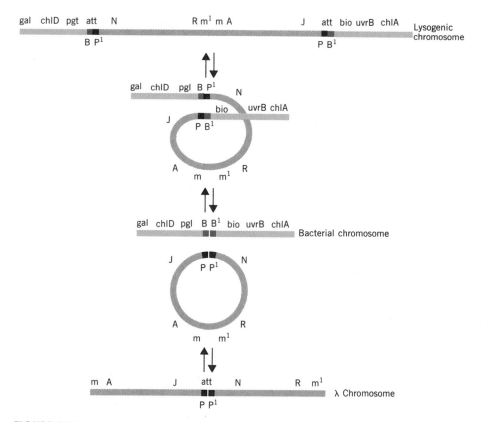

FIGURE 22.3 Insertion of *λ* into the bacterial chromosome. A, J, N, and R are phage genes. attP.P^1 and attB.B^1 are the phage and bacterial sites of insertional recombination. m and m^1 are the ends of the *λ*-DNA molecule. The bacterial genetic symbols are: gal, a cluster of three genes that determine enzymes of galactose catabolism; ch1D, ch1A, two genes that determine chlorate sensitivity; pg1, structural genes for phosphogluconolactonase; bio, a cluster of at least five genes that determine enzymes of biotin biosynthesis; uvr B, a gene that determines resistance to ultraviolet light. From A. M. Campbell, The Bacteriophage Lambda, Cold Spring Harbor Laboratory, Cold Spring Harbor, N.Y., 1971 (© Cold Spring Harbor Laboratory).

FIGURE 22.4   A schematic representation of the λ genetic map. Not all genes are represented nor are the distances precise. The genes in the left and right arms control DNA synthesis and code for structural components of the phage. The regulatory region determines whether the phage will integrate with the host chromosome or enter into the lytic cycle. From A. M. Campbell, *The Bacteriophage Lambda,* Cold Spring Harbor Laboratory, Cold Spring Harbor, N.Y., 1971 (© Cold Spring Harbor Laboratory).

The genes for the various functions are clustered in three parts of the λ-chromosome (Figure 22.4). The replication functions for the productive cycle of infection are located on the left and right arms, while the regulatory functions concerned with creating lysogeny (a lysogenic condition exists when a bacterium possesses and transmits to its progeny the power to produce phage particles and is therefore prone to lysis) are located in the center of the λ-chromosome between the left and right arms.

The genes in the left and right arms of λ code are involved in DNA replication and the production of head and tail proteins. Only these genes are necessary and sufficient for the replication of an active mature λ-phage particle. The genes in the central regulatory region are involved in the insertional recombination of the λ-DNA to the host chromosome, excision of the prophage DNA from the chromosome, and determination of whether the lysogenic or productive cycle will be activated. Since the genes in the regulatory region are not necessary for

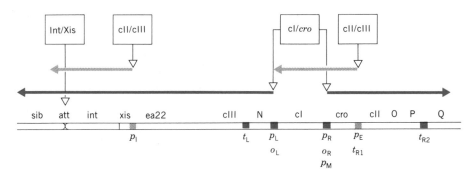

FIGURE 22.5   Regulation of transcription during the early stage of development. Immediately after infection, transcription from p_L and p_R is terminated mainly at t_L and t_{Rl}, with some RNA chains continuing to t_{R2}. The N protein eliminates these termination events and provides for transcription of the remainder of the early gene region (solid arrows). The cII and cIII proteins partition λ development toward the lysogenic response because cII stimulates RNA synthesis from p_E and p_I (hatched arrows) and delays the expression of late lytic functions; cIII stabilizes cII. The cI protein binds to the operators o_L and o_R, shutting off the early λ transcripts from p_L and p_R; cI also maintains its own further synthesis by regulating the p_M promoter for the cI gene. The alternative late stage of productive development utilizes the cro and Q proteins; Q turns on the genes for head, tail, and lysis proteins, and cro turns off transcription of the cI gene from p_M and reduces the synthesis of the p_L and p_R transcripts for early proteins. From H. Echols and G. Guarneros, *Lambda II,* Cold Spring Harbor Laboratory, Cold Spring Harbor, N.Y., 1983 (© Cold Spring Harbor Laboratory).

phage replication, they can be replaced by fragments of DNA that are "cloned" during phage replication (see Section 22.7.1). The complex regulatory region controlling lysogeny is diagrammed in Figure 22.5. In the lysogenic state, the production of repressor from the *cI* gene prevents transcription from the promoters P_L and P_R that control the transcription of the *N* gene and *Q* gene, respectively. Repression by the *cI* gene product (λ-repressor) maintains the prophage state. The expression of the *N* and *Q* genes is necessary for the expression of the delayed early and late genes, respectively, during the productive lytic cycle. The *N* gene codes for an antitranscription terminator that allows transcription of delayed early genes controlled by P_L and P_R. The *Q* gene product activates the late genes involved in DNA replication and the synthesis of head and tail proteins.

The maintenance of the prophage state is delicately balanced with the potential for conversion to the lytic state by the presence of two sets of operators and the repressor that control the expression from P_L and P_R. The organization of this regulatory region is shown in Figure 22.6. There are three tandem operator sites that control the expression of the *N* and *cro* genes called O_L1, O_L2, and O_L3 and O_R1, O_R2, and O_R3, respectively. The repressor binds most strongly to O_L1 and O_R1 and progressively less to O_L2 and O_L3 and O_R2 and O_R3. The repressor (*cI*) gene is located between the operator regions. The concentration of repres-

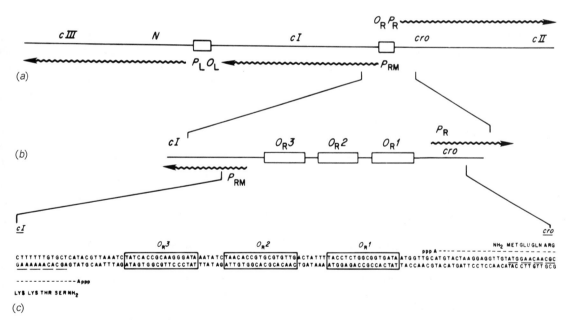

FIGURE 22.6 λ genes and regulatory elements. (*a*) A portion of the λ genome. The arrows indicate the directions and startpoints of transcription of various genes. $O_L P_L$ and $O_R P_R$ are the left and right operator and promoter regions. P_{RM} is the cI promoter active in lysogen. (*b*) Expanded diagram of the λ O_R region. O_R^1, O_R^2, and O_R^3 are repressor and *cro* binding sites, each 17 bp long. The startpoints of transcription from P_R and P_{RM}, which are located outside the operators, are indicated. As described in the text, polymerase bound to either promoter overlaps repressor binding sites in the operator. (*c*) Nucleotide sequence of the λ O_R region. The startpoints of the *cI* and *cro* mRNAs are shown, as are the amino terminal portions of the corresponding protein sequences. From Ptashne, et al., *Cell* 19:1, Massachusetts Institute of Technology, 1980 (© Massachusetts Institute of Technology).

sor that determines whether the lysogenic or lytic cycle is to occur is itself controlled by an autogenous mechanism involving the expression of the repressor gene. When the concentration of repressor is low, the repressor will bind to O_L1 and O_R1 to prevent expression of the N and cro genes, thus preventing the lytic cycle. When the concentration of repressor is high, O_L2, O_L3, O_R2, and O_R3 will bind repressor. Repressor bound to O_R3 will repress the synthesis of the repressor itself from gene cI. Thus, an excess of repressor can lead to the lytic state.

In order to release the phage from its lysogenic state, the concentration of repressor is reduced to just the level that will allow expression of the cro gene. The cro gene product will then bind to the O_R3 and prevent repressor synthesis. O_R3 has a higher affinity for cro product than O_R1, and therefore, repressor synthesis is inhibited and cro gene expression not inhibited. Once the repressor gene is repressed by cro product, then a reduced concentration of repressor will lead to transcription from P_L and P_R and set off a whole series of genes involved in λ-DNA replication and head and tail synthesis. Finally, about 100 new λ particles are produced and cell lysis occurs.

22.2.4   BACTERIOPHAGE M13

M13 is a small filamentous phage that infects *E. coli* bearing F pili. It is 895 nm long and 9 nm in diameter and contains a single-stranded circular DNA molecule that extends throughout the length of the filament (the DNA strand contained in the phage particle is called the "plus strand"). The DNA contains 6408 nucleotides and codes for 10 genes. The phage infects the cell by binding to the F pili, but the exact mechanism of DNA injection is unknown. M13 is unique among bacteriophages, since infection does not result in a lytic cycle. After replication, virus particles are secreted from the host cell with up to 1000 particles being produced during one cell generation.

After injection of the single-stranded DNA (ssDNA) into the host cell during infection, the ssDNA is converted into a circular supercoiled double-stranded DNA called the parental replicating form (RF) by the host's replication system (Figure 22.7). Thus, a complementary DNA strand (the "minus strand") is synthesized for the injected ssDNA. Most interestingly, the host RNA polymerase makes the RNA primer for synthesis of the "minus strand."

In the second stage of DNA replication, the parental RF replicates with the help of a phage gene-encoded function to make additional RFs to ensure an adequate rate of transcription. The host RNA polymerase can only utilize double-stranded DNA as a template. The production of additional RFs results in the adequate transcription of the DNA replication protein genes and coat protein genes.

In the final stage of replication, ssDNA (plus strand) is again produced from the progeny RFs with the assistance of another phage-encoded gene product and these progeny ssDNA molecules are packaged into phage filaments. The complementary DNA strand (minus strand) made during RF formation is never incorporated into progeny phage particles.

Thus, the DNA of M13 phage undergoes a cycle in which a single-stranded circular DNA is converted to a circular double-stranded DNA molecule on infection. Then after a few cycles of replication as a double-stranded molecule, single-stranded circular DNA molecules are again produced for incorporation

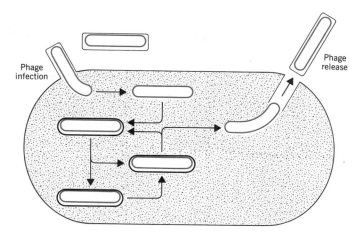

FIGURE 22.7 Schematic representation of the life cycle of phage M13. The plus strand (color) in the phage is injected into the host *E. coli* cell. A replicating form (RF) forms with a complementary minus (black) strand. After replication of the RF, plus strands (color) are formed that are coated and released from the cell.

into the phage particle. The incorporation of only the "plus" strand into the phage particle makes this a useful system for cloning only one of two complementary strands of DNA during genetic engineering experiments (see Section 22.6).

22.3 POLIOVIRUS

In the previous section we described several DNA viruses of bacteria, called bacteriophages. Although bacteriophages with genomes of single-stranded or double-stranded RNA are known, they are rare. In contrast, the viruses of animals and of plants more commonly have genomes of single-stranded RNA. Many animal viruses and about 90% of the known plant viruses have genomes of RNA that is messenger RNA. That is, the RNA that is isolated from virus particles will direct the synthesis of one or more proteins in the properly constructed cell free translation system (Section 20.9). These viruses are regarded as (+)RNA viruses in recognition of the (+) messenger polarity of the genomic RNA.

Regardless of the kind of nucleic acid that forms the genome in the virus particle, the action of the virus is similar; it subverts the machinery of the cell at the molecular level using the protein synthetic apparatus and other metabolic systems to eventually produce new virus particles. This obviously requires replication of the virus nucleic acid. As we indicated in Section 19.1, for most (+)RNA viruses this replication is accomplished without involvement of DNA. Rather, the (+)RNA of the infecting virus particle is transcribed into the complementary (−)RNA strand, which in turn serves as the template for more (+)RNA synthesis.

Poliovirus is an example of a (+)RNA virus that infects human and primate cells. Poliovirus has been studied in the laboratory for 25 years and, because of the great medical importance it once had, it is one of the most well-researched animal viruses. Poliovirus is a member of the *picornavirus* group of animal viruses, the term being derived from *pico,* for small, and *RNA.* Among the members of the picornavirus group are hepatitis A virus, foot-and-mouth disease virus of cattle, the rhinoviruses that are among the causes of human colds, and mengovirus (see cover of this book).

The poliovirus particle is composed of a single RNA molecule and 60 copies each of four different protein molecules. Eucaryotic cells are generally equipped to translate monocistronic mRNAs, mRNAs with a single functional AUG initiation codon near the 5′ end and a single termination codon near the 3′ end of the mRNA (Section 20.7). This might pose a problem for a (+)RNA animal virus such as poliovirus, which must express its several genes, the genes for four coat proteins, and the proteins needed for RNA replication, in the environment of the primate cell. The (+)RNA viruses generally follow one of two strategies for expressing several protein molecules from the genes encoded on a single, long (+)RNA molecule.

One approach, not employed by poliovirus, is to initiate transcription of the virus (−)RNA at more than one site. Initiation at the 3′ end of the (−)RNA template produces new, full-length virus genomic (+)RNA suitable for translation from close to its 5′ end or for encapsidation in the virus particles. Internal initiation at one or more sites can produce other RNAs that can serve as mRNAs for internally coded virus proteins. These latter *subgenomic* mRNAs are the mRNAs for the genes that are not translated from the full length (+)RNA because of the limitations of the eucaryotic translation systems.

Poliovirus proteins are expressed in a different way. The poliovirus (+)RNA from virus particles is, for the purposes of translation, a monocistronic mRNA. The theoretical coding capacity of poliovirus RNA, if every nucleotide residue were part of one open reading frame, is for a very large protein of about molecular weight 275,000 [(approx. 7500 nucleotide residues) × (1 amino acid/3 nucleotide residues) × (MW 110 per amino acid residue) = approx. 275,000]. The following diagram locates the single initiation codon beginning at residue 743 and single termination codon at residue 7392.

A **polyprotein** of molecular weight about 245,000 is translated from this long open reading frame. As is indicated by Figure 22.8, the polyprotein contains the amino acid sequences of 11 different proteins covalently connected in one polypeptide chain. These are specified as regions 1A, 1B, . . . , 3D of the polyprotein and, similarly, as the individual proteins. Thus, the poliovirus system fits the general picornavirus pattern, the 4-3-4 arrangement of polypeptides, referring to the peptide sequence 1ABCD-2ABC-3ABCD. The synthesis of the polyprotein is actually terminated by a pair of stop codons, but, of course, there are no other stop codons occurring in the long open reading frame (ORF).

The functional proteins of poliovirus are generated from the polyprotein by protein processing, the cleavage of the polyprotein to produce intermediate polypeptides and finally the functional proteins. With two exceptions, processing cleavages of the poliovirus polyprotein occur at glutamine–glycine bonds. One of these is the asparagine–serine bond that, when split, generates 1A and

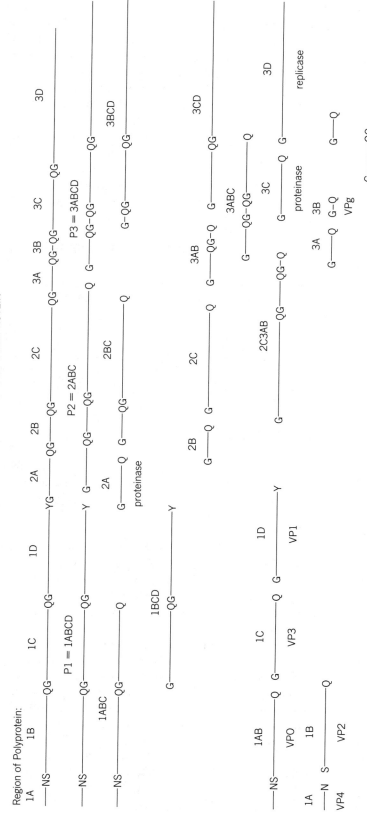

PROCESSING OF POLIOVIRUS POLYPROTEIN

FIGURE 22.8   The poliovirus polyprotein and the polypeptides and functional proteins derived from it by specific proteolytic processing.

G = glycine, N = asparagine, Q = glutamine, S = serine, Y = tyrosine

1B from 1AB. Proteins 1A, 1B, 1C, and 1D are coat proteins and the asparagine–serine cleavage reaction occurs only after the virus particle has been assembled. The other exception is the tyrosine–glycine bond that appears to be the first bond to be split in the polyprotein.

As shown in Figure 22.8, proteins 2A and 3C are proteinases. Antibody against 3C was able to interfere with the processing of the polyprotein *in vitro*. Only the cutting of the glutamine–glycine bonds was prevented. This evidence points to 3C as the principal, perhaps the only, virus specific proteinase that catalyzes the hydrolysis of specific Gln–Gly bonds. Very likely, this proteinase is able to function, though perhaps only inefficiently, while still embedded in the polyprotein as region 3C. This would allow the first polyprotein molecules that were synthesized to be cleaved. The proteinase for the tyrosine-glycine bond is 2A.

Protein 3B is also designated VPg, for *virion protein, genome-linked.* The VPg is found in virus particles linked covalently to the 5′ end of the RNA by an unusual phosphodiester bond between the 5′-uridylate residue of the RNA and a tyrosine phenolic hydroxyl group of the VPg. Protein 3D is an RNA-dependent RNA polymerase that is required for poliovirus RNA replication. Thus, the poliovirus genes are expressed into a series of proteins, after polyprotein processing, that meet the requirements for replication of poliovirus that are not met by the machinery of the cell. Note that the cleavage of a polyprotein precursor assures the production of the four proteins of poliovirus particles, 1A, 1B, 1C, and 1D (also known as coat proteins VP4, VP2, VP3, and VP1, Figure 22.8) in equimolar amounts, as required by the structure of the virus particle.

22.4 EXTRACHROMOSOMAL DNA OF CELLS: PLASMIDS

Many cells contain extrachromosomal DNA that can replicate autonomously of the chromosomal DNA. These DNA-containing elements are called plasmids. Plasmids contain their own origin of replication (*ori*) region (see Section 19.4) that allows them to replicate independently. The *ori* is usually quite host specific, although there are so-called promiscuous plasmids that can replicate in a number of different host cells. The number of copies of a plasmid within a cell can vary from one to a thousand depending on the plasmid and the conditions for growth of the cell. When only a small number (two to five copies) of a particular plasmid is normally present in a cell, it is called a "low copy number" plasmid. When 20 to 100 copies are normally present per cell, it is called a "high copy number" plasmid. Cells may also contain more than one type of plasmid; these plasmids are called "compatible plasmids." If two plasmids cannot exist simultaneously within a host, they are called "incompatible plasmids." Some bacteria may contain several different types of plasmid per cell. Some plasmids are capable of transferring from one cell to another and are called **self-transmissible** plasmids. Many plasmids cannot be transmitted from one cell to another and are called **nontransmissible** plasmids.

Plasmids can vary in size from two to three kb (kilobase pairs) or about 1.25 megadaltons (1 megadalton = 10^6 daltons) to 300 kb or about 200 megadaltons. Thus, the smallest plasmids can code for only 2 or 3 genes, whereas the largest plasmids can code for about 300 genes. Self-transmissible plasmids tend to be larger (about 40 kb in size) than nontransmissible plasmids, since the self-transmissible plasmids require additional genetic information for the transfer mechanism. Plasmids or plasmidlike elements have been found in bacteria, plants, and animal cells.

22.4.1 BACTERIAL PLASMIDS

The use of antibiotics became widespread after 1940 and in the early 1960s Watanabe reported that certain species of *Shigella* were simultaneously resistant to two or more antibiotics. It was also shown that these properties could be transferred to nonresistant cells. It turned out that these cells contained self-transmissible plasmids that conferred this drug resistance property to the bacterium. It is possible that the indiscriminate use of antibiotics led to the evolution of plasmids containing genes for multiple resistance to antibiotics. At the practical level, this is causing serious health problems in hospitals that tend to harbor highly resistant bacterial strains. A typical bacterial plasmid DNA is illustrated in Figure 22.9.

What type of genetic information is present on bacterial plasmids? Many of the plasmid genes confer antibiotic or drug resistance to the cell that harbors the plasmid, since the genes code for enzymes that either destroy or modify antibiotics in such a way to make them ineffective. Thus, these cells become resistant to antibiotics. If the plasmid is removed from these cells they become sensitive to the antibiotics once more, thus demonstrating the plasmid origin of their resistance.

Many other phenotypes have also been attributed to plasmid-encoded genes in bacteria. These include sexuality, production of toxins, possession of restriction endonuclease function, capability for toluene degradation, ability to cause plant tumours, resistance to heavy metals, and drug production. When no phenotype can be attributed to the plasmid, it is called a "cryptic" plasmid. This does not mean that the plasmid does not contain genetic information. It just means that a phenotype has not been identified for the plasmid.

22.4.2 REPLICATION OF BACTERIAL PLASMIDS

The replication of plasmids is dependent on the replication machinery of the host cell and the property of the *ori* region and can occur in two modes:

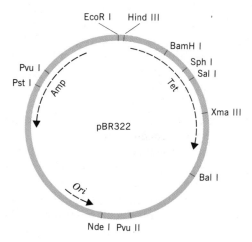

FIGURE 22.9 An *E. coli* plasmid showing its origin of replication (*Ori*), many restriction sites, tetracycline resistance (Tet), and ampicillin resistance (Amp) genes.

stringent replication or **relaxed replication.** The stringent replication-controlled plasmids are found in one or few copies per cell, whereas the relaxed replication plasmids are present in many copies (20–100) per cell. In the "relaxed" state, replication is not obligatorily coupled to chromosome replication; a plasmid may replicate twice before the chromosome has finished replicating even once and not all plasmids in the cell are replicating at the same rate. In the "stringent" state, each plasmid usually replicates once per generation and all plasmid DNA in a cell population would be replicated once before any are replicated a second time. Plasmids that replicate in the relaxed mode are used in genetic engineering experiments, since DNA molecules from various sources can be integrated into the plasmid DNA and high copy numbers of the recombinant DNA may be obtained. Also, if the inserted DNA contains a gene that can be expressed in the host cell, then a large amount of a specific protein can be produced from the gene.

One of the important control features of replication concerns the segregation of the replicated plasmids into the daughter bacterial cells during cell division. This is particularly important in the case of stringently replicating plasmids, since there may be only two copies of the plasmid per cell and equal distribution of the plasmid copies is required to ensure that each bacterial progeny maintains the plasmid-infected state. The regulatory features for ensuring proper segregation are also encoded into the *ori* region of the plasmid.

The replication of the *E. coli* plasmid ColE1 has been studied extensively. The *ori* region of this plasmid contains two sites that are recognized by RNA polymerase. The RNA polymerase synthesizes RNA in a convergent manner from these two sites on the complementary DNA strands to form two primers for the DNA synthesis machinery of the host. RNase H, a ribonuclease that hydrolyzes RNA present in an RNA–DNA hybrid structure, cleaves the complementary primers at precise locations in the *ori* at which DNA synthesis is initiated from the primers by DNA polymerase III. The initiation step is not understood fully as yet. During elongation, one strand may be synthesized in a leading-strand manner, whereas the other strand may be synthesized in the lagging-strand mode (see Section 19.6). The usual host enzymes for DNA replication are used to replicate the DNA during elongation.

22.5   BASIS OF RECOMBINANT DNA METHODOLOGIES

The first recombinant DNA experiments were carried out in 1973 by Boyer and Cohen. These studies were based on a number of earlier fundamental biological and biochemical findings that laid the foundation for the field of genetic engineering. The recombinant DNA technology now allows the recombination of DNA molecules that would never or very rarely occur in nature. It is a powerful tool that is currently being utilized to manipulate the genetic makeup of plants (Box 22.A), animals, and microorganisms. This should lead to improvements in several major areas of human endeavour, such as agriculture, medicine, and industry. In agriculture, we should see new strains of plants that are disease resistant, provide higher yields, and are salt tolerant, and animals that are more disease resistant and provide more protein. In medicine rare hormones, antibiotics and important enzymes will be produced more readily. In industry, fermentation processes will become much more efficient and provide higher yields

BOX 22.A INTERKINGDOM TRANSFER OF GENETIC INFORMATION: NATURAL TRANSFER OF GENES FROM BACTERIUM TO PLANT

Various strains of the crown gall bacterium, *Agrobacterium tumefaciens,* can invade a wound site of a plant and there induce the formation of a gall, which is an outgrowth of disorganized tissue. The "crown" of a tree is the region at which the trunk and root system join. Since *A. tumefaciens* is a soil-inhabiting organism, it is frequently at the crown that a gall is induced. Crown galls frequently grow larger than the original diameter of the tree and may result in the death of the tree. The crown gall disease has been studied for more than 75 years, and it is not limited to trees. Most dicotyledenous plants are susceptible to experimental inoculation with the bacterium, as are several gymnosperms and a few monocots, which develop small swellings at the site of inoculation rather than galls.

The interest in crown gall increased greatly when gall tissue was experimentally freed of detectable bacteria. The bacteria-free gall tissue, when cultured on a nutrient-containing agar substrate, was able to proliferate without certain plant growth substances, auxins and cytokinins, that are necessary for the increase of normal plant tissue on the same medium. That is, the plant cells seemed to have been **genetically transformed** at some point by exposure to *A. tumefaciens.* An intense search was then initiated for a tumor-inducing principle, that is, the substance or substances that mediate this transformation from normal growth to tumor, from auxin- and cytokinin-dependent growth to independent growth.

Eventually, an association was made between the presence of a very large plasmid in *A. tumefaciens* and its ability to induce galls. This plasmid, which in some variants exceeds 200,000 base pairs, is circular and is designated pTi. pTi is now known to have at least three regions that are important to the ability of *A. tumefaciens* to induce tumors, in addition to the nucleotide sequences necessary to regulate and specify the structural genes for replication of the plasmid. The **T-DNA** region of pTi, which is about 25,000 base pairs, is transferred to th plant cell nucleus where it becomes integrated into the nuclear DNA at one or a few apparently randomly selected locations. The T-DNA specifies genes for auxin and cytokinin production, and these genes have the proper signal sequences to facilitate their expression in the plant cell. The subsequent synthesis of auxin and cytokinin, catalyzed by the enzymes translated from mRNA that is transcribed from the T-DNA, explains the independence of gall cells from a supply of exogenous growth substances.

The T-DNA also specifies genes for the synthesis of the unusual amino acids known as opines, such as octopine (Section 3.5). A second region of pTi, outside of the T-DNA region, specifies bacterial enzymes that degrade the same opine that is synthesized under the control of the T-DNA. In fact, pTi plasmids may be classified according to the type of opine that can be both synthesized (plant genes) and degraded (bacterial genes) by the gene products of the pTi plasmid. The bacterium seems to have evolved a mechanism by which it can genetically engineer the plant to produce a specific opine. This is indeed a subtle and effective subversion of the plant's metabolism. The opine that the plant synthesizes amounts to the bacterium's private stock, a supply of carbon and reduced nitrogen for which the bacterium has developed a specialized metabolism.

The third important region of pTi, for tumor induction, is the *Vir* region. Functions controlled by the *Vir* region include the recognition of chemical signals from the wounded plant, which are or are related to the aromatic ketone acetosyringone (2,6-dimethoxy-4-acetophenol). These chemical signals apparently activate other *Vir* genes. The products of these other *Vir* genes mediate the transfer and integration of the T-DNA.

The *Vir* and T-DNA regions need not be on the same plasmid, and the opine degradation genes, as important as presumably they are to the survival of the bacterium, are not needed for tumor formation. Thus, tumorigenicity *A. tumefaciens* is maintained even when the *Vir* and T-DNA regions are moved to separate, relatively small and easily manipulated plasmids. That is, the full size and difficult-to-manipulate pTi plasmid is not required in its entirety. The development of such binary systems, with two plasmids, allows Nature's genetic engineer of plants to be used by researchers to transform plants with other genes.

The general approach is to substitute the auxin and cytokinin genes of the T-DNA region, or the entire, central T-DNA, with other DNA of interest. *A. tumefaciens* that bears the modified T-DNA plasmid and a plasmid with the *Vir* genes, can transform plants. At first the substituted DNA was one that encodes a drug resistance gene, such as a gene for kanamycin resistance. Since the T-DNA genes for auxin and cytokinin synthesis, but not the bordering sequences necessary for DNA transfer, had been removed, the plasmid is said to be "disarmed." The plant cells transformed with a disarmed plasmid system, unlike crown gall cells, are dependent upon auxin and cytokinin for growth in culture. For several plants, including tobacco, petunia, and tomato, transformed cells could be "regenerated" into plants of normal appearance on a selective medium that contained kanamycin. The resulting plants were kanamycin resistant. This new character of the **transgenic** plants was shown to be the result of a true genetic change. The kanamycin resistance was inherited as a simple dominant character in crosses of drug-resistant and nonresistant plants.

The transformation system becomes more versatile when both a drug resistance gene and other genes are placed in the T-DNA region. Plants regenerated on drug-containing medium after transformation have a high probability of having been transformed not only with the drug resistance gene but also with the other genes. Successful production of new proteins in such multiply transformed plants usually requires the presence of the proper promoter and 5′ and 3′ flanking sequences. Thus the plant transformation technology based upon the pTi plasmid derivatives provides an opportunity to test, for a given gene, the effects of nucleotide sequences that lie outside of the coding sequences of the open reading frame.

Using transformed plants, legume seed promoters have been shown to limit the production of legume seed proteins to the seeds of transformed tobacco plants. The promoter of the small subunit of pea ribulose-1,5-*bis*phosphate carboxylase places the expression of this gene under light control in transformed petunia, and the small subunit is synthesized and accumulates in the chloroplasts of the transformed petunia plants, as might be expected. Tobacco plants have been made resistant to an herbicide by transforming them with the gene for an herbicide-resistant bacterial enzyme of aromatic amino acid biosynthesis. The potential of the transformation technology for creating crop plants with new and valuable properties is obvious.

No system similar to the pTi system is available for animal cells. However, animal cells in culture have been transformed at low efficiency by simply supplying them with DNA in a precipitated form, and animal germ cells have been transformed by the microinjection of foreign DNA.

for the same input of energy and growth medium. Energy sources that are too expensive relative to current energy supplies may become more competitive and be safe and renewable energy resources. What were the fundamental discoveries that had a profound impact on the initiation of genetic engineering? They include studies on bacterial genetics, bacteriophages, plasmids, nucleic acids, and various enzymes that utilized nucleic acids as substrates. Some of these fundamental studies will be discussed in the following sections.

22.5.1   TYPE I RESTRICTION ENDONUCLEASES

Restriction enzymes are as a class one of the most important tools available for recombinant DNA technology. These endonucleases are found in bacteria and have the unique property of recognizing specific short DNA sequences and cleaving the phosphodiester bonds of both strands of the DNA at this recognition site. Thus, a bacterium can protect itself from invasion by a foreign DNA by cleaving the incoming DNA with its restriction endonuclease. By using a battery of different restriction enzymes, the scientist is able to cleave a chromosome into various discrete DNA fragments. This has allowed the preparation of two distinct DNA fragments from two different sources that can then be ligated together to form a recombinant DNA molecule. Also, these enzymes do not remove any bases from the DNA during cleavage. Therefore, the cleavage site can be reclosed to restore the original restriction site. How were these useful enzymes found? As described in Box 6.D, the discovery of these enzymes was based on the early observation that the efficiency of infection of *E. coli* by λ-bacteriophage depended on the host in which λ had been propagated. Phage grown on *E. coli B* infected *E. coli B* with high efficiency (1/1), but infected *E. coli K* with low efficiency (1/10^4) and vice-versa. The restriction of λ was based on the presence of a host-specific restriction/modification system.

This system includes a restriction endonuclease and a modifying enzyme (a DNA methylase) that both recognize a specific sequence in the DNA. If the DNA sequence is unmethylated, then the restriction enzyme will cleave the DNA. If the DNA is methylated at the recognition site by the modifying enzyme, then the restriction enzyme is not able to cleave the DNA. Thus in the aforementioned case with λ, the restriction/modification system of *E. coli B* would not restrict λ-DNA that had been replicated and modified in *E. coli B*, and therefore, the λ would be able to replicate efficiently in this host. However, if this same bacteriophage λ were used to infect *E. coli K*, the restriction/modification system of *E. coli K*, which has a different sequence recognition specificity from that found in *E. coli B*, would quickly restrict the incoming λ-DNA and destroy its infectivity. Thus, the restriction/modification system is a mechanism to destroy foreign DNA that may invade a bacterium. As it turned out, the restriction systems of *E. coli B* and *E. coli K* were not suitable for genetic engineering, since their restriction enzymes did not cleave the unmodified DNAs at specific sites. Their restriction/modification systems are classified as Type I.

22.5.2   TYPE II RESTRICTION ENDONUCLEASES

The more commonly used restriction/modification systems are classified as Type II. These systems usually contain separate restriction and modification enzymes that cleave and methylate DNA at the same recognition site. Thus, cleavage of DNA with these restriction enzymes results in the production of specific DNA fragments. The specificity of several of these enzymes is illustrated in Table 22.1.

The restriction enzymes usually recognize a short palindromic DNA sequence containing four to six bases. After cleavage, three types of end groups are generated: 5' tails, 3' tails, and blunt ends.

HindIII	5'–A↓AGCTT–	5' – – – –A	
	3'–TTCGA↑A–	3' – – – –TTCGA	5' Tail
PstI	5'–CTCGA↓G–	5' G– – – –	
	3'–G↑AGCTC–	3' AGCTC– – – –	3' Tail
HpaI	5'–GTT↓AAC–	5' – – – –GTT	
	3'–CAA↑TTG–	3' – – – –CAA	Blunt end

When two different restriction enzymes produce similar tails—for example, two 5' tails, these tails may be partially or totally complementary or compatible—for example, BamHI and BglII.

BamHI   ———G↓GATCC———   ———A↓GATCT———   BglII
         ———CCTAG↑G———   ↓   ———TCTAG↑A———

BamHI   ———G   GATCT———   BglII
         ———CCTAG   A———

         ligate   ↓

              ———GGATCT———
              ———CCTAGA———

This ligated product cannot be restricted by either BamHI nor BglII but can be cleaved by MboI or Sau3A.

When compatible end groups are ligated, they may or may not regenerate a restriction site. In some cases, a new restriction site may be made on ligation of compatible ends.

The DNA-modifying enzymes use *S*-adenosylmethionine as the methyl donor for methylating the recognition sites. In the host cell, the modifying enzyme methylates the DNA during replication. Thus, the DNA is protected from its own restriction enzyme.

Restriction endonucleases are very useful in recombinant DNA studies in several ways. They can generate DNA fragments with complementary tails from two different chromosomes. This permits the ready ligation of these molecules, for example, if two chromosomes are treated with EcoRI, then all the DNA fragments will have EcoRI 5′ tails and any two EcoRI fragments can be ligated with DNA ligase.

TABLE 22.1   Restriction Enzymes and Their Recognition Sites

ENZYME	RECOGNITION SEQUENCE	TERMINI
AvaI	C↓PyCGPuG	5′
BamHI	G↓GATCC	5′
BglII	A↓GATCT	5′
ClaI	AT↓CGAT	5′
EcoRI	G↓AATTC	5′
HaeIII	GG↓CC	Blunt
HindIII	A↓AGCTT	5′
HpaI	GTT↓AAC	Blunt
KpnI	GGTAC↓C	3′
MboI	↓GATC	5′
PstI	CTCGA↓G	3′
PvuII	CAG↓CTG	Blunt
RsaI	GT↓AC	Blunt
SalI	G↓TCGAC	5′
Sau3A	↓GATC	5′
SstI	GAGCT↓C	3′
TaqI	T↓CGA	5′
XbaI	T↓CTAGA	5′
XhoI	C↓TCGAG	5′

Restriction maps can be made of chromosomes or DNA fragments by cleaving the DNA with several different restriction enzymes. The process of preparing restriction maps is described in Box 6.E. This process is very important, since many techniques used in recombinant studies—for example, DNA sequencing, cloning, and so on—require a knowledge of the restriction sites available on a DNA fragment or cloning vehicle. Unique restriction sites on a cloning vehicle are very important as cloning sites, for example, if there is a unique EcoRI site on a plasmid, EcoRI fragments from another chromosome can be inserted into it and cloned. In DNA sequencing techniques (Box 6.C), relatively small DNA fragments are sequenced. These small fragments are obtained from larger fragments by restriction endonuclease treatment and subcloning of these smaller fragments.

22.6 CLONING DNA FRAGMENTS

Techniques are now available for cloning DNA fragments from all sources. DNA fragments are usually obtained by restriction endonuclease treatment of chromosomes. However, they can also be prepared by chemical synthesis, random shearing of DNA followed by tailing, and cDNA synthesis from mRNA. Each method has its advantages and shortcomings. Chemical synthesis of a gene requires prior knowledge of the amino acid sequence of the protein product and the method is quite laborious. Chemical synthesis is more frequently used to synthesize a short deoxyoligonucleotide based on the N-terminal amino acid sequence of a protein to form a molecular hybridization probe to detect the complete gene in a gene bank. Random shearing of DNA is a practical method to obtain a gene bank in which every gene in the chromosome will be represented (Figure 22.10). Tailing can now be accomplished by filling the tails of sheared DNA by use of the Klenow fragment (DNA polymerase I from *E. coli* can be cleaved with a protease to yield two fragments; the Klenow fragment is capable of synthesizing DNA, but does not contain the 5′ to 3′

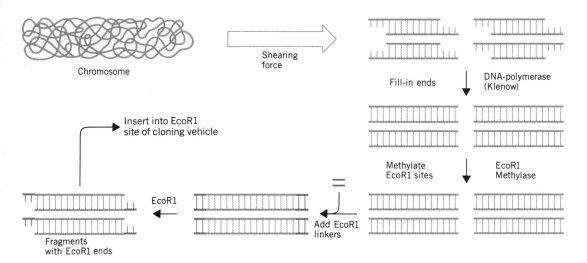

FIGURE 22.10 Construction of a gene bank by random shearing of the chromosome.

exonuclease function) of DNA polymerase I to provide blunt-ended molecules. These blunt-ended DNA fragments are then treated with a specific DNA-modifying enzyme (e.g., EcoRI methylase) to protect specific sites from the coupled DNA restriction endonuclease (i.e., EcoRI). Polylinkers (short deoxyoligonucleotides that contain several restriction endonuclease sites) are then ligated to these protected fragments and then restricted with the specific restriction enzyme (EcoRI) to provide cohesive tails for linkage to cloning vehicles.

In order to obtain large amounts of a particular fragment of DNA, a cloning vehicle is necessary. The most frequently used cloning vehicles are either plasmids, phages, or cosmids, which can produce large copies of a particular DNA fragment.

22.7 PLASMIDS AS CLONING VEHICLES

Plasmids that are useful for cloning have the following properties:

1. They are small and contain 3 to 5 kb of DNA.
2. They are present in high copy number, for example, 50 copies per cell.
3. They contain a suitable marker or markers, for example, antibiotic resistance markers.
4. They are not transmissible, that is, they are not transferred from one cell to another.
5. They contain suitable restriction sites that can be used for inserting DNA fragments for cloning — this has become less of a problem, since polylinkers (short oligonucleotides) containing several unique restriction sites can be inserted into plasmids.
6. They are stably maintained during replication in the host, that is, during cell division, the plasmids are segregated evenly into the daughter cells.
7. They can replicate in a suitable host or hosts; for example, one can insert an *ori* from a yeast into an *E. coli* plasmid and that plasmid will now be able to replicate in both *E. coli* and yeast.

Usually, *ori* sites are quite host specific. A number of plasmids have been isolated from different hosts and several have the properties just listed and are suitable cloning vectors. Also, a large number of specialized plasmids have been constructed by genetic engineering that facilitate the cloning of certain DNA fragments — for example, DNA fragments that contain regulatory elements such as promoter and terminators, and that allow the selection of clones by their color or their resistance to antibiotics.

Gene banks can be made in plasmids. However, since the maximum size of the DNA fragment that can be readily recombined and replicated with plasmids is about 10 kb, gene banks would be limited to smaller chromosomes, such as those found in procaryotes. For constructing gene banks from larger eucaryotic chromosomes, phages and cosmids are more suitable cloning vectors.

22.7.1 BACTERIOPHAGE λ AS A CLONING VEHICLE

The *E. coli* phage λ (see Section 22.2.3) has been used as a cloning vehicle and a system for creating gene banks of various organisms. λ can be used as a cloning system, since its genome has a regulatory region that is not essential for replication of the phage particle. This region (see Figure 22.4), which lies between the left and right arms of the phage genome, can be removed and replaced by a similar size fragment of DNA that is then replicated during the replication cycle of the phage (Figure 22.11).

In practice, λ-phage have been constructed that have two specific restriction endonuclease sites (e.g., EcoRI) flanking or partially within the regulatory region. When the phage is treated with EcoRI, the regulatory region is removed and two phage fragments, the left and right arms, are obtained. DNA fragments equal or nearly equal in size to the excised regulatory region can then be inserted between the left and right arms, ligated, and encapsulated in phage coat and tail proteins to form an infective λ-phage particle. The size of the insert can be as large as 20 kb. When cells are infected with this now virulent recombinant phage, up to 200 phage particles per cell can be formed, thus resulting in a high copy of the DNA insert. If the insert contains a gene that can be expressed in *E. coli,* it could also result in the production of a large amount of the gene product. The large size of the insert makes λ-phage gene banks more practical for large genomes than plasmid gene banks, which can readily accommodate only about 10-kb fragments.

22.7.2 COSMIDS

Cosmids are also used for cloning. This system takes advantage of the properties of both the plasmid and λ systems. A cosmid consists of a circular DNA molecule containing a plasmid origin of replication (*ori*), which allows autonomous replication, a suitable antibiotic marker, a λ *cos* site (these are the cohesive

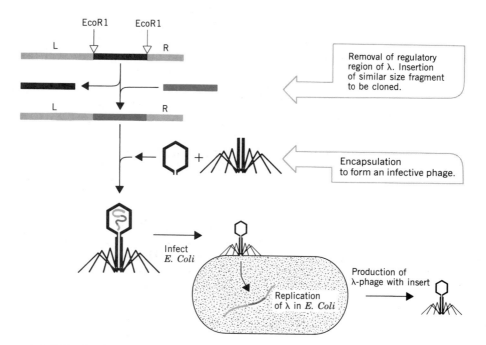

FIGURE 22.11 Construction of a gene bank in phage λ. Treatment of the λ phage with EcoRI removes the regulatory region (black bar) of the phage. EcoRI fragments of similar size are inserted between the left (L) and right (R) arms of the phage. The recombined molecules are packaged *in vitro* into phage heads and tails to make infective phage particles.

ends of the linear λ chromosome that allow circularization of the molecule after infection; the *cos* site also is required for ligating and packaging the right amount of DNA into a λ head), and a unique restriction site (e.g., EcoRI) that can be used for inserting a DNA fragment. The cosmid can replicate like a plasmid. The strategy for using the cosmid as a cloning vehicle is illustrated in Figure 22.12.

The cosmid can be used to make gene banks by partial endonuclease restriction (e.g., EcoRI) of chromosomes that will result in fairly large DNA fragments. These fragments will be mixed with the EcoRI restricted cosmids and joined between two linear cosmid molecules. When the proper size DNA fragment (these can be up to 35 kb long) is inserted between two cosmid molecules, the distance between the resulting two *cos* sites will determine whether the recombinant DNA can be properly packaged into the λ head to form a λ-like particle (Figure 22.12). When these particles are used to infect *E. coli,* the recombinant cosmid DNA will be injected into the host and replicated by use of the plasmid *ori* present on the molecule. Proper selection techniques will allow the detection of specific recombinant DNA molecules in the clones.

22.8 GENE LIBRARIES

Construction of chromosomal gene banks or libraries has been very useful for isolating specific genes. For instance, DNA obtained from human placental tissue has been used to construct human gene banks. Two general types of gene banks have been constructed: (1) mRNA-directed DNA gene banks and (2) chromosomal DNA gene banks.

The general strategy of constructing a gene bank from mRNA is to partially purify the mRNA population of a cell, convert the mRNA sequences to double-stranded DNA sequences, and insert the DNA sequences into a suitable cloning vector. This bank should contain at least one DNA copy of each of the mRNA species. Usually, these banks are constructed to obtain a specific gene whose mRNA is a major fraction of the total mRNA population. The general strategy for constructing a chromosomal gene bank is to obtain large fragments of the chromosome that will ensure the presence of all the genes in an intact form and insert these fragments into a suitable cloning vector and host that will allow the subsequent selection of a specific gene from this gene bank.

22.8.1 LIBRARIES OF mRNA SEQUENCES

For the construction of a cDNA (cDNA is DNA that has a complementary sequence to mRNA) gene bank from mRNA, mRNA is isolated from a differentiated eucaryotic cell that is synthesizing a large amount of a specific protein. Usually, this indicates that the cell is actively transcribing the gene for this protein and that a relatively high concentration of the mRNA is present in the cell. Most eucaryotic mRNAs have a long poly(A) tail at the 3′ end and can be separated from the tRNA and rRNA, which comprise the bulk of cellular RNA, by hybridizing the poly(A)-tailed mRNA with a separation column containing poly(dT) residues. All RNAs that do not contain poly(A) tails (i.e., rRNA and tRNA) will not hybridize with the poly(dT) residues in the column and will flow through the column; the poly(A)-containing mRNA will hybridize with

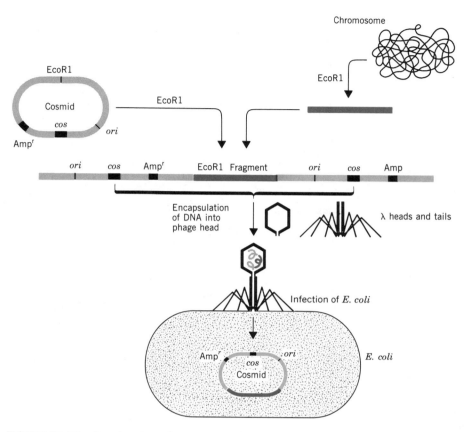

FIGURE 22.12 Construction of a gene bank with cosmids. See text for details.

the poly(dT) residues and therefore be attached to the column. The mRNAs can be eluted from the column by raising the temperature and reducing the salt concentration of the eluting buffer. The poly(A)-containing mRNA is then incubated with **reverse transcriptase,** which copies the mRNA into cDNA and finally into double-stranded DNA. This double-stranded DNA can then be treated with the Klenow fragments of DNA polymerase I to form blunt-ended molecules. These blunt-ended molecules can then be "tailed" by the use of **deoxynucleotidyl terminal transferase** that can add short homopolymers to 3′ ends of double-stranded DNA that are complementary to a plasmid DNA that also has been tailed with a complementary homopolymer (Figure 22.13). The complementary tails of the DNAs allow them to be ligated to form a recombinant DNA and the recombinant plasmid will then be able to clone the inserted DNA derived from mRNA.

The use of cDNA-constructed eucaryotic gene banks is particularly helpful when expression of eucaryotic genes is desired in a procaryotic system such as *E. coli.* The presence of introns in eucaryotic genes and the absence of hnRNA-processing enzymes in procaryotes prevent the translation of the RNA produced from native eucaryotic genes into functional protein molecules. Thus, the conversion of the mature eucaryotic mRNA into DNA partially overcomes this problem, since the mature eucaryotic mRNA does not contain introns but

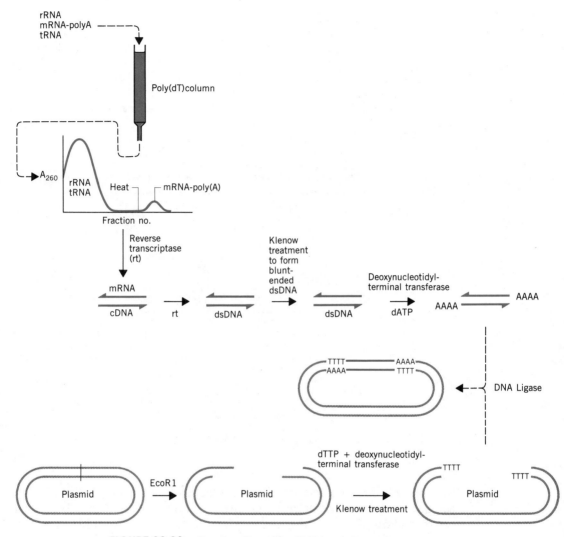

FIGURE 22.13 Construction of a cDNA bank from mRNA.

does contain the codon sequences that can be translated by the procaryotic translation system into a protein sequence. The DNA derived from eucaryotic mRNA will still have to be inserted behind a suitable procaryotic promoter and Shine–Dalgarno sequence in proper translation phase to obtain transcription and translation products.

22.8.2 LIBRARIES FROM CHROMOSOMAL DNA

For the construction of a chromosomal gene bank, the chromosomal DNA is cleaved into relatively large fragments by random shearing or partial restriction endonuclease treatment (see Section 22.6). The large fragments of DNA will ensure the presence of all genes in intact form and also reduce the number of clones that will have to be examined in order to find a particular gene. These fragments are suitably treated and inserted into cloning systems that can tolerate large DNA fragments (see Section 22.7). Usually, gene banks have been made with use of λ-phage and cosmids.

22.9 SCREENING LIBRARIES FOR SPECIFIC SEQUENCES

How does one find a specific gene in a gene bank? Various methods have been developed to find a specific gene. These methods are based on whether the gene in question is or is not being expressed in the cloning system. The immunological screening methods are used when there is a high probability that the gene is being expressed in the cloning system. The hybridization probe methods can be used to screen gene banks in which a gene is not being expressed. This method

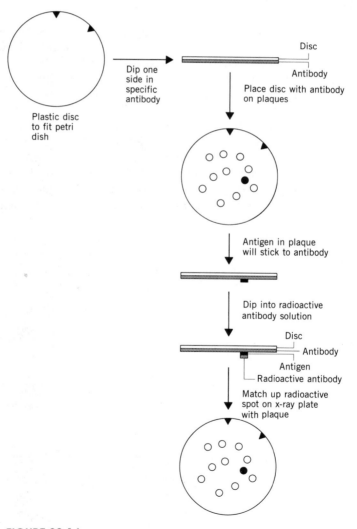

FIGURE 22.14 Immunological screening of a gene bank.

depends on identifying a gene with the use of a radioactive RNA or deoxyoligonucleotide probe that is complementary to the sequence of the gene being sought (see Box 6.F).

22.9.1 IMMUNOLOGICAL SCREENING

In immunological screening methods (see Box 4.E), a gene is sought for a protein or enzyme that has been purified and for which an antibody has been produced. This antibody is used to screen a gene bank for a clone that is producing the protein (Figure 22.14). Since antibody–antigen interactions are very specific, this screening method will identify clones that are producing the protein of interest or some very closely antigenically related protein. It is possible to screen tens of thousands of clones very rapidly and find a specific clone. After the clone is detected, further work is necessary to confirm that the gene is actually present in the clone.

22.9.2 SCREENING BY MOLECULAR HYBRIDIZATION

For screening gene banks by the hybridization probe method, two approaches are generally used. If a mRNA is available to probe for a gene, the mRNA itself can be labeled or a radioactive cDNA can be synthesized from it for use as a probe. If these probes are not available, then a small deoxyoligonucleotide probe can be synthesized chemically, based on the amino acid sequence of the protein product of the gene (Figure 22.15).

Since there is ambiguity in amino acid codons (e.g., leucine has six possible codons), several deoxynucleotide sequences are made for a stretch of five to seven amino acids usually at or near the amino terminal end of the protein. There are now commercial DNA synthesizers (Box 6.G) that can simultaneously make multiple deoxynucleotide sequences, taking into account ambiguous amino acid codons. The use of deoxyoligonucleotides containing 15 to 20

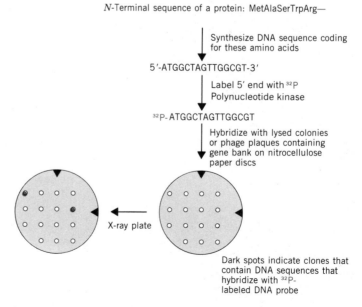

N-Terminal sequence of a protein: MetAlaSerTrpArg—

Synthesize DNA sequence coding for these amino acids

5'-ATGGCTAGTTGGCGT-3'

Label 5' end with ^{32}P Polynucleotide kinase

^{32}P-ATGGCTAGTTGGCGT

Hybridize with lysed colonies or phage plaques containing gene bank on nitrocellulose paper discs

X-ray plate

Dark spots indicate clones that contain DNA sequences that hybridize with ^{32}P-labeled DNA probe

FIGURE 22.15 Hybridization probe of a gene bank.

bases (15-mer to 20-mer) is necessary for forming stable complementary base pairings or hybrids between the probe and the sequence in the gene bank.

The radioactively labeled probes are used to screen a gene bank containing DNA fragments from all parts of the chromosome, using techniques similar to those described in Box 6.F. If a particular clone contains the gene for the probe, the probe will hybridize with the DNA. The clone is then purified and the gene obtained from the cloning vehicle. This method has several advantages, since it can detect genes that are not being expressed in the cloning system and it can detect the presence of the complete eucaryotic gene containing introns and exons in a procaryotic system. This latter fact is becoming less important, since cloning systems in eucaryotic systems, for example, yeasts, are being developed. These systems can transcribe and translate complete eucaryotic genes and may be the better system for cloning and expressing eucaryotic genes.

22.10 SEQUENCE SPECIFIC EFFECTS IN CLONING AND EXPRESSION

There are several factors to consider when carrying out recombinant DNA experiments. If the sole purpose is to clone and amplify a DNA fragment and not to obtain expression of the genes contained in the fragment, then it is usually quite feasible to do so, since the DNA replication mechanism usually cannot distinguish between a "foreign" DNA versus a "native" DNA. The identification of bacteria that bear the recombinant clone can be facilitated by having suitable genetic markers on the vector.

The more difficult type of experiment is to obtain not only cloning, but the expression of a gene in a foreign environment, for example, the expression of a eucaryotic gene in a procaryotic cell and vice-versa. Several factors must be considered, since barriers for expression of foreign genes exist at several levels.

1. The foreign gene may have sequences that make it highly unstable in the host; it may recombine with the host's chromosome or it may be deleted by the host's endonucleases.

2. The promoter sequence of the foreign gene may not be recognized by the host RNA polymerase, a transcription barrier.

3. The promoter may require a positive effector that is lacking in the host cell, a regulatory transcription barrier.

4. The RNA that is synthesized may need processing prior to translation, a post-transcription barrier.

5. The foreign RNA may be highly susceptible to the host's ribonucleases.

6. The Shine–Dalgarno or translation initiation signal on the foreign mRNA may not be recognized by the host translation system, a translation barrier.

7. The frequency of codon use of the foreign gene may be so different from the host that the mRNA is translated very slowly, a translation barrier. This may also lead to a high turnover rate of the foreign mRNA.

8. The primary protein product may require covalent modification prior to forming the active conformation; the host may not have the proper modification enzyme, a post-translation barrier.

9. The foreign protein may be attacked by host proteases that can recognize "abnormal" proteins.

Thus, recombinant DNA experiments can be planned and executed with great care, but there are unknown and unpredictable circumstances that could lead to failure. However, rapid advances are being made currently to overcome most of these difficulties.

REFERENCES

1. *Playfair,* Immunology at a Glance, *Blackwell, London.*
2. *B. Lewin,* Genes, *2nd ed. New York: Wiley, 1985.*
3. *G. E. Russell, ed.,* Biotechnology and Genetic Engineering Reviews, *Vols. 1 (1984), 2 (1984), 3 (1985). Newcastle-upon-Tyne: Intercept, Ltd.*

REVIEW PROBLEMS

1. Design an experiment to isolate the human insulin gene. Assume you know the amino acid sequence of the preproinsulin molecule and that you have a source of insulin. How would you identify the DNA fragment containing the insulin gene (use placental DNA as the source of human DNA)?

2. The Gram positive bacterium *Bacillus subtilis* can secrete several proteins into the medium. Assume a high copy *B. subtilis* plasmid is available with suitable restriction sites. Assume that there is a high degree of specificity at the transcription and translation levels (i.e., the transcription and translation machineries are highly specific for *B. subtilis* signals). How would you construct a secretion vector to secrete human insulin from *B. subtilis.*

Pentose Phosphate Metabolism

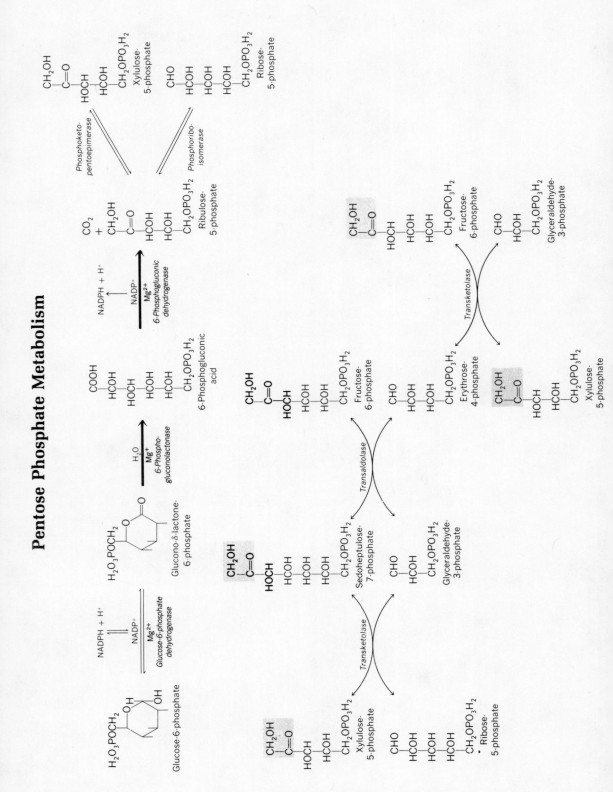